Creating and Transforming the Twentieth Century, Revised and Expanded

Creating and Transforming the Twentieth Century, Revised and Expanded

Technical Innovations and Their Lasting Impact

VACLAV SMIL

OXFORD
UNIVERSITY PRESS

OXFORD
UNIVERSITY PRESS

Oxford University Press is a department of the University of Oxford.
It furthers the University's objective of excellence in research, scholarship,
and education by publishing worldwide. Oxford is a registered trade mark of
Oxford University Press in the UK and in certain other countries.

Published in the United States of America by Oxford University Press
198 Madison Avenue, New York, NY 10016, United States of America.

© Oxford University Press 2025

All rights reserved. No part of this publication may be reproduced, stored in a retrieval system, transmitted, used for text and data mining, or used for training artificial intelligence, in any form or by any means, without the prior permission in writing of Oxford University Press, or as expressly permitted by law, by license or under terms agreed with the appropriate reprographics rights organization. Inquiries concerning reproduction outside the scope of the above should be sent to the Rights Department, Oxford University Press, at the address above.

You must not circulate this work in any other form
and you must impose this same condition on any acquirer

Library of Congress Cataloging-in-Publication Data
Names: Smil, Vaclav, author.
Title: Creating and transforming the twentieth century, revised and expanded :
technical innovations and their lasting impact / Vaclav Smil.
Description: New York, NY : Oxford University Press, 2025. |
Includes bibliographical references and index.
Identifiers: LCCN 2024034838 (print) | LCCN 2024034839 (ebook) |
ISBN 9780197784648 (hardback) | ISBN 9780197784655 (updf) |
ISBN 9780197784662 (epub) | ISBN 9780197784679 (digital-online)
Subjects: LCSH: Technological innovations—Economic aspects. |
Technological innovations—Economic aspects—History—19th century. |
Technological innovations—Economic aspects—History—20th century. |
Technological innovations—History—19th century. |
Technological innovations—History—20th century.
Classification: LCC HC79.T4 S645 2025 (print) | LCC HC79.T4 (ebook) |
DDC 338/.064—dc23/eng/20241227
LC record available at https://lccn.loc.gov/2024034838
LC ebook record available at https://lccn.loc.gov/2024034839

DOI: 10.1093/9780197784679.001.0001

Printed by Sheridan Books, Inc., United States of America

Contents

Preface   vii

PART I. CREATING THE TWENTIETH CENTURY: TECHNICAL INNOVATIONS OF 1867–1914 AND THEIR LASTING IMPACT

1. The Great Inheritance   3
2. The Age of Electricity   37
3. Internal Combustion Engines   107
4. New Materials and New Syntheses   163
5. Communication and Information   215
6. A New Civilization   277
7. Contemporary Perceptions   325

PART II. TRANSFORMING THE TWENTIETH CENTURY: TECHNICAL INNOVATIONS AND THEIR CONSEQUENCES

1. Transforming the Twentieth Century: Debts and Advances   339
2. Energy Conversions: Growth and Innovation   363
3. Materials: Old Techniques and New Solutions   425
4. Rationalized Production: Mechanization, Automation, Robotization   475
5. Transportation, Communication, Information: Mass and Speed   531
6. New Realities and Counterintuitive Worlds: Accomplishments and Concerns   585

7. A New Era or an Ephemeral Phenomenon? Outlook for
   Technical Civilization   641

   PART III. LOOKING BACK, LOOKING AHEAD:
            TWO DECADES LATER

References   675
Index   717

Preface

The first book of this historical duo was—using the words that my favorite composer chose when dedicating his quartets to Joseph Haydn—the result of *lungha e laboriosa fattica*, and yet, at that time, more than two decades ago, I wished that the task could continue. I had a selfish and an objective reason for this: further immersion in the world of pre-World War I ideas, inventions, and innovations would bring more revelations, surprises, and confirmations, and I would have also liked more space because many topics were addressed only cursorily, many reflections and considerations had to be left out.

At the same time, I have always followed Faraday's great dictum—work, finish, publish—and so, however incomplete and imperfect, *Creating the Twentieth Century: Technical Innovations of 1867–1914 and Their Lasting Impact*, the story of one of the greatest adventures in history, my homage to the creators of a new world, was published by Oxford University Press in June 2005. By that time, I was almost done with writing the inevitable second book. As soon as I began to prepare the first volume, it was obvious that it would need a successor to complete the first book's mission. The companion volume explained how the epoch-making pre-World War I inventions and innovations were developed, diffused, commercialized, and improved—and it also traced how these developments combined with new fundamental advances to transform the twentieth century. Oxford University Press published *Transforming the Twentieth Century: Technical Innovations and Their Consequences* before the end of 2006.

Three commonalities are evident. Scores of historical engravings, photographs, and patent drawings are an integral part of the two books: images transport us into the past in a way words cannot do. To appreciate the transformative nature of technical advances I had to introduce scientific and engineering terms and concepts that are not part of common vocabulary. And some readers might feel that the text contains too many numbers, but without adequate quantification there is no real appreciation of the true magnitude of accomplishments that created and transformed the twentieth century. Metric system and scientific units and prefixes are used throughout.

My 2005 preface also contained a brief and clear summary of what not to expect. To repeat, this book duo is neither a brief but comprehensive summation of world history of the two pre-World War I generations and of the twentieth century seen through a prism of technical innovations, nor an economic or social history of the modern era written with an engineering slant. These are not books by a historian for historians who have spent lifetimes studying the era and who expect to find some discipline-specific framing. The books are not intended to be either extended arguments for technical determinism in human affairs or uncritical exaltations of the era. My aim was not to offer any new socioeconomic explanations of the origins of the age, to follow any particular methodology, or to derive any political lessons.

I am quite content to leave the genre of these two books undefined: *they are simply an attempt to tell a story of amazing changes, of the greatest discontinuity in history, and of its extended aftermath*, and to do so from a multitude of perspectives. Their aim is to bring out the uniqueness of the first period and remarkable transformations of the second one and to be reminded of the lasting debt we owe to those who invented the fundamentals and created the structure of the modern world. Or, to paraphrase Braudel's (1950) remarks offered in a different context, I do not seek a philosophy of this great discontinuity but rather its multiple illumination.

I thank Molly Balikov and Oxford University Press for giving me the opportunity to revisit and revise my favorite book duo two decades after its publication. Cuts were done to make room for updates and for a new closing chapter containing additional evaluations and observations concerning the unique era that created the twentieth century and of the innovations that transformed it and provided the foundations for the latest advances. The result is a new single-volume, well-illustrated, and comprehensively referenced edition that will, I hope, illuminate the complex technical origins of the modern world and its prospects.

PART I

CREATING THE TWENTIETH CENTURY

*Technical Innovations of 1867–1914 and
Their Lasting Impact*

Contents
1. The Great Inheritance 3
2. The Age of Electricity 37
3. Internal Combustion Engines 107
4. New Materials and New Syntheses 163
5. Communication and Information 215
6. A New Civilization 277
7. Contemporary Perceptions 325

Frontispiece I.1 Technical advances that began to unfold during the two pre-World War I generations and that created the civilization of the twentieth century resulted in the first truly global human impacts. Some of these are detectable from space, and nighttime images of Earth (here, the Americas in 2000) are perhaps the most dramatic way to show this unprecedented change. Before 1880, the entire continents were as dark as the heart of Amazon remains today.

Image based on NASA's composite available at http://antwrp.gsfc.nasa.gov/apod/ap001127.html.

1

The Great Inheritance

> "You must follow me carefully. I shall have to controvert one or two ideas that are almost universally accepted."
> "Is not that a rather large thing to expect us to begin upon?' said Filby, an argumentative person with red hair."
> "I do not mean to ask you to accept anything without reasonable ground for it. You will soon admit as much as I need from you."
> —H. G. Wells, *The Time Machine*

Imagine an exceedingly sapient and durable civilization that began scanning a bubble of space, say 100 light years in diameter, for signs of intelligent life about half a billion years ago. Its principal surveillance techniques look for any emissions of organized electromagnetic radiation as opposed to the radio frequencies emitted from stars or light that originates from natural combustion of carbon compounds or from lightning. For half a billion years its probes that roam the interstellar space have nothing to report from nine planets that revolve around an unremarkable star located three-fifths of the way from the center of an ordinary looking spiral galaxy that moves inexorably on a collision course with its nearest neighbor. And then, suddenly, parts of that star system's third planet begin to light up, and, shortly afterward, they begin to transmit coherent signals as the two kinds of radiation emanating from the Earth's surface provide the evidence of intelligent life.

Less than a century later a closer approach of the probes would reveal an organized pattern of nighttime radiation in the visible (400–700 nanometers [nm]) and near-infrared (700–1,500 nm) parts of the electromagnetic spectrum produced by electric lights whose density is highest in the planet's most affluent and heavily populated regions (see the Frontispiece). And then the probes could detect a growing multitude of pulsed, modulated signals in frequencies ranging from less than 30 kilohertz (kHz) (very-low-frequency radio band) to more than 1 gigahertz (GHz) (radar bands). These signals have their origins in fundamental scientific and technical advances of the 1880s and 1890s: that is, in the invention of durable incandescent electric lights, in the introduction of commercial generation and transmission of electricity, in the production and detection of Hertzian waves, and in the first tentative wireless broadcasts—and these capabilities became considerably developed and commercialized before World War I.

That was the time when the modern world was created, when the greatest technical discontinuity in history took place. This conclusion defies the common perception that sees the twentieth century as the period of unprecedented technical advances that originated in systematic scientific research and whose aggressive deployment and commercialization brought profound economic, social, and environmental transformations. The last two decades of the twentieth century witnessed an enormous expansion of increasingly more affordable and more powerful computing, of instantaneous access to the globe-spanning World Wide Web, and of rising ownership of portable electronics led by mobile phones: as a result, they have been singled out as a particularly remarkable break from the past.

This was as expected according to those who maintain that the evolution of our technical abilities is an inherently accelerating process (Vinge 1993; Coren 1998; Moravec 1999; Kurzweil 2005). One reality has been critical for creating this impression of accelerating technical innovation: the fact that the number of transistors crowded on a microchip has been doubling approximately every two years since 1972. This trend, known as Moore's Law, was predicted in 1965, by Intel's co-founder, and it has continued despite repeated forecasts of its imminent demise (Moore 1965; Intel 2024). In 1972, Intel's 8008 chip had 2,500 transistors; a decade later there were more than 100,000 components on a single memory microchip; by 1989, the total surpassed 1 million; by 2000, the Pentium 4 processor had 42 million transistors. And, so far, the trend had continued in the twenty-first century: by 2023, the highest component count on a consumer microprocessor was 134 billion (Figure I.1.1).

Indubitably, the twentieth century was exceedingly rich in innovations, and microelectronics has expanded immensely our capacities for problem analysis and information transfer, to the point where the promoters of artificial intelligence see it set to dominate our decision-making before 2050. The idea of accelerating evolution implies the existence of a grand evolutionary trend as well as a tacit assumption of a purpose and a goal—and none of these conclusions can be justified by studying the evolutionary record. Moreover, there has been nothing inevitable about the course of the biosphere's

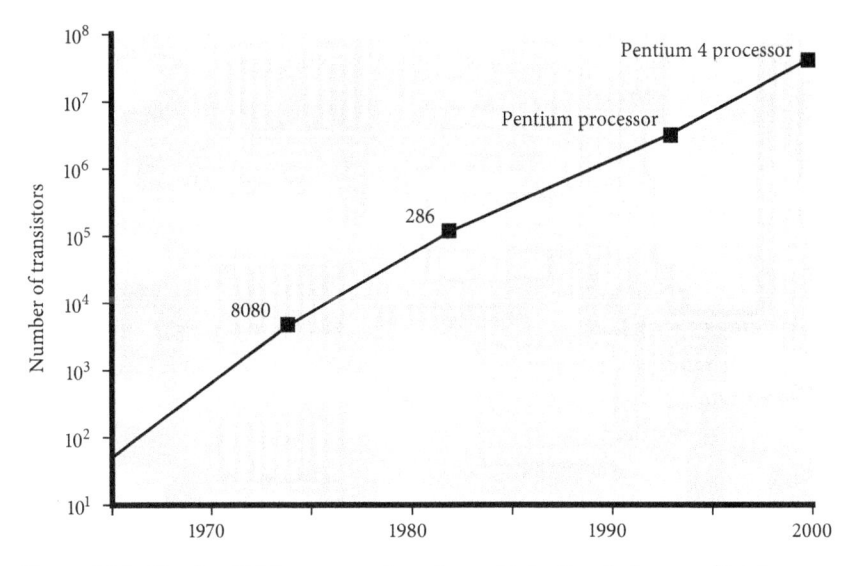

Figure I.1.1 Gordon E. Moore, a co-founder and later the chairman of Intel Corporation, predicted in 1965 that the number of transistors per integrated circuit would double every year. Later he revised the rate to every 18 months, and the actual doubling time between 1971 and 2001 was almost exactly 2 years. This rate is not based on any laws of physics, and it can continue for years, but not for decades.
Plotted from data in Moore (1965) and Intel (2024).

evolution during the past 4 billion years (Smil 2002). From a paleontologist's perspective (Gould 1989: 14),

> evolution has been a staggeringly improbable series of events, sensible enough in retrospect . . . but utterly unpredictable and quite unrepeatable. . . . Wind back the tape of life . . . let it play again from an identical starting point, and the chance becomes vanishingly small that anything like human intelligence would grace the replay.

Moreover, a closer examination of quotidian realities reveals that many techniques whose everyday use keeps defining and shaping modern civilization had not undergone any fundamental change during the twentieth century. Their qualitative gains (higher efficiency, increased reliability, greater convenience of use, lower specific pollution rates) took place without any change of basic, long-established concepts and practices. I will demonstrate that the fundamental means to realize nearly all twentieth-century accomplishments were put in place even before the century began, mostly during the three closing decades of the nineteenth century and in the years preceding World War I. That period ranks as history's most remarkable discontinuity not only because of the extensive sweep of its innovations, but no less so because of the rapidity of fundamental advances that were achieved during that time.

And that transformation is particularly impressive when contrasted with the long span of slow, incremental technical advances of the antique, medieval, and early modern (1500–1800) eras, with examples ranging from the improving efficiency and power of waterwheels and greater maneuverability of sails to higher productivity and lower use of fuel in metal smelting (Smil 2017a). By 1850, their performances were distinctly better than those of their late medieval predecessors, and the difference was even wider when comparing them to their original antique models—but even so theirs were unmistakably just slowly rising cumulative trends. In technical terms, there are only two periods of human history that stand apart as times of two remarkable, broad, and rapid innovation spurts. The first technical saltation, purely oriental, took place during the Han Dynasty, in China (207 BCE–9 CE); the second one, entirely occidental in both its genesis and its nearly instant flourish, unfolded in Europe and North America during the two generations preceding World War I.

In both instances, those widespread and truly revolutionary innovations changed not only the course of the innovating societies, but they were eventually translated into profound global impacts. The concatenation of Han advances laid strong technical foundations for the development of the world's most persistent empire (and, until the eighteenth century, also the world's richest economy) and for higher agricultural and manufacturing productivities far beyond its borders (Needham et al. 1965, 1971; Wang 1982; Temple 1986). The dynasty's most important innovations were in devising new mechanical devices and advancing the art of metallurgy.

The most remarkable new artifacts included the breastband harness for horses and prototypes of an efficient collar harness, wooden moldboard plows with curved shares made from non-brittle iron, multitube seed drills, cranks, rotary winnowing fans, wheelbarrows, and percussion drills (Figure I.1.2). Metallurgical innovations included the use of coal in ironmaking, production of liquid iron, decarburization of iron to make steel, and casting of iron into interchangeable molds. But this innovative period was spread over two centuries, and some of its products were not adopted by the rest of the

Figure I.1.2 Sichuanese salt well made with a percussion drill, one of the great inventions of the Han dynasty in China. The same technique was used to drill the first US oil well in Pennsylvania in 1859.
Reproduced from a Qing addition to Song's (1673) survey of China's techniques.

Old World for centuries, or even for more than a millennium after their initial introduction (Smil 2017a).

The Unprecedented Saltation

In contrast, impact of the late nineteenth and the early twentieth century advances was almost instantaneous because their commercial adoption and widespread diffusion were very rapid. Analogy with logic gates in computing captures the importance of these events. *Logic gates* are fundamental building blocks of digital circuits that receive binary inputs (0 or 1) and produce outputs only if a specified combination of inflows takes place. A great deal of potentially very useful scientific input that could be used to open some remarkable innovative gates was accumulating during the first half of the nineteenth century. But it was only after the mid-1860s, when so many input parameters began to come together, that a flood of new instructions surged through Western society and our civilization ended up with a very different program to guide its future.

The most apposite evolutionary analogy of this great technical discontinuity is the Cambrian eruption of highly organized and highly diversified terrestrial life. This great evolutionary saltation began about 540 million years ago, and it had produced—within a geologically short spell of less than 25 million years, or only about 0.5% of the evolutionary span—virtually all the animal lineages that are known today (McMenamin and McMenamin 1990; Zhuravlev and Riding 2000). Many pre-World War I innovations were patented, commercialized, and ready to be diffused in just a matter of months (telephone, light bulbs) or a few years (gasoline-fueled cars, synthesis of ammonia) after their conceptualization or experimental demonstration. And, as they were built on fundamental scientific principles, it is not only their basic operating modes that have remained intact but often many specific features of their pioneering designs are still very much recognizable among their most modern upgrades.

The era's second key attribute is the extraordinary concatenation of many scientific and technical advances. The first category of these scientific advances embraces those fundamentally new insights that made it possible to introduce entirely new industries, processes, and products. Certainly, the most famous example of this kind is a fundamental extension of the First Law of Thermodynamics that was formulated by Albert Einstein as a follow-up of his famous relativity paper: "An inertial mass is equivalent with an energy content μc^2" (Einstein 1907). By 1943, this insight was converted into the first sustained fission reaction; 1945 saw the explosions of the first three fission bombs (Alamogordo, Hiroshima, Nagasaki); and, by 1956, the first commercial fission reactor began generating electricity (Garwin and Charpak 2001; Smil 2017a).

But during the twentieth century everyday lives of billions of people were much more affected by Heinrich Hertz's discovery of electromagnetic waves much longer than light but much shorter than sound.

This adventure started in 1886, when Hertz detected the spark-generated waves just across a lecture room in Karlsruhe. Soon the reach of these waves progressed to Marconi's Morse signals on land, to communication between ships and across the Atlantic, then to Fessenden's pioneering radio broadcasts, and, after World War I, to television—and to portable electronic devices. By 2000, Hertzian waves made possible such wonders as pinpointing locations with Global Positioning Systems or sending messages, data, and images from mobile phones—be it in Asia's packed subways or on mountain peaks in Europe.

And then there were new scientific insights that did not launch new products or entirely new industries but whose broad theoretical reach has clarified a variety of everyday challenges and has been used to construct better devices and more efficient machines. An outstanding example was the realization that a ratio calculated by multiplying the characteristic distance and velocity of a moving fluid by its density and dividing that product by the fluid's viscosity yields a dimensionless number whose magnitude provides fundamental information about the nature of the flow. Osborne Reynolds (1842–1912), a priest in the Anglican Church and the first professor of engineering in Manchester, found this relationship in 1883, after experiments with water flowing through glass tubes (Rott 1990).

Low Reynolds numbers correspond to smooth laminar flow that is desirable in all pipes, as well as along the surfaces of ships or airplanes. Turbulence sets in with higher Reynolds numbers, and completely turbulent flows are responsible for the cavitation of propellers used on ships (in Chapter I.2, we will see how Charles Parsons solved this very challenge) and nuclear submarines, for vibration of structures, erosion of materials, and relentless noise. A great deal of the cockpit and cabin noise in a cruising jetliner does not come from the powerful gas turbines propelling the aircraft: it is from the turbulent boundary layer that generates wall pressure fluctuations along the aluminum alloy or carbon fiber fuselage, causing vibrations and interior noise.

The period's fundamental technical advances include, above all, large-scale electricity generation and transmission and the inventions of new prime movers and energy converters. Internal combustion engines and electric motors have eventually become the world's most common mechanical prime movers. Transformers and rectifiers ensure the most efficient use of electricity in energizing many specialized assemblies and machines whose sizes range from microscopic to gargantuan, from stationary designs (in manufacturing, commerce, and households) to the world's fastest trains. Yet another group of technical advances includes production processes whose commercialization put on the market new, or greatly improved, products or procedures whose use, in turn, boosted other technical capabilities and transformed economic productivities and private consumption alike.

By far the most important new materials were inexpensive, high-quality steels and aluminum produced electrolytically by the Hall-Héoult process.

Moving assembly lines and the liquefaction of air are excellent examples of new processes whose adoption changed the nature of industrial production. The last process is also a perfect illustration of how ubiquitous and indispensable the synergies of these inventions are. Less than two decades after its discovery the liquefaction of air found one of its most massive (and unanticipated) uses as the supplier of nitrogen for the Haber-Bosch synthesis of ammonia. High crop yields and the deep greens of suburban lawns are thus linked directly to air liquefaction through nitrogen fertilizers. And, most remarkably, in many cases, the ingredients necessary for completely new systems fell almost magically into place just as they were needed. The most notable concatenation brought together incandescent filaments, efficient dynamos and transformers, powerful steam turbines, versatile polyphase motors, and reliable cables and wires for long-distance transmission to launch the electric era during a mere dozen or so years.

The third remarkable attribute of the pre-World War I era is the rate with which all kinds of innovations were promptly improved after their introduction: made more efficient, more convenient to use, less expensive, and hence available on truly mass scales. For example, as I will detail in the next chapter, the efficiency of incandescent lights rose more than six-fold between 1882 and 1912, while their durability was extended from a few hundred to more than 1,000 hours. There were similarly impressive early gains in the efficiency of steam turbines or electricity consumption of aluminum electrolysis. The fourth notable characteristic of the great pre-World War I technical discontinuity is the imagination and boldness of new proposals. There is no better testimony to the remarkable pioneering spirit of the era than the fact that so many of its inventors were eager to bring to life practical applications of devices and processes that seemed utterly impractical, even impossible, to so many of their contemporaries.

Three notable examples illustrate these attitudes of widely shared disbelief. On March 29, 1879, just nine months before Thomas Edison demonstrated the world's first electrical lighting system, American Register concluded that "it is doubtful if electricity will ever be used where economy is an object" (cited in Ffrench 1934: 586). In the same year, the Select Committee on Lighting by Electricity of the British House of Commons heard an expert testimony that there is not "the slightest chance" that electricity could be "competing, in general way, with gas," and *The Engineer* wrote on November 9, 1877, that "electricity for domestic illumination would never, in our view, prove as handy as gas. An electric light would always require to keep in order a degree of skilled attention which few individuals would possess" (quoted in Beauchamp 1997: 136).

Henry Ford reminisced that the Edison Company objected to his experiments with internal combustion and that its executives offered to hire him "only on the condition that I would give up my gas engine and devote myself to something really useful" (Ford 1922: 34). And three years before the Wright brothers took off above the dunes at Kitty Hawk in North Carolina on December 17, 1903 (Figure I.1.3), Rear Admiral George W. Melville (1901: 825) concluded that "outside of the proven impossible, there probably could be found no better example of the speculative tendency carrying man to the verge of chimerical than in his attempts to imitate the birds."

Finally, there is the epoch-making nature of these technical advances, the proximate reason for writing this book: most of them are still with us not just as inconsequential survivors or marginal accoutrements from a bygone age but as the very foundations of modern civilization. Such profound and abrupt discontinuity with such lasting consequences has no equivalent in history. Perhaps the closest analogy would be the emergence of the first

Figure I.1.3 Orville Wright is piloting while Wilbur Wright is running alongside as their machine lifts off briefly above the sands of the Kitty Hawk, North Carolina, at 10:35 A.M. on December 17, 1903.
Library of Congress image (LC-W86-35) is reproduced from the Wrights' glass negative.

settled agricultural societies nearly 10,000 years ago. But the commonly used term of "Agricultural Revolution" is a misnomer because during that gradual process foraging continued to coexist first with incipient and then with slowly intensifying crop cultivation and animal domestication (Smil 2017a). In contrast, the pre-World War I innovations tumbled in at a frenzied pace.

When seen from the vantage point of the early twenty-first century there is no doubt that the two generations between the late 1860s and the beginning of World War I remain the greatest technical watershed in human history. Moreover, as stressed at the outset, this was the first advance in nearly 4.5 billion years of the planet's evolution that led to the generation of cosmically detectable signals of intelligent life on the Earth: a new civilization was born, one based on a synergy of scientific advances, technical innovation, aggressive commercialization, and intensifying, and increasingly efficient, conversions of energy.

The Knowledge Economy

Technical advances of the antiquity, Middle Ages, and the early modern era had no scientific foundation. They had to be based on observations, insights, and experiments, but they were not guided by any coherent set of accumulated understanding that could at least begin to explain why some devices and processes work while others fail. They involved an indiscriminate pursuit of ideas that opened both promising paths of gradual improvements, be it of waterwheels or sails, as well as cul de sacs of *lapis philosophorum* or *perpetuum mobile*. Even the innovations of the early decades of the Industrial Revolution conformed to this pattern. Writing at the very beginning of the nineteenth century Joseph Black noted that "chemistry is not yet a science. We are very far from the knowledge of first principles. We should avoid everything that has the pretensions of a full system" (Black 1803: 547). Mokyr's (1999) apt characterization is that the first Industrial Revolution created a chemical industry without chemistry, an iron industry without metallurgy, and power machinery without thermodynamics.

In contrast, most of the technical advances that appeared during the two pre-World War I generations had their basis in increasingly sophisticated scientific understanding, and, for the first time in history, their success was shaped by close links and rapid feedback between research and commercialization. Naturally, other innovations that emerged during that period still owed little to science because they resulted from random experimenting or serendipity. This is not surprising as we must keep in mind that the new process of scientifically based technical developments was unfolding along with the underlying trend of traditionally incremental improvements.

The first foundations of a new knowledge economy appeared during the seventeenth century, their construction accelerated during the eighteenth century, and the process matured in many ways before 1870. Its genesis, progress, grand features, and many fascinating details are best presented by Mokyr (2002). The most fundamental component of the subsequent change was the development and commercialization of new *prime movers* (Smil 2008, 2017a). Prime movers are those energy converters whose capacities and efficiencies determine the productive abilities of societies as well as their tempo of life. They are also critical for energizing chemical syntheses whose accomplishments help to form our surroundings as well as to expand the opportunities for feeding and healing ourselves. They include (listed in historical sequence) human and animal muscles, water wheels, windmills, external and internal combustion engines, turbines, nuclear reactors, and devices that convert other energies directly to electricity (photovoltaic and fuel cells).

At the beginning of the eighteenth century both the dominant and the largest prime movers were the same ones as in late antiquity. Muscles continued to be the most common prime movers: humans could sustain work at rates of just between 50 and 90 watts (W), while draft mammals could deliver no more than 300 W for small cattle, 400–600 W for smaller horses, and up to 800 W for heavy animals. Capacity of European waterwheels, the most powerful prime movers of the early modern era, averaged less than 4 kilowatts (kW). This mean-s that it took until 1700 to boost the peak prime mover ratings roughly 40-fold (Figure I.1.4), and there was no body of knowledge to understand the conversion of food and feed into mechanical energy and to gauge the efficiency of this transformation.

By 1800, there was still no appreciation of thermal cycles, no coherent concept of energy, no science of thermodynamics, no understanding of metabolism. Antoine Lavoisier's (1743–1794) suggestion of the equivalence between heat output of animals and men and their feed and food intake was only the first step on a long road of subsequent studies of heterotrophic metabolism. But there was important practical progress as James Watt (1736–1819) converted Newcomen's steam engine from a machine of limited usefulness and very low efficiency to a much more practical device capable

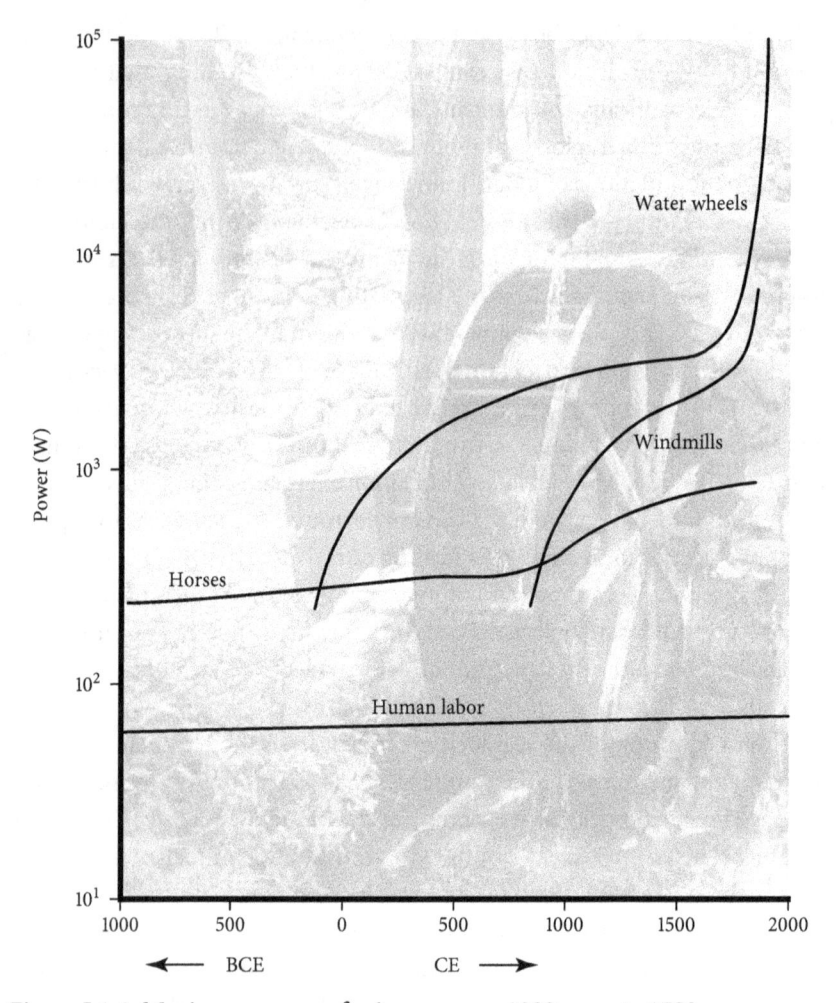

Figure I.1.4 Maximum power of prime movers, 1000 B.C.E. to 1700 C.E. Waterwheels became the most powerful inanimate prime movers of the preindustrial era and lost this primacy to steam engines after 1730.

of about 20 kW that began revolutionizing many tasks in coal mining, metallurgy, and manufacturing (Thurston 1878; Dalby 1920; Rosen 2012). Watt also invented a miniature recording steam gauge, and this indicator made it possible to study the phases of engine cycles.

By 1870, thermodynamics was becoming a mature science whose accomplishments helped to build better prime movers and design better industrial processes. This transformation started during the 1820s, with Sadi

Carnot's (1796–1832) formulation of essential principles that he intended to be applicable to all imaginable heat engines (Carnot 1824). This was a bold claim but one that was fully justified: thermodynamic studies soon confirmed that no heat engine can be more efficient than a reversible one working between two fixed temperature limits and that the highest theoretical efficiency of this Carnot cycle cannot surpass 65.32%. Another important insight published during the 1820s was Georg Simon Ohm's (1789–1854) explanation of electricity conduction in circuits and the formulation of what later became the eponymous law relating current, potential, and resistance (Ohm 1827). Correct understanding of this relationship was a key to devising a commercially affordable system of electric lighting because it minimized the mass of expensive conductors.

Understanding of energy conversions progressed rapidly during the 1840s and 1850s, with the brilliant deductions of Robert Mayer (1814–1878) and James Prescott Joule (1818–1889). When Mayer worked as a physician on a Dutch ship bound for Java, he noted that sailors had a brighter venous blood in the tropics. He correctly explained this fact by pointing out that less energy, and hence less oxygen, was needed for basal metabolism in warmer climates. Mayer saw muscles as heat engines energized by oxidation of blood, and this led to the establishment of a mechanical equivalent of heat and hence to a formulation of the law of conservation of energy (Mayer 1851), later known as the First Law of Thermodynamics. A more accurate quantification of the equivalence of work and heat came independently from Joule's work: in his very first attempt he was able to come within less than 1% of the actual value (Joule 1850).

Soon afterward William Thomson (Lord Kelvin, 1824–1907) described a universal tendency to the dissipation of mechanical energy (Thomson 1852). Rudolf Clausius (1822–1888) formalized this insight by concluding that the energy content of the universe is fixed and that its conversions result in an inevitable loss of heat to lower energy areas: energy seeks uniform distribution, and the entropy of the universe tends to maximum (Clausius 1867). This Second Law of Thermodynamics—the universal tendency toward disorder—became perhaps the most influential, as well as a much misunderstood, cosmic generalization. And although its formulation did not end those futile attempts to build perpetuum mobile machines it made it clear why that quest will never succeed.

There is perhaps no better illustration of the link between the new theoretical understanding and astonishing practical results than Charles Parsons's invention and commercialization of the steam turbine, the most powerful commonly used prime mover of the twentieth century (Figure I.1.5). Parsons, whose father was an astronomer and a former president of the Royal Society, received mathematical training at the Trinity College in Dublin and at Cambridge, joined an engineering firm as a junior partner, and proceeded to build the first model steam turbine because thermodynamics told him it could be done. He prefaced his Rede lecture describing his great invention by noting (Parsons 1911: 1) that "the work was initially commenced because calculation showed that, from the known data, a successful steam turbine ought to be capable of construction. The practical development of this engine was thus commenced chiefly on the basis of the data of physicists." In 1885, Parsons designed his first prototype turbine rated at 7.5 kW, and, in 1899, Parsons's company delivered its first 1 MW turbine, a more than 130-fold increase in power in just 14 years.

Pre-1870 gains in chemical understanding were, comparatively, even greater as science started from a lower base. Brilliant chemists of the late eighteenth century—Antoine Lavoisier, Wilhelm Scheele, Joseph Priestley—began to systematize the fragmentary understanding of elements and compounds, but there was no unifying framework for their efforts. Early nineteenth-century physics had a solid grasp of mechanics, but large parts of chemistry still had the feel of alchemy. Then came a stream of revolutionary chemical concepts. First, in 1810, John Dalton put atomic theory on a

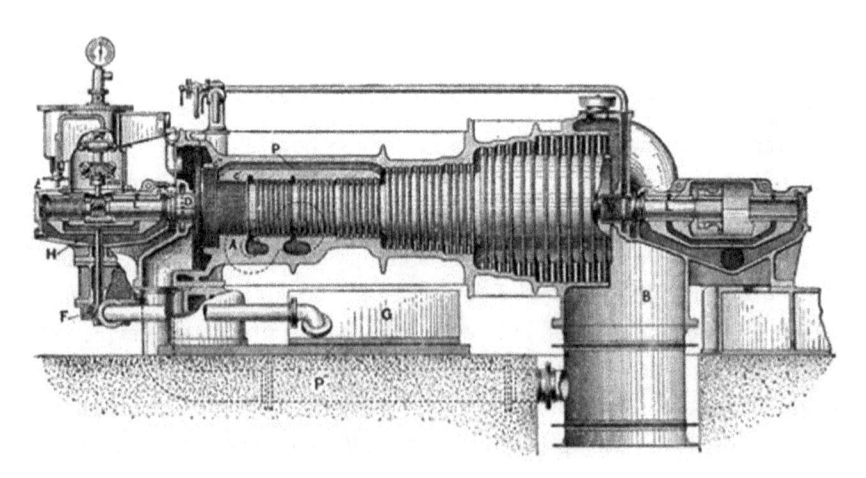

Figure I.1.5 Longitudinal cross-section through the casing of 1-MW-capacity Parsons's steam turbine designed in 1899.
Reproduced from Ewing (1911).

quantitative basis, and, in 1828, Friedrich Wöhler (1800–1882) synthesized the first organic compound (urea). Beginning in the 1820s, Justus von Liebig (1803–1873; Figure I.1.6) established standard practices of organic analysis and attributed the generation of carbon dioxide (CO_2) and water to food oxidation, thus providing a fundamentally correct view of heterotrophic metabolism (Liebig 1840).

After 1860, Friedrich August Kekulé (1829–1896) and his successors made sense of the atomic structure of organic compounds. In 1869, Dimitrii Mendeleev (1834–1907) published his magisterial survey of chemistry and placed all known—as well as yet unknown—elements, in their places in his periodical table. This achievement remains one of the fundamental pillars of the modern understanding of the universe (Mendeleev 1891). These theoretical advances were accompanied by the emergence of chemical engineering, initially led by the British alkali industry, and basic research in organic synthetic chemistry brought impressively rapid development of the

Figure I.1.6 What Justus von Liebig (1803–1873) helped to do so admirably for chemistry, other early nineteenth-century scientists did in their disciplines: made them the foundations of a new knowledge economy.
Photograph from author's collection.

coal-tar industry. Its first success was the synthesis of alizarin, a natural dye derived traditionally from the root of the madder plant (*Rubia tinctorum*).

Badische Anillin- & Soda-Fabrik (BASF) synthesized the dye by using the method invented by Heinrich Caro (1834–1910), the company's leading researcher; his method was nearly identical to the process proposed in England by William Henry Perkin (1838–1907). Synthetic alizarin has been used ever since to dye wool and to stain microscopic specimens. German and British patents for the process were filed less than 24 hours apart, on June 25 and 26, 1869 (Brock 1992). This was not as close a contest as the patenting of the telephone by Alexander Graham Bell and Elisha Gray that occurred just a few hours apart. These two instances illustrate the intensity and competitiveness of the era's quest for innovation.

Remarkable interdisciplinary synergies combining new scientific understanding, systematic experiments, and aggressive commercialization can be illustrated by developments as diverse as the birth of the electric era and the synthesis of ammonia from its elements. Edison's light bulb was not a product of intuitive tinkering by an untutored inventor. Incandescent electric lights could not have been designed and produced without combining deep familiarity with the state-of-the-art research in the field, mathematical and physical insights, a punishing research program supported by generous funding from industrialists, a determined sales pitch to potential users, rapid commercialization of patentable techniques, and the continuous adoption of the latest research advances.

Fritz Haber's (1868–1934) discovery of ammonia synthesis was the culmination of years of research effort based on decades of previously unsuccessful experiments, including those done by some of the most famous chemists of that time (among them two future Nobelians, Wilhelm Ostwald and Walter Nernst). Haber's success required familiarity with the newly invented process of air liquefaction, willingness to push the boundaries of high-pressure synthesis, and determination, through collaboration with BASF, not to research the process as just another laboratory curiosity but to make it the basis of commercial production (Stoltzenberg 1994; Szöllösi-Janze 1998; Smil 2001; Johnson 2022).

And Carl Bosch (1874–1940), who led BASF's development of ammonia synthesis, made his critical decision as a metallurgist and not as a chemist. When, at the crucial management meeting in March 1909, the head of BASF's laboratories heard that the proposed process would require pressures of at least 100 atmospheres, he was horrified. But Bosch remained confident: "I believe it can go. I know exactly the capacities of the steel industry. It should be risked" (Holdermann 1954: 69). And he did: Bosch's confidence challenged the German steelmakers to produce reaction vessels of unprecedented size capable of operating at previously unheard-of pressures—but less than five years later these devices were in commercial operation at the world's first ammonia plant at Oppau, north of Ludwigshafen.

The Age of Synergy

At this point I must anticipate, and confront, those skeptics and critics who would insist on asking—despite all arguments and examples given so far in support of the Age of Synergy—how justified is my conclusion to single out this era, how accurate is my timing of it? As De Vries (1994: 249) noted, history is full of "elaborate, ideal constructions that give structure and coherence to our historical narratives and define the significant research questions." And while these historiographical landmarks are based on events that indubitably took place, these concepts tend to acquire life of their own, and their interpretations tend to shift with new insights. And so, the concept of the Renaissance, the era seen as the opening chapter of modern history, came to be considered by some historians as little more than an administrative convenience, "a kind of blanket under which we huddle together" (Bouwsma 1979: 3).

Even more germane for any attempt to delimit the Age of Synergy is the fact that some historians have questioned the very concept of its necessary precursor, the Industrial Revolution. Its dominant interpretation as an era of broad economic and social change (Ashton 1948; Landes 1969; Mokyr 1990; Allen 2017) has been challenged by views that see it as a much more restricted, localized phenomenon that brought significant technical changes only to a few industries (cotton, ironmaking) and left the rest of the economy in premodern stagnation until the 1850s: Watt's steam engines notwithstanding, "the British economy was largely traditional 90 years after 1760" (Crafts and Harley 1992).

Or, as Musson (1978: 141) put it, "the typical British worker in the mid-nineteenth century was not a machine-operator in a factory but still a traditional craftsman or labourer or domestic servant." At the extreme, Cameron (1982) argued that the change was so small relative to the entire economy that the very title of the Industrial Revolution is a misnomer, and Fores

(1981) went even further by labeling the entire notion of a British industrial revolution a myth. Temin (1997) favors a compromise, seeing the Industrial Revolution's technical progress spread widely but unevenly. I am confident that the concept of the Age of Synergy is not a mental construct vulnerable to devastating criticism but an almost inevitable identification of a remarkable reality.

The fact that the era's epoch-making contributions were not always recognized as such by contemporary opinion is not at all surprising. Given the unprecedented nature of the period's advances, many commentators simply did not have the requisite scientific and technical understanding to appreciate the reach and the transforming impact of new developments. During the early 1880s, most people thought that electricity would merely substitute faint light bulbs for similarly weak gas lights: after all, the first electric lights were explicitly designed to match the luminosity of gas jets and gave off only about 200 lumens, or an equivalent of 16 candles (compared to 1,600 lumens for standard twentieth-century 100 W incandescent bulb). And even one of the era's eminent innovators and a pioneer of electric industry did not think that gas illumination was doomed.

William Siemens (see Figure I.4.6) reaffirmed in a public lecture on November 15, 1882 (i.e., after the first two Edison plants began producing electricity) his long-standing conviction that gas lighting "is susceptible of great improvement and is likely to hold its own for the ordinary lighting up of our streets and dwellings" (Siemens 1882: 69). A decade later, during the 1890s, people thought that gasoline-fueled motor cars were just horseless carriages whose greatest benefit would be to rid the cities of objectionable manure. And many people also hoped that cars would ease the congestion caused by slow-moving and often uncontrollable horse-drawn vehicles (Figure I.1.7).

And given the fact that such fundamental technical shifts as the widespread adoption of new prime movers have invariably long lead times—for example, by 1925, England still derived 90% of its primary power from steam engines (Hiltpold 1934)—it is not surprising that even H. G. Wells, the era's most famous futurist, maintained in his first nonfictional prediction that if the nineteenth century needs a symbol then that symbol will be almost inevitably a railway steam engine (Wells 1902a). But, in retrospect, it is obvious that, by 1902, steam engines were a symbol of a rapidly receding past. Their future demise was already irrevocably decided as it was only a matter of time before the three powerful, versatile, and more energy-efficient prime movers

Figure I.1.7 A noontime traffic scene at the London Bridge as portrayed in *The Illustrated London News*, November 16, 1872.

that were invented during the 1880s—steam turbines, gasoline-powered internal combustion engines, and electric motors—would completely displace wasteful steam engines.

I also readily concede that, as is so often the case with historical periodization, other and always somewhat questionable brackets of the era that created the twentieth century are possible and readily defensible. This is inevitable given the fact that technical advances have always some antecedents and that many claimed firsts are not very meaningful. Patenting dates also make dubious markers as there have been many cases when years, even more than a decade, elapsed between the filing and the eventual grant. And too many published accounts do not specify the actual meanings of such claims as "was invented" or "was introduced." Consequently, different sources will often list different dates for their undefined milestones, and entire chains of these events may be then interpreted to last only a few months, many years, or even decades.

Petroski (1993) offered an excellent example of a dating conundrum by tracing the patenting and eventual adoption of the zipper, a minor but ubiquitous artifact that is now produced at a rate exceeding 15 billion units every year. US patent 504,038 for a clasp fastener was granted to Whitcomb

L. Judson of Chicago in August of 1893, putting the invention of this now universally used device squarely within the Age of Synergy. But an examiner overlooked a very similar idea that was patented in 1851, by Elias Howe, Jr., the inventor of sewing machine. And it took about 20 years to change Judson's idea of the slide fastener to its mature form when Gideon Sundback patented his "new and improved" version in 1913, and about 30 years before that design became commercially successful. In this case, there is a span of some seven decades between the first, impractical invention and widespread acceptance of a perfected design. Moreover, there is no doubt that patents are an imperfect measure of invention as some important innovations were not patented, and many organizational and managerial advances are not patentable.

Turning once again to the Industrial Revolution, we see that its British, or more generally Western phase, has been dated as liberally as 1660–1918, or as restrictively as 1783–1802. The first span was favored by Lilley (1966) who divided the era into early (1660–1815) and mature (1815–1918) periods; the second one was the time of England's economic take-off as defined, with a specious but suspect accuracy, by Rostow (1971). Other dates can be chosen, most of them variants on the 1760–1840 span that was originally suggested by Toynbee (1884). Ashton (1948) opted for 1760–1830; Beales (1928) preferred 1750–1850, but he also argued for no terminal date. Leaving aside the appropriateness of the term "revolution" for what was an inherently gradual process, it is obvious that because technical and economic take-offs began at different times in different places there can be no indisputable dating even if the determination would be limited to European civilization and its overseas outposts.

This is also true about the period that some historians have labeled the Second Industrial Revolution. This clustering of innovations is dated to between 1870 and 1914 by Mokyr (1999) and to 1860–1900 by Gordon (2000). But in Musson's (1978) definition the Second Industrial Revolution was most evident between 1914 and 1939, and King (1930) was certain that, at least in the United States, it started in 1922. And, in 1956, Leo Brandt and Carlo Schmidt described to the German Social Democratic Party Congress the principal features and consequences of what they perceived was the just unfolding Second Industrial Revolution that was bringing a reliance on nuclear energy (Brandt and Schmid 1956). Three decades later Donovan (1997) used the term to describe recent advances brought by using the Internet during the 1990s, and e-Manufacturing Networks (2001) called

John T. Parsons—who patented the first numerical control machine tool—the "father of the Second Industrial Revolution."

Moreover, just before the end of the twentieth century, some high-tech *enamorati* began to reserve the term only for the ascent of future nanomachines that will not only cruise through our veins but will be eventually able to build complex self-replicating molecular structures—even as others have already labeled the ascent of semiconductors, computing (mainframe and personal), and the internet as the Third Industrial Revolution and claimed that we are "poised at the beginning of the Fourth Industrial Revolution . . . powered by cloud, social, mobile, the Internet of things, and artificial intelligence" (Trailhead 2024). Consequently, I am against using this constantly morphing and ever-advancing term.

In contrast, constraining the singular Age of Synergy can be done with greater confidence. My choice of the two generations between the late 1860s and the beginning of World War I is bracketed at the beginning by the introduction of the first practical designs of dynamos and open-hearth steelmaking furnaces (1866–1867), the first patenting of a sulfite pulping process (1867), the introduction of dynamite (1866–1868), and the definite formulation of the Second Law of Thermodynamics (1867). Also, in 1867 or 1868, the United States, the indisputable overall leader of the great pre-World War I technical saltation, became the world's largest economy by surpassing the British gross domestic product (GDP) (Maddison 1995).

Late in 1866 and in the early part of 1867, three engineers concluded independently that powerful dynamos can be built without permanent magnets by using electromagnets: Alfred Varley got his patent on December 24, 1866, and this idea was publicly presented for the first time in January 1867 by Charles Wheatstone and Werner von Siemens (Schellen and Keith 1884; Thompson 1901). These new dynamos, after additional improvements during the 1870s, were essential to launching the electric era during the 1880s. The open-hearth furnace was another fundamental innovation that made its appearance after 1867: in 1866, William Siemens and Emile Martin agreed to share the patent rights for its improved design, the first units became operational soon afterward, and the furnaces eventually became the dominant producers of steel during most of the twentieth century (Dichman and Reynolds 1911; Buell 1936; Smil 2016).

Tilghman's chemical wood pulping process (patented in 1867) opened the way for mass production of inexpensive paper. And, by 1867, Alfred Nobel was ready to produce his dynamite, a new, powerful explosive that proved to be another epoch-making innovation, both in destructive and constructive senses. The tangled history of the Second Law of Thermodynamics does not allow us to cite a single year as the date of its clear formulation (Wolfram 2023), but it was in 1867 when Rudolf Clausius published the book summarizing his work on what he initially referred to as "the second fundamental theorem" (Clausius 1867).

The year 1867, the beginning of my preferred time span of the Age of Synergy, also saw the design of the first practical typewriter and the publication of Marx's *Das Kapital*. Both accomplishments—an ingenious machine that greatly facilitated information transfer and a mass of dense ideological writing—had an enormous impact on the twentieth century that was suffused with typewritten information and that experienced the prolonged rise and sudden fall of Marx-inspired communism. Inventors of the new technical age and the entrepreneurs who translated the flood of ideas into new industries and thus formed new economic and social realities (sometimes they were the same individuals) came from all regions of Europe and North America, but, in national terms, there can be no doubt about the leading role played by the United States. Surpassing the United Kingdom to become the world's largest economy during the late 1860s was only the beginning. Just 20 years later Mark Twain's introduction of a Yankee stranger (preparatory to his imaginary appearance at King Arthur's court) sounded as a fair description of the country's technical prowess rather than as an immodest boast of a man who

> "learned to make everything; guns, revolvers, cannons, boilers, engines, all sorts of labor-saving machinery. Why, I could make anything a body wanted—anything in the world, it didn't make any difference what; and if there wasn't any quick new-fangled way to make a thing, I could invent one—and do it as easy as rolling off a log." (Twain 1889: 20)

The most prominent innovations that marked the end of the era just before World War I were the successful commercialization of Haber-Bosch ammonia synthesis, the introduction of the first continuously moving assembly line at the Ford Company, Irving Langmuir's patenting of a coiled tungsten filament that was found in all incandescent light bulbs until they were taken off the market, and the first vacuum tubes (Figure I.1.8). The first accomplishment was perhaps the most important technical innovation of the modern era: without it the world could not support (depending on the prevailing diets) more than about half of its recent population (Smil 2001, 2011).

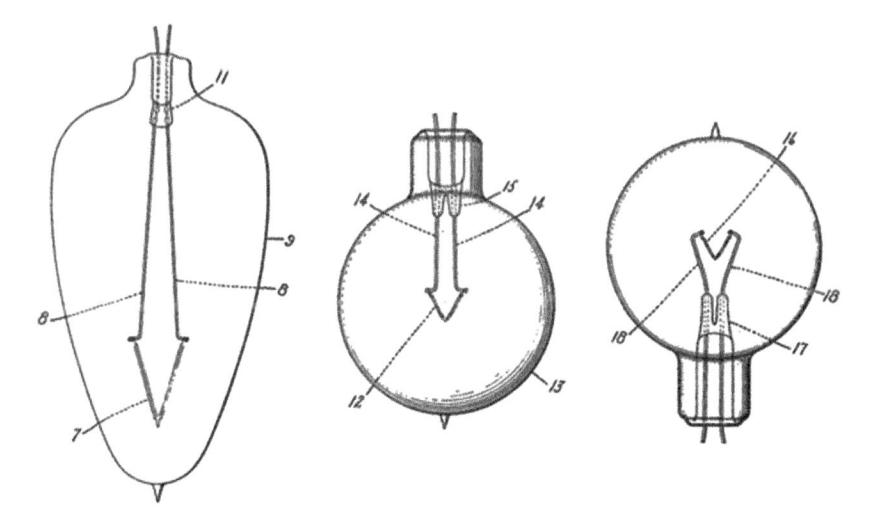

Figure I.1.8 Three illustrations that accompanied Irving Langmuir's US Patent 1,180,159 (1913) for an incandescent electric lamp show different kinds of coiled filaments that resulted in higher conversion of electricity to light.
Drawings and patent specification available at the US Patent and Trademark Office.

The second invention revolutionized mass production far beyond car assembly: every mobile phone, now the most owned electronic device, comes off Asian line assemblies. The third one kept brightening dark hours during the entire twentieth century until its inherently low efficiency led to its displacement (first by fluorescent, then by halogen lights and light-emitting diodes). And the fourth one provided the hot-glass foundations of a new electronic era. The last pre-World War I years also brought some fundamental scientific insights whose elaboration led to key innovations of the twentieth century. Robert Goddard's concept of a multiple-stage rocket (patented in 1914) led eventually to communication and Earth observation satellites and to the exploration of space. Niels Bohr's model of the atom was the conceptual harbinger of the nuclear era (Bohr 1913).

I will readily concede that there are good arguments for both extending and trimming my timing of the Age of Synergy. Extending the 1867–1914 span just marginally at either end would make it identical to the Age of Energy (1865–1915), which Howard Mumford Jones defined for the United States simply as the half century that followed the end of the Civil War (Jones 1971). He used the term "energy" not in its strict scientific sense but to describe a period of extraordinary change, expansion, and mobility. Trimming the span to 1880–1910 would shorten the era to the three exceptionally inventive decades that were marked by an unprecedented concatenation of innovation, incipient mass production, and shifting consumption patterns. Other choices are possible, even if less defensible: for example, opting for the three decades between 1875 and 1905 would cover the time between the introduction of Otto's four-stroke internal combustion engine and the publication of Einstein's first relativity paper (Einstein 1905) that some interpretations see as the start of a new era.

But one choice I am deliberately ignoring is the dating of well-known long-wave economic cycles, whose idea was originally formulated by Kondratiev (1935), elaborated by Schumpeter (1939), and embraced by many economists after World War II (Freeman 1983; Van Duijn 1983; Vasko, Ayres, and Fontvieille 1990; Tylecote 1992; Keklik 2003). Dating of these waves varies. The second Kondratiev cycle commenced between 1844 and 1851 and ended between 1890 and 1896, which means that my preferred dating of the Age of Synergy (1867–1914) would be cut into two segments. Perez (2002) distinguished three stages of economic growth associated with technical advances during that period: the Age of Steam and Railways (the second technical revolution) that lasted between 1829 and 1874; the Age of Steel, Electricity, and Heavy Engineering (the third revolution, 1875–1907); and the Age of Oil, the Automobile, and Mass Production (the fourth revolution, 1908–1971).

The search for long waves is a fascinating enterprise, but the historical evidence forces me to conclude that the Age of Synergy was a profound technical singularity, a distinct discontinuity, and not just another (second, or second and third) instalment in a series of regular repetitions, a period that belongs to two or three debatably defined economic cycles. My concern is not with the timing of successive sequences of prosperity, recession, depression, and recovery but with the introduction and rapid improvement and commercialization of fundamental innovations that defined a new era and influenced economic and social development for generations to come.

Even a rudimentary list of such epoch-defining artifacts must include telephones, sound recordings, light bulbs, practical typewriters, chemical pulp, and reinforced concrete for the pre-1880 years. The astonishing 1880s brought reliable incandescent electric lights, electricity-generating plants, electric motors and trains, transformers, steam turbines, gramophones, popular photography, practical gasoline-fueled four stroke internal combustion engines, motorcycles, cars, aluminum production, crude oil tankers, air-filled rubber tires, first steel-skeleton skyscrapers, and prestressed concrete. The

1890s saw Diesel engines, x-rays, movies, the liquefaction of air, wireless telegraphy, the discovery of radioactivity, and the synthesis of aspirin.

And the period between 1900 and 1914 witnessed mass-produced cars, the first airplanes, tractors, radio broadcasts, vacuum diodes and triodes, tungsten light bulbs, neon lights, the common use of halftones in printing, stainless steel, the synthesis of ammonia, the hydrogenation of fats, and air conditioning. Removing these items and processes from our society would deprive it not only of a very large part of its anthropogenic environment we now take for granted but it would also render virtually all twentieth-century inventions useless as their production and/or functioning depend on an uninterrupted supply of electricity and the use of many high-performance materials, above all on structural concrete and steel and on aluminum.

Although my dating (1867–1914) of the Age of Synergy differs from Gordon's (2000, 2016) timing of the Second Industrial Revolution (1860–1900), we share the conclusion that the development and diffusion of technical and organizational advances introduced during those years created a fundamental transformation in the Western economy because it ushered the world into the golden age of productivity growth that lasted until 1973, when quintupling of oil prices by the Organization of the Petroleum Exporting Countries (OPEC) led to years of high inflation and low growth (Figure I.1.9). Neither the pre-1860 advances nor the recent diffusion and enthusiastic embrace of computers and the internet are comparable with the epoch-making sweep and lasting impacts of that unique span of innovation that dominated the two pre-World War I generations.

Or, as Gordon (2016; 601) put it,

> The digital Third Industrial Revolution (IR #3), though utterly changing the way Americans obtain information and communicate, did not extend across the full span of human life as did IR #2, with the epochal changes it created in the dimensions of food, clothing, housing and its equipment, transportation, information, communication, entertainment, the curing of diseases and conquest of infant mortality, and the improvement of working conditions on the job and at home.

An excellent confirmation of the period's epochal nature comes from a list of the twentieth century's greatest engineering achievements that the US National Academy of Engineering released to mark the end of the second

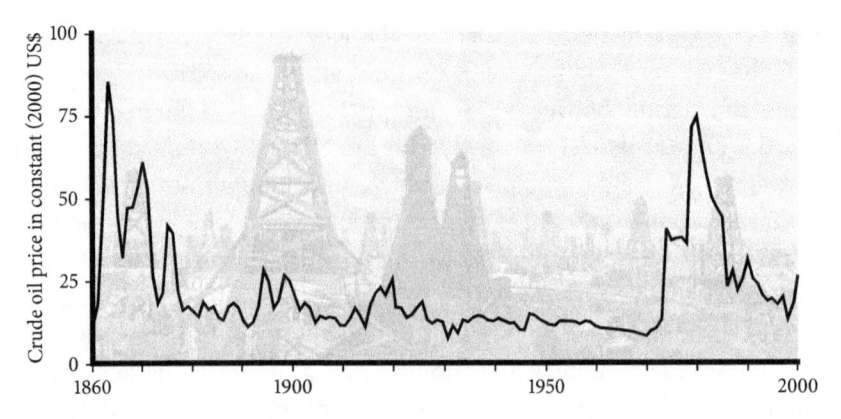

Figure I.1.9 A century-long declining trend of crude oil prices (expressed in constant monies) ended with OPEC's first round of sudden price increases in 1973–1974.
Based on a graph in Energy Institute (2023).

millennium (NAE 2024). No fewer than 16 of the 20 listed categories of engineering achievements had not just their genesis but often also a considerable period of rapid pioneering development during the two pre-World War I generations. These achievements are headed by electrification, automobiles, and airplanes (numbers 1–3 on the NAE's list); include the telephone, air conditioning, and refrigeration (numbers 9 and 10); and conclude with petrochemical techniques and high-performance materials (numbers 17 and 20).

Another list, commissioned by the Lemelson-MIT Prize Program (that recognizes and rewards innovation) and assembled by polling 1,000 American adults in November 1995, was topped solely by the inventions that were introduced and considerably developed before World War I. In 34% of all responses, the automobile ranked as the most important invention of modern times, followed by the light bulb (28%), telephone (17%), and aspirin which was tied, at 6%, for the fourth place with the PC, the highest-ranking of all post-World War I innovations (Lemelson-MIT Program 1996). Finally, in van Duijn's (1983) list of 38 major innovations that shaped the course of six key twentieth-century growth sectors (steel, telecommunication, cars, aircraft, illumination, and photography) 23, or 60%, were invented between 1867 and 1914.

A few revealing comparisons are perhaps the best way to impress an uninitiated mind on the scope and impacts of the period's remarkable innovations. During the first generation of the nineteenth century the everyday life of most people was not significantly different from that of the early eighteenth century. Even a well-off New England farmer plowed his fields with a heavy wooden plow pulled by slow oxen. Poorly sprung carriages and fully rigged sail ships were the fastest means of transport, and information—leaving aside the limited network of optical telegraphs built in France and a few neighboring countries (Holzmann and Pehrson 1994)—traveled only as fast as people or animals did.

Illumination came in meagre increments of tallow candles or smoky oil lamps. Recycling of organic wastes and planting of legumes were the only manageable sources of nitrogen in agriculture, and average staple grain yields were only marginally above the late medieval levels. One of the key inescapable social consequences of these realities was the fact that even a very large agricultural labor force was unable to feed adequately slowly growing populations. Even though more than four-fifths of the work force were required to produce food through taxing labor, large shares of the rural population did not have even enough bread, meat was an occasional luxury for most people, and recurrent spells of serious food shortages were common in Europe while frequent famines were affecting Asia (Kiple and Ornelas 2000; Smil 2017a).

A hundred years later a single small tractor provided traction equal to that of a dozen large horses, and the Haber-Bosch synthesis of ammonia made it possible to supply optimum amounts of the principal macronutrient to crops. Highest crop yields rose by roughly 50%, and the average food supply in all industrializing countries was more than even liberally defined nutritional requirements. The age of steam locomotives was about to give way to more efficient and more powerful diesel engines and electric motors, ship makers turned to new powerful steam turbines, and the luminosity of Edison's original light bulb was greatly surpassed by new lights with incandescing tungsten wires.

I must close this brief recounting of fundamental pre-World War I advances by emphasizing that I have no interest in forcing the concept of the Age of Synergy or exaggerating the era's importance (although that would be hard to do). Also, I am not promoting any simplistic, deterministic, techniques-driven interpretation of modern history. I would never argue that the adoption of a particular technical advance determines the fortunes of individual societies or that it eventually dictates a convergence of their behavior. Identical machines and processes deployed as a part of a democratic, transparent society will have very different effects compared to those they could bring when operating as part of a dictatorship where autarky and the lack of accountability foster mismanagement.

Nor am I claiming that all the twentieth-century technical accomplishments were either direct derivatives (albeit improved and mass-produced in unprecedented quantities) or at least indirect descendants of innovations introduced during the two pre-World War I generations. But it is indisputable that these enduring innovations were indispensable for creating the twentieth century, and it is also possible to forecast with a high degree of confidence that many of them will not lose this importance for generations to come. And so, to understand better where we have come from and where we are heading, I will portray the genesis of those enduring innovations and appraise their legacy. But, before I begin this account of the pre-World War I advances and of their defining importance for the twentieth century, I will take one more brief, comparative look at the era's accomplishments and their enduring quality.

The Distance Traveled

An effective way to appreciate the distance traveled between 1867 and 1914 is to contrast the state of our understanding of the world in the early nineteenth and the early twentieth centuries with the realities at the very beginning of the twenty-first century. This could be done perhaps most impressively by contrasting the degree of comprehension that a competent scientist of one period would have when, just for the sake of this thought experiment, we would transport him 100 years into the future. If Antoine Lavoisier (Figure I.1.10), one of the founders of modern chemistry, were not guillotined during the French Revolution's Age of Terror in 1794 (for being one of the 27 tax collectors working on behalf of the king), he would have been 70 years old in 1813. A few men alive at that time could have equaled his comprehensive understanding of natural sciences and technical advances—but for him the world of the year 1913 would hold countless inexplicable wonders.

Steam turbines in large power plants, high-voltage transmission lines, electric lights and motors, oil drilling rigs, refineries, internal combustion engines, cars, radio broadcasts, x-rays, high explosives, high-performance steels, synthetic organic compounds and fertilizers, aluminum, and airplanes—all of these, and scores of other machines, materials, objects, and processes would stun and puzzle him, and most of them would be utterly

Figure I.1.10 Antoine Lavoisier (1743–1794)—the greatest chemist of the eighteenth century and an accomplished polymath—would view most of today's technical achievements with utter incomprehension.
Portrait from E. F. Smith Collection, Rare Book & Manuscript Library, University of Pennsylvania, Philadelphia.

incomprehensible to his lay contemporaries. In contrast, were one of the accomplished innovators of the early twentieth century—Edison, Fessenden, Haber, or Parsons—be transported to the early twenty-first century he would have a deep understanding of most of them (as he created some of them) and at least a highly competent familiarity with the rest of the items listed in the preceding paragraph. Moreover, he would not need a great deal of explanation to understand many devices and processes that he never saw at work but whose operation is so clearly derived from the foundations laid down before World War I.

This legacy of the pre-World War I era is most obvious as far as energy sources and prime movers are concerned. As already emphasized, no two physical factors are of greater importance in setting the pace and determining the ambience of a society than its energy sources and its prime movers. The global fossil fuel era began sometime during the 1890s, when coal, increasing volumes of crude oil, and a small amount of natural gas began supplying more than half of the world's total primary energy needs by displacing traditional biomass energies of wood, charcoal, crop residues, and dried dung (Smil 2017b). By the late 1920s, biomass energies provided no more than 35% of the world's fuels, and, by 2000, their share was about 10% of global energy use, falling to about 5% by 2020.

The two prime movers that dominate today's transportation and fossil-fueled electricity-generating plants—internal combustion engines and steam turbines—were also invented and rapidly improved before 1900. And an entirely new system for the generation, transmission, and use of electricity—by far the most versatile form of energy—was created in less than 20 years after Edison's construction of the first installations in London and New York, in 1882. The only new primary energy source that has made a substantial commercial difference during the twentieth century was nuclear fission (in 2000, the world derived from it about 16% of electricity), but, due to its arrested development, its share of the global primary energy supply has been slowly falling.

The only new and important twentieth-century prime mover that entered everyday use after 1914 was the gas turbine. Its development took off during the 1930s, and it led to both stationary machines for electricity generation and to the era of jet-powered flights. Naturally, the stories of nuclear fission and gas turbines will feature prominently in the second section of this book. Only the first decades of the twenty-first century saw a rapid rise of two new renewable energy conversions, wind turbines and solar photovoltaic cells, but, by 2023, they still supplied less than 8% of the global primary energy whose provision remains dominated (80% in 2023) by fossil fuels still overwhelmingly converted into useful energies in processes introduced before 1914.

To appreciate the legacy of the epoch-making innovations introduced between 1867 and 1914, I will concentrate on four classes of fundamental advances: formation, diffusion, and standardization of electric systems; invention and adoption of internal combustion engines; the unprecedented pace of introducing new, high-performance materials and new chemical syntheses; and the birth of a new information age. This will be followed first by introducing some additional perspectives on that eventful era and then by a brief restatement of its lasting legacies. In closing, I will look at two major trends that had governed the twentieth century, the creation of high-energy societies and mechanized mass production aimed at raising the standards of living—and at how these achievements were perceived by those who lived through those remarkable decades.

And although this is a book about inventions, improvements, and applications of techniques and about the power of applied scientific understanding, I wish I could have followed concurrently those fascinating artistic developments that took place during the two pre-World War I generations. Not only were they remarkable by any historic standards, but, much like their technical counterparts, they had set in place many tastes, preferences, and sensibilities that could be felt during the twentieth century. Just think of an extraordinary concatenation of artistic talents in fin de siécle Paris—where Émile Zola's latest instalment of his expansive and gripping Rougon-Macquart cycle could be read just before seeing Claude Monet's canvases that transmuted the landscape around Giverny into images of shimmering light and, later the same day, one could go to a concert presenting Claude Debussy's intriguing *L'Apres-midi d'un faun.*

Moreover, the world of technique left many brilliant imprints on the world of arts. One of my great favorites is *Au Bonheur des Dames*, Zola's (1883) masterful portrait of the new world of spreading affluence and frenzied mass consumption, set in the late 1860s, that was closely modeled on his thorough studies of large Parisian department stores of the early 1880s (Figure I.1.11). And the restlessly kinetic pre-World War I futurist paintings by Giacomo Balla, Umberto Boccioni, or Marcel Duchamp evoke the speed and dynamism of a new age driven by electricity and motors better than do a list of technical specifications (Humphreys 1999; Lord 2014).

What legacies of those two incomparable generations have we taken with us into the twenty-first century? I keep pointing out repeated failures of enthusiastic forecasts that have been predicting the demise of some well-established techniques and the imminent adoption of new technical and managerial approaches. Nuclear fission should have become the dominant method of electricity generation by 2000, but, after its retreat, we are still waiting for its renaissance. In the new century, we have been promised the end of all internal combustion engines, perhaps as early as in 2035 and of all fossil-fueled conversions by 2050—even as the global combustion of fossil carbon (and the rising concentrations of atmospheric CO_2) have kept setting new annual records.

Realities of energy and material supply, food and industrial production, construction, transportation, and communication prevailing during the second decade of the twenty-first century only strengthen my conviction that many epoch-making innovations that were introduced before World War I and that had served us so well, and often with no fundamental

Figure I.1.11 Grandiose, elegantly domed, and richly decorated central hall of *Grands Magasins du Printemps* (owned by Jules Jaluzot & Cie.) on Boulevard Haussman in Paris, the city's premiere luxury department store. Its three wings and eight floors of diverse merchandise reflected the rise of affluent middle class during the two pre-World War I generations.
Reproduced from *The Illustrated London News*, October 18, 1884.

modifications, during the twentieth century have much greater staying power than may be commonly believed. We will continue to rely on them for much of the twenty-first century, and this realization makes it even more desirable, as well as more rewarding, to understand the genesis of these innovations.

Frontispiece I.2 Cross-section of Edison's New York station (thermal capacity, 93 MW) completed in 1902. Boilers are on the right; steam engine and dynamo hall on the left. Four steel-plate stacks were 60 m tall.

Reproduced from the cover page of *Scientific American*, September 6, 1902.

2
The Age of Electricity

> There is a powerful agent, obedient, rapid, easy, which conforms to every use, and reigns supreme on board my vessel. Everything is done by means of it. It lights it, warms it, and is the soul of my mechanical apparatus. This agent is electricity.
> —Captain Nemo to Pierre Aronnax in Jules Verne's *Twenty Thousand Leagues Under the Sea*

I do not know of any better description of electricity's importance in modern society than taking this quotation from Jules Verne's famous science fiction novel and substituting "in modern civilization" for "on board my vessel." In 1870, when Verne set down his fictional account of Captain Nemo's global adventures, various electric phenomena had been under an increasingly intensive study for more than a century (Figuier 1888; Fleming 1911; MacLaren 1943; Dunsheath 1962). Little progress followed the pioneering seventeenth-century investigations by Robert Boyle (1627–1691) and Otto von Guericke (1602–1686). But Pieter van Musschenbroek's (1692–1761) invention of the Leyden jar (a condenser of static electricity) and Benjamin Franklin's (1706–1790) bold and thoughtful experiments in Boston (beginning in 1749) and later in Paris revived interest in the properties of that mysterious force (Priestley 1768; Park 1898).

Electricity ceased to be a mere curiosity and became a subject of increasingly systematic research with the work by Luigi Galvani (1737–1798, of twitching frog legs fame, during the 1790s), Alessandro Volta (1745–1827; his pioneering paper that described the construction of the first battery was published in 1800), Hans Christian Ørsted (1777–1851, who uncovered the magnetic effect of electric currents in 1819), and André Marie Ampère (1775–1836, who contributed the concept of a complete circuit and quantified the magnetic effects of electric currents). Their research opened many new experimental possibilities, and, already in 1820, Michael Faraday (1791–1867), using Ørsted's discovery, built a primitive electric motor, and, before 1830, Joseph Henry's (1797–1878) experimental electromagnets became powerful enough to lift briefly loads of as much as 1 ton (t).

But it was Michael Faraday's discovery of the induction of electric current in a moving magnetic field, in 1831, that eventually led to the large-scale conversion of mechanical energy into electricity (Faraday 1839). Its revelation is easily stated: the magnitude of the electromotive force that is induced in a circuit is proportional to the rate of change of flux. This discovery of how to generate alternating current (AC) was the logic gate that opened the route toward the practical use of electricity that was not dependent on bulky, massive, low-energy-density batteries. Eventually, this route led to three classes of machines whose highly reliable and remarkably efficient work makes it possible to permeate the modern world with inexpensive electricity: turbogenerators, transformers, and electric motors.

Shortly after Faraday's fundamental discoveries came the invention of the telegraph, which, by the 1860s, evolved into a well-established, globe-spanning network of wired communication. But, by 1870, there were no commercially viable electric lights, and, as the puzzled Aronnax noted when he was told about electric propulsion of the *Nautilus* by Nemo, until that time electricity's "dynamic force has remained under restraint and has only been able to produce a small amount of power"—because there were no reliable means of large-scale electricity generation. Experiments with electricity and the expanding telegraphy field were energized by batteries, commonly known as *Voltaic bimetallic piles* whose low-energy density (<10 Wh/kg) was suited only for applications that required limited power.

Then, after millennia of dependence on just three basic sources of energy—combustion of fuels (biomass or fossil fuel), animate metabolism (human and animal muscles), and conversion of indirect solar flows (water and wind),—everything changed in a single decade. During the 1880s, the combined ingenuity of inventors, support of investors, and commercially viable designs of enterprising engineers coalesced into a new energy system without whose smooth functioning there would be no modern civilization. Practical carbon-filament electric lights—soon supplanted by incandescing finely drawn wires—illuminated nights. Parsons's invention of the steam turbine created the world's most powerful prime mover that made the bulky and inefficient steam engines obsolete and allowed the inexpensive generation of electricity on large scales. Transformers made long-distance electricity transmission possible. Efficient induction motors converted the flow of electrons into mechanical energy, and innovative electrochemistry began producing new materials at affordable prices.

The economic and social transformations brought by electricity were so profound because no other kind of energy affords such convenience, flexibility, and instant and effortless access to consumers: a flip of a switch, the push of a button, or now even just a preprogrammed order. In households, electricity had eventually eliminated daily chores that ranged from tiresome—drawing and hauling water, washing and wringing clothes by hand, ironing them with heavy wedges of hot metal—to light but relentlessly repetitive tasks (trimming wicks and filling oil lamps). Electricity on farms did away with primitive threshing of grains, hand-milking of cows, preparation of feed by manual chopping or grinding, and pitchforking of hay into lofts. Electric pumps eliminated slow and laborious water-raising powered by people and animals. In workshops and factories, electricity replaced poorly lit premises with bright lights, while convenient, efficient, and precisely adjustable electric motors did away with transmission belts driven by steam engines. On railways, electricity supplanted inefficient, polluting steam engines with faster means of clean conveyance.

No other form of energy can equal electricity's flexibility: it can be converted to light, heat, motion, and chemical potential and used extensively in every principal energy-consuming sector except commercial flight. Until 1998, the last sentence would have said simply "with the exception of flight"—but in that year AeroVironment's unmanned Pathfinder aircraft rose to 24 kilometers (km) above sea level. In August 2001, a bigger Helios, a thin, long, curved, and narrow flying wing (its span of just over 74 meters (m) is longer than that of a Boeing 747) driven by 14 propellers powered by 1 kilowatt (kW) of bifacial solar cells, became the world's highest-flying plane as it soared to almost 29 km (Figure I.2.1; NASA 2021).

Besides this all-encompassing versatility, electricity is also a perfectly clean as well as silent source of energy at the point of consumption, and its delivery can be easily and precisely adjusted to provide desirable speed and accurate control for flexible industrial production (Nye 1990; Schurr et al. 1990). Electricity can be converted without virtually any losses to useful heat, and large electric motors can transform more than 90% of electricity into mechanical energy; it can also generate temperatures higher than the combustion of any fossil fuel, and, once the requisite wiring is in place, any

Figure I.2.1 Helios prototype flying wing, powered by photovoltaic cells, during its record-setting flight above the Hawaiian Islands on July 14, 2001. NASA photograph ED 010209-6.

new converters can be just plugged in. Perhaps the best proof of electricity's importance comes from simply asking what we would not have without it. The answer is just about everything in the modern world.

We use electricity to power our lights, electronic devices, a panoply of converters ranging from handheld hair dryers to the world's fastest trains, almost every life-saving measure (modern synthesis and production of pharmaceuticals is unthinkable without electricity, vaccines need refrigeration, hearts are checked by electrocardiograms and during operations are bypassed by electric pumps), and most of our food is produced, processed, distributed, and cooked with the help of electric machines and devices. This chapter describes the genesis and evolution of electric systems that took place between the 1870s—the last pre-electric decade in history—and 1914. During that span of less than two generations, we made enormous progress as we put in place the foundation of a new energy system whose performance is now far ahead of anything we had before World War I but whose basic features remained remarkably constant throughout the twentieth century.

Fiat Lux: Inventing Electric Lights

Thomas Edison was accustomed to keeping a brutal work pace and demanded others to follow suit, and his search for a durable filament that would produce more than an ephemeral glow was particularly frustrating (Josephson 1959; Israel 1998; Freeberg 2013; Morris 2020). But the often-repeated dramatic story of continuous and suspenseful "death-watch" that took 40 hours of waiting for the first successful filament to stop incandescing is just a legend, derived from later reminiscences of Edison's assistant (Jehl 1937). Edison's laboratory records indicate that the lamp that made history by passing the 10-hour mark—the one with a very fine carbonized cotton filament, a piece of six-cord thread fastened to platinum wires—was attached to an 18-cell battery at 1:30 A.M. on October 22, 1879, and that it still worked by 3 P.M. and continued to do so for another hour after increased power supply overheated and cracked the bulb (Friedel and Israel 1986).

A filament incandescent for nearly 15 hours represented a major advance in the quest for electric light, but it was a step clearly retracing the previous research of other inventors. Edison's principal, and lasting, contribution to the development of electric light was in changing the basic operating conditions, not in discovering new components. And the initial success had to be followed by numerous improvements and modifications, in Edison's laboratory and by others, before reliable and reasonably long-lasting light bulbs were ready for mass marketing. What is so remarkable is that so many basic components and procedures that were finessed during the 1880s remained in place during the entire twentieth century because many subsequent innovations had greatly improved electric lighting's performance without changing its basic operating principles.

Early Electric Lights

The quest to use electricity for lighting began decades before Edison was born. The possibility to do so was first realized at the very beginning of the nineteenth century. In 1801, Humphry Davy (1778–1829) was the first scientist to describe the electric arc that arises as soon as two carbon electrodes are slightly separated. In 1808, Davy publicly demonstrated the phenomenon on a large scale at the Royal Institution in London by using 2,000 Voltaic cells to produce a 10-centimeter (cm)-long arc between electrodes of willow charcoal (Davy 1840). The first trials of arcs preceded by decades the introduction of large electricity generators. The world's premiere of public lighting took place in December 1844, when Joseph Deleuil and Léon Foucault briefly lit Place de la Concorde by a powerful arc (Figuier 1888).

The first dynamo-driven outdoor arc lights were installed during the late 1870s: starting in 1877, P. N. Yablochkov's (1847–1894) much admired electric candles were used to illuminate downtown streets and other public places in Paris (Grand Magasins du Louvre, Avenue de l'Opéra) and London (Thames Embankment, Holborn Viaduct, the British Museum's reading room). By the mid-1880s, arc lamps were common sights in many Western cities, but they were massive and complicated devices that required skilled installation and frequent maintenance. Because a continuous arc wears away the electrodes, a mechanism was needed to move the rods to maintain a steady arc and to rejoin them and separate them once the current was switched off and on.

Many kinds of self-regulating mechanisms were invented to operate arc lamps, but that still left considerable costs associated with the carbons and the labor for re-carboning the devices. A typical 10-amp (A) arc lamp using, respectively, carbons of 15 and 9 millimeters (mm) for its positive and negative electrodes and operating every night from dusk to dawn would have consumed about 180 m of carbon electrodes a year (Garcke 1911a). Placing these 500-watt (W) lamps 50 m apart for basic street illumination would have required annual replacement of 3. 6 km of carbons for every kilometer of road, a logistically cumbersome and costly proposition precluding the adoption of arc lamps for extensive lighting of public places.

Glass-enclosed arcs extended the lamp's useful life to as much as 200 hours, but they did not lower the cost of re-carboning enough to make arcs a better choice than advanced gas lighting, and they also reduced the maximum luminosity from as much as 18,000 to no more than 2,500 lumens (lm). As for the electricity supply, a typical 10-A arc consuming 500–600 W was connected by a line of just 50–60 volts (V), a voltage too low for efficient large-scale electricity distribution. Obviously impractical for most indoor and for all household uses and uneconomical for outdoor applications, arc lights could not become the sources of universal illumination. Less powerful and much more practical incandescent lights were the obvious choice to fill this enormous niche—and, in 1802, Humphry Davy demonstrated their possibility when he placed a 45-cm-long piece of platinum wire (diameter of 0.8 mm) in a circuit between the bars of copper and induced first red heat and then brilliant white light (Davy 1840).

A tedious account of inventors and their failed or promptly superseded attempts would be needed to review all activities that took place between 1820, when William de La Rue experimented with a platinum coil inside a bulb, and January 27, 1880, when Edison was granted his basic patent (US Patent 223,898) for a carbon-filament electric lamp (Edison 1880a; Figure I.2.2).

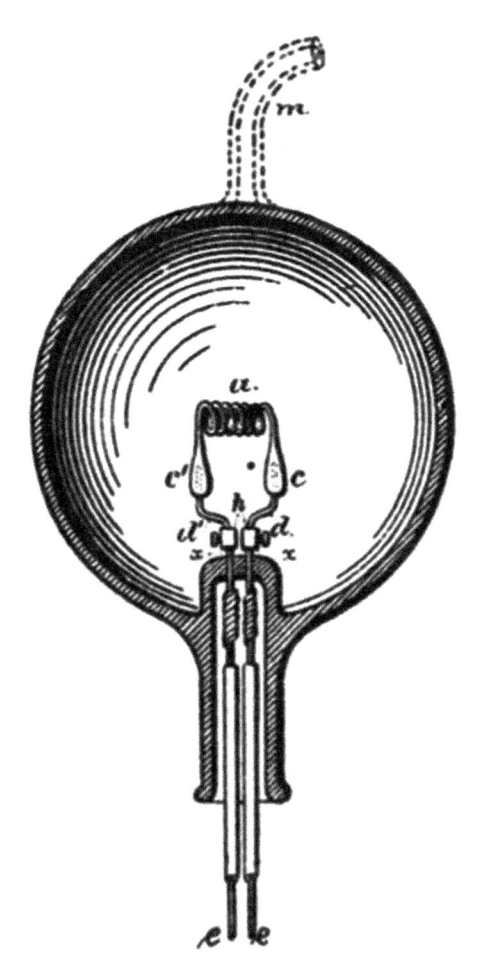

Figure I.2.2 Drawing attached to Edison's fundamental US Patent 223,898 for the incandescent light shows a tightly coiled filament, but carbonized materials used by Edison in 1879–1880 would have made that impossible, and the first lamps had simple loop filaments. The coiled metallic filament was patented only in 1913 by Irving Langmuir (see Figure I.1.8).
Reproduced from Edison (1880a).

More than a score of inventors in the United States, the United Kingdom, France, Germany, and Russia sought patents for the "subdivided" electric light for nearly four decades before Edison began his experiments (Pope 1894; Hammer 1913; Howell and Schroeder 1927; Bright 1949; Friedel and Israel 1986; Bowers 1998).

Their names have been forgotten, and even the historians of lighting mention some of them primarily because the men were involved in complex litigations with Edison's company during the 1880s (Pope 1894; Dyer and Martin 1929; Covington 2002). They sought materials with very high melting points because the proportion of light to heat radiated by a filament rises with the increasing temperature. Carbon, with the highest melting point of all elements (3,650°C), is an excellent incandescing material, and they used it as solid rods or charcoal tubes but could not produce sustained light. Platinum has a relatively low melting point of (1,772°C), which can be raised by alloying. In addition, unsuccessful early lamp designs had an imperfect vacuum or their imperfectly sealed glass bulbs were filled with nitrogen.

Edison's R&D Dash

Edison (Figure I.2.3) began his lighting experiments—in August 1878, after a memorable trip to the Rockies and California, when he rode most of the way from Omaha to Sacramento on a cushioned cowcatcher of a locomotive (Josephson 1959)—by alloying platinum with iridium and testing various coatings, and his first patent application for the incandescent platinum-iridium light was made on October 5, 1878 (US Patent 214,636, granted on April 22, 1879). Ten days later, on October 15, 1878—after the initial capital was raised from investors by Edison's long-time friend, patent attorney Grosvenor P. Lowrey—an agreement was reached to set up Edison Electric Light Company with the objective "to own, manufacture, operate and license the use of various apparatus in producing light, heat or power by electricity" (Josephson 1959: 189). In December 1878, Edison boasted to a *New York Sun* reporter that "I am all right on my lamp. I don't care anything more about it" (quoted in Friedel and Israel 1986: 42), but there was little progress during the winter months of 1878–1879.

At that time, Edison was only one of many inventors (and a late starter) racing to develop practical incandescent lighting. What eventually set Edison apart was not his legendary perseverance in pursuing a technical

Figure I.2.3 Thomas A. Edison in 1880 at the time of his work on incandescent lights and electric systems.
Library of Congress photograph by Emil P. Spahn (LC-USZ62-98067).

solution or the financing he received from some of the richest men of his time, but the combination of a winning conceptualization and a rapid realization of an entire practical commercial system of electric lighting (Friedel and Israel 1986; Israel 1998). In that sense, he deservedly ranks ahead of even Joseph Wilson Swan (1828–1914), an English chemist and physicist who remains credited in the United Kingdom as the inventor of the light bulb.

Swan rejected platinum as the best incandescent substance already during the 1850s and concentrated instead on producing suitable carbon filaments. By 1860, he made a lamp that contained the key ingredient of Edison's early promising models of 1879: a carbonized paper strip mounted in an (imperfectly) evacuated glass vessel and connected to a battery. These short-lived lights could reach only red glow, but Swan resumed his experiments during the late 1870s, when better vacuum pumps became available.

A working lamp—demonstrated for the first time at a meeting of the Newcastle-on-Tyne Chemical Society on December 18, 1878—had platinum wires and a carbon filament, the same components as Edison's first longer-lasting lamp revealed 10 months later (Bowers 1998).

But Swan's lights operated, as did those of all other inventors before 1879, with filaments whose resistance was very low, ranging from less than 1 ohm (Ω) to no more than 4–5 Ω. Mass deployment of such lights would have required very low voltages and hence impracticably high currents, resulting in a ruinously large mass of transmission wires. Moreover, these lamps were connected in series and supplied with a constant current from a dynamo, making it impossible to switch on the lights individually and forcing a shutdown of the entire system because of a single interruption. Edison's key insight was that any commercially viable lighting system must minimize electricity consumption and hence must use high-resistance filaments with lights connected in parallel across a constant-voltage system.

This concept, so contrary to the prevailing wisdom of the time, was at first questioned even by Francis R. Upton (1852–1921), the only highly trained scientist (mathematics and physics, from Princeton and Berlin University) in Edison's laboratory, who was hired specifically to formulate and buttress the inventor's ideas in rigorous scientific terms. Forty years later, Upton recalled how eminent electricians of that time maintained that the subdivision of the electric current was commercially impossible and how muddled was their understanding of the very concept (Jehl 1937). A few simple calculations illustrate the difference between the two approaches.

An incandescent lamp whose filament had resistance of just 1 Ω and operated at 10 V would have required the current of 10 A (V/Ω), and, as Upton calculated, a copper conductor 12,000 inches long would have required a massive cross-section of 12,000 square inches. In contrast, voltage of 100 V and a light bulb whose filament had a resistance of 100 Ω—these being parameters chosen by Edison and Upton to reduce the cost of copper conduits and to make filaments stable (Martin 1922)—needed just 1 A and a conduit with a cross-section of just 120 square inches (Jehl 1937). This means that, for the same transmission distance, a system composed of low-resistance lamps would have required 100 times the mass of identical conducting wires to distribute electricity than would the high-resistance arrangement, obviously a nonstarter from the economic point of view.

As Edison noted in his specification for the first patent concerning high-resistance lamps (US Patent 227,229 submitted on April 12, 1879),

By the use of such high-resistant lamps I am enabled to place a great number in multiple arc without bringing the total resilience of all the lamps to such a low point as to require a large main conductor; but, on the contrary, I am enabled to use a main conductor of very moderate dimensions. (Edison 1880b: i)

Edison's actual choice was 0.75 A, 110 V, and 147 Ω. This was a winning concept, but the patented lamp itself was only an improved, but impractical, version of his older platinum lights.

The lamp had an alloyed platinum wire inside a small vacuum tube (his technicians succeeded in creating a vacuum of nearly 1-millionth of the atmospheric pressure) that was placed inside a glass cover. A flexible metallic aneroid chamber underneath the vacuum tube would expand and briefly interrupt the electrical circuit and prevent the filament from overheating. This was an ingenious design, but the complex arrangement of connections, wires, and magnets and the placement of the vacuum bulb within another glass container were not a basis for a sturdy and practical lamp.

Edison's Success and Competitors

At that time Edison was also designing a new telephone and building his first dynamo, the largest electricity generator of that time, and he returned to his search for a better filament only in August 1879. Many different substances were subsequently carbonized in a small furnace, and Edison's first success with carbonized cotton sewing thread in October 1879 was followed by intense activity to improve the design. A filament made of carbonized cardboard (Bristol board) proved to be a much better choice than cotton thread, and new lamps lasted easily more than 100 hours. But before the demonstration, scheduled for December 31, 1879, took place, the *New York Herald* scooped the event by publishing a full-page article in its December 21 issue.

Authored by Marshall Fox, who spent two weeks in the Menlo Park laboratory preparing the piece, the lengthy article was informative, accurate, and well-written (it is reproduced in Jehl 1937). Fox conveyed well not only the incredulity of the event (electric light being produced from "a tiny strip of paper that a breath would blow away") and the glory of its final product ("the result is a bright, beautiful light, like the mellow sunset of an Italian autumn") but also the basic scientific preconditions and challenges of Edison's quest. Not surprisingly, it was later judged by Edison as "the most accurate story of the time concerning the invention" (Jehl 1937: 381).

The first public demonstration of incandescent lighting took place, as planned, at Menlo Park on December 31, 1879. Edison used 100 cardboard filament lamps—long-stemmed and topped by an onion-shaped dome, each consuming 100 W and producing about 200 lm—to light the nearby streets, the laboratory buildings, and the railway station. In so many ways, this was not an end but merely a frantic beginning of the quest for incandescent light. Other inventors rapidly switched to high-resistance designs, and new patents were filed at a dizzying rate. In 1880, Edison alone filed nearly 30 patents dealing with electric lamps and their components (TAEP 2024). Swan's 1880 British patents incorporated high resistance and the parallel arrangement, and, after they were upheld in British courts, Edison decided not to challenge them. Instead, a joint Edison & Swan United Electric Light Company was set up in 1883, to manufacture Swan's lamps in the United Kingdom.

In the United States, Edison had to embark on a many-sided defense of his patents while also defending himself against numerous claims of patent infringement. Between 1880 and 1896, more than $2 million was spent in prosecuting more than 100 lawsuits of the latter kind (Dyer and Martin 1929). Edison's greatest loss came on October 8, 1883, when the US Patent Office decided that his light bulb patents rested on the previous work of William Sawyer. The US Circuit Court of Appeals made its final decision in favor of the Edison lamp patent only on October 4, 1892, more than 12 years after the filing of the patent itself, leaving the inventor to complain that he had never enjoyed any benefits from his lamp patents.

But protracted litigations did not stop the search for better filaments. Edison's laboratory tested more than 6,000 specimens of any available plant fibers, ranging from rare woods to common grasses, and Edison finally settled on a Japanese bamboo filament that lasted about 600 hours, a great improvement on the 150 hours obtained with the cardboard lamps of December 1879. But cellulose filaments, introduced by Swan in 1881, proved to be the most popular carbon choice, and their improved versions became the industry standard by the early 1890s. Glass bulbs were initially free-blown from 2.5-cm tubing, first in short and later in longer pear shapes, and the glass widened gently after it emerged from the base collar. Mold-blown bulbs, first introduced by the General Electric Company during the early 1890s, had a short but almost perpendicular step just above the collar and only then did they widen into a rounded shape. Early bulbs came in a

Figure I.2.4 Different kinds of Edison's light bulbs from the 1880s.
Reproduced from Figuier (1888).

profusion of shapes, all of them with a short tip as the glass was nipped off while mouth-blowing the bulb, and Edison himself designed many of these shapes (Figure I.2.4).

The first electric decade also saw the first colored, ground glass, opal, and etched lamps, and ornamental lamps of various shapes. The earliest bases were wooden or, between 1881 and 1899, plaster of Paris. Porcelain was introduced in 1900, and glass insulation, standard in later bulbs, came a year later. Platinum was used for lead wires, and carbon filaments were connected first with clamps and then, after 1886, with carbon paste. Edison's bamboo filament was abandoned after Edison General Electric merged with Thomson-Houston in 1892, to form the still operating and now highly diversified General Electric Company. But subsequent lamp designs incorporated more than the great inventor's high-resistance, parallel-connection design; they also perpetuated his patented mounting.

A simple screw base—whose initial inspiration came to Edison early in 1880, as he unscrewed the cover of a kerosene can, and whose design was done by two of Edison's long-time associates, Edward H. Johnson and John Ott during the fall of 1880 (Friedel and Israel 1986)—was borne by about 70% of all lamps sold in 1900, and it became an industry standard by 1902. Nothing demonstrated better the state of incandescent lamps after more than a dozen years of their rapid evolution than did the lighting at Chicago's Columbian Exposition of 1893, the first such event illuminated solely by electricity (Bancroft 1893). Arc lights were still used for some outdoor locations, but the Westinghouse Electric & Manufacturing Company was the principal lighting contractor, and it used 90,000 "Stopper" lamps operating at 105 V. Inside the exposition's Electrical Building were more than 25,000 incandescent lights, including 10,000 General Electric lamps.

All these lamps shared one undesirable property: a very low efficiency of converting electricity into visible light. The earliest designs produced just over one lm/W, which means (converting with the average of 1.46 mW/lm) that they turned no more than 0.15% of electric energy into light (Figure I.2.5). Still, this was an order of magnitude better performance than for paraffin candles, whose combustion converted a mere 0.01% of energy in the solid hydrocarbon into light, and was nearly four times the rate for gas lights: using typical performance data from Paton (1890), I calculated that gas jets converted less than 0.04% of the manufactured fuel to visible light.

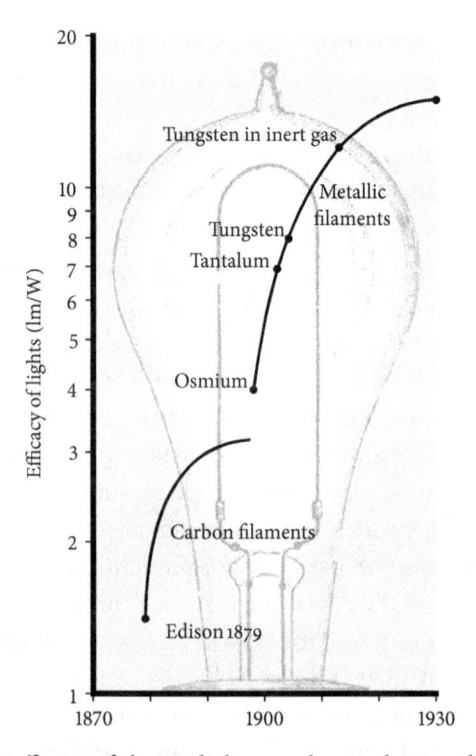

Figure I.2.5 The efficacy of electric lights rose by nearly an order of magnitude between 1879 (Edison's first long-lasting carbon filament) and 1912 (Langmuir's tungsten filament in inert gases). Calculated and plotted from data in a variety of contemporary and retrospective publications.

Improved filaments raised the typical performance to 2.5 lm/W (0.37%) by the mid-1890s (Figure I.2.5), and it was obvious that the ongoing large-scale diffusion of electric lighting would greatly benefit from replacing inefficient and fragile bits of carbonized cellulose with more luminous, as well as sturdier, metallic filaments.

Metallic Filaments

By 1898, light bulbs reached an efficacy of 3.5 lm/W (0.51%) and, in that year, Carl Auer von Welsbach (1858–1929)—who in 1885 patented the incandescent mantle, a delicate gauze cylinder impregnated with thorium and cerium oxides whose greater luminosity delayed the demise of gas lighting (Welsbach 1902)—introduced the first working metal filament made of osmium. This metal's melting point (2,700°C) is 1,000°C above that of platinum, and osmium lights had an efficacy of 5.5 lm/W. But the rarity, and hence the high cost of the element prevented their commercial use. Lamps with tantalum filaments (melting point of 2,996°C) made from a drawn wire reaching efficacies up to 7 lm/W were patented by Werner von Bolton and Otto Feuerlein in 1901 and 1902 and were used until 1911. In 1904, Alexander Just and Franz Hanaman patented the production of tungsten filaments; the metal's high melting point (3,410°C) allowed an efficacy of at least 8 lm/W, but preparing bendable wire from this grayish white (hence its German name, *Wolfram*, and symbol W) lustrous metal was not easy (MTS 2024).

Several methods were tried during the following years before William David Coolidge (1873–1975) found that high temperature and mechanical treatment make tungsten so ductile that it can be drawn into a fine wire even at ambient temperature; his patent application for making tungsten filaments was filed on June 19, 1912 (Figure I.2.6). General Electric's first tungsten lamp was introduced in 1910, and it produced no less than 10 lm/W (efficiency of about 1.5%). The metal became the dominant incandescent material by 1911, and the last cellulose lamps were made in 1918. When incandescing inside lamps, both filaments must be kept at temperatures far below their respective melting points to avoid rapid vaporization. Consequently, their energy output peaks in the near infrared (at about 1 micrometer [μm]) and then falls through the visible spectrum to about 350 nm (into the ultraviolet A range), producing most of the visible light in red and yellow wavelengths,

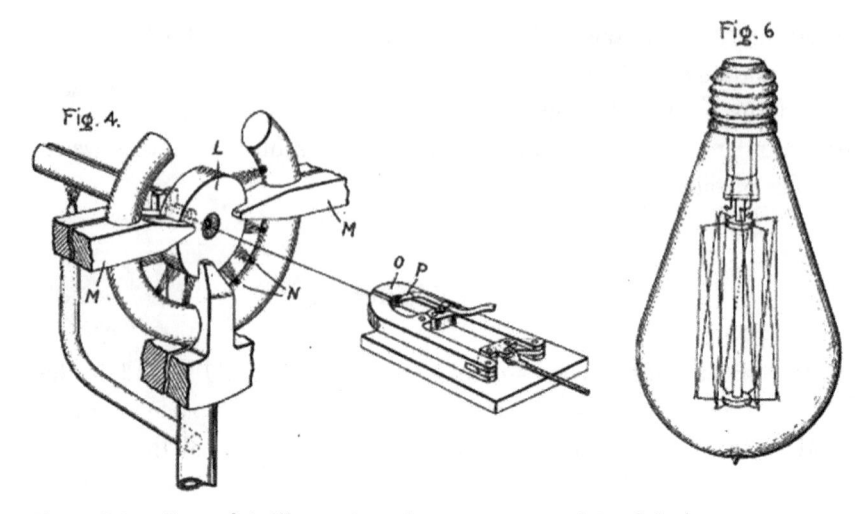

Figure I.2.6 Two of six illustrations that accompanied Coolidge's US Patent 1,082,933 for making tungsten filaments show his wire-drawing apparatus and an incandescent lamp made with the ductile metal.

unlike daylight, which peaks at 550 nanometers (nm) and whose intensity declines in both infrared and ultraviolet directions.

One more step was needed to complete the evolution of standard incandescent light: to eliminate virtually all evaporation of the filament and thus to prevent any black film deposits inside a lamp. That step was taken in April 1913, by Irving Langmuir (1881–1957), who discovered that, instead of maintaining a perfect vacuum, it is more effective to place tungsten filaments into a mixture of nitrogen and argon and, to reduce the heat loss due to convection, to coil them (US Patent 1,180,159; see Figure I.1.8). These measures raised the efficacy to 12 lm/W for common light bulbs and more for lamps of high power. Less than a quarter-century after Edison's 1879 experiments, the production of incandescent lamps was thus a mature technique, and gradual gains during the twentieth century—such as tungsten coils (introduced in 1934) that further reduced heat loss from convection—brought further gains. Efficacy rose to more than 15 lm/W for 100-W lamps, and rated life spans increased to 1,000 hours for standard sizes, but no fundamental changes were ahead (Figure I.2.5).

More efficacious lights, based on entirely different principles, also trace their origins to the pre-World War I period, but they became commercially successful only after 1950. The first discharge lamps were built by Peter

Cooper Hewitt (mercury vapor in 1900), and Georges Claude (1870–1960) demonstrated neon discharge (bright orange red) in 1910 (first neon lights were used in 1912, at a West End cinema in London). Further development of these ideas led to commercial discharge lamps: low-pressure sodium lamps were introduced in 1932, and low-pressure mercury vapor lamps, commonly known as fluorescent lights, were first patented in 1927 and came on the market during the late 1930s (Bowers 1998).

In fluorescent lamps, phosphorous compounds that coat the inside of their glass absorb ultraviolet rays generated by excitation of low-pressure mercury vapor and re-radiate them as illumination approximating daylight. The best fluorescent lights are now producing nearly 110 lm/W, converting about 15% of electricity into visible radiation—and they also last about 25 times longer than does a tungsten filament. In 1912, Charles Steinmetz experimented with metal halide compounds in mercury lamps to correct their blue-green color—and exactly 50 years later General Electric revealed its first commercial metal halide light. Although the earliest light bulbs were only as luminous as the gas jets that they were designed to replace (typically equivalent to just 16 candles or about 200 lm), their light was safer, more reliable, quiet, and less expensive, which is why they rapidly displaced even Welsbach's greatly improved gas mantle.

By 1914, 100-W tungsten light bulbs were able to produce more than 1,000 lm, and, despite the inroads made by more efficient sources, they continued to dominate the lighting market throughout the twentieth century: in the United States of the early 1990s, 87% of all lights used one or more hours per day (453 million out of a total of 523 million) were incandescent (EIA 1996). Those smallish fragile glass containers with glowing filaments were the first source of convenient and affordable light that gave us, finally, an easy mastery over darkness: they were undoubtedly the most common, most recognizable, and most beneficial inventions of the pre-World War I period that helped to create the twentieth century and keep it brightly lit.

Edison's System

The invention of commercially viable light bulbs had a great practical and symbolic importance in ushering in the electric era, but the bulbs were just at the end of the line. Edison's lasting contribution is not that he invented a light bulb (in that quest he was, as we have seen, preceded by a score of other inventors). The fundamental importance of Edison's multifaceted and, even for him, frenzied activity that took place between 1879 and 1882 is that he put in place the world's first commercial system of electricity generation, transmission, and conversion. In Hughes's (1983: 18) words, "Edison was a holistic conceptualizer and determined solver of the problems associated with the growth of systems."

Few complex technical systems have seen such a rapid transformation of ideas into a working commercial enterprise. But the first step was taken reluctantly. One of the businessmen present at the Menlo Park demonstration on December 31, 1879, was Henry Villard, the chairman of Oregon Railway & Navigation Company, who became an instant convert and persuaded a reluctant Edison, who preferred to concentrate on the development of a larger-scale urban system, to install electric lights in his latest steamship *Columbia*. The first lighting system was installed in March 1880; the ship left for Portland in May, and it arrived in Oregon after a 20-week journey around South America with its lighting in excellent condition. But this success was not followed immediately by any larger land-based installations.

This is understandable given the amount of technical, economic, and managerial challenges. Electricity had to be made cheaper than gas lighting, a mature, well-established industry with extensive generation and transmission infrastructures in place in every major Western city and with profitable operations owned by some of the leading investors of that time.

By the early 1880s, customers were used to the gently hissing sound of burning gas, but the experience was made less comfortable by the evolved heat and emissions of water vapor and carbonic acid. While electric light of 100 candles generated just 1.2 megajoules (MJ) of heat an hour and released no emissions, Argand gas burners of equivalent luminosity warmed the surroundings at the rate of about 20 MJ/hour and emitted nearly 1 kg of water vapor and nearly half a cubic meter of carbonic acid per hour of use (Anonymous 1883; Paton 1890). Moreover, there were dangers of asphyxiation or explosion from leaking gas, and, obviously, the burner could be used only in an upright position.

The greater convenience of electric lights was obvious: steady, noiseless, nonpolluting, and flexible, ready to be used in all positions. But (recall Siemens's opinion cited in Chapter I.1) not everybody agreed that the days of gas lighting were numbered, and it was very difficult to project the costs and reliability of future electric systems. That is why Edison set aside his work on incandescent filaments and began to design not only larger dynamos but also to make detailed studies of operating costs, profits, and pricing of the coal gas industry to set the goals that his new system had to meet to prevail.

Dynamos, Engines, Fixtures

After his return from a memorable trip to California in August 1878, Edison began testing two of the most successful dynamos of the day, machines by Werner Siemens (1816–1892) and Zénobe-Théophile Gramme (1826–1901) that were used to power arc lights. By that time dynamo design advanced far beyond the first, hand-cranked, toylike generator built by Hypolite Pixii in 1831 (MacLaren 1943). By far the most critical gain came in 1866–1867, with the realization that residual magnetism in electromagnets makes it possible for the generators to work even from a dead start, that is, without batteries and permanent magnets. In his memoirs, Werner Siemens (1893; Figure I.2.7) recalled his first presentation of the idea of what he named a "dynamo-electric machine" to a group of Berlin physicists in December 1866 and its publication on January 17, 1867—as well as the fact that the priority of his

Figure I.2.7 Werner Siemens, a founder of one of the world's leading makers of electric (and now also electronic) equipment, inventor of self-exciting dynamo and builder of first long-distance telegraph lines.
Portrait reproduced from Figuier (1888).

invention was immediately questioned but later confirmed. There is, however, enough evidence to conclude that Charles Wheatstone (1802–1875), Henry Wilde, and Samuel Varley discovered the same phenomenon independently during the same time.

In an 1883 lecture, Siemens's brother William (1823–1883) noted that "the essential features involved in the dynamo-machine . . . were published by their authors for the pure scientific interest attached to them without being made subject matter of letters patent," and that this situation "retarded the introduction of this class of electrical machine" because nobody showed sufficient interest in their requisite commercial development (Siemens 1882: 67). That is why Siemens gave great credit to Gramme for his initiative in using Antonio Pacinotti's 1861 idea to build a ring armature wound with many individual coils of wire insulated with bitumen and to introduce a new type of commutator.

Gramme's invention of a new machine, the *magneto-electrique produisant de courant continu*, was presented to the Académie des Sciences in Paris in July 1871 (Chauvois 1967). In contrast to Gramme's hollow cylinder, Siemens's improved dynamo had windings crossing near the center. Both machines could supply continuous current without overheating, were used to energize arc lights in an increasing number of European cities of the 1870s and were later replaced by more powerful dynamos designed by Charles Brush and Elihu Thomson. Edison began his systematic work on producing better dynamos in February 1879 by examining closely these existing designs.

Design of large dynamos needed for the contemplated urban system capable of powering more than a thousand lights began in the summer of 1880, and its most distinct feature was the direct connection to a Porter-Allen steam engine, obviating the inefficient belting. The prototype proved the basic viability of the concept. The first installation of an underground distribution network on the grounds of the Menlo Park laboratory took place in spring and summer of 1880, and the first streetlights supplied by underground conduits were lit on November 1, 1880. Edison's house was connected to the network a week later. Optimistic as ever, Edison now predicted he would have a working central station in lower Manhattan before May 1, 1881. He needed to move fast because, by December 1880, no fewer than six companies were installing arc lights in the city and because Hiram Maxim put in the first incandescent lights in the vaults and reading rooms of the Mercantile Safe Deposit Company (Bowers 1998).

But first, a new Edison Electric Illuminating Company of New York, set up by Edison's attorney Grosvenor Lowrey, had to be formed (in December 1880, with initial capital of $1 million) to conform to a state law that restricted the use of streetlights to enterprises incorporated under gas statutes. Then a deal had to be made between Edison and the company, controlled by its investors, to share the eventual profit, and the system's components had to be assembled. Armington & Sims supplied the steam engines, and Babcock & Wilcox provided the boilers, but everything else—switch boxes, fittings, sockets, wall switches, safety fuses, fuse boxes, consumption meters, and insulated underground conductors—had to be designed, tested, improved, and redesigned.

Once new factories began turning out large numbers of electric components, Edison abandoned his resistance to install small, isolated systems in individual plants or offices. In February 1881, the first installation of this kind was completed in the basement of Hinds, Ketcham & Company, a New York lithography shop. In November of that year, the Edison Company for Isolated Lighting was organized to handle new orders coming from universities, hotels, steamships, and textile mills, where the electric lights were particularly welcome to reduce the risk of fire. By the beginning of 1883, there were more than 150 isolated systems in the United States and Canada and more than 100 in Europe. But Edison's main goal was a large centralized system in Manhattan, the realization of which was falling behind the previous year's expectation. Finally, the contract for this project was signed on March 23, 1881.

The First Central Plants

Edison initially wanted to supply the entire district from Canal Street on the north to Wall Street on the south but had to scale the coverage to about 2.5 square kilometers (km^2) between Wall, Nassau, Spruce, and Ferry Streets, Peck Slip, and the East River. The location, the First District in lower Manhattan, was selected above all because of its high density of lighting needs and its proximity to New York's financial and publishing establishments. Edison's crews had surveyed the area's lighting needs as well as its requirements for mechanical power already during 1880, and then produced large maps annotated in color inks that showed the exact number of gas jets in every building, their hours of operation, and their cost (Dyer and Martin 1929).

The laying of more than 24 km of underground conductors began soon after the contract was signed, and it proceeded according to Edison's feeder-and-main system, for which he was granted a patent (US Patent 239,147) in March 1881 (Figure I.2.8). This arrangement reduced the cost of copper from the originally calculated $23.24 per lamp to just $3.72. Meanwhile, the work progressed on a larger dynamo intended first for powering Edison's exhibits at the forthcoming Paris International Electrical Exposition (Beauchamp 1997; Figure I.2.9). The massive machine (nearly 30 t) of unprecedented power (51.5 kW at 103 V) achieved public notoriety because of its nickname, Jumbo, derived from the fact that it was shipped to France on the same vessel that, on a previous voyage, transported P. T. Barnum's eponymous elephant from the London Zoo (Beauchamp 1997). By the end of 1881, the second Jumbo was installed at Edison's first operating central station in London, which began generating electricity on January 12, 1882.

Its location was chosen by Edison's London representatives to circumvent the lengthy application process for digging up the streets by running electric lines under the Holborn Viaduct (Hughes 1962). Machinery for this temporary installation was in a four-story row house (No. 57), with a 93-kW generator supplying initially 938 lamps, including 164 streetlamps along 800 m of the viaduct and lights in the General Post Office, the City Temple, and businesses along the street with direct current of 110 V. A second dynamo was added by April 1881, and the station remained in operation until 1884. But Edison's plans for a large central plant in London were thwarted by the Electric Lighting Act of the British Parliament, which made such installations impossible until it was revised in 1888.

Just as the Manhattan station was undergoing a series of final tests, Edison's first American small hydroelectric station, powered by a 107-cm waterwheel, was readied for service on the Fox River in Appleton, Wisconsin. Two small dynamos rated at a total of 25 kW were housed in a wooden shed, and they powered 280 weak lights, but the sturdy installation, ordered by the town's paper manufacturer H. F. Rogers, was in full operation by September 30, 1882, and it worked until 1899 (Dyer and Martin 1929). The first small English hydro-generating plant was opened during the previous year in Godalming by the Siemens brothers, whose company installed a small water turbine in the River Wey that operated only until 1884 (Electricity Council 1973).

The Manhattan station was to be in an entirely different class. Edison's property search ended with two houses, 255 and 257 Pearl Street, both much

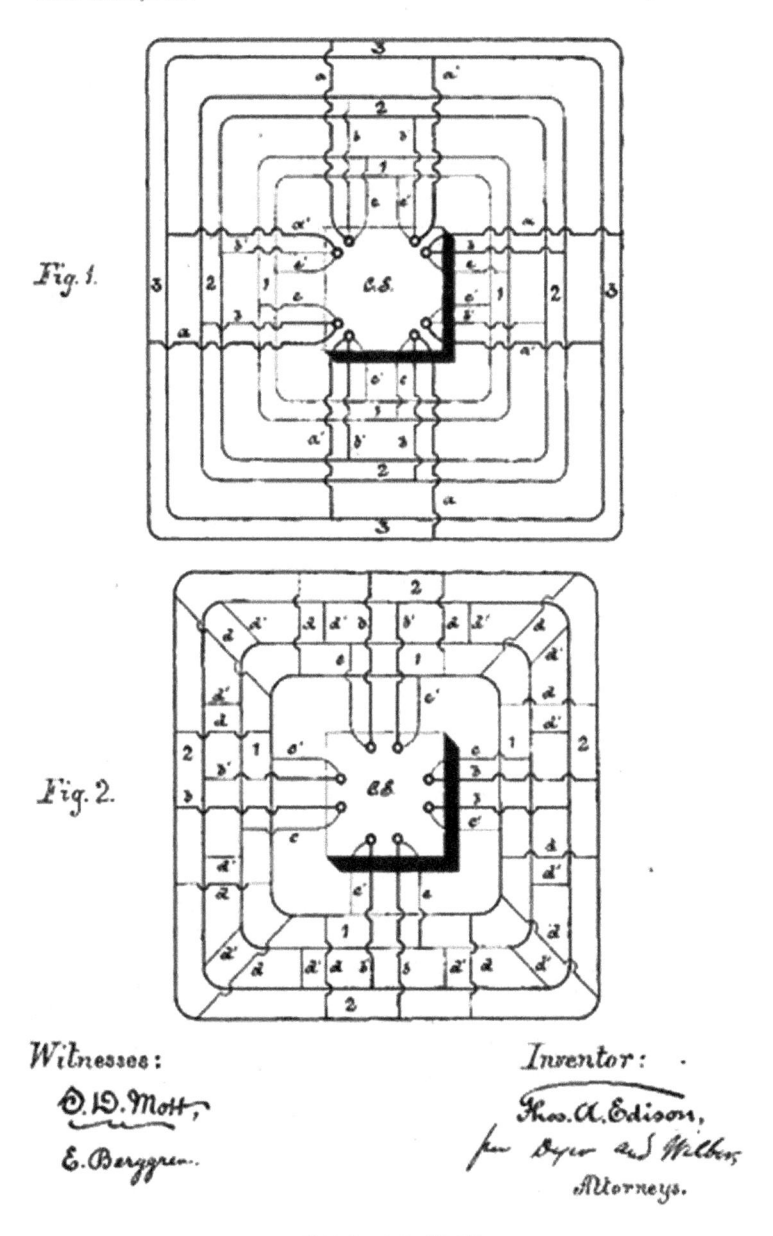

Figure I.2.8 Drawing attached to Edison's US Patent 239,147 for a system of electric lighting (feeder and main arrangement).
Reproduced from Edison (1881).

more expensive than he anticipated. The station's heavy equipment was put in No. 257, coal (and ash) was stored in the basement, four Babcock & Wilcox boilers (about 180 kW each) sat on the ground floor, and six Porter-Allen engines (each rated at 94 kW) and six large direct-connected Jumbo dynamos sat on the reinforced second floor (Martin 1922; Figures I.2.9 and I.2.10). The neighboring house, No. 255, was used for material storage and as a repair shop and, frequently, as a dormitory for Edison and his crew working on the station (Israel 1998).

Even before all connections were completed, Edison turned on the first light in J. P. Morgan's office at 3 p.m. on September 4, 1882: just a single dynamo was online, and it supplied about 400 lights. When the time came to engage the second machine, the usually confident inventor admitted that he was extremely nervous as he scheduled a Sunday test—and, as he recalled later, his concerns were justified: "One engine would stop and the other would run up to a thousand revolutions; and then they would seesaw. The trouble was with the governors. . . . I grabbed the throttle of one engine and E. H. Johnson, who was the only one present to keep his wits, caught hold of the other, and we shut them off" (Edison, cited in Martin 1922: 56).

Figure I.2.9 Edison's Jumbo, a massive dynamo that was first displayed in Paris in 1881.
Reproduced from *Scientific American*, December 10, 1881.

Figure I.2.10 Dynamo room of Edison's New York Pearl Street Station with six directly driven Jumbos.
Reproduced from *Scientific American*, August 26, 1882.

By the end of 1882, three more Jumbos were added at the Pearl Street station, whose output was lighting more than 5,000 lamps, and, in January 1883, the company began charging for its electricity using Edison's first ingenious but cumbersome electrolytic meters. Edison had every reason to be pleased. As he noted in 1904, in an article he wrote for *Electrical World and Engineer*, "As I now look back, I sometimes wonder at how much was done in so short a time" (cited in Jehl 1937: 310). The scope of his early work during those years is perhaps best revealed by the patents he obtained between 1880 and 1882, in addition to nearly 90 patents on incandescent filaments and lamps: 60 patents dealing with "magneto or dynamo-electric machine" and its regulation, 14 patents for the system of electric lighting, a dozen patents concerning the distribution of electricity, and 10 patents for electric meters and motors (TAEP 2024).

Higher than anticipated costs and frequent dynamo outages meant that the company lost money in 1883 ($4,457.50), but it made a good profit ($35,554.79) in 1884 (Martin 1922). Its operation provided a unique learning opportunity and served as an irreplaceable advertisement for the new system. One of the first adjustments Edison made soon after starting

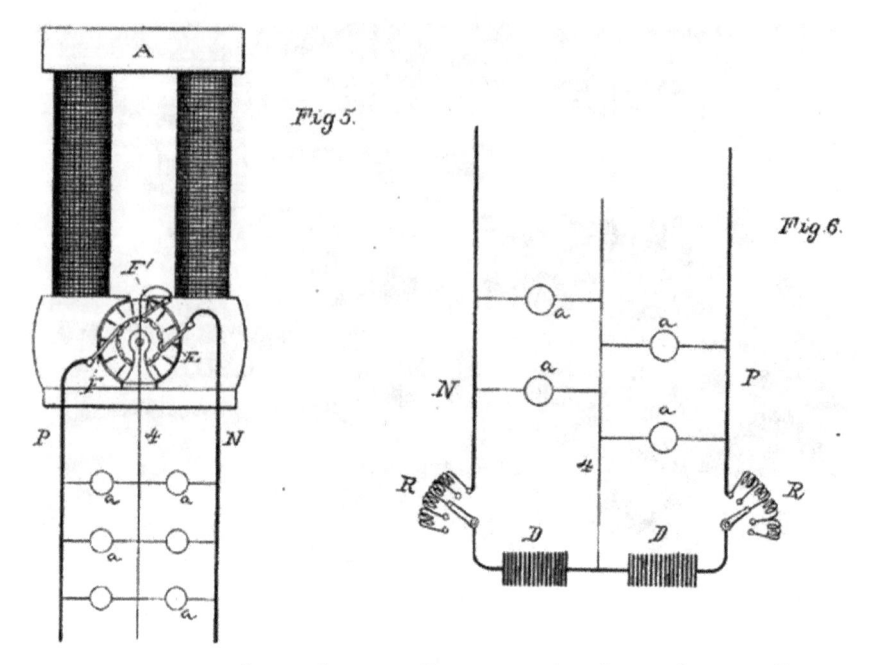

Figure I.2.11 Two of seven drawings illustrating Edison's specification of his US Patent 274,290 for three-wire electrical distribution. The left image shows wiring attached to a dynamo; the right, to secondary batteries.
Reproduced from Edison (1883).

the Pearl Street station, on November 27, 1882, was to file a patent for the three-wire distribution system (US Patent 274,290), an arrangement that was independently devised in England by John Hopkinson (1849–1898) a few months earlier and that remains the standard in our electric circuits.

Edison's first central station with three-wire distribution was completed in October 1883, in Brockton, Massachusetts. This configuration saved about two-thirds of the copper mass compared to the two-wire conduit used in Manhattan (Figure I.2.11). While the feeder-and-main system is still with us, three-wire DC transmission was soon superseded, both in its maximum spatial reach and its unit cost, by the transmission of three-phase and single-phase AC, which was initially shunned by Edison but strongly favored by George Westinghouse, Nikola Tesla, Sebastian de Ferranti, and Steinmetz. By 1884, the Pearl Street station was serving more than 10,000 lamps, and its success led to a rapid diffusion of similar installations for which the Edison Company had a virtual monopoly during most of the 1880s.

By 1891, more than 1,300 central Edison plants were in operation in the United States, supplying about 3 million lights. Fire that damaged the Pearl Street station in 1890 destroyed all but one of the Jumbos, No. 9; in 1893, it was moved to the Columbian Exposition in Chicago, and it was eventually rebuilt and ended up in the Ford Museum in Greenfield Village near Dearborn, Michigan (Sinnott and Bowditch 1980; ASME 2024). Edison's role in the genesis of the electric era cannot be overestimated. He was able to identify key technical challenges, resolve them by tenacious interdisciplinary research and development, and translate the resulting innovations into commercial use.

There were other contemporary inventors of light bulbs and dynamos, but only Edison had the vision of a complete system as well as the determination and organizational talent to make the entire system work, and he and his coworkers translated his ideas into realities in an astonishingly short period. One of the greatest tributes paid to this work came from Emil Rathenau, one of the creators of Germany's electric industry and the founder of Allgemeine Elektrizitäts Gesselschaft. Rathenau (quoted in Dyer and Martin 1929: 318–319) recalled his impressions after seeing Edison's display at the Paris Electrical Exhibition of 1881.

> The Edison system of lighting was as beautifully conceived down to the very details, and as thoroughly worked out as if it had been tested for decades in various towns. Neither sockets, switches, fuses, lamp-holders, nor any of the other accessories necessary to complete the installation were wanting; and the generating of the current, the regulation, the wiring with distribution boxes, house connections, meters, etc., all showed signs of astonishing skill and incomparable genius.

This informed praise alone negates many derisory remarks that Edison's antagonists and critics produced since the very beginning of the electric era. Edison was, without any doubt and to resort to modern parlance, an extraordinary systems thinker, but it seems to me that Friedel and Israel (1986: 227) captured best the essence of his achievements by noting that "the completeness of that system was more the product of opportunities afforded by technical accomplishments and financial resources than the outcome of a purposeful systems approach." This more subtle interpretation does not change the basic facts. Edison was an exceptionally inventive, ambitious, and confident man who awed, motivated, inspired, and alienated his coworkers and financial backers. The combination of unusual insights, irrepressible enthusiasm, and the flair for (self)promotion led him often to voice exaggerated expectations and to make impossible promises.

But obstacles and setbacks that derailed his overoptimistic schedule served only to fortify his determination to overcome them and to come up with better solutions. His devotion to the pursuit of invention was legendary, his physical stamina incredible. What is not perhaps stressed enough is that the financial support he received from his backers (including some of the richest men of his time) and the skills and talents of his craftsmen made it possible for him to explore freely so many ideas and possibilities. In that sense, the Menlo Park laboratory was a precursor of the great corporate R&D institutions of the twentieth century.

In retrospect, there is no doubt that the combination of Edison's inventions or radical improvements of several key components of the emerging electric system and his indefatigable push for its commercialization have been his greatest legacy. His were epoch-making achievements, but while all electric networks are conceptual descendants of Edison's system of centralized electricity supply, their technical particulars as well as their operational arrangements had changed substantially before the end of the nineteenth century. Unlike the remarkably durable specifications and external features of Edison's incandescent light bulbs, Edison's original electric system was not such a resilient survivor—and it did not deserve to be.

Above all, Edison's first station was a very inefficient operation: its heat rate was about 146 MJ/kWh, which means that it was converting less than 2.5% of the coal it burned into electricity. By 1900, the two key components of the electricity-generating system were different: steam turbines connected directly to alternators, rather than steam engines coupled with dynamos, became the preferred prime movers, and increasingly higher voltages of AC, rather than relatively low voltages of DC, distributed the generated electricity. On the consumption side, improved incandescent lights still used a large share of the generated electricity, but electric motors in factories and in urban transportation were rapidly becoming the largest consumers.

Turbines, Transformers, Motors

During the remainder of the 1880s, the new electric industry saw advances that were no less fundamental than those made during the first three years of its development, when it was dominated by Edison's quest to introduce a profitable commercial system. These changes affected every stage of the innovation process as they transformed the generation of electricity, its transmission, and its final uses. Edison became an increasingly marginal player, and the greatest acclaim must deservedly go to two engineers born in the mid-1850s: Charles Algernon Parsons (1854–1931) for his invention of steam turbogenerator, and Nikola Tesla (1856–1943) for his development of polyphase electric motor. Although William Stanley's (1858–1916) contributions to the design of transformers stand out, the genesis of that device was much more a matter of gradual refinements that were introduced by nearly a dozen engineers in several countries.

Every one of these three great innovations made a fundamental difference for the future of large-scale electricity generation, and their combination amounted to a fundamental transformation of what was still a nascent industry. By 1900, these advances came to define the performance of modern electric enterprises, determining the size and efficiency of turbogenerators, the location and layout of stations, the choice of switchgear, and transmission and distribution arrangements. Without these innovations, it would have been impossible to reach the magnitude of generation and hence the economies of scale that made electricity one of the greatest bargains of the twentieth century.

These techniques were later transformed by a series of incremental innovations whose pace had notably accelerated during the post-World War II economic expansion and then began to approach some inevitable performance plateaus. Consequently, all components and all processes of the modern electric industry became more efficient and more reliable than they were in 1914, yet they demanded less material per unit of installed capacity, were more economical, and their operation has reduced impacts on the environment. At the same time, there is no mistaking their 1880s pedigree.

When Edison began to outline his bold plan for centralized electricity generation for large cities, the steam engine was the only available prime mover. During the 1880s, design of these machines was a mature art (Thurston 1878; Ewing 1911; Dickinson 1939), but they were hardly the best prime movers for large-scale electricity generation. After more than a century of development, their best efficiencies reached about 15%, but the typical sustained performances of large machines were much lower. And although the mass-to-power ratio for the best designs fell from more than 500 g/W in 1800 to about 150 g/W by the 1880s, they remained very heavy.

In addition to relatively low efficiency, large mass, and restricted capacity, the top speeds of steam engines were inherently limited due to their reciprocating motion. Common piston speeds were below 100 m/min, and, while the best marine engines of the 1880s could reach up to 300 m/min, they worked normally at 150–180 m/min (Ewing 1911). Edison's first generators installed in the experimental station in Menlo Park in 1880 were driven, awkwardly and unreliably, by belts attached to steam engines. This arrangement was soon substituted by directly connected high-speed engines used in Edison's first installations in 1881. A Porter-Allen machine could run at 600 revolutions per minute (rpm); an Armington & Sims engine for the Paris exhibition was rated at about 50 kW and ran at 350 rpm (Friedel and Israel 1986).

Simple calculations reveal the limitations inherent in these maxima. Imagine an urban system that does not serve any electric motors but only 1 million lights (merely six lights per household for 100,000 families and another 400,000 public, office, and factory lamps) whose average power is just 60 W. Such a system would require (assuming 10% distribution losses) 66 MW of electricity. With the best dynamos converting about 75% of mechanical energy into electricity, this would call for steam engines rated at 88 MW. Those at Edison's Pearl Street station had a total capacity of nearly 3.4 MW, which means that such a dim and frugal lighting system for a city of 400,000 people would have required 26 generating stations. Their load factor would have been low (no more than 35%) because they had only one kind of final use, but even with a relatively high efficiency of 20% they would have consumed annually about 250,000 t of steam coal.

This was, indeed, the inevitable setup of the earliest urban electric systems: they served just limited parts of large cities, and hence the generating companies had to bring large quantities of coal into densely populated areas and burn the coal there. Switching from DC to AC would have allowed for efficient transmission from peri-urban locations, but it would not have changed the required number of generating units. Moreover, this setup was so inefficient that, were the entire US population (about 57 million people in 1885) to be served by such low intensity systems, they would have consumed the country's total 1885 production of bituminous coal! Obviously, a more powerful but a much less massive and more efficient prime mover to turn the dynamos was needed.

Steam Turbines: Laval, Parsons, Curtis

Water-driven turbines offered such an option since the 1830s, when Benoit Fourneyron introduced his effective designs, but, before the 1880s, there were not even any serious attempts made to develop an analogical machine driven by steam. The first successful design was introduced by a Swedish engineer, Carl Gustaf Patrick de Laval (1845–1913), whose most notable previous invention was a centrifugal cream separator that he patented in 1878 (Smith 1954). His impulse steam turbine, introduced in 1882, extracted steam's kinetic energy by releasing it from trumpet-shaped nozzles on a rotor with appropriately angled blades. Such a machine is subjected to high rotational speeds and huge centrifugal forces. Given the materials of the 1880s, it was possible to build only machines of limited capacity and to run them at no more than two-thirds of the speed needed for their best efficiency. Even then, helical gearing was needed to reduce the excessive speed of rotation.

Blades in Laval's turbine are turned solely by impulse: there is no drop in pressure as the steam passes the moving parts, and its velocity relative to the moving surfaces does not change except for the inevitable friction. Charles Parsons (Figure I.2.12) chose a very different approach in designing his successful steam turbine. As he explained in his Rede lecture,

Figure I.2.12 There is no portrait of Charles Algernon Parsons dating from the early 1880s when he invented his steam turbine. This portrait (oil on canvas) was painted by William Orpen in 1922.
Reproduced here courtesy of the Tyne and Wear Museums, Newcastle upon Tyne.

It seemed to me that moderate surface velocities and speeds of rotation were essential if the turbine motor was to receive general acceptance as a prime mover. I therefore decided to split up the fall in pressure of the steam into small fractional expansion over a large number of turbines in series, so that the velocity of the steam nowhere should be great.... I was also anxious to avoid the well-known cutting action on metal of steam at high velocity. (Parsons 1911: 2)

Consequently, the steam moved in an almost nonexpansive manner through each one of many individual stages (there can be as many as 200), akin to water in hydraulic turbines whose high efficiency Parsons aspired to match. Parsons filed the key British patent (GB Patent 6,735) on April 23, 1884; the US application for a rotary motor was filed on November 14; and US Patent 328,710 was granted in October 1885. He proceeded immediately to build his first compound turbine. This machine (now preserved in the lobby of the Parsons's Building at Trinity College in Dublin) was rated at just 7.5 kW, produced DC at 100 V and 75 A, ran at 18,000 rpm, and had an efficiency of a mere 1.6%. Many improvements followed during the decades before World War I (Parsons 1936).

Improvements concentrated above all on the design of blades and overcoming the challenges of building a machine composed of thousands of parts that had to perform faultlessly while moving at high speeds in a high-pressure and high-temperature environment and do so within extremely narrow tolerances as clearances over the tips of the moving blades are less than 0.5 mm (see Figure I.1.5). Parsons also had to design a dynamo that could withstand high speeds and resist the great centrifugal force and solve problems associated with lubrication and controls for the machine. The first small single-phase turbo-alternators for public electricity generation were ordered in 1888, and the first two of the four 75-kW, 4,800-rpm machines began operating at Forth Banks station, in Newcastle upon Tyne, in January 1890. Their consumption was 25 kg of steam per kilowatt-hour at a pressure of 0.4 MPa (or a conversion efficiency of just over 5%), and they supplied AC at 1 kV and 80 Hz.

In 1889, Parsons lost the patent right to his parallel flow turbine when he left Clarke, Chapman & Company, established C. A. Parsons & Company, and turned to designs of radial flow machines that proved to be less efficient. A 100-kW, 0.69-MPa turbo-alternator that his new company supplied to Cambridge Electric Lighting in 1891 was Parsons's first condensing turbine.

Figure I.2.13 Engraving of Parsons's first 1-MW turbogenerator (steam turbine and alternator) installed in 1900 at Elberfeld plant in Germany.
Reproduced from *Scientific American*, April 27, 1901.

Parsons's small turbines of the 1880s were noncondensing, exhausting steam against atmospheric pressure and hence inherently less efficient. The condensing unit was also the first one to work with superheated steam (its temperature rose above that corresponding to saturation at the actual pressure), albeit only by 10°C. In 1894, Parsons recovered his original patents (by paying £1,500 to his former employers) and proceeded to design not only larger stationary units but also turbines for marine propulsion.

A rapid increase of capabilities then led to the world's first 1-MW units: two of them, with the first tandem turbine-alternator arrangement, were built by C. A. Parsons & Company in 1899, for the Elberfeld station in Germany, to generate single-phase current at 4 kV and 50 Hz (Figure I.2.13). The first 2-MW (6 kV, three-phase AC at 40 Hz) turbine was installed at the Neptune Bank station in 1903, and Parsons's designs reached a new mark with a 5-MW turbine connected in 1907, in Newcastle upon Tyne (Parsons 1911). That turbine converted coal to electricity with about 22% efficiency. Parsons largest pre-World War I machine was a 25-MW turbo-alternator, installed in 1912, at the Fisk Street station of the Commonwealth Edison Company in Chicago.

American development of steam turbines did not lag far behind the British advances (MacLaren 1943; Bannister and Silvestri 1989; GE 2024). George Westinghouse bought the rights to Parsons's machines in 1895, built his company's first turbine in 1897, and, by 1900, delivered a 1.5-MW machine to the Hartford Electric Light Company. In 1897, General Electric made an agreement with Charles Curtis (1860–1953), who patented a new turbine concept a year earlier. This turbine could be seen as a hybrid of Laval's and Parsons's designs (Ewing 1911). Curtis machines eventually reached capacities of up to 9 MW, and they were installed at many smaller plants throughout the United States during the first two decades of the twentieth century.

General Electric's first turbine, a 500-kW unit installed in Schenectady, New York, in 1901, had a horizontal shaft, but its second machine of the same size was vertical, and the company soon began offering vertical Curtis turbines with capacities of up to 5 MW. The third American company that advanced the early development of steam turbine was Allis-Chalmers, originally also a licensee of Parsons's machines. And during the first decade of the twentieth century, Swiss Brown Boveri Corporation became the leading producer of turbines in Europe (Rinderknecht 1966). Consequently, it took less than two decades after the first commercial installation to establish steam turbines as machines that were in every respect superior to steam engines. Their two most desirable features were a large range of sizes and unprecedented generation efficiencies. In just 20 years, the size of the largest machines rose from 100 kW to 25 MW, a 250-fold increase, and, by 1914, it was clear that it would be possible to eventually build sets of hundreds of megawatts (MW).

Parsons's (1911) figures show rapid gains of thermal efficiency, from about 2% for his 1884 model turbine, to about 11% for the first low-superheat 100-kW machine in 1892, and to nearly 22% for the 5-MW, higher superheat design in 1907. The efficiency for a 20-MW turbine installed in 1913 was just over 25% (Dalby 1920): an order of magnitude gain in efficiency in three decades was clearly a very steep learning curve. In contrast, British marine engine trials of the best triple- and quadruple-expansion steam engines conducted between 1890 and 1904 showed maximum thermal efficiencies of 11–17% (Dalby 1920). And even Parsons's pioneering 100-kW turbine, built in 1891, weighed only 40 g/W, or less than 20% of the best comparable steam engines; that ratio fell by 1914 below 10 g/W, and it was just over 1 W/g for the largest machines built after the mid-1960s (Hossli 1969).

Naturally, these declining power-to-mass ratios, especially pronounced with larger units (Hunter and Bryant 1991), translated into great savings in metal consumption and in much lower manufacturing costs per unit of installed power. Giant steam engines also required additional construction costs for the enormous buildings needed to house them. Direct delivery of rotation power rather than awkward arrangements for converting reciprocating motion was the most obvious mechanical advantage, one that was

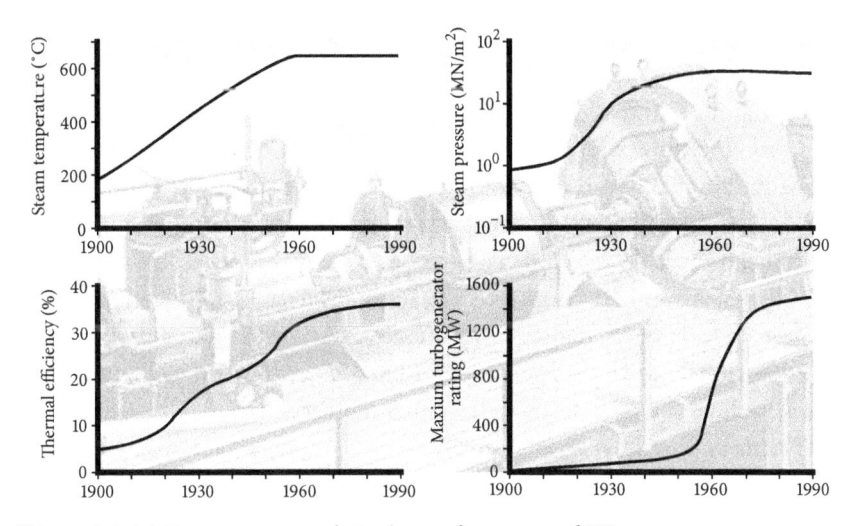

Figure I.2.14 Long-term trends in the performance of US steam turbogenerators show higher steam temperature and pressure, improved average efficiency, and rising unit power.

also largely responsible for lack of vibration. The twentieth century had witnessed a remarkable rise of turbine specifications very much along the lines that were set by Parsons during the first two decades of turbine development (Termuehlen 2001).

Technical advances have been impressive. By 1940, steam pressures rose by an order of magnitude, steam temperatures nearly tripled, and the highest ratings reached 1 gigawatt (GW) by the late 1960s and eventually leveled off at 1.5 GW (Figure I.2.14). By 2000, large turbines rotated at up to 3,600 rpm and worked under pressures of 14–34 megapascals (MPa) and temperatures up to just above 600°C. Three-phase alternators were the world's largest continuous energy converters, and the highest conversion efficiencies of the late twentieth-century fossil-fueled plants were up to 40–43% (compared to just 4% in 1890). In 2000, about three-quarters of the world's electricity was generated by steam turbines. Consequently, even with relatively rapid additions of new photovoltaics and wind-turbine capacities, there is little doubt that Parsons's machines will remain indispensable producers of the world's electricity at least throughout the first half of the twenty-first century.

Before leaving the story of steam turbines, just a few paragraphs on their marine applications. Parsons's 1894 prospectus seeking the necessary capital for the new Marine Steam Turbine Company claimed, quite correctly, that the new system would revolutionize the current method of using steam as a motive power. The experimental vessel *Turbinia*—30 m long, displacing 40 t, and driven by a single radial flow turbine capable of 715 kW—became the fastest vessel in the world, reaching 34.5 knots, 4 knots above the fastest British destroyers powered by compound steam engines (Osler 1981). A public display of *Turbinia*'s speed at a grand Naval Review at Spithead on June 26, 1897, convinced the Royal Navy to order its first turbine-driven destroyer in 1898. Six years later, 26 naval ships, including the massive *Dreadnought*, had Parsons direct-drive turbines, as did soon four of the world's largest passenger ships—*Mauretania, Lusitania, Olympic*, and *Titanic*, icons of the Golden Age of trans-Atlantic shipping of the early twentieth century.

Diesel engines or gas turbines now power most cargo and passenger ships, but steam turbines can still be found on vessels ranging from the largest aircraft carriers to tankers carrying liquefied natural gas. Vessels of the Nimitz class, the latest series of the US nuclear carriers, have two pressurized water reactors that power four geared steam turbines capable of 194 MW (DCS Documentation 2024). Steam turbines also drive centrifugal pumps and compressors, including those employed during the Haber-Bosch synthesis of ammonia. And the commercial success of steam turbines led to the search for practical gas turbines whose remarkable rise will be detailed in this book's companion volume.

Transformers

I cannot think of another component of the electric system—indeed, of another device—whose ubiquitous service is so essential for the continuous functioning of modern civilization, yet that would be so absent from public consciousness as the transformer. Coltman (1988) is correct when he ascribes this lack of recognition to the combination of the transformer's key outward attributes: it does not move, it is almost completely silent, and it is usually hidden (underground, inside buildings, behind screens, or boxed in: plates protruding from large transformer boxes are external radiators used for cooling the device). Yet without these ingenious devices, we would be stuck in the early Edisonian age of electricity with limited transmission distances and with a rather high density of generating stations in urban areas and with smaller places relying on isolated generating systems.

This prospect clearly worried the early proponents of electricity. William Siemens (1882: 70) was concerned that "the extension of a district beyond the quarter of a square mile limit would necessitate an establishment of unwieldy dimensions." This would mean not only a large increase in the total cost of electric conductors per unit area but also that a "great public inconvenience would arise in consequence of the number and dimensions of the electric conductors, which could no longer be accommodated in narrow channels placed below the kerb stones but would necessitate the construction of costly subways—veritable cava electrica."

Electricity is easiest to generate and most convenient to use at low voltages, but it is best transmitted over long distances with the least possible losses at high voltages because the power loss in transmission varies as the square of the current transmitted through the wires. Consequently, switching to the higher voltages of AC reduced the dimensions of conductors, but such high voltages could not have been produced for transmission and then reduced for final use in households and offices without transformers. These devices convert one electric current into another by either reducing or increasing the voltage of the input flow. What is most welcome from an engineering point of view is that transformers do this with virtually no loss of energy and that the conversion works effectively across an enormous range of voltages (Coltman 1988; Del Vecchio et al. 2017).

Basic equations quantify the advantage of a system with transformers. The rate of transmitted electricity is equal to the product of its current and voltage (W = AV), and because voltage equals current multiplied by resistance (Ohm's law, V = AΩ), power is the product of $A^2Ω$. Consequently, transmitting the same amount of power with 100 times higher voltage will result in cutting the current by 99% and reducing the resistance losses also by 99%: the combination of low current and high voltage is always the best choice for long-distance transmission. Transformers step up the generated current for transmission and then step it down for local distribution and actual use. At that point, many gadgets may need further voltage reduction and current conversion; for example, 120 V AC must be transformed into 20 V DC to run the Dell laptop on which this book was revised.

All transformers work by electromagnetic induction, whose existence was demonstrated for the first time in 1831, by Michael Faraday. A loop of wire carrying AC (the primary winding) generates a fluctuating magnetic field that will induce a voltage in another loop (the secondary winding) placed in the field. In turn, the current flowing in the secondary coil will induce a

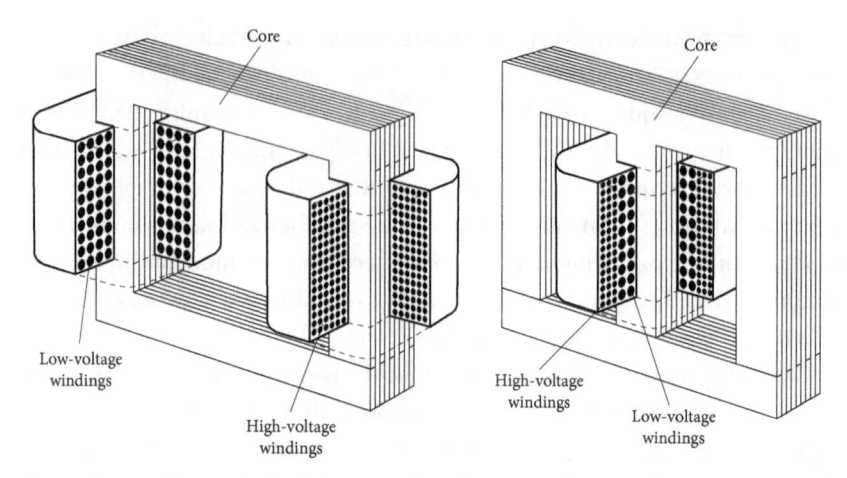

Figure I.2.15 Two principal designs of modern transformers that are made with laminated iron sheets. In the core form, on the left, the two circuits enclose separate arms of the transformer, while in the E-shaped shell form both the primary and the secondary coils envelope the central bar.

voltage in the primary loop, and both loops will also produce self-induction (Figure I.2.15). Because the total voltage induced in a loop is proportional to the total number of its turns (if the secondary has three times the number of turns of the primary, it will supply tripled voltage), and because the rate at which the energy is being transformed (W) must equal the product of its voltage and its current (VA), the product of the secondary voltage and the secondary current (negligible losses aside) must equal the product of the primary voltage and the primary current. When Faraday wound two coils on an iron ring, rather than on a wooden one as in his initial induction experiments, he brought together the essential elements of the modern transformer.

The other important reason that transformers do not occupy a prominent position in the history of nineteenth-century invention is the absence of a single protean mind behind their introduction—no eureka-type discovery—just gradual improvements based on Faraday's experiments (Fleming 1901; MacLaren 1943). The need for transformers arose with the first tentative steps toward the AC system. Lucien H. Gaulard (1850–1888) and John D. Gibbs got the first patents for its realization in 1882, and they displayed their transformer (they called it a "secondary generator") in 1883, at the Electrical Exhibition in London. To add incandescent light bulbs to an AC system that supplied arc lights connected in series (turned on and off at the same time), they used a two-coil transforming device.

Its two windings were made of flat copper sheet rings inserted on a cast iron wire core and insulated one from another by paper and varnish. A year later Gaulard and Gibbs staged another demonstration of this inefficient model at the international exhibition in Torino, Italy. In 1884, three Hungarian engineers employed at Ganz Works in Budapest improved the Gaulard-Gibbs device by designing two kinds of transformers for parallel connection to a generator and making them more efficient (Németh 1996). The key innovation—using closed iron cores that work much better than do the open-ended bundles—was suggested by Otto Titusz Blathy (1860–1939), the youngest engineer in the group. Shunt connection was the idea of Karoly Zipernowsky (1853–1942), head of the electrical section of Ganz Works; and the experiments were performed by Miksa Déri (1854–1938). These transformers (later called "ZBD transformers") were shown first in May 1885, at the Hungarian National Exhibition in Budapest.

George Westinghouse purchased the rights to the Gaulard-Gibbs design and took options on ZBD transformers as well. Edison's distrust of AC systems led him to ignore these developments, and transformers were the only fundamental electric device to whose design he did not make any significant contribution. In contrast, William Stanley, a young engineer working for Westinghouse, began designing improved transformers (he called them "converters") in 1883. By December 1885, he had a model that was much less expensive to make than the ZBD device and that became the prototype of devices we still use today (Stanley 1912; Hawkins 1951). In the first patented version (US Patent 349,611 of September 21, 1886; three drawings on the left in Figure I.2.16), the soft iron core, encircled by primary (b1) and secondary (b2) coils, was either annular or rectangular with curved angles and could consist either of a single piece of metal or wires or strips.

But, as shown in a patent drawing filed by George Westinghouse in December 1886 (US Patent 366,362), the preferred composition of the core soon changed to thin plates of soft iron constructed with two rectangular openings and separated from each other, individually or in pairs, by insulating material (two drawings on the right in Figure I.2.16). From this design, it was a short step to stacked laminations stamped out from iron sheets in the form of the letter E, making it easy to slide pre-wound copper coils in place and then to lay straight iron pieces across the prongs to close the magnetic circuit; these became the most obvious marks of Stanley-type transformers

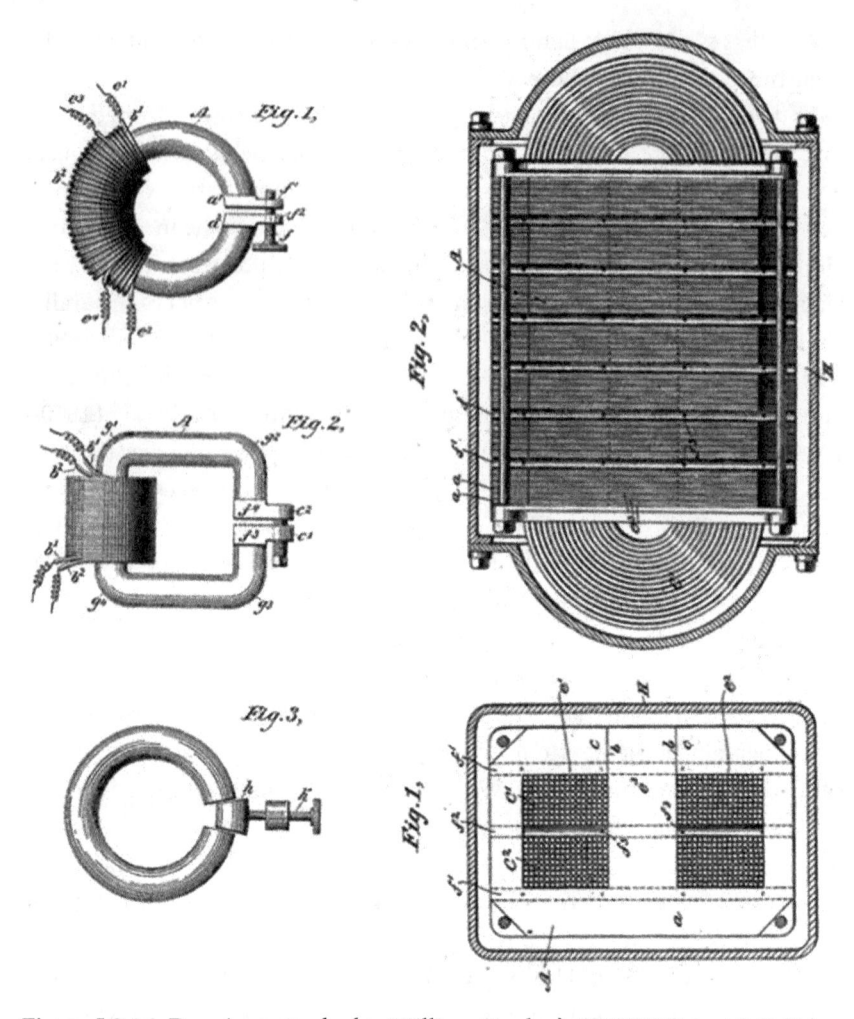

Figure I.2.16 Drawings attached to William Stanley's 1886 US Patent 349,611 for induction coil, a prototype of modern transformers (left), and to US Patent 366,362 filed by George Westinghouse in December 1886. Stanley's device has an annular core; the later design has the core (A) made of thin insulated plates of soft iron with two rectangular openings.

(Coltman 1988; Figure I.2.15). Soon after the Westinghouse Electric Company was incorporated in January of 1886, Westinghouse himself took out patents for improved designs of induction coils and transformers, including both the air-cooled and oil-insulated devices (both are still the standard practice today).

In 1890, Sebastian Ferranti designed the largest devices in operation for London's Deptford station. Electricity was generated at 2.5 kV, stepped up to 10 kV for more than 11 km transmission by underground cables, and then stepped down by transformers in the central London area to the 2.4 kV required for the distribution system (Electricity Council 1973). A big increase in transformer performance was needed by 1895, to accommodate transmission from the Niagara Falls hydroelectric station. Electricity generation, based on Tesla's polyphase AC concept, was at 5 kV and 25 Hz, and three-phase transmission to Buffalo carried the current at 11 kV. General Electric built two 200-kVA transformers for the aluminum plant of the Pittsburgh Reduction Company and a 750-kVA transformer for the Carborundum Company (MacLaren 1943). By 1900, the largest transformers were rated at 2 MVA and could handle inputs of 50 kV. But these devices operated with considerable hysteresis and eddy current losses.

Hysteresis is the memory effect of magnetic materials that weakens the transforming capacity by delaying magnetic response, and *eddy currents* are induced in metals with low resistivity. Both phenomena dissipate large amounts of energy, and they made the early transformer cores, made of pure iron, relatively inefficient. Eddy currents can be reduced by laminating the core, and, in 1903, English metallurgist Robert A. Hadfield (1858–1940) discovered that silicon steel greatly increased its resistivity while leaving the magnetic properties largely intact. Moreover, with the same maximum flux density silicon steel has hysteresis losses 75% lower than does sheet iron, and it is not subject to aging, whereby the hysteresis loss can double with time when the metal is exposed to higher temperatures (>65°C) or mechanical fatigue. All transformer cores, as well as parts of rotating machinery subject to alternating fields, are made of silicon steel.

Nearly 150 years after its invention, the transformer remains little more than an artfully assembled and well-cooled bundle of iron sheets and copper coils. But, as with so many other pre-World War I inventions, this conservation of a fundamentally ideal design has been accompanied by some very impressive gains in typical performance, including better cooling and insulation introduced gradually since the 1930s: low-power devices (up to 50 kVA) are still cooled by natural air flow, larger ones have forced air circulation, and those above 200 kVA are usually immersed in mineral oil in a steel tank (Wildi 1981). Just before World War I, the largest transformers could work with inputs of up to 150 kV and power of 15 MW; by 2025, the largest transformers rated at 587.1 MVA and could accommodate currents up to 1,100 kV.

But Stanley's (1912: 573) appraisal is as correct today as it was when he confessed in 1912 to a meeting of American Institute of Electrical Engineers to

> a very personal affection for a transformer. It is such a complete and simple solution for a difficult problem. It so puts to shame all mechanical attempts at regulation. It handles with such ease, certainty, and economy vast loads of energy that are instantly given to or taken from it. It is so reliable, strong, and certain. In this mingled steel and copper, extraordinary forces are so nicely balanced as to be almost unsuspected.

First Electric Motors

The history of electric motors resembles that of incandescent lights: in both cases, the eventual introduction of commercially viable devices was preceded by a long period of experimental designs and, in neither case, did this unsystematic effort lead directly to a successful solution of an engineering challenge. Tesla's contributions to the introduction and rapid diffusion of AC electric motors were no less important than Edison's efforts to commercialize incandescent light. The two most common converters of electricity, lights and motors, supply two very different final demands, but, as far as economic productivity is concerned, there is no doubt that during the twentieth-century it was revolutionized even more by electric motors than it was by electric lights.

The electric motor is fundamentally a generator working in reverse, and so it is not surprising that the first attempts to harness the changing electromagnetic field for motion date to the 1830s, to the years immediately following Faraday's fundamental experiments. There are many studies describing these earliest attempts as well as later developments preceding the first DC motors during the 1870s and the invention of induction machines during the latter half of the 1880s (Bailey 1911; Hunter and Bryant 1991; Pohl and Muller 1984). The most remarkable pioneering efforts were those of Thomas Davenport in Vermont, in 1837 (US Patent 132) and contemporaneous designs (beginning in 1834) by M. H. Jacobi in St. Petersburg. Davenport used his motors to drill iron and steel parts and to machine hardwood in his workshop, while Jacobi used his motors to power the paddle wheels of a boat carrying 10–12 people on the Neva River.

Both inventors had high hopes for their machines, but the cost and durability of the batteries used to energize those motors led to early termination of their trials. A similar experience met Robert Davidson's light railway car that traveled on some British railroads in 1843, and heavy motors built during early 1850s (with considerable congressional funding) by Charles Page in Massachusetts. None of these battery-powered devices could even remotely compete with steam power: in 1850, their operating cost was, even under the most favorable assumptions, nearly 25 times higher. As a result, during the 1850s, two of England's most prominent engineers held a very low opinion of such machines. Isambard K. Brunel (1806–1859), perhaps the most famous engineer of his time, was against their inclusion in the Great Exhibition of 1851 because he considered them mere toys, and William Rankine (1820–1872), an outstanding student and popularizer of thermodynamics, wrote, in 1859, that the true practical use of electromagnetism was not to drive machines but to make signals (Beauchamp 1997).

The first electric motor that was sold in quantity was Edison's small device mounted on top of a stylus and driving a needle in rapid (8,000 punctures a minute) up-and-down motion. This stencil-making electric pen, energized by a bulky two-cell battery, was the first device to allow large-scale mechanical duplication of documents (Pessaroff 2002). Edison obtained a patent in August 1876, and eventually thousands of these devices were sold to American and European offices (Edison 1876). And it was also during the late 1870s that the first practical opportunities to deploy DC motors not dependent on batteries arose with the commercial introduction of Gramme's dynamo. This came about because of a fortuitous suggestion made in 1873, at the industrial exhibition in Vienna.

Several versions of the story were published during subsequent decades, but the most authentic account comes from a letter sent in 1886 by Charles Félix, at that time the director of a sugar factory in Sermaize, to *Moniteur Industriel* and reproduced in full by Figuier (1888: 281–282). Gramme Company was exhibiting one dynamo powered by a gas engine and another one connected to a Voltaic pile. During his visit of the exposition Félix suggested to Hippolyte Fontaine, who oversaw Gramme's display, "Since you have the first machine that produces electricity and the second one that consumes it, why not let electricity from the first one pass directly to the second and thus dispense with the pile?" Fontaine did so, and the reversibility of machines and the means of retransforming electrical energy to mechanical energy were discovered. Electricity produced by the gas engine-powered dynamo was led by a 100-m cable to the second dynamo, and this machine operating in reverse as a motor was used to run a small centrifugal pump.

Before the end of the decade came the first demonstration of a DC motor (at 130 V) in traction when Siemens & Halske built a short (300 m) circular narrow-gauge railway (*Bahn ohne Dampf und ohne Pferde*—railway without steam or horses—as an advertisement had it) at the Berlin Exhibition of 1879, with a miniature locomotive electrified from the third rail. But as the first central electricity-generating stations were coming online during the 1880s, such motor-driven devices began their transformation from display curiosities to commercial uses. By 1887, the United States had 15 manufacturers of electric motors producing about 10,000 devices a year, most of them with ratings of only a small fraction of one horsepower, many of them inefficient and unreliable (Hunter and Bryant 1991). Leading innovators included Sprague Electric Railway & Motor Company (with the first powerful mine hoist in Aspen, Colorado, the precursor of electric elevators for tall buildings), the Thomson-Houston Company, and Eickemeyer Motors.

Rudolf Eickemeyer's contribution was the winding of armature coils in a form to mold them to the exact shape, and to insulate them thoroughly before emplacing them in the armature; both procedures became the industry standard (MacLaren 1943). But the company that made the greatest difference was, thanks to its rights to Tesla's patents, Westinghouse Electric. Nikola Tesla (1857–1943)—a largely self-educated Serbian engineer who, before his emigration to New York in 1884, worked with new electric systems in Budapest and Paris (Figure I.2.17)—was not actually the first inventor to reveal publicly a discovery of the principle of the rotating electromagnetic field and its use in an induction motor. Galileo Ferraris (1847–1897) presented the same insight to the Royal Academy of Science in Torino, on March 18, 1888, and he published his findings in April, a month before Tesla's May 16, 1888, lecture at the American Institute of Electrical Engineers (Martin 1894; Popović, Horvat, and Nikić 1956).

But Tesla, who claimed that he got the idea for his device one afternoon in 1882, while walking in a park in Budapest with a friend and reciting a stanza from Goethe's *Faust*, was far ahead in developing the first practical machine (Ratzlaff and Anderson 1979; Cheney 1981; Carlson 2015). He built its first models while working in Strasbourg in 1883, and, after immigrating to the

Figure I.2.17 Nikola Tesla: a portrait from the late 1880s when he worked on polyphase electric motors.
Image available from Tesla Society.

United States, he hoped to develop the design for Edison's company. He was hired by Edison immediately after presenting the letter of introduction he got in Paris from Charles Batchelor, Edison's old associate, but the two incompatible men soon parted. Tesla, a gifted scientist and mathematician, was appalled by what he felt to be Edison's brute force approach to problem-solving and believed that a bit of theory and a few calculations would have eliminated a great deal of wasteful searching for technical solutions—and Edison, with his new central electric system built on low-voltage DC, was not ready to embrace the radical ideas of an avid high-voltage AC promoter.

After he secured generous financing, Tesla set up his electric company in April 1887, and he proceeded to build single-phase, two-phase, and three-phase AC motors and design appropriate transformers and a distribution system. Almost immediately, he also began a spate of patent filings (totaling 40 between 1887 and 1891). The two key patents for his polyphase motor (US Patents 391,968 and 391,969, filed on October 12, 1887) were granted on

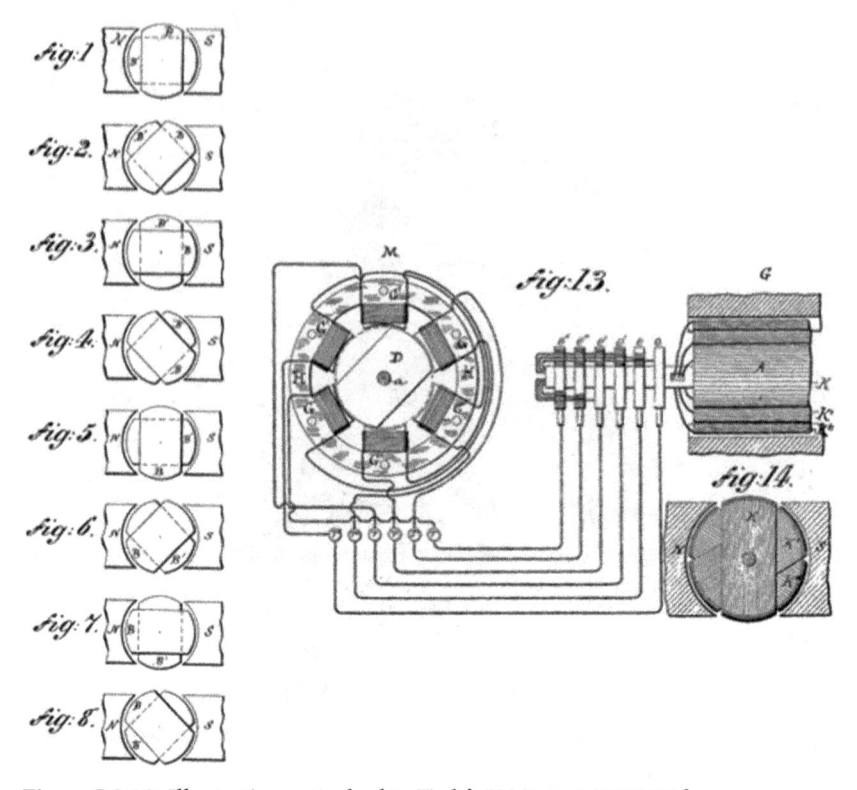

Figure I.2.18 Illustrations attached to Tesla's US Patent 391,968 for an electromagnetic motor. Drawings 1–8 show the principle of a two-phase motor's action and drawing 13 shows the connections of a motor to a generator.
Reproduced from Tesla (1888).

May 1, 1888. After many challenges, they were finally upheld by the courts in August 1900. The first patent specification illustrates the principle of a polyphase motor's operation (Figure I.2.18), and it gives clear indications of the simplicity and practicality of the device. That was Tesla's foremost goal, not an interesting gadget but

> a greater economy of conversion than has heretofore existed, to construct cheaper and more reliable and simple apparatus, and, lastly, the apparatus must be capable of easy management, and such that all danger from the use of currents of high tension, which are necessary to an economical transmission, may be avoided. (Tesla 1888: 1)

Elegant design is one of the most highly valued achievement in engineering, and Tesla's AC motor was an outstanding example of this accomplishment. By using the rotating magnetic field (with two or more alternating currents out step with each other) produced by induction, he had eliminated the need for a commutator (used to reverse the current's direction) and for contact brushes (allowing for the passage of the current). Westinghouse acquired all of Tesla's AC patents in July 1888, for $5,000 in cash and 150 shares of the company's stock, as well as for the royalties for future electricity sales (the last being obviously the most rewarding deal). Tesla's motors were first shown publicly in 1888, and Westinghouse Company produced its first electrical household gadget, a small fan powered by a 125 W AC motor, in 1889. This was a modest beginning of a universal conquest, but, by 1900, there were nearly 100,000 such fans in American households (Hunter and Bryant 1991).

As with other inventions of the remarkable 1880s, subsequent improvements further raised the already high efficiencies of induction motors (the largest machines now convert up to 96% of electricity into kinetic energy), lowered their production costs, and vastly extended their uses without departing in any fundamental way from Tesla's basic designs. Tesla's first patent was for a two-phase machine. Mikhail Osipovich Dolivo-Dobrowolsky (1862–1919) built the first three-phase induction motor in Germany, in 1889, while he was the chief electrician for the AEG, and three-phase machines soon became very common in industrial applications: the phase offset of 120 degrees means that, at any given moment, one of the three phases is near or at its peak, and this assures a more even power output than with two phases.

Adoption of Electric Motors and Their Impact

Only a few other inventions have had such a transforming impact on industrial productivity and quality of life. In less than two decades after their commercialization, electric motors surpassed lights to become the single largest consumer of electricity in the United States, and, after 1900, they were rapidly adopted in industry. In 1899, only 22% of about 160,000 motors produced in the United States (only about 20% of them powered by AC) were for industrial uses; a decade later the share was nearly 50% of 243,000 (more than half of them run by AC), and, before the end of the 1920s, electric motors became by far the most important prime mover in Western industrial production (Schurr et al. 1990).

Many reasons explain this dominance. Induction motors are among the most rugged and most efficient of all energy-converting machines (Behrend 1901; Anderson and Miller 1983; Andreas 1992; Hughes and Drury 2019). Some of them are totally enclosed for operation in extremely dusty environments or work while completely submerged. All of them can be expected to deliver years of virtually maintenance-free operation, can be mass-produced at low cost, and their unit costs decline with higher capacities. They are also compact, with large machines weighing less than 5 g/W (somewhat heavier than gasoline engines). By 1914, they were made in an impressive variety of sizes ranging from the smallest units driving dental drills to nearly 1,000 machines of 20–52 kW that were used to operate the lock gates, rising-stem gate valves, and chain fenders of the newly built Panama Canal (Rushmore 1912; Figure I.2.19). Today's motors come in a much wider range of capacities, ranging from a fraction of a watt in electronic gadgets to 80 MW for the largest two-pole induction motor for the liquid natural gas (LNG) industry (GE 2018).

And their scope of applications has increased to the point that everything we eat, wear, and use has been made with their help (Smil 2017a). Electric motors mill our grain, weave our cloth, roll out our steel, and mold our plastics. They power diagnostic devices and are installed aboard cars, planes, and ships; even larger numbers of small and tiny electric motors go every hour into new electric appliances, tools, and consumer electronics. They do work ranging from frivolous (opening car sunroofs) to life-saving (moving blood in heart-lung machines), from essential (distributing the heat from hydrocarbons burned by household furnaces) to entertaining (powering all kinds of amusement park rides). They also lift an increasingly urbanized humanity to high-rise destinations, move parts and products along assembly lines, and make it possible to micromachine millions of accurate components for devices ranging from giant turbofan jet engines to implantable heart pacemakers.

Perhaps the best way to appreciate their impact is to realize that modern civilization could have access to all fuels and even to generate all electricity—but it could not function so smoothly and conveniently without electric motors, these new alphas (in baby incubators) and omegas (powering compressors in morgue coolers) of our world. Not surprisingly, electric motors now consume more than two-thirds of all electricity produced in the

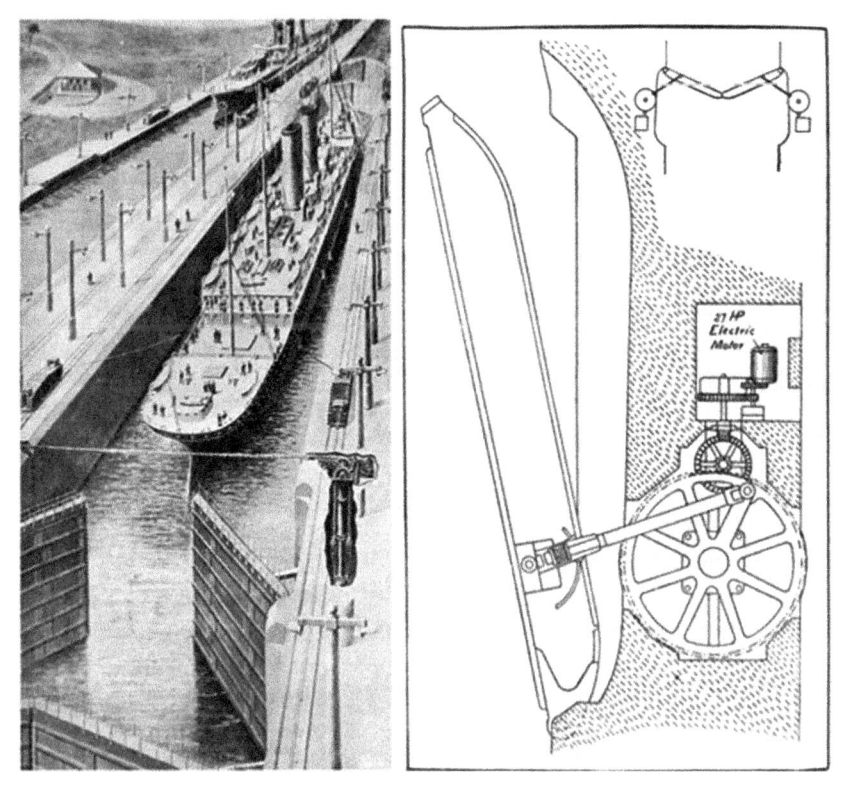

Figure I.2.19 A set of lock gates of the Panama Canal and a plan view of mechanism for their operation. One 20-kW electric motor was needed for each leaf of 46 gate pairs, and 5.6-kW motors were used for the miter-forcing machines that made gates to come together with a tight seal and lock them in that position.
Reproduced from *The Illustrated London News*, March 21, 1914, and from Rushmore (1912).

United States, and they are doing so with increasing efficiencies (Hoshide 1994; Hill 2018). Because one design cannot accommodate all these demands, induction motors have evolved to include several basic kinds of machines, with three-phase and single-phase motors being by far the most common choices.

Three-phase motors have the most straightforward construction. Their stators are made of a steel frame that encloses a hollow cylindrical core composed of stacked laminations. Two major types of three-phase induction motors are distinguished by their rotor wiring: squirrel-cage rotors (introduced by Westinghouse Company during the early 1890s) are made up of bars and end rings; wound rotors resemble stators. Because the three-phase current produces a rotating magnetic field, there is no need for additional windings or switches within the motor to start it. And three-phase motors have also the highest efficiencies, are the least massive per unit of installed power, and are the least expensive to make. Not surprisingly, three-phase induction motors are the norm for most permanently wired industrial machinery as well as for air- and liquid-moving systems.

The diffusion of three-phase induction motors in industrial production was a much more revolutionary step than the previous epochal transition that saw waterwheels replaced by steam engines in factories. That substitution did not change the basic mode of distributing mechanical energy that was required for countless processing, machining, and assembling tasks: factory ceilings continued to be clogged by complex arrangements of iron or steel line shafts that were connected by pulleys and belts to parallel counter-shafts and these, in turn, were belted to individual machines. With such arrangements, a prime mover outage or an accident at any cracked shaft or a slipped belt stopped the whole assembly. Conversely, even if most of the machines were not needed (during periods of low demand), the entire shaft assemblies were still running.

Electric motors were used initially to drive relatively short shafts for groups of machines, but, after 1900, they were increasingly used as unit drives. Their eventual universal deployment changed modern manufacturing by establishing an enduring pattern of industrial production (Schurr et al. 1990; Hunter and Bryant 1991). While the total installed mechanical power in US manufacturing had roughly quadrupled between 1899 and 1929, the capacity of electric motors grew nearly 60-fold and reached more than 82% of the total available power compared to a mere 0.3% in 1890 and less than 5% at the end of the nineteenth century (USBC 1954; Figure I.2.20). In the United States, the substitution of steam and direct water-powered drive by motors was practically complete just three decades after it began during the late 1890s.

The benefits of three-phase motors are manifold. There are no friction losses in energy distribution, an individualized power supply allows optimal machine use and maximum productive efficiencies, they open the way for flexible plant design and easy expansion, and they enable precise control and any desired sequencing of tasks. Plant interiors look different because the unit electric drive did away with the overhead clutter, noise, and risks brought by rotating shafts and tensioned belts, and the ceilings were freed for installation of better illumination and ventilation, steps that further helped

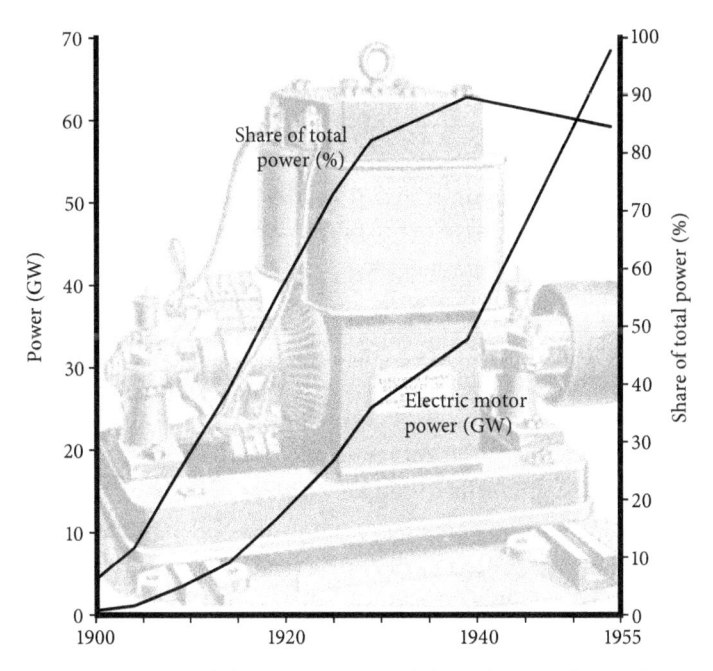

Figure I.2.20 Capacity of electric motors and their share in the total power installed in the US manufacturing.
Plotted from data in USBC (1954).

to boost productivity. But three-phase wiring is not normally supplied to homes and offices, and hence a large variety of machines of smaller capacity draw on single-phase supply.

Tesla filed his application for a single-phase motor on October 20, 1888, which was granted on August 14, 1894 (Tesla 1894), and Langdon Davies commercialized his single-phase design in 1893. Construction of these machines is very similar to that of three-phase motors: they have the same squirrel-cage rotor and a stator whose field alternates poles as the single-phase voltage swings from positive to negative. Efficiencies of single-phase motors are lower than in three-phase motors, but they, too, can deliver years of reliable work with minimal maintenance. Induction motors using single-phase 120-V AC are most common in household and office appliances, including ceiling and table fans, kitchen appliances, and air conditioners. The principle of synchronous motors was first demonstrated by Gramme's 1873 Vienna displays: they are simply alternators connected to a three-phase supply. Because its frequency is fixed, the motor speed does not vary with load or voltage. Their principal use is at low speeds (below 600 rpm) where induction motors are heavy, expensive, and inefficient. Westinghouse deployed the first synchronous motor in 1893, in Telluride, Colorado, to run an ore-crushing machine. Modern electronic converters can produce very low frequencies, and synchronous motors can be thus run at the very low speeds desirable in rotary kilns in cement plants or in ore or rock milling and crushing. And synchronous motors are the best choice to power clocks and tape recorders. And DC motors, going back to the 1870s, remain ubiquitous. Some of them get their supply from batteries; many others receive DC from an electronic rectifier fed with AC. They have been entirely displaced by AC machines in nearly all common stationary industrial applications, but because of their very high starting torque, they have been always the motors of choice for electric trains. Between 1890 and 1910, before they were displaced by internal combustion engines, they were relatively popular as energizers of electric cars.

There is no shortage of electric motors in road and off-road vehicles. Their total global number in 2020 was in the billions: in addition to their starter motors, modern cars (total of about 1.5 billion vehicles in 2024) also have an increasing variety of small DC servomotors that operate windshield wipers, windows, locks, sunroofs, or mirrors by drawing on battery-supplied DC. DC motors are also common in chairs for invalids, as well as in treadmills, garage-door openers, heavy-duty hoists, punch presses, crushers, fans, and pumps used in steel mills and mines. Another advantage of DC motors is that their power tends to be constant because changes in torque automatically produce a reverse change in speed; consequently, these motors slow down as a train starts going uphill, and, in cranes and hoists, they lift heavy loads more slowly than they do light ones. I will comment on the latest mass-market for electric motors—to power electric cars—in the book's closing chapter.

Systems Mature and (Not Quite) Standardize

Remarkable accomplishments of the 1880s refashioned every original component of the new electrical system and introduced machines, devices, and converters whose performance had soon greatly surpassed the ratings, efficiencies, and reliabilities of initial designs. But this extraordinarily creative period was not conducive to standardization and optimization. This dissipative design is a phenomenon that recurs with rapidly moving innovations: before World War I it was also experienced by the nascent automotive industry. As a result, 10 years after the first central plants began operating in the early 1880s, one could find many permutations of prime movers, generators, currents, voltages, and frequencies among the operating coal-fired and hydro-powered stations.

Prime movers were still dominated by steam engines, but steam turbines were on the threshold of rapid commercial gains. Generators were mostly dynamos coupled indirectly or directly to reciprocating engines, but again, unit arrangements of turbines and alternators were ascendant. The earliest systems, isolated or central, produced and transmitted DC, commonly at 105–110 V or at 200 V with a three-wire arrangement. But the first AC systems were installed within months after the commercial availability of AC transformers, such as Ferranti's small installation in London's Grosvenor Gallery in 1885, and Stanley's town plant in Barrington, Massachusetts, which led George Westinghouse, who came to see it on April 6, 1886, to enter the AC field (Stanley 1912). And so, inconspicuously, began the famous battle of the systems.

AC Versus DC

Edison, the leading proponent of DC, was not initially concerned about the competition, but he soon changed his mind and embarked on what was certainly the most controversial and most bizarre episode of his life. His intensive anti-AC campaign was not based on either scientific and economic arguments (on both counts DC had its advantages in particular settings) or on pointing out some indisputable safety concerns with high-voltage AC (HVAC). Instead, he deliberately exaggerated the life-threatening dangers of AC, during 1887 was himself involved in cruel demonstrations designed to demonize it (electrocutions of stray dogs and cats coaxed onto a sheet of metal charged with 1 kV from an AC generator), and made repeated personal attacks on George Westinghouse (1846–1914), a fellow electric-systems pioneer, inventor (his fame began with his railway air brake patented in 1869), and industrialist and, thanks to Stanley's influence, an early champion of AC (Figure I.2.21).

Perhaps worst of all, in 1888, Harold Brown, one of Edison's former employees, became a very active participant in a movement to convince the state of New York to replace hanging with electrocution by AC. Brown was

Figure I.2.21 George Westinghouse had a critical role in the development of electric industry thanks to his inventiveness and entrepreneurial abilities, his unequivocal advocacy of alternating current, and his support of Tesla's work.
Photograph courtesy of Westinghouse Electric.

eventually selected to be the state's electric chair technician, and he made sure that the AC was supplied by Westinghouse's alternators (Essig 2003). Edison's war on HVAC lasted about three years. In 1889, he was still writing that laying HVAC lines underground would make them even more dangerous and that

> [m]y personal desire would be to prohibit entirely the use of alternating currents. They are as unnecessary as they are dangerous... and I can therefore see no justification for the introduction of a system which has no element of permanency and every element of danger to life and property. (Edison 1889: 632)

Edison's vehement, dogmatic, and aggressive attitude has been painful to accept and difficult to explain even by some of his most admiring biographers (Dyer and Martin 1929; Josephson 1959). But his opposition to HVAC stopped suddenly in 1890, and soon he was telling the Virginia legislature that debated an AC-related measure that "you want to allow high pressure wherever the conditions are such that by no possible accident could that pressure get into the houses of the consumers; you want to give them all the latitude you can" (quoted in Dyer and Martin 1929: 418). David (1991), in a reinterpretation of the entire affair, argues that Edison's apparently irrational opposition had a very rational strategic goal and that its achievement explains the sudden cessation of his anti-AC campaign. By 1886, Edison became concerned about the financial problems of the remaining electric businesses he still owned (by that time he had only a small stake in the original Edison Electric Light), and hence he welcomed a suggestion to consolidate all Edison-related enterprises into a single new corporation.

In return for cash ($1.75 million), 10% of the shares, and a place on the company's board, Edison's electric businesses were taken over by the Edison General Electric Company organized in January 1889, and, a year later, Edison liquidated his shares and stopped taking any active role on the company's board. David (1991) argues that the real purpose of Edison's anti-AC campaign was to support the perceived value of those Edison enterprises that were entirely committed to the DC-based system and thus to improve the terms on which he, and his associates, could be bought out. Once this was accomplished, the conflict was over.

The battle of the systems thus ended abruptly in 1890, although some accounts date it as expansively as 1886–1900 (Winchell 2019). The latter delimitation is correct if one wants to include the period when AC became virtually the only choice for new central electric systems. And it must be also noted that Edison's reasoning had its vigorous British version as the DC side—led by two luminaries of the new electric era, R. E. B. Crompton and John Hopkinson—was joined by one of the world's foremost physicists, Lord Kelvin, while Sebastian Ziani de Ferranti, W. M. Mordey, and Sylvanus Thompson were the leading British experts behind HVAC (Flood, Cartney, and Whitaker 2008). While in 1886 one could see the DC–AC competition as something of a standoff, by 1888, the choice was fundamentally over thanks to the rapid advances of three technical innovations that assured the future AC dominance: the already described invention of induction motors, the commercialization of an accurate and inexpensive AC meter, and the introduction of the rotary converter.

In April 1888, Westinghouse engineer Oliver Shallenberger designed the first reliable meter for AC, and, the very next year, Westinghouse Electric began to manufacture it. But as David (1991) points out, neither the AC motors nor the AC meters could have dislodged the entrenched DC systems (by 1891, they accounted for more than half of all urban lighting in the United States) without the invention of a rotary converter that made it possible to connect old DC central stations, as well as already extensive DC transmission networks, to new long-distance HVAC transmission lines. The first rotary converter was patented in 1888, by another of Edison's former employees, Charles S. Bradley. The device combined an AC induction motor with a DC dynamo to convert HVAC to low-voltage DC suitable for distribution to final users.

When the original single-phase AC systems began to be replaced by more efficient polyphase transmission, new converters were eventually developed to take care of that transposition, and its reverse proved very useful during the 1890s, when it made it possible to use the existing DC-generating equipment to transmit polyphase HVAC over much larger areas than could be ever reached by a DC network. And so the only matter that was unclear by 1890 was how fast the AC systems would progress and how soon the superior polyphase AC generation and its high-voltage distribution would become the industry standard.

A few bold projects accelerated this inevitable transition.

The first one was the already mentioned Deptford station planned by the London Electric Supply Corporation, a new company registered in 1887, to provide a reliable electricity supply to central London (Figure I.2.22). The site, about 11 km from downtown on the south bank of the Thames, was chosen due to its location on the river and easy access to coal shipments by barges. Sebastian Ziani de Ferranti (1864–1930), the company's young chief electrical engineer, conceived the station on an unprecedented scale: large steam engines were to drive record-size dynamos whose shafts weighed almost 70 t (Anonymous 1889). Because of the magnitude of the service (more than 200,000 lights) and the plant's distance from the city's downtown, Ferranti decided to use the unprecedented voltage of 10 kV. The company's financial difficulties eventually led to his dismissal from the project's leadership in 1891 and to the scaling-down of the Deptford operation, but Ferranti's unequivocal conviction about the superiority of high voltage was fully vindicated.

America's first AC transmission over a longer distance took place in 1890, when Willamette Falls in Oregon was linked by a 20-km, 3.3-kV line with Portland, where the current was stepped down first to 1.1 kV and

Figure I.2.22 The massive exterior of London Electric Supply Corporation's Deptford station and one of its record-size Ferranti-designed dynamos under construction.
Reproduced from *The Illustrated London News*, October 16, 1889.

then to 100 V and 500 V for distribution (MacLaren 1943). A more decisive demonstration of long-distance HVAC transmission took place at the Frankfurt Electrotechnical Exhibition in 1891 (Beauchamp 1997). Oscar Muller built a 177-km-long line to a generator at Lauffen, on the upper Neckar River, to transmit 149 kW of three-phase AC, initially at 15 kV. The first small three-phase system in the United States was installed by the newly organized General Electric in 1893, in Concord, New Hampshire, and there is no doubt that the tide was completely turned when Westinghouse Company and General Electric made designs for the world's largest AC project at the Niagara Falls (Passer 1953; Hunter and Bryant 1991).

On August 26, 1895, some of the water was diverted to two 3.7-MW turbines to generate electricity at 2.2 kV and use it at nearby plants for electrochemical production of aluminum and carborundum. In 1896, part of the output was stepped up to 11 kV and transmitted more than 30 km to Buffalo for lighting and streetcars. By 1900, 10 Westinghouse two-phase 25-Hz generators of the first powerhouse (total of 37.3 MW) were in operation, and General Electric built 11 more units (total of 40.9 MW) for the second Niagara station, completed by 1904. The project's aggregate rating of 78.2 MW accounted for 20% of the country's installed generating capacity (Figure I.2.23). Successful completion of this project confirmed the concept of large-scale HVAC generation and transmission as the basic paradigm of future electric systems.

General Electric's thrust into polyphase AC systems was greatly helped by the insights and discoveries of Charles Proteus Steinmetz (1865–1923), a brilliant mathematician and engineer who emigrated from Germany to New York in 1889, where he worked first for Rudolf Eickemeyer's company and was hired from there in 1893 by General Electric (Kline 1992). Steinmetz's work ranged from finding solutions to high transmission losses (his first patents at General Electric) to highly theoretical mathematical techniques used to analyze AC circuits. Between 1891 and 1923, he received more than 100 AC-related patents and published a dozen books. After 1900,

Figure I.2.23 Generator hall of Niagara Falls power plant.
Reproduced from *Scientific American*, January 25, 1896.

Steinmetz had a critical role in setting up a new science-based laboratory at General Electric (one of the precursors of large-scale corporate R&D institutions) and guiding its early progress.

After the mid-1890s, it was only a matter of time before all necessary technical capabilities were put in place to support the expected expansion of AC systems. Parsons built the world's first three-phase turbo-alternator in 1900, for the Acton Hall Colliery in Yorkshire, where the 150-kW machine was to energize coal-cutting machinery. The first public supply of three-phase current (at 6.6 kV and 40 Hz) came from the Neptune Bank station (still powered by steam engines) that was officially opened by Lord Kelvin, a former opponent of AC, in 1901 (Electricity Council 1973). But, long after the dominance of AC was assured, the inertia of complex techniques made DC linger. In the United Kingdom, DC still held 40% of all generating capacity in 1906 (Fleming 1911). In the United States, DC's share in the total end-use capacity was 53% in 1902 and still 26% by 1917 (David 1991).

The trend toward increasing transmission voltages that was set during the first decade of AC development continued, albeit at a slower pace, for nearly 100 years. The highest American ratings rose from 4 kV during the late 1880s to 60 kV by 1900 and to 150 kV in 1913; the first 300 kV lines were installed in 1934, and, by the 1970s, AC was transmitted by lines of 765 kV (Smil 2008). By the 1980s, HVAC interconnections tied together most of the European continent as well as the eastern and western part of the United States and Canada, while east–west links across the continent remained very weak. And, by the early 1970s, transmission techniques made a full circle by returning to DC, but this time at high voltages (HVDC).

HVDC, made possible by using mercury arc rectifiers that convert AC to DC and then reconvert it back to AC for distribution, was pioneered in Canada to transfer large blocks of electricity from hydro-projects in northern regions of Manitoba, British Columbia, and Quebec (Arrillaga 1983). HVDC has been also used in submarine transmission cables, first pioneered on a smaller scale in 1913, when the British Columbia mainland was linked to Vancouver Island. In 1954 came the first large-scale application when the Sweden–Gotland cable carried 20 MW at 100 kV over 96 km, but, a decade later, New Zealand's two islands were connected in 1965 by a 250-kV tie that carries 600 MW.

Expanding Generation and Consumption

Proliferation of electric supply was very rapid: as already noted, in 1891, the United States had 1,300 central electric systems and many more isolated generation units (Hunter and Bryant 1991). The Edison Electric Illumination Company of New York, whose northernmost service boundary in 1883 was Nassau Street, reached 59th Street by 1890 and 95th Street by 1898, and, by 1912, it served practically every street in Manhattan and the Bronx (Martin 1922). In terms of 50-W light bulb equivalents, its connections rose from 3,144 lights on December 1, 1882, to nearly 1.5 million lights by 1900, and to more than 12 million by the end of 1913. By 1900, every major city in Europe and North America, and many on other continents, had various systems of public electric supply from central stations. This electricity was used initially for lighting and transportation and soon also for a variety of household and industrial tasks (Wordingham 1901; Gay and Yeaman 1906).

During the first decade of the twentieth century many of these small stations located in or near downtowns were shut down and new, larger plants were built either on the outskirts of urban areas or directly in such coal-producing districts as the Midlands and Yorkshire in England and the Ruhr in Germany. In 1904, the Newcastle upon Tyne Electric Supply Company commissioned its Carville station, whose design set the basic pattern for all subsequent large generating stations (Electricity Council 1973). The initial version had two 3.5-MW and two 1.5-MW units (combinations of boilers, turbines, and alternators that became later known as a turbogenerator) operated from a control room.

For large coal-fired stations the mine-mouth locations remained a preferred choice ever since. Plants of increasingly larger capacities were built until the 1960s in the United Kingdom, primarily in the Midlands, and in the United States were concentrated in Pennsylvania, the Ohio River Valley, and the Tennessee Valley Authority region. These stations were designed to burn finely pulverized coal blown into their boilers to maximize combustion efficiency. The first experiments with this technique began at Willesden station in England, in 1903. By 1914, the use of powdered coal was developed into a successful commercial practice (Holmes 1914), but most electricity-generating plants adopted it only after World War I.

The two pre-World War I generations also set the course for the construction of large-scale waterpower projects thanks to the introduction of two new turbines that eventually came to dominate hydroelectric generation. Several new water turbine designs appeared during the two decades after Benoit Fourneyron built his first reaction turbines during the late 1820s and the 1830s (Smith 1980). The most successful design was the one patented first in the United States, in 1838, by Samuel B. Howd and then improved and commercially deployed by James B. Francis (1815–1892). This machine is known as the Francis turbine, but it is clearly a product of multiple contributions (Hunter 1979). During the 1880s, Lester Allen Pelton (1829–1908) developed and patented an impulse turbine driven by water jets discharged into peripheral buckets, a design for very high-water heads. In contrast, in 1913, Viktor Kaplan patented a reaction turbine with adjustable vertical-flow propellers that has become a preferred choice for low dams.

Before World War I more than 500 hydroelectric stations were in operation around the world (mostly in Europe and North America), and the world's first pumped storage stations were also built. These stations use electricity produced during low-demand periods (mostly during the night) to pump water into a nearby high-lying reservoir to be released into downpipes and converted almost instantly into hydroelectricity during periods of peak demand. In 2025, this remains the most economical way of indirect electricity storage. But the period of hydroelectric megaprojects—exemplified in the United States by the Tennessee Valley Authority and by the Hoover and Grand Coulee dams—came only during the 1930s (ICOLD 2024).

Expanding the availability of electricity led almost immediately to applications beyond household and street lighting. Few facts illustrate the rapid pace of post-1879 developments better than the frequency of major national and international exhibitions that were devoted exclusively to that newly useful form of energy or where electricity played a major part (Beauchamp 1997). Of 38 international expositions that took place during the exceptionally inventive 1880s, seven were devoted solely to electricity, including the already noted Paris Electrical Exhibition of 1881 and similar displays in London in the same year, Munich (1882), Vienna (1883), and Philadelphia (1884). Electricity was a major attraction at British Jubilee Exhibitions in 1886–1887, and even more so at the Chicago Columbian World Fair in 1893 (Bancroft 1893).

But already in Paris in 1881, several exhibitors showed—still as a curiosity rather than as a commercial proposition—sewing machines, lathes, and a printing press powered by electric motors. By 1890, General Electric was selling electric irons, fans, and an electric rapid cooking apparatus that could boil a pint of water in 12 minutes (Electricity Council 1973). Household

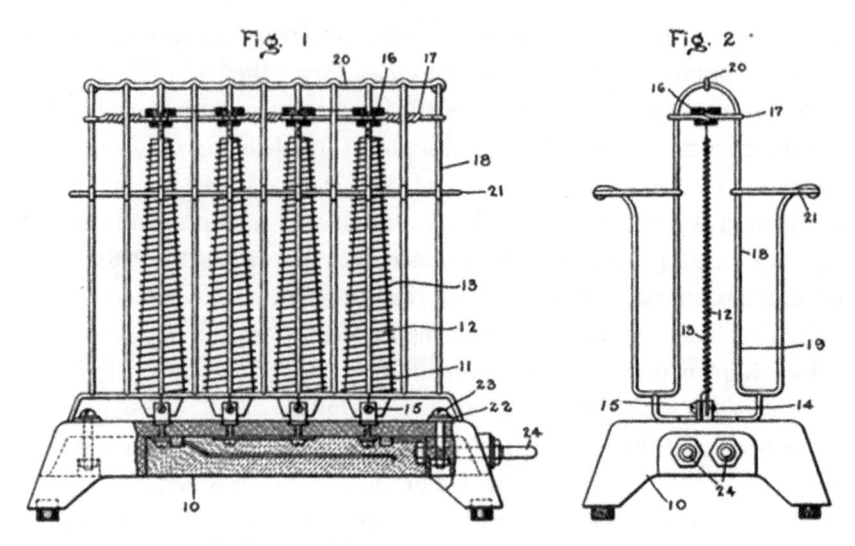

Figure I.2.24 Frank Shailor's "electric heater," the first practical toaster whose US Patent 950,058 was assigned to General Electric, which began its production in 1909. Its construction was made possible by a new nickel-chromium alloy; Shailor used John Dempster's (US Patent 901,428) combination of 62% Ni, 20% Fe, 13% Cr, and 5% Mn for resistance wires.

appliances were introduced to the British market during the Crystal Palace Electrical Exhibition in 1891, and they included irons, cookers, and electric fires with the heating panels set behind wrought-iron screens. Two years later hot plates, fans, bells, bed warmers, radiators, and a complete Model Electric Kitchen with a small range, saucepan, water heater, broiler, and kettle were featured at a Chicago exhibition (Bancroft 1893; Giedion 1948). Early electric ranges resembled the well-established gas appliances, and the first toaster had only one set of resistance wires (Figure I.2.24).

Chicago displays also contained an enormous array of industrial applications of electricity ranging from ovens and furnaces to metal- and woodworking machines (lathes, drills, presses) and from railway signaling equipment to an electric chair. Among the notable pre-World War I commercial introductions were (with the dates of initial appearance in the United States) the electric washing machine (1907, featured soon afterward in the Sears & Roebuck Company's catalog), vacuum cleaners (originally called electric suction sweepers), and electric refrigerators (1912). Willis Haviland Carrier (1876–1950) designed his first air conditioning system in 1902, after he realized that air could be dried by saturating it with chilled water to induce condensation. His first installation, with the cooling capacity of nearly 50 t of ice a day, was to control temperature and humidity for the Brooklyn printing company Sackett-Williams. These controls eliminated slight fluctuations in the dimensions of their printing paper that caused the misalignment of colored inks (Ingels 1952).

Carrier's first patent (US Patent 808,897) for the "Apparatus for Treating Air" was granted in 1906, and, in 1911, he completed the *rational psychrometric formulae*, which has become a lasting foundation of basic calculations required to design efficient air conditioning systems (ASHRAE 2001). Textile, film, and food processing industries were among the first commercial users, while cooling for human comfort began only during the 1920s, with department stores and movie theaters. The first air-conditioned American car was engineered in 1938, and, in the same year, Philco-York marketed the first successful model of a window air conditioner (NAE 2024). Widespread use of household air conditioning, and with it the large-scale migration to a suddenly comfortable Sunbelt, had to wait until the 1960s.

By the end of the twentieth century almost 70% of all US households had either central or window air conditioning, and central units were being installed in virtually all new custom-built US houses, regardless of the latitude.

While mechanical refrigeration for commercial operations (cold storages, ice making, meatpacking, dairy, breweries) was perfected during the 1880s, its large equipment, the need for manual operation by skilled personnel, and the use of potentially dangerous ammonia as the refrigerant did not make the technique readily transferable to household use. The first practical design for a kitchen unit, patented by Fred Wolf in 1913, used copper tubing with flared joints to reduce the risk of leaks—and it even produced ice cubes (Nagengast 2000).

The first affordable household refrigerator, General Motor's Frigidaire, became a common possession only during the late 1920s. The cost of refrigeration fell and its safety rose with the replacement of dangerous refrigerants by chlorofluorocarbons in 1931 (NAE 2024). During the 1980s, these gases were incontrovertibly implicated in the destruction of stratospheric ozone. The genesis—and solution—of this environmental concern will be covered in this book's companion volume. By 2000, only color televisions rivaled refrigerators as virtually universal household possessions in affluent countries.

The first portable electric vacuum cleaner, brought out in 1905 by Chapman & Skinner Company in San Francisco, was a heavy (more than 40 kg) machine with a fan nearly half a meter across to produce suction. The familiar upright machine (patented in 1908) was the invention of James W. Spangler. William Hoover's Suction Sweeper Company began its production in 1907 and introduced its 0 model, weighing about 18 kg, a year later; a high-speed universal motor for vacuum cleaners became available in 1909. The key features of the early Hoover arrangement are still with us: long handle, disposable cloth filter bag, and cleaning attachments. Electrolux introduced its first canister model in 1921. The first radical departure from the basic pre-1910 design came only with James Dyson's invention of the non-clogging Dual Cyclone vacuum during the 1980s (Dyson 1998).

The age of electricity also brought new diagnostic devices and medical equipment whose basic modes of operation remain unchanged but whose later versions became more practical or more reliable to use (Davis 1981). The first dental drill powered by a bulky and expensive DC motor (replacing previous hand cranks, springs, and foot treadles) was introduced in 1874, and plug-in drills became available by 1908. A small battery powered the first endoscope, a well-lit tube with a lens for viewing the esophagus, designed in 1881 by Polish surgeon Johann Mickulicz (1850–1905). At the Buffalo exhibition in 1901, visitors could see for the first time an incubator for premature babies; two years later Willem Einthoven (1860–1927) designed the first electrocardiograph (he was awarded the 1925 Nobel Prize for Physiology or Medicine), and the device has been manufactured for diagnostic use since 1909 (Davis 1981).

By far the most important diagnostic contribution that would have been impossible without electricity began with Wilhelm Conrad Röntgen's (1845–1923) experiments with electric discharges in highly evacuated glass tubes. That led, on November 8, 1895, to the discovery of x-rays (Brachner 1995). Their value as a diagnostic tool was appreciated immediately, and consequently, in 1901, when the President of the Royal Swedish Academy of Sciences presented Rontgen with the first Nobel Prize in Physics, he noted that they were already in extensive use in medical practice (Odhner 1901). By 1910, their use was standard in medical diagnostics in large Western cities, and they also found many applications in science and industry.

Pre-World War I decades also set several basic management practices in the electric industry. Already in 1883, John Hopkinson (1849–1898) had introduced the principle of the two-part tariff method of charging for consumed electricity, and the combination of a fixed (basic service) charge and a payment for the used quantity. And almost immediately, the industry had to deal with the challenge of variable demand. While electricity demand was dominated by lighting, the daily load factors of generating stations were very low, often no more than 8–10%; in 1907, the average nationwide load for the United Kingdom was just 14.5%. That is why utilities were eager to see the installation of more electric motors and looked for opportunities to reduce the large disparities between peak and average loads.

Electricity use for traction had begun on a small scale already in the early 1880s, with the first electric tramway built by Werner Siemens in Berlin's Lichterfelde in 1881, and all major modes of electrically driven urban and intercity transport were introduced before 1913 (Figure I.2.25). The relevant first British dates were as follows (Electricity Council 1973): a short DC railway line in 1883 (on the beach at Brighton), tramway in 1885, underground railway 1890 (in London), main-line railway electrification 1895, AC railway line in 1908, and trolley buses in 1911 (but Werner Siemens had built the first trolley bus in Berlin already in 1882).

Figure I.2.25 Electric streetcar between Frankfurt-am-Main and Offenbach had conductors of wrought iron (diameter of 3 cm) suspended by iron wires that were attached to telegraph poles. Identical arrangement was used in many other European cities.
Reproduced from *Scientific American*, August 26, 1882.

In the United States, the first electric streetcars were introduced by Frank J. Sprague (1857–1934) in Richmond, Virginia, in 1888 (cars lit with electric lights traveled at 24 km/h), and soon afterward in Boston (Cambridge). The subsequent diffusion of this innovation was very rapid: within five years 14 of the 16 US cities with more than 200,000 people and 41 of 42 cities with populations between 50,000 and 200,000 had electric streetcars. By 1900, there were 1,200 such systems in operation or under construction; by 1902, 99% of the old horse-drawn systems were converted, and, by 1908, more than 1,200 companies owned about 62,000 km of electrified track and elevated railways (Dyer and Martin 1929; Sharlin 1967).

These developments brought electricity's benefits to more people (lighting excepted) than any other application at that time. Electric traction eventually spread from subways and commuter trains to high-speed rail links. Advances were rapid: already by 1902, Siemens & Halske was running an experimental four-motor polyphase car of 2.2 MW at speeds exceeding 160 km/h (Perkins 1902). All of today's fastest trains are electric. The French TGV, current speed record holder, draws 25 kV at 50 Hz from a fixed overhead wire; the current is then stepped down by massive transformers to 1,500 V, and it is supplied to synchronous 1.1 MW AC traction motors (High Speed Rail Alliance 2024).

In addition to transportation and industrial motors, electric welding, electrometallurgy, and electrochemical industries (particularly after the Hall-Héroult process for producing aluminum was introduced in 1886) emerged as major industrial users of electricity. Electric arc welding with a carbon electrode was patented in the United Kingdom in 1885, by Nikolai Benardos and Stanislaw Olszewski, and, seven years later, Charles L. Coffin (1844–1926) got the US patent for a metal electrode (Lebrun 1961). Advances in electrometallurgy were made possible by William Siemens's invention of the electric furnace in 1878. A quantitatively small but qualitatively immensely important use of electricity is in the enormous variety of monitoring and analytical devices used in industry, science, and healthcare.

As is commonly the case with technical innovations, diffusion of electricity was affected by national idiosyncrasies. Inertial reliance on the long-established steam power using cheap coal slowed down the British embrace of electricity, and the unfortunate Electric Lighting Act of 1882, which limited the operating licenses to seven years and gave local authorities the right to take over company assets after 21 years, created another disincentive that was remedied only by new legislation in 1888. These two factors explain why the United States and Germany pioneered widespread applications of electricity. By 1900, the national totals were (in GWh) 0.2 in the United Kingdom, 1.0 in Germany, and nearly 5 in the United States; by 1914, the respective figures rose to 3.0, 8.8, and 24.8, which means that the United States per capita generation reached 250 Wh/year, compared to about 135 Wh in Germany and to just 70 Wh in the United Kingdom (USBC 1975; Mitchell 1998).

American and German cities were also far ahead of English urban areas in developing the symbiotic relationship between electricity and industrial production (Martin 1922; Hughes 1983; Nye 1990; Platt 1991). By 1911, Chicago used more than 80% of its electricity for industrial and traction power; the analogical share was 66% in Berlin, but only 39% in London, where lighting still dominated the demand. And in both the United States and Germany, developments were disproportionately driven by just two companies: after the early 1890s, General Electric and Westinghouse introduced most of the technical advances in American electric industry, and AEG and Siemens played the same role in Germany (Feldenkirchen 1994; Strunk 1999).

American electrification efforts paid off rapidly in cities: by 1906, Boston's average per capita output of electricity was nearly 350 Wh/year, New York's almost 300 Wh/year, and Chicago's 200 Wh/year, compared to just over 40 Wh/year in London (Fleming 1911). But the country's large size (and hence the cost of transmission) delayed access to electricity in smaller towns and in rural areas. Only about 5% of potential customers had electricity in 1900; 10 years later still only 1 in 10 American homes was wired. The share surpassed 25% by 1918, and most urban households became connected only by the late 1920s (Nye 1990).

In technical terms, European and North American developments proceeded along very similar lines as generating machinery was designed for more efficient performance with higher pressures and greater steam superheat, as established central stations grew by adding larger turbogenerators (often produced in large series of standard sizes), and as long-distance transmission steadily progressed toward higher AC voltages. Electric motors surpassed lights as the principal users of electricity in all industrializing countries (e.g., in the United Kingdom, the crossover had taken place already before 1910), but Europe and North America retained distinct operating parameters for electricity distribution and its final use in households.

Early polyphase systems operated with frequencies ranging from 25 to 133.33 Hz, but, already by 1892, Westinghouse Company had adopted just two rates, 25 Hz for large-load supply and 60 Hz for general distribution.

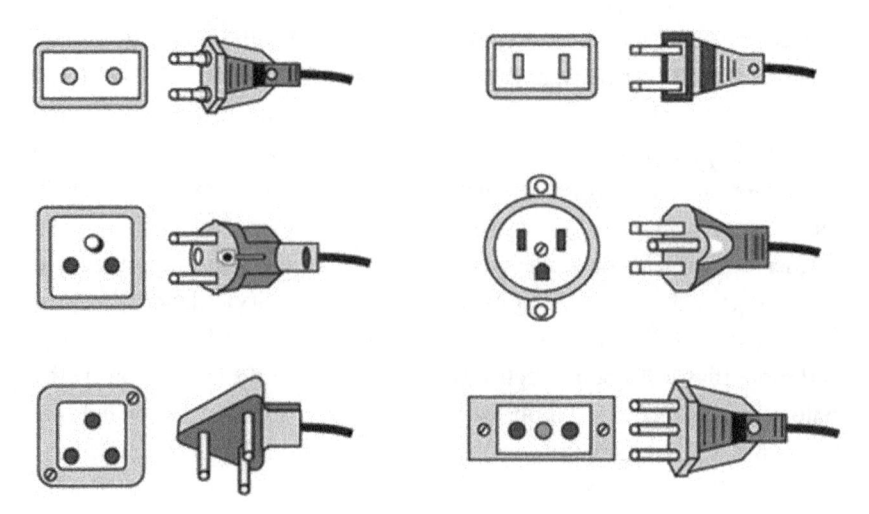

Figure I.2.26 Major types of electric plugs used around the world.
Simplified from a more detailed illustration in USDC (2002).

Accordingly, household and office supply in North America is now single-phase AC of 120 V (maximum of 125 V) at 15 A and 60 Hz; the current oscillates 60 times every second between ±170 V, averaging 120 V, with the high voltage supplied to the smaller prong. In contrast, the pre-World War I trend in the United Kingdom was toward increasing variation rather than standardization, and, in 1914, London had no less than 10 different frequencies and a multitude of voltages (Hughes 1983). The European standard eventually settled on single-phase AC of 220 V at 50 Hz, but amperage is not uniform; for example, Italy uses 10 A or 16 A (BIA 2001). Japanese ratings conform to the US standard; Chinese ratings, to the European one.

But these disparities are nothing compared to the bewildering variety of sockets (shaped as circles, rectangles, or squares), plugs (with flat or rounded blades, and with or without grounding pins), and connectors used by households: worldwide there are 15 basic types of electric plugs and sockets with different arrangements of flat and round pins (ITA 2002; Figure I.2.26). Unfortunately, there is little hope that this unhelpful diversity will be eliminated any time soon; the only solution when using incompatible devices and plugs is to buy appropriate transformers and adapters.

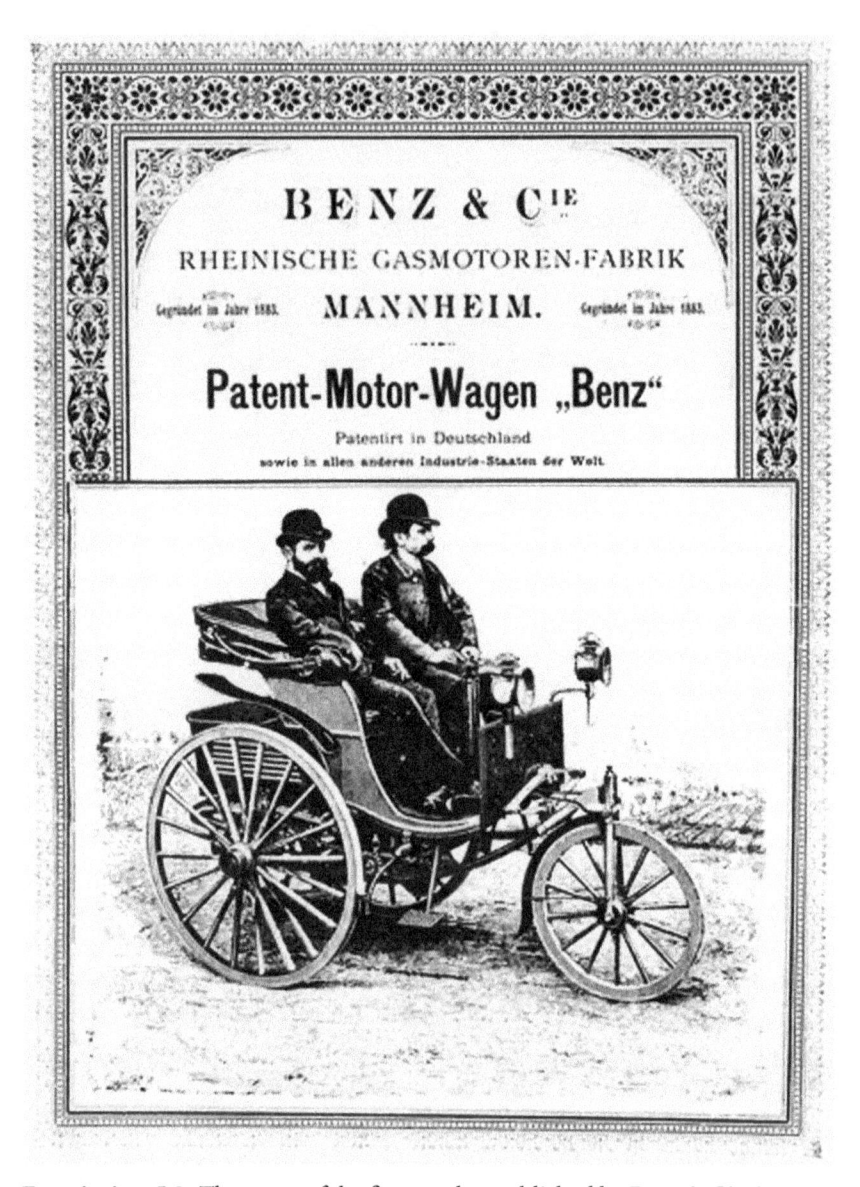

Frontispiece I.3 The cover of the first catalog published by Benz & Cie. in 1888 shows a slightly modified version of the three-wheel vehicle patented and publicly driven for the first time in 1886.

Courtesy of Mercedes-Benz Konzernarchiv, Stuttgart.

3
Internal Combustion Engines

After supper my wife said, "Let's go over to the shop and try our luck once more...." My heart was pounding. I turned the crank. The engine started to go "put-put-put," and the music of the future sounded with regular rhythm. We both listened to it run for a full hour, fascinated, never tiring of the single tone of its song. The longer it played its note, the more sorrow and anxiety it conjured away from the heart.
—Karl Benz, recalling the first successful run of his two-stroke engine on New Year's Eve 1879

Karl Benz (1844–1929) had a very good reason to be pleased as he listened to his two-stroke engine: his contract machine-building business was failing and a new gasoline-fueled engine, one that he designed and built in his free time, was to solve his difficulties. But neither he nor his wife Bertha (1849–1944)—who had more faith in his inventions than he did himself—could have foreseen the consequences of this experiment. By July 1886, Benz mounted a successor of that engine—still a low-powered but now four-stroke machine—on a three-wheel chassis that was surmounted by two high-perched seats, an innovation that made him eventually famous as the "inventor" of automobile (Figure I.3.1). Quotation marks are imperative because the motor car is one of the least appropriate artifacts whose genesis can be ascribed to a single inventor (Beaumont 1902/1906; Walz and Niemann 1997).

But Benz was surely right about the music of the future. In a few generations, its rhythm became ubiquitous, and there are now few other human creations that have been so admired and cherished and yet so disparaged and reviled as internal combustion engines. In these ingenious devices, steel pistons are tightly confined in cylinders as they are rapidly driven by mini explosions of compressed fuel, and their frenzied reciprocating motion is then converted

Figure I.3.1 Karl Benz.
Photograph courtesy of Mercedes-Benz Konzernarchiv, Stuttgart.

into smooth rotation that is transmitted to wheels or propellers. Attitudes aside, their regular rhythm has come to define much of the tempo of modern civilization. Installed in cars, they bring choices and opportunities of personal freedom and mobility as well as environmental degradation and traffic congestion. They enlarge the world of commerce, powering trucks, ships, and planes as well as many kinds of off-road machines that are now indispensable in agriculture, mining, and construction.

The history of internal combustion engines is an even better example of the multifaceted, complex, synergistic origins of a major technical advance than the formation of electric systems. The first working engines were relatively inefficient coal gas–fueled sources of stationary power, and, as their performance improved, liquid-fueled engines diffused to dominate road, off-road, water, and air transportation. Consequently, it is useful to separate the history of internal combustion engines into five stages, some consecutive or partially overlapping, others concurrent.

The first period embraces failed efforts to build explosively powered machines, a quest going back to the seventeenth century. The pace picked up during the first half of the nineteenth century, culminating with the first serious attempts to build steam-powered powered vehicles. Walter Hancock's (1799–1852) series of quaintly named machines—"Infant," "Enterprise," "Autopsy," "Era," and "Automaton"—built during the early 1830s, was particularly notable (Beaumont 1902). The steam engine has obvious disadvantages in road transportation, but it was not eliminated as a serious contender for powering passenger vehicles until the very end of the nineteenth century. Even some of the early proponents of powered flight tried to use it despite its inherently high mass-to-power ratio.

The second stage in the development of internal combustion engines, during the 1850s, saw the construction of the first commercially promising but heavy stationary machines powered by coal gas. Their subsequent improvement, particularly by Nicolaus Otto and Eugen Langen, led eventually to producing thousands of small horizontal engines for workshop and industrial uses. Initially, they were non-compressing and mostly double-acting machines, but Otto's pioneering 1876 design of a compression four-stroke engine proved immediately popular. This machine embodied the key operating features of all modern engines, but it was not suitable for transportation, and, despite its prompt patenting, its priority was later disputed. Practical automotive designs emerged during the third stage, when gasoline-powered engines were used in the first carriage-like vehicles of the 1880s. This period is usually portrayed as dominated by the largely German invention of the automobile beginning with motorized carriages built by Gottlieb Daimler and Wilhelm Maybach and, independently, by Karl Benz.

But better engines are just one precondition of a practical and reliable machine: a lengthy list of other requirements ranges from the overall design, easy starting, and good transmission to cushioned wheels. These improvements began during the early 1890s, with a fundamental redesign of four-wheeled vehicles by Emile Levassor and continued with innovations made in France, Germany, the United States, and the United Kingdom. Another notable event of that intensively innovative period was Rudolf Diesel's (1858–1913) patenting of a new kind of engine, one that needed neither carburetor nor a sparking device and that could use heavy oils rather than gasoline. The fourth stage includes the rapid maturation of high-performance four-stroke engines and gradual emergence of vehicles whose overall design set the trends of automotive developments for much of the twentieth century. This stage also included the development and maturation of reciprocating engines for the first generations of airplanes.

Motorized vehicles could not become a mass market item if their production continued the practices prevailing during the first two decades of their existence. Artisanal manufacturing, either as unique items or in very small series by skilled (originally mostly bicycle) mechanics, could not deliver either the numbers needed or lower prices. Hence, the fifth key stage of early development (and one that was largely concurrent with the fourth) was marked by the first steps toward highly efficient mass production of automotive engines and other car parts that eventually made the car industry the leading manufacturing activity of every large modern economy. Only a short lag separated these advances from a no less impressive large-scale production of aircraft.

Remarkably, nearly all basic challenges of durable design and affordable car manufacturing were resolved in a highly effective fashion before World War I. Obviously, the late twentieth-century cars did not share outward features with their early twentieth-century predecessors: low profile versus a tall, boxy appearance; enclosed versus open or semi-open bodies; curved versus angular body shapes; sunken versus free-standing headlights; and thick versus thin tires were just the most obvious differences. But, as I will show, the evolution has been much more conservative under the hood. And while commercial long-distance flight ceased to rely on reciprocating engines during the 1960s, these efficient and highly reliable machines keep powering large numbers of smaller planes used for short-haul transport, pastime flying, and tasks ranging from spraying crops to water-bombing burning forests.

Taken for granted because of their ubiquity, internal combustion engines should still be an object of admiration if only because these affordable machines, now made up of many thousands of precision-machined (or cast or extruded or molded) parts, work so reliably for such long periods of time (Taylor 1984; Heywood 1988; Stone 1993; van Basshuysen and Schäfer 2007). What is even more remarkable is that they do so requiring relatively little service while subject not only to environmental extremes but also to abusive handling by users, most of whom know nothing about their design or operating requirements. The reliability, durability, and affordability of internal combustion engines are major factors that work against their rapid replacement with alternatives that may be less noisy and much less polluting. This process is now underway with passenger vehicles as cleaner but costlier and heavier electric cars are gaining new market share (particularly in China and the European Union), but electrifying heavy-duty road (and off-road) transportation is a more demanding task.

The importance of internal combustion engines for the smooth functioning of modern societies is taken entirely for granted in countries where their numbers come close, match, and even surpass the total numbers of population. There were 340 million people in the United States in 2023, but more than 290 million automotive and close to 100 million other internal combustion engines. The latter category included about 50 million engines in lawnmowers and other garden machines, about 12 million outboard and inboard engines in recreational boats, 9 million motorcycles, more than 1 million snowmobiles, with emergency electricity generators being among the latest popular additions. While the ubiquity of internal combustion engines is obvious, few people realize that in affluent countries their aggregate installed power has been, for generations, considerably greater than that of any other prime mover.

Reliable US data show that the total power of automotive engines had surpassed that of all prime movers used in electricity generation (steam engines, steam, and water turbines) already before 1910 (USBC 1975). During the last decade of the twentieth century, the total capacity of the US vehicular internal combustion engines was more than 20 terawatts (TW) compared to less than 900 gigawatts (GW) installed in steam and water turbines, or more than a 20-fold difference (Figure I.3.2). And a worldwide comparison shows that while, in 2000, the total installed power of electricity-generating turbines was about 3.2 TW, the global fleet of road vehicles alone had installed power of at least 60 TW.

Multiple Beginnings

Cummins (1989: 1) opens his meticulous history of internal combustion engines by noting, half seriously, that their development began with the invention of the cannon, a device whose key drawback was "that it threw away the piston on each power stroke." But this very action clearly demonstrated the potential utility of tamed explosion, and the first sketchy suggestions of engines exploiting that principle had appeared already in the late seventeenth century. But the first serious attempts to use hot gas rather than steam as a working medium date to the very end of the eighteenth century, and many non-compression (and hence inefficient) engines using mixtures of

112   THE TWENTIETH CENTURY

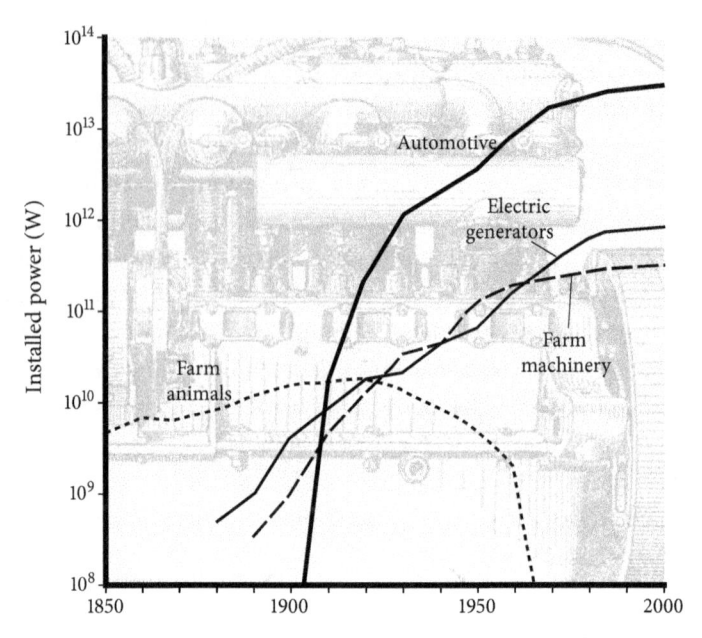

Figure I.3.2 Long-term trends of the total power of main categories of the US prime movers show that automotive internal combustion engines surpassed the aggregate power of both draft animals and electricity-generating equipment before 1910.
Plotted from data in USBC (1975).

coal gas and air were built before 1860. The development of internal combustion engine thus resembles the early history of incandescent light, with more than a dozen inventors—from Philippe Lebon in 1801 to Eugenio Barsanti and Felice Mateucci in 1854—patenting and demonstrating their designs.

The first engine that caused a great deal of public interest and that was manufactured for sale was conceived only in 1858 and patented in 1860, by Jean Joseph Etienne Lenoir (1822–1900). Like a steam engine, after which it was patterned, it was a stationary, horizontal, double-acting machine with slide valves to admit the mixture of illuminating gas and air and release the burned and expanded gas. Ignition was with an electric spark without any fuel compression. This slow (about 200 rpm) engine had a thermal efficiency of less than 4%, and fewer than 500 units (rated at about 2 kilowatts [kW]) were manufactured during the 1860s. In 1862, a liquid-fuel version of the engine, mounted on a three-wheel cart and using a primitive carburetor, propelled the vehicle from Paris to Joinville-le-Point for a total distance of 18 kilometers (km). But there was no follow-up because Lenoir did not pursue any further development of his engine.

Otto and Langen

Nicolaus August Otto (1832–1891; Figure I.3.3) was an unlikely candidate for revolutionizing the design of internal combustion engines (Langen 1919; Sittauer 1972). He was a traveling salesman for a wholesale food company (what Germans call *Kolonialwarengrosshandlung*)—but what he really wanted to do was to design a better engine. His experiments with Lenoir's engine led to the first Otto engine fueled with an alcohol–air mixture, but a better machine emerged only after Otto approached Eugen Langen (1833–1895), the well-off owner of a sugar-refining business, to invest in a newly established company (N. A. Otto & Cie.) that was formed in 1864.

Figure I.3.3 Nicolaus Otto, the inventor of eponymous two- and four-stroke gas-fueled internal combustion engines.
Reproduced from Abbott (1934).

The new company soon produced a two-stroke engine that resembled the Barsanti-Mateucci machine, but some of whose features were sufficiently different to be patented in 1866 in Germany and a year later in the United States. Its tall cylinder surmounted by a large five-spoked flywheel was heavy and noisy, but, when it was displayed in 1867 at the Paris Exhibition, it was found to be more than twice as efficient as other featured gas engines. This earned it a Grand Prize, and the small company, suddenly beset with orders, began serial production in 1868. Once a new substantial investor was found, the whole enterprise was expanded and then in 1872 reorganized as Gasmotorenfabrik Deutz AG, named after a suburb of Cologne, where it was relocated. Today's Deutz AG remains one of the world's leading engine makers (Deutz 2024).

Gottlieb Daimler (1834–1900), an experienced engineer who worked previously also in France and England, was appointed production manager. Daimler brought along Wilhelm Maybach (1846–1929)—a draftsman whom he met for the first time in the machine workshop of the Bruderhaus orphanage in Reutlingen, where Maybach lived since the age of 10—as a new head of the design department (Rathke 1953). The company's improved non-compression engine, launched in 1874, had an efficiency of about 10%, but this performance was achieved by installing a nearly 4 meter (m) tall cylinder and letting the gas expand to 10 times its charged volume (compared to just 2.5 times in Lenoir's engine). Moreover, at about 900 g/W, the engine was much heavier than the best steam engines of the day.

But it was a commercial success, and hundreds of units were built every year during the latter half of the 1870s in Germany and under licenses in France, the United Kingdom, and the United States. The company was engaged in numerous patent litigation trials, and Otto felt that a much better machine was needed to preempt the competition. By the spring of 1876, he had such a design: a horizontal four-stroke compression engine that he conceived and had built without Daimler's help. The four-stroke operation was not the only innovation: Otto's patents stressed more the novelty of stratified combustion, whereby the gas and air mixture is preceded by air alone to create a lean charge near the piston and to produce a smoother, gradual burning of the fuel.

Nothing came out of this proposition during Otto's time and for decades afterward, but the concept was revived during the 1970s as one of the means to lower the emissions of combustion gases that act as precursors

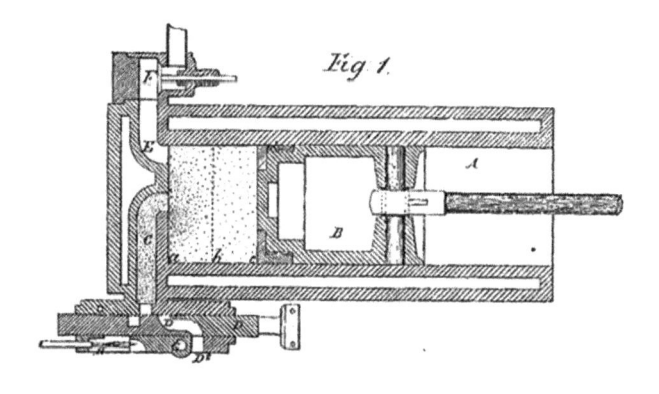

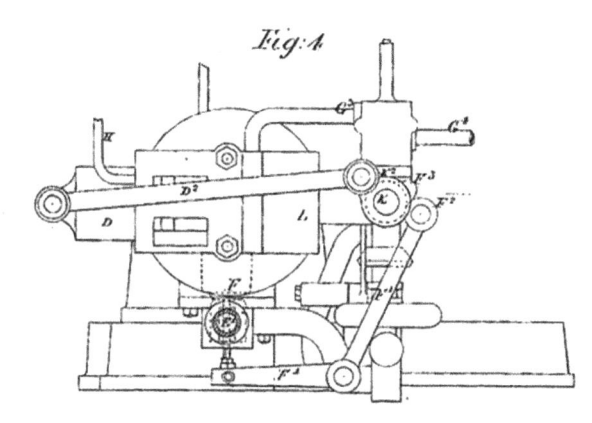

Figure I.3.4 Cross-section of a cylinder (with continuous ignition flame on the lower right) and overall arrangement of Otto's gas-fueled four-stroke engine patented in the United States in 1877. This illustration accompanies US Patent 194,047.

of photochemical smog. Otto's distrust of electric ignition led him to retain a small continuous ignition flame that was controlled by a sliding valve (Figure I.3.4). This is how Otto summarized the engine's key feature in his US patent (Otto 1877: 1):

> According to my present invention an intimate mixture of combustible gas or vapor and air is introduced into the cylinder, together with a separate charge of air or other gas, that may or may not support combustion, in such a manner and in such proportions that the particles of the combustible

gaseous mixture are more or less dispersed in an isolated condition in the air or other gas, so that on ignition, instead of an explosion ensuing, the flame will be communicated gradually from one combustible particle to another, thereby effecting a gradual development of heat and a corresponding gradual expansion of the gases, which will enable the motive power so produced to be utilized in the most effective manner.

Under Maybach's direction the new design was improved and turned rapidly into a series of production models as both the company's tests and independent appraisals confirmed the engine's superior performance. Although its overall thermal efficiency was basically the same as for the noncompressing engine (about 17%), the new design reduced the piston displacement by 94% and the mass-to-power ratio by nearly 70%. And although the engines were still quite heavy (at best, 200 g/W), they were lighter than comparably sized small steam engines, they could be built with much higher maximum capacities than the original Otto & Langen machines, and they were also much quieter and vibrated much less. Otto's horizontal four-stroke engines were targeted for workshops that were too small for installing their own steam engine. Eventually, nearly 50,000 units with a combined capacity of about 150 megawatts (MW) (averaging about 3 kilowatts [kW] per engine) were made over the period of 17 years.

Specifications for Otto's typical early four-stroke machine were a compression ratio of about 2.6, an air-to-fuel ratio of 9:1, a bore of 210 millimeters (mm), and a stroke length 350 mm for a 6-kW engine running at 160 revolutions per minute (rpm), and they were still quite heavy at more than 250 grams (g) per watt (W) (Clerk 1909). The fundamental operating principle of Otto's invention has not changed, but the inventor would be impressed by the much improved performance of today's identically powerful engines. A horizontal Honda model GX240, a versatile small engine, provides an excellent example (Honda Engines 2024). This engine runs much faster (3,600 rpm), has a much higher compression ratio of 8.5:1 (hence a much smaller bore and stroke, 77 mm and 58 mm), and it is much lighter (slightly more than 4 g/W). And, of course, it is fueled by gasoline and not by illuminating gas made from coal. But before I describe the rapid improvements of gasoline engines during the 1880s, I must clarify the origins of the four-stroke engine by looking at the role of Beau de Rochas.

Beau de Rochas and Otto: Ideas and Machines

The success of Otto's four-stroke engine led to many attempts to challenge its patent rights, but all of them were unsuccessful until the chance discovery of a forgotten unpublished patent. This claim was filed by a French engineer Alphonse Eugène Beau (1815–1893, who later in life styled himself Beau de Rochas) with the Société de Protection Industrielle on January 16, 1862 (No. 52,593) under an expansive title *Nouvelles recherches et perfectionnements sur les conditions pratiques de la plus grande utilisation de la chaleur et en général de la force motrice, avec application aux chemins de fer et à la navigation* (Payen 1993). The patent contained no drawings, and it was not based on any experimental model— but it detailed the principle of a four-stroke engine powered by a gas–air mixture that is compressed before its combustion.

Beau de Rochas did so by prescribing first the four parameters "for perfectly utilising the elastic gas in an engine" (citing from Donkin's [1896: 432–433] translation).

1. The largest possible cylinder volume with the minimum boundary surface.
2. The greatest possible working speed.
3. The greatest possible number of expansions.
4. The greatest possible pressure at the beginning of expansion.

As for the sequence (shown in Figure I.3.5 by both piston movement and pressure-volume diagram), "the following operations must then take place on one side of the cylinder, during one period of four consecutive strokes:

1. Drawing in the charge during one whole piston stroke
2. Compression during the following stroke.
3. Inflammation at the dead point, and expansion during the third stroke.
4. Discharge of the burnt gases from the cylinder during the fourth and last stroke.

Because Beau de Rochas failed to pay the requisite fee, the patent was never published, and it circulated in just a few hundred copies distributed by the inventor. He also did not try to assert his rights when Otto's engine first appeared on the market or for many years afterward. Only in 1884, after a

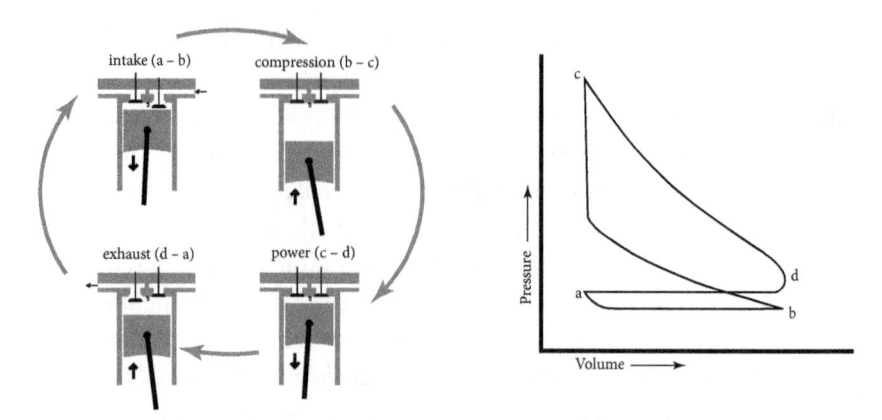

Figure I.3.5 Idealized Otto cycle (with heat supplied and exhausted at constant volume) and piston movements corresponding to the four phases of the pressure-volume diagram.

patent attorney called attention to Beau's old private pamphlet, did the Körting brothers, builders of large gas engines in Hannover, contest the originality of Otto's idea. Although it was obvious that Otto's engine was designed without any reference to Beau's obscure description, Otto eventually lost the legal fight in Germany in 1886, but his rights were upheld in the United Kingdom. Not surprisingly, German engineers, as well as British and American automakers, talk about the Otto cycle, but in France it remains *cycle de Beau de Rochas*. Whatever the name, the four strokes—aspiration, compression, inflammation, refoulement in Beau's language, *Ansaugen, Verdichten, Arbeiten, Ausschieben* in Otto's tongue—have become one of the ruling rhythms of modern civilization.

But the Beau–Otto affair goes far beyond the common patent interferences and usual complications and revisions in the dating of inventions: it goes to the very core of the concept of the Age of Synergy that I advocate in this book. Beau de Rochas was content to set down his ideas on paper and then walk away from them, although he had opportunities to do otherwise: he was Lenoir's friend, since 1852 he lived and worked in Paris in the government's transportation department, and he was engaged in many projects that involved propulsion machinery. Moreover, he could have constructed at least a working prototype—but he simply chose not to do so. He made designs for railways and ship docks and planned the use of Lake Geneva for hydroelectric generation, but he did not try to convert his 1862 idea of a four-stroke engine even into a toylike model.

In contrast, Otto's approach was among the best examples of the advances characteristic of the Age of Synergy. First, he developed a non-compression engine better than Lenoir's inefficient machine; then, working not only against the dominant consensus but even against the judgment of his experienced production manager, he developed a much better four-stroke compression engine; he had the design patented properly and promptly; and, without delay, he then committed resources to its further improvement, started its manufacturing in large series, and licensed its production widely in his country and abroad. This is the very procedure whose numerous repetitions eventually created the civilization of the twentieth century. Not surprisingly, Otto's inventive elan was sapped by the German denial of his four-stroke patent. He died before his fifty-ninth birthday in Cologne, the town where he spent nearly all his adult life.

German Trio and the First Car Engines

The next critical stage in the history of internal combustion engines continued to unfold in Germany, and we have already met all three protagonists, Benz, Daimler, and Maybach. Karl Friedrich Benz (1844–1929; Figure I.3.1) was born in Karlsruhe to a family whose men were smiths for several generations. After his technical studies, Benz set up an iron foundry and a mechanical workshop, and, in 1877, with his main business struggling, he began to build stationary engines. Gottlieb Daimler (Figure I.3.6) worked for Otto & Langen (later Deutz AG) between 1872 and 1882, and he left the company after disagreements with Otto. And Wilhelm Maybach, whom Daimler recruited to N. A. Otto & Cie in 1873, improved the designs of both Otto's two patented engines and brought them to commercial production. The efforts of these three men eventually created in the first practical high-speed, four-stroke, gasoline-fueled engines.

After leaving Deutz at the end of 1881, Daimler, soon joined by Maybach, set up a shop in the Gartenhaus on his comfortable property in Canstatt, Stuttgart's suburb, and set out to develop a powerful yet light gasoline-fueled engine (Walz and Niemann 1997; Adler 2006). Gasoline, by that time readily available as the first volatile product of simple thermal refining of American and Russian crude oil, was an obvious choice. With about 33 megajoules per liter (MJ/L) (nearly 44 MJ/kg), the fuel has about 1,600 times the energy density of illuminating gas, and its low flashpoint (−40°C) makes it ideal for

Figure I.3.6 Gottlieb Daimler.
Photograph courtesy of Mercedes-Benz Konzernarchiv, Stuttgart.

easy starting (but also hazardous to use). By 1883, the two engineers had the prototype of a high-revolution (about 600 rpm) gasoline-fueled engine with a surface carburetor and inflammation-prone hot-tube ignition (heated by an external gasoline flame) that caused repeated flare-ups.

In 1885, Daimler and Maybach fastened a small (about 350 W) and light, air-cooled version on a bicycle, creating a prototype of the motorcycle, which was driven for the first time on November 10, 1885, by Daimler's son Paul just 3 km from Cannstatt to nearby Untertürkheim (Walz and Niemann 1997). Figure I.3.7 shows this—still rather unwieldy—machine as well as its subsequent transformation into a modern-looking motorcycle with electric engine starter that became popular just before World War I (O'Connor 1913). Going a step further, in March 1886, Daimler ordered a coach with four wooden wheels from a Stuttgart coachmaker and mounted on it a larger (0.462 liters [L], 820 W), water-cooled version of the engine (Figure I.3.8).

Daimler and Maybach's inelegant vehicle, with passengers precariously perched high above the ground, made its first rides in Daimler's garden, then a trip from Cannstatt to Untertürkheim sometime during the fall of 1886, and, in 1887, it was driven from Cannstatt to Stuttgart at about 18 km/h. Their engine had only a single opening for inlet of the charge and the

INTERNAL COMBUSTION ENGINES 121

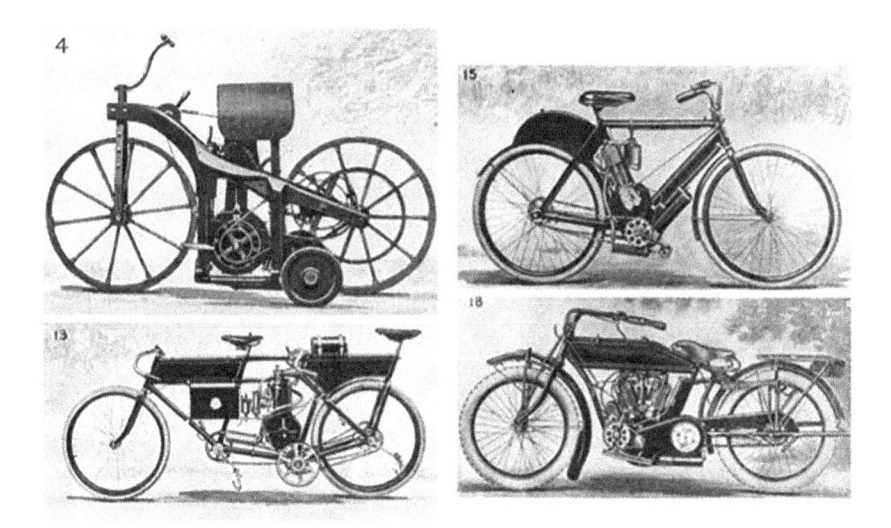

Figure I.3.7 Daimler and Maybach's unwieldy 1885 motorcycle with subsidiary wheels compared to a tandem machine of 1900 and to standard motorcycles of 1902 and 1913.
Reproduced from O'Connor (1913).

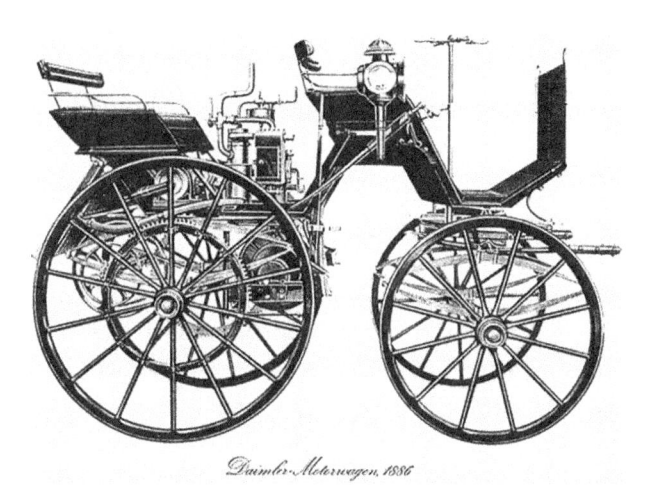

Figure I.3.8 Daimler and Maybach's first motor car. Their engine was installed in a coach ordered from Wimpf & Sohn in Stuttgart for the forty-third birthday of Daimler's wife, Emma Pauline.
Image courtesy of Mercedes-Benz Konzernarchiv, Stuttgart.

exhaustion of gases into the cylinder. Gasoline was supplied from the float chamber to a carburetor, and the engine was cooled by water. Remarkably, their achievements were being independently anticipated and duplicated by Karl Benz working in Mannheim, about 120 km northwest of their Cannstatt workshop (Walz and Niemann 1997; Grünewald 2013).

On January 29, 1886, Karl Benz was granted the German patent (DR Patent 37,435) for a three-wheeled vehicle powered by a four-stroke, single-cylinder gasoline engine: this date is often considered as the origin of the first automobile. This was Benz's goal all along: he decided to concentrate on developing a two-stroke engine intended from the very beginning to be used in vehicles. As described in this chapter's epigraph, he succeeded in doing that just before the end of 1879; by 1882, he had a reliable two-stroke gasoline-fueled, water-cooled horizontal engine with electric ignition, and, in 1883, after securing an investor, a new company was finally set up. After the expiry of Otto's patent, Benz began designing four-stroke gasoline engines.

The first three-wheeled motorized carriage was driven publicly for the first time on July 3, 1886, along Mannheim's Friedrichsring. The light vehicle (total weight was just 263 kg, including the 96-kg single-cylinder four-stroke engine) had a less powerful (0.954 L, 500 W) and a slower-running (just 250 rpm) engine than did Daimler and Maybach's carriage, and it could go no faster than about 14.5 km/h. The vehicle's smaller front wheel was first steered by a small tiller and then by a horizontal wheel; the engine was placed over the main axle under the narrow double seat, and drive chains were connected to gears on both back wheels (see the frontispiece to this chapter). The car had to be started by turning clockwise a heavy horizontal flywheel behind the driver's seat, but water cooling, electric coil ignition (highly unreliable), a spark plug (just two lengths of insulated platinum wire protruding into the combustion chamber), and differential gears were key components of Benz's design that are still standard features in modern automobiles.

The first intercity trip of Benz's three-wheeler took place two years later, in August 1888. Benz's wife Bertha took the couple's two sons and, without her husband's knowledge, drove the three-wheeler from Mannheim to Pforzheim to visit her mother, a distance of about 100 km (Mercedes-Benz Group 2024). After their arrival, they sent a telegram telling Benz about the completion of their pioneering journey. That trip was made in a vehicle with a backward facing third seat that was placed above the front wheel, and, by 1889, other additions included body panels, two lamps, and a folding top. These machines were seen largely as curious, if not ridiculous, contrivances, and, before 1890, only a few three-wheelers were made for sale; the same was true about the 3-horsepower (hp) Victoria, Benz's first four-wheeler, which went on sale only in 1893.

Meanwhile, in 1889, Daimler and Maybach introduced a new two-cylinder V engine (angled at 17 degrees) that displaced 0.565 L and ran at 920 rpm. Although they mounted the engine on a better chassis and used steel wire rather than wooden spoke wheels, the vehicle with larger back wheels still retained a decidedly carriagelike look. But the design pioneered the concept of power transmission through a friction clutch and sliding-pinion gears. A year later, in November 1890, when new partners were brought in and the Daimler Motoren Gesselschaft (DMG) was set up, they produced their first four-cylinder engine. But neither Benz nor Daimler and Maybach designed the first vehicle that was not merely a horseless carriage: that was done by a French engineer, Emile Levassor (1844–1897).

As a partner in Panhard & Levassor, Levassor was introduced to the German V engine by Daimler's representative in Paris, and, in 1891, working for Armand Peugeot (1849–1915), he designed an entirely new chassis. Cars with Daimler and Maybach's engines took four out of the first five places in the world's first car race in July 1894 (a steam powered De Dion & Bouton tractor beat them). A year later, when Levassor himself drove his latest model to victory in the Paris–Bordeaux race (a round trip of nearly 1,200 km) at average speed of 24 km/h (Beaumont 1902/1906), the car's Daimler-Maybach 4.5-kW engine weighed less than 30 g/W, an order of magnitude lighter than Otto's first four-stroke machines.

A momentous event in the company's history took place on April 2, 1900, when Emil Jellinek (1853–1918), a successful businessman and the Consul General of the Austro-Hungarian Empire in Monaco, set up a dealership for Daimler cars (Robson 1983). He placed the initial order for 36 vehicles worth 550,000 Goldmarks and then doubled it within two weeks. His condition for this big commitment: exclusive rights for selling the cars in the Austro-Hungarian Empire, France, Belgium, and the United States under the Mercédès (the name of Jellinek's daughter) trademark. Maybach responded to this opportunity by developing a design that Mercedes-Benz (2024) persists in describing as the first modern automobile.

Being basically a race car, the Mercedes 35 had an unusually powerful, 5.9-L, 26-kW (35 hp) four-cylinder engine running at 950 rpm with two carburetors and with mechanically operated inlet valves; the vehicle had a lengthened profile and a very low center of gravity. Shifting was done with a gear stick moving in a gate. What was under the hood was even more important: Maybach reduced the engine's overall weight by 30% (to 230 kg) by using an aluminum engine block, and the much greater cooling surface of his new honeycomb radiator (standard designs that are still with us) made it possible to reduce the coolant volume by half. Consequently, the mass-to-power ratio of this powerful engine was reduced to less than 9 g/W, 70% below the best DMG engine in 1895, and the vehicle's total body mass was kept to 1.2 tons (t).

Within months the new car broke the world speed record by reaching 64.4 km/h, and a more powerful Mercedes 60, with unprecedented acceleration and more elegant bodywork, was introduced in 1903. This combination of performance and elegance, of speed and luxury, has proved to be the most endurable asset of the marque: the company advertised it as much in 1914 as it does today, and it continues to charge the buyers a significant premium for this renown (Figure I.3.9). Introduction of the Mercedes line could be seen as the beginning of the end of Germany's automotive *Gründerzeit*. During the first decade of the twentieth century, the center of automotive development

Figure I.3.9 Ludwig Holwein's elegant pre-World War I advertisement for a Mercedes.
Courtesy of Mercedes-Benz Konzernarchiv, Stuttgart.

shifted to the United States as all three German protagonists left the automobile business. Daimler died in 1900, Benz left his company in 1906, and Maybach left DMG in 1907.

Before leaving this early history of German internal combustion engines, I should clarify the subsequent joining of Daimler's and Benz's names. In 1926, Benz & Cie and DMG joined to form Daimler-Benz AG, uniting the names and traditions of the two great pioneers of internal combustion engines and automaking: the two men who lived for decades in cities that were just a bit more than 100 km apart never met during their long lives. In 1998, Daimler-Benz AG bought Chrysler, America's fourth largest automaker (ranking behind General Motors, Ford, and Honda) to form DaimlerChrysler but that merger ended in 2007. Then, until 2022, the company was known as Daimler AG and now it is the Mercedes-Benz Group.

That I have said so far nothing about the US automakers has a simple explanation. As the leading British expert put it at the beginning of the twentieth century (Beaumont 1906: 268), by 1906 "progress in the design and manufacture of motor vehicles in America has not been distinguished by any noteworthy advance upon the practice obtaining in either this country or on the Continent." Matters changed just two years later with the arrival of Ford's Model T—but before taking up that epoch-making event I must describe the invention of the most efficient internal combustion piston engine by yet another German engineer.

The Diesel Engine

Rudolf Diesel (1858–1913), the man who invented a different internal combustion engine, set out to do so as a 20-year-old student. He succeeded (although not in the way he initially envisaged) nearly 20 years later, but ended his life before he could witness his machine's rise to global prominence (Figure I.3.10). Diesel's account of his invention, written in 1913, explains how the idea of a better engine dominated his life since he was a university student (Diesel 1913: 1–2).

> When my esteemed teacher, Professor Linde, explained to his audience at the Munich Polytechnic in 1878, during his lecture on thermodynamics, that the steam-engine transforms into effective work only 6–10% of the heat value of its fuel, when he explained Carnot's theorem, and pointed out that with isothermic changes of condition all heat conducted to a gas would

Figure I.3.10 Rudolf Diesel failed to design a near-Carnot cycle engine, but his new machine became the most efficient means of converting chemical energy of liquid fuels into motion through internal combustion.
Reproduced from *Scientific American*, October 18, 1913.

be converted to work, I wrote in the margin of my college notebook: "Study whether it is not possible to realize in practice the isotherm?" Then and there I set myself the task! That was still no invention, not even the idea of one. Thereafter the wish to realize the ideal Carnot cycle dominated my existence.

Diesel's invention belongs to the same category as Charles Parsons's steam turbine: both had their genesis in understanding the laws of thermodynamics and hence both are excellent examples of fundamental innovations that appeared for the first time only during the Age of Synergy: inventions driven and directed by practical applications of advanced scientific knowledge, not products of stubborn experimenting or chance outcomes. There is, however, an important difference: Parsons had realized his initial goals almost perfectly. Diesel's main objective, inspired by Carl Linde's (1842–1934) lectures, proved to be technically impossible, and he had to depart fundamentally from his initial idea of a near-ideal engine, but he still designed a prime mover of unprecedented efficiency.

Diesel also had an explicit social goal for his machine that he described in his famous programmatic publication (Diesel 1893): he wanted to change the world by introducing a prime mover whose affordable operation would make small businesses competitive. After his studies in Munich, Diesel became employed by Linde's ice-making company in Paris in 1880, and he spent most of the 1880s perfecting and developing refrigeration techniques, including an ammonia absorption motor (Diesel 1937). After several years of experiments, he filed a patent application for a new internal combustion engine in February 1892, and, in December of the same year, he was granted the first German patent (Patent 67,207) for the machine. Thanks to the interest by Heinrich Buz (1833–1918), director of the Maschinenfabrik Augsburg, and with financial support by Friedrich Krupp, the difficult work of translating the idea into a working device began soon afterward.

The operating mode of this four-stroke engine was to be unique. Air alone was to be drawn into the cylinder during the charging stroke and then compressed to such an extent (initial specifications called for as much as 30 megapascals [MPa]) that its temperature alone (between 800°C and 1,000°C) would ignite a liquid fuel as it entered the cylinder. But, as Diesel was eager to stress, to claim that the self-ignition of the fuel is the most important attribute of his design would be an incorrect and superficial view: it would be absurd to build a much heavier engine merely for the sake of ignition when the machine is cold because, once the engine is warmed up, self-ignition takes place at lower pressures. The engine's principal goal was the highest practical efficiency, and this necessitated the use of compressed air (Diesel 1913). Diesel's initial goal was to achieve isothermal expansion: after adiabatic compression of the air, the fuel should be introduced at such a rate that the heat of combustion would just replace the heat lost due to gas expansion; hence, all thermal energy would be converted to kinetic energy and the temperature inside the cylinder would not surpass the maximum generated during the compression stroke.

As the conditions close to isothermal combustion could not be realized in practice, Diesel concluded in 1893 that combustion under constant pressure was the only way to proceed (Diesel 1893). But even then, it took more than three years of intensive development and steady downward adjustments of operating pressure before a new engine (German Patent 86,633, US Patent 608,845) was ready for official acceptance testing. This took place in February 1897, and although the pressure–volume diagram bore little resemblance to Carnot's ideal cycle, the engine was still far more efficient than any combustion device (Figure I.3.11). Official testing results showed that

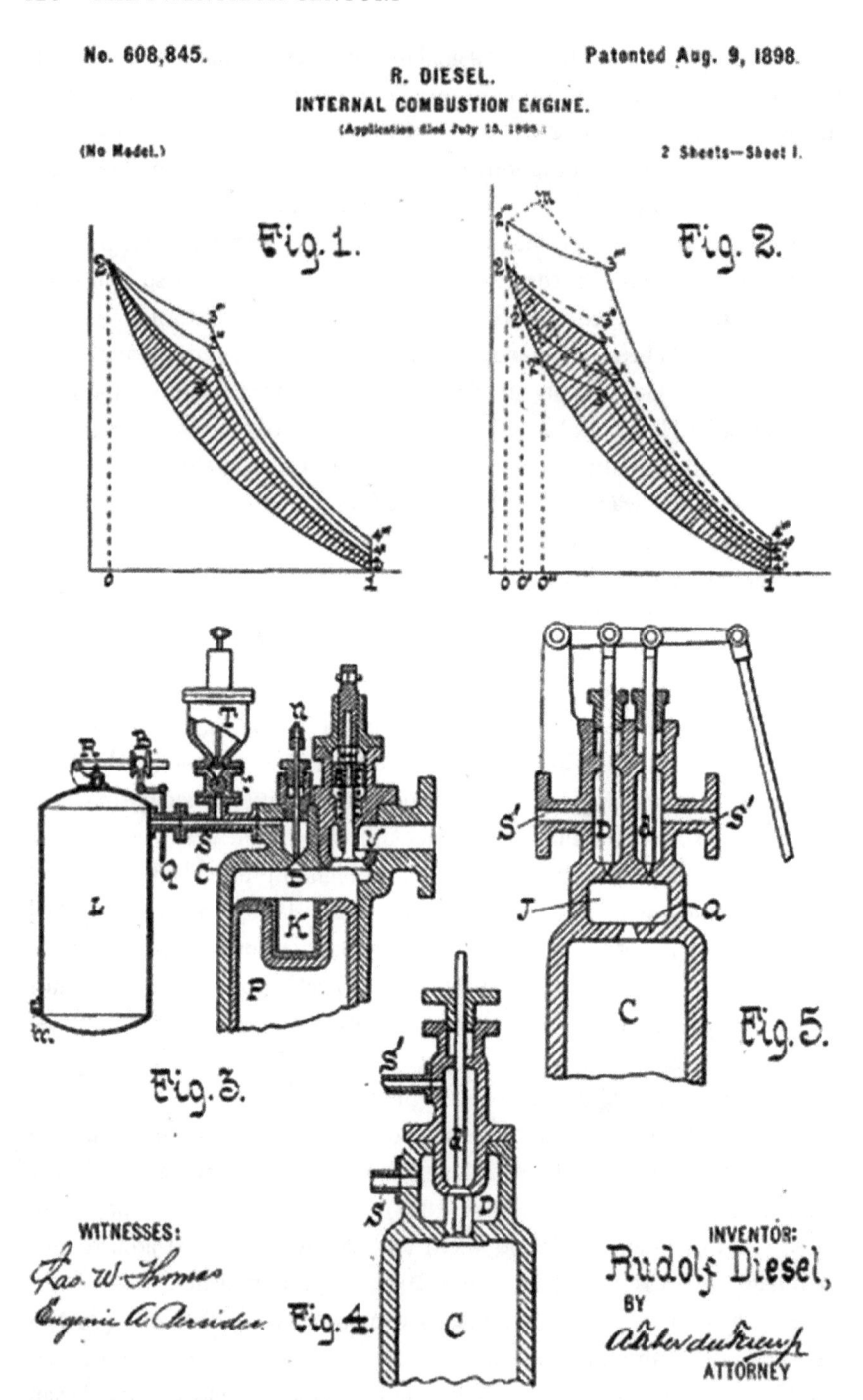

Figure I.3.11 Pressure-volume diagrams and details of cylinder heads from Diesel's US Patent 608,845 (1898) of a new type of internal combustion engine.

the 13.5-kW engine, whose top pressure of 3.4 MPa was an order of magnitude below Diesel's initial maximum specification, had a thermal efficiency of 34.7% at full load, corresponding to 26.2% of the brake thermal efficiency (Diesel 1913).

By 1911, a much larger (445 kW) diesel engine had a mechanical efficiency of 77% and a brake thermal efficiency of 31.7% (Clerk 1911). In contrast, typical brake efficiencies of commonly deployed Otto engines at that time were between 14% and 17%. Diesel's dream of a Carnot-like engine was gone, but expectations for a machine whose efficiency was still twice as high as that of any other combustion device were high, licensing was done swiftly, but actual commercialization of new engines ran into problems associated above all with the maintenance of high pressure and with the timing of fuel injection. The latter challenge was solved satisfactorily only in 1927 when, after five years of development, Robert Bosch (1861–1942) introduced his high-precision injection for diesel engines (Smil 2009).

Moreover, Diesel disliked being engaged in the commercialization phase of his inventions, and he also judged many producers to be incapable of building his splendid machine properly. And, like other inventors, he had to face the inevitable lawsuits contesting his invention, the first one already in 1897. Given the engine's inherently higher mass-to-power ratio, its first applications were in stationary uses (pumps, oil drills) and in heavy transport. A French canal boat was the first diesel-powered vessel in 1903; a French submarine followed in 1904. A small electricity-generating plant to supply streetcars in Kiev was commissioned in 1906, and the first ocean-going vessel with diesel engines was the Danish *Selandia*, in 1912. The first diesel locomotive, with Sulzer engines able to sustain a speed of 100 km/h, was built in 1913, for the Prussian Hessian State Railways (Anonymous 1913a). By that time there were more than 1,000 diesel engines in service, most of them rating 40–75 kW. A small diesel truck was built in 1908, but passenger diesel vehicles had to wait for Bosch's injection pump.

In 1913, Johannes Lüders, a professor at the Technische Hochschule in Aachen, published his *Dieselmythus*, a work highly critical of Diesel's accomplishments (Lüders 1913). That attack, the less than enthusiastic commercial acceptance of his engines, the slow pace of their adoption by the automotive market, prolonged anxiety and illness (including a nervous breakdown), and excessive family spending and resulting financial problems—any one of these factors makes it most likely that Diesel's disappearance was not an accident but a suicide. On September 29, 1913, he boarded a ship from Belgium to England, where he was to attend the opening of a new factory to produce his engines. Next morning, he was not aboard the ship.

Benz died at 85, Maybach at 83: if Diesel had lived so long, he would have witnessed a universal triumph of his machine (Smil 2009). Diesel's conviction that his engine would eventually become a leading prime mover of road vehicles began to turn into reality just 10 years after his death. But his accomplishment was indisputable even at the time of his death. As Suplee (1913: 306) noted in a generous obituary, Diesel's engine "is now known all over the world as the most efficient heat motor in existence, and the greatest advance in the generation of power from heat since the invention of the separate condenser by Watt."

Creating Car Industry and Car Culture

The slow pace of diffusion and acceptance of automobiles is reflected in the small numbers of vehicles in use in the United States. A mere 300 of them were on the roads in 1895; in 1900, 17 years after Daimler and Maybach built their first gasoline engine, the total was just 8,000, which means that only one of out of every 9,500 Americans owned a motor vehicle; by 1905, the nationwide registration total stood at just short of 78,000 cars (USBC 1975). During the 1890s, *Scientific American*, a leading source of information on technical advances, kept paying a great deal of attention to what were by that time only marginal improvements in steam engine design, while its coverage of emerging car engineering was rather unenthusiastic. Similarly, Byrn's (1900) systematic review of nineteenth-century inventions devoted only as many pages (7 out of 467) to automobiles as it did to bicycles. Cost and lack of infrastructure had much to do with this: the first generation of automobiles manufactured in small series by artisanal methods were expensive, driving on unpaved roads was difficult, and the unreliability of early vehicle designs and a near complete absence of any emergency services turned these dusty (or muddy) trips into unpredictable adventures.

Three cartoons from the British magazine *Punch* capture the essence of these experiences (Figure I.3.12). Travails of automotive pioneers of the 1890s were also vividly captured by Kipling's recollection of "agonies, shames, delays, rages, chills, parboilings, road-walkings, water-drawings, burns and starvations" (quoted in Richardson 1977: 27). The matter of

INTERNAL COMBUSTION ENGINES 131

Figure I.3.12 In the first year of the twentieth century, *Punch* carried many cartoons spoofing the new automotive experience. In the top image, a farmer calls, "Pull up, you fool! The mare's bolting!" "So's the car!" cries the motorist. The bottom left illustration shows "the only way to enjoy a motor-car ride through a dusty country," and the bottom right is not "a collection of tubercular microbes escaping from the congress but merely the Montgomery-Smiths in their motor-car, enjoying the beauties of the country."
Reproduced from *Punch* issues of June 12, June 19, and July 31, 1901.

public safety was addressed for the first time in 1865, in relation to heavy steam-powered traction by the justly ridiculed British legislature, which required that three people driving any "locomotives" on highways and the vehicles, traveling at the maximum speed of 6.4 km/h, be preceded by a man with a red flag (Rolls 1911). Incredibly, these restrictions were extended in

1881 to every type of self-propelled vehicle, and they were repealed only in 1896. No other country had such retarding laws but concerns about the speed of new machines and accidents were seen everywhere. At the same time, nothing could erase a growing feeling that something profoundly important was getting under way.

In November 1895, the opening editorial of the inaugural issue of *American Horseless Age* had no doubts that "a giant industry is struggling into being here. All signs point to the motor vehicle as the necessary sequence of methods of locomotion already established and approved. The growing needs of our civilization demand it; the public believe in it" (cited in Flink 1975: 13). But first the slow, fragile, unreliable, and uncomfortable vehicles with obvious carriage and bicycle pedigrees had to evolve into faster, sturdier, more reliable, and much more comfortable means of passenger transportation. Although European engineers had accomplished some of this transformation by the beginning of the twentieth century, only mass production could make the new vehicles more affordable. Henry Ford took that critical step beginning in 1908, and, by 1914, the two factors that were so important in shaping the twentieth century—car making as a key component of modern economies and car ownership as a key factor of modern life—were firmly in place.

Technical Challenges

The virtually unchallenged dominance that road vehicles powered by internal combustion engines enjoyed during the twentieth century was achieved within a single generation after their first demonstrations—but their fate appeared unclear as late as 1900. Decades of experience with high-pressure steam engines made it possible to build some relatively lightweight machines that won several car races during the 1890s. Among new steam-powered vehicles that were designed after 1900 was Leon Serpollet's beak-shaped racer that broke the speed record for 1 km in 29.8 seconds (s) (equivalent to 120.8 km/h) in Nice, in 1902, and Francis and Freelan Stanley's steam car that set a new world speed record for the fastest mile (205.4 km/h) in 1906. But after that steam-powered passenger cars went into a rapid decline, although heavy commercial vehicles stayed around for several more decades (Flink 1988). And Edison was far from being alone in believing that electric cars would soon prevail. The late

1890s and the first years of a new century looked particularly promising for those vehicles.

In 1896, at the first US track race at Narragansett Park, in Rhode Island, a Riker electric car decisively defeated a Duryea vehicle, and, three years later, in France, another electric car, a bullet-shaped "Jamais Contente" driven by Camille Jenatzy, reached a speed of 100 km/h. And, of course, the electrics were clean and quiet, no high-pressure boilers and hissing hot steam, no dangerous cranking and refills with flammable gasoline. Their commercial introduction began in 1897, with a dozen Electric Carriage & Wagon Company's taxicabs in New York. in 1899, US production of electric cars surpassed 1,500 vehicles, compared to 936 gasoline-powered cars (Burwell 1990). Two years later Pope's Electric Vehicle Company was both the largest maker and the largest owner and operator of motor vehicles in the United States—but, by the end of 1907, it was bankrupt (Kirsch 2000). The combination of technical improvements, ease of use, and affordability shifted the outcome in favor of gasoline-powered vehicles.

There was not a single component of engine and car design that remained unimproved. While fundamental operating principles remain intact even today, the first quarter-century of automotive engineering was remarkable not only for the breadth of its innovations but also for the rapidity with which important substitutions and improvements took place. First to be resolved was the problem of convenient steering. In horse carriages, the front axle with fixed wheels swiveled on a center pivot, and, while it could be turned easily by a horse, it was a challenge for a driver. Benz avoided the problem by opting for a single front wheel governed by a tiller. But there was an effective old solution to front-wheel steering: Rudolph Ackermann's 1818 invention that linked wheels by a rod so that they could move together while turning on independent pivots at the ends of a fixed front axle. England's first gasoline car, built by Edward Butler in 1888, was the first four-wheeler with Ackermann steering, setting a standard for nearly all future vehicles.

As already noted, in 1891, Emile Levassor of the Parisian company Panhard & Levassor, originally makers of woodworking machinery, began to reconfigure the motorized vehicles pioneered by Benz, Daimler, and Maybach in a radical and enduring fashion (Figure I.3.13). He moved the engine from under the seats and placed it in front of the driver, a shift that placed the crankshaft parallel with the car's principal axis rather than parallel with the axles and made it possible to install larger, more powerful engines.

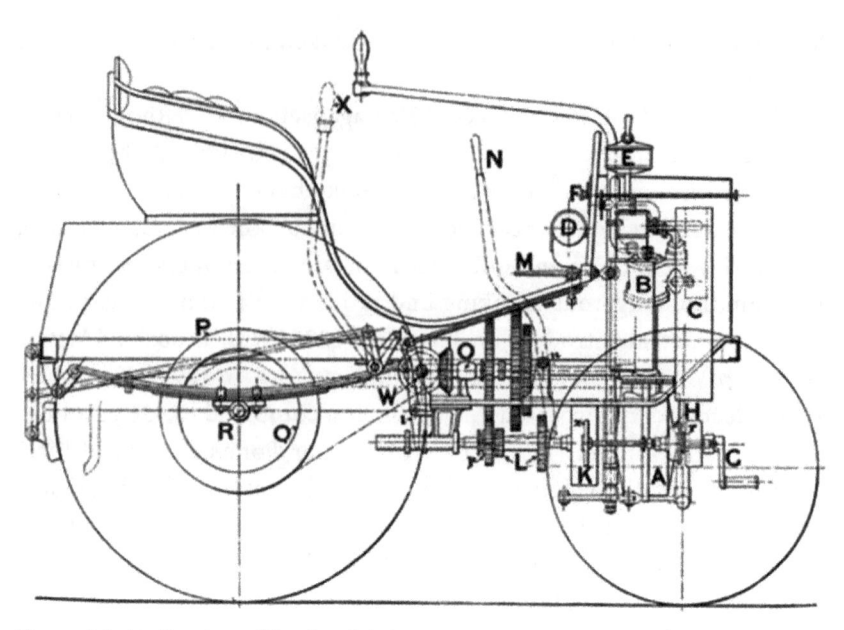

Figure I.3.13 Section of Panhard & Levassor 1894 car equipped with Daimler's motor. A is the crank box of the inclined cylinder, B the carburetor, and C the exhaust silencer. Gasoline flowed from D to E and F (ignition tube and mixture regulator). Three shifting wheels (L) were operated by the lever N.
Reproduced from Beaumont (1902/1906).

This also led inevitably to the design of a protective, and later also aerodynamic, hood. The engine had a friction clutch and sliding-pinion gears, and he replaced the primitive leather drive belt with a chain drive.

As one of the founders of the most illustrious British car marque put it in his encyclopedic survey of motor vehicles, "with all the modifications of details, the combination of clutch, gearbox and transmission remains unaltered, so that to France, in the person of M. Levassor, must be given the honour of having led the development of the motor-car" (Rolls 1911: 915–916). Although the car had Ackermann steering, it was controlled by a long horizontal lever that was finally replaced by 1898 by a modern wheel at the end of an inclined column. Even more important, ignition had to be made much less dangerous as well as much more reliable. The risky open-flame hot tube was first replaced by low-voltage magnetos (generators producing a periodic electric pulse) whose design was improved in 1897 by Robert Bosch

(1861–1942). Bosch's magneto was successfully used for the first time in a de Dion-Bouton three-wheeler that was then able to reach an unprecedented speed of 50 km/h. But the real breakthrough came in December 1901, when Gottlob Honold showed his employer a new high-voltage magneto with a new type of spark plug.

The patent was granted on January 7, 1902, and the timing could not have been better because the high-voltage magneto and spark plug that made high-speed engines a reality became available just as serial car production was finally taking off. Fouling was a recurrent problem with the early designs, but gradual improvements raised the life span of spark plugs to more than 25,000 km. Nickel and chromium were used until the 1970s; a copper core was introduced during the early 1980s, and, shortly afterward, a center electrode was made of 99.9% platinum whose use allows for the maintenance of optimal operating temperature and for reaching a self-cleaning temperature within seconds (Bosch Global 2024; Kanemitsu 2001). In 1902, Bosch produced a total of about 300 spark plugs; a century later the worldwide annual production topped 350 million.

Various carburetor designs of early years eventually gave way to the spray-nozzle device, developed by Maybach in 1893, which became the basis for modern devices that introduce a fine jet of gasoline into the air that is entering a cylinder. Mechanical, cam-operated inlet valves replaced the automatic devices that opened late on the induction stroke and closed prematurely, thus resulting in a weakened charge. Renault was the first car maker to introduce a transmission that was directly connected to the engine: it had three forward and one reverse gears selected with a gear shift. Engine control switched from cutting out impulses to throttling—that is, to reducing the volume of fuel charged into a cylinder at one stroke.

The overall engine design benefited from a greater variety of new high-performance steel alloys whose use made cylinders, pistons, valves, and connecting rods both lighter and more durable, and engine performance was greatly enhanced by improvements in lubrication. Finally, chain transmission to wheels running on a fixed axle was replaced by propeller drive on a rotating axle. Four- and even six-cylinder cars (the latter arranged in a V) eliminated the need for a substantial flywheel, while the honeycomb radiator, patented by DMG in 1900, doubled the efficiency of cooling, and a single lever became the standard for engaging both forward and reverse gears. As in all initial stages of a prime mover's development, installed power kept increasing not only in high-performance vehicles but also in cars designed for the mass market. In 1904, Maybach completed the first six-cylinder Mercedes with 53 kW, and, in 1906, he introduced a 90-kW race engine with overhead intake and exhaust valves and with dual ignition. For comparison, today's remaining small passenger cars (Honda Civic, Toyota Corolla) rate typically between 79 and 90 kW, and the now dominant mid-size SUVs have engines capable of 180–220 kW.

There were also innovations that made cars easier to start, drive, and service in emergencies. All early vehicles were started by turning a hand crank, a demanding and sometimes a dangerous task: a premature engine firing due to an advanced spark setting would rotate the starting crank violently, and it could easily break a wrist or thumb. Early automakers tried to design self-starting engines with acetylene, compressed air, gasoline vapors, and springs (Bannard 1914). The solution arrived in 1911, when Charles Kettering, at his Dayton Engineering Laboratories (the company, now Delco Remy, continues to make starters and other electric parts for cars), succeeded in designing the first practical electric starter, for which he received a US patent in 1915 (Jeffries 1960).

Laborious hand-cranking remained common on cheaper models, but, by 1920, nearly every US car was available with at least optional electric starting. The three-speed and reverse gearbox (the H slot) that remains a standard in today's passenger cars was first offered with a Packard in 1900, the year when Frederick Simms also built the first car fender (NAE 2024). Electric lights and electric horns gradually replaced paraffin and acetylene lanterns and bulb horns, and mechanically operated windshield wipers kept the windows clean in semi-enclosed automobiles. Detachable wheels and a spare wheel were first introduced in the United Kingdom, as was the first speedometer (made in 1902, by Thorpe and Salter with the range of 0–35 mph or 0–55 km/h). The inflatable rubber tire, patented in the United Kingdom in 1888 (also US Patent 435,995 in 1890) by Scottish veterinary surgeon John Boyd Dunlop (1840–1921) and almost immediately used on newly popular bicycles, was superior to solid tires but because it was glued solidly to the wheel rim, it was not easily repaired.

The first detachable rubber tire was produced by the Michelin brothers, Andre (1853–1931) and Edouard (1859–1940), in 1891. The older brother studied engineering and architecture, and the younger one painting, but both returned home to manage their small family business in Clermont-Ferrand, which they eventually transformed into a very successful multinational corporation (Michelin 2024). Their tire enabled Charles Terront to win the Paris–Brest–Paris bicycle race despite a puncture. In 1900, Michelin tires were on Jenatzy's vehicle when it broke the 100 km/h barrier—and a year before that, the company adopted Bibendum, the tire-bulging Michelin Man, as its symbol.

Once the early cars shed their carriage form and began to reach higher speeds, the detachable tires had a huge new market. The company's notable pre-World War I innovations included a dual wheel for buses and trucks that allowed heavier loads (in 1908) and the removable steel wheel that made it possible to use a ready-to-ride spare in 1913. That same year, Michelin released the first detailed map of France that was specifically designed with motorists in mind (today their maps cover all continents). Safer rides were made possible by replacing band brakes with drum brakes that were introduced by Renault in 1902. In December of the same year, Frederick William Lanchester (1868–1946) patented disk brakes, whose clamps hold onto both sides of a disk that is attached to the wheel hub. But as these brakes required more pedal pressure to operate, they became common only with the power-assisted braking that became standard on many American cars during the 1960s.

Yet another important improvement to increase the safety of motoring arrived in 1913, with William Burton's (1865–1914) introduction of thermal cracking of crude oil, initially at the Standard Oil of Indiana's Whiting refinery (Sung 1945). This procedure increased the yield of gasoline in refining, and its product was a fuel that contained only about 40% volatile compounds, in contrast to almost purely volatile natural gasolines that were used since the beginning of the automotive era and that were a frequent cause of accidental flare-ups during hot summer months. Still, the less volatile fuel alone was not good enough to eliminate violent knocking that came with higher compression. That is why all pre-World War I engines worked with compression ratios no higher than 4.3:1 and why the ratio began to rise to modern levels (between 8 and 10) only after the introduction of leaded gasoline (Smil 2023a).

Perhaps no single achievement conveys more impressively the early advances in automotive design than improvements in the key engine parameter, its mass-to-power ratio. I already noted that when Daimler and Maybach set out to develop a new engine, one of their key goals was to make it considerably lighter than Otto s four-stroke gas-fueled machine that weighed about 270 g/W. By 1890, the best DMG engine was down to about 40 g/W; in 1901,

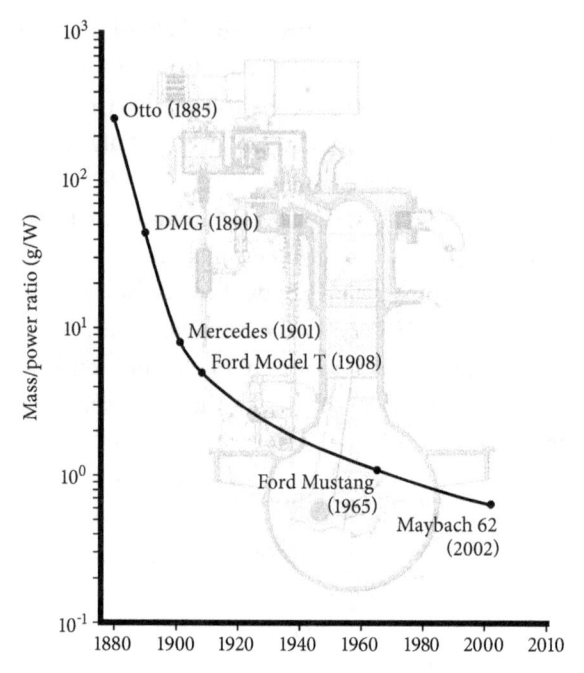

Figure I.3.14 The rapid decline of mass-to-power ratios of automotive engines during the first three decades of their development was followed by a slower rate of improvement after World War I. Plotted from a wide variety of engine specifications.

Maybach's famous first Mercedes weighed 8.5 g/W and Ford's Model T engine needed eventually less than 5 g/W (Figure I.3.14). As shown in Figure I.3.14, the mass-to-power ratio continued to decline after World War I before it stabilized around 1 g/W for typical passenger cars, which means that about 98% of the entire 1880–1960 decrease took place before 1913.

By the end of the first decade of the twentieth century, the car was in many ways a rapidly maturing machine. Looking back, Flower and Jones (1981) noted that the performance of faster cars available by 1911 was comparable not just to that of a 1931 model but even to a typical car built immediately after World War II because many changes took place without any radical modifications of its basic design and of its mode of operation. But a radical price cut was needed if car ownership were to diffuse widely among average urban and rural families. In this respect, one car will always have a very special place in the annals of automotive history: Ford's Model T. Because of its low price, remarkably sturdy construction, and reliable service, the combination had a profound socioeconomic impact. And the latter effect refers not only to how the car was promoted, marketed, and used but also in how it was made: Ford's Model T was a key catalyst for the development of the modern auto industry.

Ford's Model

The infatuation with machines that led Otto from grocery sales to four-stroke engines was also the beginning of Henry Ford's (1863–1947) achievement as the world's unrivaled producer of passenger cars (Figure I.3.15). While Ford worked at the Edison Illuminating Company in Detroit, where he was the chief engineer, he built his first motorized vehicle, a small quadricycle, in 1896 (Ford 1922). But his entrepreneurial beginnings were not particularly auspicious. In 1898, he became the chief engineer and manager of Detroit Automobile Company, which was, after he left it, reconstituted in 1902 as Cadillac. His second, short-lived company did not produce any cars.

Third time lucky, Ford Motor Company (FMC) was set up in 1903, with Alexander Malcolmson, a Detroit coal dealer, as the principal investor

Figure I.3.15 Henry Ford's portrait taken in 1934.
Library of Congress photograph (LC-USZ62-78374).

(Brinkley 2003). The company was at first quite successful with its models A and B, but its later upscale models—including Model K, the Gentleman's Roadster, able to go 112 km/h but retailing for $2,800 (FMC 1908)—were not selling well. Model N, introduced in 1906, was the first car aimed at a mass market, a goal that was finally achieved with Model T, which was launched on October 1, 1908. This was an unabashedly programmatic car, a vehicle that was intended, in Ford's own words, for the great multitude, but one that was built of good-quality materials and sold at a price affordable for anybody with a good salary. Unlike so many previous, and too many subsequent, advertising claims produced by the car industry, these statements were quite accurate: this was a resolutely down-market vehicle designed to fill a large waiting market niche.

America needed a car not just for urban residents but for settlers on vast expanses of newly cropped land, which is why Ford—himself a former farmer—saw it as a farmer's car. He gave it a generous clearance to make it go not just along muddy roads (at that time fewer than 10% of all US roads were surfaced) but, if need be, also across a plowed field, and it had a combination of simplest design (so any machine-minded owner could fix it) and good materials. Although the wide-tracked wheels, high seats, and small, initially wooden body (the combination that led to nicknaming the car a "spider") conveyed a sense of overall fragility, this impression was false.

Heat-treated vanadium steel gave the car a ruggedness that is amply attested by contemporaneous photographs of Model Ts carrying and pulling some heavy loads and running all kinds of attachments. Ford stressed this quality in his advertisements, claiming that "the Model T is built entirely of the best materials obtainable ... axles, shafts, springs, gears, in fact a vanadium steel car is one evidence of superiority" (FMC 1909). The first water-cooled, four-cylinder engines had a capacity of 2,760 cubic centimeters(cm^3), ran at 1,600 rpm, rated about 15 kW, and (including the transmission) weighed about 12 g/W. The touring version of the car weighed 545 kilograms (kg) and could reach speeds up to 65 km/h.

Of the five versions introduced during the first model year, the touring car for five passengers was, and remained, by far the most popular, while the coupe, looking much like a top hat, continued to be produced in only small numbers (Figure I.3.16).

Ts were the first Fords with left-hand drive (a logical feature for easy passing), and none of the early models had any battery: headlamps were lit by carbide gas, running and rear lights by oil, a magneto produced electricity

Figure I.3.16 Four Model Ts (clockwise starting from the upper left corner) as they appeared in the company s catalog (FMC 1909): touring car, by far the most popular version of the vehicle; roadster with a vertical windshield; a town car; and a coupe that looks much like a top hat.

for the four spark plugs, and starting was by a hand crank. The company's catalog reserved the greatest praise for springs that "no conceivable accident could possibly affect" and that made it "one of the easiest riding cars ever built" (FMC 1909). Mass production was the company's aim from the very beginning. The first Model T catalog boasted that Ford had already worked out the "quantity production system" that allowed it to build "a hundred cars a day, a system whereby overhead expense is reduced to minimum. . . . We know how to build a lot of cars at a minimum cost without sacrificing quality" (FMC 1909). But that was just a prelude.

A much more efficient system, introduced at the Highland Park factory in 1913, has been described, analyzed, praised, and condemned many times (Flink 1988; Ling 1990; Batchelor 1994; Brinkley 2003; Nye 2013). Whatever the verdict, the system was widely copied, fine-tuned, and adjusted, and it remained the dominant means of large-scale automobile production until the combination of widespread robotization of many basic tasks and changes on assembly lines that were pioneered by Japanese automakers began its transformation during the 1970s (Womack, Jones, and Roos 1990). The leading goal of Ford's assembly line was to do away with the artisanal mode of car manufacturing that saw small groups of workers assembling entire vehicles from parts they brought themselves (or which were carted or trucked to them) from assorted stores and that often had to be adjusted to fit.

As Ford expanded his production, this form of operation eventually required more than 100 chassis stations that were placed in two rows stretching over nearly 200 m, an arrangement that created logistic problems and floor congestion. The other option, whereby fitters and mechanics were walking around the assembly hall to perform their specialized tasks on each vehicle, was no more effective. As a result, productivity was low, with most European manufacturers requiring 1.5–2 workers to make one car per year; Ford was able to make nearly 13 cars a year per worker already in 1904.

Ford's quest for high labor productivity relied on the use of identical parts to be used in identical ways in assembling identical-looking vehicles by performing repetitive identical tasks—and eventually doing so along endlessly moving belts whose speed could be adjusted to the point where rapid assembly did not compromise the desired quality of construction.

This manufacturing process began at the EMF Company in 1911, when two of Ford's former employees, W. E. Flanders and Max Wollering, installed a mechanized assembly line to speed up chassis production. Chassis were first pushed, and then pulled, by cables actuated by electric motors, and Ford's first moving line, installed at Highland Park in 1913, was hardly more impressive: just 45 m long, with some 140 assemblers working on five or six chassis (Lewchuk 1989).

There was only a marginal increase in labor productivity at Ford during the early part of 1913, but then the idea of putting labor along a moving line began to bring the expected profits. The rope-drawn chassis assembly line cut the time for finishing a car from 12 hours and 30 minutes in October to 2 hours and 40 minutes in December (Flink 1988). The next year, the introduction of a chain-driven conveyor cut the assembly time to just 90 minutes.

The system was used first to assemble magnetos, where the time was cut from 20 to 5 minutes. By 1914, the new plant was producing a thousand automobiles a day and

> the work has been analyzed to the minutest detail with a view to economizing time. Men are given tasks that are very simple in themselves,

and, by dint of repetition day in and day out, acquire a knack that may cut the time of the operation in two. A man may become a specialist in so insignificant an operation, for instance, as putting in a certain bolt in the assembling of the machine. (Bond 1914: 8)

This method of production was undoubtedly exploitative, stressful, and—perhaps the strongest charge leveled against it—dehumanizing. And there is no simple explanation for why it was accepted by so many for so long.

Opportunity for unskilled workers was an obvious draw, and relatively high wages had clearly an important effect. In 1914—in what *The Economist* called "the most dramatic event in the history of wages"—Ford more than doubled the assembly-line pay to $5/hour and cut the working day from nine to eight hours. This reduced high turnover and delayed the unionization of labor (Lewis 1986). Ford passed most of the savings achieved by his streamlined assembly to consumers: lower prices led to increasing sales, which led to higher production at lower costs. Consequently, Ford's approach was not one of marketing but of true economies of scale: he kept increasing the volume of production and converting the higher productivity into lower prices. The introductory price (for red or gray vehicles) was $850 to $875, but, by 1913, mass production of more than 200,000 vehicles (with bodies of sheet steel) lowered it by 35% to $550. A year later it was down to $440, and, after World War I, the car retailed for as little as $265 for the runabout model and $295 for the touring car (MTFCA 2024). Ford's great downmarket move paid off.

During the first fiscal year of Model T production (1908–1909), it shipped nearly 18,000 vehicles, about 15% of US passenger vehicles; by 1914, Ford's share was about 44%, and three years later it rose to 48%. The famous standard black took over by 1915, and no other colors were produced until 1925: the choice was dictated simply by the cost and durability of black varnishes, which had little resemblance to modern spray finishes (Boggess 2000). By the time the model was discontinued in 1927, the company's official data sheet shows that 14,689,520 Model Ts were made (Houston 1927). The durability of these vehicles and hardships of the deepest economic crisis of the twentieth century combined to keep a few millions of Models T on American roads throughout the 1930s. More than a century after their introduction, there are still many (rebuilt and reconditioned) Model Ts in perfect running order (MTFCA 2024).

The First Two Decades

The early car industry was much beholden to bicycle making. This is hardly surprising given the history of that simple machine (Whitt and Wilson 1982; Herlihy 2006). Incredibly, the modern bicycle, an assembly that consists of nothing more than a two equally sized wheels, a sturdy frame, and a chain-driven back wheel, emerged only during the 1880s, a century after Watt's improved steam engines, half a century after the introduction of locomotives, and years after the first electric lights! Previous bicycles were not just clumsy but often very dangerous contrivances whose riding required either unusual dexterity and physical stamina or at least a great deal of foolhardiness. As such, they had no chance to be adopted as common means of personal transport (Figure I.3.17).

Only after John Kemp Starley and William Sutton introduced bicycles with equal-sized wheels, direct steering, and a diamond-shaped frame of

Figure I.3.17 The evolution of bicycles from a dangerous vertical fork of 1879 and Starley's 1880 Rover (wheels are still of unequal size) to modern-looking "safety" models of 1890s with diamond frame of tubular steel and pneumatic tires.

Images reproduced from *Scientific American*, April 19, 1892, and July 25, 1896.

tubular steel—their Rover series progressed to direct steering and acquired a frame resembling the standard diamond by 1886 (Figure I.3.17)—did the popularity of bicycles take off. And they soon became even easier to ride with the addition of pneumatic tires and backpedal brake (US Patent 418,142), both introduced in 1889. As with so many other techniques of the astonishing 1880s, the fundamental features of those designs have been closely followed by nearly all subsequent machines for most of the twentieth century. Only in the late 1970s came new high-tech designs that introduced expensive alloys, composite materials, and unorthodox upturned handlebars.

Early bicycle makers found the task of assembling small cars a naturally kindred enterprise, and bicycles contributed to the birth of the automotive era because they provided ordinary people with their first experience of individualized, long-distance transportation initiated at one's will and because both of these new speedy machines created demand for paved roads, road signs, traffic rules and controls, and service shops (Rae 1971; Flink 1975). Early cars benefited from these developments in terms of both innovative construction (steel-spoked wheels, welded tubes for bodies, rubber tires), convenient roadside service, and the experience of free, unscheduled travel. But too many former bicycle mechanics switched to car making and only some, including Opel, Peugeot, Morris, and Willys, prospered by producing either unique items or very small series of expensive yet often unreliable vehicles. That is why Ford's Model T was such a fundamental breakthrough.

Car races played an important role both in bringing the new machines to widespread public attention and in improving their design and performance. The first one was organized by *Le Petit Journal* in 1894, between Paris and Rouen (126 km), and the very next race in 1895 was ambitiously extended to a return trip between Paris and Bordeaux, nearly 1,200 km that the winning vehicle completed with an average speed of 24 km/h (Flower and Jones 1981). In 1896, the Paris–Marseille–Paris race, beset by heavy rains, covered more than 1,600 km: Emile Levassor was among many injured drivers, and his injuries caused his death a few months later. The first decade of the twentieth century saw more intercity races, beginning with the Thousand Miles Trial in the United Kingdom in 1900, the Paris–Berlin race in 1901, and the Paris–Madrid race in 1903 that ended prematurely in Bordeaux after many accidents and 10 fatalities. Louis Renault, who won the first leg of the race in the light car category (Figure I.3.18), learned of the death of his brother Marcel only after arrival at Bordeaux.

Figure I.3.18 Louis Renault winning the first leg of the ill-fated Paris-Madrid race in May 1903.
Reproduced from the cover of *The Illustrated London News*, May 30, 1903.

But soon even longer races were held, such as the first long-distance rally from Paris to Constantinople in 1905, and, audaciously, the 14,000-km Beijing–Paris run in 1907, when cars often had to be towed or even carried over rough, roadless terrain. This was followed by inauguration of the world's two most famous speedways at Le Mans in 1906 and Indianapolis in 1909 (Boddy 1977; Flower and Jones 1981). Today's variety of car races (from Formula 1 to monster drag cars, from NASCAR to dune buggy competitions) and the magnitude of annual spectator pilgrimages demonstrate how enormously the interest in speed and performance has grown (Hamilton 2022).

The international character of the emerging car industry was seen in both licensing agreements and sales. French winners of car races had German motors in their vehicles, DMG licenses launched the car-making industry in the United Kingdom (1893), and exports were important for most of Europe's pioneering car producers as well as for Ford's company. For example, in 1900, when Benz & Cie was still the world's largest automaker, it sold 341 out of the total of 603 vehicles abroad. As already noted, there were only some 8,000 cars in the United States in 1900, and fewer than 2,400 in France, the country that pioneered modern car design during the 1890s. In 1905, Germany and France each had fewer than 20,000 passenger vehicles, the United Kingdom's total was about 30,000, and US registrations reached 77,000.

As is so often the case in the early stages of a new industry, too many companies competed in the market. By 1899, there were more than 600 car manufacturers in France; nearly 500 car companies were launched in the United States before 1908, and about 250 of them were still in operation by the time Ford began selling his Model T (Byrn 1900; Flink 1988). Post-World War I consolidation reduced these numbers to fewer than 20 major companies, and, by the end of the twentieth century, about 70% of all passenger cars made in North America were assembled by just three auto makers: General Motors, Ford, and Honda. In 2023, the three big US brands (GM, Ford, and Stellantis) sold about 43% of all vehicles (US Auto Manufacturer Sales Data 2024).

Americans were late starters compared to Germans and the French. The first machine that attracted public attention was designed by Charles and Frank Duryea in Springfield, Massachusetts. These bicycle mechanics decided to build it after reading about Benz's latest vehicle in *Scientific American* in 1889. Their machine was ready in 1893; they won America's first (88 km) car race, held in Chicago, in November 1895 (beating a Benz, the only other car that finished the course) and sold their first car in February 1896. By the century's end, there were at least 30 American car makers, and they produced 2,500 vehicles in 1899 and slightly more than 4,000 in 1900 (Flink 1975).

All but one of the great enduring names in the American car industry appeared before 1905, with production conspicuously concentrated in Michigan (May 1975; Kennedy 1941). Ransom Olds (1864–1950), financed by S. L. Smith, began making his Oldsmobiles, which were nothing but buggies with an engine under the seat, in 1899. In 1901, Olds introduced America's first serially produced car, the Curved Dash, with a single-cylinder 5.2 kW engine (May 1977). The Cadillac Automobile Company began selling its vehicles in 1903, the same year that David D. Buick (1854–1929) sold his first car. In 1908, these brands became divisions of General Motors under

William Durant (1861–1947). Only Walter Chrysler (1875–1940) did not begin making his cars before World War I.

Once the gasoline-powered engine prevailed, the speed of post-1905 American automobilization was unparalleled. By 1913, Americans could choose among cars of more than 120 manufacturers, ranging from Ford's cheapest Model T (at $525 it was not the cheapest car available, as Metz and Raymond were selling inferior vehicles for, respectively, $395 and $445) to "America's Foremost Car," the Winton Six, whose "freedom-from-faults" had made it the leader with "up-to-the-minute in everything that makes a high-grade car worth having." France had the largest motor vehicle industry until 1904, when its output of nearly 17,000 cars was surpassed by the US production of just over 22,000 vehicles (Kennedy 1941; Flink 1988). Then the gap widened rapidly: by 1908, US output reached 60,000, and, by 1913, it approached half a million (USBC 1975).

At that time US car production was a quasi-monopolistic endeavor, a licensed enterprise with companies belonging to the Association of Licensed Automobile Manufacturers (ALAM) and paying a fee for every vehicle produced. This peculiar situation arose from an even more peculiar patent application. George B. Selden, a Rochester patent lawyer, filed a claim on May 8, 1879, but the patent was not issued until 16 years later, on November 5, 1895 (US Patent 549,160) because Selden kept filing amendments and thus deliberately deferring the start of the 17-year protection period (Selden 1895; Kennedy 1941). By that time its design, shown in derisory drawings of an awkward looking carriage, was left far behind by intervening engineering advances. Selden never built any vehicle, and the automotive pioneers were not aware of the pending patent's existence until Albert A. Pope of the Electric Vehicle Company bought the rights to Selden's patent in November 1899. He promptly filed and won an infringement suit against the Winton Company (Flink 1975).

As the wording of Selden's claim was so broad—applying to any vehicle with "a liquid hydrocarbon gas-engine of the compression type comprising one or more cylinders" (Selden 1895: 3)—a group of 32 automakers formed ALAM in 1903 and agreed to pay 1.25% of the retail value of their cars to the Electric Vehicle Company (Kennedy 1941). Producers soon succeeded in reducing the rate, but ALAM continued the legal fight against those manufacturers that refused to pay the fees as a means of regulating competition and keeping prices high. Henry Ford's new company was sued in October 1903 and, after a fight resembling the contested invention of incandescent lights, lost the case completely in 1909.

This put the Ford Company in a very precarious position just as it was entering the mass-production stage with the Model T. Ford appealed, maintaining in his advertisements that the Selden patent did not cover a practical machine and that his company was the true pioneer of the gasoline automobile. The dubious patent was finally invalidated in January 1911, the year when other monopolies were outlawed with the antitrust verdicts that dissolved the Standard Oil Company of New Jersey and the American Tobacco Company (Kennedy 1941). And so, the passenger car in America, though still adventuresome to drive, not particularly comfortable, and still rather expensive, was on the verge of becoming a mass property as it began to cast a widening spell.

Soon even the author of what is now seen as one of the great classics of English children's literature, *The Wind in the Willows*, wrote admiringly of it. As Kenneth Grahame's animal friends, traveling in a horse-drawn gypsy wagon, were swept into a deep ditch by a rapidly passing magnificent car, the Toad did not complain. Instead, they

> found him in a sort of a trance, a happy smile on his face, his eyes still fixed on the dusty wake of their destroyer.... "Glorious, stirring sight!" murmured Toad, never offering to move. "The poetry of motion! The real way to travel! The only way to travel! Here to-day—in next week tomorrow! Villages skipped, towns and cities jumped—always somebody else's horizon! O bliss! (Grahame 1908: 40–41)

Toad's monologue ("O what a flowery track lies spread before me, henceforth! What dust-clouds shall spring up behind me as I speed on my reckless way!") prefigured the infatuation of billions of drivers and passengers that followed during the twentieth century.

Expanding Applications

Not surprisingly, it did not take long for European and American engineers to begin mounting new gasoline engines on chassis other than flimsy buggies or sedate carriages. Some of these applications, including small delivery vehicles and taxicabs, did not call for any special engines or body designs, but many other uses required more powerful engines and sturdier bodies.

Heavier designs began appearing during the late 1890s, and, by 1913, just about every conceivable type of heavy-duty commercial vehicle that is now in service was present in forms ranging from embryonic to accomplished. More than a century later we rely on a great variety of internal combustion engines whose ratings span five orders of magnitude. The smallest and lightest engines (<1 kW) are installed in lawnmowers, outboard motors, and ultralight airplanes. Most passenger cars do not have engines more powerful than 150 kW, off-road machines can rate above 500 kW, airplane engines more than 1 MW, and the largest marine engines and stationary machines for electricity generation have power ratings in tens of MW.

After 1903, another process began unfolding in parallel with the development of specialized engines for vehicles, locomotives, and ships: efforts to use new engines in flight, to become finally airborne with a machine heavier than air. A different type of engine was needed to meet this challenge, a light and powerful one with a very low mass-to-power ratio. The Wright brothers achieved that goal on December 17, 1903, when their airplane briefly lifted above the dunes at North Carolina's Kitty Hawk (Wright 1953; McCullough 2016). Rapid improvements of post-1905 designs put in place all the key features that dominated aeroengines until gas turbines (jet engines) took over as the principal prime movers in both military and commercial planes about half a century later.

But before turning to the heaviest and to the lightest machines I should mention small engines that are used wherever low weight, small size, and low-cost operation are needed. Their development began during the late 1870s, with Dugald Clerk's (1854–1932) two-stroke gas-fueled compression engine designed in 1879, and with Benz's first two-stroke gasoline-fueled machines. Until the 1990s, they were common in lawnmowers, grass trimmers, and snow and leaf blowers, in chain saws, in outboard engines for motorboats, and in motorcycles, but, in 2002, Honda, the world's leading maker of motorcycles, announced plans to use only four-stroke engines because they produce fewer emissions and consume less fuel. But two-stroke engines are still around, in the United States mostly in outboard motors.

Heavy-Duty Engines

After 1905, it was obvious that steam-powered wagons, trucks, and tractors faced an inevitable extinction. Benz & Cie built the world's first gasoline-powered bus, able to carry just eight passengers in 1895. London got its first famous red double-deckers, built on a chassis virtually identical to that of a 3-ton (t) truck, in 1904, and, by 1908, there were more than 1,000 of them in service. They displaced some 25,000 horses and 2,200 horse-drawn omnibuses, and Smith (1911) concluded that their higher speed had an effect of more than doubling the width of the city's main thoroughfares. Horse-drawn omnibuses disappeared from London by 1911, from Paris in January 1913. As much as we complain about congestion caused by motor vehicles, the equivalent amount of traffic powered by horses would be unmanageable (see Figure I.1.7).

DMG built its first gasoline-powered truck, a 3-kW vehicle, in 1896; it could carry up to 5 tons at speeds not surpassing 12 km/h. In the United States, Alexander Winton began manufacturing his first delivery wagons, powered by a single-cylinder, 4.5-kW engine, two years later. Two-ton and three-ton trucks were the first common sizes, with smaller machines popular for city and suburban delivery (Figure I.3.19). Fire engines were the

The ¾ Ton Utility Truck–$1250

Figure I.3.19 A 1913 advertisement for a small (four-cylinder) utility truck (maximum carrying capacity of 0.9 t) made by the Gramm Motor Truck Co. This "most practical and serviceable truck of its size ever built" was designed for city and suburban delivery.
Reproduced from *Scientific American*, January 18, 1913.

most frequently encountered special truck applications before World War I (Smith 1911), and heavier 5-t trucks that could haul gross trailer loads of up to 40 t became standard in the United States for long-distance transport before World War I (Anonymous 1913b). The preferred US practice that emerged before World War I is still with us. The semi-trailer is made up of the tractor unit (now normally with two-wheel steer axle and two rear drive axles with a pair of double wheels on each side) that carries a turntable on which is mounted the forward end of the trailer (initially just a two-wheeled, and now typically eight-wheeled), hence named a standard 18-wheeler.

By 1913, the area served daily by a truck was more than six times larger than the one served by horse-drawn wagons, the total workload was nearly four times greater, garaging was done with less than 15% of space required for stabling horses, and the overall cost was some 60% lower (Perry 1913). Massive deployment of trucks began only with US involvement in World War I: by 1918, the country's annual truck production was nearly 230,000 units, more than nine times the 1914 total (Basalla 1988). Postwar production continued at a high rate, but truck capacities remained relatively low until after World War II, and eventually they reached maximum gross weights of as much as 58 t (tractor and two trailing units). But most US states limit the 18-wheelers (with engines of about 250 kW) used in highway transport to about 36 t. Where these restrictions do not apply, trucks have grown much bigger: a fully loaded Caterpillar 797F, the largest US dump truck powered by a 2.98-MW engine and used in surface mining, weighs 624 t (Caterpillar 2024).

The use of internal combustion engines in field machinery began almost as soon as the first small-scale sales of passenger cars (Dieffenbach and Gray 1960). The first, and excessively heavy gasoline-powered tractor was built by John Froehlich, in Iowa, in 1892, and sold to Langford, in South Dakota, where it was used just in threshing for a few months. Froelich formed the Waterloo Gasoline Tractor Engine Company, which was later acquired by the John Deere Plow Company, whose successor remains a leader in producing tractors and other self-propelled agricultural machinery. By 1907, there were no more than 600 agricultural tractors in the United States, and most of these early models resembled more the massive steam-powered machine rather than today's tractors: their mass-to-power ratios were between 450 and 500 g/W, compared to just 70–100 g/W for modern machines.

By 1912, the number of tractor makers rose above 50; the first international trials had taken place (Ellis 1912); and smaller, more practical, and more affordable machines began to appear (Williams 1982). Given the climate and terrain of California's Central Valley, it is not surprising that this region pioneered the widespread use of gasoline-fueled tractors and combines (Olmstead and Rhode 1988). By 1912, the United States had some 13,000 working tractors, and 80 companies produced more than 20,000 new machines in 1913 as the massive replacement of draft animals by internal combustion engines was finally under way (Rose 1913). But it was only during the 1920s that the power of agricultural machinery surpassed that of draft animals (see Figure I.3.2).

Shortly after World War I diesel engines finally began their conquest of the heavy automotive market. Besides the engine's inherently higher conversion efficiency, diesel fuel is also cheaper than gasoline, yet it is not dangerously flammable. Moreover, high engine efficiency and low fuel volatility mean that these diesel-powered vehicles have a much longer range per tank than do equally powerful gasoline-fueled machines. Additional mechanical advantages include diesel engine's high torque, its resistance to stalling when the speed drops, and its sturdiness (well-maintained engines can go 500,000 km without an overhaul).

In 1924, Maschinenfabrik Augsburg-Nürenberg (MAN; the successor of the company where Diesel worked between 1893 and 1897) was the first engine maker to build a direct-injection diesel engine, and, eight years, later the company stopped producing gasoline-fueled engines and concentrated on diesel engines, which remain one its major products (MAN 2024). Both Benz and Daimler produced their first diesel trucks in 1924, and, after they merged in 1926, they began developing a diesel engine for passenger cars; it was ready in 1936, when model 260-D, a rather heavy 45-hp saloon that became a favorite taxicab, was displayed at the Berlin car show (Williams 1972; Davis 2011). By the late 1930s, most of the new trucks and buses built in Europe had diesel engines, a dominance that was extended after World War II to North America and Asia and, with exported vehicles, to every continent. The mass-to-power ratios of automotive diesels eventually declined to less than 5 g/W, and today's lightest units in passenger cars are not that different from gasoline-fueled engines.

The use of steam engines in shipping continued well after World War I, while steam turbines were gaining an increasing share of the market. But by 1939, 25% of the world's merchant marine ships were propelled by diesel engines. More than a century after the first vessel was equipped with a small diesel engine, that triumph is even more evident, with about 90% of the world's largest freight ships, including the crude oil supertankers, powered by diesel engines. MAN, Mitsui, and Hyundai are their leading producers. The maximum size of these large machines keeps increasing. In 1996, the world's largest engine rated about 56 MW, but just five years later Hyundai built a 69.3 MW engine, and, in 2006, a WärtsiläSulzer RTA96-C aboard the container ship *Emma Maersk* reached 80.08 MW (Wärtsilä 2006). Dimensions of the largest machines are best illustrated by their cylinder bores and piston strokes: their maxima are now, respectively, almost 1 m and more than 2.5 m, dimension an order of magnitude larger than those of automotive diesels (Calder 2006; Smil 2009).

Low-rpm diesel engines of different sizes are also deployed in many localities far from centralized electrical supplies—be it in low-income countries of Asia, Africa, and Latin America or in isolated places in North America and Australia—to provide light and mechanical energy for refrigeration and crop processing. Other market niches where diesel engines are either dominant or claim large shares of installed power include heavy construction (cranes, and earth-moving, excavating, and drilling machines), tractors and self-propelled harvesters, locomotives (for both freight and passenger service, but the fastest ones are electric), and main battle tanks (although the best one, the US Abrams M1/A1, is powered by a gas turbine).

Engines in Flight

At the close of the nineteenth century, the old ambition to fly in a heavier-than-air machine appeared as distant as ever. Utterly impractical designs of bizarre flying contraptions were presented as serious ideas even in technical publications. One of my favorites is a new aerial machine that was pictured and described in all seriousness in the May 9, 1885, issue of *Scientific American* (Figure I.3.20), but I could have selected other preposterous designs from the last third of the nineteenth century. The goal was elusive, but some innovators were determined: Otto Lilienthal (1848–1896), the most prominent German aviation pioneer, completed more than 2,500 short flights with various gliders before he died in 1896 when his glider stalled. So did (fiction anticipating reality) a man obsessively driven to invent flying machines in Wells's *Argonauts of the Air* that was published a year before Lilienthal's death.

Building the heavier than air flying machines during the last decade of the Age of Synergy was about much more than just mounting a

Figure I.3.20 Ayres's bizarre new aerial machine, as illustrated in *Scientific American* of May 9, 1885.

reciprocating engine on wings or on fuselage. Airplanes fit perfectly into the class of achievements that distinguish that period from anything that preceded it: they came about only because of the combination of science, experimentation, testing, and the quest for patenting and commercialization. The two men who succeeded where so many other failed—Wilbur (1867–1912) and Orville (1871–1948) Wright (Figure I.3.21)—traced their interest in flying to a toy, a small helicopter powered by a rubber string they were given by their father as children when they lived in Iowa. This toy was invented by Alphonse Pénaud in 1871, and the two boys built several copies.

Years later, their interest in flying was rekindled by Lilienthal's gliding experiments, and, after reading a book on ornithology in the spring of 1899, Wilbur sent a letter to the Smithsonian Institution inquiring about publications on flight. The letter was answered in a matter of days by a package that contained reprints of works by Lilienthal and Langley. The brothers also ordered Octave Chanute's (1832–1910) book that described progress in flying machines (Chanute 1894) and began correspondence with this pioneer of theoretical manned flight. Soon they augmented the engineering experience from their bicycle business with numerous tests as they embarked on building and flying a series of gliders (Figure I.3.22). Beginning in the fall of 1901, these experiments also included a lengthy series of airfoils and wing shapes in a wind tunnel (Culick 1979).

Figure I.3.21 Wilbur and Orville Wright on the rear porch steps of their house, 7 Hawthorne Street, in Dayton, Ohio, in 1909.
Library of Congress portrait (LOT 11512-A).

Figure I.3.22 Wilbur Wright piloting one of the experimental gliders above the dunes of Kill Devil Hills at Kitty Hawk, North Carolina, in October 1902.
Library of Congress images (LC-W861–11 and LC-W861-7).

While gliding attempts have a long history, the first successful trial of an unmanned powered plane took place on October 9, 1890, when Clément Ader's *Eole*, a bat-winged steam-driven monoplane, was the first full-sized airplane that lifted off under its own power. In 1896, Samuel Pierpoint Langley (1834–1906), astronomer, physicist, and the secretary of the Smithsonian Institution, received a generous US government grant ($50,000) to build a gasoline-powered aircraft (NASM 2024a). By 1903, his assistant, Charles M. Manly, designed a powerful (39 kW, 950 rpm) five-cylinder radial engine that he mounted on *Aerodrome A*; on October 7, 1903, he launched it, with himself at the controls, near Widewater, in Virginia, by a catapult from a barge. The plane immediately plunged into the river, and it once more failed to fly during the December 8 test on the Potomac River, in Washington, DC, when it reared and collapsed onto itself. Manly was pulled unhurt from the icy water, but the collapse spelled the end of Langley's project.

Just nine days after Manly's mishap, on December 17, 1903, Orville Wright piloted the first successful flight at Kitty Hawk, near Kill Devil Hills, in North Carolina: it was more of a jump of 37 m, with the *Flyer* pitching up and down, staying airborne for just 12 seconds, and damaging a skid on landing (Wright 1953; USCFC 2003; see Figure I.1.3). Their second flight, after repairing the skid, covered about 53 m, and the third one 61 m. During the fourth flight Wilbur covered nearly 250 m in level flight, then the plane began plunging and crash-landed with a broken front rudder frame—but in 59 seconds, it had traveled 260 m. When it was carried back to its starting point, it was lifted by a sudden wind gust, turned over, and destroyed. The first *Flyer*, a fragile machine with a wingspan of 12 m and weighing just 283 kg, thus made only those four flights totaling less than 2 minutes. The brothers telegraphed their father in Ohio with their news and took their broken plane to Dayton.

Why did the Wrights, bicycle makers from an Ohio town, succeed in less than five years after they ordered information on flights from the Smithsonian? Because they started from the first principles and studied in detail the accomplishments and mistakes of their immediate predecessors. Because they combined a good understanding of what was known at that time about aerodynamics with practical tests and continuous adjustments of their designs. Because they aimed to produce a machine that would not only lift off by its own power but that would be also properly balanced and could be flown in a controlled manner. There is no doubt that the matters of aerodynamic design and flight control were a much greater challenge than coming up with an engine to power the flight, although they did that, too (Wright 1953; McCullough 2016).

What is so remarkable about the Wrights' accomplishment is that they designed the plane's every key component—wings, balanced body, propellers, engine—and that to do so they prepared themselves by several years of intensive theoretical studies, calculations, and painstaking experiments with gliders that were conducted near Kill Devil Hills between 1901 and 1903. And, after getting negative replies from engine manufacturers whom they contacted with specifications required for their machine, the Wrights designed the engine themselves, and it was built by their mechanic, Charles Taylor, in just six weeks. This was by no means an exceptional engine. Indeed, as Taylor noted later, it "had no carburettor, no spark plugs, not much of anything... but it worked" (cited in Gunston 1986: 172).

The four-cylinder 3.29-L engine laid on its side, its body was of cast aluminum, and its square (10 × 10 cm) steel cylinders were surrounded by water jackets, but the heads were not cooled, so the engine got progressively hotter and began losing power. The crankshaft was made from a single piece of steel; one of its ends was driving the camshaft sprocket and the other one the flywheel and two propeller chain sprockets. The Wrights' initial aim was to get 6 kW, but they did better: the finished 91-kg engine first developed 9.7 kW, and, at Kitty Hawk, it was rated as high as 12 kW: this unexpected performance gave it a mass-to-power ratio as low as 7.6 g/W. The brothers applied for a patent on March 23, 1903, nine months before the *Flyer* took off and received the standard reply that the US Patent Office was sending to many similar applications: an automatic rejection of any design that had not already flown. Their patent (US Patent 821,393) was granted in May 1906, and, expectedly, it was commonly infringed and ignored. The key patent drawing shows no engine, just the detailed construction of wings, canard, and the tail (Figure I.3.23).

Afterward, as with the development of automobiles, there was a pause in development. The Wrights did not do any flying in 1906 and 1907 while they were building several new machines. But, starting in 1908, the tempo of aeronautic advances speeded up noticeably. In September 1908, Wilbur Wright stayed over Le Mans for 91 minutes and 25 seconds, covering nearly 100 km, and, during that year's last day, he extended the record to 140 minutes. Louis Blériot (1872–1936)—an engineer whose first unsuccessful flying machine, built in 1900, was an engine-powered ornithopter designed to fly by flapping its wings—crossed the English Channel by flying from Les

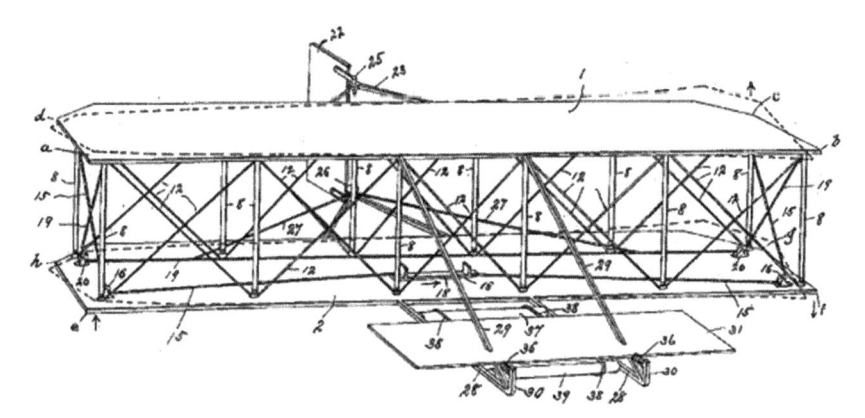

Figure I.3.23 Drawing of the Wrights' flying machine that accompanied their US Patent 821,393 filed on March 23, 1903.
Drawing available at http://www.uspto.gov.

Baraques to Dover on July 25, 1909, in 37 minutes. This flight, worth £1,000 from London's *Daily Mail*, was done in a monoplane of his own design, the fourth such machine since he built the world's first single-winged aircraft in 1907 (see Figure I.7.1).

Everything of importance was improving at the same time. The Wrights' first plane was a canard (tail-first) and a pusher (rear engine) with skids, and its engine was stationary, with in-line cylinders. But soon there was a variety of propulsion, tail, and landing gear designs, as well as a tendency toward a dominant type of these arrangements, and in-line stationary engines were replaced by radial machines (Gunston 1999). The Antoinette—an engine designed by Lèon Levavasseur and named after the daughter of his partner—was originally built for a speedboat, but, between 1905 and 1910, it became the most popular radial engine with a very low mass-to-power ratio (less than 1.4 g/W) unsurpassed for 25 years.

The first successful rotary engine (with the crankshaft rigidly attached to the fuselage, and cylinders, crankcase, and propeller rotating around it as one unit) was the Gnome, a 38-kW machine designed in France by the Seguin brothers in 1908, and whose improved versions powered many fighter planes during World War I (Gunston 1986). Ovington (1912: 218) called it "theoretically one of the worst designed motors imaginable, and practically the most reliable aeroplane engine I know of." Other rotary engines followed soon afterward, including the Liberty, the most popular engine of World War I, whose mass-to-power ratio was just above 1 g/W. Before 1913, there were also planes with simple retractable landing gear, as well as the first machines (in 1912) with a monocoque (single-shell) fuselage that was required for effective streamlining and was made of wood, steel, or aluminum. Hoff (1946: 215) captured the speed of these advances and the driving force behind them by noting that in the early 1900s "the only purpose of the designer was to build an airplane that would fly, but by 1910 military considerations became paramount."

France—where Alberto Santos Dumont (1873–1932), a Brazilian airship pioneer, made his first flight in a biplane of his own design in 1906—led this military development. By 1914, the country had about 1,500 military and 500 private planes, ahead of Germany (1,000 military and 450 private). Much as with cars, early development of American military airpower was slow. The US Army bought a Wright biplane in 1908, but little progress was achieved even by April of 1917, when the United States declared war on Germany: at that time the country had only two small airfields, 56 pilots (and 51 students), and fewer than 300 second-rate planes, none of which could carry machine guns or bombs on a combat mission (USAF Museum 2024). Only in July 1917 did the Congress appropriate the largest-ever sum ($640 million, or about $15.4 billion in 2023 US$) to build 22,500 Liberty engines; by October 1918, more than 13,000 of them were built.

But it was an American pilot who demonstrated for the first time a capability that eventually became one of the most important tools in projecting military power. On November 14, 1910, Eugene Ely (1886–1911) took off with a Curtiss biplane from the *USS Birmingham* to complete the first ship-to-shore flight, and less than 10 weeks later, on January 18, 1911, he took off from the Tanforan racetrack and landed on an improvised deck of the cruiser *Pennsylvania*, anchored in the San Francisco Bay (see Figure I.6.8). Thirteen years later the Japanese built the world's first aircraft carrier, and it was the destruction of their carrier task force by three American carrier air groups during the battle of Midway between June 4 and 7, 1942, that is generally considered the beginning of the end of Japan's Pacific empire (Smith 1966; Parshall and Tully 2006).

Of course, both the pre-World War I and the dominant World War I airplane designs were constrained by many limitations. Most obviously, to minimize drag (an essential consideration given the limited power of early aeroengines), structural weight had to be kept to a minimum, and thin wings (maximum thickness-to-length ratio of 1:30), be they on mono- or biplanes, had to be braced. This requirement disappeared only with the adoption of light alloys (Hoff 1946), and today's largest passenger planes have unbraced cantilevered wings (the longest, 80 m, on the Airbus 380). These structures are designed to withstand considerable vertical movements: even seasoned air travelers may get uncomfortable as they watch a wingtip rising and dipping during spells of high turbulence.

American airpower lagged in World War I, but an unprecedented mobilization of the country's resources made it globally dominant during World War II. Some P-51 Mustangs, perhaps the best combat aircraft of World War II, with engines that rated as high as 1.1 MW, had a maximum level speed of just over 700 km/h (Taylor 1989). That was the pinnacle of reciprocating engines. Jet propulsion eliminated them first from combat and long-distance passenger routes and soon afterward from all longer (generally more than two hours) flights. But reciprocating engines still power not only those nimble, short-winged machines that perform amazing feats of aerial acrobatics at air shows but also airplanes that provide indispensable commercial and public safety services (Gunston 2002). These range from suppressing forest fires and searching for missing vessels to seeding Californian or Spanish rice fields, from carrying passengers from thousands of small airports to applying pesticides to food and feed crops.

Frontispiece I.4 The foreground of this dramatic photograph, taken in April 1889, shows the bottom member of the unfinished North Queensferry cantilever of the Forth Bridge and Garvie main pier in the background.
Reproduced from the cover of *The Illustrated London News*, October 12, 1889.

4
New Materials and New Syntheses

> The careful text-books measure
> (Let all who build beware!)
> The load, the shock, the pressure
> Material can bear.
> So, when the buckled girder
> Lets down the grinding span,
> The blame of loss, or murder,
> Is laid upon the man.
> Not on the Stuff—the Man!
> —Rudyard Kipling, *Hymn of Breaking Strain*

Motion and speed—rotation of steam turbines and electric motors, reciprocation of internal combustion engines, travel by new automobiles and bicycles, promise of airplanes—were the obvious markers of the epoch-making pre-World War I advances. At the same time, another category of less flamboyant innovations was changing the societies of Europe and North America. None of those new engines, turbines, and conveyances would have been possible without new superior materials whose applications also revolutionized construction. The bold cantilever of the Firth of Forth bridge, shown in the frontispiece to this chapter, was thus no less a new departure and an admirable symbol of the 1880s than were Benz's engines or Tesla's motors. The bridge was the first structure of its kind to be built largely of steel: only that alloy could make this defiant design possible because only that material could bear the loads put on those massive cantilevers.

This was not the first use of metal in bridge building: wrought-iron chains were used for centuries to suspend walkways across deep chasms in China, and the era of cast iron structures began in 1779, when Abraham Darby III completed the world's first cast iron bridge across the Severn's Shropshire Gorge. But all of those were expensive exceptions as wood continued to rule even the most advanced pre-industrial societies, be it Qianlong's China or Enlightenment France. Wood was the dominant material in construction, farm implements, artisanal machinery, and household objects. If these items, and other objects made of natural fibers, were to be excluded from a census of average family possessions, then the total mass of materials that were not derived from biomass amounted to as little as a few kilograms (kg) and typically to no more than 10 kg per capita. Moreover, those artifacts were fashioned from low-quality materials, with cast iron pots and a few simple clay and ceramic objects dominating the small mass of such possessions.

In contrast, by the end of the twentieth century an average North American family owned directly more than a ton of metals and alloys in cars, appliances, furniture, tools, and kitchen and garden items. In addition, two-thirds of these families owned their house, and its construction needed metals (iron, steel, copper, aluminum), plastics, glass, ceramics, paper, and other materials. Any reader who has never tried to visualize this massive increase of material possessions will enjoy looking at Menzel's (1994) photographs that picture everything that one family owned in 30 different countries (from Iceland to Mongolia) during the early 1990s. Differences between possessions in some of the world's richest and poorest countries captured in Menzel's book are good indicators of the material gain that distinguished modern economies from pre-industrial societies.

And, of course, total per capita mobilization of materials in modern economies, including a large mass of by-products and hidden flows (e.g., coal mining overburden, ore and crop wastes, or agricultural soil erosion), is vastly greater than direct consumption of finished products. Studies of such resource flows showed that annual per capita rates during the mid-1990s added up to 84 tons (t) in the United States, 76 t in Germany, and 67 t in the Netherlands (Adriaanse et al. 1997). When the US total is limited only to materials used in construction and to industrial minerals, metals, and forestry products, its annual rate shows increase from about 2 tons per capita in 1900 to about 12 t per capita in 2000 (Smil 2023b).

This great mobilization of non-biomass materials began during the two pre-World War I generations. In 1800, steel was a relatively scarce commodity, aluminum was not even known to be an element, and there were no plastics. By 1870, the world's annual output of steel still prorated to less than 300 grams (g) per capita (yes, grams); there was no mass production of cement, and aluminum was a rarity used to make jewelry. And then, as with electricity and internal combustion, great discontinuity took place. By 1913, the world was producing more than 40 kilograms (kg) of steel per capita, a jump of two orders of magnitude in half a century, and the American average was more than 200 kg per person. The cost of aluminum fell by more than 90% compared to 1890, and concrete—a mixture of cement, aggregates (sand and gravel), and water—was a common, although an oft-reviled, presence in the built environment.

But the material revolution did not end with these newly ubiquitous items. Maturing synthetic inorganic chemistry began producing unprecedented volumes of basic chemicals (led by sulfuric acid), and, by 1909, Fritz Haber synthesized ammonia from its elements, an achievement that opened the way for not only feeding the growing world population but also securing unprecedented quantity and quality of average food intakes (Smil 2001, 2022). On the destructive side of technical advances, societies moved from gunpowder to nitroglycerine and dynamite and beyond to even more powerful explosives. Dynamism of the Age of Synergy thus did not come only from new energies and new prime movers; it was also fashioned by new materials whose quantity and affordability combined with their unprecedented qualities to set the patterns whose repetition and elaboration still shape our world.

Steel Ascendant

Steel is, of course, an old material, one with a long and intricate history that is so antithetically embodied in the elegant shapes and destructive power of the highest quality swords crafted in elaborate ways in such far-flung Old World locations as Damascus and Kyoto (Verhoeven, Pendray, and Dauksch 1998). But it was only during the two pre-World War I generations that steel became inexpensive, its output truly massive, and its use ubiquitous. Unlike in the case of electricity, where fundamental technical inventions created an entirely new industry, steelmaking was a well-established business long before the 1860s—but one dominated by artisanal skills and hence not suited to the high-volume, low-price production that was to become a hallmark of late nineteenth-century ferrous metallurgy.

Lowthian Bell (1884: 435–436), one of the century's leading metallurgists, stressed that by 1850 "steel was known in commerce in comparatively very limited quantities; and a short time anterior to that period its use was chiefly confined to those purposes, such as engineering tools and cutlery, for which high prices could be paid without inconvenience to the customer." But by 1850, at least one cause of the relative rarity of steel was removed as the production of pig (cast) iron, its necessary precursor, expanded thanks to taller,

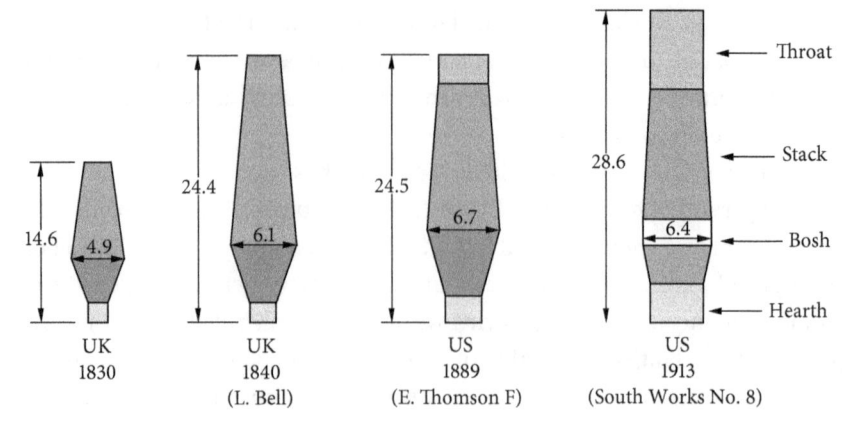

Figure I.4.1 Increasing height and volume of blast furnaces, 1830–1913.
Based on data in Bell (1884) and Boylston (1936).

and hence much more voluminous and more productive, coke-fueled blast furnaces. Bell concluded that the typical furnaces were too low and too narrow for the efficient reduction of iron ore, and his design increased the overall height by 66%, the top opening by more than 80%, and the hearth diameter by 33% (Bell 1884; Figure I.4.1).

During the latter half of the nineteenth century, the technical primacy in ironmaking passed from the British to the Americans (Hogan 1971). Before 1870, the world's most advanced group of tall blast furnaces operated in England's Northeast, using Cleveland ores discovered in 1851 (Allen 1981). Pennsylvanian ironmakers then took the lead, especially with the furnaces built at Carnegie's Edgar Thomson Works: their hearth areas were more than 50% larger; their blast pressures and rates were twice as high. As a result of these cumulative advances, the global output of pig iron roughly doubled to 10 megatons (Mt) between 1850 and 1870, and then it jumped to more than 30 Mt by 1900, and reached 79 Mt by 1913 (Kelly and Fenton 2003; Figure I.4.2).

This growth was driven by rapidly rising demand for steel whose inexpensive production became a reality thanks to new processes that could convert pig iron into an alloy with superior qualities and hence with a much larger scope of applications. Both pig iron and steel are alloys whose differences in carbon content translate into special physical, and hence structural, properties. Cast iron contains between 2% and 4.3% carbon, while steel has merely 0.05% to 2% of the element (Bolton 1989; Smil 2016).Cast iron has

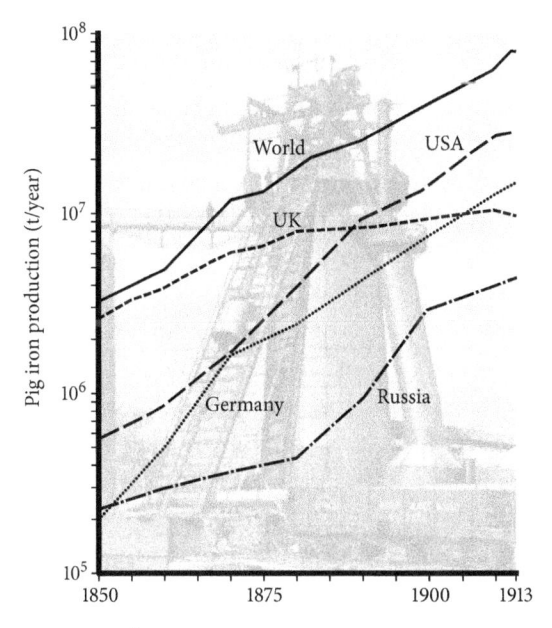

Figure I.4.2 Pig iron smelting, 1850–1913.
Plotted from data in Campbell (1907) and Kelly and Fenton (2003).

very poor tensile strength (much less than bronze or brass), low impact resistance, and very low ductility. Its only advantage is good strength in compression, and its widest uses have been in such common artifacts as water pipes, motor cylinders, pistons, and manhole covers. In contrast, the best steels have tensile strengths an order of magnitude higher, and they can tolerate impacts more than six times greater.

Steels remain structurally intact at up to 750°C, compared to less than 350°C for cast iron. Addition of other elements—including, singly or in combination, aluminum (Al), chromium (Cr), cobalt (Co), manganese (Mn), molybdenum (Mo), nickel (Ni), titanium (Ti), vanadium (V), and tungsten (W), in amounts ranging from less than 2% to more than 10% of the mass—produces alloys with a variety of desirable properties. Low-carbon sheet steel goes into car bodies; highly tensile and hardened steels go into axles, shafts, connecting rods, and gear. Stainless steels are indispensable for medical devices as well as for chemical and food-processing equipment, and tool steels are made into thousands of devices ranging from chisels to extrusion dies.

Pre-industrial societies were producing limited amounts of steel primarily by *cementation*, that is, by adding the desired amount of carbon to practically carbon-free wrought iron. The alloy remained a commodity of restricted supply even by the mid-eighteenth century, when cast iron production began rising substantially. Benjamin Huntsman (1704–1776) began producing his crucible cast steel by carburizing wrought iron during the late 1740s, but that metal was destined only for such limited-volume applications as razors, cutlery, watch springs, and metal-cutting tools. As advancing industrialization needed more tensile metal, particularly for the fast-growing railways, wrought iron filled the need.

This iron had to be produced in puddling furnaces by the sequence of reheating the brittle pig iron in shallow coal-fired hearths and pushing and turning it manually with long rods to expose it to oxygen and produce an alloy with a mere trace (0.1%) of carbon. Later this extraordinarily hard labor—which involved manhandling iron chunks of nearly 200 kg for longer than an hour in the proximity of very high heat—was done mechanically. The puddled material was then, after another reheating, rolled or hammered into desired shapes (rails, beams, plates). The first innovation that made large-scale steelmaking possible was introduced independently and concurrently by Henry Bessemer (1813–1898) in England and by William Kelly (1811–1888) in the United States. Bessemer revealed his process in August 1856 and patented it in England (GB Patent 2219) in 1856, and in the United States (US Patent 16,083) in 1857 (Bessemer 1905).

Molten pig iron was poured into a large pear-shaped tilting converter lined with acid material, and subsequent blasting of cold air through "tuyeres" would decarburize the molten metal and drive off impurities. Kelly rushed to file his patent (US Patent 17,628 in 1857) only after he learned that Bessemer filed his in the United Kingdom, and Kelly was declared the inventor of the process by US courts. The blowing period lasted as little as 15 and usually less than 30 minutes, and it produced spectacular displays of flames and smoke issuing from the converter's mouth (Figure I.4.3). But the process proved to be a great disappointment: while the air forced into molten iron would burn off carbon and silicon, phosphorus and sulfur were left behind.

This drawback was not discovered during Bessemer's pioneering work because, by chance, he did his experimental converting with pig iron made from Blaenavon ore that was virtually phosphorus-free. The simplest solution to the phosphorus problem was to use iron ore of great purity, but their supply was obviously limited. A technical fix for the sulfur problem that made the Bessemer process much more widely applicable was discovered by Robert Forester Mushet (1811–1891). This experienced metallurgist added

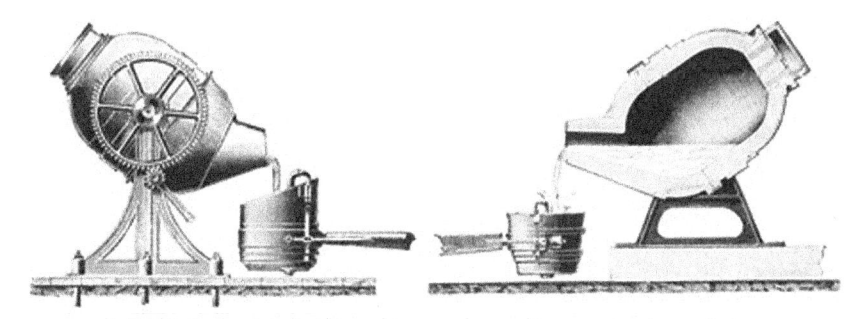

Figure I.4.3 Turning gears and a cross-section of Bessemer converter pouring the molten metal and a converter in operation at John Brown & Co. Foundry.

Top image reproduced from Byrn 1900); bottom image reproduced from *The Illustrated London News*, July 20, 1889.

small amounts of *spiegel* iron, a bright crystalline iron ore with about 8% Mn and 5% carbon (C), to the decarbonized iron to partially deoxidize the metal (Mn has a great affinity for oxygen) as well as to combine with some of the sulfur and remove the resulting compounds in slag. Patent specifications for this process were filed in 1856 (Osborn 1952).

By 1861, Bessemer steel was being rolled into rails in several mills around England, and it was first produced in the United States in 1864. One more essential step was needed to make the process universal: to remove phosphorus from pig iron to be able to use many iron ores that contain the element. Many engineers tried for several decades to solve this challenge, and the solution was found by two young metallurgists, Sidney Gilchrist Thomas (1850–1885) and his cousin Percy Carlyle Gilchrist (1851–1935). Their reasoning was not new: to use a basic material that would react with the acidic phosphorus oxides present in the liquid iron and remove them in slag (Almond 1981).

After several years of experiments with different linings, the cousins produced hard, dense, and durable blocks from impure limestone and sealed the joints with a mixture of tar and burned limestone (or dolomite). As the basic lining would be insufficient to neutralize the phosphorus compounds, they also added lime to the charged ore. Thomas and Gilchrist took out a dozen British and foreign patents between 1878 and 1879, and their innovation was finally acclaimed at the Iron and Steel Institute meeting in May 1879. This technique produced most of the world's steel between 1870 and 1910. Its share in the American output peaked at 86% of the total in 1890 (Hogan 1971). But this success was short-lived because a different procedure, one whose first practical application was one of the markers of the beginning of the Age of Synergy, that proved to be an epoch-making innovation and produced most of the world's steel for more than two-thirds of the twentieth century.

Open-Hearth Furnace

The story of open-hearth steelmaking is one of the best illustrations of the difficulties with attributing the origins of technical inventions and dating them accurately. The principal inventor of the furnace was one of four brothers of the German family that has no equals as far as the inventiveness of siblings is concerned. We have already met Werner, the founder of Siemens & Halske Company and one of the inventors of self-excited dynamos (see Figure I.2.8). The adaptation of the open-hearth furnace for more efficient steelmaking was the idea of Carl Wilhelm, known as William, after he became a British subject in 1859 (Figure I.4.4). Wilhelm, an inventor and promoter in the fields of thermal and electrical engineering, elaborated the idea in cooperation with his younger brother Friedrich (1826–1904).

The idea of open-hearth melting itself was not new; it was the heat economy introduced into the process by Siemens's regenerative furnace

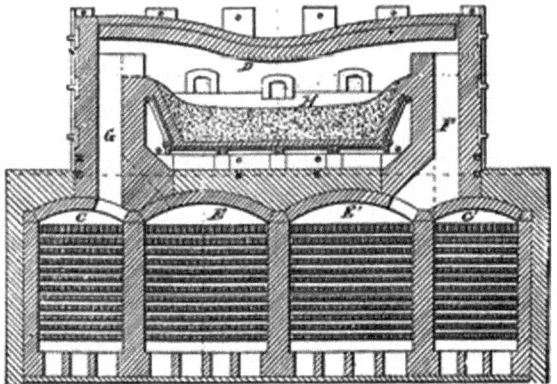

Figure I.4.4 William Siemens (reproduced from *Scientific American*, December 22, 1883) and a section through his open-hearth steelmaking furnace. Gas and air are forced through chambers C and E, ascend separately through G, ignite in D, and melt the metal in hearth (H); hot combustion gases are led through F to preheat chambers E' and C'. Once these are hot, they begin receiving gas and air flows and the entire operation is reversed as combustion gases leave through G to preheat C and E.
Reproduced from Byrn (1900).

that made the commercial difference. The furnace was a simple rectangular brick-lined chamber with a wide, saucer-shaped and shallow hearth whose one end was used to charge pig iron as well a small quantity of iron ore or steel scrap and to remove the finished steel (Riedel 1994). Unlike in ordinary furnaces, where much of the heat generated by fuel combustion escaped with hot gases through a chimney, Siemens's furnace led the hot gases first through a regenerator, a chamber stacked with a honeycomb mass of bricks that absorbed a large share of the outgoing heat.

As soon as the bricks were sufficiently heated, the hot gases were diverted into another regenerating chamber, while the air required for combustion was preheated by passing through the first, heated, chamber. After its temperature declined to a predetermined level, the air flow was reversed, and this alternating operation guaranteed the maximum recovery of waste heat. Moreover, the gaseous fuel used to heat the open-hearth furnace (usually produced by incomplete combustion of coal) was also led into a regenerative furnace that then required four brick-stacked chambers (Figure I.4.4). This energy-conserving innovation (fuel savings amounted to as much as 70%) was first used in 1861, at glassworks in Birmingham.

In 1867, after two years of experiments, Siemens was satisfied that high temperatures (between 1,600°C and 1,700°C) generated by this process would easily remove any impurities from mixtures of wrought iron scrap and cast iron charged into the furnaces. As Bell (1884: 426) put it, "the application of this invention to such a purpose ... is so obvious, that its aid was speedily brought into requisition in what is now generally known as the Siemens-Martin or open-hearth process." The double name is justified: although the Siemens brothers did their first tests in 1857, and then patented the process in 1861, it took some time to perfect the technique, and a French metallurgist Emile Martin (1814–1915) succeeded in doing so first and filed definitive patents in summer 1865. Concurrent trials by Siemens were also promising, and, by November 1866, Siemens and Martin agreed to share the rights; the new process was commercialized for the first time in 1869 by three British steelworks.

In the Bessemer converter, the blowing process was over in less than half an hour, whereas open-hearth furnaces needed commonly half a day to finish the purification of the metal. Hot metal was transferred into a giant ladle, which was then lifted by a crane and its contents poured into molds to form the desired ingots. Between 1886 and 1890, the process spread to US steelmaking (Almond 1981), and, in 1890, Benjamin Talbot (1874–1947) devised a tilting furnace that made it possible to tap slag and steel alternatively and hence to turn basic open-hearth steelmaking from a batch into a virtually continuous process.

As in so many other cases of technical advances, the initial design was kept largely intact, but the typical size and average productivity grew. During the late 1890s, the largest plant of the US Steel Corporation had open hearths with areas of about 30 square meters (m^2); by 1914, the size was up to 55 m^2, and, during World War II, it reached almost 85 m^2 (King 1948).

Heat sizes increased from just more than 40 t during the late 1890s to 200 t after 1940. This tiltable, basic Siemens-Martin furnace came to dominate American steelmaking during the first two-thirds of the twentieth century. In the United States, its share rose from just 9% of all steel production in 1880 to 73% by 1914, and it peaked at about 90% in 1940; similarly, in the United Kingdom, its share peaked just above 80% in 1960 (Figure I.4.5). These peaks were followed by rapid declines in the technique's importance as basic oxygen furnaces became the dominant means of modern steelmaking during the last third of the twentieth century, with electric arc furnaces not far behind.

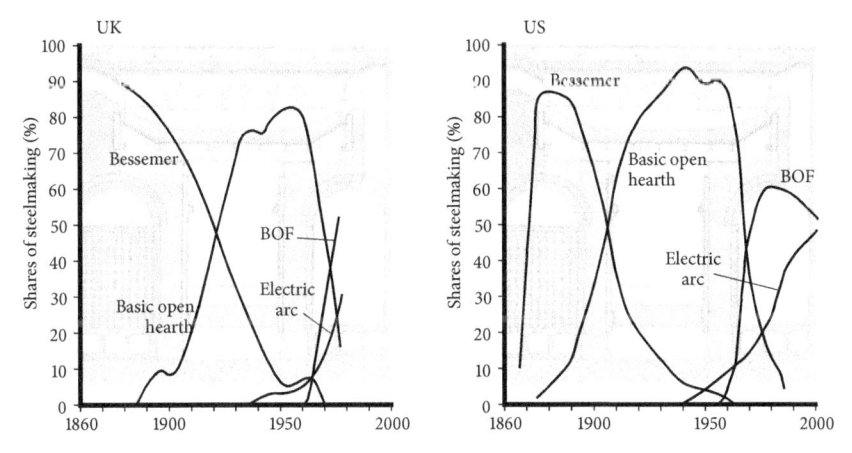

Figure I.4.5 Shares of leading methods of steelmaking in the United Kingdom (1880–1980) and the United States (1860-2000).

Plotted from data in Almond (1981) for the United Kingdom and in Campbell (1907) and USBC (1975) for the United States.

The last smelting innovation that was introduced before 1914 was the electric arc furnace. William Siemens built the first experimental furnaces with electrodes placed at the top and bottom of a crucible or opposed horizontally (Anonymous 1913c). Paul Héroult commercialized the process in 1902 to produce steel from scrap metal. These furnaces operate on the same principles as the arc light: the current passing between large carbon electrodes in the roof melts the scrap, and the metal is produced in batches. By 1910, there were more than 100 units of different designs operating worldwide, with Germany in the lead, far ahead of the United States. The subsequent combination of decreased electricity costs and increased demand for steel during World War I transformed the electric arc furnaces into major producers of high-quality metal, with nearly 1,000 of them at work by 1920 (Boylston 1936). Their rise to dominance in modern steelmaking will be described in the companion volume.

Steel in Modern Society

All these metallurgical innovations meant that the last quarter of the nineteenth century was the first time in history when steel could be produced not only in unprecedented quantities, but also to satisfy specific quality demands and to be available in the large batches needed to make very large parts. Global steel output rose from just half a million tons in 1870 to 28 Mt by 1900 and to about 70 Mt by 1913. Between 1867 and 1913, US pig iron smelting rose 21-fold, while steel production increased from about 20,000 t to nearly 31 Mt, a more than 1,500-fold rise. Consequently, steel's share in the final output of the metal rose precipitously. In 1868, about 1.3% of US pig iron was converted to steel; by 1880, the share was slightly less than a third, and it reached about 40% by 1890, almost 75% by 1900, and 100% by 1914 (Figure I.4.6; Hogan 1971; Kelly and Fenton 2003).

As the charging of scrap metal increased with the use of open-hearth and, later, electric and basic oxygen furnaces, steel production began surpassing pig iron output: by 1950, the share was 150%, and, by 2000, US steel production was 2.1 times greater than the country's pig iron smelting (Kelly and Fenton 2003). Initially, nearly all steel produced by the open-hearth furnaces was rolled into rails. But soon the metal's final uses became much more diversified as it filled many niches that were previously occupied by cast and wrought iron and, much more important, as it found entirely new markets with the rise of many new industries that were created during the two pre-World War I generations.

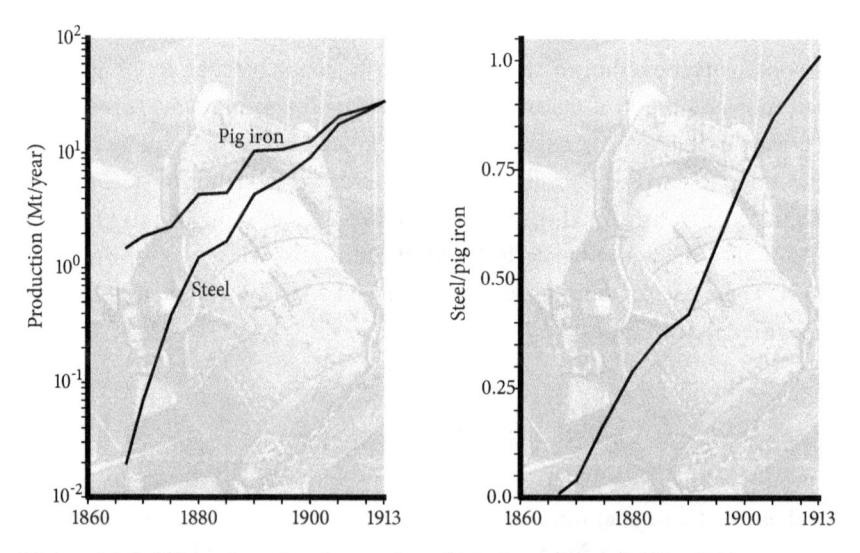

Figure I.4.6 US steel production and steel/pig iron shares, 1867–1913.
Plotted and calculated from data in Temin (1964) and Kelly and Fenton (2003).

Substitutions that required increasing amounts of steel included energy conversions (boilers, steam engines), land transportation (locomotives, rolling stock), and, after 1877, when the Lloyd's Register of Shipping accepted steel as an insurable material, the metal rapidly conquered the shipping market. Steel substitutions also took place in the production of agricultural implements and machinery as well as in textile and food industries, in industrial machines and tools, and in the building of bridges. Undoubtedly the most remarkable, elegant, and daring use of steel in bridge building was the spanning of Scotland's Firth of Forth to carry the two tracks of the North British Railway (see frontispiece to this chapter).

This pioneering design by John Fowler and Benjamin Baker was built between 1883 and 1890, and it required 55,000 t of steel for its 104-meter (m)-tall towers and massive cantilevered arms: the central span extends more than 105 m, and the bridge's total length is 2,483 m (Hammond 1964). The structure—both reviled (William Morris called it "the supremest specimen of all ugliness") and admired for its size and form—has been in continuous use since its opening. Today's longest steel bridges use the metal much more sparingly by hanging the transport surfaces on steel cables: the central span of the world's longest suspension bridge, the 1915 Çanakkale across the Dardanelles Strait (completed in 2022), measures 2,023 m (Dezeen 2022).

New markets that were created by the activities that came into existence during the last third of the nineteenth century—and that produced enormous demands for steel—included most prominently the electrical industry with its heavy generating machinery, and the oil and gas industry, dependent on drilling pipes, well casings, pipelines, and complex refinery equipment. An innovation that was particularly important for the future of the oil and gas industry was the introduction of seamless steel pipes. The *pierce rolling process* for their production was invented in 1885, by Reinhard and Max Mannesmann, at their father's file factory in Remscheid, Germany (Mannesmann RöhrenwerkAG 2024). Several years later they added *pilger rolling*, which reduces the diameter and wall thickness while increasing the tube length. The combination of these two techniques, known as the *Mannesmann process*, produces all modern pipes.

The first carriagelike cars had wooden bodies, but a large-scale switch to sheet steel was made with the onset of mass automobile production that took place before 1910. Steel consumed in US automaking went from a few thousand tons in 1900 to more than 70,000 t in 1910, and then to 1 Mt by 1920 (Hogan 1971). Expanding production of motorized agricultural machinery (tractors, combines) created another new market for that versatile alloy. Beginning in the 1880s, significant shares of steel were also used to make weapons and heavy armaments for use on both land and sea, including heavily armored battleships.

The pre-World War I period also saw the emergence of many new markets for specialty steels, whose introduction coincides almost perfectly with the two generations investigated in this book (Law 1914). Re-enter Bessemer, this time in the role of a personal savior rather than an inventor. After Mushet lost his patent rights for perfecting the Bessemer process, his mounting debts and poor health led his 16-year-old daughter Mary to make a bold decision: in 1866, she traveled to London and confronted Bessemer in his home (Osborn 1952). We will never know what combination of guilt and charity led Bessemer to pay Mushet's entire debt (£377 14s 10d) by a personal check and later to grant him an allowance of £300 a year for the rest of his life. Mushet could thus return to his metallurgical experiments, and, in 1868, he produced the first special steel alloy by adding a small amount of tungsten during the melt.

Mushet called the alloy "self-hardening steel" because the tools made from it did not require any quenching to harden. This precursor of modern tool steels soon became known commercially as *RMS* (Robert Mushet's Special Steel), and it was made, without revealing the process through patenting, by a local company and later, in Sheffield, under close supervision of Mushet and his sons. One of its most notable new applications was in the mass production of steel ball bearings, which were put first into the newly introduced safety bicycles and later into an expanding variety of machines; eventually (in 1907) the Svenska Kullagerfabriken (SKF) company introduced the modern method of ball-bearing production.

Manganese—an excellent deoxidizer of Bessemer steels when added in minor quantities (less than 1%)—made brittle alloys when used in large amounts. But in 1882, when Robert Abbot Hadfield (1858–1940) added more of the metal (about 13%) to steel than did any previous metallurgist, he obtained a highly wear-resistant and nonmagnetic alloy that found its most important use in toolmaking. Other alloys followed during the following decades. Between 3% and 4% molybdenum was added for tool steels and for permanent magnets; as little as 2–3% nickel sufficed to make the best steel for cranks and shafts steel and, following James Riley's 1889 studies, also for armor plating of naval ships. Between 0.8% and 2% silicon was used for the manufacture of springs, and niobium and vanadium were added in small amounts to make deep-hardening steels (Zapffe 1948; Smith 1967).

Nickel and chromium alloy (Nichrome), patented by Albert Marsh in 1905, provided the best solution for the heating elements in electric toasters: dozens of designs of that small household appliances appeared on the US market in 1909, just two months after the alloy became available (NAE 2024; see Figure I.2.25). Finally, in August 1913, Harry Brearley (1871–1948) produced the first batch of what is now an extended group of stainless steels by adding 12.7% chromium, an innovation that was almost immediately used by Sheffield cutlers. Some modern corrosion-resistant steels contain as much as 26% chromium and more than 6% nickel (Bolton 1989), and their use ranges from household utensils to acid-resistant reaction vessels in chemical industries. The introduction of new alloys was accompanied by fundamental advances in understanding their microcrystalline structure and their behavior under extreme stress. Discovery of x-rays (in 1895) and the introduction of x-ray diffraction (in 1912) added new valuable techniques for their analysis.

During the twentieth century, steel's annual worldwide production rose 30-fold to nearly 850 Mt, and then China's enormous expansion of steelmaking raised the 2022 total to 1 gigatons (Gt), accounting for 54% the global output (Kelly and Fenton 2003; WSA 2024). Annual per capita consumption of finished steel products is a good surrogate measure of economic advancement. The global average in 2022 was more than 130 kg, and the rate ranged from highs of between 250 and 600 kg for affluent Western economies to about 160 kg in China, less than 40 kg in India, and just a few kilograms in the poorest countries of sub-Saharan Africa (WSA 2024). Exceptionally high steel consumption rates for South Korea and Taiwan (around 1 ton per capita) are anomalies that do not reflect actual domestic consumption but rather the extensive use of the metal in building large cargo ships for export.

The Age of Synergy also introduced the use of the most versatile alloy in a hidden and, literally, supporting role. Although normally painted, steel that forms sleek car bodies, massive offshore drilling rigs, oversize earth-moving machines, or elegant kitchen appliances is constantly visible, and naked steel that makes myriads of household utensils and industrial tools is touched every day by billions of hands. But one of the world's largest uses of steel is normally hidden from us: steel in buildings, where it is either used alone in forming the structure's skeleton and bearing its enormous load or where it reinforces concrete.

Steel in Construction

Structural steel in the form of I beams, which had to be riveted together from smaller pieces, began to form the skeletons of the world's first skyscrapers (less than 20 stories tall) during the 1880s. Its advantages are obvious. When tall buildings are built with solid masonry, their foundation must be substantial, and load-bearing walls in the lower stories must be very thick. Thinner walls were made possible by using cast iron (good in compression) with masonry, and later cage construction used iron frames to support the floors. The 10-story (42 m high) Chicago's Home Insurance Building, designed by William Le Baron Jenney (1832–1907) and finished in 1885 (the Field Building now occupies its site), was the first structure with a load-carrying frame of steel columns and beams. The total mass of this structural steel was only a third of the mass of a masonry building and resulted in increased floor space and larger windows (Figure I.4.7).

As electric elevators became available (America's first one was used in Baltimore, in 1887, and the first Otis installation was done in New York, in 1889), and as central heating, electric plumbing pumps, and telephones made it more convenient to build taller structures, skyscrapers soon followed, in both Chicago and New York. The early projects included such memorable (and no longer standing) Manhattan structures as the World Building (20 stories, 94 m in 1890) and the Singer Building (47 stories, 187 m in 1908). Henry Grey's (1849–1913) invention (in 1897) of the universal beam mill made it possible to roll sturdy H beams. In the United States, they were made for the first time by the Bethlehem Steel's mill at Saucon, Pennsylvania, in 1907, and their availability made the second generation of skyscrapers possible. Rolled in a single piece directly from an ingot, H beams had a substantially higher tensile strength (Cotter 1916; Hogan 1971). An undisputed paragon of structures using this new material is New York's Woolworth Building at 233 Broadway, designed by Cass Gilbert (1859–1934) and finished in 1913.

The building used for the first time the key techniques that still characterize the construction of skyscrapers. Its 57 stories (241 m high) are founded on concrete piers that reach Manhattan's bedrock; sophisticated wind bracing minimizes its swaying, and high-speed elevators provide local and express service. A new, elite group of fearless construction -ironworkers emerged to assemble these massive and tall steel skeletons—and more than a hundred years later they still form a special professional caste (Figure I.4.7).

Figure I.4.7 The beginnings of the skyscraper era. The Home Insurance Building in Chicago, completed in 1885 (and demolished in 1931), was the first structure supported by steel beams and columns. The engraving of New York's ironworkers building an early skyscraper and taking "coolly the most hazardous chances" is reproduced from *The Illustrated London News* of December 26, 1903. A century later, safety harnesses were compulsory, but the basic assembly procedure and the need for casual tolerance of great heights and an uncommon sense of balance and physical dexterity have not changed.

The Woolworth Building remained the city's tallest structure until 1930, when it was surpassed by the Bank of Manhattan and Chrysler buildings and, a year later, by the Empire State Building (Landau and Condit 1996). These structures, and most of their post-1970 companions—with the Petronas Towers in Kuala Lumpur becoming the world's tallest building (452 m) when opened in 1999—share the same fundamental structural property, as they hide variably shaped steel skeletons adorned with ornamental cladding of stone, metal, plastic materials, or glass (Figure I.4.8).

The other major use of steel in construction, burying it inside concrete, also emerged during the pre-World War I period. This technique, whereby concrete gets its tensile strength from embedded steel bars or

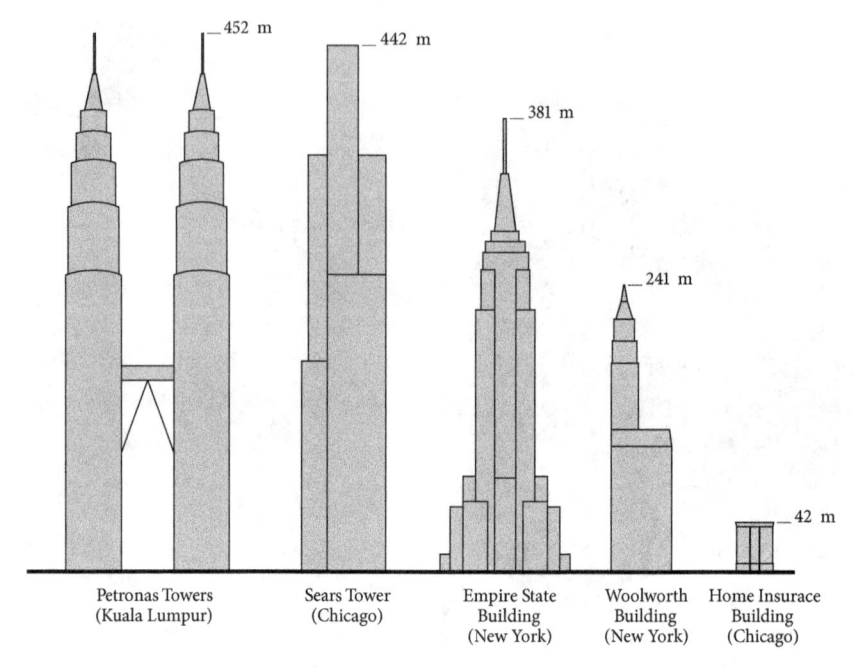

Figure I.4.8 Height comparison of five notable steel-skeleton skyscrapers built between 1885 and 1999.

steel lattices, created much of the built environment of the twentieth century. Concrete—a mixture of cement, sand, gravel, and water—is of course an ancient material, the greatest structural invention of the Roman civilization. The Pantheon, the pinnacle of its concrete architecture, built between 126 and 118 BCE, still stands in the heart of Rome's old city: its bold dome, spanning 43.2 m, consists of five rows of square coffers of diminishing size converging on the stunning unglazed central oculus (Lucchini 1966; Marder and Jones 2018).

Better cement to produce better concrete has been available only since 1824, when Joseph Aspdin began firing limestone and clay at temperatures high enough to vitrify the alumina and silica materials and to produce a glassy clinker (Shaeffer 1992; Courland 2011). Its grinding produced a stronger Portland cement (named after limestone whose appearance it resembled when set) that was then increasingly used in many compressive applications, mainly in foundations and walls. *Hydration*, a reaction between cement and water, produces tight bonds and a material that is very strong in compression but has hardly any tensile strength; hence it can be used in beams only when reinforced: such a composite material has monolithic qualities, and it can be fashioned into countless shapes.

Three things had to happen to engender the widespread use of reinforced concrete: a good understanding of tension and compression forces in the material, the realization that hydraulic cement protects iron from rust, and understanding that concrete and iron form a solid bond (Peters 1996). But even before this knowledge was fully in place, there were isolated trials and projects that embedded cast or wrought iron into concrete beginning during the 1830s, and the first proprietary system of reinforcement was introduced in England and France during the 1850s (Newby 2001). In 1854, William Wilkinson patented wire rope reinforcement in coffered ceilings, and, a year later, François Coignet introduced a system that was used in 1868 to build a concrete lighthouse at Port Said. Fireproofing, rather than the monolithic structure, was seen as the material's most important early advantage.

Between 1867 and 1878, a Parisian gardener, Joseph Monier (1823–1906), patented reinforced concrete troughs, pipes, panels, footbridges, and beams, an outgrowth of his production of garden tubs and planters that he strengthened with simple metal netting to make them lighter and thinner (Marrey 2013). Many similar patents were granted during the following decades, but there was no commensurate surge in using the material in construction. The pace picked up only after 1880, particularly after Adolf Gustav Wayss (1851–1917) bought the patent rights to Monier's system for Germany and Austria (in 1885) and after a Parisian contractor, François Hennebique (1842–1921), began franchising his patented system of reinforced construction, particularly for industrial buildings (Straub 1996; Delhumeau 1999). Inventors designed a variety of shaped reinforcing steel bars to increase the surfaces at which metal and concrete adhere (Figure I.4.9).

And one additional requirement had to be solved: the large-scale production of cement. The patenting of inclined rotary kilns that process the charge at temperatures of up to 1,500°C began in the United Kingdom during the 1870s; the first commercially successful patents were granted in the United Kingdom 1885 (to Fredrick Ransome) and 1889 (to Frederick Stokes). The early kilns (starting in the United States in 1892) were small (less than 8 m long and 1.5 m in diameter) but American producers scaled-up these designs rapidly, and, by 1907, the country produced half of the world's cement in kilns up to about 20-m long and 2 m in diameter (Moore 2011). Concurrently, engineers had also developed standard tests of scores of various cement formulas and concrete aging.

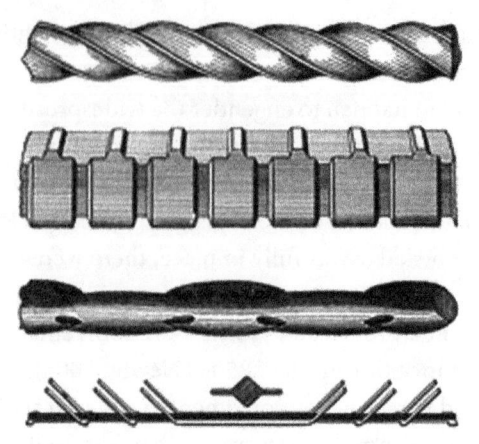

Figure I.4.9 Shaped steel bars for reinforcing of concrete: spiral twist by E. L. Ran-some, a square design with projections by A. L. Johnson, alternating flat and round sections by Edwin Thacher, and a bar with bent protrusions by Julius Kahn and the Hennebique Co.
Reproduced from Iles (1906).

The Ingalls Building (16 stories, 54 m) completed in Cincinnati, Ohio in 1903, was the world's first reinforced concrete skyscraper (Condit 1968), and, in 1906, Thomas Edison got into the act with his cast-in-place concrete houses that he built in New Jersey. Most of the 86 original houses still stand, but such utilitarian structures did not make concrete a material of choice for residential construction (Atlas Obscura 2024). Much more creatively, two architects, Auguste Perret and Robert Maillart (1872–1940), made the reinforced concrete esthetically acceptable. Perret did so with his elegant apartments, including the delicate façade of 25 Rue Franklin in Paris, and with public buildings, most notably the Theatre des Champs-Elysées. Maillart designed and built (in Switzerland between 1901 and 1940) more than 40 elegant concrete bridges (Billington 1989).

By the 1960s, new high-strength reinforced concrete began to compete with the steel frame in the building of the tallest skyscrapers (Shaeffer 1992). The world's tallest building, the 828-m-high Burj Khalifa completed in Dubai in 2009 (United Arab Emirates), is a combination of reinforced concrete and structural steel (Aldred 2010; Burj Khalifa 2024). Its Y-shaped, 586-m-high reinforced concrete core is supported by a large reinforced concrete mat (supported by massive bored reinforced concrete piles), and it is surmounted by a steel superstructure, including a telescopic steel spire.

After 1950 also came widespread commercial applications of *pre-stressing* concrete, another fundamental pre-World War I invention involving concrete and steel. Its origin can be dated exactly to 1886, when Carl Dochring came up with an ingenious idea to stretch the reinforcing bars while the concrete was wet and release the tension after the material had set (Abeles 1949). This puts the individual structural members, which are usually pre-cast offsite, into compression and makes it possible to use about 70% less steel and 40% less concrete for the same load-bearing capacity and hence to build some amazingly slender structures.

During the first decades of the twentieth century, Eugene Freyssinet (1879–1962) developed much of the underlying technical understanding of pre-stressing, and he also came up with the idea of *post-stressing* by tensioning wires that are threaded through ducts formed in precast concrete (Grotte and Marrey 2000; Ordóñez 1979). Reinforced concrete is simply everywhere—in buildings, bridges, highways, runways, and dams (Courland 2011). The material assumes countless mundane and often plainly ugly shapes (massive gravity dams, blocky buildings) as well as daring and visually pleasing forms (thin shells). Its unmatched accretion is the world's largest dam—China's Sanxia Dam on the Yangzi, 185 m tall and 2.3 kilometers (km) long—which contains nearly 28 million cubic meters of concrete reinforced with nearly half a million tons of steel bars (Power-Technology 2020).

In terms of both global annual production and cumulatively emplaced mass, reinforced concrete is now the leading man-made material that has been used both for the shoddily built apartment blocks of Beijing's Maoist period (they looked dilapidated even before they were completed) and for the elegant sails of Sydney Opera House, which looks permanently inspirational. Unadorned reinforced concrete forms the world's tallest free-standing structure, Toronto's oversize CN tower, as well as Frank Lloyd Wright's perfectly proportioned cantilevered slabs that carry Falling Water over a cascading Pennsylvanian stream. As Peters (1996: 58) noted, "[R]einforced concrete has been variously attacked as the destroyer of our environment and praised as its savior," but whatever the reaction may be, his conclusion is indisputable: "Our world is unthinkable without it." Yet only history-minded engineers know its origins, and so reinforced concrete may be perhaps the least appreciated of all fundamental pre-World War I innovations, an even more obscure foundation of modern civilization than the electrical transformer described in Chapter I.2.

Aluminum

This light metal is much more common in nature than is iron: this most common of all metallic elements is, after oxygen and silicon, the third largest constituent of Earth's crust, amounting to about 8% by mass (Press and Siever 1986). But because of its strong affinity for oxygen, the metal occurs naturally only in compounds, never in a pure state. And, unlike iron, which has been known and worked for more than 3,000 years, aluminum was identified only in 1808, by Humphry Davy (of the first electric arc fame); isolated in a slightly impure form as a new element only in 1825, by Hans Christian Ørsted (of the magnetic effect of electric currents fame); and two years later produced in powder form by Friedrich Wohler (of the first-ever urea synthesis fame). It remained a rare commodity until the late 1880s: Napoleon III's infant son got a rattle made of it, and 2.85 kg of pure aluminum were shaped into the cap set on the top of the Washington Monument in 1884 (Binczewski 1995).

The quest for producing the metal in quantity was spurred by its many desirable qualities. The most notable attribute is its low density: with merely 2.7 g/cm3, compared to 8.9 g/cm3 for copper and 7.9 g/cm3 for iron, it is the lightest of all commonly used metals. Only silver, copper, and gold are better conductors of electricity, and only a few metals are more malleable. Its tensile strength is exceeded only by the best steels, and its alloys with magnesium, copper, and silicon further enhance its strength. Moreover, it can be combined with ceramic compounds to form composite materials that are stiffer than steel or titanium. Its high malleability and ductility mean that it can be easily rolled, stamped, and extruded, and it offers attractive surface finishes that are highly resistant to corrosion. The metal is also generally regarded as safe for use with food products and for beverage packaging. Finally, it is easily recyclable, and its secondary production saves 95% of the energy needed for its production from bauxite.

Bauxite, named after the district of Les Baux in southern France where Pierre Berthier first discovered the hard reddish mineral in 1821, is the element's principal ore, and it contains between 35% and 65% of alumina (aluminum oxide, Al_2O_3). But the first attempts to recover the metal did not use this abundant ore. Henri Saint-Claire Deville (1818–1881) changed Wohler's method by substituting sodium and potassium and began producing small amounts of aluminum by a cumbersome and costly process of reacting metallic sodium with the pure double chloride of aluminum and sodium at high temperature. In 1854, Robert Bunsen separated aluminum electrolytically from its fused compounds, and, shortly afterward, Deville revealed his apparatus for reducing aluminum from fused aluminum-sodium chloride in a glazed porcelain crucible that was equipped with a platinum cathode and carbon anode.

Although this process did not lead to commercial production, Borchers (1904: 113) summed up Deville's contribution by noting that he introduced the two key principles that were then "repeatedly re-discovered and patented, viz: 1. The use of soluble anodes in fused electrolytes; and 2. The addition of aluminum to fused compounds of the metal during electrolysis, by the agency if alumina." Worldwide production of the metal was just 15 tons in 1885, and, even after Hamilton Y. Castner's new production method cut the cost of sodium by 75%, aluminum still cost nearly US$20/kg in 1886 (Carr 1952).

A Duplicate Discovery

When the commercially viable solution finally came—during that extraordinarily eventful decade of the 1880s—it involved one of the most remarkable instances of independent concurrent discovery. In 1886, Charles Martin Hall and Paul Louis Toussaint Héroult (Figure I.4.10) were two young chemists of the same age (both born in 1863, and both died in 1914), working independently on two continents (the first in a small town in Ohio, the second near Paris) to come up with a practically identical solution of the problem during the late winter and early spring months of 1886. Just two years later, their discoveries were translated into the first commercial enterprises producing electrolytic aluminum.

Hall's success was the result of an early determination and a critical chance encounter: his aspiration to produce aluminum dated to his high school years, but after he enrolled at the Oberlin College in Ohio, in 1880, he had the great luck of meeting and working with Frank F. Jewett, an accomplished scientist who came to the college as a professor of chemistry and mineralogy after four years at the Imperial University in Tokyo and after studies and research at Yale, Sheffield, Gottingen, and Harvard (Edwards 1955; Craig 1986). Jewett provided Hall with laboratory space, materials, and expert guidance, and Hall began intensive research efforts soon after his graduation in 1885.

Figure I.4.10 Portraits of Charles Hall (left) and Paul Héroult from the late 1880s, when the two young inventors independently patented a nearly identical aluminum reduction process.
Photographs courtesy of Alcoa.

After a few false starts, he began to experiment with cryolite, the double fluoride of sodium and aluminum (Na_3AlF_6), and, on February 9, 1866, he found that this compound with a melting point of 1,000°C was a good solvent of Al_2O_3. Hall had to energize his first electrolysis of cryolite, which he did on February 16, 1886, with inefficient batteries. Graphite electrodes were dipped into a solution of aluminum oxide and molten cryolite in a clay crucible, but, after several days of experiments, he could not confirm any aluminum on his cathode. Because he suspected that the cathode deposits originated in the crucible's silicon lining, he replaced it with graphite.

On February 23, 1886, after several hours of electrolysis in a woodshed behind Hall's family house, he and his older sister Julia found deposits of aluminum on the cathode. Hall's quest is a near-perfect example of a fundamental technical advance that was achieved during the most remarkable decade of the Age of Synergy by the means so emblematic of the period: the combination of good scientific background and a readiness to experiment, persevere, and make necessary adjustments, and all of this followed by prompt patenting, fighting off the usual interference claims, and no less persistent attention to scaling-up the process to an industrial level and securing financial backing.

Hall's first American specification was received in the Patent Office on July 9, 1886—but it ran immediately into a problem as the office notified him of a previous application making a very similar claim: Héroult filed his first

French patent (Patent 175,711) on April 23, 1886, and by July 1887, he formally contested Hall's priority. But as Hall could prove the date of his successful experiment on February 23, 1886, he was eventually granted his US rights, divided among two patents issued on April 2, 1889. US Patents 400,664 and 400,766 specified two immersed electrodes, while US Patents 400,667 specified the use of cryolite (Figure I.4.11). While his patents were pending, Hall looked for financial backing, and he finally found committed support from Pittsburgh metallurgist Alfred E. Hunt. The new Pittsburgh Reduction Company was formed in July 1888, and it began reducing alumina before the end of the year (Carr 1952). This was a small establishment powered by two steam-driven dynamos and producing initially just slightly more than 20 kg of aluminum per day, and 215 kg/day after its enlargement in 1891 (Beck 2001).

Its electrolytic cells (60 cm long, 40 cm wide, and 50 cm deep) were made of cast iron, held up to about 180 kg of cryolite, and had 6–10 carbon anodes (less than 40 cm long) suspended from a copper bus bar, with the tank itself acting as the cathode. A new plant at New Kensington near Pittsburgh reached an output of 900 kg/day by 1894, and, in the same year, an even larger facility was built at Niagara Falls to use the inexpensive electricity from what was at that time the world's largest hydro station. In 1907, the Pittsburgh Reduction Company was renamed Aluminum Company of

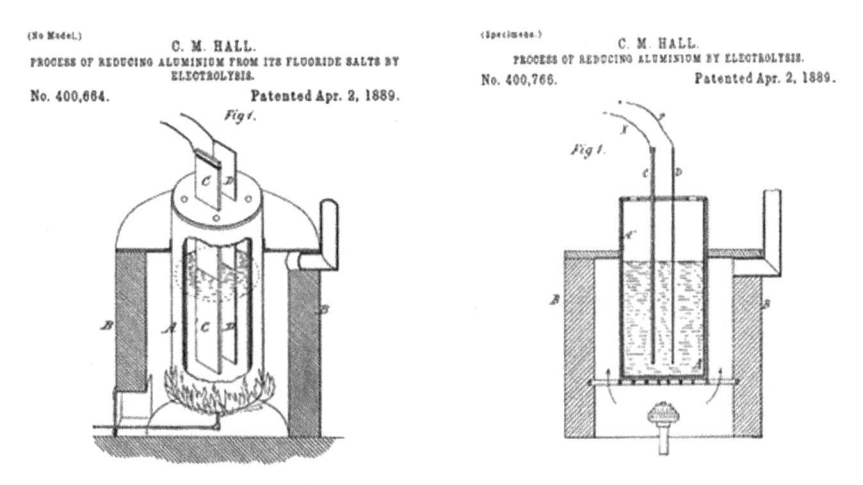

Figure I.4.11 Illustrations of aluminum reduction process in Hall's two US patents (400,664 on the left, 400,766 on the right) granted on April 2, 1889. In both images, A is a crucible, B a furnace, and C the positive and D the negative electrode. The first patent proposed $K_2Al_2F_8$ as the electrolyte; the second one, $Na_2Al_2F_8$.

America (Alcoa). Alcoa, now a multinational corporation with operation in 39 countries, is still among the metal's leading global producers, now led by China's Chinalco and Hongqiao Group and Russia's Rusal. Similarities between Hall's and Héroult's independent specifications are striking.

Héroult, who conducted his experiments on the premises of his father's small tannery and who did not rely on batteries but on a small steam-powered Gramme dynamo to produce a more powerful current, described his invention as a "process for the production of aluminum alloys by the heating and electrolytic action of an electric current on the oxide of aluminum, Al_2O_3, and the metal with which the aluminum shall be alloyed" (cited in Borchers 1904: 127). Like Hall, he also made a provision for external heating of the electrolytic vessel, and his aluminum furnace also had a carbon lining and a carbon anode. Héroult's technique was commercialized for the first time in 1888, by the Aluminium Industrie Aktiengesellschaft in Neuhausen, Switzerland, and by the Société Electrométallurgique Française (Ristori 1911). Larger works, using inexpensive hydroelectricity, were soon established in France, Scotland, Italy, and Norway.

Mature Hall-Héroult Process

The Hall-Héroult process remains the foundation of the world's aluminum industry. Nothing can change the fact that the abundant Al_2O_3 is nonconducting and that molten cryolite, in which it is dissolved, is an excellent conductor and hence a perfect medium for performing electrochemical separation. Heat required to keep the cryolite molten comes from the electrolyte's resistance and not from any external source; during the electrolysis, graphite is consumed at the anode with the release of carbon dioxide (CO_2), while liquid aluminum is formed at the cathode and collects conveniently at the bottom of the electrolytic vessel, where it is protected from reoxidation because its density is greater than that of the supernatant molten cryolite.

Another important step toward cheaper aluminum that took place in 1888, and on which we continue to rely more than a century later, was Karl Joseph Bayer's (1847–1904) patenting of an improved process for alumina production. After washed bauxite is ground and dissolved in NaOH, insoluble compounds of iron, silicon, and titanium are removed by settling, and fine particles of alumina are added to the solution to precipitate the compound, which is then calcined (converted to an oxide by heating) and shipped as a white powder of pure Al_2O_3 (Lumley 2010). Typically, 4–5 t of bauxite will produce 2 t of alumina, which will yield 1 t of the metal.

As with so many other late nineteenth-century innovations, the electrolytic process was greatly scaled up while its energy consumption fell considerably. During the twentieth century, average cell size in Hall-Héroult plants had doubled every 18 years, while energy use decreased by nearly half between 1888 and 1914 and then dropped by nearly as much by 1980 (Beck 2001; Figure I.4.12). By 1900, the worldwide output of aluminum reached

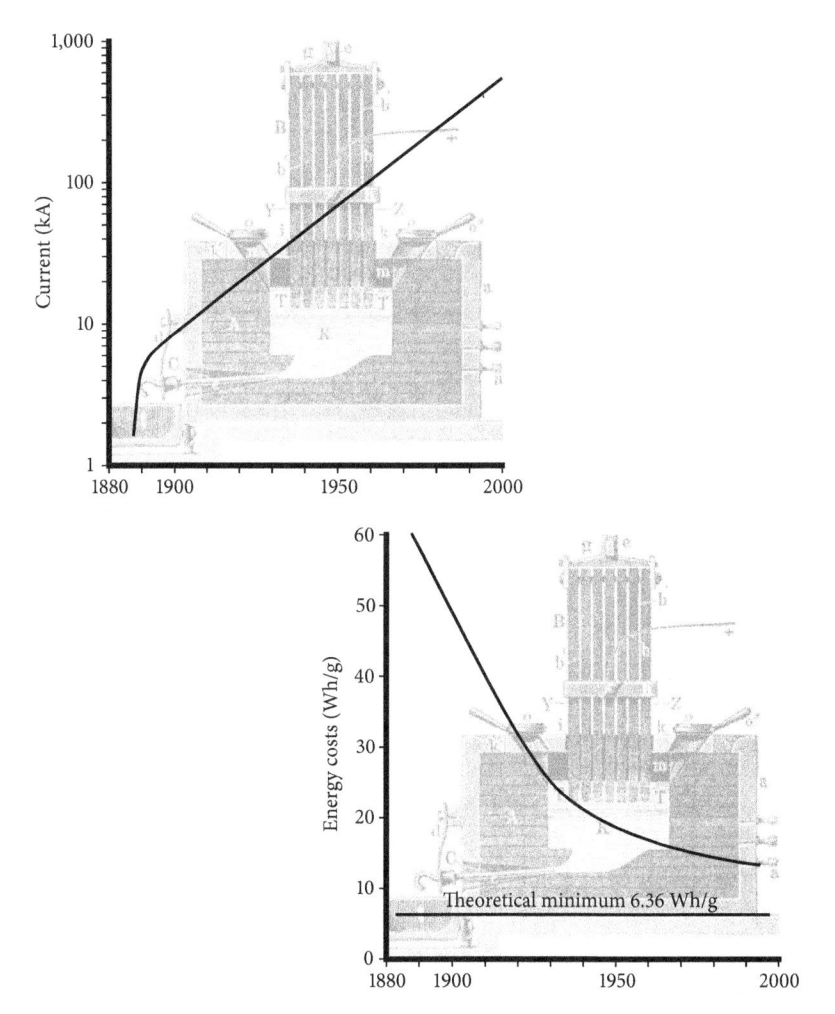

Figure I.4.12 Increasing cell size (indicated by higher current) and declining energy cost of the Hall-Héroult aluminum production, 1888–2000.
Based on Beck (2001).

8,000 t, and its price fell by more than 95% compared to the cost in 1885 (Borchers 1904). By 1913, its global production was 65,000 t, and the demand rose to nearly 700,000 t by the end of World War II. By the end of the twentieth century, annual aluminum smelter production surpassed 24 Mt by 2022 it rose to 69 Mt (USGS 2024). The metal is produced both in pure (99.8%) form and as various alloys, and most of it is sent for final processing in rolls of specified thickness.

Inexpensive electrolytic aluminum was a novelty that had to take away market shares from established metals and create entirely new uses (Devine 1990a). During the early 1890s, the metal's main use was for deoxidation of steel, but a much larger market soon emerged with aluminum's use for cooking utensils and then with parts for bicycles, cameras, cars, and many other manufactures. No machine has been as defined by aluminum and its alloys as twentieth-century aircraft: steel alloys are the quintessential material in land and sea transport, but they are too heavy to be used for anything but some special parts in the air.

Aluminum bodies began displacing wood and cloth in aircraft construction during the late 1920s, after the mid-1930s demand for the metal was driven mainly by the need for large numbers of military planes (Hoff 1946). After World War II the market for aluminum alloys expanded rapidly with the large-scale adoption of military jet airplanes during the late 1940s and of commercial jets during the late 1950s. High-strength aluminum alloys (mainly alloy 7075 made with copper and zinc) made up 70–80% of the total airframe mass of modern commercial planes, which means that for the Boeing 747, this metal added up to more than 100 t per plane (Figure I.4.13). The rest was accounted for by steel alloys, titanium alloys (in engines, landing gear), and, increasingly, composite materials. The Boeing 787, launched in 2007, was the first commercial airplane with composite materials (carbon fiber) accounting for half of the plane's mass (Boeing 2024). Space flight and Earth-orbiting satellites have been other important markets for aluminum alloys.

But the metal's light weight and resistance to corrosion guarantees its expanding uses in construction (from window and door frames to roofing and cladding), and aluminum extrusions are widely used in railroad rolling stock (large hopper cars that carry bulk loads), in automobiles and trucks (in 2000, a typical sedan contained twice as much aluminum, mostly in its engine, as it did in 1990), in both recreational boats and commercial shipping,

NEW MATERIALS AND NEW SYNTHESES 191

Figure I.4.13 The Boeing 747, the largest passenger aircraft of the late twentieth century whose construction required more than 100 t of aluminum alloys.
Photograph from author's collection.

and in food preparation (cooking and table utensils) and packaging (ubiquitous carbonated beverages and beer cans, foils). And aluminum wires, supported by steel towers, are now the principal long-distance carriers of both high-voltage AC (HVAC) and DC (HVDC).

The Hall-Héroult process is not without its problems. Even after all those efficiency gains, it is still highly energy intensive: at the beginning of the twenty-first century primary aluminum production in the United States averaged about 15.2 megawatt hours (MWh) of electricity per ton of the metal, or about 55 gigajoules per ton (GJ/t); two decades later it declined only slightly to 14.9 MWh/t (IAI 2024). Adding the inevitable electricity losses in generation and transmission and the energy costs of bauxite mining, production of alumina and electrodes, and the casting of the smelted metal raises the total rate to almost 200 GJ/t, and well above that for less efficient producers. This is an order of magnitude more than producing steel from blast-furnace smelted pig iron (Smil 2023b). But as there are no obvious alternatives, it is safe to conclude that our reliance on those two great 1888 commercial innovations in metal industry—Bayer's production of alumina and Hall-Héroult's electrolysis of the mixture of cryolite and alumina—will continue during the twenty-first century.

Industrial Gases

Separation of gases from the atmosphere is a nearly perfect example of advances that characterized the Age of Synergy but these techniques, so indispensable for shaping the modern world, have been often neglected in standard accounts of inventions and innovations or treated much more cursorily than other age-defining materials. For example, Williams, in *The History of Invention* (1987), a treatment of exemplary breadth based on the multivolume Oxford set, does not devote even a single sentence to air liquefaction, cryogenics, or mass production of industrial gases. And industry is also a perfect example of technical innovations whose basic processes were developed, and to a considerable degree improved, during the two pre-World War I generations but whose true importance became clear only decades later.

Scientific foundations for extracting oxygen and nitrogen from the atmosphere were laid by the gradual advances of nineteenth-century physics and chemistry, with the single most important contribution made by Thomas Joule and William Thompson (Lord Kelvin) in 1852 (Almqvist 2003). They demonstrated that when highly compressed air flowing through a porous plug (a nozzle) expands to the pressure of the ambient air, it will cool slightly. The first experimental success in liquefying oxygen—yet another example of virtually simultaneous inventions whose frequency was one of the markers of innovative intensity during the Age of Synergy—came just before the end of 1877, when Louis-Paul Cailletet, a metallurgist working in ironworks at Chatillon-sur-Seine, got an oxygen mist in his glass tube, and when Raoul Pictet's better apparatus in Geneva produced a limited but clearly visible flow of liquid oxygen.

Six years later in Kraków, Sigmund von Wroblewski and Carl Olszewski did a better job by liquefying oxygen at a pressure of less than 2 megapascals (MPa) and used the same apparatus to liquefy nitrogen. In 1879—a year after his lectures at Munich's Technische Hochschule inspired young Rudolf

Diesel to devote himself to the design of a superior internal combustion engine (see Chapter I.3)—Carl von Linde (1842–1934) established the Gesselschaft für Linde's Eismaschinen in Wiesbaden and began to work on improving the methods of refrigeration (Figure I.4.14). On June 5, 1895, he was granted German Patent 88,824 for his process for the liquefaction of atmospheric and other gases, which combined the Thomson-Joule effect with what Linde termed "countercurrent cooling" (Linde 1916). Compressed air (at 20 MPa) was expanded through a nozzle at the bottom of an insulated chamber (achieving a drop of about 40°C), and it was used to precool the incoming compressed air in a countercurrent cooler. As the cycle is repeated, the temperature of the air expanded at the nozzle progressively declines, and the air eventually liquefies.

In 1902, Linde received German Patent 173,620 for the method used to separate the principal gases from the liquefied air, and he set up the first commercial air separator at Höllriegelskreuth near Munich. The key feature of Linde's patent was rectification (i.e., purification by repeated distillation) through self-intensification: nearly pure oxygen vapor rising from the

Figure I.4.14 Carl von Linde (1842–1934), the inventor of artificial cooling systems and founder of Linde AG.
Photograph from the Linde corporation.

bottom of the rectification column was liquefied by encountering the cooler downward-flowing liquid mixture of oxygen and nitrogen. Georges Claude (1870–1960) introduced his new air liquefaction design also in 1902. Unlike Linde's liquefaction (with its simple nozzle expansion), his process used an external works (an expansion engine) for refrigeration of the compressed air (the challenge of lubricating the piston at low temperatures had to be solved first), and his rectification method incorporated *retour en arrière*, whereby the liquefied stream was first divided into oxygen-rich and oxygen-poor fractions, and the latter was used for a more efficient washing out of the oxygen from the ascending vapor.

By 1910, Linde perfected this approach by introducing a double rectification, with the two streams first formed in the lower pressure column and then the final concentration of the lower boiling component (nitrogen) in the vapor and the higher boiling gas (oxygen) in the liquid done in the upper column. A century later these basic solutions devised by Linde and Claude are still the foundation of the large-scale production of gases (and Linde AG, headquartered in Wiesbaden, remains the leading supplier of cryogenic processes), but incremental improvements had resulted not only in much higher production volumes but also in lower capital and energy costs. Securing large volumes of hydrogen was a greater challenge.

Badische Anillin- & Soda-Fabrik's (BASF) first ammonia pilot plant in Oppau, near Ludwigshafen, used hydrogen produced from chlorine-alkali electrolysis, a process unsuitable for large-scale commercial operations. Other known methods to produce hydrogen were either too expensive or produced excessively contaminated flow. Given BASF's (indeed, the country's) high dependence on coal, Carl Bosch (whose leading role in the commercialization of ammonia synthesis is described later in this section) concluded that the best solution was to react the glowing coke with water vapor to produce water gas (about 50% hydrogen [H_2], 40% carbon monoxide [CO], and 5% each of nitrogen [N_2] and CO_2) and then remove CO cryogenically (Smil 2001). But, in the summer of 1911, Wilhelm Wild (1872–1951) began removing CO by a catalytic process during which the reaction of the gas with water shifted CO to CO_2 and produced additional H_2. This *Wasserstoffkontaktverfahren*, patented by Wilhelm Wild and Carl Bosch, became the dominant technique of mass-producing hydrogen for the next four decades.

New Syntheses

Even reasonably well-informed people do not usually mention explosives when asked to list remarkable innovations that helped to create, rather than to imperil, the modern world: compared to the destructive power of modern explosives, their constructive roles have received a disproportionately small amount of attention. Yet their constant use is irreplaceable not only in construction projects but also in the destruction of old buildings and concrete and steel objects and, above all, in ore and coal mining. How many Americans would even guess that during the late 1990s coal extraction (mostly in surface mines) accounted for 67% of all industrial explosives used annually in the United States (Kramer 1997)?

There was only a brief delay between the patenting of the first modern explosive and its widespread use. Dynamite patents were granted to Alfred Nobel (1833–1896; Figure I.4.15) in 1867 (GB Patent 1345) and in 1868 (US Patent 78,317), and, by 1873, the explosive was such a great commercial success that he could settle in Paris in a luxurious residence at 59 Avenue Malakoff, where a huge crystal chandelier in his large winter garden lit his prize orchids and where a pair of Russian horses were waiting to take his carriage for a ride in the nearby Bois de Boulogne (Fant 1993). In contrast, the first new synthetic materials—nitrocellulose-based Parkesine (1862) and Xylonite (1870) used for combs, shirt collars, billiard balls, and knife handles—made little commercial difference after their introduction. And the first commercial product made from the first successful thermoset plastic—a Bakelite gear-lever knob for a Rolls Royce—was introduced only in 1916, nearly a decade after Leo Baekeland filed his patent for producing the substance (Bijker 1995). Numerous (and always only black) Bakelite moldings, sometimes elegant but more often either undistinguished or outright ugly-looking (and still too brittle)—ranging from the classic rotary telephone to cigarette holders, and from early radio dials to electric switch covers—began appearing only after World War I.

So did products made of polyvinyl chloride, now encountered in cable and wire insulation as well as in pipes and bottles, a polymer that was first produced in 1912. Many new plastics were introduced during the 1930s, but their widespread commercial use came only after 1950. Modern plastics are thus much less beholden to the pre-World War I period than electricity or steel, and they fill the image of quintessential twentieth-century

Figure I.4.15 Alfred Nobel. His wealth and fame did little to ease his near-chronic depression and feelings of worthlessness (see Chapter I.6). Photograph © The Nobel Foundation.

materials: their introduction, proliferation, and impact will be examined in this book's companion volume. But there is another chemical synthesis, one involving a deceptively simple reaction of two gases, whose laboratory demonstration was followed by an unusually rapid commercialization that established an entirely new industry just before the beginning of World War I. This accomplishment ranks as one of the most important scientific and engineering advances of all time, and that is why this chapter closes with a detailed narrative and appraisal of the Haber-Bosch synthesis of ammonia from its elements.

Powerful Explosives

Introduction of new powerful explosives during the 1860s ended more than half a millennium of gunpowder's dominance. Clear directions for preparing different grades of gunpowder were published in China by 1040, and its European use dates to the early fourteenth century (Needham et al. 1986). The mixture of the three components capable of detonation typically included about 75% saltpeter (potassium nitrate, KNO_3), 15% charcoal, and 10% sulfur. Rather than drawing oxygen from the surrounding air, ignited gunpowder used the element present in the saltpeter to produce a very rapid (roughly 3,000-fold) expansion of its volume in gas. Historians have paid a great deal of attention to the consequences of gunpowder use in firearms and guns (Kelly 2004), particularly to its role in creating and expanding European empires after 1450 (Cipolla 1966; McNeill 1989).

Gunpowder's destructive capacity was surpassed only in 1846, when Ascanio Sobrero (1812–1888), a former student of Justus von Liebig, who also worked with Alfred Nobel when they both studied with Jules Pelouze in Paris in the early 1850s, created nitroglycerin. This pale yellow, oily liquid is highly explosive on rapid heating or on even a slight concussion, and it was produced by replacing all the hydroxide (–OH) groups in glycerol—$C_3H_5(OH)_3$,—with NO_2 groups (Escales 1908; Marshall 1917). Glycerol, a sweet syrupy liquid, is readily derived from lipids or from fermentation of sugar; among its many commercial uses are antifreeze compounds and plasticizers. Nitroglycerol is the most common of nitroglycerin's many synonyms, but the proper chemical designation is 1,2,3-propanetriol trinitrate (Figure I.4.16). The compound is best known for its use as an antianginal agent and coronary vasodilator.

After an explosion at the Torino armory, Sobrero stopped all experiments with the dangerous substance and did not become famous for his invention: who knows his name today besides the historians of chemistry? Enter Nobel, whose great fame does not rest on discovering a new explosive compound but, first, on a process to ignite nitroglycerine in a controllable and predictable fashion and then on the preparation and marketing of the substance in a form that is relatively safe to handle and practical to use for industrial blasting. Nobel's nitroglycerin experiments began in 1862, when he was detonating the compound by using small amounts of black powder attached to a slow fuse. In 1863, he got his first patent for using nitroglycerin in blasting and introduced an igniter (a wooden capsule filled with black powder) to trigger the compound's explosion.

In November of the same year, he invented dynamite, a new explosive that, to quote his US Patent 78,317 granted on May 26, 1868 (Nobel 1868: i),

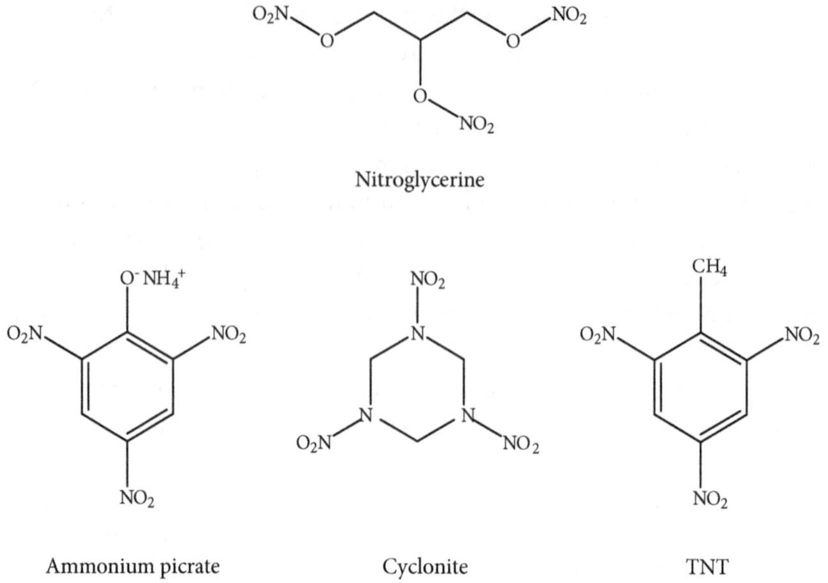

Figure I.4.16 Structures of 1,2,3-propanetriol trinitrate (nitroglycerine), ammonium picrate (2,4,6-trinitrophenol ammonium salt), cyclonite (hexahydro-i,3,5-trinitro-i,3,5-triazine), and TNT (i-methyl-2,4,6-trinitrobenzene).

consists in forming out of two ingredients long known, viz. the explosive substance nitroglycerin, and an inexplosive porous substance, hereafter specified, a composition which, without losing the great explosive power of nitroglycerine, is very much altered as to its explosive and other properties, and which, at the same time, is free from any quality which will decompose, destroy, or injure the nitroglycerine, or its explosiveness.

Nobel decided to give this combination a new name, dynamite, "not to hide its nature, but to emphasize its explosive traits in the new form; these are so different that a new name is truly called for" (cited in Fant 1993: 94). Nobel's dynamite was made with highly porous Kieselguhr, a mineral that was readily available along the banks of the Elbe and the Alster (the Elbe's small northern tributary) near Hamburg, where Nobel moved from Sweden in 1865. The mineral is diatomaceous earth formed of siliceous shells of unicellular microscopic diatoms, aquatic protists found in both fresh and marine waters that often display nearly perfect radial symmetry. Diatomaceous earth—also used as an insulator, an abrasive in metal polishes, an ingredient in toothpaste, to clarify liquids, and as a filler for paper, paints, and detergents—can absorb about three times its mass of nitroglycerine.

A 1:3 mixture of diatomaceous earth and nitroglycerine produces a reddish yellow, crumbly mass of plastic grains that is slightly compressible and safe from detonation by ordinary shocks (Shankster 1940). Today's standard dynamite composition is basically the same as Nobel's ratio: 75% nitroglycerol, 24.5% diatomaceous earth, and 0.5% sodium carbonate, the last compound being added to neutralize any acidity that might be formed by the decomposition of nitroglycerin. Nobel also came up with a detonator, without which the new explosive would be nearly impossible to use. This igniter was a metal capsule filled with $Hg(CNO)_2$ (mercury fulminate): a strong shock produced by this primary explosive, rather than high heat, set off the nitroglycerin.

The Nobel igniter, patented in 1867, opened the way for widespread use of dynamite, making it one more critically important item to mark the beginning of the Age of Synergy, whose ethos Nobel embodied as much as did Edison, Hall, Parsons, or Daimler: he wanted to see his inventions commercialized and diffused as rapidly as possible. By 1870, he had five nitroglycerine factories in Europe; established the first dynamite factory in Paulilles, in southern France in 1871; and watched the explosive conquer the blasting market worldwide. By 1873, there were 15 dynamite factories in Europe and North America, and, between 1871 and 1874, the total output of the explosive had quadrupled to just more than 3,000 t.

As the most powerful explosive yet available—its velocity of detonation is as much as 6,800 meters per second (m/s), compared to less than 400 m/s for gunpowder (Johansson and Persson 1970)—the compound was more effective in breaking rocks apart rather than just displacing them. Consequently, blasting for mining operations, railroad construction, and other building projects became not only safer but also more efficient and less expensive. Nobel had tried unsuccessfully to gelatinize nitroglycerin already in 1866, and he finally succeeded in 1875, by preparing a gelatinous dynamite, the combination of nitroglycerine (about 92%) and nitrocellulose (cotton treated with nitric acid, also known as nitrocotton or gun cotton). This tough yet plastic and water-resistant compound became the explosive of choice for packing boreholes in rock-blasting operations. Cheaper ammonium nitrate was later used to replace a relatively large share of nitroglycerin and to produce many kinds of plastic and semiplastic ammonia gelatin dynamites of different explosive intensity.

These substances contain 50–70% explosive gelatin (a mixture of nitroglycerine and nitrocotton), 24–45% ammonium nitrate, and 2–5% wood pulp (Shankster 1940). Ammonium nitrate—a compound that was first made by J. R. Glauber in 1659 by reacting ammonium carbonate with nitric acid—is difficult to detonate, but it is a powerful oxidizing agent, and its heating under confinement results in highly destructive explosions. Mixtures of ammonium nitrate and carbonaceous materials (charcoal, sawdust) were patented as explosives by C. J. Ohlsson and J. H. Norbin in 1867, but mixtures of the compound with fuel oil (generally referred to as ANFO) became commonly used only during the 1950s (Marshall 1917; Clark 1981). Since that time, explosives based on ammonium nitrate that are inexpensive and safe to use have captured most of the blasting market by replacing dynamite; some 2.5 Mt of these mixtures accounted for 99% of US industrial explosives sales during the late 1990s (Kramer 1997).

Unfortunately, ANFO has been used not only as a common industrial explosive but also in powerful car and truck bombs by terrorists. The two worst terrorist attacks that took place in the United States before September 11, 2001—the bombing of the World Trade Center in New York City on February 26, 1993 (when, fortunately, only six people were killed), and the destruction of the Alfred Murrah Federal Office Building in Oklahoma City on April 19, 1996 (169 adults and children killed)—involved vehicles packed with ANFO mixtures. The bombing also resulted in a renewed interest in additives able to desensitize ammonium nitrate, and new tests showed that this option does not work very well with larger volumes of explosives that are used in terrorist acts: the Oklahoma bomb contained about 1.8 t of nitrate (Hands 1996).

Although Nobel was frequently reproached for inventing such powerful means of destruction, dynamite did not take the place of gunpowder for military uses. While it made it easier to commit terrorist attacks—in 1882, Czar Alexander II was among the first victims of a dynamite bomb—slower acting, and preferably smokeless propellants were needed for guns. Such explosives were produced during the 1880s: *poudre B* (nitrocellulose is first gelatinized in ether and alcohol and then extruded and hardened to produce a propellant that could be handled safely) by Paul Vieille (1854–1934) in 1884, Nobel's own Ballistite (using nitroglycerine instead of ether and alcohol) in 1887, and cordite, patented in England by Frederick Abel and James Dewar in 1889. The combination of these new powerful propellants and better guns made from new alloys resulted in considerably longer firing ranges.

The destructiveness of these weapons was further increased by introducing new explosive fillings for shells. Ammonium picrate was prepared in 1886; trinitrotoluene (TNT), which was synthesized by Joseph Wilbrand in 1863 and which must be detonated by a high-velocity initiator, was first manufactured in Germany in 1891—by 1914, it became a standard military compound; and the most powerful of all prenuclear explosives, cyclotrimethylenetrinitramine, known either as cyclonite or RDX (for royal demolition explosive), was first made by Hans Henning in 1899, for medicinal uses by treating a formaldehyde derivative with nitric acid. Its widespread use came only during World War II.

Structural formulas of these three powerful explosives clearly show that they are variations on the common theme, namely, a benzene ring with three nitro groups (Figure I.4.16). Cyclonite production took off only after cheap supplies of formaldehyde became available, and its combination of high velocity of detonation (8,380 m/s) and large volume of liberated gas make it an unsurpassed choice for bursting charges. Consequently, this substance was produced and used on a massive scale during World War II, as was TNT, which was deployed in more than 300 million antitank mines.

Ammonia from Its Elements

Inclusion of this item may seem puzzling because the synthesis of the simplest triatomic nitrogen compound does not figure on commonly circulating lists of the greatest inventions in modern history. Whether they were assembled by the canvassing of experts or of the public, these compilations (a few of them were noted in Chapter I.1) almost invariably contain automobiles, telephones, airplanes, nuclear energy, space flight, television, and computers as the most notable technical inventions of the modern era. And yet, when measured by the most fundamental of all human yardsticks, that of physical survival, none of these innovations has had such an impact on our civilization as did the synthesis of ammonia from its elements (Smil 2001, 2022). The two keys needed to unlock this puzzle are a peculiar attribute of human metabolism and the evolution of nitrogen's presence and cycling in the biosphere.

In common with other heterotrophs, humans cannot synthesize amino acids, the building blocks of proteins, and must ingest them (eight essential amino acids for adults, and nine for children) in plant or animal foods to grow all metabolizing tissues during childhood and adolescence and to maintain and replace them afterward. Depending on the quality of their diets, most adults need between 40 and 80 g of food proteins (which contain roughly 6–13 g of nitrogen) each day. Crops, or animal foods derived from crop feeding, provide nearly 90% of the world's food protein, with the rest of the nutrient coming from grasslands, fresh waters, and the ocean. To produce this protein, plants must have an adequate supply of nitrogen—and this macronutrient is almost always the most important factor that limits yields in intensive cropping.

Unfortunately, the natural supply of nitrogen is restricted. Nitrogen makes up nearly 80% of Earth's atmosphere, but the molecules of dinitrogen gas (N_2) are nonreactive, and only two natural processes can split them and incorporate the atomic nitrogen into reactive compounds: lightning and biofixation. Lightning, producing nitrogen oxides, is a much smaller source of reactive nitrogen than biofixation. In biofixation, free-living cyanobacteria, bacteria, and, more importantly, bacteria symbiotic with legumes (above all, those belonging to genus *Rhizobium*) use their unique enzyme, nitrogenase, to cleave atmospheric dinitrogen's strong bond and incorporate the element first into ammonia and eventually into amino acids. In addition, reactive nitrogen is easily lost from soils through leaching, volatilization, and denitrification.

As a result, shortages of nitrogen limit crop yield more than shortages of any other nutrient. Traditional crop production could improve the nutrient's supply only by cultivating leguminous crops or by recycling crop residues and animal and human wastes. But this organic matter has only very low concentrations of nitrogen, and even its (utterly impracticable) complete recycling would be able to support only a finite number of people even if they were to subsist on a largely vegetarian diet. In contrast, it is much easier to cover any shortages of phosphorus and potassium. The latter nutrient is not commonly deficient in soils, and it is easily supplied just by mining and crushing potash. And the treatment of phosphate rocks by diluted sulfuric acid, the process pioneered by John Bennett Lawes (1814–1900) in England during the 1860s, yields a concentrated source of soluble phosphorus in the form of ordinary superphosphate.

But at the close of the nineteenth century, there was still no concentrated, cheap, and readily available source of fertilizer nitrogen needed to meet the higher food requirements of expanding populations. And not for lack of effort.

Two new sources were commercially introduced during the 1840s—guano (bird droppings that accumulated on some arid islands, mainly in the Pacific) and Chilean nitrate ($NaNO_3$) extracted from near-surface deposits in the country's northern desert regions—but neither had a high nitrogen content (15% in nitrate; up to 14% in the best and less than 5% in typical guanos), and both sources offered only a limited supply of the nutrient. So did the recovery of ammonium sulfate from coking ovens, which was introduced for the first time in Western Europe starting in the 1860s: the compound contained only 21% nitrogen. As the nineteenth century was coming to its close, there was a growing concern about the security of the future global food supply, and no other scientist summarized these worries better than William Crookes (1832–1919), a chemist (as well as physicist) whom we will again encounter in Chapter I.5 in relation to his comments on wireless transmission. Crookes based his presidential address to the British Association for the Advancement of Science, which he delivered in September 1898 (Crookes 1899,) on "stubborn facts"—above all, the limited amount of cultivable land, growing populations, and relatively small number of food-surplus countries. He concluded that the continuation of yields that prevailed during the late 1890s would lead to a global wheat deficiency as early as in 1930. Higher yields, and hence higher nitrogen inputs, were thus imperative, and Crookes saw only one possible solution: to tap the atmosphere's practically unlimited supply of non-reactive nitrogen to produce fixed nitrogen that could be assimilated by plants.

He made the existential nature of this challenge quite clear (Crookes 1899: 45–46).

> The fixation of nitrogen is vital to the progress of civilised humanity. Other discoveries minister to our increased intellectual comfort, luxury, or convenience; they serve to make life easier, to hasten the acquisition of wealth, or to save time, health, or worry. The fixation of nitrogen is a question of the not far-distant future.

And he had no doubt as to how this will happen: "It is the chemist who must come to rescue.... It is through the laboratory that starvation may ultimately be turned into plenty" (Crookes 1899: 3).

But in 1898, there was no technical breakthrough in sight. The German cyanamide process—in which coke was reacted with lime and the resulting calcium carbide was combined with pure nitrogen to produce calcium cyanamide (CaCN$_2$)—was commercialized in that year, but its energy requirements were too high to become a major producer of nitrogen fertilizer. That the sparking of nitrogen and oxygen can produce nitrogen oxides (readily convertible to nitric acid) was known for more than a century, but the very high temperatures needed for this process (up to 3,000°C) could be produced economically only with very cheap hydroelectricity.

The first commercial installation of this kind (the Birkeland-Eyde process) was set up in Norway, in 1903, and it was enlarged before 1910, but the total output remained small. The breakthrough came 11 years after Crookes's memorable speech, when Fritz Haber (1868–1934; Figure I.4.17) demonstrated the process of ammonia synthesis from its elements. And just four years after that, his small bench-top apparatus was transformed into a large-scale commercial process by a dedicated engineering team led by another outstanding German chemist, Carl Bosch (1874–1940; Figure I.4.17).

Figure I.4.17 Fritz Haber's portrait (left) taken during the early 1920s. Carl Bosch's photograph (right) from the early 1930s.

Haber photograph courtesy of Bibliothek und Archiv zur Geschichte der Max-Planck-Gessselschaft, Berlin. Bosch photograph courtesy of BASF Unternehmensarchiv, Ludwigshafen.

Genesis of the Haber-Bosch Process

Fritz Haber was born to a well-off merchant family (dyes, paints, chemicals) in Breslau; he studied in Berlin, where Carl Liebermann (1842–1914), best known for the synthesis of alizarin (a red dye), was one of his supervisors, and in Heidelberg under Robert Wilhelm Bunsen. Afterward, he went through several jobs before Hans Bunte offered him an assistantship at the Technische Hochschule in Karslruhe, where Haber stayed for the next 17 years, working initially in electrochemistry and then in thermodynamics of gases (Szöllösi-Janze 1998).

His first experiments with ammonia took place in 1904, at a request from the Österreichische Chemische Werke in Vienna, and, as noted in his 1919 Nobel lecture, the challenge was only seemingly simple (Haber 1920: 326).

> We are concerned with a chemical phenomenon of the simplest possible kind. Gaseous nitrogen combines with gaseous hydrogen in simple quantitative proportions to produce ammonia. The three substances involved have been well known to the chemist for over a hundred years. During the second half of the last century each of them has been studied hundreds of times in its behaviour under various conditions during a period in which a flood of new chemical knowledge became available.

But none of the many illustrious chemists who attempted to perform that simple reaction—$N_2 + 3H_2 \leftrightarrow 2NH_3$—during the nineteenth century had succeeded in the task, and, by the beginning of the twentieth century, it appeared that such a synthesis may not be possible, a perception that deterred many new attempts. Haber's first experiments to synthesize ammonia at 1,000°C and at atmospheric pressure were soon abandoned due to very low yields obtained with an iron catalyst, which he chose after testing more than 1,000 materials. He returned to the task only three years later and, with his English assistant Robert Le Rossignol, began new experiments: under a pressure of 3 MPa they obtained much higher, although still clearly non-commercial, yields (Smil 2001; Sheppard 2017).

New calculations showed that a pressure of 20 MPa, at that time the limit he could obtain in his laboratory, and a temperature of 600°C would yield about 8% ammonia—but he had no catalyst that would work at such a relatively low temperature. By 1908, his search for ammonia synthesis received assistance from BASF, at that time the world's largest chemical company, which was interested in advancing research on nitrogen fixation by the Birkeland-Eyde process. As already noted, this technique used a high-voltage electric arc to produce nitrogen oxide that could be later converted to nitric acid. In 1907, Haber and his pupil A. König published a paper on that topic, and their work attracted the attention of BASF. On March 6, 1908, the company concluded two agreements with Haber: the first one provided generous financial assistance to do more work on the electric arc process; the second one, a smaller sum of money to continue his previous work on high-pressure ammonia synthesis.

Nothing pathbreaking ever came out of arc experiments, while the high-pressure catalytic synthesis, thanks to Le Rossignol's skills in constructing the necessary apparatus (including a small double-acting steel pump) and Haber's perseverance in searching for better catalysts and for the best combination of pressure and temperature, began soon yielding promising results. Haber's first German patent for ammonia synthesis (Patent 235,421, valid since October 13, 1908) is generally known as the "circulation patent," and its basic principle is still at the core of every ammonia plant today (BASF 1911). Commercialization of the process could not be done without catalysts that could support rapid conversion at temperatures of 500–600°C.

Initially Haber believed that osmium would be the best choice, and the finely divided metal supporting ammonia yields of about 6% and higher was clearly of economic interest; hence, he advised BASF to buy up all the osmium that was at that time in the possession of a company that used it to produce incandescent gas mantles. Haber filed his osmium catalyst patent on March 31, 1909, and continued to search for other catalysts; uranium nitride looked particularly promising. But above all, he concentrated on perfecting his laboratory apparatus built to demonstrate the potential of high-pressure synthesis with gas recirculation; by the end of June, the setup was finally ready.

A convincing demonstration of ammonia synthesis is one of those rare great technical breakthroughs that can be dated with great accuracy. On July 3, 1909, Haber sent a detailed letter to the BASF directors (now in the company's Ludwigshafen archives) that described the events of the previous day, when Alwin Mittasch and Julius Kranz witnessed the demonstration of Haber's synthesis in his Karslruhe laboratory (Haber 1909).

> Yesterday we began operating the large ammonia apparatus with gas circulation in the presence of Dr. Mittasch, and we were able to keep its

uninterrupted production for about five hours. During this whole time it had functioned correctly and it produced liquid ammonia continuously.... The steady yield was 2 cm³/minute, and it was possible to raise it to 2.5 cm³. This yield remains considerably behind the capacity for which the apparatus has been constructed because we have used the catalyst space very insufficiently.

The apparatus used in the experiment is shown in Figure I.4.18. Inside an iron tube kept under a pressure of 20.3 MPa was a nickel heating coil used to raise the gas temperature to a desired level; the synthesized ammonia was separated from the flowing gas at a constant high pressure, and the heat from the exothermic reaction could be removed from exhaust gases and used to preheat the freshly charged gas replacing the synthesis gas removed by the conversion. Recycling the gases over the catalyst made it possible to sustain relatively high production rates. Haber's third key German patent application (Patent 238,450, submitted on September 14, 1909, and issued on September 28, 1911) detailed the synthesis under very high pressure (Haber 1911).

Once there was no doubt about the feasibility of a high-pressure catalytic synthesis of ammonia from its elements, BASF moved swiftly to commit its resources to the commercialization of the process.

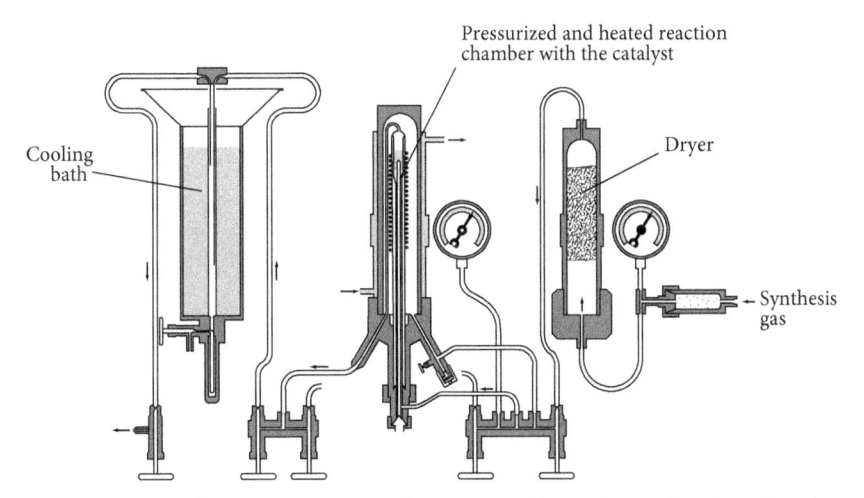

Figure I.4.18 Laboratory apparatus that was used by Haber and Le Rossignol in their successful experiments to synthesize ammonia from its elements.
Reproduced from Smil (2001).

This challenge called for scaling up Haber's bench-top apparatus—a pressurized tube that was just 75 centimeters (cm) tall and 13 cm in diameter—into a large-throughput assembly operating under pressures that were at that time unprecedented in industrial synthesis and filling it up with a relatively inexpensive yet highly effective catalyst. And it also called for, first and foremost, producing economically large volumes of the two feedstock gases. Rapid success could not have happened without Carl Bosch's decisive managerial leadership and his technical ingenuity. Bosch came to BASF in 1899, directly after graduating from Leipzig University and eventually was put in charge of BASF's nitrogen fixation research. That goal eluded him, as it did so many other outstanding synthetic chemists of the time. But when Haber demonstrated his bench-top process to BASF, it was Bosch who oversaw turning it into a profitable commercial operation.

I have already related in Chapter I.1 how, in March 1909, Bosch's understanding of steel metallurgy persuaded the BASF leadership to proceed with the commercialization of Haber's process. But before any large-scale production of ammonia could begin, Bosch and his team had to solve several unprecedented engineering challenges, including the failure of the first scaled-up converters. These 2.5-m-long experimental tubes (heated electrically from the outside and filled with the catalyst) exploded after about 80 hours of operation under high pressure. Fortunately, Bosch anticipated such a mishap and placed them in reinforced concrete chambers. Bosch's solution was to contain high temperature within an inner wall of soft (low carbon) steel and to force a mixture of pressurized cold hydrogen and nitrogen into the space between the inner and outer walls.

Consequently, the inner shell was subject to equal pressure from both sides, and the strong outer steel shell was under high pressure but remained much cooler. With this solution, converter sizes increased to 4 m and 1 t in the first pilot plant in 1912, and units 8 m long weighing 3.5 t were installed in the first commercial process in 1913. Perhaps the most notable challenge among other engineering problems that had to be overcome before setting up the entire operation was the production of hydrogen (nitrogen was derived from Linde's air liquefaction). In 1912, Bosch and Wilhelm Wild patented their *Wasserstoffkontaktverfahren*, a catalytic shift reaction that transforms CO and steam into CO_2 and hydrogen.

The second major set of challenges involved the search for the best and most inexpensive catalyst. Alwin Mittasch (1869–1953) undertook systematic testing of all metals known to have catalytical abilities in pure form or as binary (e.g., Al-Mg, Ba-Cr, Ca-Ni) or ternary catalysts, as well as more complex mixtures (Mittasch 1951). Magnetite (Fe_3O_4) from the Gallivare mines in northern Sweden supported the highest yield, and Mittasch then sought the best possible combination with catalytic promoters. The first patent for such mixed catalysts was filed on January 9, 1910. Nothing illustrates the efficacy of Mittasch's exhaustive search better than the fact that most commercial catalysts used in ammonia synthesis during the twentieth century were just slight variations on his basic combination that relied on magnetite with additions of Al_2O_3, potassium oxide (K_2O), calcium oxide (CaO), and magnesium (Mg). These promoted iron catalysts are also extraordinarily stable, able to serve for up to 20 years without deactivation.

Construction of the first ammonia plant at Oppau, near Ludwigshafen, began on May 7, 1912, and production started on September 9, 1913, with the gas used as a feedstock for the synthesis of ammonium sulfate fertilizer. But less than a year after the plant's completion, Oppau's ammonia was diverted from making fertilizer to replacing Chilean nitrate, whose imports were cut off by the British naval blockade. There is no doubt that the Haber-Bosch synthesis was one of the factors that helped to prolong World War I. Blockaded Germany could do without Chilean nitrates that were previously used for producing explosives, and its troops had enough ammunition to keep launching new offensives for nearly four years, until the spring of 1918. To satisfy this large new wartime demand, a second, much larger ammonia plant was completed at Leuna, near Halle, in 1917. when the war ended Germany was the only country with a considerable capacity to produce inexpensive inorganic nitrogen fertilizer.

But the synthesis of ammonia had a limited impact on the world's food production before the 1950s because only a few countries were relatively intensive users of inorganic fertilizers and as the economic downturn of the 1930s and World War II set back the industry. Global synthesis of ammonia remained below 5 Mt until the late 1940s, and, while several European countries had high average rates of fertilizer applications already before World War I, in the United States more than one-third of all farmers did not use any fertilizer nitrogen even by the late 1950s. But then the situation changed rapidly, particularly as nitrogen fertilizers became the key nutrient that unlocked the yield potential of new cultivars everywhere except in the conflict-torn and mismanaged economies of sub-Saharan Africa.

Importance of Synthetic Ammonia

While the importance of electricity or internal combustion engines for modern society is obvious, even reasonably educated urbanites do not appreciate the essential role played by nitrogen fertilizers, and even most scientists are not aware of the extent to which global civilization depends on the Haber-Bosch synthesis of ammonia. Global production of ammonia nearly doubled during the 1950s, further quadrupled by 1975, when, after a brief period of stagnation that began during the late 1980s, it rose to nearly 130 Mt by the end of the twentieth century. During the late 1990s, the global output of ammonia and sulfuric acid was virtually identical, but because of ammonia's much lower molecular weight (17 vs. 98 for H_2SO_4), the gas is the world's leading chemical in terms of synthesized moles.

During the late 1990s, about two-thirds of ammonia, or an equivalent of about 85 Mt of nitrogen a year, were used by the fertilizer industry (Figure I.4.19), but, in contrast to the pre-1960 period, when the gas was a feedstock for making a variety of compounds, most of it now is used for the synthesis of urea. Unlike gaseous ammonia, urea is a solid compound, containing 45% nitrogen, and it is produced in small granules that are easily stored, shipped, and applied to fields. Most nitrogenous fertilizers are applied to cereals and to oil and tuber crops. A detailed balance of nitrogen flows in the global agroecosystems showed that, at the end of the twentieth century, synthetic nitrogenous fertilizers supplied about half of the nutrient available to the world's crops, with the other half coming from leguminous crops, organic recycling, and atmospheric deposition (Smil 1999a).

Does this also mean—with nitrogen being the leading limiting input in food production—that roughly one-half of humanity is now alive thanks to the Haber-Bosch synthesis? This question cannot be answered without referring to a specific food supply: how many people would not be alive without ammonia synthesis depends on prevailing diets. Detailed accounts of the global nitrogen cycle indicate that, by 2000, about three-quarters of all nitrogen in food proteins available for human consumption came from arable land (with the rest coming from pastures and aquatic species). If synthetic fertilizers provided about half of all nitrogen in harvested crops, then at least every third person, and more likely two people out of five, got the protein in their diets from the Haber-Bosch synthesis. As with most global averages, such a revelation both overestimates and underestimates the importance of this great invention.

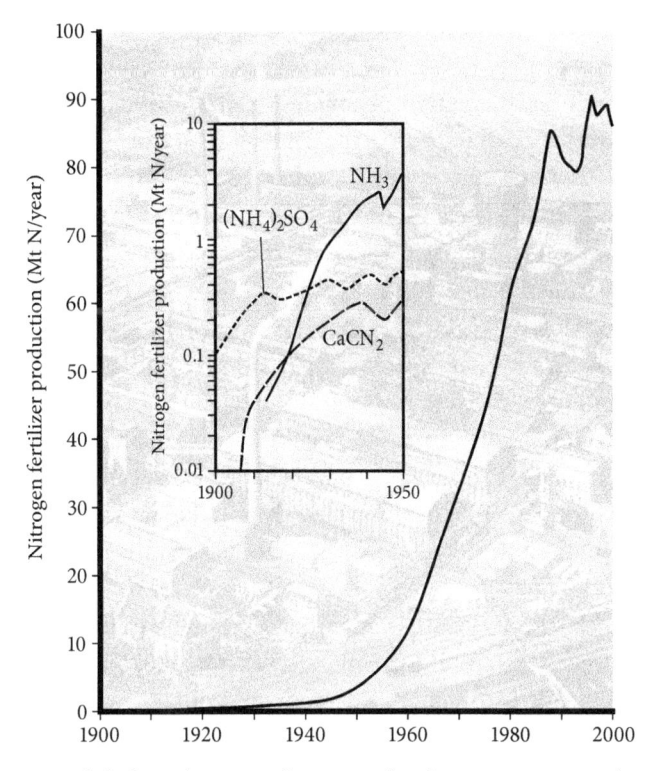

Figure I.4.19 Global production of nitrogen fertilizers, 1900–2000 (in megatons of nitrogen per year). Pre-1913 output includes ammonium sulfate from coking and cyan-amide and nitrate from the electric arc process.
Based on data in Smil (2001).

In affluent nations, fertilization helps to produce an excess of food in general and assures high animal food intakes in particular (most of those countries' crops are fed to animals rather than being eaten directly by people), and it also helps to produce more food for export. Consequently, significant cuts in fertilizer use would result in less meaty diets and in lower exports but would not at all imperil the overall adequacy of the food supply. In contrast, rising fertilizer applications have been essential for lifting populous low-income countries from the conditions of bare subsistence and widespread malnutrition. Norman Borlaug, one of the leaders in the development and diffusion of new high-yielding varieties of crops, captured the importance of nitrogen fertilizers in his speech accepting the Nobel Prize for Peace in 1970 (Borlaug 1970): "If the high yielding dwarf wheat and rice varieties are the catalysts that have ignited the Green Revolution, then chemical fertilizer is the fuel that has powered its forward thrust."

Perhaps the best illustration of different degrees of dependence on nitrogen fertilizers is a comparison of the United States and Chinese situations. In the late 1960s, Americans applied less than 30 kg of nitrogen per hectare of their farmland, while the Chinese rate was less than 5 kg/hectare (ha). Then came a stunning reversal of China's international posture: the country's return to the United Nations in 1971 was followed by Nixon's state visit in 1972. The consensus opinion was that this historic shift was motivated almost solely by the need for a strategic partner to counter the expansionist USSR: Realpolitik made yesterday's despised aggressors raining bombs on Vietnam today's valued allies as Mao chatted with Nixon in Zhongnanhai. But such major policy shifts rarely have a single cause, and, as far as I am aware, at the time of China's opening nobody pointed out a reason ultimately more powerful than the fear of the Soviet hegemony: the need to avert another massive famine.

The world's greatest and largely man-made famine claimed at least 30 million lives between 1958 and 1961 (Smil 1999b; Yang 2008). In its wake, China's population expanded at an unprecedented rate: with no population controls in place, it grew from 660 million people in 1961 to 870 million by 1972. This addition of more than 200 million people in a single decade represented the fastest population growth in China's long history and the highest ever national increment in global terms. At the same time, slowly rising crop yields could not keep up even with basic food needs: by 1972, China's average per capita food supply was below the levels of the early 1950s! The only effective solution was to increase rapidly the synthesis of nitrogenous fertilizers—and soon after Nixon's visit China placed orders for 13 of the world's largest and most modern ammonia-urea complexes.

Such an order could not be filled without turning to M. W. Kellogg, America's and the world's leader in ammonia synthesis: 8 of the 13 ammonia plants came from Kellogg, and they fed their product to urea plants delivered by Kellogg's Dutch subsidiary. Was China's opening to the world a matter of grand politics and strategic alliances? Undoubtedly—but it was also a matter of basic survival. More purchases of ammonia-urea complexes followed, as did more fertilizer imports: by the early 1980s, China became the world's largest consumer and, a decade later, also the world's largest producer of fertilizer nitrogen. During the late 1990s, American applications averaged about 50 kg of nitrogen per hectare—but the Chinese rate surpassed 200 kg/ha and, in five of the most intensively cultivated provinces with a total population equal to that of the United States, topped 300 kg/ha! By the end of the twentieth century fertilizer nitrogen provided about 60% of the nutrient in China's crops, and because more than 80% of the country's protein was derived from crops, roughly half of all nitrogen in China's food came from inorganic fertilizers.

Similarly high degrees of high existential dependence have been evolving, nationally or regionally, in other land-scarce countries. When these countries reach the limit of their cultivated area, then they must turn, even if they remain largely vegetarian, to higher nitrogen inputs. During the last two generations of the twentieth century, India increased its total applications of nitrogenous fertilizers by roughly the same rate as did China (a more than 40-fold rise), while Indonesia applied during the late 1990s nearly 25 times as much synthetic fertilizer as it did in 1960 (FAO 2024). Because during the twenty-first century most of the population growth will take place in Asia and Africa, this dependence is here to stay because there is still no practical alternative to supply the needed mass of this most important macronutrient.

A world that will eventually function without the two great mainstays of modern civilization—internal combustion engines and electricity generated by the burning of fossil fuels—is not difficult to foresee. During the early 2020s, electric vehicles were claiming rising shares of the global car market as more new capacity was installed every year in solar and wind conversions than in steam turbogenerators and gas turbines. In contrast, there is no substitute for dietary proteins whose production requires adequate nitrogen supplied to crops. Conferring the fixation capability possessed by legumes on wheat and rice has remained, much like the promise of early commercial nuclear fusion, elusive (Smil 2023a), and the dependence on the Haber-Bosch synthesis of ammonia will likely extend far into the twenty-first century, a legacy of one the most important, yet so inexplicably little appreciated, pre-World War I technical advances.

Frontispiece I.5. Alexander Graham Bell lecturing on his telephone to an audience in Salem, Massachusetts, on February 12, 1877, and a group in the inventor's study in Boston listening to his explanations. The early apparatus was housed in an oblong box about 18 cm wide and tall, and about 32 cm long (inset on the right).

Reproduced from the cover of *Scientific American*, March 31, 1877.

5
Communication and Information

> The apparatus ... is all contained in an oblong box about 7 inches high and wide, and 12 inches long. This is all there is visible of the instrument, which during the lecture is placed on a desk at the front of the stage, with its mouthpiece toward audience. Not only was the conversation and singing of the people at the Boston end distinctly audible in the Salem Hall, 14 miles away, but Professor Bell's lecture was plainly heard and applause sent over the wires by the listeners in Boston.
> —*Scientific American* of March 31, 1877, reporting on A. G Bell's latest telephone demonstration

Less than a year after Alexander Graham Bell exhibited the first version of his telephone at the Centennial Exposition in Philadelphia in June 1876, he was demonstrating a better design during public lectures. Although much admired, the first device had a limited capacity as it could transmit only over short distances and did so with a much-diminished signal; in contrast, the improved device made the first intercity audio communication possible. By the time of the Salem lecture, pictured in the frontispiece to this chapter, the record distance was about 230 kilometers (km), from Boston to North Conway, in New Hampshire. Bell's telephone was only the first of several fundamental inventions that eventually revolutionized every aspect of modern communication. Almost exactly 10 years after Bell's first intercity telephone calls came a discovery with even farther-reaching consequences.

What Heinrich Hertz succeeded in doing between 1886 and 1888 was to generate, send, and for the first time to receive electromagnetic waves whose frequencies "range themselves in a position intermediate between the acoustic oscillations of ponderable bodies and the light-oscillations of the ether" (Hertz 1887: 421). He generated sparks by an induction coil, sent electromagnetic waves across a room where they were reflected by a metal sheet, and then measured their frequency (distance between their crests) by observing, under microscope, tiny sparks that could be seen at the other end of the room across the small gap of a receiving wire loop. Expressed in units that honor his name, these broadcasts had frequencies of 50–500 megahertz (MHz), or 6 meters (m) to 60 centimeters (cm) (Aitken 1976). Lower frequencies of this range are now used for FM radio and broadcast TV, the higher ones for garage door openers, medical implants, and walkie-talkies.

Has any other physical experiment that was so simple—a few batteries, coils, and wires positioned in a 14-m-long lecture room—eventually brought such a wealth of amazing spin-offs? Hertz pursued this research only to confirm Maxwell's theoretical conclusions about the nature of electromagnetic waves, and he could not foresee any use for these very rapid oscillations. That perception changed soon after his premature death in 1894, and his fundamental discovery was first transformed into wireless telegraphy and soon afterward into broadcasts of voice and music. And more wonders based on those rapid oscillations were to follow during the twentieth century: television, radar, satellite telecommunication, and the beginnings of mobile telephony followed (starting in the 1990s) by wireless internet.

But the Age of Synergy revolutionized every kind of communication, a term that I use in the broadest sense to embrace all forms of printed, visual, and spoken information as well as the means and nontechnical requirements of its large-scale diffusion. Consequently, in this chapter, I examine first not only the progress in flexible and rapid printing of large editions of books and periodicals but also the new techniques that were introduced to meet the rising demand for printing paper. When describing the invention of the telephone, I first explain its telegraph pedigree, and, before outlining the early history of wireless communications, I look more closely at Hertz's experimental breakthrough. These techniques created the new economies and new social realities of the twentieth century, but only the American origins of telephone—or at least Bell's famous imploration, "Mr. Watson—come here—I want to see you" (Bell 1876a)—are more widely known.

And there are some major misunderstandings: neither Thomas Edison nor the Lumière brothers invented moving pictures, nor did Marconi make the world's first radio broadcasts (Garratt 1994). Moreover, even many informed people would not rank (or even be aware of) sulfate pulp or linotypes being among the key innovations of the two pre-World War I generations. Because of its topical sweep, this is inevitably the most heterogeneous chapter of this book, yet several critical commonalties qualify all these achievements as quintessential innovations of the Age of Synergy. Once introduced in their basic functional form during the Age of Synergy, all of them became subject to intensive refinement and redesign that boosted their performance and brought them close to technical maturity often in less than a generation after their initial patenting. And all of them deserve attention because of their combined impact on civilization.

Sulfite and sulfate pulp continue to furnish most of our paper needs, and there are no radical innovations on the papermaking horizon to change that. Before they were displaced by photo typesetting, linotype machines were used to set all books, newspapers, and merchandise catalogs during the first two-thirds of the twentieth century. Cellular phones have now largely done away with the wired touch-tone telephones that were introduced in 1963—but until that time the classical rotary telephone, whose diffusion began in the early 1920s, was the standard equipment, and landlines are not about to disappear. Development and commercialization of radio and television, as well as of sonar and radar, were only a matter of time after Hertz demonstrated the generation and reception of high-frequency waves. The age of inexpensive and simple-to-operate cameras using roll film started with the first of George Eastman's boxy Kodaks in 1888, exactly 75 years before Kodak Instamatics were introduced in 1963.

But before I start the topical surveys of major communication techniques, I must mention one of their forms that has been inexplicably neglected in standard accounts of modern logistics: parcel deliveries. There is no justification for this omission either on commercial or on emotional grounds. Few people are aware of the relatively late origins of postal parcel deliveries. British parcel post began operating in August 1883 (Samuelson 1896). In the United States, where international parcel post has been available since 1887, private companies did very good business by delivering mail orders to millions of rural customers, and their opposition to postal parcel service delayed its enactment until 1912. The service began on January 1, 1913, and the enthusiastic response translated into impressive economic benefits; for example, during the first five years of parcel post delivery Sears, Roebuck & Company, the second of the two giant mail-order retailers (established in 1893, 21 years after Montgomery Ward), tripled its earnings.

The high intensity of parcel deliveries (by trucks, vans, motorcycles, bicycles and airplanes) is now evident in all affluent societies. United Parcel Service, the world's largest parcel delivery company, was established in 1907, in Seattle, as the American Messenger Company and got its present name in

1919. In 2022, its capitalization peaked at nearly $190 billion and its annual revenues are close to $100 billion (Trading Economics 2024), the total higher than the gross domestic product of about 60 countries in 2023 (World Bank 2024). And with the internet, millions of consumers became new converts to the convenience of receiving goods at (or selling them from) home; their repeat orders (and sales) have been driving the expansion of parcel deliveries, and, by 2020, Amazon was out-shipping FedEx.

Printed Word

I will start with the printed word simply because of the longevity of the art. Gutenberg's movable type (first used in 1452) led to a well-documented explosion of printing. By 1500, more than 1,700 presses were at work across Europe, and they produced more than 40,000 separate titles or editions amounting to more than 15 million incunabula copies (first printed editions) (Johnson 1973). Subsequent centuries brought slow but cumulatively important improvements that resulted in such remarkable typographic achievements as the first comprehensive encyclopedia edited by Denis Diderot and Jean le Rond D'Alembert (1751–1777). The mechanization of printing advanced during the first half of the nineteenth century with the introduction of high-volume presses.

A new press built in 1812, by Friedrich Koenig, still had a flat type of bed, but the paper was pressed onto it by a revolving cylinder: when powered by a steam engine, this machine could print more than 1,000 sheets per hour. Introduction of a durable and flexible papier-mâché matrix made it possible to cast curved stereos once the cylindrical rollers replaced the flat printing bed and the type itself was locked onto a circular surface. This took some time to accomplish, and Richard Hoe (in 1844) and William Bullock (in 1865) were the two principal designers of practical rotary machines (Sterne 2001). Hoe's revolving press was fed sheets of paper by hand, and Bullock's roll-fed rotary could produce 12,000 newspapers per hour. Mechanization of folding, stitching, and binding helped to make progressively larger runs of newspapers and books cheaper.

But there were still obvious and irksome complications and self-evident limits. Setting the text with movable type was a highly time-consuming process that cried for mechanization. And the collecting of rags—to be converted, after sorting and cleaning, by boiling in alkaline liquor into new paper—was imposing obvious limits on the total mass of printed matter. Societies where used textiles were the only feedstock for papermaking could never hope to become both extensively and deeply literate and well informed. These challenges were resolved by new inventions and by their rapid improvement.

Typesetting and Typewriting

Setting texts with movable type was a highly repetitive, monotonous job but also one that required a high degree of alertness, combining persistence and patience. Every character had to be individually selected from a distribution box and lined up backward with appropriate spacing (achieved by inserting faceless pieces of type) in a composing stick. Once a stick was filled, the type was lifted onto a shallow tray (galley); finished galleys were used to produce proofs, marked proofs were used to correct any typesetting errors, and pages were then imposed (arranged in numerical sequence) and fastened into a "forme" that could be easily moved around, inked, and printed. When the printing was done, the type had to be disassembled, laboriously cleaned, and returned to distribution boxes.

There were many attempts to mechanize this task, and many typesetting machines were invented and patented during the nineteenth century. John Southward, writing in the late 1880s for the ninth edition of *Encyclopaedia Britannica*, noted that fewer than half a dozen of these machines had stood the test of practical experience, that a few relatively successful designs were confined to special classes of work, and that "it is open to doubt whether the nimble fingers of a good compositor, aided by the brains which no machinery can supply, do not favourably compare on the ground of economy with any possible mechanical arrangement" (Southward 1890: 700–701).

But even as Southward was expressing his doubts about the future of practical typesetting machines, their tentative and limited success was turning into a commercial triumph. Indeed, his very words (and more than 10,000 pages of the nineth edition of the world's most famous encyclopedia) were typeset by an improved version of a relatively simple machine introduced in 1872 by Charles Kastenbein. But Southward's encyclopedic entry does not contain any mention of the machine whose basic design was completed by 1884, that had been used to set newspapers since 1886, and that eventually proved the superiority of mechanical arrangement over nimble fingers.

Fingers, and brains, were still needed—not to pick, place, and space the letters but simply to type: the machine was Mergenthaler's linotype.

Ottmar Mergenthaler (1854–1899; Figure I.5.1) belonged to that large group of German craftsmen who made up a notable part of the multimillion emigration from Western Europe's largest nation to the New World and whose mechanical skills were behind many innovations introduced in the United States during the two pre-World War I generations. He was apprenticed to a watchmaker, and, after his arrival in the United States, he began working in his cousin's engineering workshop in Washington, D.C. Among other things, the shop produced models to accompany patent applications, and that is how Mergenthaler met, in 1876, Charles T. Moore, who brought in plans of a machine designed to transfer a page for printing by lithography. Building its model led Mergenthaler to the idea of designing a composing machine, a task that preoccupied him for the next 15 years (Mengel 1954; Kahan 2000).

Technical requirements for a successful composing machine were very demanding. The tasks of composing, justifying, casting, and distributing the

Figure I.5.1 Ottmar Mergenthaler, the designer of the most successful commercial linotype.
Reproduced from Coraglia (2024).

type had to be automated, and the machine had to be operated by a single person and perform at a rate sufficiently faster than hand composing to justify its cost. Mergenthaler's solution was patented in 1885 (Figure I.5.2). The operation began when a keystroke let a small thin bar (matrix) containing the female mold of a character descend from a vertical storage bin (magazine). Matrices, and appropriate space bands, were then placed in alignment in a line, a task that required an accuracy of 1/100 of a millimeter (mm) and uniform depth to print every letter and number clearly (Scott 1951).

Mergenthaler had to solve first an unprecedented problem of precision manufacturing of the matrices: the process had to be relatively inexpensive to produce the 1,200 matrices needed for every machine, and they had to be strong to withstand repeated gripping, lifting, pushing, pressing, and spacing. This required building a matrix factory and designing complex machines to mill, grind, emboss, cut, stamp, and finish the blanks punched from sheets of brass. At least one critical task could use an already commercialized process: cutting of characters was done by a punch machine invented by Linn Boyd Benton (1844–1932). And Mergenthaler used

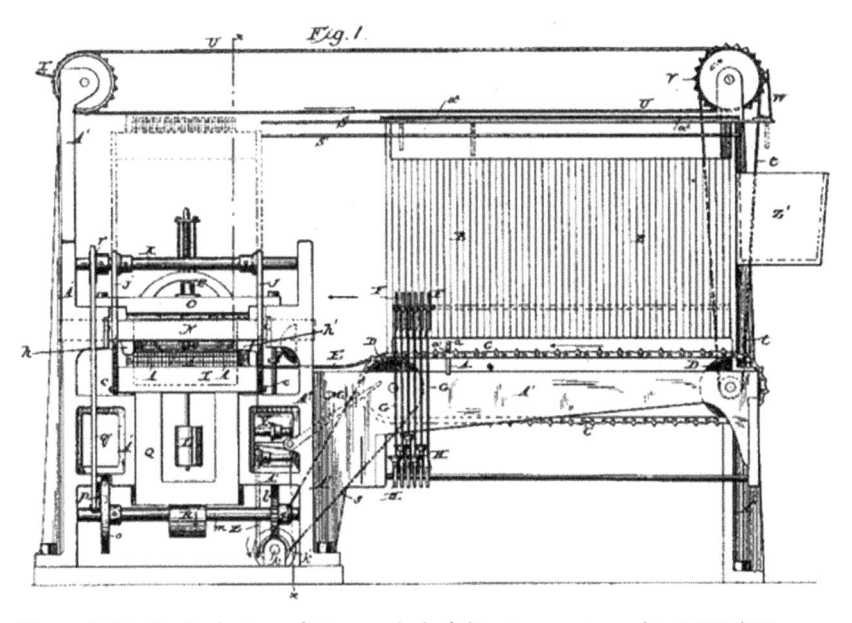

Figure I.5.2 Basic design of Mergenthaler's linotype patented in 1885 (US Patent 317,828). A few notable parts include matrix magazines (B), matrix distributor (U), separable mold (N), and melting pot (O).

another already available invention, from J. D. Schuckers, for the automated spacing of matrices.

Completed matrices were released from the assembler (so the operator could proceed to set the next line) and were transferred by the first elevator for justification and casting with hot (285°C) metal (an alloy of 85% lead and 12% antimony). After casting, the matrix line was carried upward by the second elevator, space bands were separated first and returned to the appropriate box for reuse, and then the matrix was lifted to the level of the distributor bar that was suspended over the magazine and passed along it until their unique combinations of keys caused them to fall into their original compartments to be ready for forming new matrix lines. Cast slugs were pushed into trays or galleys to be used for direct printing or for making a stereotype for a rotary press; afterward they could be melted, and the metal could be reused.

The first production version of a linotype was used to set the *New York Tribune* on July 3, 1886. More than 50 machines were at work the very next year, and, in a fashion emblematic of the period's innovation, the invention became almost immediately the subject of numerous improvements; eventually there were more than 1,500 patents related to its design and operation. Mergenthaler put his Simplex linotype in production by 1890, and two years later came Model 1, the first truly successful design. By 1895, there were already more than 1,000 Mergenthaler linotypes around the world, and about 6,000 copies of Model 1 were made before more reliable and more complex designs took over. The machine, whose invention has been called the second greatest event in the history of printing, and which admiring Edison saw as the eighth wonder of the world, was used to set all the news of defeats and victories that took place during the first six decades of the twentieth century (Figure I.5.3).

During the 1950s, photo typesetting and photo-offset printing began a new trend, and the last of the Linotype company's nearly 90,000 machines was produced in 1971. But thousands of linotypes were still used during the 1980s, and many survived into the 1990s. As linotypes began disappearing, some typesetters, while appreciative of the abilities of new computerized equipment, missed the sound, smell, and feel of those complex, clanging, heat-radiating machines. I also remember them fondly, from Prague of the 1960s, and I agree with a retired Italian linotype operator that there is something magic about *l'arte di fondere i pensieri in piombo* (Coraglia 2024).

Figure I.5.3 Mergenthaler's 1890 Simplex linotype design (reproduced from Scientific American, August 9, 1890) and composing room of the *New York Times* full of linotype machines.
Photograph taken in 1942 by Marjory Collins.

The letters of the two leftmost vertical rows of keys on the Linotype console that were used to test the machine spell a magical "etaoin shrdlu." Those on the top row of typewriters give another enigmatic line: "qwertyuiop"—but the first successful machine could type it only in capitals. This machine had an even more extensive pedigree of failed designs than practical incandescent light. The first real breakthrough came in 1867, when Christopher Sholes, Carlos Glidden, and Samuel Soule filed a patent for a type writing machine, which was granted in June of the following year (Figure I.5.4). The trio sold the rights for $12,000 to James Densmore and George W. N. Yost, who in turn arranged for production of the machine by E. Remington & Sons, a renowned maker of guns and sewing machines (Zellers 1948).

The design of the first commercial typewriter betrays its origins: it was done by William Jenne from Remington's sewing machine division, and this treadle typewriter of 1874 clearly resembled a sewing machine. Its sturdy upright body decorated with gold and bold flower decals sat on a four-legged table with a foot treadle used to advance the paper. The platen was made of vulcanized rubber; there was a wooden space bar and a four-line keyboard of capital letters starting with the QWERTY sequence. The type bars with

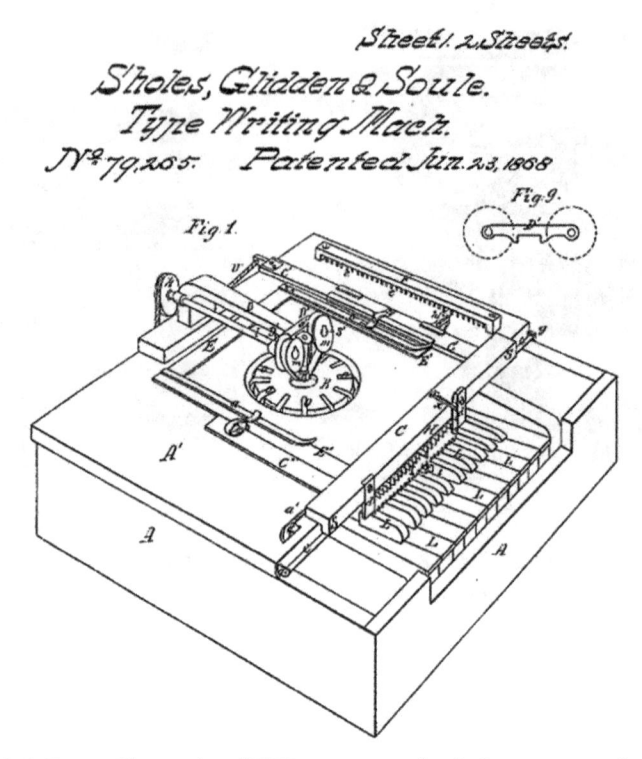

Figure I.5.4 Patent illustration (US Patent 79,265) of a "type-writing" machine designed in 1867 by Sholes, Glidden, and Soule.

raised-relief letters and numbers hit the inked ribbon against the underside of the platen (up-strike) so the typist could not see the text. A design that made it possible to print both capital and small letters by depressing the same key was patented (US Patent 202,923) by Byron A. Brooks in 1878.

But the so-called visible machines, where the typist could see the text, became common only by the late 1890s, and portable typewriters (led by 1906 Corona) appeared before World War I, when standard machines acquired many noise-reducing features and other mechanical improvements (Typewriter Topics 1924; Figure I.5.5). On such machines were most of the twentieth century's great novels and stories written. Hemingway and Faulkner used Underwoods, and Hemingway also had Royals (both desk and portable models), Arthur Koestler had a portable Remington, Primo Levi an Olivetti (ClassicTypewriter Page 2024). But the typewriter became more than an indispensable tool of trade for novelists and journalists. During World War II, typewriters came under the control of the War

Figure I.5.5 Four notable early typewriters: Remington's Standard No. 6, an "invisible" design of 1894 (upper left); a Remington-Sholes Visible from 1908 (upper right); a portable Senta, which came out first in 1912 (lower left); and a standard Royal from the early 1920s (lower right).
Reproduced from Typewriter Topics (1924).

Production Board as tens of thousands of them were taken along by armies into battle (a ship sunk off Normandy during D-Day carried 20,000 Royals and Underwoods) together with tripod stands; thousands of these machines were fitted with non-Latin alphabets and special characters to be used in different parts of the world (Frazier 1997).

Alternatives to type bars appeared even before the first visible typing. The first electric typewriter, with a typewheel printhead, was sold in 1902, but IBM's Selectric became the best-selling model only during the 1970s, just before typewriters gave way to personal computers (PCs). And the QWERTY arrangement? Sholes chose this irrational order deliberately to slow down the speed of typing and hence to minimize the jamming of type bars. Rational keyboard designs—for English language typewriters, August

Dvorak's famous statistics-based rearrangement speeded up typing by more than a third (Dvorak et al. 1936)—have been available for decades, but the universal design was not only maintained by virtually all twentieth-century typewriters but also has been transferred, without any second thoughts, to computer keyboards.

Papermaking

By the beginning of the nineteenth century, steadily rising demand for paper was putting greater strain on the supply of rags. In different places and at different times, the paper making feedstock included cotton, linen, flax, hemp, and even silk fibers. Small runs of artisanal handmade sheets are still made from rags, while the paper for the US currency is composed of 75% cotton and 25% linen (Bureau of Engraving and Printing 2024). Demand for larger volumes of fibrous feedstock further increased after the introduction of a new papermaking technique that produced long strips rather than single sheets of paper. The machine for this semicontinuous production was invented in 1798, by Nicolas Louis Robert, a clerk at the Essonne Paper Mills in France, and the English rights for its use were bought by Henry Fourdrinier (1766–1854), the owner of a mill in Kent.

By 1803, the Fourdrinier brothers and Bryan Donkin had constructed the first practical machine at a mill in Frogmore, in Hertfordshire (Wyatt 1911). This machine produced strips of paper up to 3.6 m wide in lengths of 15 m, and the strips had to be taken off wet and hung to dry. The basic principle remains. Pulp is laid on a continuous wire mesh at the wet end of the machine, most of the water is expelled in the felt press section, and the process is finished by passing paper over a series of heated cylinders. Performance is a different matter: by 1910, the largest fourdriniers were 5 m wide and could produce rolls of paper nearly 13 km long at the speed of 240 m/minute (min). Today's large fourdriniers can produce rolls at speeds of up to 450 m/min.

During the first half of the nineteenth century, paper recycling became more common, and the search for new feedstocks led to the use of jute sacking, cereal straws, Manila hemp, and esparto grass, but wood was obviously the most abundantly available source of fiber. Mechanical pulping was developed for practical use by Heinrich Volter in Bautzen by 1845, but it has been used commercially on a larger scale only since the early 1870s.

The process had high yield (up to 90% of wood becomes pulp), but it removed hardly any lignin. This polymer normally constitutes between 30% and 40% of wood, but its presence in the pulp makes for an inferior paper. Furthermore, wood resins in mechanical pulp contribute to paper's rapid deterioration, betrayed by distinct yellowing with age. These drawbacks make little difference for ephemeral paper uses, and that is why mechanical wood pulp still accounts for about 15% of all pulp produced worldwide (FAO 2024). Its most important use is for newsprint, and it also goes into toilet paper, towels, cardboard, and building board.

The soda (or alkaline) process, patented by Charles Watt and Hugh Burgess in 1854, was introduced in the United States for the treatment of deciduous woods, mainly for poplars, aspen, beech, birch, maple, and basswood. The first American mill using the process was in Pennsylvania, in 1863. Chipped wood was boiled under pressure with sodium hydroxide for up to nine hours, and the bleached pulp was used for book and magazine papers. The soda process was gradually displaced by the sulfite (or acid) process that eventually produced most of the world's paper during the first third of the twentieth century.

Benjamin Chew Tilghman (1821–1901), better known for his patenting of sandblasting, began his work on this treatment during the 1850s. His US Patent 70,485 was for treating wood with acid, boiling it under pressure, and adding a suitable base substance for easier bleaching (Tilghman 1867). An indirect boiling process, with steam at 125–135°C circulating through copper or lead coils to boil the mixture for at least 20–30 hours, was patented in Sweden, in 1871, by Carl Daniel Ekman (1845–1904) and in Germany, in 1874, by Alexander Mitscherlich (1836–1918). Ekman produced the first sulfite pulp at the Bergvik pulp mill (Edge 1949; Steenberg 1995).

With its longer boiling at lower temperature this process yields stronger pulps than does the direct boiling method that used hot steam (140–160°C) blown into the boiler for 8–10 hours per batch. In Europe, this new method became generally known as the Ritter-Kellner process. Karl Kellner (1851–1905) developed it in 1873, when he worked for Baron Hector von Ritter-Zahony's paper factory in Görz; he had it patented in 1882 and tried subsequently, in cooperation with North American manufacturers, to monopolize (unsuccessfully) the world cellulose market. America's sulfite pulp production began only in 1883, at the Richmond Paper Company's plant in Rumford, Rhode Island (Haynes 1954). Figure I.5.6 shows an early design of American pressurized tanks used for pulp production.

Figure I.5.6 Apparatus for treating wood with bisulfite of lime.
Reproduced from *Scientific American*, March 29, 1890.

Feedstock for both methods is ground coniferous wood that is boiled under pressures of 480–550 kilopascals (kPa) in a solution of bisulfites of calcium or magnesium until all resins and gums are dissolved and cellulose fibers can be recovered, washed, and bleached (Edge 1949). Just before World War I, sulfite pulp accounted for about half of all wood pulp produced in the United States (Keenan 1913). The sulfite process produces strong pulp, but its acidity eventually embrittles the paper, whose disintegration is accelerated by air pollution and high humidity. Most of the paper produced between 1875 and 1914 has, long ago, entered the stage of advanced disintegration, and even when carefully handled, the brittle pages of those old publications tear easily, and some literally crumble. And the same fate awaits most of the pre-1990 publications because the first official paper standard for permanent papers was adopted in the United States only in 1984, and the international norm, ISO 9706, came a decade later.

The second pulp-making innovation, and the one that has dominated global papermaking since the late 1930s, came shortly afterward: in 1879, Swedish chemist Carl F. Dahl invented a pulping process that used sulfates rather than sulfites to boil coniferous wood in large upright pressurized boilers for about four hours (Biermann 1996). The sulfate process produces much stronger paper, hence the use of the Swedish (or identical German) word for strength to label the process *kraft pulping*. The residual lignin colors the paper brown, and so bleaching is required to produce white paper, and additives are used to control pH, strength, smoothness, and brightness.

While the kraft process yields superior pulp at a lower price, it also produces highly offensive hydrogen sulfide, and sulfate pulps are also more difficult to bleach. Advantages of sulfate pulping—inexpensive chemical; a large amount of energy produced during the recovery process; and pine trees, abundant worldwide, are well suited for it—made it the world's leading method of papermaking. By 2020, the sulfate process accounted for nearly 70% of the world's wood pulp production; mechanical pulping produced about 15% of the total, and the rest comes from sulfite and semi-mechanical treatment (FAO 2024).

The importance of these relatively simple chemical processes—pressurized boiling of ground wood in acid solutions—that were commercialized before World War I is not generally appreciated. They made it possible to convert the world's largest stores of cellulose sequestered in tree trunks into high-quality paper and ushered in the age of mass paper consumption. In 1850, near the end of the long epoch of rag paper, there were fewer than 1,000 papermaking machines in Europe and about 480 in the United States; annual per capita output of paper amounted to less than 5 kilograms (kg). By 1900, Germany alone had 1,300 machines and the worldwide total surpassed 5,000. US paper production doubled during the 1890s to 2.5 megatons (Mt), and, by 1900, it surpassed 30 kg per capita. Paper became one of the cheapest mass-produced commodities of industrial civilization.

The consequences of this change could be seen everywhere. Picture postcards, invented in Austria in 1869 and available in France and the United States by 1873, became very cheap and highly popular. In 1867, Margaret Knight (1838–1914) designed the first flat-bottom paper bag and soon afterward also a machine to make such bags; better machines, by Luther Childs Crowell in 1872 and by Charles Stillwell in 1883, made their production even cheaper (Stilwell 1889). By the mid-1890s the largest mail order business in the United States, Montgomery Ward & Company, whose first catalog in 1872 was a single page, was distributing seasonal catalogs containing more than 600 pages (Montgomery Ward 1895). Writing slates disappeared even from the poorest schools: the just cited catalog was selling 160 cream-colored pages of its School Spelling Tablet for three cents.

The combination of cheaper paper and halftone reproduction (see the next section) led to an enormous expansion of publishing, particularly in the United States with its new 10-cent illustrated magazines (Phillips 1996), and the printing of large editions of books, ranging from bibles and historical novels to how-to manuals, became affordable. As books became cheaper, the total number of public libraries also increased rapidly; in the United Kingdom, the Local Government Act of 1894 made it possible to have them even in remote rural parishes (Fleck 1958). And the rising demand for paper continued even during the age of electronics.

Global consumption of paper (and paperboard) had more than sextupled during the second half of the twentieth century, and it has continued to increase, albeit at a somewhat lower rate, reaching a new record (nearly 420 Mt in 2021). North American consumption has declined since 2000, the European Union mean has changed little, but, as expected, Asian per capita rates have been up (FAO 2024). These realities expose the myth of the paperless office that should have become, according to the forecasts of electronic techno enthusiasts, the norm before the end of the twentieth century (Sellen and Harper 2002). And the creation and expansion of e-commerce has led to rising demand for cardboard and packaging paper.

Accessible Images

There was no shortage of excellent images in the pre-industrial world, and some of them were reproduced in relatively large number of copies; they ranged from exquisite copper engravings of Renaissance and Baroque masters to fascinating multicolored *ukiyo-e* prints that depicted everyday life of Tokugawa Japan. But all these images had two things in common: they were produced in artisanal fashion, one at a time and at a relatively high cost. The invention of lithography—by Alois Senefelder (1771–1834)—made the reproduction process much cheaper but prints still had to be pulled off one by one. The introduction of steam-driven flat bed and, later, rotary presses speeded up the printing process, and electric motors made it easier—but none of these innovations changed the painstaking process of cutting, engraving, or lithographing the originals.

Both engravings and woodcuts were inherently labor-intensive interior techniques unsuited for any rapid capture of images in the open. Only the development of photography changed that, but, despite many impressive improvements, the scope of this new art was limited for decades by cumbersome and inconvenient procedures as well as by the absence of simple techniques to reproduce these images and to make inexpensive copies for personal use and distribution. In 1826, Joseph Nicéphore Niépce (1765–1833) produced the world's first saved photograph, a very crude image of nearby buildings seen from an upper window of his home that was captured on light-sensitive bitumen and fixed on a pewter plate (Brown 2002).

During the late 1830s, the first *nature morts* and photographs of buildings and people appeared on polished silver-plated copper plates made by Louis Jacques Mandé Daguerre (1789–1851). And William Henry Fox Talbot (1800–1877) and John Herschel (1792–1871) independently invented the process of photography on sensitized paper (Schaaf 1992; Frizot 1998). In 1839, Herschel used for the first time the terms "negative" and "positive"; in 1847, Claude Niepce, Joseph's cousin, invented the glass plate; and, four years later, Frederick Scott Archer introduced a cumbersome wet collodion process that left behind a remarkable record of portraits, documents of daily life, and landscape images (Brown 2002).

Wet plates were dominant for about three decades. The first dry, silver-bromide sensitized, gelatin-emulsion plates were produced in 1864, by W. B. Bolton, and B. J. Seyce, and many inventors, including Joseph Wilson Swan of incandescent light fame, improved them during the 1870s (Abbott 1934). An easily produced and highly sensitive plate, prepared by Charles Bennett, reached the market only in 1878. Still, taking pictures required patience (long exposures), skill, and considerable expense (heavy, bulky, and fragile plates). Four kinds of innovations revolutionized photography: a reduction in exposure time, the development of convenient photographic film to replace heavy glass plates, the availability of affordable handheld cameras, and access to inexpensive custom film development that obviated the need for an in-house darkroom and knowledge of photographic chemistry.

Cameras, Films, Photographs

The earliest photographs required very long exposure times. Niepce's faint view from his window took eight hours to capture; two decades later typical daguerreotypes required 30 seconds when the plate was exposed in a studio. Talbot took the first high-speed flash image in 1851, using a spark from the discharge of a Leyden jar, but that was clearly impractical for casual use. Progressive improvements that began during the 1850s reduced the typical exposure times, and, by 1870, Talbot was able to expose his plates for just 1/100 of a second. The most revealing breakthrough in high-speed photography came when Eadweard Muybridge (1830–1904), an Englishman who started his American photographic career with the images of Yosemite in 1867, developed a technique to capture motion in rapid sequences with a camera capable of shutter speeds of as much 1/1,000 of a second (Hendricks 2001).

This work was prompted by Leland Stanford—at that time a former governor of California, the president of the Central Pacific, the future founder of the eponymous university, and an avid horse breeder—who wanted to confirm his belief that a galloping horse will, at one point, have all its legs off the ground simultaneously. In 1878, Muybridge proved this to be correct: his most famous photo sequence confirmed that all four hooves of a galloping horse are momentarily off the ground (Figure I.5.7). After his return from a European tour, where he showed his Zoopraxiscope, a view box that displayed a rapid series of stills that conveyed motion (the principle used later in animated cartoons), Muybridge made an agreement with the University of Pennsylvania to prepare a comprehensive series of sequences of animal and human locomotion.

He used a fixed battery of 24 cameras positioned parallel to the 36-m-long track with a marked background: these cameras captured the motion of

Figure I.5.7 Four of the 12 silhouetted frames of Muybridge's famous sequence of "automatic electro-photographs" showing a trotting horse (*Sally Gardner*, owned by Leland Stanford) at the Palo Alto track on June 15, 1878.

animals and people as they broke strings stretched across the track and activated the shutters in turn. The complete set of 781 sequences was published in 1887 (Muybridge 1887), and it includes such predictable images as running at full speed and throwing a baseball, as well as such oddities as a woman pouring a bucket of water over another woman. Truly instantaneous photography was made possible only with highly sensitive dry plates, which reduced exposures to fractions of a second in well-lit environments. Even amateur photographers could now capture movement, and, by 1888, their task was made easier by new devices that measured the time required for exposure.

George Eastman (1854–1932), the man who removed the other two obstacles that precluded the popular diffusion of photography, developed his interest in taking pictures when he was a poorly paid clerk at Rochester Savings Bank. He bought a heavy camera and a wet-plate outfit for a planned trip to Santo Domingo in 1877. The trip fell through, but Eastman found a new calling: he experimented with coatings, began making Bennett's dry plates, and, in 1879, he devised and patented a new coating machine to do the job (Brayer 1996). Instead of glass, in 1884, he introduced a three-layered negative stripping film (paper, soluble gelatin, gelatin emulsion), and, a year later, in cooperation with camera maker William H. Walker, he invented a roll holder that could be attached to a standard plate apparatus. All that was needed was to put the smaller size of his film into a handheld, easy-to-use camera.

His solution was a simple and heavily advertised camera whose principal selling slogan was "You press the button, we do the rest." The US patent for the camera was issued in September 1888, and Kodak's first version was in production between June and December of that year (Coe 1988). This design did not include any technical innovations as it used only already tested components, but it produced the first handheld camera suitable for amateur photography. Kodak's operation was not quite as simple as the slogan had it: the film had to be advanced and rewound, and the first 13 Kodak models had a string connected to the shutter mechanism whose protruding top had to be pulled two or three times to wind up the barrel shutter spring to its full tension (Figure I.5.8).

The camera's wooden box (a rectangular prism, 16.3 cm long) was covered with smooth black leather; the 57 mm lens was centered on one narrow face. Because the lens had a great depth of field, it produced focused images from as close as 1.2 meters, and its angle (60 degrees) made it possible to dispense

234   THE TWENTIETH CENTURY

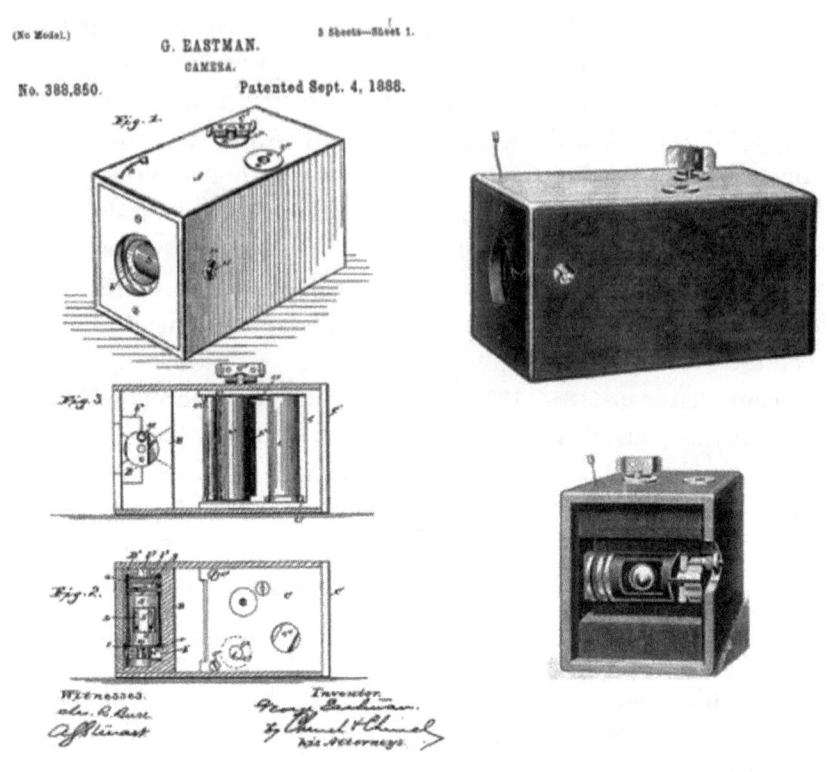

Figure I.5.8 Patent drawings for Eastman's Kodak camera and its actual appearance and a detail of its shutter mechanism.
Reproduced from *Scientific American*, September 15, 1888.

with a viewfinder. Once photographs were taken, the entire camera was returned to the Rochester, New York, factory where the film was removed, developed, and printed. The camera took circular photographs (so the users would not have to worry about keeping the camera level) with a diameter of 6.25 cm, and photographers had to keep track of the pictures taken. The original list price was $25, a considerable amount in 1888, and about 5,200 units were produced before new models were introduced in 1889.

That was also the year when the stripping film, whose development required time- and labor-intensive operations, was replaced by the precursor of modern films made of modified celluloids. This nitrated cellulose was first made by Alexander Parkes (1813–1890) in 1861, and it was used to make a variety of small items such as combs, knife handles, pens, and boxes. Henry M. Reichenbach, a chemist working for Eastman, discovered that by dissolving the cellulose nitrate in alcohol and by using several additives, he could produce thin, flexible, and perfectly clear films by casting the mixture on glass plates (Coe 1977). Once set, this celluloid was coated with gelatin emulsion to make the world's first transparent roll film, whose production was patented in April 1889, and which was used in a new Kodak camera released in October 1889.

Several technical advances of the 1890s made amateur photography even easier. They included the easy-to-load cartridge film that was patented by Samuel N. Turner in 1895. This film was backed by black paper marked with numbers that could be read through a small window and thus allowed for advancing a precise length of a film for each exposure. The Folding Pocket Kodak, the obvious ancestor of all roll-film cameras marketed during the next six decades, was introduced in 1897. Kodak's subsequent most notable pre-World War I innovation was the world's first practical safety film that used cellulose acetate instead of the highly flammable cellulose nitrate (Kodak 2024).

By the beginning of the twentieth century, amateur photography was thus one of the most accessible, if often disparaged, arts. Alexander Lamson's reaction in *The Beacon* of July 1890 expressed the feeling of "real" amateurs: "Photographers who merely 'press the button' and leave someone else to 'do the rest' are not really entitled to the honorable name" (Lamson 1890: 154). The 1890s were also the first years when just about everybody could also enjoy inexpensively reproduced photographs thanks to halftone printing. Previous reproduction techniques could either produce limited numbers of excellent (continuous tone) and costly copies, while several processes aimed at larger print runs (most notably photogravure) were cumbersome as well as very expensive. Because the most common relief printing processes could render only solid blacks and whites, large numbers of skilled artists were engaged during the second half of the nineteenth century in converting photographs into reproducible engravings.

Only halftone reproduction could reproduce photographs by breaking up the continuous tones of the photographic image into a pattern of tiny dots of variable size that could be printed using the inexpensive relief technique (Phillips 1996). Because of the limited resolution of our vision, these variably sized dots blend and produce the illusion of shaded grays from an image that is composed entirely of black and white. Moritz and Max Jaffé used a gauze screen as early as 1877, and Charles-Guillaume Petit and Georg Meisenbach were other inventors credited with commercialization of the process whose first versions were quite complicated and relatively slow.

By 1885, Frederic Eugene Ives (1856–1937) perfected the process by inventing a screen that was made from two sheets of glass with a fine series of etched lines (six per millimeter is now standard for fine resolution) forming a crossline grid that was filled with an opaque substance (Ives 1928). Rephotographing a full-tone photograph through the screen produces the halftone effect through a sensitized copper plate marked with a dotted pattern. Ives never patented his crossline design, and halftone printing diffused rapidly once the Levy Company began producing crossline screens in 1892. The new technique cut the cost of newspaper illustrations by about 95% and made illustrated printing commonplace. During the late 1880s, full-page engraving in *The Illustrated London News* or *Harper's Magazine* cost about $300; a few years later it was an inexpensive option (less than $20 for a full page) for every printer (Mott 1957).

Halftones introduced an unprecedented realism to printed illustration and merged the market for photographs with the market for the printed word; this combination brought an enormous expansion in the publishing industry after 1890. And because it was cheaper to print halftones than text, the new technique made illustrated newspapers and magazines not only less expensive to produce but also more attractive, further stimulating their sales (Phillips 1996). Photographs became an indispensable tool for advertising, and nothing has changed in this respect in the early twenty-first century: the halftone technique remains as essential today to convert photographs into advertising images as when it was introduced during the 1890s.

Ives's other great printing invention also dates to the early 1880s: producing primary-color separations (Ives used blue, red, and green filters), making their halftones, and then overprinting precisely aligned layers of yellow, cyan, and magenta ink to get a full-spectrum color reproduction of the original image. We still use three colors of ink (cyan, magenta, yellow) and black, and you can see their overlaps by examining newspaper or magazine illustrations with a magnifying glass or by noticing their alignment markers, often left near the edges of color-printed pages.

Movies

After Edison refused Muybridge's offer to transform Zoopraxiscope into a more sophisticated device, he set out to develop a device that, in his own words, would do for the eye what his phonograph did for the ear. But, unlike in the case of incandescent light, most of the work on Edison's concept to combine a camera and a peephole viewer was done by his assistant, William Kennedy Laurie Dickson (1860–1935). The development of Edison's Kinetograph and Kinetoscope is a perfect example of the complexity of technical inventions and of the inappropriateness of narrow attributions. Edison's initial idea was to follow his phonograph design and have tiny photographic images affixed to a rotating cylinder. But after his visit to Paris in 1889—where he met Etienne-Jules Marey (1830–1904), professor of physiology, whose Chronophotographe used a continuous roll of film to produce a sequence of still images for research in cardiology—he abandoned that impractical approach and directed Dickson to pursue the roll design.

The US patent for the Kinetograph (a camera) and Kinetoscope (the viewer) was filed in August 1891 (Figure I.5.9). Dickson ordered 35-mm film from Eastman (whose cameras used 70-mm film) and fed it vertically through the Kinetoscope. Edison's first film studio was completed in May 1893, and on May 9 was the first public showing of an electrically powered Kinetoscope at the Brooklyn Institute of Arts and Science. The reward for lining up and peering into the machine was to see three Edison employees hammering on an anvil and passing around a bottle of beer. Such was the sublime beginning of American motion pictures.

Soon, with Dickson and William Heise in charge of the motion picture production, Edison's studio began releasing an increasing variety of full-length (in 1894. that meant less than a minute) films. The first copyrighted product, a would-be comical clip that showed Fred Ott, an Edison employee, sneezing into a handkerchief, was followed by pictures of an amateur gymnast, of a lightning shave in a barbershop, and of Eugene Sandow, an Austrian strong man, flexing his upper torso. Nearly 80 pictures were made in 1894, and they included wrestlers, cock fights, terriers attacking rats, Buffalo Bill and an Indian war council, and the gyrations of skimpily dressed dancers. Comparisons with some leading subjects of mainstream US filmmaking a century later (during the 1990s) are unavoidable: wrestling, violence, Westerns, sex, even another muscle-displaying Austrian (Arnold Schwarzenegger, of course).

The pattern of moviemaking was set right at the beginning; all that remained was to embellish it and make it gradually more provocative and—technical advances permitting—much more elaborate. Although the resulting images could be seen only by one person at a time, Kinetoscope parlors (the first one opened in New York in April 1894) showing short clips

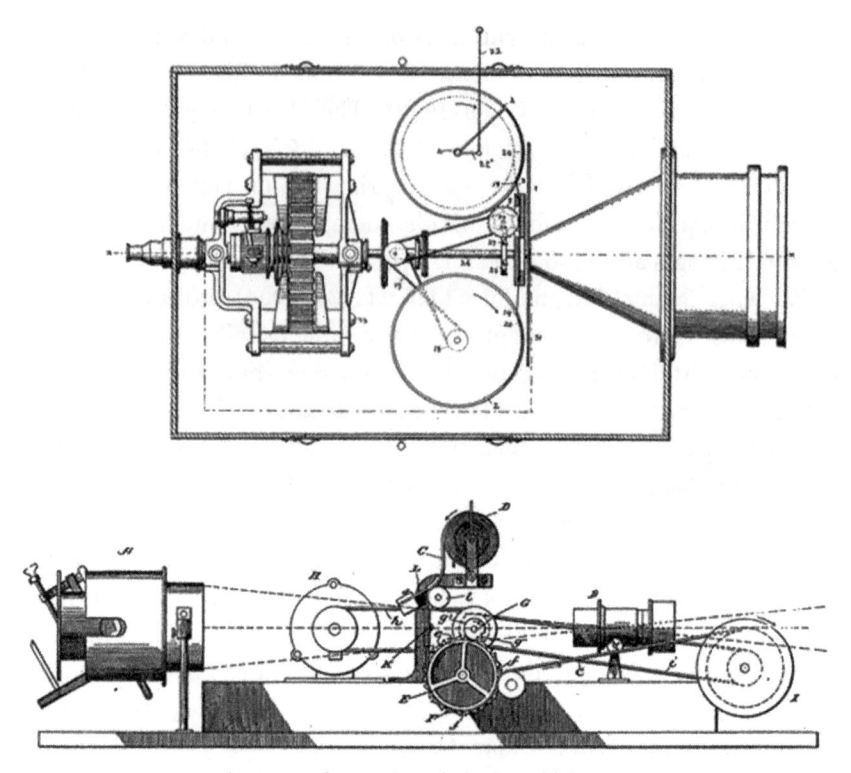

Figure I.5.9 Patent drawings for Dickson's (Edison's) kinetographic camera (US Patent 589,168) filed in August 1891 and granted 6 years later (top), and for Jenkins and Armat's Phantoscope (US Patent 586,953), whose slightly altered version was produced as Edison's Vitascope (bottom).

became popular in many large US cities. Real movies were born and greatly improved between 1895 and 1910 (Mitry 1967; Musser 1990; Rittaud-Hutinet 1995; Nowell-Smith 1996). Precise attributions of beginnings are questionable: moving pictures have no precise moment of origin, nor can any country or any individual claim a clear-cut priority. But the cinematographic protagonists clearly fall into two groups: the largely forgotten ones, and a few iconic names.

On the technical side, the first group includes about a dozen Frenchmen, Germans, and Britishers. Louis Aimé Augustin Le Prince (1842–1890) patented his 16-mm lens "apparatus for producing animated pictures of natural scenery and life" in January 1888 (US Patent 376,247), and the images of his in-laws and his son he took with it in October of that year in the garden of their house in Leeds were the first successfully photographed and projected motion pictures. Leon Bouly designed his Cinématographe in 1892 and patented an improved version a year later. Max and Emil Skladanowsky held the first public show of their 15-minute movie program produced with their Bioskop on November 1, 1895, in Berlin. And the first portable apparatus to make films was the Kineopticon made by Bert Acres, who began his work on a kinetic camera in 1893 (de Vries 2017).

The best-known names are, of course, those of Edison and the Lumière brothers, but the invention attributed to Edison was a case of outright deception and the Lumières' principal contribution was skillful marketing. As the Kinetoscope was enjoying increasing popularity, Edison was in no rush to develop a competing device. But when Francis Jenkins and Thomas Armat introduced their Phantoscope (Figure I.5.9), Edison's company bought the rights to the machine and agreed to manufacture the projector on the condition that it would be advertised as Edison's invention, the Vitascope. The promotional brochure claimed that Edison invented "a new machine ... which is probably the most remarkable and startling in its results of any that the world has ever seen" (quoted in Musser 1990: 113).

Edison's role in all of this was indisputably unethical: he appeared at the first private screening of the Vitascope, on March 27, 1896, and acted for the benefit of assembled reporters as the inventor of yet another magical machine, assuring widespread press coverage of the event. The first Vitascope theater opened its doors on April 23, 1896, in a New York City music hall, and a year later there were several hundred of these projectors across the country. Standard film histories concur that all those inventive Americans pioneering their competing scopes were not the creators of cinema and that that primacy belongs to Louis Lumière (1864–1948), the younger son of Lyon photographer Antoine Lumière (Mitry 1967; Conreur 1995; Rittaud-Hutinet 1995).

At the age of 17, Louis improved the preparation of a gelatin-bromide plate used for snapshots, and the family's company began producing these plates in 1882. His brother Auguste (1862–1954) left a clear account that credits Louis with the key advances. But Auguste also conceded that Louis did not construct an entirely new machine from fundamental principles. Origins of the Lumières' machine (French Patent 245,032) go directly to Edison's, Marey's and Bouly's designs. After Antoine Lumière saw the display of Edison's new device on a visit to Paris in 1894, he bought it and had one of his workers disassemble it. Not only was he familiar with Marey's work, but he also picked up Bouly's expired patent and with it the name of the device: Bouly, much like Beau de Rochas, did not pay his annual patent fees. Perhaps the most important advantage of the Cinématographe Lumiére was that it combined a camera and projector in a small box that weighed less than 5.5 kg and hence could be easily used for filming in plein air (Institute Lumiére 2024).

But if the Lumières' apparatus was highly derivative, their promotion effort was very effective. Indeed, Sauvage (1985) saw it as the biggest stunt in film history. In any case, by the mid-1890s, public interest was piqued by the Kinetoscope, and it was clear that the market was ready for a new commercial display of moving images. The first demonstration of the Cinématographe took place on March 22, 1895, in a lecture at the Societé d'Encouragement l'Industrie Nationale in Paris. Three days before that, the brothers shot their first, 52-second movie, *La sortie des usines*, which showed workers passing through the front gate of the Lumière et fils factory in Lyon. The first public screening of 10 short clips filmed between March and December 1895 was in the Salon Indien in the basement of the Grand Café, on the Boulevard des Capucines on December 28, 1895, and it was attended by 33 people.

The show lasted only 20 minutes, and it consisted mainly of scenes of everyday Parisian life. By January 1896, up to 2,500 people a day were paying for the show, whose most impressive clip was "L'arrivée d'un train en gare de La Ciotat," filmed in June 1895. This consisted of nothing but a steam locomotive pulling carriages and moving diagonally across the screen as it approached the platform in a town on the Cote d'Azur, where the Lumières had a summer residence. Many spectators felt that the machine would leave the screen, and O'Brien (1995: 34) rightly argued that ever since the filmmakers have tried to emulate, with more difficulty and at a much higher cost "the same visceral surprise that the Lumières accomplished less strenuously by planting a camera on the platform of La Ciotat Station as a train was arriving."

London saw the first Lumière movies on February 20, 1896; New York on June 29; and, by the end of summer, shows took place in cities from Madrid to Helsinki. Cinématographe and Vitascope movies were incompatible, as the former used film with a single pair of round perforations for each frame, while the latter had two perforations. The Lumières eventually made more than 1,500 films, each lasting about 50 seconds. Many Lumière vignettes can be seen on the Web, and the Library of Congress now has 341 early Edison motion pictures on its Inventing Entertainment website (LOC 2024).

Movie shows eventually ceased to be dependent on saloons, theaters, and arcades as specialized movie theaters proliferated by the turn of the century

(LOC 2024). By 1908, there were about 8,000 of these establishments in the United States (nickelodeons, usually for fewer than 200 people), and this created such a huge demand for new films that it could be satisfied only by imports, mainly from France: by the end of 1906, at least one new film was finished every day, with the Pathé-Frères studio in Vincennes being the leading producer. Georges Méliès (1861–1938), a magician and theater impresario, became the first great innovator of cinematographic art (Conreur 1995). He pioneered narrative cinema, fantastic subjects, and science fiction, and he also developed such special effects as double exposure (in *La caverne maudite* in 1898), actors performing with themselves on a split screen (in *Un homme de tête* in the same year), and the dissolve (in *Cendrillon* in 1899).

Among the technical innovations tried before World War I were wide screens (75 mm film), which became popular only during the 1950s, and the first talking movies (Edison's Kinetophone in 1913). The earliest short films included most of the subjects that became staples of twentieth-century cinematography: mundane events, trivia, beauty, conflicts. As the movies lengthened (to 5–10 minutes by 1905, and to about 15 minutes by 1910) storytelling became common. In 1906, the world's first feature film (*The Story of the Kelly Gang*) that lasted more than one hour was not made either in France or in New York but in Australia by the J. & N. Tait theatrical company (Vasey 1996).

By the end of the 1900s, several film companies opened their studios in the western suburb of Los Angeles, and, within a decade, these Hollywood newcomers dominated not only America's but the world's moviemaking (Gomery 1996). Their success broke up the power of the Motion Pictures Patents Company, a combination of leading US and European equipment and moviemakers that was formed to charge inflated prices for cameras, projectors, and movies. But that short-lived monopoly group had at least one salutary effect: in 1909, it adopted 35-mm film as its standard-size stock, and, a century later, after every aspect of moviemaking had changed drastically, 35-mm film is still with us (Rogge 2022).

Only historians of cinema are now aware of the trust's brief existence, but every movie fan recognizes the name Paramount, the company established by Adolph Zukor in July 1912. And so, the pattern of twentieth-century Hollywood moviemaking was essentially set before World War I: use newspapers and pulp magazines, classical novels whose copyright had expired, and successful theatrical plays for suitable plots; integrate vertically all the activities from production to promotion; and go after foreign markets.

More than a century later, they are still playing the same show, except that the potential for success and failure has been greatly magnified. And just one last notable pre-World War I movie event before I turn to sound: during his tour of the United States, Charlie Chaplin was eventually persuaded to join Mack Sennett's independent Keystone Company late in 1913.

Reproducing Sound

Invention of the first of two distinct ways of reproducing sound—its conversion into electromagnetic waves, virtually instantaneous transmission across increasing distances, and reconversion to audible frequencies by the means of telephony—led in a matter of months to the success of the second method whereby words or music were stored ingeniously for later replay by the first of many possible recording techniques. The first invention, as already noted in Chapter I.1, is certainly the best-known instance of nearly simultaneous independent filing in the history of patenting. The second invention, finalized just before the end of 1877, further increased Edison's fame, built on his improved stock ticker and quadruplex telegraph, and only his demonstration of practical incandescent light received a greater acclaim when it was unveiled two years later.

The appearance of practical telephony is yet another example of success preceded by several decades of interesting experiments and proposals. These developments failed to reach the commercial stage but provided essential steppingstones toward an invention whose many impacts have not been dulled nearly 150 years later. The latest stage in the evolution of telephones—the worldwide diffusion of wireless devices that can also receive and transmit internet messages and take and send digital images—has rejuvenated and reconfigured this now classic technique. As a result, telephones have become, once again, one of the most desirable—and highly addictive—personal possessions.

The phonograph, despite Edison's determined effort, was never a great commercial success, and it was the gramophone, Emile Berliner's most famous invention, which became the standard means of music and voice reproduction for much of the twentieth century. Its dominance began to decline only during the 1960s, with the introduction of compact audiocassettes (by Philips in 1963), and its nearly complete demise came with the introduction of compact discs in the early 1980s. Gramophones are still around, but the entire field of recorded sound is one of the best examples of how some key innovations of the pre-World War I period were transformed during the twentieth century into new and qualitatively superior systems.

Wireless transmission of sound and faithful reproduction of words and music at receiving points that were increasingly distant from transmitters rank among the most significant innovations of the Age of Synergy. Although regular radio broadcasts began to take place only during the early 1920s, all key components and procedures for widespread diffusion of wireless transmission were put in place before 1914. And it is a much less appreciated fact that the same was true as far as television was concerned: nearly all critical components for the wireless transmission of images were either proposed or invented by 1912.

Inventing Telephones

The gestation period of devices that could successfully transmit human voice was less than half a century, a span shorter than the time that elapsed between the first failed trials of incandescent lights and Edison's 1879 success. The first known rudimentary demonstration of telephony was a magic lyre that Charles Wheatstone (1802–1875) showed in 1831: when he connected two instruments by a rod of pine wood, a tune played on one was faithfully reproduced by the other. Wheatstone did not pursue the implications of this short-distance nonelectric telephony, but, in 1837, he became famous as one of the co-inventors of the telegraph. The first possibility of electric telephony was noted in 1837, by Charles G. Page, who heard the sound produced by an electromagnet at the instant when the electric circuit is closed and described the phenomenon as "Galvanic music" (Gray 1890).

The first explicit formulation of the idea of electric telephony came in 1854, when Charles Bourseul concluded that a flexible plate that would vibrate in response to the fluctuating air pressure generated by a voice could be used to open and close an electric circuit and that the transmitted signals could be reproduced electromagnetically by a similar plate at the receiving end. Philip Reis was the first experimenter who tried to make the idea work (and who coined the term "telephony"). The quest was renewed by two Americans during the 1870s, by Alexander Graham Bell (Figure I.5.10), an amateur inventor, and Elisha Gray, a partner in the Western Electric Manufacturing Company in Chicago.

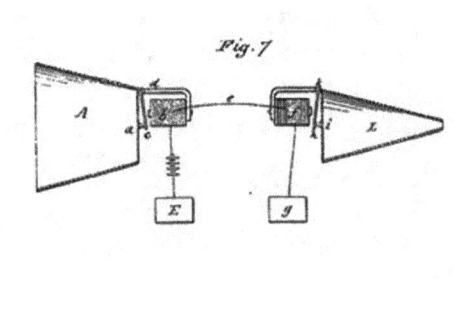

Figure I.5.10 Alexander Graham Bell and the key figure of his patent application for "improvement in telegraphy." The photograph was taken in 1904 when Bell was 57 years old.
Patent drawing reproduced from Bell (1876b).

Bell and Gray eventually found out about each other, and Bell's triumph was not a result of superior design and, fundamentally, not even of luck in filing the patent application just hours ahead of Gray's caveat, but of a very different perception of the importance of telephony (Gray 1890; Garcke 1911b; Hounshell 1981). Both inventors were initially motivated by coming up with a multiplex system for telegraphic transmissions: after decades of sending just one message in one direction at a time, a duplex system, introduced in 1872, effectively doubled the capacity of telegraphic networks, and multiplexing could multiply it even more dramatically and become one of the most lucrative inventions.

Gray began by investigating the possibility of a telegraph wire carrying several frequencies; this musical telegraph could become a multiple transmitter, given a receiver that could segregate the individual tones. And if sending several tones at once, why not human voice? By 1874, reports on Gray's experiments anticipated that, once perfected, the invention would do away with telegraph keys because the operators would simply talk over the wire. At the same time, there was widespread skepticism about the usefulness of voice communication: after all, Reis's experiments led nowhere, and Gray's own patent attorney believed that telephony would be just a toy.

Bell was approaching the task from an acoustic point of view: his appreciation of speech properties came from his father, a professor of elocution and a creator of a system for teaching the deaf. After he learned about Gray's work, in the spring of 1874, he was able to get financial support from Gardiner G. Hubbard, whose deaf daughter he was tutoring, and, by July of 1875, he was able to transmit sounds by shouting into the diaphragm of the transmitter (Rhodes 1929; Bruce 1973). This caused the attached reed to vibrate and generate weak currents in the transmitting electromagnet, which were then reproduced by the receiving diaphragm. Gray, still pursuing his multiplex goal, finally decided to file at least a formal notice of the telephone's basic concept: he did so on February 14, 1876, and was preceded by just two hours by Bell's full patent application. Comparison of the two concepts, both using a liquid transmitter, shows how nearly identical they were (Figure I.5.11).

But there was no immediate legal contest: as soon as Gray learned that Bell anticipated him by just hours, he agreed with his attorney and with his financial sponsor against going ahead with an immediate filing of a patent and contesting Bell's claim. Less than a month after the patent filing, and on the evening of the day when the first sentences were exchanged telephonically between Bell and Watson, an exultant Bell wrote to his father: "I feel that I have at last struck the solution of a great problem—and the day is coming when telegraph wires will be laid on to houses just like water or gas—and friends converse with each other without leaving home" (Bell 1876a: 3).

Bell's specification for "improvement in telegraphy" described "undulatory currents" that are produced "by gradually increasing and diminishing the resistance of the circuit" and that, unlike the intermittent ones, can transmit simultaneously a much larger number of signals and hence can be used to construct an apparatus for "transmitting vocal or other sounds telegraphically" (Bell 1876b). The last of seven figures of Bell's patent application sketched the simple operating principle of such a device (Figure I.5.10).

You could build your own basic telephone following these simple instructions—but it would be a very primitive electromechanical device. You would have to shout into it, and a single apparatus at each end would have the double duty of both transmitter and receiver. Still, this great novelty got an enthusiastic reception at the 1876 Centennial Exposition held in Philadelphia, and, a few months later, Bell was demonstrating an improved design (see the frontispiece to this chapter). But it was Edison, rather than

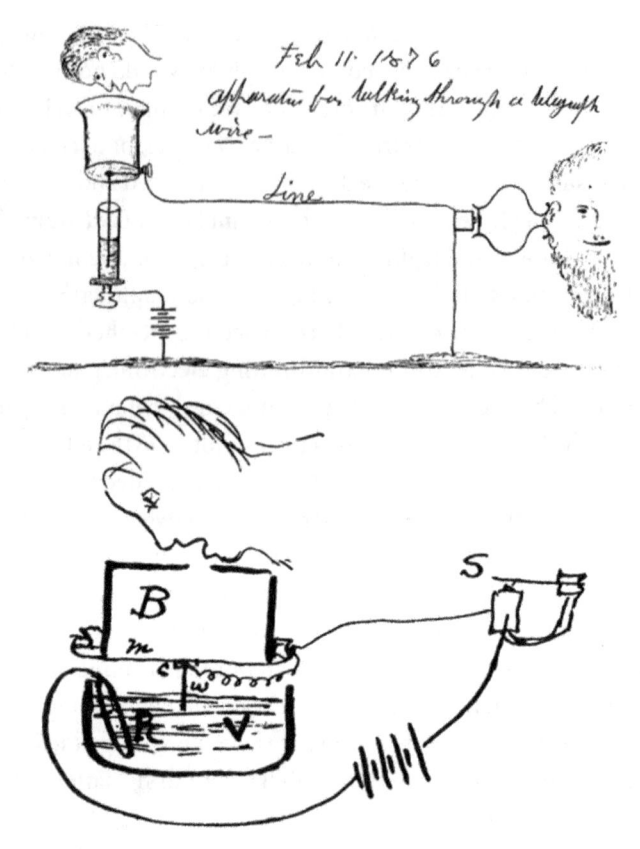

Figure I.5.11 Comparison of Gray's (top) and Bell's concepts of the telephone shows two near-identical designs with liquid-bath transmitters. Gray's sketch was done on February 11, 1876; Bell's on March 9, 1876, more than three weeks after the two men filed their patent applications.

"The armature c is fastened loosely by an extremity to the uncovered leg d of the electro-magnet b, and its other extremity is attached to the center of a stretched membrane, a. A cone, A, is used to converge sound-vibrations upon the membrane. When a sound is uttered in the cone the membrane a is set in vibration, the armature c is forced to partake of the motion, and thus electrical undulations are created upon the circuit A b e f g.... The undulatory current passing through the electromagnet f influences its armature h to copy the motion of the armature c. A similar sound to that uttered into A is then heard to proceed from L." (Bell 1876b: 3)

Reproduced from Hounshell (1981).

Bell, who came up with the first practical transmitter. Edison's involvement began after Western Union had second thoughts about entering the telephone business once it saw the attention Bell's device got in Philadelphia.

Initially, this leader of the telegraph industry declined to invest in a new form of telecommunication and rejected Bell's offer to sell his patents to the company and thus obtain the telephone monopoly. Once Western Union reconsidered its telephone policy, it turned to Edison to come up with a commercial device (Josephson 1959). Belatedly, Edison, who built himself a replica of Reis's device in 1875, re-entered the field of telephony. After several months of intensive experimentation (in which he was greatly disadvantaged due to his poor hearing), he was able to replace Bell's liquid transmitter with a superior variable-resistance carbon device, essentially a microphone controlled by the received sound as the vibration of a diaphragm modulated an externally supplied electric current. This was achieved by changing the resistance of carbon granules enclosed in a small button due to the changing pressure applied by a plunger attached to the diaphragm (Figure I.5.12).

This design produced a much higher volume of sound and with sharper articulation than did Bell's primitive device; a caller could thus speak into what was essentially a microphone while listening at the same time through the receiver. Edison filed the basic patent in April 1877, and the next step was to insert carbon granules between the diaphragm and a metal backplate. This design, patented by Henry Hunnings in 1879, was improved first by Edison in 1886, by packing the granules in a small button container, and then, in 1892, by Anthony White, who interposed the granules between polished carbon diaphragms, with the front one of these placed against a diaphragm made of mica (Garcke 1911b). Known as the *solid-back transmitter*, this device was used basically unchanged until the mid-1920s, and its improved versions were installed in all telephones until the 1970s, when they were displaced by dynamic transducers.

The Bell Telephone Company was formed in July 1877, and, a year later, with 10,000 phones in service, Theodore Vail (1845–1920) became its new general manager; he eventually guided the corporation to become a giant monopoly (Jewett 1944). Edison's carbon transmitter became the centerpiece of a collection of other telephone patents (including Gray's receiver) that Western Union used, in November 1877, to set up its own telephone business. But it left the field entirely to Bell's company by October 1879, when it acknowledged the priority of Bell's patent and when Gray received

T. A. EDISON.
SPEAKING TELEGRAPH.

No. 474,230. Patented May 3, 1892.

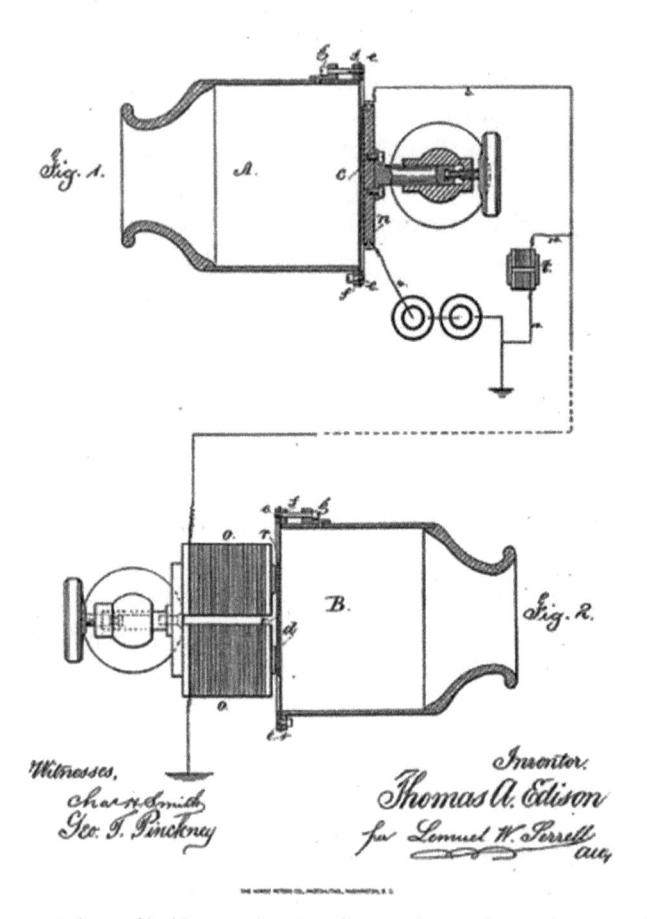

Figure I.5.12 Edison filed his application for speaking telegraph on April 27, 1877, but it was granted only on May 3, 1892. The illustration shows transmitter (top) and receiver. Resonators are marked A and B, sheet metal diaphragms c and d; clamping rings (e and f) and tightening screws (g) keep the diaphragms tensioned; the disk n, made of conducting material, was placed in front of the transmitter diagram, and the electromagnet o in front of the receiver diaphragm.

a consolation payment of $100,000. Meanwhile, Bell continued to elaborate and adjust his basic design. Almost exactly a year after filing the patent, he was able to make a call between Boston and North Conway, in New Hampshire, over nearly 230 km (Anonymous 1877), and, by October 1877,

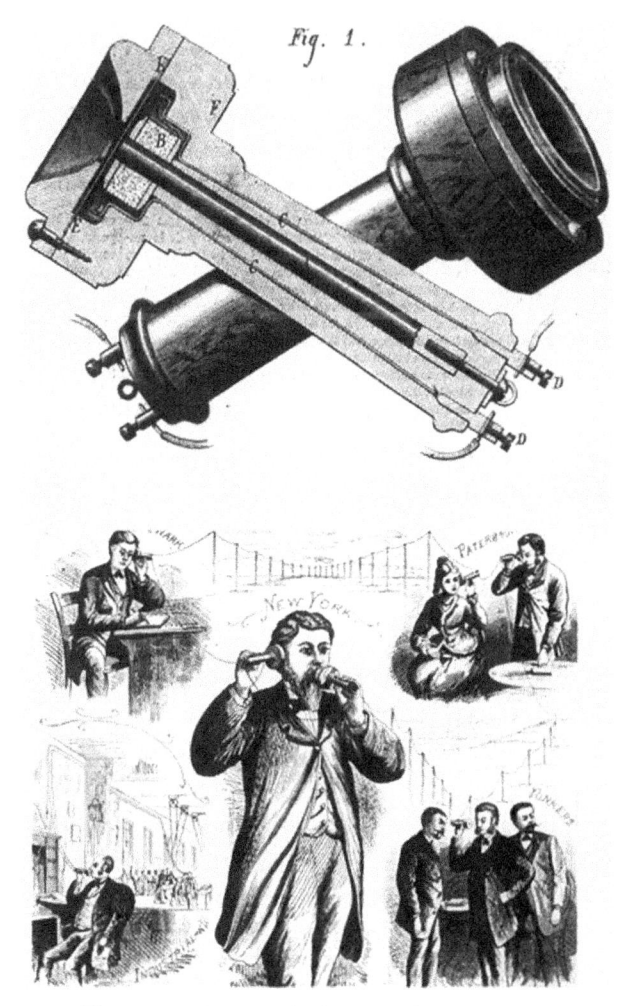

Figure I.5.13 Bell's improved 1877 telephone and some of his demonstrations of the new device.
Reproduced from *Scientific American*, October 6, 1877.

he had a simplified device in a more compact portable form, with an elongated cylinder forming a handle with a flaring mouthpiece (Figure I.5.13).

In a society where mobile phones are now used almost incessantly, it is hard to imagine that the invention was not embraced as rapidly as possible—but there were many technical, financial, and organizational challenges to overcome (Casson 1910). Getting and staying connected were major hurdles. The first manual switchboard that connected many phones through a single exchange was opened in January 1878, in New Haven, Connecticut, and a decade later more than 150,000 subscribers were served by nearly 750 main exchanges. In 1888 came the introduction of central battery supply to energize all telephones in exchange instead of relying on a subscriber's own batteries or hand cranking. This important principle assures that modern telephone systems—still powered from a large central battery (48 VDC systems) backed up by emergency generators—remain open even during electricity outages.

Calling distances continued to lengthen. In 1877, the first regular line from Boston to Somerville was less than 10 km; by 1881, the link between Boston and Providence spanned 60 km, and, by 1892, a call could be made from New York City to Chicago. But the first very expensive transcontinental call (which took more than 20 minutes to arrange) could not be made until 1915, when New York City was linked with San Francisco. The total number of American telephones rose from 48,000 in 1880 (or one phone for about every 10,000 people) to 134,000 by 1885 and to 1.35 million (or one phone for every 56 people) by 1900 (Garcke 1911b).

Trials with direct dialing began as early as 1891, when the unlikely pioneer was Kansas City undertaker Almon Strowger (US Patent 447,918), who suspected that operators were switching his business to competitors. A much better automatic system, introduced in 1901, produced "a peculiar humming sound" when the station called up was busy, and it also had an early version of a rotary dial "provided at its circumference, opposite each figure, with an aperture into which the finger may be inserted" (Anonymous 1901: 85). But none of these early automatic switching devices diffused widely before World War I, nor did the universal dial tone that was pioneered by Siemens in 1908. The first four decades of telephone service thus required large numbers of switchboard operators to run local and later long distance service, and an effective solution to this major precondition of telephone's diffusion relied on employing women.

Growth of manual telephone exchanges provided many women with their first entry into the labor force. But even the modest wages of telephone operators could not make the service cheap. By 1900, more than 20 years after the technique was invented, basic monthly telephone charge in major US cities was between a quarter and a third of the average monthly wage. This meant that most customers were businesses and that, for most Americans, telegraph remained the dominant mode of intercity communication well into the first decades of the twentieth century. And while Europe of the last two decades of the nineteenth century pioneered such critical inventions as steam turbines and internal combustion engines, its adoption of telephones lagged far behind the US pace.

In the United States, Bell's monopoly lasted until 1894, and the subsequent increase in the number of independent telephone companies was only temporary: nearly all of them were connected to Bell's long distance service, which was monopolized by the company's new parent organization, American Telephone and Telegraph. Bell's strategy of high long distance charges used to subsidize low-cost local services lasted for most of the twentieth century until the deregulation and breakup of the company on January 1, 1984.

In Samuelson's (1896: 128) words, "the telephone has been coldly received in England," and, compared to the United States, the country "has lagged behind . . . to an extent which is almost ludicrous." After the British Post Office, which had the long distance monopoly from 1896, took over all services in 1912, the United Kingdom had one phone for about 60 people, as did Germany, while the US rate fell to just below 10 people per telephone; the French ratio was still more than 150, and the Russian one dipped to just below 1,000.

The large investments of monopolistic companies in installed equipment, which they expected to last for decades, was by far the most important reason for the slow progress of telephonic innovation during most of the twentieth century. The first one-piece black set incorporating transmitter and receiver in the same unit was introduced only in the late 1920s. Subsequently, the telephone device stagnated—except for pushbutton dialing introduced in 1963—until the beginning of a rapid conversion to electronic telephony begun in the late 1970s (Luff 1978). Only then was the carbon microphone replaced by a dynamic transducer (identical device acting as a receiver), and small integrated circuits eliminated bulky bell ringer, transformer, and oscillator. Phones got smaller, lighter, and cheaper, and soon the entire innovation cycle was repeated with wireless cellular devices.

Recorded Sound

The pedigree of Edison's phonograph can be traced to both telegraph and telephone. While trying to improve his telegraph transmitter, Edison discovered that when the recorded telegraph tape was played at a high speed, the machine produced a noise that resembled spoken words (Josephson 1959).

Could he then record and replay a telephone message by attaching a needle to the diaphragm of a receiver to produce a pricked paper tape? Soon he tried a tinfoil cylinder and was surprised as he heard himself reciting "Mary had a little lamb." As this was the first device able to record and reproduce sound, it attracted a great deal of public attention. In 1878, Edison took the phonograph on a tour that included a demonstration to US President Rutherford Hayes, and he set up a new company to sell the machine that he wanted to see in every American home (Figure I.5.14).

During the early 1880s, when most of his attention was given to the development of new electric systems, Edison was looking for a substance to replace foil as the recording medium—and so were other inventors. In 1885, Chichester Bell and Charles Tainter (1854–1940) got the patent for the Graphophone, a phonograph with wax-coated cylinders incised with vertical grooves, and, by 1887, Edison patented his improved phonograph, which used wax cylinders as well as a battery-powered electric motor. Both devices gave rise to a new recording industry, but neither was a great commercial success. The phonograph was marketed as a multifunctional device: as a recorder of family's vocal mementos, a music box, a dictation machine for

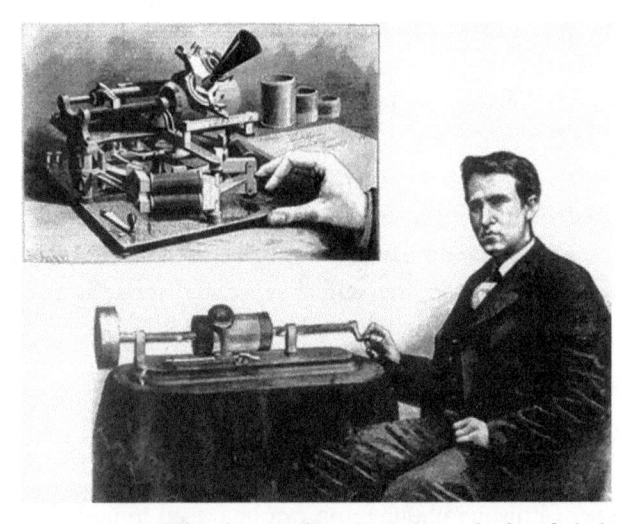

Figure I.5.14 Edison with his first perfected hand-cranked tinfoil phonograph and his improved wax-cylinder, electric motor-powered phonograph.

Edison photograph reproduced from Scientific American, July 25, 1896; inset reproduced from *Scientific American*, December 24, 1887.

businesses, an audio text for the blind, and later also in a miniature form inside Edison's talking dolls.

The two main problems with the phonograph were the recording mode and the recording medium. The vertical (hill-and-dale) cut needed a mechanism that would prevent the stylus from jumping out of a groove, and the wax cylinder was obviously too soft and too fragile to make permanent recordings; it broke easily with repeated handling, it needed a storage box, and it could not be reproduced cheaply in large quantities. But Edison—displaying his penchant for lost causes that led him later to invest years and large sums of money in fruitless efforts to beneficiate iron ore and develop a superior battery—kept improving the home phonograph for decades after its introduction, and the Edison company continued to make recorded cylinders until 1919. By that time, the era of the gramophone was in full swing (literally, given the popularity of new jazz recordings).

The origins of the gramophone can be traced directly to the failings of the phonograph: vertical-cut grooves, soft recording surface, and difficult mass production of recorded sound. All of these were successfully resolved in a relatively short time by Emile Berliner (1851–1929), a German immigrant who, after a succession of jobs in New York City, became a cleanup man in the laboratory of Constantine Fahlberg (1850–1910), the discoverer of saccharine. There he became interested in experimenting with electricity, and, in 1876, he invented a simple, loose-contact telephone transmitter (a type of microphone), which he patented in 1877, and which earned him a job with the Bell Telephone Company of Boston. Ten years later he patented (US Patent 372,786) a new system for recording sound with a lateral stylus and playing the records on a hand-cranked machine. Its first public presentation took place at the Franklin Institute on May 16, 1888 (Berliner 1888).

Berliner's solution was clearly anticipated by ideas about the "process of recording and reproducing audible phenomena" that Charles Cros (1842–1888) submitted to the French Academy of Sciences in 1877—but of whose existence Berliner, and even the US patent examiners, were ignorant (Berliner 1888). Cros's key idea, based in turn on Scott's phonautograph, was to attach a light stylus to a vibrating membrane and use it to produce undulating tracings on a rotating disk—and then, by means of a photoengraving process, to convert the undulatory spiral into relief or intaglio lines to produce a playable record. Nothing practical came out of Cros's suggestions, but when Berliner eventually learned about them, he very generously acknowledged their priority.

Berliner's first recordings were done by a stylus tracing a fine undulatory line in a thin layer of fatty ink, which he prepared by mixing one part of paraffin oil with 20 parts of gasoline and drying the deposit after the gasoline evaporated (Berliner 1888). A stylus attached to a mica diaphragm recorded the vibrations by moving laterally; the grooves were etched by acid, and a 15-cm zinc record, with 2-minute capacity, was placed on a hand-cranked turntable to be played by a steel stylus at 30 rotations per minute (rpm). Multiple reproduction was done by electroplating the master zinc record and using the metal negative to stamp out the desired number of positive copies. After trying many different substances, Berliner chose celluloid; soon after he switched to hardened rubber, which could withstand better the pressure exerted on steel needles by the cumbersome combination of the tone arm and horn.

Berliner (1888: 18) correctly anticipated that the records (he called them "phonautograms") would provide both substantial royalties for performers and a great deal of enjoyment to their collectors: "Prominent singers, speakers or performers may derive an income from royalties on the sale of their phonautograms . . . collections of phonautograms may become very valuable, and whole evenings will be spent at home going through a long list of interesting performances."

Berliner's US Gramophone Company sold 1,000 machines in 1894 and 25,000 15-cm hard rubber records. In 1895, Berliner patented (US Patent 534,543) the recording on a horizontal disk (Figure I.5.15); in 1896, he discovered that shellac (an organic compound prepared from a gummy secretion of the Asian scale insect *Coccus lacca*) from the Duranoid Company was superior to hard rubber, and he switched to it in 1897. The material remained dominant until the late 1940s, when it was replaced by vinyl.

Berliner's key recording feature, whereby the groove both vibrates laterally and propels the stylus, remained the standard audio technique until long-playing (LP) stereo records, introduced during the late 1950s, combined the lateral and vertical cut. Rising sales brought illegal competitors, and, in 1900, Berliner turned over his US patents to Eldridge R. Johnson of the National Gramophone Company in Camden, New Jersey, who formed a new business that eventually changed its name to the Victor Talking Machine Company. After its 1929 merger with RCA, it became the world's largest and best-known recording enterprise, RCA Victor. Berliner relocated his company to Montreal in 1900.

During the second year of his Montreal operation, Berliner sold 2 million single-sided records: the other side featured what was to become one of

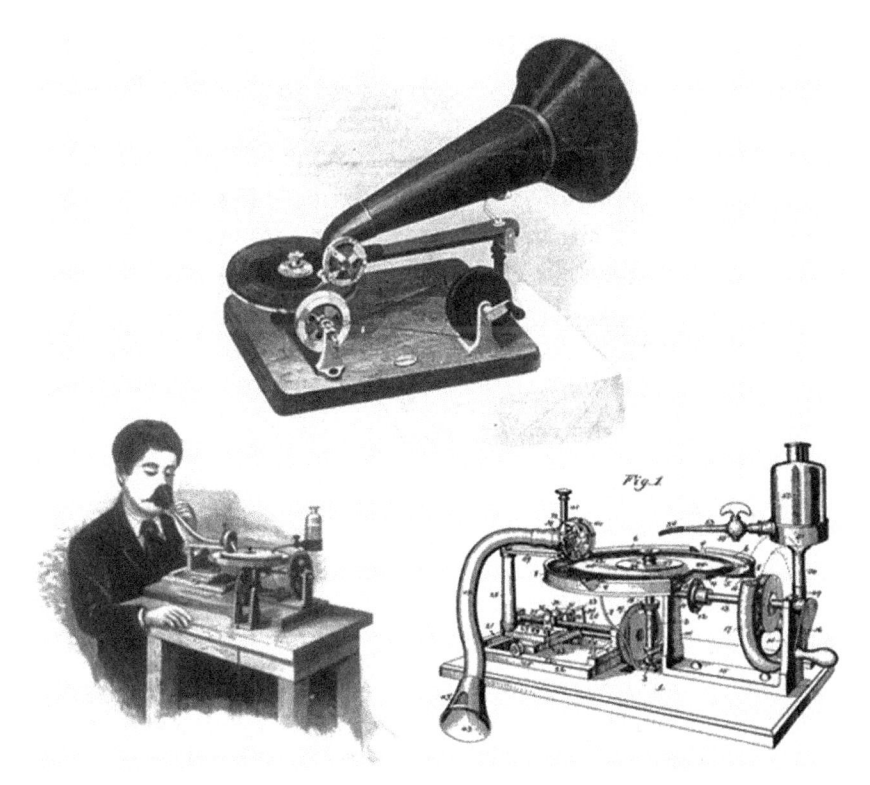

Figure I.5.15 Emile Berliner's drawing of the apparatus he invented (US Patent 534,543 filed in 1892 and granted in February 1895) to produce gramophone records (lower right) and use of the instrument for recording speech (lower left) and an early design of a hand-cranked reproducing gramophone. With Berliner's design, it was necessary that "the record tablet be covered with a thin film of alcohol, and for this purpose a thin stream of alcohol (stored in the vessel on the right) is directed upon the center of the tablet . . . from which the alcohol spreads in all directions by centrifugal force."
Top image reproduced from the cover of *Scientific American*, May 16, 1896.

the world's most famous trademarks, Francis Barraud's painting of his dog, Nipper, listening to a gramophone ("His Master's Voice," source of the company abbreviation HMV), which Berliner registered on July 10, 1900. Victor's two major milestones were the first recording that sold a million copies (the first Red Seal record with Enrico Caruso, on an exclusive contract, singing *Vesti la giubba* from Leoncavallo's I Pagliacci) in 1903, and the introduction of the Victrola (in 1906), the first entirely enclosed cabinet phonograph, which, despite its high cost ($200), became a big seller for the next two decades.

The first double-sided disks were produced in 1902, but even so the playtime remained limited to no more than 10 minutes of music. Recording complete symphonies or operas was thus still impractical, but it was done anyway: HMV's first complete opera, Verdi's *Ernani*, filled 40 single-sided disks in 1904. Prior to that, even musical aficionados might have heard a favorite composition or opera only once or twice in a live performance or had to be content just studying its score. More important, recordings introduced great music to millions of listeners who could not make it to a concert hall. On the pop side, the earliest hits included folk songs, band music, and, beginning in 1912, the craze for ballroom dancing led to a proliferation of dance band records (but the first Dixieland Jazz record, which sold 1.5 million copies, came in 1917).

Before leaving the subject of recorded sound, I must note the genesis and delayed commercialization of magnetic recordings. Valdemar Poulsen (1869–1942), a Danish telephone engineer, patented his system of sound recording and reproduction on steel piano wire in 1898, and a working model of his Telegraphone was one of the electrical highlights at the Paris Exposition in 1900 (Daniel, Mee, and Clark 1999). An improved design, able to record 30 minutes of sound, became available in 1903, but better established and cheaper Edisonian phonographs and dictaphones delayed commercial development of the device, and the first Magnetophon recordings were made only in 1935.

Hertzian Waves

There were no fundamental obstacles to prevent the discovery of high-frequency electromagnetic waves years, even decades, before Hertz did so. Already in 1864, James Clerk Maxwell (1831–1879; Figure I.5.16) had presented to the Royal Society of London a dynamical theory of the electromagnetic field (Maxwell 1865), and his ideas generated a great deal of interest and disbelief, particularly after they were published in the two-volume *Treatise on Electricity and Magnetism* (Maxwell 1873). Implications of Maxwell's theory were clear: the existence of waves of varying length that would propagate at the speed of light, curl around sharp edges, and be absorbed and reflected by conductors; moreover, it was known how to calculate their length and how to produce them (Lodge 1894). But Maxwell himself made no attempts to test the wavelike nature of electromagnetic radiation,

Figure I.5.16 James Clerk Maxwell (left) postulated the existence of electromagnetic waves longer than light but shorter than sound more than two decades before Hertz's experiments confirmed the theory. This portrait of David Edward Hughes (right) was taken during the mid-1880s, just a few years after his pioneering experiments with "aerial electric waves."
Maxwell photograph from author's collection; Hughes photograph courtesy of Ivor Hughes.

and it took more than two decades after the publication of his paper before somebody investigated their existence.

Or, more precisely, before somebody did so and published the results, because here I must make an important detour to describe one of the most remarkable cases of lost priority and misinterpreted invention. Sometime in December 1879, seven years before Hertz began his experiments, David Edward Hughes (1831–1900), a London-born physicist who returned to the city from Kentucky, where he was a professor of music as well as physics, began sending and receiving electromagnetic pulses over distances of as much as several hundred meters (Hughes 1899; Figure I.5.16). As already noted, Hughes was, in 1878, one of the inventors of a loose contact microphone, and experiments with this device led him to suspect for the first time the existence of what he called "extra current" produced from a small induction coil.

He found that microphones made sensitive receivers of these waves, and he tested the transmission first indoors over distances of up to 20 m; then, walking up and down Great Portland Street with the receiver in his hand and the telephone to his ear, he could receive signals up to 450 m away.

Between December 1879 and February 1880, eight people, most of them members of the Royal Society, witnessed these experiments, but George Gabriel Stokes (1819–1903), famous Cambridge mathematician and a future president of the society, concluded that everything could be explained by well-known electromagnetic induction effects. Although Hughes continued his experiments, he was so discouraged by his failure to persuade the Royal Society experts "of the truth of these aärial electric waves" that he refused to write any paper on the subject until he had a clear explanation of their nature, a demonstration that came from Hertz's experiments.

Crookes, who witnessed the December 1879 experiment, wrote two decades later to Fahie that "it is a pity that a man who was so far ahead of all other workers in the field of wireless telegraphy should lose all credit due to his great ingenuity and prevision" (Fahie 1899: 305). How far ahead? A writer in *The Globe* of May 12, 1899 (cited in Fahie 1899: 316), summed up this best when he said that the 1879 experiments "were virtually a discovery of Hertzian waves before Hertz, of the coherer before Branly, and of wireless telegraphy before Marconi and others." Incredibly, a few years after Hughes's first broadcasting experiments, there was another near discovery by none other than Thomas Edison, and it went public. One of the world's great missed fundamental discoveries was a small item displayed in 1884 in a corner of the Philadelphia Exhibition's largest electrical exhibit, which belonged, predictably, to Thomas Edison: it was an "apparatus showing conductivity of continuous currents through high vacuo." In its patent application, filed on November 15, 1883, Edison (1884: 1) noted that

> if a conducting substance is interposed anywhere in the vacuous space within the globe of an incandescent electric lamp, and said conducting substance is connected outside of the lamp with one terminal, preferably the positive one, of the incandescent conductor, a portion of the current will, when the lamp is in operation, pass through the shunt circuit thus formed, which shunt includes a portion of the vacuous space within the lamp.

This device became known as a *tripolar incandescent lamp*, but the observed phenomenon, called by Edison "etheric force" and commonly known as the *Edison effect*, remained just a curiosity without any practical applications. Another 12 years had to pass after Edison secured this useless patent before the current passing through space was recognized as electromagnetic oscillations at wavelengths far longer than light but far shorter than audible sound. Interestingly, neither Edison, who through his long career experienced a full measure of triumphs and failures stemming from his obsessive inventive drive, nor Hughes, who ended his life as one of the most honored inventors of his generation, regretted their near misses.

Hughes's 1899 correspondence with Fahie reveals a man who looked back without bitterness, crediting Hertz with "a series of original and masterly experiments." And these, done between 1886 and 1889 without any knowledge of Hughes's work, were exactly that. Heinrich Rudolf Hertz (1847–1894; Figure I.5.17) was steered in the direction of these fundamental experiments by his famous teacher, Hermann von Helmholtz (1821–1894). In 1879, Helmholtz made an experimental validation of Maxwell's hypotheses the subject of a prize by the Berlin Academy of Sciences, and he believed that Hertz would be the best candidate to solve the problem. Although Hertz soon abandoned this line of inquiry, his "interest in everything connected with electric oscillations had become keener" (Hertz 1893: 1).

Hertz finally began investigating what he termed "*sehr schnell elektrische Schwingungen*" (very rapid electric oscillations) in 1886, at the Technische Hochschule in Karlsruhe. His experimental setup was very simple, and his work is well documented in his diaries, letters, and papers (Hertz 1893; Hertz and Susskind 1977). He produced electromagnetic waves by using a large induction coil, basically a transformer that received pulsed voltage from batteries to primary winding and produced a much higher voltage in the secondary winding. In his first set of experiments, he used a straight copper wire with a small discharge gap for the inducing circuit and a rectangle of insulated wire for the induced circuit (Figure I.5.17).

Subsequently, his discharger consisted of a straight copper wire 5 mm in diameter with a 7.5 mm spark gap in the middle that was attached to spheres 30 cm in diameter made of sheet zinc and placed 1 m apart. Once this simple dipole antenna was connected to an induction coil, its alternate charging and discharging produced sparks across the gap. His secondary circuit was even simpler, what we came later to call a *loop receiving antenna*, without any rectifier or amplifier, just a coil of wire 2 mm thick formed in a circle of 35 cm in radius left with just a small spark gap that could be regulated by a micrometer screw (Hertz 1887). Hertz set up all the experiments in such a way that the spark of the induction coil was visible from the place where the spark in the micrometer took place.

The most important reason for Hertz's success was his choice of frequencies. His experiments were done in a lecture room 15 m long and

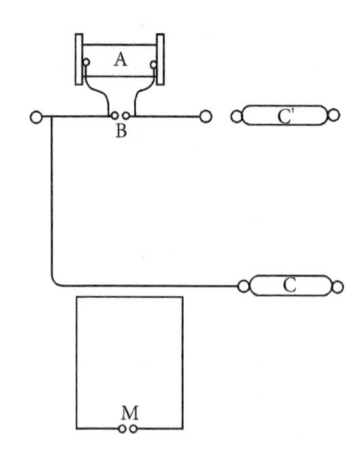

Figure I.5.17 Heinrich Hertz's experiments opened the way for the entire universe of wireless communication and broadcast information. The simplicity of his epochal discovery is best illustrated by one of his early instrumental arrangements (right). An inducing circuit (top) contained the induction coil (A) and straight copper wire conductors (C and C') with the discharger (B) at the center; the induced circuit was a rectangular wire with the gap adjustable by a micrometer (M).
Reproduced from Hertz (1887).

14 m wide, but rows of iron pillars reduced the effective area to about 12 × 8 m. This meant that to radiate the waves from one end of the room, bounce them off a metal sheet at the other end, and measure the crests or nodes of the standing waves (by the strength of sparks), at least one-half wavelength (and preferably several) had to fit into the length of the room. Hertz understood this requirement, and the best reconstruction of his experiments indicates that he was using wavelengths between 6 m and 60 cm—that is, frequencies of 50–500 megahertz (MHz).

His ingenious experiments proved that these invisible electromagnetic waves behave much as light does as they are reflected and refracted by surfaces and that they travel through the air with finite, light-like velocity. In retrospect, the discovery of the waves that "range themselves in a position intermediate between the acoustic oscillations of ponderable bodies and the light-oscillations of the ether" (Hertz 1887: 421) was obviously one of the most momentous events in modern history. Hertz's feat was akin to identifying a new, enormous continent superimposed on a previously well-explored planet, accessing an invisible but exceedingly bountiful realm of the universe, the existence of an entirely new phenomenon of oscillations.

Less than a decade after Hertz's experiments came the first wireless telegraph transmissions, and, less than two decades later, the first radio broadcasts. By the late 1930s, radio was the leading means of mass communication and entertainment, the first scheduled television broadcasts were taking place, and radar was poised to become a new, powerful tool in warfare and—after World War II—also in commercial aviation. Hertzian waves had changed the world, and there is yet no end to their impact: their subdivision, modulation, transmission, and reception for the cellular telephony and wireless Internet are just the latest installments in the still unfolding story. But Hertz did not foresee any practical use for his invention, and, because of his premature death, he never got a chance to revise this conclusion once research into communication with high-frequency waves got under way.

Wireless Communication

Guglielmo Marconi's (1874–1937; Figure I.5.18) fame rests on filing the first wireless telegraph patent in 1896 (GB Patent 7,777; US Patent 586,193) and being the first sender of wireless signals over increasingly longer distances and finally across the Atlantic in 1901 (Marconi 1909; Jacot and Collier 1935; Boselli 1999). David Hughes aside, there are at least four other pioneers of wireless telegraphy whose work made Marconi's achievements possible or preceded his first broadcasting demonstrations. In 1890, Nikola Tesla's invention of his eponymous coil, a device that could step up ordinary currents to extremely high frequencies, opened the way for the transmission of radio signals (Martin 1894). In 1890, Edouard Branly (1844–1940) noticed that an ebonite tube containing metal filings that would not normally conduct, when placed in a battery circuit, became conductive when subjected to oscillatory current from a spark generator. This device made a much better detector of electromagnetic waves than did Hertz's spark gap.

In June 1894, at the Royal Institution, and again in August in Oxford, Oliver Joseph Lodge (1851–1940; see Figure I.6.3) used an improved version of Branly's tube, which he named a "coherer" to demonstrate what might have been the world's first short-distance wireless Morse broadcasts (Jolly 1974). *Nature* reported on the first event, and, in 1897, Lodge recalled that, on the second occasion, when signals were sent some 60 m (and through two

Figure I.5.18 Guglielmo Marconi, who, without a formal scientific or engineering education, put into practice what several much more experienced men only contemplated doing: sending and receiving wireless signals by using a simple patented apparatus. This photograph was taken in 1909, when Marconi's work was rewarded by a Nobel Prize in physics.
Photograph © The Nobel Foundation.

stone walls) from the Clarendon Laboratory to the Oxford Museum, he used Morse signals. On balance, Aitken (1976) believes that Lodge, not Marconi, was the inventor of wireless telegraphy. But Lodge himself acknowledged that "stupidly enough no attempt was made to apply any but the feeblest power so as to test how far the disturbance could really be detected" (Lodge 1908: 84). He, much like Hertz, had a purely scientific interest in the matter and took steps toward the first commercial uses only after Marconi's 1896 patent.

And yet already, two years before Lodge's London and Oxford demonstrations, William Crookes—who also witnessed Hughes's 1879 experiment, and whose concerns about world food production that he expressed in 1898 were noted in Chapter I.4—spelled out (albeit still timidly, as his vision was limited to Morse signals) the commercial potential of Hertzian waves. He noted that an almost infinite range of ethereal vibrations or electrical rays... un-folded to us a new and astonishing world—one which it is hard to conceive should contain no possibilities of transmitting and receiving intelligence. . . . Here, then, is revealed the bewildering possibility of telegraphy without wires, posts, cables, or any of our present costly appliances. . . . This is no mere dream of a visionary philosopher. All the requisites needed to bring it within the grasp of daily life are well within the possibilities of discovery.... (Crookes 1892: 174, 176)

Russian historians routinely claim that the inventor of wireless telegraphy is Alexander Stepanovich Popov (1859–1906), who began his work with electromagnetic waves as he tried to detect approaching thunderstorms (Radovsky 1957). He designed an improved version of Lodge's coherer and used a vertical antenna to pick up the discharges of atmospheric electricity. In his lecture to the Russian Physicist Society on May 7, 1895, Popov reported that he transmitted and received wireless signals over a distance of 600 m (Constable 1995). Popov's first public demonstration of wireless transmission took place in March 1896, and, a year later, he installed the first ship-to-shore link between the cruiser *Africa* and Russian Navy headquarters in Kronstadt.

While none of these four inventors pushed for commercialization of their discoveries, 22-year-old Marconi came to England from Italy in 1896 with the determination to patent the achievements of his Italian work and to commercialize his system of wireless telegraphy (Jacot and Collier 1935; Boselli 1999). He arrived in February 1896, and, on March 30 (helped by his cousin Henry Jameson-Davis: Marconi's mother was Annie Jameson, a daughter of the famous Irish whiskey maker who married a well-off Bolognese merchant and landowner), he got a letter of introduction to William Preece, chief engineer of the British Post Office (Constable 1995). What Marconi brought to England was nothing fundamentally new: his transmitter was a version of high-frequency oscillator originally developed by his mentor Augusto Righi (1850–1920) in Bologna, his receiver was an improved version of Branly's coherer, and his antenna was of the grounded vertical type.

What was decisive was his confidence that this system would be, with improvements, eventually able to send signals over long distances and his determination to achieve this goal in the shortest possible time. As Preece correctly observed in an 1896 lecture, many others could have done it, but none of them did (Aitken 1976). Marconi's preliminary British patent application was filed in London, on June 2, 1896, and the first field tests sponsored by the Post Office were done on the Salisbury Plain, where, in September 1896, he transmitted 150 MHz signals over the distance of up to 2.8 km. His US patent application was filed on December 7, 1896 (Figure I.5.19). In the same month, after his signals crossed 14 km of the Bristol Channel, Marconi incorporated the Wireless Telegraph and Signal Company. In September 1899, his signals spanned 137 km across the English Channel between Wimereux and Chelmsford. This achievement convinced Marconi that the signals follow Earth's curvature, and, by July 1900, the directors of his company approved the ultimate trial of trans-Atlantic transmission.

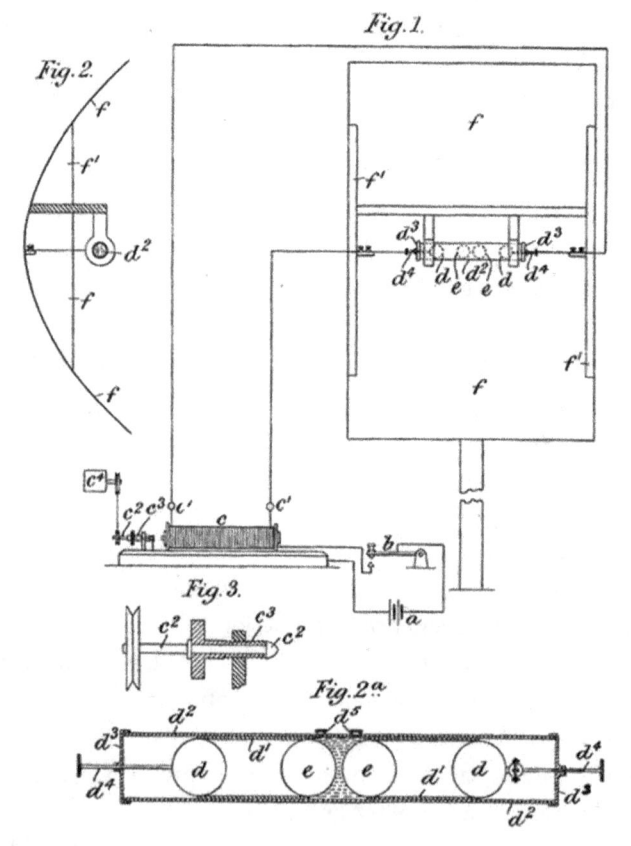

Figure I.5.19 Illustration of the transmitting component of Marconi's pioneering patent (US Patent 586,193) for sending and receiving electrical signals. Fig. 1 is a front elevation and Fig. 2 a vertical section of the transmitter with parabolic reflector (a is a battery, b a Morse key, c an induction coil, and d metallic balls). Fig. 2a is a longitudinal section of the oscillator.

John Ambrose Fleming (1849–1945) became the scientific adviser of the project, and he built a new, more powerful transmitter while Marconi himself designed the inverted wire cone antenna with a 60-meter diameter to be suspended in a ring of 20- to 60-m-tall wooden masts. The site selected for the transmitter, with alternator driven by an 18.7-kilowatt (kW) oil engine whose output was transformed to 20 kilovolts (kV), was at Poldhu, in Cornwall, where construction began in October 1900, and Marconi's mast at Cape Cod was to be the receiver. Setbacks delayed the first attempt: a nearly completed Poldhu aerial collapsed on September 17, 1901, and, on November 26, 1901, a gale took down the Cape Cod aerial as well. But Marconi quickly redesigned the Poldhu transmitter, an inverted wire pyramid anchored by four 60-m wooden towers and decided to shorten the distance and build a new receiving station on Signal Hill that overlooks St. John's, Newfoundland, 2,880 km from Poldhu.

The first trans-Atlantic signal, sent by the most powerful spark transmitter of that time, was picked up on December 12, 1901, on the simplest untuned receiver (Figure I.5.20): Marconi heard faint triple dots of Morse "S" at 12:30 and then again at 1:10 and 2:20 P.M., but nothing afterward. He hesitated before releasing the information to the press four days later. The news was greeted with enthusiasm, but there was also skepticism about the claim, and some of it remains even today. Major uncertainty surrounds the actual wavelength of the first transmission. Immediately after the event, Marconi claimed it was about 800 kHz; he did not quote any figure in his Nobel lecture, but in his later recollections, in the early 1930s, he put it at about 170 kHz (Bondyopadhyay 1993).

MacKeand and Cross tried to settle this uncertainty by using the best possible information to reconstruct Marconi's complete system and then to model its most likely performance. They concluded that the transmission centered on 475 and 540 kHz, that its total power at Poldhu reached about 1 kW and at St. John's was about 50 picowatts (pW), and that it is likely that Marconi received high-frequency wide-band signals, "spurious components of the spark transmitter output, propagated across the Atlantic by sky waves near the maximum usable frequency" (MacKeand and Cross 1995: 29). In contrast, Belrose (1995, 2001) found it difficult to believe that Poldhu signals could have been heard on Signal Hill because the broadcast took place during the day (hence their heavy attenuation), during a sunspot minimum period, and because the untuned receiver used by Marconi had no means of amplification.

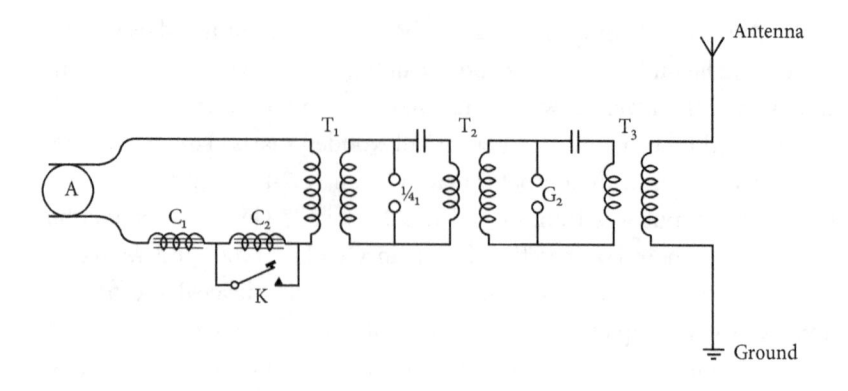

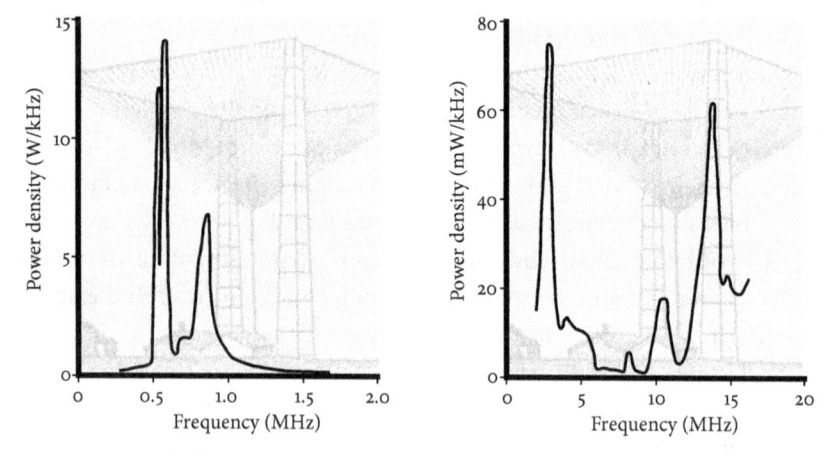

Figure I.5.20 The basic circuit of Marconi's Poldhu transmitter contained an alternator (A), chokes (C1 and C2), low-frequency (T1) and high-frequency (T2 and T3) transformers, spark gaps (G1 and G2), capacitors (P1 and P2), and telegraph key (K). Transmitted power at low frequencies had the highest density at about 0.6 MHz (bottom left); at high frequencies, at about 3 MHz (bottom right).
Based on Fleming's original drawing in Bondyopadhyay (1993) and on Aitken (1976) and Belrose (1995).

Definitive demonstration of long-range wireless transmission took place in February 1902, when Marconi and a group of engineers sailed from Southampton to New York on the *Philadelphia*, fitted with a four-part aerial: coherer-tape reception was up to 2,500 km from Poldhu, and the audio signal reached as far as 3,360 km (McClure 1902; Marconi 1909). Although Marconi was completely surprised by the difference between the maximum daylight reception (1,125 km) and the nighttime maxima (he was not aware of the ionospheric reflection of radio waves during the night), he was now confident that as soon as he set up an American station like that at Poldhu, he would be able to transmit and receive easily across the Atlantic. The Canadian government provided financial assistance to set up a high-power station at Glace Bay, Nova Scotia, and the exchange of messages with Poldhu began in December 1902. In 1907, an enlarged Glace Bay station and a new European station at Clifden, in Ireland, were used for the first commercial signaling across the Atlantic (Marconi 1909).

Marconi's first major customer was the British Royal Navy, which made a substantial purchase of radio sets in 1900, and wireless communication was used (unsuccessfully) for the first time during the Boer War (1899–1902) and shortly after that routinely during the Russo-Japanese War of 1904–1905. Commercially much more important was Marconi's contract with Lloyds of London, the world's premier shipping insurer, which committed itself in 1901 to 14 years of exclusive use of Marconi's wireless telegraph system. But the diffusion of wireless telegraphy in merchant shipping was not a rapid affair. Only in 1914—two years after the signals sent on the night on April 14, 1912, by the senior radio operator Jack Phillips from the *Titanic* (he drowned, and several nearby ships did not have their receivers on) called for help for the sinking passenger liner—did an international conference mandate the presence of wireless on all ships carrying more than 50 people (Pocock 1995).

Marconi's aggressive advances had other and much more experienced engineers playing catch-up. Lodge filed his syntony (two circuits tuned to the same frequency) patent in 1897 (Lodge 1908). Adolf Slaby (1849–1913), the first professor of electro-technology at Berlin's Technische Hochschule, was present at Marconi's May 1897 experiments, and, after his return to Germany, he and Georg Wilhelm Alexander von Arco (1869–1940), working for Allgemeine Elektrizitätsgesselschaft (AEG), began to patent various improvements to receivers and antennas. So did, independently, Karl Ferdinand Braun (1850–1918), working for Siemens & Halske in Strassburg. After the two rival companies joined to set up Gesselschaft für drahtlose Telegraphie (much better known as Telefunken), those inventions provided the foundation for Germany's leading role in the early development of wireless broadcasting.

In the United States, Tesla filed his two key radio patent applications (US Patents 645,576 and 649,621) in 1897. He specified an apparatus that would produce "a current of excessively high potential" to be transmitted through the air, and he also recognized that the receiving coils may be moveable, "as, for instance, when they are carried by a vessel floating in the air or by a ship at sea" (Tesla 1900: 2; Figure I.5.21). As these filings preceded by more three years Marconi's US patent application for an apparatus for wireless telegraphy (US Patent 773,772, filed on November 10, 1900), Tesla's priority seemed secure: his reaction to the news of Marconi's trans-Atlantic broadcast was to note that that achievement was based on using 17 of his patents.

But, in 1904, the US Patent Office reversed itself and recognized Marconi as the inventor. This only aggravated Tesla's already precarious financial position, but worse was to come in 1909, when Marconi and not Tesla got the Nobel Prize in physics. This bitter story had its unexpected ending in 1943 when, a few months after Tesla's death, the US Supreme Court finally ruled in his favor (Cheney 1981). But this was no righting of intellectual wrongs, merely a way for the court to avoid a decision regarding Marconi Company's suit against the US government for using its patents. Marconi's success is a perfect illustration of the fact that no fundamental inventions are needed to become a much-respected innovator.

Being first to package and slightly improve what is readily available, being aggressive in subsequent dealings, and making alliances with powerful users can take an entrepreneur and his company a lot further than coming up with a brilliant new idea: many of today's very successful businesses have followed that precept. The fact that Marconi was not a great technical innovator is best illustrated by the fact that he considered Morse signals quite adequate for the shipping business and that he, unlike Tesla and Fessenden, did not foresee the multifaceted development of the radio and broadcasting industry (Cheney 1981; Belrose 1995).

Spark generators, favored by Marconi, could only transmit Morse code. The first continuous wave signals that could be modulated by audio frequencies for the transmission of voice or music were produced by arc transmitters and high-frequency (HF) alternators. Valdemar Poulsen designed the first effective arc transmitter in 1902, and the largest devices were patterned on Kristian Birkeland and Samuel Eyde's process, which was commercialized in the same year to produce nitrogen oxides as feedstocks for the synthesis of inorganic fertilizers. These transmitters became eventually truly giant: Telefunken's Malabar, in the Dutch East Indies, commissioned in 1923, had input of 3.6 megawatts (MW) and a 20-ton (t) electromagnet

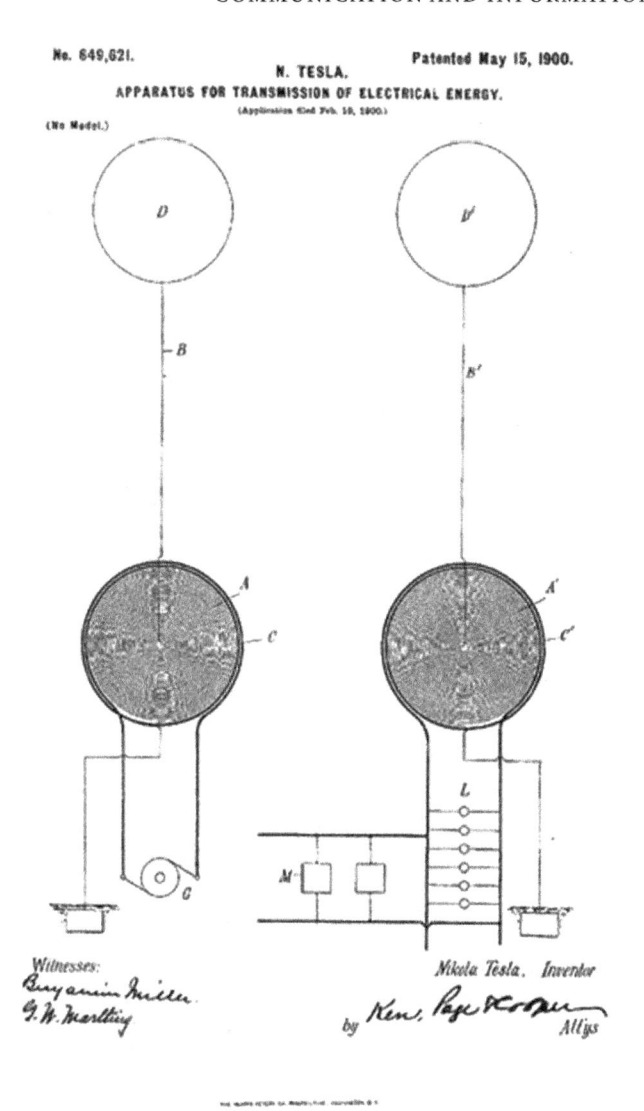

Figure I.5.21 Tesla filed his application for an apparatus for transmission of electrical energy on September 2, 1897, and received US Patent 649,621 in May 1900. The transmitter consists of a suitable source of electric current (G), transformer (A and C being, respectively, high-tension secondary and lower voltage primary coils), conductor (B), and terminal (D), "preferably of large surface, formed or maintained by such means as balloon at an elevation suitable for the purposes of transmission." In the receiving station, the signal is led from the elevated terminal (D') via a conductor (B') to the transformer, whose coils are reversed, with A being the primary, and the secondary circuit contains "lamps (L), motors (M), or other devices for utilizing the current."

to quench the arc. Europe's largest arc transmitter, completed in 1918, in Bordeaux, had eight 250-m masts and covered an area of nearly 50 hectares.

Tesla, already famous because of his electric motor and alternating current (AC) inventions, built the first low-frequency radio alternator (operating at 30 kHz) in 1899, but two men whose work made the first radio broadcasts possible—a Canadian, Reginald Aubrey Fessenden (1866–1932), and a Swede, Ernst Frederik Werner Alexanderson (1878–1975)—are well known only to students of radio history. Fessenden (Figure I.5.22) was first to conceive and Alexanderson first to build and then to perfect high-frequency (HF) alternators that could produce continuous wave signals. Fessenden's early accomplishments included work in Thomas Edison's laboratory and professorships of electrical engineering at Purdue and Pittsburgh University.

Figure I.5.22 In contrast to Tesla and Marconi, only radio experts and historians of invention now know the name of Reginald Fessenden, a Canadian working in the United States, who made the world's first radio broadcast in December 1906.
Photograph courtesy of the North Carolina State Archives.

In 1900, he began to work for the US Weather Bureau and pursued his idea that HF well above the voice band should make wireless telephony possible.

His first success came on December 23, 1900, when he transmitted a couple of sentences—"One-two-three-four, is it snowing where you are Mr. Thiessen? If it is, would you telegraph back to me?"—over 1.6 km between two 15-m masts built on Cobb Island, in the Potomac River, in Maryland (Raby 1970). After he left the bureau, he set up his broadcasting hut and antenna at Blackmans Point, on Brant Rock in Massachusetts, and its transAtlantic counterpart at Machrihanish, in Scotland, so his challenge was to span a distance longer than did Marconi's wireless telegraphy experiments. His eventual success was made possible by a new 50-kHz alternator that was built by General Electric according to the designs of Ernst Alexanderson, a young Swedish engineer who was hired by General Electric in 1902 (Nilsson 1987). Alexanderson's machine had the periphery of its tapered disk rotating at 1,100 km/h (i.e., the speed of sound) and yet wobbling less than 0.75 mm.

This was an inherently wasteful, and hence costly, way to broadcast: due to high heat losses in winding and armature and to high frictional losses arising from up to 20,000 rpm, the efficiency of smaller units was less than 30%. Moreover, the technique was limited to relatively low frequencies of no more than 100 kHz, compared to 1.5 MHz for spark generators, but it was the first means of generating true continuous sine waves. After World War I, vacuum tube transmitters began replacing the pioneering transmitting devices, but some spark generators remained in operation until after World War II, long after arc transmitters and most of the alternators were gone.

The first long-distance broadcast using Alexanderson's alternator took place accidentally in November 1906, when an operator at Machrihanish clearly overheard the instructions relayed from Fessenden's Brant Rock headquarters to a nearby test station at Plymouth. Encouraged, Fessenden prepared the world's first programmed radio broadcast: on Christmas Eve of 1906, he gave a short speech and then played a phonographic recording of Handel's *Largo* followed by his own short violin solo of Gounod's *O Holy Night*, a reading from the Bible, and wishes of merry Christmas (Fessenden 1940). This broadcast was heard by radio operators on ships of the US Navy along the Atlantic coast and on the vessels of the US Fruit Company as far away as the Caribbean, and its variant was repeated on New Year's Eve.

Fessenden also developed *heterodyning*—a way to transfer a broadcast signal from its carrier to an intermediate frequency in the receiver to avoid retuning the receiver when changing channels—and amplitude modulation

(AM). AM frequencies, now used in the range of 540–1,600 kHz, remained the best way to broadcast voice and music until the diffusion of frequency modulation (FM, now in the band between 88 and 108 MHz) that was invented by Edwin Armstrong (1890–1954). But these admirable achievements required further development and modification to make broadcasting a commercial reality (White 2024).

Contributions by John Fleming, Edwin Armstrong, Lee De Forest, and others were essential in that respect. The year of the first broadcast brought another milestone in the history of radio because of Lee De Forest's (1873–1961) invention of the *triode* or, as he called it, the "Audion tube." In 1904, Fleming invented the *diode* (a device based on Edison's unexploited effect), which was essentially an incandescent light bulb with an added electrode and could be used as a sensitive detector of Hertzian waves as well as a converter of AC to direct current (DC) (Figure I.5.23).

But, as he later recalled in his biography,

> Sad to say, it did not occur to me to place the metal plate and the zig-zag wire in the same bulb and use an electron charge, positive or negative, on the wire to control the electron current to the plate. Lee de Forest, who had been following my work very closely, appreciated the advantage to be so gained, and made a valve in which a metal plate was fixed near a carbon filament in an exhausted bulb and placed a zig-zag wire, called a grid, between the plate and the filament. (Fleming 1934: 144)

The added grid electrode, interposed between the incandescing filament and the cold plate, acted as a modulator of the flowing current and made the Audion the first highly sensitive practical amplifier (Figure I.5.23). In 1912, when Armstrong introduced the first acceptable AM receiver, de Forest began using a series of his Audions to amplify HF signals, and he also discovered that feeding part of the output from the triode back into its grid produces a self-regenerating oscillation in the circuit that can be used, when fed to an antenna, for broadcasting speech and music. Triodes could be adapted for reception, amplification, and transmission, and for more than half a century, all electronics—radios, televisions, and the first computers—depended on an increasing variety of vacuum tubes. Tetrodes and pentodes (with four and five electrodes in vacuum) and cathode ray tubes (for TV screen and display monitors) were eventually added to diodes and triodes. Domination of

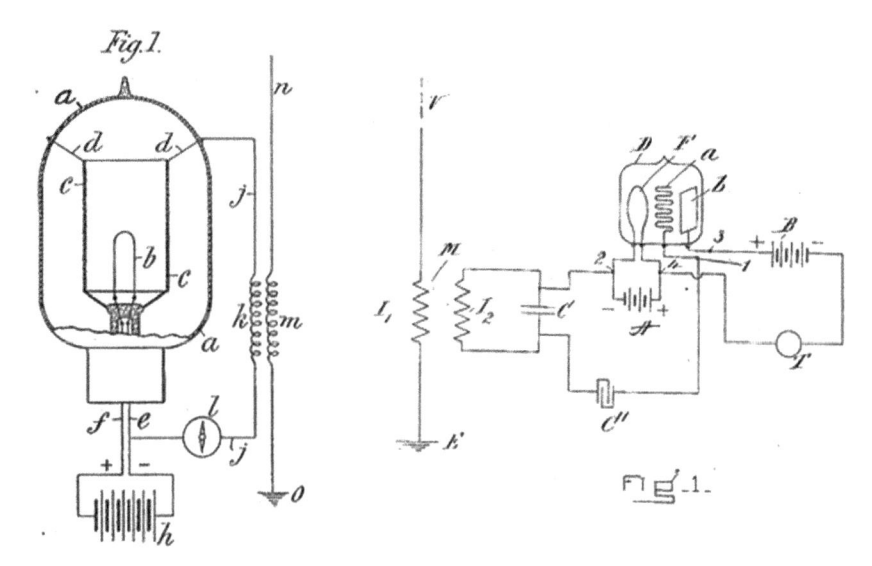

Figure I.5.23 Detail of John A. Fleming's patent drawing (US Patent 803,684 filed on April 10, 1905) for the diode (left), an "instrument for converting alternating electric currents into continuous currents." A glass bulb (a) contains a carbon filament (b) operating at 6–8 V and 2–4 A and connected to leads (e and f) by platinum wires. An aluminum cylinder (c), suspended by platinum wires (d), is open at the top and bottom, and it surrounds the filament without touching it. Lee De Forest's diagram (right) of "a wireless telegraph receiving system comprising an oscillation detector constructed and connected in accordance with the present invention," namely, his triode (US Patent 841,387), shows an evacuated glass vessel (D) containing three conducting members: a metallic electrode (F) connected in series to a source of electricity that can heat it to incandescence; a conductor (b) made of platinum and, interposed between these two members, a grid-shaped platinum wire (a).
Both drawings available at http://www.uspto.gov.

vacuum tubes ended only when transistors, superior amplifiers that operate on an entirely different principle, began taking over during the 1950s.

By 1913, the combination of better generators, antennas, amplifiers, and receivers provided a good foundation for the commercial development of radio broadcasting, but large-scale diffusion of the new communication medium took place only after World War I (White 2003). Several reasons explain this delay: only limited numbers of radio amateurs and military and ship operators had requisite transmitters and receivers, public access to radio was delayed by restrictions imposed on the use of airwaves during World

War I, and it was necessary to wear headsets. That is why radio's expansive years came between the two world wars as transmission distances increased, the numbers of radio stations multiplied, and every aspect of reception was greatly improved thanks to new tuning circuits, capacitors, microphones, oscillators, and loudspeakers. Radio's second great transformation came during the 1950s, with the deployment of transistors that made it inexpensive and easily portable.

Frontispiece I.6. An image that embodies aspirations of a new era: "The Progress of the Wheel: The Ousting of the Horse from London Thoroughfares," portrayed in *The Illustrated London News,* June 27, 1903.

6
A New Civilization

> Every now and again something happens—no doubt it's ultimately traceable to changes in industrial technique, though the connection isn't always obvious—and the whole spirit and tempo of life changes, and people acquire a new outlook which reflects itself in their political behaviour, their manners, their architecture, their literature and everything else. . . . And though, of course, those black lines across the page of history are an illusion, there are times when the transition is quite rapid, sometimes rapid enough for it to be possible to give it a fairly accurate date.
> —George Orwell, in a BBC broadcast on March 10, 1942

Orwell's observations make a perfect epigraph to fortify this book's arguments about the unequaled pre-World War I technical saltation. Their discovery was serendipitous: I began reading Orwell's wartime essays (Orwell 1942) a few months before I was to start writing this chapter. As soon as I came across the quoted passage, I felt as if this book had been distilled, 60 years before it was written, into a paragraph. Besides capturing the very intent and essence of this book, by asserting matter-of-factly that great civilizational shifts are ultimately traceable to technical changes, Orwell also believed that some of these momentous transformations could be rather accurately dated. And although these might be seen as only inconsequential stylistic preferences, I was pleased that Orwell used three specific terms that are not commonly encountered in writings about epochal shifts. Rather than using the inappropriate, but now generally accepted term "technology," he wrote correctly about "technique" (a perceptive reader might have noticed that this is the only instance when the word "technology" appears in this book).

Orwell also talked about the spirit and tempo of life, two elusive but critical concepts that I feel are essential for real understanding of civilizations. And, serendipity squared, within days after reading Orwell's essay I came across a fitting frontispiece. I realized that, unlike with specific topical chapters, it would not be easy to choose an appropriate illustration for these closing reflections. I rejected dozens of possibilities until I came across a 1903 engraving in *The Illustrated London News* depicting a "continual procession of motors and cycles" that could be seen "any fine holiday afternoon and Sunday morning" along London's Kensington High Street.

This picture appears entirely unremarkable when seen from the vantage point of the early twenty-first century, yet it would have been quite unimaginable in the mid-1860s, and so it conveys perfectly the great technical saltation accomplished during the two pre-World War I generations. And the two portrayed conveyances and their users also tell us much about the attendant socioeconomic impacts. Phrases that come to mind capture the tempo and spirit of the modern age: mobility (both in the physical and in the social sense), mass consumption, democratization, opportunity, gains in equality, emancipation, rise of the middle class, recreation, and leisure. Also, aspiration and anticipation: so much had changed, but so many changes were ahead as those determined riders proceeded on their way.

The principal subject of this chapter is to explain, in a terse and resonant manner, those social, economic, and behavioral waves that began to radiate across modern societies after new techniques had disturbed and transformed the traditional surface. Doing this in a comprehensive way would require another book, and that is why my survey of new realities and enduring trends will concentrate only on two conjoined fundamental attributes and legacies of the era: the increase of energy consumption and mechanized mass production and the delivery of new services.

Missing Perspectives

Writing this book was a constant exercise in restraint. Again and again, I wanted to supply more technical details regarding inventions because those numbers and their improvement over time tell best the stories of unprecedented achievements and of the continuous quest for technical perfection. At the same time, I would have preferred to prepare the ground for these details with generous explanations of conditions that prevailed before the great wave of the late nineteenth-century innovations and to enliven the technical presentations by supplying much more personal information regarding the background, effort, and expectations of the era's leading inventors. And so, exercising restraint once again, I chose only two subjects for this section. I will first describe the origins and early development of modern data processing, and my second choice is to reflect a bit more systematically on the personal triumphs, trials, and tragedies, as well as on the notable idiosyncrasies, of some of the era's leading protagonists.

BC (Before Computers)

I did not set out to write a comprehensive history of technical advances brought by the Age of Synergy; instead, I concentrated on the four classes of fundamental innovations that have created and largely defined the twentieth century. Even within those confines, there are many omissions and unavoidable simplifications. Undoubtedly, the most important set of innovations that I left out deals with what could be broadly described as modern information management. Seen from the vantage point of the early twenty-first century, this may not seem to be an important omission. After all, the enormous post-1970 advances in computing—exemplified by Moore's law (see Figure I.1.1)—have greatly overshadowed even those accomplishments that were achieved during the first three decades of computer evolution before 1971 (the year of the first microprocessor).

Those post-1971 developments—so promptly translated into widespread ownership of affordable personal machines and leading to the emergence and, after 1993, to very rapid adoption of the Internet and mobile telephony—appear to relegate pre-computer data management to the category of inconsequentially primitive tinkering. This would be a wrong interpretation of historic reality. Although semiconductors and microchips, the key components of modern computers, were invented only after World War II, we should not forget that, for the first two-thirds of the twentieth century, information management was directly beholden to innovations introduced well before 1914.

The two pre-World War I generations were not as epoch-making for data management as they were for other advances described in this book, but they were much more important than the perspective skewed by recent accomplishments would lead us to believe. This conclusion is justified not only because of the fundamental fact that the Age of Electronics could not arise without reliable and affordable electricity generation. Two important pre-World War I advances included automated handling and evaluation of massive amounts of statistical information and the invention and widespread diffusion of mechanical and electromechanical calculating devices. Both retained their importance until the early 1970s, and I have vivid memories on both accounts.

The first new skill I had to learn after we arrived in the United States in 1969 was to program in FORTRAN. At that time, Penn State was a proud owner of a new IBM 360/67, and there was only one way to talk to the machine: to spend hours at a keypunch producing stacks of programming and data input cards, a technique introduced for the first time by Herman Hollerith (1860–1929) during the 1890s (Hollerith 1894). And a few years later, when I needed to do some weekend calculations at home, I was still lugging from the university one of those portable yet not-so-light Burroughs machines. So here are at least brief reminders of the pre-World War I origins of two modern techniques that were instrumental in ushering in the computing age.

The unfinished construction of Charles Babbage's analytical engine has been the best-studied advance of the early history of automated calculation: more than a dozen books were written about it and its creators (Babbage and Augusta Ada Byron, Countess of Lovelace); a partial prototype was completed and is now displayed in the Science Museum in London. Curiously, a much more successful effort by George (1785–1873) and Edvard (1821–1881) Scheutz, who succeeded in finishing and selling (with difficulty) two of their machines, which could not only do complex calculations but also print the results, remains generally unknown (Lindren 1990).

Both Babbage and the Scheutzes aimed too high, but many simple calculating machines were designed and offered for sale throughout the nineteenth century (Chase 1980; History-Computer 2024). Some of them were admirable examples of clever mechanical design, but none of the devices that were offered before 1886 was easy to use, and hence none of them was in demand by the growing office market. The necessity to enter numbers with levers and the absence of printers were the most obvious drawbacks. The first key-operated calculator was the Comptometer designed by Dorr Eugene Felt (1862–1930). Its prototype was housed in a wooden macaroni box and held together by staples and wire, and its production by Felt & Tarrant Manufacturing began in 1886. More than 6,000 units were sold during the subsequent 15 years (Damer 2006). By 1889, Felt added a printer to make the first Comptograph, but the machine never sold well.

A prototype of the first very successful adding and listing device was completed by 1884, by William Seward Burroughs (1857–1898), whose work as a bank clerk motivated him to ease the repetitiveness of tasks and improve the accuracy of calculations. In 1886, Burroughs and his three partners formed the American Arithmometer Company (renamed Burroughs Adding Machine Company in 1905); its first adding machines had a printing mechanism activated by pulling a lever (Figure I.6.1). Initially modest sales rose rapidly after 1900, thanks to an expanded product line, and their cumulative total reached 50,000 machines by 1907 (with more than 50 models); two decades later it surpassed 1 million units (Cortada 1993).

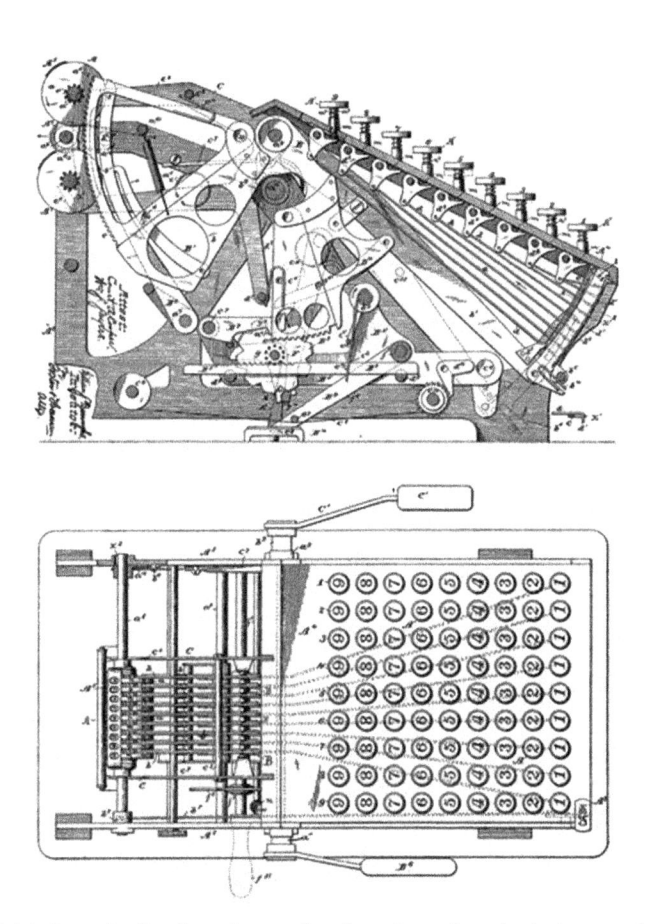

Figure I.6.1 Longitudinal section and a plan view of a calculating machine invented by William S. Burroughs and granted US Patent 388,116 in 1888.
Images available at http://www.uspto.gov.

Although there was no shortage of competition, Burroughs machines were the world's dominant calculators throughout the first half of the twentieth century. Class 1 models, introduced in 1905 and weighing nearly 30 kilograms (kg), were famous because of their glass side walls showing the mechanism inside the machine. Subsequent important milestones included the first portable device (9 kg) in 1925, the first electric key-actuated machine three years later, and the first account machine with programmed control panel in 1950 (History-Computer 2023). Two years later, a new era began as Burroughs built an electronic memory system for the ENIAC computer, but the company kept producing portable calculators throughout the 1960s. For more than eight decades, complex calculations, from astronomical ephemerids to artillery tables, were done by hundreds of different models of Burroughs machines and their competitors.

Hollerith's inventions were both derivative and original. His challenge was to find a practical automatic substitute for the highly labor-intensive, expensive, and increasingly protracted manual tabulation of US Census data, which depended on marking rolls of paper in appropriate tiny squares and then adding them up. Hollerith began to work on a solution in 1882 (after he joined MIT), and, in 1884 (after he moved to the US Patent Office), he filed the first of his more than 30 patent applications, which detailed how the information recorded on punch cards would be converted into electrical impulses, which in turn would activate mechanical counters (Hollerith 1894; Figure I.6.2). The storage medium chosen by Hollerith was first introduced in 1801, when Joseph-Marie Jacquard (1752–1834) programmed complex weaving patterns by means of stiff pasteboard cards with punched holes. Hollerith's punched cards could store an individual's census data on a single card, and his prototype used a tram conductor's ticket punch to make the holes.

His great innovation was not just to have large numbers of these cards read by an automatic machine but to devise means of tabulating the results and extracting information on any specific characteristics or on their revealing combinations. Reading was done by a simple electromechanical device, as spring-mounted nails passed through the holes and made contacts. Punched cards were used to process a large variety of statistical information and to keep records that ranged from the US censuses to particulars of prisoners in Hitler's Germany, where IBM's subsidiary Deutsche Hollerith Maschinen was taken over by the Nazis before World War II (Black 2001). The age of encoded paper ended with the rise of magnetic memories, whose advantage

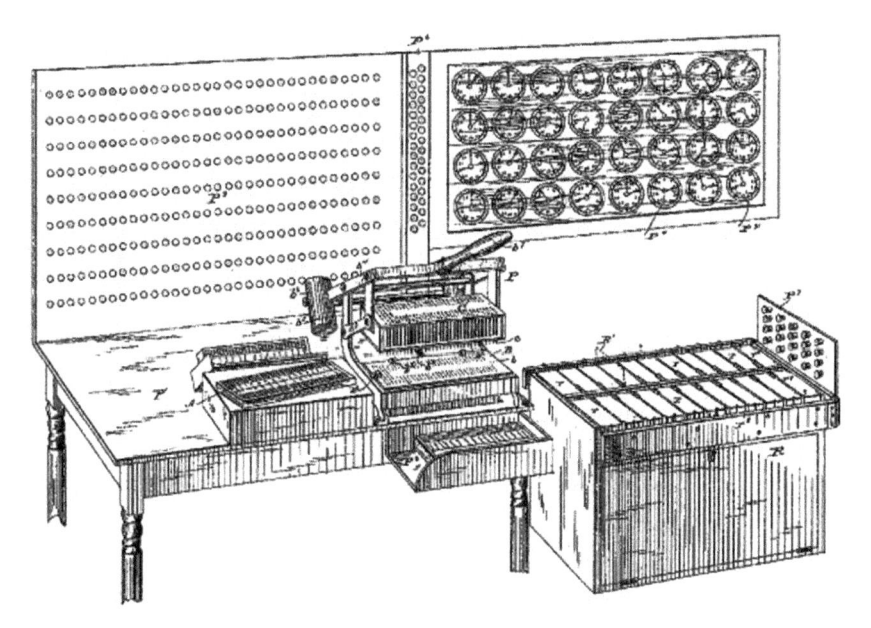

Figure I.6.2 Holerith's 1889 machine for compiling statistics (US Patent 395,781). Punched paper strips or cards placed on a stationary nonconducting bed-plate (B) with holes that contain embedded wires are read by lowering a movable platen of pins (C). The circuit wires from the bed-plate are connected to the switchboard (P_3). Counters (P_4, upper right hand) and sorting box (R) were other prominent features of the apparatus.

includes not only enormous storage capacity per unit volume but also the ease of reuse by erasing the stored information, an obvious impossibility with punched cards.

Triumphs, Tragedies, Foibles

Although I did sprinkle the text of topical chapters with brief references to some biographical facts, much more information could have been shared about the fascinating or mundane lives of great innovators and about their fate following their (often singular and brief) periods of fame. Not surprisingly, there were quite a few deserved triumphs as well as personal and family tragedies. And, as is rather common with creative individuals, many of them behaved in highly idiosyncratic ways, and, while some of their beliefs, quirks, and foibles were quaint or amusing, others made them appear unbalanced and even psychotic.

One of the most admirable qualities of many creators of a new technical age, and one of a few readily quantifiable marks of their intellectual triumphs, is their prodigious inventiveness. Edison's record was unparalleled during the Age of Synergy: 1,093 US and 1,239 foreign patents were granted between 1868 and 1931, with an additional 500–600 applications that were unsuccessful or abandoned (TAEP 2024). Nearly 40% of his US patents pertained to incandescent lights and to the generation and transmission of electricity, with recorded sound, telegraphy and telephony, and batteries being the other major activities, and dozens of patents were obtained for ore mining and milling and cement production.

Tesla's worldwide patent count surpassed 700. Frederick Lanchester, builder of the first British car and the inventor of disk brakes, held more than 400 patents. George Westinghouse, one of the creators of the electric era, amassed 361 patents, the most famous of which was the one for the compressed air brake (US Patent 88,929 in 1869) he designed after witnessing a head-on collision of trains between Schenectady and Troy. His other notable railroad-related inventions included automatic signaling and an automatic air and steam coupler; as already noted, he also had numerous patents in the field of new alternating current (AC) generation and transformation (Prout 1921). Among Nobel's 355 patents are substitutes for rubber and leather, perforated glass nozzles, the production of artificial silk, and the world's first aluminum boat. Ernst Alexanderson had 344 patents, the final one granted in 1973, when he was 95 years old!

These examples show that the triumphs of many great innovators of the Age of Synergy were not limited to a single class of devices or to a group of kindred advances within a particular field. Many multitalented thinkers and experimenters left behind entire catalogs of inventions and improvements strewn over their specialty and frequently extending over several fields. The Siemens brothers contributed to such disparate advances as regenerative furnaces, dynamos, and intercontinental telegraphs. Werner will be always best known for his dynamo, but his inventions also included electric rangefinders, mine exploders, and a continuous alcoholmeter that was used by the Russian government to levy taxes on the production of vodka (Siemens 1893). Tesla's first US patent, in 1888, was for an electric motor (US Patent 391,968; see Figure I.2.19); his last, in 1928 (US Patent 1,655,114), was for what we would call today a vertical short takeoff and landing aircraft.

Before he designed the first practical electric starter for cars, Charles Kettering patented a driving mechanism for cash registers (US Patent 924,616 in 1909) and later investigated the most suitable anti-knocking additives for gasoline, refrigeration, and humidity control (Jeffries 1960). Emile Berliner, the inventor of a loose-contact transmitter and gramophone, also patented a parquet carpet, porous cement tiles designed to improve the acoustics of concert halls, a lightweight internal combustion engine, and a tethered helicopter. Besides amplitude modulation and heterodyne, Reginald Fessenden also patented an electric gyroscope, sonar oscillator, and a depth finder.

Despite the keen perception and clever thinking that they displayed in technical matters, many inventors turned out to be dismal businessmen. Edison spent a fortune created by his numerous electrical inventions on his futile enterprise of mining and enriching iron ore in New Jersey. Lee De Forest was not the only inventor who spent a great deal of money on protracted law suits, but he was an exceptionally poor businessman: several of his companies failed; in 1903 he was accused of stealing one of Fessenden's inventions (and eventually found guilty of patent infringement); and, a decade later, he was tried with two of his business partners for misleading stock offerings (and found not guilty).

Other inventors were surprisingly susceptible to the fraudulent claims of assorted spiritualist movements that were much in vogue during the late nineteenth century. Two of the era's leading British scientists—William Crookes and Oliver Joseph Lodge—became committed spiritualists. Lodge (Figure I.6.3), who shared his spiritualistic enthusiasm with his friend Arthur Conan Doyle, believed that the unseen universe is a great reality to which we really belong and shall one day return (Lodge 1928), and he hoped that after his death he would be able to send a message from "the other side" by using a medium. Carl Kellner, one of the inventors of the sulfite papermaking process, was not only a devoted spiritualist but also a student of Asian mysticism (OTO 2024).

The political and social judgments of many famous inventors and industrialists of the era were questionable, some reprehensible. Henry Ford's anti-Semitism was of such an intensity that, in the early 1920s, his framed photograph hung on the wall of Hitler's Munich office and copies of the German edition of *The International Jew*, a series of articles that appeared in Ford's newspaper *The Dearborn Independent* (1922), were displayed on the future Führer's table (Baldwin 2002). And in 1935, before the Fascist invasion of Ethiopia, Louis Lumière dedicated his photograph to *il Duce*

Figure I.6.3 Oliver Joseph Lodge, one of the leading pioneers of radio age who promised to send signals from "the other side" after his death.

"*avec l'expression de ma profonde admiration*," while, in July 1941, the name of his brother Auguste appeared on the list of French ultra collaborators who created the Légion Volontaires Français to fight along with the Nazis (Davey 1971).

And while it is not surprising that a large and disparate group of pre-World War I innovators included individuals with dubious or even reprehensible beliefs, one of the last attributes one would associate with the creators of a new era would be their lack of imagination. Counterintuitively, this was not an uncommon failing. Recognizing excellence in a particular field is no guarantee that every technical judgment would be acute and rewarding; it is perhaps the most fascinating idiosyncrasy of the creative process to see that many minds that were so inventive and so open to radical experimentation could be surprisingly resistant to other and not even so shocking ideas. One demonstration of this conservative behavior is the reluctance, or an outright refusal, to carry some inventions even a step further. There is no special term for this attitude, but "Watt's syndrome" may be a good description.

Watt's experiments and insights turned an extremely inefficient Newcomen's machine with a very limited range of applications into the first useful mechanical prime mover energized by fuel combustion—but his 1769 patent and its subsequent 25-year extension impeded the next step of steam-driven innovation. Because Watt was afraid to work with high-pressure steam, he not only made no attempts to develop steam-driven transportation but also actively discouraged William Murdock, the principal erector of his engines, from doing so (Robinson and Musson 1969). Watt's patent expired in 1800, and, just a few years later, Richard Trevithick and Oliver Evans had high-pressure boilers ready, and the diffusion of railways, steamships, and higher efficiency stationary engines was on the way. Some 80 years later, Karl Benz displayed a classic Watt syndrome by refusing to have anything to do with either very fast-running engines or vehicles other than motorized horse carriages.

Marconi's reluctance to broadcast anything but Morse signals let others take the lead in developing radio. And I have also noted, respectively, in Chapters I.2 and I.1, Edison's two famous failures of imagination: his militant rejection of AC and his belief that electricity, rather than internal combustion engines, would be the dominant automotive prime mover of the twentieth century. In the first instance, Edison reversed himself quickly, but the second conviction gripped him for most of the pre-World War I period. Instead of trying to invent, for example, a reliable and lucrative electric starter for internal combustion engines (the device that was eventually designed in 1911 by Kettering), he persevered for years in his futile quest for a superior battery.

Another display of failed imagination is a surprisingly common lack of confidence regarding the potential importance of one's own inventions. Combative confidence and exaggerated expectations might be expected as a part of aggressive inventive process—but not an almost inexplicable diffidence and technical timidity. Among the most famous examples are Elisha Gray's initial decision not to pursue his rights to patenting a telephone because he believed that the device was just an interesting toy, Hertz's complaints to his students that the electromagnetic oscillations he just discovered could not possibly have any practical use (think of today's electronic universe!), and Louis Lumière's initial conviction that the cinema was an invention without any future as he expected that people would get bored watching his images of city scenes, trains, workers, and landscapes, most of which they could see just by walking around.

Besides these surprising failures of imagination, there were also some notable personal tragedies. While most of the era's inventors had fulfilling careers and led interesting lives, there were also suicides, premature deaths, and, more frequently, convictions of being unjustly treated. I have described Diesel's feelings of failure and his (almost certain) suicide (Chapter I.3). Ironically, by the time of Diesel's disappearance in 1913, his great invention was well set on its road to a commercial triumph, while an earlier sudden death may have changed the history of the world's most popular entertainment. In September 1890, Louis Le Prince boarded a train in Dijon, after a visit to his brother, and never made it to Paris. We will never know why, but we could ask: If Le Prince had lived for at least another five years, could he had perfected the camera with which he took, in 1888, what are arguably the world's first moving pictures, and would he be known today as the inventor of cinema?

Alfred Nobel did not commit suicide and did not die a mysterious death, but one of his early experiments with nitroglycerine killed his youngest brother, Emil, and four other people, and his near-chronic depression, despondency, and searing self-appraisal would have made Freud shudder. This is how, at age 54, when refusing his brother's request to write an autobiography, Nobel saw himself (cited in Fant 1993: 1):

Alfred Nobel—a pitiful creature, ought to have been suffocated by a humane physician when he made his howling entrance into his life. Greatest virtues: keeping his nails clean and never being a burden to anyone. Greatest weaknesses: having neither wife and kids nor sunny disposition nor hearty appetite. Greatest single request: to not be buried alive. Greatest sins: not worshiping Mammon. Important events in his life: none.

Two years later, in a letter to Sofie Hess, an Austrian flowers salesgirl whom he met as a 20-year-old in 1876 and who was his mistress for the next 18 years, he wrote (quoted in Fant 1993: 318): "[W]hat a sad end I am going toward, with only an old servant who asks himself the whole time if he will inherit anything from me. He cannot know that I am not leaving a last will." But, of course, he did, and probably the world's most famous one: it divided his diminished but still considerable riches among 19 relatives, coworkers, and acquaintances (20% of the total), various institutions (16%), and the Swedish Academy of Sciences in order to "constitute a fund, the interest on which shall be annually distributed in the form of prizes to those who, during the preceding year, shall have conferred the greatest benefit on mankind" (Nobel 1895).

While the aging Nobel was depressed but lucid, the last two decades of Tesla's life were marked by a deepening psychosis. He told journalists of his great love for a white pigeon ("Yes, I loved her as man loves a woman"), and, finally, a few days before his death on January 7, 1943, he sent a messenger with a sealed envelope containing money and addressed to Samuel Clemens (Mark Twain's real name) at 35 South Fifth Avenue. That was the address of Tesla's first New York laboratory of the late 1880s, and, in 1943, Mark Twain, whom Tesla wanted to help ("He was in my room last night. . . . He is having financial difficulties") had been dead for 33 years (Cheney 1981).

Mergenthaler found he had tuberculosis even before he was 40; he moved from Baltimore to New Mexico, but there his house and his papers were destroyed by fire, and he died in Baltimore before his forty-sixth birthday (Kahan 2000). Hertz died at the age of 36, after he became ill with an infection of the mouth and ears and underwent several unsuccessful surgeries. Wilbur Wright died of typhoid at age 45. Daimler's long and accomplished life had an ironic ending. He did not like to drive (and may have never driven at all), and his death came after he insisted, with his health failing, on being taken in poor weather in an open vehicle to inspect a possible site for a new factory. On the return trip he collapsed, fell out of the car, and died soon afterward.

But what beset his fellow inventor's family was infinitely more tragic. As Maybach's new powerful and elegant Mercedes 35 was breaking the world speed records in 1901 and 1902, his adolescent son Adolf (born in 1884) was succumbing to schizophrenia. His condition later deteriorated to such an extent that he could not be kept at home and had to be cared for in sanatoriums. Yet the worst was to come decades later: Adolf was murdered in 1940, 11 years after his father's death, as one of the thousands of victims of the Nazi *Euthanasieprogramm.* Who remembers this as the company is advertising the pride of German engineering, new luxury models of a great marquee selling for hundreds of thousands of dollars?

Not a few inventors discovered that even an extensive record of remarkable accomplishments did not automatically translate into widespread recognition and financial rewards. Fessenden felt slighted in both his native Canada and in the United States (Raby 1970). He got his first substantial reward only for his sonar patent and received a large settlement for his pioneering radio inventions only in 1928, after years of litigation and just four years before his death. Much as today, the willingness to engage in self-promotion and a dose of unorthodox behavior helped to capture press attention and to create an erroneous but greatly appealing image of a prototypical heroic inventor who appears to base his work on nothing but acute intuition and succeeds due to his uncommon perseverance.

Edison was the master of this game, while Tesla was too eccentric to play it successfully. This difference helps to explain why Edison remains, more than 70 years after his death, widely admired by a public that craves such suitably heroic figures and why Tesla's following—although highly, and some of it even fanatically, devoted to the memory of that master electrician—is much more cultlike, and why, outside of his native Serbia (where he will always be a national hero), it is largely limited to scientists and to individuals who are intrigued by his research into ultrahigh-frequency discharges, large-scale energy transmissions without wires, and death rays. However, Tesla got one honor that eluded Edison (and Carnot, Mayer, or Maxwell): the unit of magnetic flux density carries his name, and so he joins the most select company of scientists and engineers—including Ampère, Coulomb, Faraday, Hertz, Ohm, and Volta—whose names were chosen for international scientific units.

But public admiration and official honors went not only to such skilled self-promoters as Edison or Marconi but also to inventors whose work is now known only to historians of science and engineering. In the words of his brother, William Siemens (see Figure I.4.5) "forced the public opinion of England to honour him in his lifetime, and in a still more striking manner after his death" (Siemens 1893: 270). He was knighted, got honorary degrees from both Oxford and Cambridge, and, on November 21, 1883, the London *Times* obituary spoke of his "singularly powerful and fertile mind." His funeral service took place in Westminster Abbey, and later a window in the cathedral was dedicated to his memory: England's highest honors for an immigrant German engineer.

I end these reflections on personalities by noting a few idiosyncrasies that have been a surprisingly frequent accompaniment of the creative process. Edison could not just nap but sleep deeply just about anywhere—fully dressed in a much-crumpled three-piece suit lying on wooden desks and benches (Figure I.6.4) and on bare floors, much like an ascetic Chinese sage with only a bent arm for his pillow. During the construction of the world's first electric network, his company kept a large stock of tubes in the cellar of the station at Pearl Street. Edison recalled that "as I was on all the time, I would take a nap of an hour or so in the daytime—and I used to sleep on those tubes in the cellar. I had two Germans who were testing there, and both of them died of diphtheria, caught in the cellar, which was cold and damp. It never affected me" (quoted in Dyer and Martin 1929: 400).

Cherish the thought of perhaps the most influential inventor of a new era reposing on iron pipes in a dingy cellar—and try to imagine one of today's CEOs getting involved in the same—pun unintended—down-to-earth fashion. In contrast, there were Tesla's obsessive neuroses about germs and cleanliness: he eventually required even his closest friends to stand at a

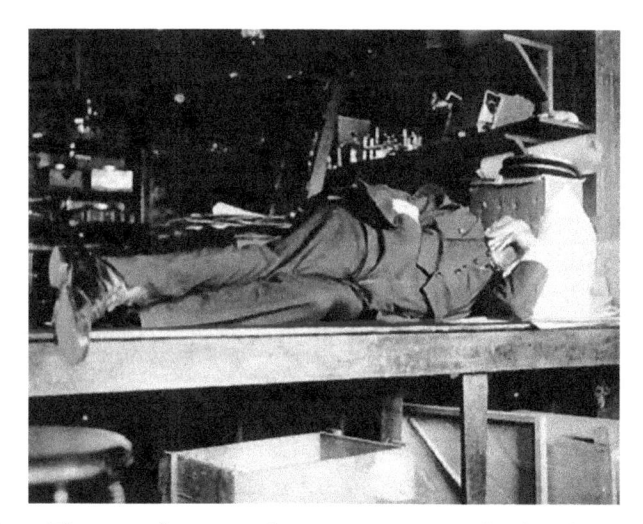

Figure I.6.4 Thomas Edison snatching a nap on a wooden bench.
Image available from the National Park Service.

distance lest they contaminate him—although he did not worry at all about close contact with thousands of pigeons, including many in his hotel rooms, that he cared for throughout his life (Cheney 1981). The Wright brothers were unsurpassed workaholics ("two of the workingest boys I ever saw," as one acquaintance put it) who had no friends outside their immediate family (Tobin 2003). In 1926, 14 years after Wilbur's death, their sister Katharine, 52 years old at that time, decided to marry Henry Haskell. Orville thought this a betrayal of the family, and he not only refused to attend the wedding but also cut all contacts with her until shortly before her death in 1929.

So Much Has Changed

Byrn (1896: 82) noted in his essay on technical progress that

> [i]t is so easy to lose sight of the wonderful, whence familiar with it, that we usually fail to give the full measure of positive appreciation to the great things of this great age. They burst upon our vision at first like flashing meteors: we marvel at them for a little while, and then we accept them as facts, which soon become so commonplace and so fused into the common life as to be only noticed by their omission.

And what an omission that would be. The best way to appreciate the enormity of technical advances that took place between 1867 and 1914 is to try to construct a modern world devoid of just 10 major achievements introduced during that era.

No *electricity* and hence no nonpolluting, convenient, and inexpensive lights and no electric motors to power myriads of precisely controllable machines, appliances, and trains; no *internal combustion engines* and hence no fast and affordable motorized vehicles, no freedom of personal movement, no inexpensive ocean transport, and no long-distance flight; no *reproduction of sound* and hence no telephones and music recordings; no *photographic film* and hence no convenient cameras and no movies; no way to generate, broadcast, and receive *electromagnetic waves* and hence no radio or TV; no *steel alloys* and no *aluminum* and hence no skyscrapers, no affordable machines and appliances; no easy way to produce *inexpensive paper* or to *reproduce images* and hence no mass publication of books and periodicals; and no *nitrogen fertilizers* and hence very widespread malnutrition, shortened life spans, and a world that could not feed its population.

Byrn (1896: 6) thought that this would leave "such an appalling void that we stop short, shrinking from the thought of what it would mean to modern civilization to eliminate from its life these potent factors of its existence." The void would be so profound that what would exist would be only a prelude to modern civilization that was already in place by the middle of the nineteenth century: a society relying on steam engines and draft horses, whose nights would be sparsely and dimly lit by kerosene and coal gas, whose dominant metal would be brittle cast iron, and whose best means of long-distance communication would be a telegraphic message.

And what was no less remarkable than this wealth of innovations was their lasting quality. Inevitably, not a few technical advances that were introduced during the Age of Synergy and that later came to dominate their respective markets were eventually replaced by new, superior twentieth-century designs (e.g., internal combustion engines displaced by gas turbines in long-distance flights) or were relegated to small niche markets (the gramophone is still preferred only by some audiophiles). But the opposite is true in a remarkably large number of cases where not only the basic design, but even particular forms and modes of operation have endured with minimal number of adjustments or even survived in virtually unchanged form.

Enduring Artifacts and Ubiquitous Innovations

Perhaps the most impressive way to illustrate the enduring qualities of so many pre-World War I artifacts is a do-it-yourself exercise of comparing many objects and machines that we use today with the designs introduced, and rapidly improved, four to five generations ago. Most people will never have a chance to compare an 1880s Thomson dynamo, or a large diesel engine built before 1910 with their modern counterparts. Consequently, such comparisons are made most conveniently with smaller objects of everyday use with which everybody is familiar. Light bulbs and spark plugs are excellent examples. A century later Edison could almost mistake a standard incandescent bulb for one of his own design.

During the 1990s, light bulbs were machine-made rather than mouth-blown, their filaments were different, and Edison's lamps were not filled with inert gases. But when hidden by frosting (which was applied externally to some early lamps since the 1880s, internally by General Electric as early as 1903), the only outward feature that would give an old lamp away immediately would be the glass tip surmounting the globe. The shape, size, and proportions of General Electric's basic 100-watt (W) Soft White bulb made in 2000 were very similar to Edison's lamps made during the 1890s (Figure I.6.5). A gently angled neck widens into a spherical top whose diameter

Figure I.6.5 Comparison of two light bulbs made a century apart: General Electric's standard frosted 100-W bulb made in 2000 (left) and Edison's low-voltage carbon-filament lamp from the 1890s (its flat buttonlike contact is a key dating feature).

accounts for just more than half of the light bulb's overall length; a metal screw with four turns is another shared feature.

More important, the lamp's basic operational ratings are virtually identical to specifications set down by Edison and Upton in 1879: consuming 100 W of electricity at 120 volts (V), the 2000 GE light drew the current of 0.83 amps (A) and had a resistance of 144.5 ohms (Ω). And, if we move to 1913 and compare that year's tipless, internally frosted, gas-filled, coiled tungsten-filament lamp with a 100-W Soft White made in 2000, we have two almost identical items. There may be no better example of a relatively complex artifact that has remained basically unchanged during the century of rapid technical innovation and that has been produced in billions of copies to become one of the most common possessions of the twentieth century.

Spark plugs made a century apart share every key component: a terminal nut that connects to a spark plug wire, the metal-core center electrode that projects from the porcelain insulator nose, the ground electrode welded to the threaded part of the shell that forms the plug's reach, and the hexagonal section in the upper part of the shell used to tighten the device with a spark plug wrench. One more example of an enduring design that is decidedly low-tech (Figure I.6.6). In 1892, William Painter invented a clever way

> for the sealing of bottles by using compressible packing disks and metallic caps, which have flanges bent into reliable locking engagement with annular locking shoulders on the heads of bottles, while the packing-disk is in each case under heavy compression and in enveloping contact with the lip of the bottle. (Painter 1892: 1)

And crimped metal bottle caps, still found on billions of beverage bottles, are also an excellent example of innovations that are not examined in this book, which concentrates on fundamental, first-order advances and on their direct derivatives that often followed the primaries with an admirable speed. Electricity generation, internal combustion engines, inexpensive steels, aluminum, ammonia synthesized from its elements, and transmission of electromagnetic waves are the key primaries; electric motors, automobiles, airplanes, skyscrapers, recorded sound, and projected moving images are the ubiquitous derivatives. But there was much more to the period than introducing, rapidly improving, and commercializing those epoch-making inventions. There were many new inventions of simple objects of everyday use whose mass manufacturing was made possible, or commercially

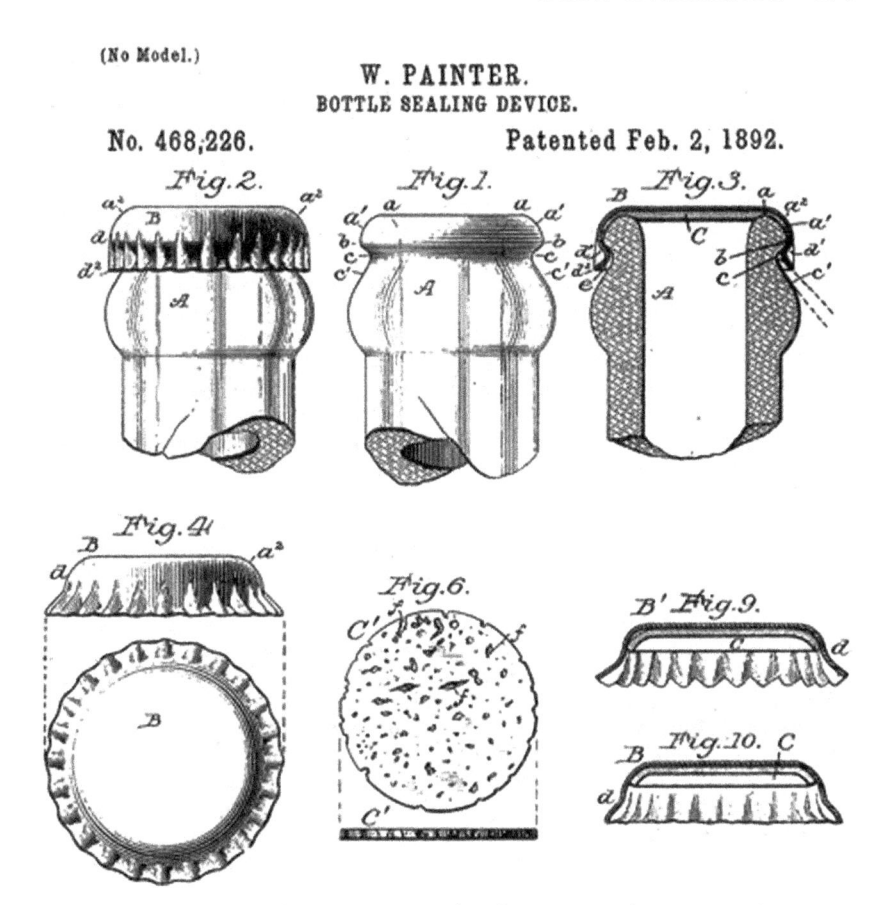

Figure I.6.6 Drawings that accompanied William Painter's patent application for a bottle-sealing device.

viable, thanks to the availability of new or better, and cheaper, materials. Consequently, those inventions can be classed as secondary or perhaps even tertiary derivations.

Their realization is based on one or more preceding fundamental innovations, but their commercial success owes no less to their ingenious designs. Many items belong to a large category of simple objects made from cheaper metals and better alloys. These include the following enduring classics, depicted in Figure I.6.7: the Gem paper clip (actually never patented) and other (patented) clip designs whose production required cheap steel wire and machines to bend it; scores (and eventually hundreds) of kinds of barbed wire (beginning with Joseph Glidden's pioneering 1874

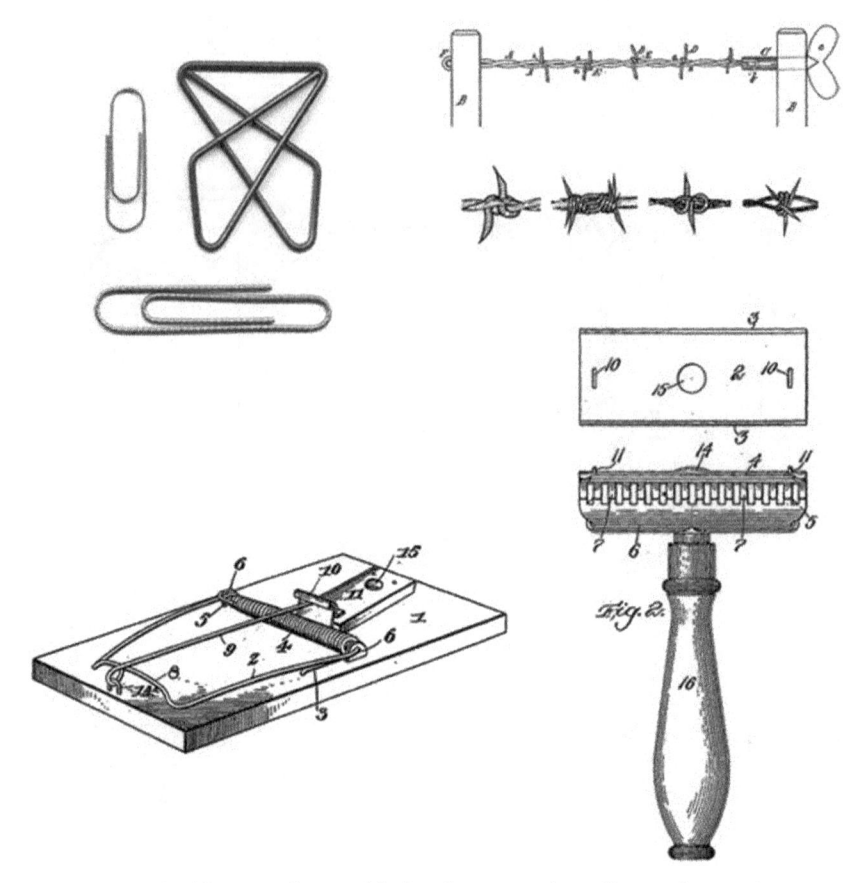

Figure I.6.7 Artifacts made possible by cheap, good-quality steel: two sizes of Gem paperclips introduced during the 1890s and a large Ideal clip from 1902; Glidden's 1874 barbed wire and details of several other elaborate twists; Hooker's "animal trap" of 1894; and King Gillette's razor designed in 1901. Gems and Ideal are scans of actual clips, other drawings are reproduced from their respective patent applications.

twist) made of galvanized steel (which, unlike round or oval iron wire, was extremely durable because of its high homogeneity and tensile strength); William Hooker's 1894 spring mouse trap; and King Gillette's 1904 razor blades made from tempered steel a mere 0.15 millimeters (mm) thick. This metallic group also included staplers (Charles Henry Gould, in 1868), sprinkler heads on showers (Harry Parmelee of New Haven, in 1874), Swiss Army knives (Victorinox company, set up by Karl Elsener in 1891), and zippers (plastic teeth came decades later).

In addition, many other products of persistent tinkering had little or nothing to do with the epoch-shaping primary innovations that were taking place during those eventful decades. The genesis of these numerous items of everyday use introduced during the Age of Synergy should be seen as further proof of the period's enormous wave of inventiveness. A reasonably complete list of these often very simple but enduring novelties would be tediously long. A few prominent examples include flat-bottom (instead of V-shaped) paper bags (1867), spring tape measures (1868), drinking straws (1888), and ice cream cones (1904). The last two items also point to a large category of dietary and culinary innovations whose introduction created entirely new consumption habits.

Now world-famous branded foodstuffs that emerged in the United States during the two pre-World War I generations range from cheese to chocolate. Empire Cheese Company began selling Philadelphia Cream Cheese in 1880. In 1906, Will Keith Kellogg, brother of the notorious electric shock and enema lover John Harvey Kellogg, added sugar to his sibling's ascetic corn flakes (patented in 1896) and began marketing them as breakfast cereal (Powell 1956). Campbell's soup trademark was registered in the same year; Milton Hershey began selling his Milk Chocolate Bar in 1894, and his rival Frank Mars opened for business in 1911 (Brenner 1999). Any list of widely consumed new generic foodstuffs should be headed by two leading contributors to the late twentieth-century increase of obesity: hamburgers, whose US debut is variously dated between 1885 and 1904 (McDonald 1997), and American-style (sausage- and cheese-loaded) pizza: first New York pizzerias date to 1890s (Myhrvold and Migoya 2021).

The world of drink was enriched (if you share the taste for thirst-inducing, syrupy, excessively sweetened, and explosively carbonated liquids that leave peculiar aftertastes) or burdened (if your preferences run more into good tea, mineral water, or fruit juices) by Coca-Cola and Pepsi Cola. Coca-Cola was invented in 1886, by Atlanta physician John S. Pemberton. Pemberton sold rights to this "intellectual beverage and temperance drink" to a pharmacist, Asa Briggs Candler, in 1891, and, by 1894, the first bottled cola was available in US drugstores (Hoy 1986). Caleb Bradham's concoction of water, sugar, vanilla, essential oils, and cola nut extract was renamed Pepsi Cola in 1898. Yet another notable beverage brand that has endured is Canada Dry Ginger Ale, formulated in 1904 by Toronto pharmacist John McLaughlin. That brand is now owned by Keurig Dr. Pepper, and that excessively sweetened liquid is the oldest major US soft drink, created in 1885 by

Charles Alderton at Morrison's Old Corner Drug Store in Waco, Texas (Dr Pepper Museum 2024)

And there were also new pastimes ranging from a variety of board games (Hofer and Jackson 2003) made affordable by cheaper paper and printing to new sports. This included basketball (James Naismith's 1891 invention) and two of my favorites, cross-country skiing and tennis. Primitive skis, descendants of snowshoes, were around for millennia, but the sport of ski-running (or Nordic skiing) began in 1879, with the Huseby races near Christiania (today's Oslo), and 13 years later the first Holmenkollen festival attracted more than 10,000 spectators. Similarly, tennis has a venerable pedigree (medieval *jeu de paume*), but the modern version was played for the first time in December 1873, in north Wales, and patented shortly afterward (GB Patent 685/1874) by Walter Clopton Wingfield. The only major modification came in 1877, at the first All-England Lawn Tennis Championships at Wimbledon, when the original 26-meter (m)-long hourglass field became a 24 × 17 m rectangle; later the net height was lowered by 10 centimeters (cm) to 90 cm (Alexander 1974).

Given the multitude of innovations that flooded in during the Age of Synergy, it is not difficult to construct narratives of today's life whose minor material ingredients and mundane actions originated not just during that inventive period but, more restrictively, even just within a single eventful decade. Here is an example constrained to the 1880s, one that does not refer directly to the decade's most important classes of inventions (electricity generation, transmission and conversion, new materials, internal combustion engines). A man wakes up late one day in one of America's large East Coast cities. First, he makes a cup of Maxwell House coffee (a brand introduced by Joel Cheek, a Nashville hotelier, in 1886) and then he eats quickly cooked Quaker Oats (the company began selling prefilled packages in 1884). His late breakfast is interrupted by a doorbell: an Avon lady is calling (they have been doing that since 1886), but his wife is out of town.

He finds that his only clean shirt needs a bit of ironing (Henry Seely patented that useful device in 1882), uses an antiperspirant (introduced in 1888), dresses up, and finds that he is out of brown paper bags (kraft process to make strong paper was first commercialized during the 1880s) to bring his customary lunch. He takes a long walk to his office, and, once he gets downtown, he enters a multistory steel-skeleton building (Jenney completed the first structure of this kind in Chicago, in 1885) through a revolving door (Theophilus Van Kannel put the first one into a building lobby in

Philadelphia, in 1888). He stops in a small drugstore where a bored teenager is turning pages of *Cosmopolitan* (it appeared first in 1886) and buys a copy of *Wall Street Journal* (published since 1889). He must wait while the teenager fiddles with his cash register (introduced by James Ritty and John Birch in 1883).

Then he takes an elevator (Elisha Otis installed the first high-speed lift in New York, in 1889) to his floor, but, before getting to his office, he stops at a vending machine (Percival Everitt introduced it in 1881) and buys a can of Coca Cola (formulated in Atlanta, in 1886). The first thing he does in his office is make a long-distance phone call (possible since 1884), jots down a few notes with his ballpoint pen (John Loud patented the first one in 1888). We leave him at that point, the case clearly made: so many of our everyday experiences and actions were set and so many artifacts (major and minor ones) that help us to cope with them were introduced during the Age of Synergy that most people are not even remotely aware of the true scope of this quotidian debt.

Life Cycles of Innovations

Numerous as they are, those basically unchanged original designs constitute a minority of advances introduced during the two pre-World War I generations. Most of the innovations have undergone changes ranging from minor adjustments to fundamental transformations. As already stressed in Chapter I.1, this quest for improvement—motivated by factors that ranged from desire to capture larger markets to the challenges of finding the most elegant engineering solutions—was one of the key marks of the era and one of its most important bequests to the twentieth century. Its many forms took different approaches by reducing energy and material intensity and by providing longer durability, higher reliability, improved conversion efficiency, and greater versatility. As these aspects will be examined in detail in the companion volume, the only important matter I address here is the apparent inevitability and regularity of many key trends.

Perhaps the most remarkable aspect of the historical continuity of technical innovation is that so many subsequent advances appear to have the inexorability of water flowing downhill. Once the basic ideas were formulated by innovative thinking and tested by bold experiments, it was only a matter of time before they were perfected, diffused, and amplified to reach entirely

Figure I.6.8 Development of an idea. Top: Eugene Ely's Curtiss biplane nears the landing platform on the USS *Pennsylvania,* anchored in San Francisco Bay, on January 18, 1911. Arresting lines were fastened to sandbags positioned along the deck. Photograph from NASM (2024). Bottom: An F/A–18F Super Hornet nears landing aboard USS *Nimitz* in the Persian Gulf.
US Navy photograph (030331-N-9228K-008.jpg) by Michael S. Kelly.

new performance levels. Look at the photograph of a small Curtiss pusher biplane, piloted by Eugene Ely on the morning of January 18, 1911, as it is about to land on a temporary wooden platform that was laid over the deck and gun turret of the Pacific Fleet's armored cruiser *Pennsylvania* (Figure I.6.8; NASM 2024b). The ship is anchored off the San Francisco waterfront, with thousands of spectators ashore. An updraft lifts Ely's light plane just as it reaches the platform, but he compensates quickly, snags the arresting gear (just a series of ropes crossing the deck and weighed down by sandbags), and pulls to a smooth stop before reaching the safety barrier.

How that image evokes all those torpedo bombers coming back to their carriers to rearm and take off again on their missions to sink Japanese ships during World War II! How conceptually identical, despite all the enormous intervening technical advances, is this scene compared with F-18s returning to America's nuclear-powered carriers stationed in the Persian Gulf (Figure I.6.8). These constants remain: a plane, a ship, a landing deck, an arresting gear, a skillful and alert pilot. Once imagined, and once shown to be practicable, the idea—in this instance not just figuratively—takes off and reaches an execution that was entirely unimaginable by its inventors but that is nevertheless unmistakably present in the original creation.

Another convincing comparative approach is to ask a simple question: Would it have been possible to stop further developments of those newly launched techniques and freeze them at any arbitrary levels of performance and complexity? And the obvious answer is "no." Hughes (1983) conceptualized the process that follows an invention in four stages: transfer of new techniques to other places and societies, formative system growth followed by reaching a new momentum (that arises from both the accumulated mass of new machines and infrastructures and the velocity of their diffusion), and finally, a qualitative change of mature techniques.

Once Edison mastered reliable electricity generation and once Tesla and Hertz opened the ways toward innovative electricity applications, it was only a matter of time before the entire electricity-driven universe of modern machines and devices, qualitatively so superior to the initial designs, was put in place. The timing of the key advances of this great transformation was undoubtedly contingent on many external factors, but their eventual attainment was a matter of very high probability. Similarly, once Karl Benz, Gottlieb Daimler, and Wilhelm Maybach mounted their high-speed gasoline motors on light carriagelike chassis, six-lane highways clogged by millions of increasingly sleeker steel machines were only a matter of three generations away.

Historical studies repeatedly demonstrate that the evolution of individual techniques or systems frequently follows an orderly progression much as living organisms do. Initially slow growth accelerates and then slows down as it approaches a limit and eventually stops. Fittingly, it was during the Age of Synergy when Gabriel Tarde (1843–1904), a French sociologist, first described this S-shaped growth of innovations.

> A slow advance in the beginning, followed by a rapid and uniformly accelerated progress, followed again by progress that continues to slacken

until it finally stops: these, then, are the three ages of those real social beings which I call inventions or discoveries. (Tarde 1903: 127)

This kind of progression also applies to the life histories of all organisms, and the resulting patterns of development are very close to one of the growth curves: I have documented these realities with scores of examples in my comprehensive review of growth in nature and in history (Smil 2019). But this is not necessarily the case with life cycles of techniques and their performance parameters: technical advances do not have lives of their own, and their evolution is not governed by some internal calls that produce growth stages of relatively fixed proportions (Ayres 1969). This becomes particularly clear when attempts are made to fit one of the standard growth curves to a historic dataset. Some innovations that originated during the Age of Synergy followed a regular progression that conformed rather closely to a growth curve, while others departed from it quite significantly.

Efficacy of incandescent lights between 1880 and 1930 is a good example of a symmetrical S-curve, with the midpoint at around 1905, while the best fit for the highest conversion efficiency of steam turbines during the same period is obviously a straight line (Figure I.6.9). Many growth patterns of innovations introduced during the Age of Synergy, exemplified in Figure I.6.9 by the maximum US steam turbogenerator ratings and by the highest US transmission voltages, form asymmetrical curves with the central inflection points shifted to the right as the early exponential growth rates were reduced by disruptions attributable to the interwar economic crisis. Indeed, these patterns could be better interpreted as two successive S-waves, with the first one reaching its plateau during the 1930s.

Commercial penetration rates of new techniques display orderly progressions that are dictated by the necessities of developing requisite manufacturing and distribution facilities and, in many cases, also putting in place the necessary infrastructures (roads, transmission lines). Not surprisingly, costly infrastructural needs or relatively high capital costs will tend to slow down the rate of adoption; for example, it was only after World War II when virtually all of America's rural households became electrified. In contrast, some industrial adoptions proceeded quite rapidly: electric motors captured half of the prime mover capacity in US manufacturing just three decades after Tesla's first devices were made by Westinghouse (see Figure I.2.21).

A NEW CIVILIZATION   303

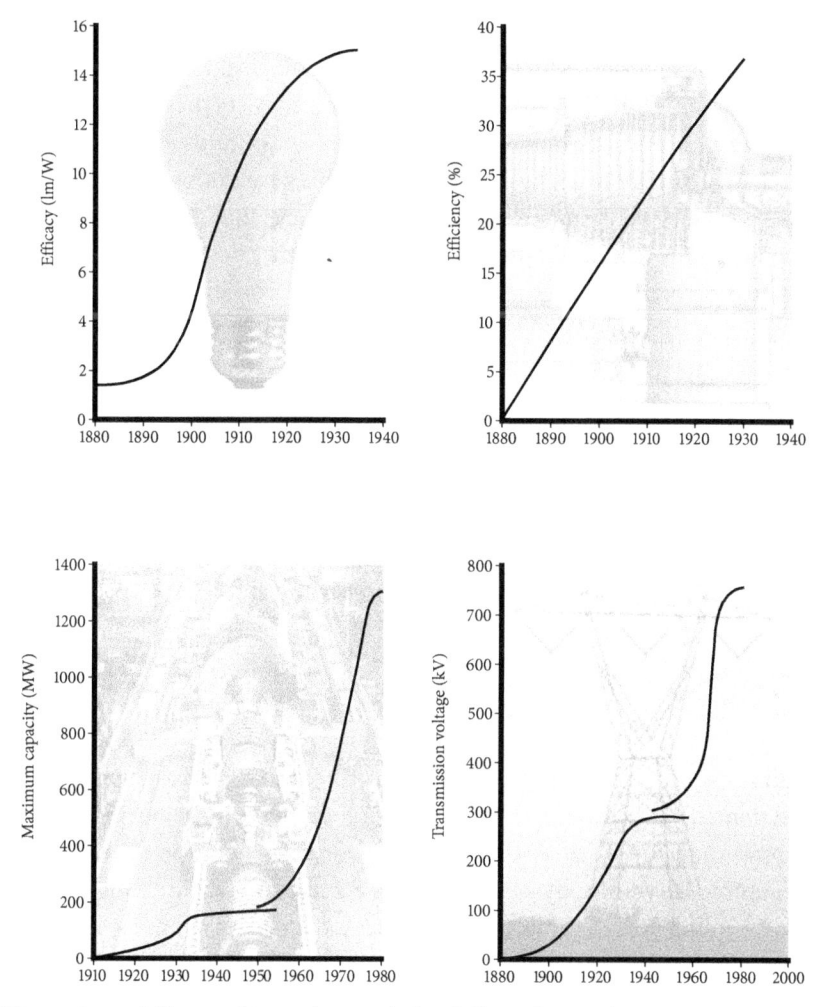

Figure I.6.9 Efficacy of incandescent lights followed a regular growth curve between 1880 and 1930 (top left), while the efficiency of best steam turbines progressed in linear fashion (top right). Histories of the highest US transmission voltages (bottom left) and the largest thermal turbogenerators (bottom right) show two successive growth curves.
Based on data and figures in Smil (2017) and Termuehlen (2001).

Once individual techniques, or their functional assemblages, reach the limits of their growth, they do not, like individual organisms, face inevitable decline and demise. Another biological analogy then becomes appropriate as they behave as mature, climax ecosystems that can maintain their performance for extended periods of time until their eventual displacement by a suite of superior techniques (a process analogical to the diffusion of new species in an ecosystem altered by climate). Some of these substitutions are gradual, even inexplicably tardy. Replacement of household incandescent lights by more efficient options is a notable case of this slow pace: the former converters, despite their inferior efficiency, continued to dominate the market throughout the twentieth century. Other substitutions, such as the adoption of color TV or compact discs (CDs), proceeded rapidly.

Technical substitutions may be accelerated or hindered by economic and political factors, but, as with the growth of individual techniques, their progress is often very orderly as inventions enter the market, come to dominate it, and then, shortly after their shares peak, begin to yield to new techniques. This wavelike progression has been also the case with transitions to new sources of primary energies and new prime movers, the two processes that have been, together with ubiquitous mechanization and mass production, among the most notable markers of a new technical era.

Markers of a New Era

Qualitative appraisals of fundamental socioeconomic changes are done by systematically presenting all relevant trends, carefully choosing the descriptive adjectives, and thoughtfully selecting noteworthy examples to capture the wholes by resorting to the specifics. Quantifying epochal changes in macro-economic terms is much more elusive as aggregate measures and reductions to growth rates hide too much and subsume too many complexities in single figures. Moreover, many pre-World War I innovations had almost immediate and far-reaching economic and social impacts, while others needed many decades before they came to dominate their respective markets.

Consequently, no comparisons of economic and social indicators of the late 1860s with those of the immediate pre-World War I years should be taken as representative measures of advances attributable solely to the great technical saltation of that era. Inevitably, most of the economic gains of the 1870s had their investment and infrastructural roots in years preceding the beginning of this period, while most of the innovations introduced after 1900 made their full socioeconomic mark only after World War I. That is why in this section I proceed along both quantitative and qualitative lines.

This book's primary concern is to detail the extent and the lasting consequences of the unprecedented number of fundamental pre-World War I technical advances. There is no adequate means of quantifying that process, but the history of patenting in the country that came to dominate this modernization process offers a valuable proxy account of these accomplishments. The first numbered US patent (for traction wheels) was issued in 1836, and the subsequent steady growth brought the total of successful applications granted to US residents to almost 31,000 by the end of 1860. After a three-year dip in the early 1860s came a steep ascent, from 4,638 patents in 1864 to 12,301 patents in 1867, and then the annual grants reached a new plateau of high sustained inventiveness at 12,000–13,000 cases a year, a fact that provides additional support for my timing of the beginning of the Age of Synergy (USPTO 2024; Figure I.6.10).

The rapid growth of patenting recommenced by 1881, and soon afterward it formed a new, and somewhat more uneven, plateau (mostly with 19,000–23,000 grants per year) that lasted until the end of the 1890s. US Patent 500,000, for a combined flush-tank and manhole, was issued in June 1893.

Figure I.6.10 US patents issued between 1836 and 1914. Notice the unprecedented rise in the annual rate of patenting that took place during the mid-1860s and the second wave of acceleration that began in 1881.
Plotted from data in USPTO (2024).

The annual rate of 30,000 grants was surpassed for the first time in 1903; US Patent 1,000,000 was issued in August 1911 (to Francis Holton for his vehicle tire), and before the beginning of World War I the total surpassed 1,100,000. Annual grants for the period 1900–1914 averaged more than 32,000, or more than three times the mean of the late 1860s, a convincing sign of intensifying inventive activity.

I do not hasten to undercut the conclusion I have just made, but I must reiterate (see Chapter I.1) that this simple quantitative focus may be somewhat misleading. The era's flood of unprecedented inventiveness also included many not just trivial but also outright ridiculous patents that should not have ever been granted (Brown and Jeffcott 1932). Among the choicest examples of the latter category are mad combinations: of match-safe, pincushion, and trap (US Patent 439,467 in 1890) and (yes, this is an exact citation) of grocer's package, grater, slicer, and a mouse and fly trap (US Patent 586,025 in 1897). Chewing-gum locket, tapeworm trap, device for producing dimples, and electrical bedbug exterminator ("electricity will be sent through the bodies of the bugs, which will either kill them or startle them, so that they will leave the bedstead," according to the US Patent 616,049 of February 7, 1898) are among my other favorites.

But there is also no doubt that many technical advances of the era had prompt and profound economic effects that were reflected in some impressive growth and productivity gains. Aggregate national accounts are the favorite indicators of economic growth, but such figures have obvious limitations of coverage and comparability. Consequently, they should be seen merely as indicators of basic trends, not as accurate reflections of all economic activity. Standard series recalculated in constant monies show that, between 1870 and 1913, the US gross domestic product grew roughly 5.3 times and that the multiples were 3.3 for Germany, 2.2 for the United Kingdom, and 2.0 for France (Maddison 1995). This growth translated, respectively, to 3.9%, 2.8%, 1.9%, and 1.6% a year. Population growth—already slow in Europe but rapid in the United States (mainly due to large immigration)—made per capita rates much more similar, with annual means of 1.8% for the United States, 1.6% for Germany, 1.5% for France, and just 1.0% for the United Kingdom.

Faster growth recorded by countries that started to industrialize aggressively only after 1850 is also shown by comparing the gains of gross domestic product per hour worked. Between 1870 and 1913, this indicator averaged 1.2% a year in the United Kingdom, while the German rate was

1.8% and the Japanese and US averages reached 1.9% (Broadberry 1992). Rising productivity was accompanied by shorter work hours: their annual total in Western economies began to decline after 1860, from nearly 3,000 to about 2,500 by 1913, and, by the 1970s, they were below 1,800 in all affluent countries (Maddison 1991). This trend was strongly influenced by declining employment in farming activities, but, just before World War I, agriculture still contributed about a third of gross domestic products in most Western countries and employed more people than did services sectors. And the term "service" called to mind the still very common household help rather than an array of activities that now account for the bulk of the Western economic product.

Average American wages rose rather slowly during the decades between the 1860s and World War I (BLS 1934), but their relatively modest progress must be seen against the falling cost of living. After a period of pronounced inflation, the American index of general price levels began to fall in 1867; then, three decades of deflation lowered it by about 45% by 1896, and renewed inflation drove it up by 40% by 1913 (NBER 2003). French industrial wages nearly doubled between 1865 and 1913, and German wages grew 2.6-fold, but British earnings went up by only 40% (Mitchell 1998). Rising economic tides of the two pre-World War I generations lifted all the boats—but even in the relatively most affluent Western countries, average disposable incomes were still very low when measured by the standards of the late twentieth century. Real incomes averaged no more than 10–15% of 1990s levels, and, despite the falling cost of food, typical expenditures on feeding a family still claimed a large share (around half) of average disposable urban income.

But none of these indicators convey adequately the epochal nature of post-1860s developments. This is done best by focusing on the two trends that became both the key drivers and the most characteristic markers of the Age of Synergy. Most fundamentally, for the first time in human history, the age was marked by the emergence of high-energy societies whose functioning, be it on mundane or sophisticated levels, became increasingly dependent on incessant supplies of fossil fuels and on a rising need for electricity. Even more important, this process entailed several key qualitative shifts, and, more than a century later, it still has not run its full course even in the countries that were its pioneers. The same conclusion is true about the other fundamental trend that distinguishes the new era: mechanized mass production that has resulted in mass consumption and in growing global interdependence.

Mass production of industrial goods and energy-intensive agricultures that yield surpluses of food have brought unprecedented improvements to the overall quality of life, whether they are judged by such crass measures as personal possessions or by such basic existential indicators as morbidity and mortality. An increasing portion of humanity has been able to live in societies where a large share, even most, of effort and time is allocated to providing a wealth of services and filling leisure hours rather than to producing food and goods. Both rising energy needs and mass production provided strong stimuli for the emergence of intensifying global interdependence, and this resulted in both positive feedback and negative socioeconomic and environmental effects.

High-Energy Societies

Thermodynamic imperatives and historical evidence are clear: rising levels of energy consumption do not guarantee better economic performance and higher quality of life—gross mismanagement of Russia's enormous energy wealth (both under the USSR and in the post-Soviet Russia) is perhaps the most obvious illustration of the fact—but they are the most fundamental precondition of such achievements. Traditional biomass energies and human and animal muscles (aided by limited exploitation of water and wind by mills) could secure the basic material necessities of life, but there was no possibility of reliable food surpluses, larger-scale industrial production, mass consumption, prolonged education opportunities, high levels of personal mobility, and increased time for leisure. A high correlation between energy use and economic performance is the norm as individual countries go through successive stages of development, and the link applies to broader social achievements as well.

New sources of primary energy and new prime movers were essential to initiate this great transition. The shift began inauspiciously in the United Kingdom during the seventeenth century and accelerated during the eighteenth century with the rising use of coal, the invention of metallurgical coke for iron smelting, and James Watt's radical improvement of Newcomen's inefficient steam engine (Smil 2017a). Even so, by 1860, coal production remained limited as the United Kingdom was the only major economy that was predominantly energized by that fossil fuel. Traditional biomass fuels continued to supply about 80% of the world's primary energy, and, by 1865, the United States still derived more than 80% of its energy needs from wood and charcoal (Schurr and Netschert 1960; Smil 2017b).

Typical combustion efficiencies of household fireplaces and simple stoves were, respectively, below 5% and 15%. Steam engines, which were diffusing rapidly both in stationary industrial applications and in land and sea transportation, converted usually less than 5% of coal's chemical energy into reciprocating motion, and small water turbines were the only new mechanical prime movers that were relatively efficient. Although English per capita consumption of coal approached 3 tons (t) per year during the late 1860s (Humphrey and Stanislaw 1979), and the US supply of wood and coal reached nearly 4 t of coal equivalent during the same time (Schurr and Netschert 1960), less than 10% of these relatively large flows were converted into space and cooking heat, light, and motion.

The Age of Synergy changed all of that, and the United States was the trendsetter. Expansion of the country's industrial production and the growth of cities demanded more coal, and in turn, industrial advances provided better means to extract more of it more productively. Internal combustion engines created a potentially huge market for liquid fuels, and newly introduced electricity generation was ready to use both coals and hydrocarbons (as well as waterpower) to satisfy a rapidly rising demand for the most convenient form of energy. Deviation-amplifying feedback of these developments resulted in an unprecedented increase of primary energy consumption, the category that includes all fossil fuels, hydroelectricity, and all biomass energies. During the two pre-World War I generations, energy consumption per capita total rose nearly twofold in England Wales, it doubled in the Netherlands—and it had nearly quintupled in Germany (Kander, Malanima, and Warde 2013).

In the United States, the total primary energy consumption rose more than fivefold during the two pre-World War I generations, but the country's rapid population growth, from about 36 million people in 1865 to just more than 97 million in 1913, reduced this to less than a twofold (1.8 times) increase in per capita terms. But these per capita primary energy multiples are very misleading because this simple quantitative contrast hides great qualitative gains that characterized energy use during the Age of Synergy. What matters most is not the total or per capita amount of available energy but the useful power that provides desired energy services. Substantial post-1870 improvements in this rate came from the combination of better performance of traditional conversions and the introduction of new prime movers and new energy sources. A few sectoral comparisons reveal the magnitude of these gains that accompanied the epochal energy transition.

Higher efficiencies in household heating and cooking did not require any stunning inventions, merely better designs of stoves made with inexpensive steel and the large-scale replacement of wood by coal. Heat recirculation and the tight structures of new coal or multifuel stoves of the early twentieth century raised efficiencies commonly 40–50% above the designs of the 1860s. Typical efficiency of new large stationary steam engines rose from 6–10% during the 1860s to 12–15% after 1900, a 50% efficiency gain, and, when small machines were replaced by electric motors, the overall efficiency gain was typically more than fourfold.

Because of transmission (shafting and belting) losses, only about 40% of power produced by a small steam engine (having 4% efficiency) would do useful work (Hunter and Bryant 1991), and another 10% of available power would be wasted due to accidental stoppages. Useful mechanical energy was thus only about 1.4% (0.04 × 0.4 × 0.9) of coal's energy content. Despite a relatively poor performance of early electricity generation (efficiencies of no more than 10% for a new plant built in 1913) and 10% transmission losses, a medium-sized motor (85% efficient) whose shafts were directly connected to drive a machine had an overall energy efficiency of nearly 8% (0.1 × 0.9 × 0.85). Coal-generated electricity for a medium-size motor in the early 1910s thus supplied at least five times as much useful energy as did the burning of the same amount of fuel to run a small steam engine of the 1860s.

Installing internal combustion engines in place of small steam engines would have produced at least two to three times as much useful energy from the same amount of fuel, and post-1910 efficiencies of steam turbines (the best ones surpassed 25% by 1913) were easily three times as high as those of steam engines of the 1860s. Higher energy efficiencies also made the enormous expansion of American ironmaking (nearly 30-fold between 1865 and 1913) possible. By 1900, coke was finally dominant, and its production conserved two-thirds of the energy present in the charged coking coal (Porter 1924). The shift from charcoal of the 1860s to coke of 1913 brought a 50% gain in energy efficiency, and better designs and heat management nearly halved the typical energy intensity of blast furnaces, from about 3 kg of coal equivalent per kilogram of pig iron in 1860s to about 1.6 kg by 1913 (Smil 2016).

This means that the overall energy costs of American pig iron production were reduced by about two-thirds. Finally, a key comparison illustrating the impressive efficiency gains in lighting: candles converted just 0.01% of paraffin's chemical energy into light, and illumination by coal gas (average yields of around 400 m³/t of coal, and typical luminosity of the gas at about 200 lumens [lm]) turned no more than 0.05% of coal's energy into light. By 1913, tungsten filaments in inert gas converted no less than 2% of electricity into light, and, with 10% generation efficiency and 10% transmission losses, the overall efficiency of new incandescent electric lighting reached 0.18% (0.1 × 0.9 × 0.02), still a dismally low rate but one nearly four times higher than for the gas lighting of 1860s!

Information available about pre-World War I sectoral energy consumption is not detailed enough to come up with an accurate weighted average of the overall efficiency gain. But my very conservative calculations, using the best available disaggregation of final energy use and the composition of prime movers, show that there was at least a twofold improvement of energy conversion in the US economy between 1867 and 1913. America's effective supply of commercial energy thus rose by an order of magnitude (at least 11-fold) during the two pre-World War I generations, and average per capita consumption of useful commercial energy had roughly quadrupled, with Germany experiencing even higher improvement

Such efficiency gains were unprecedented in history, and such rates of useful energy consumption provided the foundation for the country's incipient affluence and for its global economic dominance. In 1870, the United States consumed about 15% of the world's primary commercial energy and the country's output accounted for roughly 10% of the world's economic product; by 1913, the respective shares were about 45% and 20% (UNO 1956; Schurr and Netschert 1960; Maddison 1995; Smil 2017b). This means that the average energy intensity of US economic output rose during that period, an expected trend given the enormous investment in urban, industrial, and transportation infrastructures. Similar trends could be seen with the energy intensities of the Canadian and German economies.

No less important, this transition was coupled with qualitative improvements in the energy supply. As noted in Chapter I.1, commercial energies began supplying more than half of the world's energy use sometime during the late 1890s; for the United States that milestone was reached during the early 1880s, and, by 1914, less than 10% of the country's primary energy was from wood while about 15% of America's fossil energies came from crude oil and natural gas. The United States pioneered the transition from coal to hydrocarbons, which was driven by both a rapid diffusion of a new prime mover (power installed in internal combustion engines surpassed that in all other prime movers before 1920) and the higher quality and greater flexibility of liquid fuels. Crude oil's energy density is nearly twice as high as that of good steam coal (42 vs. 22 GJ/t), and the fuel and its refined products is easily transported, stored, and used in any conceivable conversions, including flight (Smil 2017b).

At the beginning of the twentieth century, oil resources of the pioneering fields of Pennsylvania (extraction since 1859), California (1861), the Caspian Sea (Baku 1873; Figure I.6.11), and Sumatra (1885) were augmented by new major discoveries in Texas and the Middle East (Perrodon 1981). On January 10, 1901, the Spindletop well southwest of Beaumont gave the first sign of oil production potential in Texas; by the end of the twentieth century, the state still produces a fifth of America's oil (and more comes from its offshore fields in the Gulf of Mexico). The first giant oilfield in the Middle East was discovered on May 25, 1908, in Masjid-i-Suleiman in Iran; a century later the region was producing a third of the world's crude oil, and it holds nearly two-thirds of all petroleum reserves (EI 2024).

The other key qualitative energy shift that was pioneered by the United States was the rising share of fossil fuel energy consumed indirectly as electricity. This share rose from, obviously, zero in 1881 to about 4% by 1913,

Figure I.6.11 Wooden structures of oil wells in Baku, one of the principal centers of early crude oil production. More than a century later, the still considerable untapped oil reserves of the Caspian Sea are, once again, a center of international attention.

Reproduced from *The Illustrated London News,* June 19, 1886.

surpassed 10% by 1950, and, by 2000, was nearly 35% (Smil 2017b). Finally, higher productivity in all energy industries translated into lower prices of fuels and electricity. Trends for crude oil and electricity were particularly impressive. When expressed in constant dollars, US crude oil prices in 1910 were about 90% lower than during the early 1860s (see Figure I.1.9), General Electric's household tariff for lighting fell by the same amount just during the two decades between 1892 and 1912, and English electricity prices declined by about two-thirds between 1882 and 1912 (Kander et al. 2013).

Before leaving the subject of high-energy civilization, I must stress that the trends that began during the Age of Synergy and that I describe in this section either continued for most of the twentieth century or are still very much with us. As already noted, the global shares of biomass energies fell to about 25% by 1950, to no more than 10% by 2000, and to less than 5% by 2023. Per capita energy consumption was rising until the 1970s in the United States and until the 1990s in Japan. Efficiencies of major energy conversions are still improving, although some of them (gas-fired furnaces, aluminum smelting, Haber-Bosch synthesis of ammonia) are now very close to their thermodynamic limits. The transition from coal to hydrocarbons continued as the global share of crude oil and natural gas in the world's primary energy supply rose from less than 40% in 1950 to about 65% by 2000.

Mechanization and Mass Production

Higher energy flows used with increasing efficiencies by new prime movers were the key to the sweeping mechanization of tasks that ranged from crop harvesting to office work and from manufacturing to household chores. Even in the United Kingdom, this process got fully underway only during the Age of Synergy: at the time of the 1851 census the country's traditional craftsmen still greatly outnumbered machine-operating factory workers because there were more shoemakers than coal miners and more blacksmiths than ironworkers (Cameron 1985). Mechanization made mass production the norm in all modernizing economies as it began to deliver food surpluses and affordable consumer goods while cutting the labor hours and enhancing the quality of life. Again, these trends are still very much with us, now in advanced stages in affluent countries but still in midstream in such large modernizing economies as Brazil and China.

After 1870, mechanization's reach grew ever more comprehensive, encompassing fundamental (steelmaking) as well as trivial (toy-making) procedures. The enormous expansion of machine tool manufacturing brought the mechanization of tasks that ranged from wire drawing and twisting to metal milling and shaping. Mechanization transformed activities that were both ubiquitously visible (internal combustion and electric motors displacing horses in cities) and largely hidden. An excellent example in the latter category is the speed with which electricity-powered mechanical cutting diffused in the US coal mining: from nothing in the early 1880s to about 25% by 1900 and to half of all produced coal by 1913 (Devine 1990b). The process was completed by the early 1950s, when about 95% of all underground US coal was cut from seams mechanically.

But the first area where mechanization, and other new modes of production, made the greatest difference to an average consumer was in the production of food. The combination of better machines (inexpensive steel plows, efficient harvesters and threshers), increased fertilizer use, improved storage, and cheaper long-distance transportation brought steady increases in agricultural productivity. In the United Kingdom, Samuelson (1893) calculated that between 1861 and 1881 more than 110,000 farm workers were replaced by about 4,000 skilled artisans who were making new field machines and another roughly 4,000 people who operated them. But, as with the energy transition, the United States led the way, and pages could be filled with impressive comparisons of requirements for physical labor before the Age of Synergy and by its end, when an increasingly mechanized activity was able to provide food surpluses by employing a shrinking share of the population (Schlebecker 1975; Smil 2017a, 2019).

Between 1860 and 1914, the share of the US farming population was halved to just below 30% of the total, average time required to produce a ton of wheat declined by about 45% to less than 40 hours, and the largest farms could produce it with less than 10 hours of labor (Rogin 1931; McElroy, Hecht, and Gavett 1964). New large grain mills (the first one with automatic steel rollers was built by Cadwallader Washburn, in 1879) produced superior flour, and Andrew Shriver's introduction of the autoclave (US Patent 149,256 in 1874) that used pressurized steam for the sterilization of food increased the capacity of canning 30-fold compared to the traditional method (Goldblith 1972).

New imports and mechanization made food cheaper also in Europe and in the United Kingdom. By the 1880s, even working-class English families were adding not just jam, margarine, and eggs to their regular diet but also canned sardines, coffee, and cocoa (Mitchell 1996). Bowley's (1937) detailed account shows that, compared to the 1860s, the average food basket of English families in 1913 contained three times as much meat and cheese, twice as much butter and sugar, and four times as much tea. And there were also significant increases in the consumption of fruits and vegetables, new imports of bananas and oranges, and the growing popularity of chocolate. The paragon of stores that sell gourmet food opened in Paris, at Place de la Madeleine, in 1886, when Auguste Fauchon began to offer an unmatched selection of delicacies.

The second most important area where mechanization had the greatest impact on the quality of life for the largest number of people was in providing affordable infrastructures for better household and public hygiene. Modern plumbing, whose diffusion was supported by the latest scientific discoveries of waterborne pathogens, had enormous cumulative impacts on the reduction of morbidity and on the increase in longevity. Cheaper mass-produced metals, reinforced concrete, and more powerful, more efficient electric pumps made it easier to build dams, conduits, and pipes and to bring treated drinking water (a process that became possible only with inexpensive large-scale production of chlorine) into dwellings and take the wastes out.

Continuous chlorination of drinking water began in the early years of the twentieth century. Trenton, New Jersey, had the first large US facility in 1908, and, by 1910, the country had nearly 600 treatment sites serving 1.2 million people; by 1948, some 80 million Americans drank chlorinated water supplied by nearly 7,000 utilities (Thoman 1953). Chlorination brought a rapid reduction in the incidence of waterborne infections, particularly of typhoid fever, whose pathogen, *Salmonella typhi*, was identified in 1880 by Karl Joseph Eberth. US typhoid mortality fell from about 30/100,000 in 1900 to less than 3/100,000 by 1940, and the disease was virtually eliminated by 1950 (USBC 1975).

The first wave of municipal sewage plant building dates to the 1870s, and, by 1913, most large Western European cities (North America lagged) had either sewage fields, settling tanks, grit removal, screens, or combinations of these techniques (Seeger 1999). Other environmental health gains were made possible by new techniques. Replacement of horses by engines and motors eliminated enormous volumes of excrement from cities, and this, together with paved streets, also greatly reduced the amount of airborne particulate matter. Although fuel-derived outdoor air pollution remained a serious urban problem for decades, the situation would have been much worse without more efficient stoves, boilers, and turbogenerators. Indoors, electric lights replaced air-polluting gas jets, and more affordable (and better quality) soap made more frequent washing of hands possible. As we now know, this simple chore remains the most cost-effective means of preventing the spread of many infectious diseases.

With adequate food supply and basic hygiene claiming a declining share of disposable income, rising shares of consumer expenditures began to shift first to purchases of more expensive foodstuffs (meat intake rose steadily), prepared meals, and drinks and then to acquisitions of an ever-widening array of personal and household goods whose quality was improving as mechanized mass manufacturing was turning out identical items at unprecedented speed. As Landes (1969: 289) put it, "[T]here was no activity that could not be mechanized and powered. This was the consummation of the Industrial Revolution." Ingenious machines were designed to bend leather for shoes or steel wire for paper clips, to blow glass to make light bulbs, or to shape milk and beer bottles.

The impressive rise in retail sales can be illustrated by higher revenues as well as by the number and variety of stores. Berlanstein (1964) collated interesting statistics about the growth of retail outlets for Ivry-sur-Seine, a Parisian working-class suburb: between 1875 and 1911, the number of clothing stores per 1,000 inhabitants nearly quadrupled, while the number of grocery stores quadrupled, and that of stores selling drink and prepared food rose 4.5 times. And, as already noted, consumers without access to richly stocked urban stores (see Figure I.1.11) could rely on progressively greater choice available through mail-order shopping, an innovation that was pioneered in 1872 by Aaron Montgomery Ward (1844–1913).

After 1890, mass consumption, particularly in the United States, began to embrace many nonessential manufactures whose falling prices made them soon broadly accessible. The Age of Synergy was thus the time of the democratization of possessions, habits, and tastes as machines, gadgets, and pastimes spread at an often-dizzying pace. Unfortunately, as Veblen (1902: 84) noted, there is also a near universal tendency toward excessive consumption.

> The basis on which good repute in any highly organised industrial community ultimately rests is pecuniary strength; and the means of showing pecuniary strength ... are leisure and conspicuous consumption. Accordingly,

both of these methods are in vogue as far down the scale as it remains possible.

Premiere places of conspicuous consumption were large new department stores where customers could get lost amid an artful display of goods. Zola captured this milieu with unsurpassed perfection. As Octave Mouret, director of Au Bonheur des Dames, surveyed the enormous crowd of females filling his store during the Great White Sale, he was aware that "his creation was introducing a new religion, and while churches were gradually emptied by the wavering of faith, they were replaced in souls that were now empty by his emporium" (Zola 1883/2001: 416). Nearly 150 years later, the only notable difference is that huge department stores have been displaced by much larger shopping centers sheltering scores or hundreds of smaller stores under one roof or by a prolonged inspection of online offerings. And the modern habit of walking around in visibly branded clothes or with other prominently labeled merchandise also goes back more than a century: in 1899, Louis Vuitton began to put his elegantly lettered initials on his hand-crafted products, and this practice is now imitated across all price ranges.

But, in retrospect, the most important item of conspicuous consumption and expanding accumulation was the ownership of cars. No other mass-produced item turned out to have such wide-ranging effects on the structure and performance of affluent economies, on the spatial organization of society, and on so many social and cultural habits. Only a few hundred rich eccentrics would buy the first motor cars in the early 1890s; only thousands of well-off doctors, businessmen, and engineers were eager to get one of Olds's early models a decade later—but nearly half a million individuals and families made that purchase every year just before World War I began. Extraordinarily large productivity gains were behind the car's rapid mass penetration. Data from Ford cost books show that 151 hours were needed to make a car in 1906, 39 in 1914, and 37 in 1924 (Ling 1990)—but productivity was rising in every sector of the economy.

In 1909, Thomas B. Jeffrey, a Wisconsin manufacturer of the Rambler car, noted that the "mortgage has gone from the Middle Western farm, and to take its place there is the telephone, the heating system, the water supply, improved farm machinery, and the automobile" (cited in Ling 1990: 169). By the mid-1920s, car-making became America's leading industry in terms of product value, and it has retained this primacy throughout the twentieth century. During the late 1990s, US car sales were more than 20% of all wholesale businesses and more than 25% of all retail, and automakers were the largest purchasers of steel, rubber, glass, machine tools, and robots. Adding crude oil extraction and refining, highway construction and maintenance, roadside lodging and eating, and car-dependent recreation activities leaves no doubt that automobiles have been the key factor of Western economic growth. No other machine has done so much for the still spreading *embourgeoisement* and the rise of the middle class.

By the time American car sales reached hundreds of thousands a year, many newly mass-produced goods were quite affordable—but the range of accumulated personal possession was still limited. As we have seen, the products that had eventually become such inalienable ingredients of modernity—electric lights, telephones, cars—were yet to diffuse to most of the population, and there were large differences even among neighboring countries. For example, by 1913, there were 26 people per telephone in Germany, but in France the ratio was still almost 135 (Mitchell 1998). The same was true about basic living conveniences: in 1913, fewer than 20% of Americans had flush toilets.

Regardless of the differences in the specific national rates of technical progress, cities everywhere were the greatest beneficiaries. Progress in sanitation, communication, and transportation solved or eased the three sets of key obstacles to their further growth. In turn, cities had the leading role in stimulating innovation and facilitating its diffusion (Bairoch 1991). The Age of Synergy had an overwhelmingly urban genesis, and this generalization is true not only about the outburst of technical inventiveness but also about the period's incredible artistic creativity. Just try to extend the experiment played with the absent inventions to music, literature, and painting created during the two pre-World War I generations.

Imagine that we would not have any, or most, compositions by Brahms, Bruckner, Debussy, Dvorak, Gounod, Mahler, Puccini, Rachmaninov, Ravel, and Tchaikovsky. Imagine that the novels, stories, and poems of Chekhov, Kipling, Maupassant, Rilke, Tolstoy, Twain, Verlaine, Verne, Wilde, and Zola would not exist. Remove from the world of images the paintings of French impressionists and the canvases that immediately preceded and followed that glorious era: gone are all or most of the works of Braque, Caillebotte, Cézanne, Gauguin, Manet, Matisse, Monet, Pissarro, Renoir, Rousseau, Seurat, Signac, Sisley, and van Gogh, as well as of the early Picasso.

This admirable outpouring of artistic creativity (particularly its *fin de siècle* phase) helped to widen the opportunities for visiting art exhibitions and for attending musical performances. But every pastime became more popular as leisure became a widely shared social phenomenon only during the two pre-World War I generations, when its many forms became a notable part of the process of modernization (Marrus 1974). Some of its manifestations were captured in such unforgettable impressionistic masterpieces as Pierre-August Renoir's *Déjeuner des canotiers* (1881 oil painting of a boating party gathered around a canopy-covered and wine- and fruit-laden table) or Georges Seurat's pointillistic gem *Un dimanche après-midi a l'Ille de la Grande Jatte* that was painted in 1884–1885 and depicts Parisians enjoying a summer afternoon along the banks of the Seine.

The costs of these leisure activities varied greatly. Even some simple pastimes were rather expensive. To ride the first great wheel built by George W. Ferris (1859–1896) for the 1893 Columbian Exposition in Chicago (Figure I.6.12) cost 50 cents at a time when a Chicago laborer was paid

Figure I.6.12 George Washington Ferris supported his 75-m-diameter wheel by two 42-m steel towers, and the nearly 14-m-long axle was the single largest steel forging at that time. Two reversible 750-kW engines powered the rotation of 36 wooden cars, each with 60 seats. This original Ferris wheel was reassembled at the St. Louis Exposition in 1904, and it was scrapped two years later.
Reproduced from *Scientific American* of July 1, 1893.

15 cents per hour (BLS 1934). But this was a popular, safe thrill, and, during the twentieth century, it was replicated by hundreds of eponymous structures, most recently by the world's tallest Ferris, London's Millennium Wheel. Other pastimes were cheap: Frank Brownell's 1900 Brownie camera cost a dollar, a perfect gift for children to take snapshots of their friends and pets.

And the Age of Synergy saw the birth of pastimes that targeted the largest possible passive participation as well as of elite pursuits that required not just a great deal of money but also uncommon skills. Animated cartoon films (introduced in 1906) are an excellent and enduring example of the first category, America's Cup of the other. America's unbroken string of pre-World War I Cup victories generated a great deal of public attention in its defense, and, in 1914, experts were delighted when the tradition of secrecy was lifted and technical details of competing yachts became available before the race (Anonymous 1914), which the war's outbreak postponed until 1920. And, of course, the Olympic Games were reborn in 1904.

Extensive contemporary writings dealt with many negative consequences of mechanization and mass production. People who reacted with condemnation and outrage to their often-appalling surroundings included not only social critics (from Thomas Carlyle to Karl Marx and from John Ruskin to Matthew Arnold) and perceptive novelists but also dedicated photographers who documented urban squalor and misery (Riis 1890). As Kranzberg (1982) rightly noted, this bleak interpretation of industrialization was still prevalent among many scholars a century later. Some modern critics have argued that industrialization deepened, rather than relieved, human poverty and suffering; others regretted the "irretrievable destruction of much of the beauty of the countryside" and the "loss of the peacefulness of mind which the gentle and unaltering rhythm of country life can bring" (Fleck 1958: 840).

These are indefensible views as the reality was much more nuanced. No historical study can convey it better than did Zola's monumental Rougon-Macquart cycle, perhaps the unsurpassed witness of the post-1860 era, with its astonishing sweep of desperation and hope, poverty and riches, suffering and triumphs in milieus both rural and urban and in settings ranging from an almost unimaginably brutal world of vengeful peasantry to the intrigues of Parisian *nouveaux riches*. The supposed rural harmonies hid a great deal of poverty, misery, and autocratic abuse, and many pre-industrial landscapes were neither untouched nor well cared for. And for every demonstration of undeniably reprehensible urban reality—unemployment, exploitation, low wages, unsanitary living conditions, noise, air and water pollution, illiteracy, lack of security—there are countervailing examples (Hopkins 2000).

Most significantly, millions of new jobs that paid far better than any work that could be obtained in the countryside were created by risk-taking entrepreneurs. Some of them were industrialists with exemplary social conscience. Robert Bosch instituted an eight-hour working day in 1906, four years later made Saturday afternoon free, and transferred most of the company's profits to charities (Bosch Global 2024). This charitable pattern endures: the company is 92% owned by the Robert Bosch Foundation, which funds many social, artistic, and scientific activities. And while overall social progress was not as rapid as many would have liked, it was undeniable.

Besides the gradual introduction of shorter work hours and higher purchasing power, the two pre-World War I generations also saw a diffusion of compulsory grade school attendance and the beginning of pension and health insurance schemes in many countries (first in Bismarck's Germany in the early 1880s, followed soon by Austria). For these reasons, Ginzberg (1982: 69) rightly observed that Karl Marx "was better as a critic than as a prophet" as he did not anticipate the substantial gains in the quality of life that would eventually result from mechanization and increased productivity.

Statistical bottom lines speak for themselves: in no previous period of Western history did so many people become adequately fed (e.g., British intake of high-quality animal protein roughly doubled), basically educated (as grade school attendance became compulsory: in the United Kingdom, up to age 10 in 1880, up to 12 by 1899), and able to enjoy a modicum of material affluence (more than one set of clothes and bedding, more pieces of furniture) than they did during the two pre-World War I generations. But, above all, they benefited from unprecedented declines in infant mortality, which was the main reason for rising life expectancy. Between the 1860s and 1914, Western infant mortality was roughly halved to just more than 100/1,000 newborns while life expectancy at birth increased by 20–25% to around 50 years, with the British means going from 40 years during the late 1860s to 53 years by 1913 (Steckel and Floud 1997).

Concerns about the negative consequences of globalization often sound as if the process was the invention of the closing decades of the twentieth century. But parts of Europe participated in long-range trade already during the early modern era that preceded industrialization. This process widened and intensified during the latter half of the nineteenth century to such an extent that it began to bring about a convergence of prices, perhaps the best measure of true globalization (O'Rourke et al. 1996). Between 1870 and 1913, the share of exports in the total economic product rose by 50% (to about 12%) in 10 of the most industrialized countries, and, by 1913, more than 25% of the United Kingdom's, Australia's, and Canada's gross national products originated from exports (Maddison 1995). The first era of true economic globalization peaked just before World War I: in 1913, international trade contributed 14% to the global economic product. Further gains were delayed (by two world wars and economic crisis of the 1930s), and the 1913 share of trade in the world economic product was regained only by the mid-1970s (Fouquin and Hugo 2016).

Both the epochal transition from biomass to fossil fuels and the diffusion of mechanization and mass production inevitably led to greater dependence on nonlocal resources. Coal deposits can be found in scores of countries around the world, but many coal basins contain seams of rather inferior quality, and so excellent steam and metallurgical coals became an early item of international trade. Major oil and gas fields are much less equitably distributed, and hence the universal transition from coals to hydrocarbons had to be accompanied by rising shipments of oil and—once technical progress made that possible—also by large-scale trade in natural gas (Smil 2017a). Ores, needed in increasing quantities by a rapidly expanding iron and steel industry and nonferrous metallurgy, are even more unevenly distributed; for example, only three countries (Australia, China, and Guinea) produce more than 70% of the world's bauxite (USGS 2024).

Aside from a few city states, traditional economies could produce all (or nearly all) food to assure basic nutrition for their populations, but in the new interconnected world this autarky would be very expensive, and, in most cases, it would also result in a limited choice of foodstuffs. More affluent societies with larger disposable incomes naturally created markets for a greater variety of food, and this demand was best satisfied by specialized producers who have comparative advantage thanks to their climate, long experience in particular cultivation, or low labor costs. Consequently, international trade in foodstuffs began expanding beyond the shipments of low-volume and luxury items (spices, sugar) during the 1870s as soon as long-distance transportation of bulky commodities (grains, oilseeds, refrigerated meat and butter from North America, Australia, and Argentina) could rely on large-capacity iron-hull ships.

The period also saw the rise of a new kind of enterprise as multinational corporations made their products in several countries rather than just exporting from one location. Robert Bosch was one of the first producers with a clear global aspiration: by 1913, his operations extended to 20 countries, and, still uncharacteristic for that time, his company derived more than 80% of its revenues from sales outside Germany. Several companies that were established before World War I and that eventually expanded into the world's leading multinationals have been mentioned in this book: General Electric, Brown Boveri, General Motors, Ford, Siemens, Marcon, and also the two top cola makers, Coca and Pepsi.

Remarkably, only two of the ten largest multinationals by the end of the twentieth century (when ranked by their revenues in 2000) were not set up before 1914: Walmart Stores (no. 2) and Toyota Motor Corporation (no. 10). First-ranked Exxon-Mobil originated as John D. Rockefeller's Standard Oil Trust, organized in 1882 and dissolved by the US Congress in 1911. Origins of the third, fourth, and fifth largest companies, all of them automakers (General Motors, Ford, and Chrysler), have already been described. Royal Dutch/Shell (no. 6) resulted from a 1907 merger of two young companies (Shell, from 1892; Royal Dutch, from 1903), while British Petroleum (BP) (no. 7) was set up in 1901, by William Knox D'Arcy to explore oil concessions in Persia. General Electric, whose origins go back to Edison's first electric company of 1878, ranked eighth worldwide in 2000, and Mitsubishi, set up as Tsukomo Shokai by Yataro Iwasaki in 1870 (Mitsubishi 2003), was ninth.

More than a century later, these companies continued to be at the forefront of globalization, and the interdependence generated by this now so pervasive business pattern has created many positive feedbacks as well as many regrettable trends. Principal manifestations of the process will be examined in some detail in the companion volume. There is only one more matter I want to address here: in Chapter I.7, I return briefly to those remarkable pre-World War I years to convey some of the perceptions, feelings, hopes, and fears of those who lived in those momentous times and reflected on the accomplishments, promise, and perils of technical advances that were creating a new civilization.

Frontispiece I.7 By 1914, this kind of an early-morning commuting scene—female office workers, shop assistants, and dressmakers are shown here arriving at a Paris terminus—was common in all large Western cities.
Reproduced from *The Illustrated London News,* April 11, 1914.

7
Contemporary Perceptions

> It has been a gigantic tidal wave of human ingenuity and resource so stupendous in its magnitude, so complex in its diversity, so profound in its thought, so fruitful in its wealth, so beneficent in its results, that the mind is strained and embarrassed in its effort to expand to a full appreciation of it.
> —Edward W. Byrn, *Scientific American* (1896)

This book pays homage to the astonishing concatenation of epochal innovations that were introduced and improved during the two pre-World War I generations and whose universal adoption created the civilization of the twentieth century. I have marshaled a great deal of evidence to justify this—in retrospect so inevitable—judgment. But I do not want to leave this fascinating subject without asking, and attempting to answer, one last obvious question: To what extent were the people who lived at that time aware that they were present at the creation of a new era? This question must have a complex answer that spans the entire range of responses from "not at all" through "in some ways" to "very much so." The reasons for this are obvious.

Short as it was in relative terms, the span of two generations is too long a period to be dominated by any consistent perception. Those two generations saw both the pronounced ups and downs of economic fortunes and some notable social and political upheavals, many quite unrelated to technical advances but caused by interstate rivalries or intrastate disturbances that greatly influenced the public mood. And even in the countries at the forefront of technical advances, the chores and challenges of everyday existence inevitably claimed much more attention than matters that, at least initially, appeared to have little bearing on everyday life. The earliest stages of building a new civilization were shaped more by many attributes of the preceding era that coexisted for many decades with some thoroughly modern practices. The coexistence of old and new of improvements eagerly anticipated and welcomed and widely distrusted or ridiculed, was thus entirely natural. The enormity of the post-1860s saltation was such that people alive in 1913 were further away from the world of their great-grandparents who lived in 1813 than those grandparents were from the lives of their ancestors in 1613, even in 1513. At the same time, everyday experiences of the two pre-World War I generations had inevitably much in common with the pre-industrial norms. Even in the most mechanized economies horses remained the most important prime movers in agriculture and a common sight in urban traffic: even in the United States the numbers of horses and mules peaked and began to decline only in 1922 (USBC 1975). And the first cars looked like, and were initially called, horseless carriages, and, as they rushed out to see their first airplane in the summer sky, many village children in Europe still ran barefoot during the pre-World War I years, saving their shoes for fall and winter. Many epochal inventions appeared to be just fascinating curiosities, legerdemains of scientists and engineers with little practical importance for poor families.

Some innovations had rapidly conquered many old markets or created entirely new ones. Others had to undergo relatively prolonged maturation periods because their convenience, cost, and flexibility were not immediately obvious compared to long-established devices and practices. Among the many examples in this category are Otto's massive early gas engine and awkwardly looking first cars. During the early 1870s, nobody foresaw that the gasoline-fueled successors of Otto's stationary machines would, within 50 years, energize more than 10 million vehicles, and nobody envisaged that the first automobiles of the late 1880s (expensive and unreliable toys bought by a few rich adventurers) would mutate in just two decades into the paragon of mass ownership. And, finally, the novelty of some developments was so much outside the normal frames of references that those ideas were not perceived as epochal but rather simply as mad.

This was the case, most prominently, with the first airplanes. When, in 1906, Alfred Harmsworth (Lord Northcliffe, the publisher of *Daily Mail*) offered £1,000 for the first pilot to cross the English Channel, *Punch* immediately ridiculed the challenge.

> Deeply impressed as always with the conviction that the progress of invention has been delayed by lack of encouragement, Mr. Punch has decided to offer ... £10,000 to ... the first aeronaut who succeeds in flying to Mars and back within a week. (Anonymous 1906: 380)

But the Channel prize was won by Blériot just four years later (Figure I.7.1) and, in 1913, Lord Northcliffe offered £10,000 for the first crossing of the Atlantic—this time with no parodies from *Punch*. The new challenge was taken up in early 1914 by Rodman Wanamaker, who commissioned Glenn Curtiss to build a flying boat powered by a 150-kilowatt (kW) engine that would be capable of making that trip in a single flight of 12–15 hours, but the plan had to be shelved because of World War I. If the public was confused and unsure, so were many innovators. As we have also already seen, some of them were surprisingly diffident and dismissive regarding the worth and eventual impact of their efforts. But other creators of the Age of Synergy were mindful of their place in creating a new world. As they looked back, they were proud of their accomplishments, and many of them had no doubt that even greater changes were ahead, that the worth of their inventions was more in the promise of what they would do rather than in what they had already done. Still, I suspect that most of them would be surprised to see how the great technical transformations of the twentieth century left so many of their great innovations fundamentally intact.

Figure I.7.1 Blériot's 1909 monoplane in flight and a head-on view of its propeller and three-cylinder engine.
Library of Congress images (LC-USZ62-94564 and LC-USZ62-107356).

And these feelings of confident expectations were not limited to creative and decision-making elites. The frontispiece of this closing chapter exudes the confident demeanor of young chic French ladies as they spill from a railway terminal to Parisian streets in the spring of 1914, on their way to jobs as typists, telephone exchange operators, and shop assistants. The image captures what was for them already an everyday routine but what would have been an unthinkable display of independence and opportunity in 1864, and a rare sight even in 1884. Although they were still only modestly paid, they were breaking a key feature of the traditional social order. They must have known that they were living in an era that was unlike anything else in the past, and they expected further great changes.

This shift in female work was one of the signal achievements of the age (Olivetti 2014). Several trends combined to make it possible: modern plumbing, heating, and electric appliances eased the household chores and were doing away with domestic servants, while at the same time telephones, typewriters, calculators, and keypunches were opening new opportunities for female employment, and electric trains and subways were making it possible to commute to downtowns where the new office jobs were concentrated. During the 1860s, most of the working women were employed as household servants, and, in the United Kingdom, that share was still as high as 85% by 1891 (Samuelson 1896). By 1914, household servants were a rapidly shrinking minority of the female labor force everywhere in the Western world as a new class of increasingly independent wage-earning females was expanding and creating, very visibly, a new social and economic reality.

Although in an expected minority, many astute observers were aware of the epoch-making nature of new inventions as they correctly foresaw their long-term impact. They were able to see the advantages of new ways quite clearly even decades before their general acceptance. For example, Rose (1913: 115), writing about the rapid diffusion of agricultural tractors (a process that was not completed even in the United States until the early 1960s), felt that

> there is no question but we have entered upon a new era in agriculture. The farmer desires the comforts and advantages of the city dweller, and these he gets easily and cheaply with the small gasoline engine.... It multiplies his capacity and gives him either more leisure or enables him to farm a larger area and increases his income. Power farming has just begun, and the start is encouraging.

Some innovations were so indisputably revolutionary that even the uninitiated public was made almost instantly aware of their importance years before it had any chance to buy or use these gadgets or machines or to benefit from the new processes. Although rather primitive even in comparison with designs that followed just a few years later, Bell's first telephones were greeted almost immediately with general enthusiasm. Edison's work on electric light received plenty of public attention mainly because of detailed and technically competent coverage by New York newspapers. And once electricity made industrial motors the leading prime movers, there was no doubt about the importance of that change: "Electricity came in between the big wheel at the prime mover and the little wheels at the other end, and the face of material civilization was changed almost in a day" (Collins 1914: 419).

Parsons's turbine had a key role in the swift creation of this new reality. As already described, he found a very unorthodox way to demonstrate that it was also a superior prime mover in naval propulsion by a daring public show, as his small *Turbinia* outpaced every warship at Queen Victoria's Diamond Jubilee Naval Review in 1897. This technical stunt was guaranteed to generate admiring headlines. And just a few years later, *The Illustrated London News*—a weekly that was normally much given to portraits of royalty, accounts of state visits, and engravings of angelic children—devoted an entire page of its large format to the explanation of this new system of naval propulsion (Fisher 1903). More than that, the article also included a detailed drawing of the arrangement of stationary and moving blades in Parsons's machine, a kind of illustration that is hard to imagine in today's *People* magazine.

Perhaps the best contemporary appraisal of the era's technical progress was offered by Edward Byrn in his 1896 essay written for *Scientific American*'s semicentennial contest (Byrn 1896). Although Byrn's assessment covered the years 1846–1896, most of the remarkable achievements noted in the essay originated after 1866, and his overall characterization of the period as "an epoch of invention and progress unique in the history of the world" would have been only strengthened if he were to write a similar retrospective in 1913. Byrn (1896: 82) argued that the progress of invention during the second half of the nineteenth century was "something more than a merely normal growth or natural development" and (see this chapter's epigraph) found it difficult to convey the true magnitude of the period's overwhelming contributions.

In complete agreement with my arguments (made entirely independently more than a century later: I came across his assessment in 2003), Byrn singled out the decade of 1866–1876 as "the beginning of the most remarkable period of activity and development in the history of the world" and headed the list of its great inventions with the perfection of the dynamo. Going through his long list of most-notable advances made during the subsequent three decades would be to retrace most of the ground covered in this book. What is more interesting is to list some items on Byrn's list that have not been mentioned here: compressed-air rock drills, pressed glassware, machines for making tin cans, hydraulic dredges, enameled sheet ironware for cooking, Pullman railway cars, and artificial silk from pyroxylin.

Scientific American ran another essay contest in 1913. In his winning entry, William I. Wyman ranked the 10 greatest inventions of the preceding 25 years by looking for the advances that were "most revolutionary in character in the broadest fields, which affected most our mode of living, or which opened up the largest new sources of wealth" (Wyman 1913: 337) and dated them according to their successful commercial introduction. Wyman's list—electric furnace (1889), steam turbine (1894), gasoline automobile (1890), moving pictures (1893), wireless telegraphy (1900), aeroplane (1906), cyanide process (1890), linotype machines (1890), induction motor (1890), and electric welding (1889)—contains only one item that did not make it into my selection of key pre-World War I inventions, the cyanide process of gold extraction.

The process was patented in 1888, by three Glaswegians—chemist John S. MacArthur and physician brothers Robert and William Forrest—and it revolutionized the art of color metallurgy: silver and copper could also be produced by it (Wilson 1902). Its rapid adoption brought a trebling of global gold output by 1908 and made South Africa the world's largest producer. This had major socioeconomic effects on societies whose monetary policies were based on the gold standard. More than a century after its introduction, the cyanide process—whereby ground ores are mixed with diluted cyanide [$Ca(CN)_2$, KCN, or NaCN] and the metal is then precipitated from the soluble $Au(CN)_2$ by addition of powdered zinc—remains the leading technique of gold extraction (Park 1900; Cornejo 1984), but the metal's role in the world's economy has been marginal ever since Richard Nixon took the United States off the gold standard in 1971.

Unlike Wyman and others, I refused to do any rankings in this book because the interactions of many major and minor components that drive complex dynamic systems make such orderings highly questionable. Still, a closer inspection will show that even Wyman's most debatable choices are defensible. He justified his lead ranking of the electric arc furnace because of its multiple impacts: a radical transformation of steel industry and its indispensability in producing aluminum, and, at the time of his writing, electric arc was the only promising means of fixing atmospheric nitrogen. This last role was displaced, after 1913, by the Haber-Bosch process, but the other two key contributions have become far more prominent as today's electric arc furnaces produce more than a third of the world's steel and all aluminum.

Similarly, Wyman's inclusion of electric welding was well justified by the ability of the process to join what were previously considered un-weldable metals (brass, bronze, cast iron) and to produce shapes that could be made previously only by laborious riveting. In 1964, the technique was transformed by the introduction of plasma welding that uses gas heated by an electric arc to extremely high temperatures for accurate and high-quality applications ranging from work on precision instruments to repairs of gas turbines. As a result, Wyman's identification of the most revolutionary technical advances stood the test of time quite well. His excellent essay, as well as Byrn's retrospective, shows that many well-informed observers were not only aware of the era's unique contributions but also could accurately identify its most far-reaching innovations without the benefit of longer historical perspectives that would have made it much easier to point out the cases of successful, and lasting, impacts.

In contrast, one thing that could not go unnoticed by anybody who lived during those eventful decades of rapid mechanization was the widespread demise of artisanal work. This change elicited a spectrum of feelings from enthusiasm to grudging acceptance to obvious and deeply felt regrets. A difference of opinion between the two protagonists of Zola's grand and tragic *L'Assommoir*—Gervaise, a washerwoman, and Goujet, a metalworker—captures vividly some of these emotions. Goujet, after watching silently as a bolt-making machine was churning out perfect copies, turned in a resigned way to Gervaise and said how that sight makes him feel small. His only solace was that the machines might eventually help to make everybody richer (as they indeed did). But Gervaise scoffed at this as she found the mechanical bolts poorly made: "You understand," she exclaimed with passion, "they are too well made.... I like yours better. In them one can at least feel the hand of an artist" (Zola 1877: 733).

And although we have now been living for generations in societies whose prosperity rests on the mass production of perfect copies, so many of us still repeatedly feel something of Gervaise's regret. Indeed, we are willing to pay the premium for the greatly diminished range of artisanal products, cherishing the traces left by the hands of their creators (or naively trusting fraudulent claims of some manufacturers that that indeed is the case). But we are, inescapably, ready consumers of countless low-priced mass-produced items, and we do not even seriously consider that things should be done otherwise or that there should be opposition, violent or Gandhi-like, to reverse this pattern.

This almost unquestioned acceptance became one of the surprising norms of the entire Age of Synergy. In 1811, when young Ned Ludd smashed a knitting machine and launched a rebellion against the supremacy of mechanization (Bailey 1998), the entire process of industrialization was still in its very beginnings as machines dominated only certain segments of a few industries in just a handful of countries. Two and three generations later they were ubiquitous, providing countless more reasons to oppose the diffusion of some of the dehumanizing ways of mass industrial production—but the Luddite sentiment was not overt among the men and women whose labor created the new tools of production and served them to flood the markets with new goods: its most evident champions were social critics and writers, such as John Ruskin, with his bleak view of the future (Fox 2002).

This reality reminds us how different was the entire worldview and how difficult it is to capture the prevailing attitudes even when the historical distance is just a matter of four or five generations. Above all—and in sharp contrast with the increasing frequency of immature behavior in modern society that displays so many infantile traits—most people were not burdened by unrealistic expectations. As young adults my grandparents had electric lights but no telephone, and I do not think they made a single phone call even late in their lives. Their diet was adequate but simple, their necessary daily walks long, and their schooling, much as their possessions, basic. My parents shared a great deal of these frugal realities, and I still experienced some of them as a child in post-World War II Europe: rationed food, a new book as an expensive and much prized possession, long daily walks to school, the delights of just a few weeks of ripe summer fruit and freshly picked wild strawberries.

At a time when there are more mobile phones than people (babies included), when there are fewer than two people per car ("vehicle" is a more accurate term in the world of monstrous SUVs) in affluent countries, where food costs in the richest countries are barely more than a tenth of the average disposable family income (even though so many items are carted out of season halfway around the globe), and where books have been largely displaced by screens, no amount of quiet reflection will help people growing up with these realities to grasp the modalities of pre-telephone, pre-internal combustion engine, pre-supermarket, pre-PC life. We face the same problem when trying to insert ourselves into the minds of our pre-World War I ancestors who were surrounded by the nascent manifestations of a new civilization but benefited from them only to a limited extent.

At least I am sure that my grandparents did not feel deprived because they lived and died (decades after Bell's and Benz's inventions) without telephones or automobiles. But the absence of a telephone or an automobile would be the least of it. Again, I think of my grandfathers, men who worked hard with little reward to create the modern world. My paternal grandfather helped to energize it—first as a coal miner in German deep mines (Figure I.7.2), then as a *Steiger* (foreman) in Western Bohemian hard coal pits. My maternal grandfather helped to provide its key material foundation as he built and repaired Siemens-Martin steel-making open hearth furnaces in Skoda Works, at that time one of Europe's largest industrial enterprises. Naturally, by today's standards both were poor but proud of their work, both were keen readers and independent thinkers, neither saw himself as a victim, neither looked for salvation in simplistic Marxist slogans of class struggle or in new social utopias.

As always, intellectuals were much more willing to worship these radical solutions: their convictions led soon to the three generations of the Soviet Communist empire and later to its Maoist replica, two grand "new society" experiments that were paid for with the lives of tens of millions. Others were convinced that progress, so obvious in the staggering sum of new technical advances, could lead only to better and better outcomes, with technocracy and democracy triumphant (Clarke 1985). This appeared to be self-evident to such prominent creators of America's industrial success as Thomas Edison and Andrew Carnegie, who wrote that, in America, "the drudgery is ever being delegated to dumb machines while the brain and muscle of men are directed into higher channels" (Carnegie 1886: 215). But that was not, obviously, the only—not even the dominant—perception: with gains came losses and worries, and too many everyday realities were hardly uplifting.

Figure I.7.2 My grandfather, Václav Smil, photographed at the beginning of the twentieth century in Homberg am Rhein. After generations of black coal mining east of the Rhine, in the Ruhr region, extraction began also on the river's western bank, across from Duisburg. Coal was discovered at Homberg in 1854 at a depth of 174 m, and the first shaft was opened in 1876.
Photograph by Otto Meltzer.

New machines and the rising combustion of fossil fuels meant that even rich city dwellers could not avoid ubiquitous noise and worsening air pollution. New industries and expanding cities were also polluting water and claiming fertile agricultural land at unprecedented rates. Successful innovations that created new economic opportunities were also eliminating entire classes of labor force. And high profits, including Carnegie's immense fortune, often rested not only on the deployment of new techniques but also on treating labor in ways that seem today quite intolerable. After the violent Homestead Strike of 1892, Carnegie cut the steelworkers' wages by 25–40% without changing the burden of 12-hour shifts seven days a week (Krause 1992). So much for the "higher channels" he invoked just six years earlier.

Nearly a generation later, the working conditions of new Homestead immigrants remained hard: "Their labor is the heaviest and roughest in the mill, handling steel billets and bars, loading trains, working in cinder pits; labor that demands mostly strength but demands that in large measure. . . . Accidents are frequent, promotions rare" (Byington 1910: 133). But their wages had increased, placing their households in the upper third of the US income scale, and Carnegie, retired and during the last 18 years of his life dispensing some $350 million (nearly $11 billion in 2025 monies) to charities, created a benefit fund for the employees of the company. These contrasts provide an excellent example of complexities that work against simplistic conclusions.

On an abstract level, rapid technical advances highlighted the widening gap between the impressive designs and capabilities of machines and engineering and scientific solutions on one hand and the prevailing social and economic arrangements on the other. Wells (1905: 102) believed that were our political, social, and moral devices only as well contrived to their ends as a linotype machine, an antiseptic operating plant, or an electric tramcar, there need now at present moment be no appreciable toil in the world and only the smallest fraction of the pain, the fear, and the anxiety that now makes human life so doubtful in its value.

But just three years before that, in his address to the Royal Institution on the discovery of the future, he offered another appraisal of the past and the future of a new civilization. This one left no doubt that one of the era's most provocative thinkers was acutely, and accurately, aware of its special place in human history, of its evolutionary-revolutionary nature, of its immense promise (Wells 1902b: 59–60).

> We are in the beginning of the greatest change that humanity has ever undergone. There is no shock, no epoch-making incident—but then there is no shock at a cloudy daybreak. At no point can we say, "Here it commences, now; last minute was night and this is morning." But insensibly we are in the day. . . . And what we can see and imagine gives us a measure and gives us faith for what surpasses the imagination.

PART II
TRANSFORMING THE TWENTIETH CENTURY

Technical Innovations and Their Consequences

Contents
1. Transforming the Twentieth Century: Debts and Advances — 339
2. Energy Conversions: Growth and Innovation — 363
3. Materials: Old Techniques and New Solutions — 425
4. Rationalized Production: Mechanization, Automation, Robotization — 475
5. Transportation, Communication, Information: Mass and Speed — 531
6. New Realities and Counterintuitive Worlds: Accomplishments and Concerns — 585
7. A New Era or an Ephemeral Phenomenon? Outlook for Technical Civilization — 641

Frontispiece II.1 The contrast of Mercedes models from years 1901 and 2000 exemplifies the process of technical transformations during the twentieth century: the basics remain as they were invented and improved before World War I but appearance and performance are radically different. In 1901, Wilhelm Maybach's Mercedes 35 was the most powerful and the best engineered car one could buy: its 5.9-liter four-cylinder engine developed 35 hp (26 kW) when running at 950 rpm, producing 4.4 W/cm^3 of the engine's volume. A century later, the eight-cylinder engine of a luxury Mercedes S500 displaced 5 liters, 15% less than the 1901a model, but, when running at 5,600 rpm, it developed 302 hp or 45 W/cm^3, slightly more than 10 times the performance of the 1901 model. External differences are obvious.

Photographs courtesy of Mercedes-Benz Konzernarchiv, Stuttgart.

1
Transforming the Twentieth Century
Debts and Advances

> It is to quantity, scale and magnitude, and not to change in kind, that the deepest social impacts of science and technology during the twentieth century are due.
>
> —Philip Morrison

That single sentence expresses perfectly the extended argument made in this book: transformations, not the change in kind, characterize best most of the century's technical achievements. I knew Philip Morrison (Oppenheimer's graduate student, participant in the Manhattan Project, advocate of nuclear nonproliferation) during the last 15 years of his life, when he was at the Massachusetts Institute of Technology (MIT) and lived in Cambridge. We talked and corresponded about many things, but not about the nature of twentieth-century technical advances: I discovered that perfect encapsulating quote only by accident. Developments leading to Morrisons's characterization had their beginning in Western Europe and in the United States of the 1860s: the next two generations were, by any measure, the time of technical advances unprecedented in human history.

There is nothing we can do to change the biophysical prerequisites of our existence or the recurrence of large natural catastrophes that periodically imperil all life on Earth. In this sense, our civilization is no different from the cuneiform or hieroglyphic realms of the Middle East of 5,000 years ago or from the early modern world of the seventeenth century. But in nearly all other respects our world differs fundamentally even from the realities that prevailed as recently as 1850—and the genesis of this change was the burst of technical innovations that took place during the two pre-World War I generations and that was traced and assessed in the first part of this volume.

This astonishing concatenation of technical advances created the foundations of a new civilization. These accomplishments were further refined and expanded by post-World War I developments and substantially augmented by those innovations that were commercialized after World War II and that accelerated profound economic progress and sweeping social changes during the second half of the twentieth century. As a result, by the end of the twentieth century, most of the world's 6 billion people lived in largely or overwhelmingly man-made rather than in natural or only partially modified environments (Figure II.1.1). Most of our daily interactions were with a still growing array of devices and machines rather than with soil,

Figure II.1.1 Satellite views of New Delhi and Barcelona. By the end of the twentieth century buildings and roads (concrete, steel, and asphalt) were the immediate surroundings for most of humanity.
Images from Maxar.

plants, and animals, and the energy that powered our productive activities and provided daily comforts came overwhelmingly from fossil fuels or nuclear fission rather than from conversions of renewable (direct and indirect) solar flows.

Our food, our dwellings, our material possessions, our ways of travel, and our means of production and communication depend on the incessant operation of countless devices, machines, and processes that created increasingly complex technical systems. In turn, these systems interact in complex ways, and none of them could function without uninterrupted flows of fossil energies and electricity. Affluent societies—dominated by urban populations that derive most of their livelihood from services—became the clearest embodiment of this grand transformation. But low-income countries came to be in some respects even more existentially dependent on technical advances. China and India could not produce enough food for their (in 2025 nearly 3 billion) people without massive applications of synthetic fertilizers and pesticides and without pumped irrigation, and they could not raise the living standard without producing for global markets that are accessed through telecommunication, computerization, and containerized shipping.

During the twentieth century, technical advances became the key determinants of the structure and dynamics of astonishingly productive and increasingly interdependent economies, and they brought impressive levels of affluence and a higher quality of life. At no other time did so many people enjoy such a level of well-being, such an amount and selection of inexpensive food, such a choice of goods and services, such access to education and information, and such high mobility and political freedoms. Comparisons for the Western world show that during the twentieth century life expectancies rose from the mid-40s to the late 70s; the cost of food as the share of disposable income fell by anywhere between 60% and 80%; virtually all households had an electricity supply and were saturated with telephones, radios, TVs, refrigerators, clothes washers, and microwave ovens; shares of populations with completed postsecondary degrees increased (depending on the country) three- to fivefold; and leisure travel evolved from a rarity to the world's largest economic activity (Figure II.1.2).

No less important, these accomplishments contained a tangible promise of future improvements that should eventually spread this enviable way of life to the remainder of humanity. China's enormous post-1980 economic leap—aimed at emulating the previous advances in Japan, Taiwan, and South Korea—has been perhaps the best illustration of this promise. Most

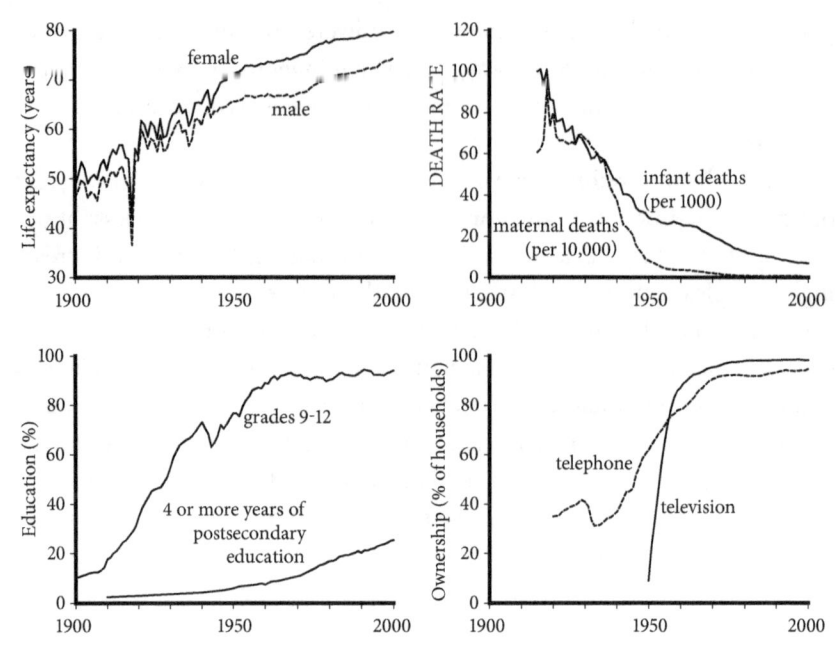

Figure II.1.2 Improvements in quality of life during the twentieth century are illustrated here by US statistics showing a gain of some three decades of average life expectancy at birth (top left), large declines of infant and maternal mortality (top right), a rise in high school enrollment and postsecondary education (bottom left), and virtually saturated household ownership of telephones and televisions (bottom right).
Plotted from data in USBC (1975).

notably, as the country's average disposable income doubled and then quadrupled, a wave of incipient affluence swept the coastal regions, where urban households became saturated first with electric fans and color TVs, later with room air conditioners and small electronic devices. Their ownership of clothes washers and refrigerators went from a negligible fraction in 1980 to more than 75% by 2000, and the rapidly advancing process of their *embourgeoisement* was perhaps best expressed by a considerable pent-up demand for those signature machines of the century—passenger cars.

This book's goal is to trace the evolution of all important techniques that were behind these accomplishments, to explain their genesis and subsequent advances, and to view the history of the twentieth century through a prism of technical innovations. The intent is to use a deliberate grinding to separate the stream of technical advances into its key constituents and to focus on every fundamental component: this includes dwelling on some inexplicably neglected developments (and forgotten innovators) whose contributions shaped the course of modern history more profoundly than some well-known advances. A similar systematic approach is impossible to use when dealing with pervasive synergies among these innovations: those links are so ubiquitous, and often so convoluted, that it would be tedious to point them out explicitly on every occasion.

For example, better-built yet more affordable cars needed stronger and less energy-intensive steels; new methods of steel production required cheaper electricity and an inexpensive large-volume supply of pure oxygen for steelmaking furnaces; more efficient electricity generation needed turbogenerators that could operate under higher pressures and temperatures and hence needed to be made of better steels; and transmitting electricity with minimal losses led to the introduction of new conductors, including aluminum wires, but aluminum smelting required roughly an order of magnitude more energy than did steel production, hence it needed additional electricity generation. And, as the following well-known car example illustrates, this pervasive interconnectedness also makes it impossible to spell out in detail the economic, social, and environmental implications of every innovation.

Mass production of cars became, first, the largest and the best-paying manufacturing industry of the twentieth century. Then it was a leader in automation and robotization that cut the labor force. Intensifying traffic necessitated large-scale construction of paved roads, overpasses, and bridges, and this was the main reason for a hugely increased extraction of sand, rock, and limestone to make cement whose annual output (above 4 gigatons [Gt] since 2013) now dominates the world's material production. Expanded road networks reordered the flows of people and goods and stimulated the growth of suburbs. Trimmed suburban, roadside, and golf-course grass became one of the largest (but purely ornamental) crops in North America, with lawns leaching fertilizer nitrogen and pesticide residues and replacing natural biodiversity. Suburban sprawl and inefficient vehicles generated photochemical smog, and this risk led eventually to emission controls, but precious metals (platinum, palladium, rhodium) used as catalysts in this process invite catalytic converter theft.

Consequently, reading this book should also be a beginning of new inquiries. I detail many links and spell out and quantify many consequences, but that will still leave many more synergies and impacts undefined and unmentioned. As you read about technical innovations, think first of their prerequisites and then begin to add many cascading links and spreading consequences. Of course, many such links and effects are obvious, perhaps even excessively researched and often in the category of worn-out clichés, while others had come as big and costly surprises, sometimes in just a matter of years, in other instances only after decades of established use. But while I strongly encourage the search for synergies, impacts, and causality, I do not interpret these links as proof of technical determinism.

Limits of Causality

My position regarding the role of technical advances in history will disappoint those who believe in the autonomous development of new techniques that inevitably impose themselves on a society and dictate its fundamental features—as well as those who see them largely as social constructs whereby cultural and political peculiarities provide their decisive stamps on initially value-neutral innovations. But rejecting these positions does not mean that truth is somewhere in the middle because shards of it can be found anywhere along the spectrum between the two extremes. As Robert Heilbroner (1967: 355) put it, "That machines make history in some sense ... is of course obvious. That they do not make all of history, however that word be defined, is equally clear."

While I have no wish to promote simplistic technical determinism, I have no doubt that many critical economic, social, and political features and trends of the twentieth century are much more explicable—especially when seen in long-term, cumulative perspectives—through the synergy of science, technical advances, and high energy use than by recourse to the traditionally dominant cultural, political, and ideological explanations. When seen through a slit of months and years, personalities and ideologies of the twentieth century loom large, but the critical impress of technical advances becomes inescapable once the national and global fortunes are viewed with a multidecadal or a centennial perspective. Can there be any doubt that Western affluence, clearly a key factor in promoting stable democratic societies and preventing violent conflicts, could not have been achieved with pre-1850 technical means? And as two very different examples—from two very different places and times but with a fascinating common denominator—illustrate, technical advances also have critical effect on the course of specific historical developments.

Historians have paid a great deal of attention to the expansionist spirit of Wilhelmine Germany and the strategic prowess of the German General Staff (Görlitz 1953; Leach 1973; Dupuy 1977; Stone 2012). But the country's endurance during World War I owed ultimately more to the rapid commercialization of a brilliant scientific discovery than to the capabilities of military leadership (Figure II.1.3). In 1909, Fritz Haber and his English assistant Robert Le Rossignol finally succeeded where many other chemists failed and demonstrated the catalytic synthesis of ammonia from its elements; in less than four years, Badische Anillin- & Soda-Fabrik (BASF), under the ingenious leadership of Carl Bosch, converted Haber's bench process into large-scale industrial synthesis in their Oppau plant (Smil 2001). When the war began, Germany's nitrate reserves, essential for explosives production, were good for only about six months, and the British naval blockade prevented the mineral's import from Chile (Stoltzenberg 1994). Only because of the

Figure II.1.3 What was a greater determinant of Germany's endurance under a three-pronged military pressure and British naval blockade during World War I: the country's military leadership (left)—or Fritz Haber's (right) invention of ammonia synthesis from its elements? The central part of the portrait of the Kaiser Wilhem II and his top military leaders is reproduced from *The Times* (1915). Eric von Falkenhayn, the Chief of the German Army's General Staff, is the first man standing on the right.

Fritz Haber's portrait by Emil Orlik is reproduced courtesy of the E. F. Smith Collection, Rare Book and Manuscript Library, University of Pennsylvania.

first Haber-Bosch ammonia plant and the construction of a much larger Leuna plant could Germany synthesize the nitrates needed for munitions and explosives and thus prolong its war-fighting capacity.

Three generations later, Deng Xiaoping's bold reform policies made the post-Mao China the world's fastest growing large economy and received an enormous amount of media coverage and analytical attention. But (as already noted in this book's first part) the reforms could not have succeeded without averting first the prospect of another famine. Between 1959 and 1961, the world's largest famine, caused largely by Mao's economic delusions, claimed the lives of 30 million people (Smil 1999b; Yang 2008), and, a generation later, the average per capita food supply stood merely at the pre-1959 level. Securing an adequate food supply was made possible only once China became the world's largest user of nitrogenous fertilizers, thanks to its import of the world's largest series of the most modern ammonia plants, which, by the late 1990s, were synthesizing enough of the key plant macronutrient to provide every citizen with an average per capita food supply that rivaled the Japanese average (Smil 2001; FAO 2024).

But these examples also illustrate the critical interplay of human and technical factors. Only strong organizational capacities and the state's commitment made it possible to complete the first converter in Germany's second large ammonia plant by April 1917, after just 11 months of construction. And only Deng Xiaoping's radical break with the communal past and his de facto privatization of farming opened the way for a much more efficient use of new technical inputs in farming. Indeed, the pervasiveness of the modern state means that most technical developments during the second half of the twentieth century were heavily promoted or retarded by massive government interventions or by a state's benign or deliberate neglect. These realities apply across an entire range of modern techniques, and well-documented developments show how the US government became involved in such different activities as promoting long-distance air transport, enabling nuclear electricity generation, and advancing telecommunications and electronic computing (Pursell 2007).

Pan American Airways became a pioneer of advanced flying—with the first pre-World War II trans-Pacific flights, the first use of jet airplanes in the United States, and its key role in developing the world's first wide-bodied jet—largely because of the provisions of the US Foreign Air Mail Act and of Juan Trippe's (Pan Am founder's) wooing of Latin American political leaders (Davies 1987; Allen 2000). The Act specified that only those corporations whose scale and manner of operation could project the dignity of the United States in Latin America could carry international mail and do so only by the invitation of the serviced countries. By orchestrating a luxurious passenger service, making Charles Lindbergh a technical advisor, and gaining the confidence of the hemisphere's leaders, the company's assured monopoly in carrying the US mail served as the foundation for its technical daring.

The wartime development of nuclear bombs cost the federal government $2.2 billion (USDOE 2024), and much larger sums were spent on amassing the post-1945 arsenal of nuclear weapons and their delivery systems. In comparison, the federal spending on nuclear electricity generation was modest, but disproportional: between 1947 and 1998, the US nuclear industry received more than 96% of $US(1998)145 billion that was disbursed by the Congress for all energy research and development (R&D) (NIRS 1999), clearly an enormously skewed distribution considering the variety of possible conversion techniques that could have been promoted. And, in 1954, the Price-Anderson Act, Section 170 of the Atomic Energy Act, guaranteed public compensation in the event of any catastrophic accident in commercial nuclear generation, a blanket protection not enjoyed by any other energy or industrial producer engaged in a risky enterprise.

Government financed not only the first large electronic computers but also the development of ground-to-air two-way radio for the military, and the postwar patents of its inventor, Al Gross, set the foundations for such ubiquitous gadgets as garage-door openers and mobile phones (Pursell 2007). Other notable US examples of government-sponsored push with an initially dominant military component include the development of space flight and communication satellites, lasers, automated fabrication, robotization, composite materials, and integrated circuits.

And the post-1973 US move toward higher average fuel efficiencies of passenger cars did not come about because the country's major automakers were vying for technical leadership, but rather from direct government intervention that legislated slowly improving performance and that raised the mean from the low of 13.5 miles per gallon (mpg) in 1973 to 27.5 mpg by 1985. The subsequent return to low oil prices sapped the government's resolve to continue along with this highly desirable and technically quite feasible trajectory. By the century's end, the mean stood still at 27.5 mpg while the continuation of the pre-1985 trend would have raised it to about 40 mpg, drastically cutting the need for crude oil imports. So much for autonomous technical advances!

Another important conclusion that weakens any simplistic claims in favor of universal technical determinism is that similar, or even superior, national accomplishments are no guarantee of specific outcomes. This point is perfectly illustrated with the Soviet achievements (Graham 1990). Basic post-World War II Soviet research was essential in developing several key advanced technical concepts, including low-temperature physics (Nobels to Landau in 1962 and to Kapitsa in 1978) and masers and lasers (Nobels to Basov and Prokhorov in 1964 and to Alferov in 2000), while Piotr Ufimtsev's equations for predicting the reflections of electromagnetic waves from surfaces were eventually used by the US designers to develop stealth aircraft.

Massive Soviet investment in military R&D produced the world's largest inventory of nuclear weapons (that peaked at 45,000 warheads in 1986 and included the most powerful fusion devices ever tested) but also some outstanding pioneering designs, and while the Soviet fighter planes had less advanced electronics than did their US counterparts, the expert judgment assessed several of them—most notably the MiG-29 and Su-25—as perhaps the world's best flying machines in their respective categories (Gunston 2002). And by the late 1980s, the USSR was the world's largest producer of crude oil (ahead of Saudi Arabia) and natural gas (ahead of the United States) and the world's second largest producer of coal (behind the United States). And yet these technical achievements could not prevent the disintegration of the Soviet empire and soon afterward the collapse of communist rule.

I revisit the limits of technical determinism in the closing chapter. Here I reiterate my respect for complex, contradictory, and counterintuitive realities and my refusal to offer any simplistic conclusions regarding the role of technical advances in creating the modern world. The most comfortable summation views this role in terms of *yin–yang* pairings: frequently dominant but most often only as *primus inter pares*; fundamental but not automatically paramount; increasingly universal but not immune to specific influences; undeniably autonomous in many respects but heavily manipulated in other ways; consequently, it is a tempting subject for what might seem to be relatively low-risk forecasts that yet turn out repeatedly to be a matter of surprising and frustratingly elusive outcomes. In contrast, there is nothing uncertain about the genesis of this grand technical transformation.

The Unprecedented Saltation

Indisputable evidence demonstrates that evolution—be it of higher life forms or of humanity's technical capabilities—has been an overwhelmingly gradual and cumulative process. Yet the record also shows that the process is punctuated, recurrently and unpredictably, by periods of sudden gains. Similarly, the history of technical advances, dominated by long periods of incremental gains or prolonged stasis, is interspersed with bursts of extraordinary creativity, but none of them even remotely approached the breadth, rapidity, and impact of the astonishing discontinuity that created the modern world during the two generations preceding World War I.

The period between 1867 and 1914, equal to less than 1% of the history of high (settled) civilizations, was distinguished by the most extraordinary concatenation of many fundamental scientific and technical advances. Their synergy produced new prime movers; new materials; new means of transportation, communication, and information processing, and they introduced the generation, transmission, and conversion of electricity, the most versatile form of energy in the modern world. No less remarkable, these bold and imaginative innovations were rapidly improved, promptly commercialized, and widely diffused, creating the fundamental reordering of traditional societies and providing the lasting foundations of modernity: not only their basic operating modes but often many of their specific features are still with us.

Even a rudimentary list of these epoch-defining innovations must include telephones, sound recordings, permanent dynamos, lights bulbs, typewriters, chemical pulp, and reinforced concrete for the pre-1880 years. The astonishing 1880s, the most inventive decade in history, brought reliable and affordable electric lights, electricity-generating plants, electric motors and trains, transformers, steam turbines, the gramophone, popular photography, practical gasoline-fueled internal combustion engines, motorcycles, automobiles, aluminum production, crude oil tankers, air-filled rubber tires, steel-skeleton skyscrapers, and prestressed concrete. The 1890s saw diesel engines, x-rays, movies, air liquefaction, and the first wireless signals. And the period between 1900 and 1914 witnessed mass-produced cars, the first airplanes, tractors, radio broadcasts, vacuum tubes, neon lights, common use of halftones in printing, stainless steel, hydrogenation of fats, air conditioning, and the Haber-Bosch synthesis of ammonia.

Genesis of these astonishing advances and their patenting, improvement, commercialization, and rapid diffusion were the subject of *Creating the Twentieth Century: Technical Innovations of 1867–1914 and Their Lasting Impact*, now the first part of this volume. I also traced some of their subsequent long-term improvements and offered many assessments of their enormous socioeconomic impacts and their lasting indispensability for the functioning of modern societies. The system of electricity generation, transmission, and industrial, commercial, and household use that was created completely de novo during the 1880s and 1890s is also a perfect illustration of my key thesis about the truly revolutionary and lasting import of the era's innovations.

Even the most accomplished engineers and scientists who were alive in 1800 would face, if teleported a century into the future, the electric system of 1900 with astonishment and near utter incomprehension. In 1900, less than two decades after the system's tentative beginnings, the world had a completely unprecedented and highly elaborate means of producing a new form of energy (by using large steam turbogenerators), changing its voltage and transmitting it with minimized losses across longer distances (by using transformers and high-voltage conduits), and converting it with increasing efficiencies using ingenious new electric motors, new sources of light (incandescent bulbs), and new industrial processes (electric arc furnaces).

In contrast, if the brilliant creators of this system, men including Thomas Edison, George Westinghouse, Nikola Tesla, and Charles Parsons (Figure II.1.4), could see the electric networks of the late twentieth century, they would be very familiar with nearly all major components because the fundamentals of their grand designs fashioned before 1900 remain unchanged. The same lack of shocked incomprehension would be experienced by the best pre-World War I engineers able to behold our automobile engines (still conceptually the same four-stroke Otto-cycle machines or inherently more efficient diesel engines), our skyscrapers (still built with structural steel and reinforced concrete), our wireless traffic (still carried by Hertzian waves), or printed images (still produced by the halftone technique).

Our quotidian debt to great innovators of the two pre-World War I generations thus remains immense, extending from complex fundamentals (electricity generation, combustion engines, wireless communication) to simple gadgets (paperclips, crimped caps on beer bottles, barbed wires, spring mouse traps). What has changed, due to the subsequent transformations, are the quality, durability, and affordability of this universe of products.

Figure II.1.4 Ranking the importance of technical advances is always arguable, but perhaps no other four innovators contributed to the creation of modern world, so utterly dependent on electricity, more than did Thomas Edison, Charles Parsons, George Westinghouse, and Nikola Tesla (clockwise from top left).

Edison's 1880 photograph is from the Library of Congress; Parsons's portrait is reproduced courtesy of the Tyne and Wear Museums, Newcastle upon Tyne; Westinghouse's portrait is reproduced courtesy of Westinghouse Electric; and Tesla's 1892 photograph is from the San Francisco Tesla Society.

Transformations and New Departures

First, a brief semantic qualifier and an outstanding illustration of a modern technical transformation: the Latin *trānsformāre* means, obviously, "changing the form," but in this book, I use the term in its widest possible meaning, one that signifies *any substantial degree of alteration* and hence spans many different final outcomes. I am applying the term not only to individual objects (devices, tools, machines) but also to processes (casting of steel, machining of parts), networks (long-distance telephony), and systems (electricity generation). Consequently, high-speed railway transport that was pioneered in Japan during the late 1950s is a perfect example of such a transformative advance.

None of the system's key ingredients was a twentieth-century invention. Steel wheels on steel rails—the arrangement that gives the advantage of firm contact, good guidance, low friction, and ability to move massive loads with low energy expenditures—was present at the very beginning of commercial railways during the 1830s. Electric motors, the most suitable prime movers for such trains, were powering trains already before the end of the nineteenth century: what took place after World War II was a critical shift from direct current (DC) to alternating current (AC), but, by that time that, too, did not require any fundamental inventions. Pantographs to lead electricity from suspended copper wires, aerodynamically styled light aluminum car bodies, regenerative brakes to slow down the trains from speeds above 200 kilometers per hour (km/h)—all these key components had to be improved, but none of them had to be invented. And the first high-speed sets ran on ballasted tracks, as did previous trains.

But when these ingredients were integrated into a new system, railway travel was transformed. No wonder that when the Japan National Railway sought a World Bank loan for its new trunk line (*shinkansen*), the bank's bureaucrats saw it as an experimental project and hence ineligible for funding. But Hideo Shima, Japan's leading railway engineer who traveled to Washington in 1960 to secure the loan, knew what he was talking about when he convinced the lenders "that *shinkansen* techniques included no experimental factor but were an integration of proven advances achieved under the 'Safety First'" (Shima 1994: 48). This is the very essence of successful transformative engineering: a system assembled from proven ingredients whose synergies change a mundane experience to such an extent that it becomes not just new but highly desirable.

When I stood for the first time at a platform of Tokyo's central station and saw the sleek *shinkansen* trains (at that time still the first, 0 series trains with their bullet noses that now look so old-fashioned) come and go at what seemed to be impossibly short intervals, it was a profoundly new experience. Many trips later this remains travel as an event because even when the rapidly receding scenery is hidden, even when standing in a crowded carriage at the end of *obon* holidays, you know that on the same track and just a few minutes behind you is another centrally controlled sleek assembly of 16 cars with 64 asynchronous motors moving in the same direction at more than 250 km/h—and that when you arrive after a journey of hundreds of kilometers, you will be within a few seconds of the scheduled time!

The twentieth century saw many other technical transformations of this kind as new developments changed the external appearance of a product, a machine, or an entire assembly (and sometimes did so beyond easy recognition) while retaining basic components and operating principles. As shown in this chapter's frontispiece, passenger cars are another prominent example of this transformation. The first automobiles of the late 1880s were open carriages with engines mounted under high-perched seats, with wooden wheels of unequal radius, dangerous cranks, and long steering columns. By 1901, Maybach's Mercedes set a new design trend, but outwardly even that pioneering vehicle had little in common with the enclosed, low-profile aerodynamic bodies of modern cars with massive rubber tires, electric starters, and power-assisted steering. But their engines run according to the same four-stroke cycle that Nicolaus Otto patented in 1877, and that Gottlieb Daimler and Wilhelm Maybach adapted for high-speed gasoline-powered automobiles starting in 1886. As Otto would have said, *Ansaugen, Verdichten, Arbeiten, Ausschieben; Ansaugen, Verdichten....*

And cars also illustrate the realities of the second major category of technical transformations, when two objects made a century apart are outwardly very similar, even identical, although their operating modes have undergone a profound transformation. The shapes and sizes of basic components of Otto-cycle engines—cylinders, pistons, valves, cranks—are constrained by their function, and these precision-machined parts must be assembled with small tolerances and run at high speeds in reciprocating motion. But unlike their pre-World War I, or for that matter nearly all pre-1970, counterparts, car engines made during the 1990s were more than complex mechanical devices. Microprocessors control and coordinate the timing of ignition, throttle position, the ratio of the fuel-to-air mixture and its metering and use dozens of sensors to monitor the engine's performance. And since 1996, all US passenger cars and light-duty trucks have been also equipped with diagnostic microcontrollers that detect possible problems and suggest appropriate action.

Other microprocessors (there were as many as 50 in a car of the late 1990s; by 2024, the typical count rose to 1,400–1,500) control automatic transmission, cruise control, and antilock brakes; actuate seat belt pretensioners; deploy airbags; optimize the operation of anti-air pollution devices; and turn the entire vehicle into a mechatronic machine with far more computing power than the Apollo 11 lunar landing module in 1969. As a result, the US car industry became the country's largest purchaser of microprocessors, and, in 2000, electronic components totaled about 8% of the average vehicle's value; by 2020 that share rose to 40% or more than $16,000 per car (Tingwall 2020). Many other industries have seen the same kind of transformation as microprocessors have become indispensable components in an expanding range of products from household appliances to industrial machinery

Analogical transformations affected the production of basic industrial materials as well as the delivery of common services. Haber-Bosch synthesis of ammonia remains the foundation of high yields in modern agriculture, but, by the late 1990s, it required only half of the energy needed in 1920 (Smil 2001, 2022). And while the sequence of dialing, waiting for a connection, and listening to a distant voice was the same with a wired telephone set in 1999 as it was in 1920, the obvious external differences between the two objects (square and bulky vs. sleek and trim shape, black vs. a choice of colors) were much less important than what was inside: all modern landline phones are fully electronic, while all pre-late-1970s devices were electromechanical.

Technical transformations of the twentieth century made a great deal of qualitative difference as they brought increased power, speed, efficiency, durability, and flexibility. Even the most imaginative innovators of the pre-World War I era who invented and commercialized the basic techniques would be impressed by these gains. Many specific comparisons of these advances are cited in the topical chapters of this book, some relatively modest (because the gains were inherently constrained by mechanical, thermodynamic, or construction requirements), others beyond the range of normal imagination (most notably the gains in computing capabilities). And, regardless of the degree of their subsequent qualitative transformations, all innovations that were introduced during the two pre-World War I generations were changed because mechanized (and later automated or robotized) mass production supplanted manual artisanal fabrication or the early assembly lines.

Again, the innovators of the pre-World War I era would be impressed by the extent of this mass production that made their inventions so much more affordable as the economies of scale helped to turn rare possessions into objects of common ownership and spread them around the world. The resulting affordability looks even more impressive when the costs are compared as shares of typical income rather than in absolute (even when inflation-adjusted) terms, and, even then, the costs undervalue the real gains because they ignore very different qualities of the products compared. A contrast of two machines in this chapter's frontispiece reminds us that, in 1900, the car was an uncomfortable conveyance that required brute force to start by cranking, offered little protection against the elements, and guaranteed frequent breakdowns; a century later, climatized interior, adjustable car seats, power-assisted steering and brakes, and audio systems offered comfortable rides, and engines ran with such a reliability that many new models require just a single annual checkup.

And the affordability, as well as quality, of services provided by complex technical assemblies and intricate engineering networks had improved to an even greater degree: American wages, electricity cost, and lighting efficiency offer perhaps the most stunning example of these gains (Figure II.1.5). In 1900, 1 kilowatt hour (kWh) of electricity (enough to light a 100-watt (W) bulb for 10 hours) in a large eastern US city cost about 15 cents (roughly $3.25 in 2000 dollars), while a century later it sold for a mere 6 cents. With hourly wages in manufacturing (expressed in inflation-adjusted monies) rising from $4 in 1900 to $13.90 in 2000, electricity as a share of average income thus became almost 190 times more affordable. But in 1900, the average efficiency of an incandescent light bulb was only about 0.6%, while a century later fluorescent light could convert as much as 15% of electricity into light. Consequently, a lumen of electric light was roughly 4,700 times more affordable in 2000 than it was 100 years earlier (or its cost was just 0.02% of the 1900 value).

But technical advances of the twentieth century went beyond the elaboration and ingenious improvements of fundamental pre-World War I innovations. They included new products whose theoretical solutions and many key components originated before 1914 (public radio, television, and gas turbines are the most prominent examples in this category) but whose practical application came only after World War I. And there were also entirely new achievements that sprang from scientific laws that were understood by 1914 but whose realization had to wait not only for further accumulation of scientific and engineering progress but also for additional fundamental discoveries: nuclear fission and solid-state electronics exemplify these innovations. Radio's pre-1914 progress has been already described, and so just a few paragraphs concerning the genesis of television and gas turbines might be in order.

356 THE TWENTIETH CENTURY

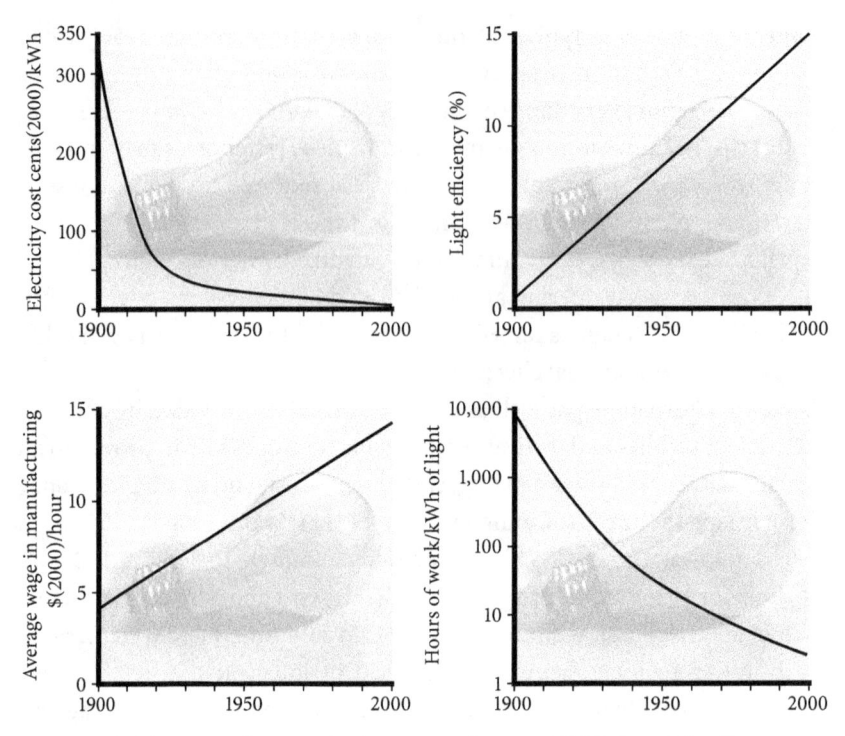

Figure II.1.5 During the twentieth century the cost of US electricity (in constant monies) fell from about $3.25 to 6 cents per kWh (top left) while the efficiency of interior lighting rose roughly 25-fold (top right) and average hourly wages in manufacturing more than tripled (bottom left). The result: largely because of technical advances, a unit of light energy became roughly 4,700 times more affordable (bottom right).
Plotted from data in Smil (2003) and USBC (1975).

The entire phenomenon of television rests on the illusion of continuous movement that is created by rapidly changing sequence of images (minimum required frequency is 25–30 times per second). The key idea—serial transmission of individual pixel signals whose rapid sequence is reconstructed in a receiver by reversing the broadcasting process—was advanced by William Edward Sawyer in 1877. Proposals for mosaic receivers and scanning with revolving disks also go back to the late nineteenth century, as does the prototype of a cathode ray tube, demonstrated in 1878 by William Crookes, and its first practical design introduced by Karl Ferdinand Braun in 1897. The basic ingredients were thus in place before World War I, and, by the mid-1920s, a Mormon teenager, Philo Farnsworth, assembled them in his mind and soon also in a working model of the first television system (see Chapter II.5).

A gas turbine does not merely substitute one working medium (gas or liquid fuel) for another (water vapor). Steam turbines are supplied from boilers by hot, pressurized fluid that is expanded in stages. Gas turbines subsume the induction of fuel, its compression, ignition, combustion, and exhaust in a single device—and unlike the internal combustion engines that do all of this in the same space (inside a cylinder) but intermittently in sequenced stages, gas turbines perform all those operations concurrently and continuously in different parts of the machine. This is why their design posed a much greater technical challenge.

Franz Stolze's designs embodied both the principle and key configurations of modern multistage machines, but his patented 1899 single-shaft turbine (driven by a primitive coal gasifier) barely produced any net power (New Steam Age 1942). Charles Curtis received his patent for a gas turbine engine (US Patent 635,919) also in 1899, but the first small experimental machine with rotary compressors was designed in 1903, by Aegidius Elling, in Norway. The Parisian Société des Turbo-moteurs built several machines between 1903 and 1906, and, in 1908, Hans Holzwarth began his long career in gas turbine design. But these machines had efficiencies below 5%, too low for any commercial use, and breakthroughs in gas turbine engineering had to wait until after World War I.

Innovations and Inventions

I use the term "innovation" in the broadest sense of introducing anything new into daily commercial use and "invention" as a more circumscribed category of devising an original device or machine or an entirely new synthetic, extractive, or fabrication process. All great inventors are innovators par excellence, but some of the twentieth century's farthest-reaching innovations—such as the Toyota system of car production with its constant quality control and just-in-time deliveries of parts (Chapter II.4)—did not require any stunning inventions, merely ingenious applications or rearrangements of established practices. Given the enormous variety of post-1914 technical advances, this book cannot be their comprehensive survey: any attempt to do so would reduce it to a briefly annotated listing of items.

Instead, I adopt the same approach I followed in explaining the epoch-making pre-World War I technical advances and pay reasonably detailed attention to the first-order innovations that remain critical for the functioning of the modern world. Chapter II.2 thus assesses the energetic foundations of the twentieth century civilization, Chapter II.3 examines the material foundation of modern societies, paying particular attention to advances in making steel (as these alloys remain the most indispensable of all metals), to the production of industrial gases (a class of fundamental innovations curiously neglected by most surveys of technical progress), and to the syntheses of plastics before ending with a brief history of silicon, the signature material of the electronic era.

Chapter II.4 is devoted to sweeping transformations brought by mechanization, automation, and robotization, and Chapter II.5 addresses those sectors with a common denominator of mass and speed: transportation, information, and communication. Besides obvious segments on cars, trains, and airplanes, the transportation section also includes a lengthy review of advances in freight shipments. This is followed by brief histories of modern computing and electronics, with a particular emphasis on rapid advances in hardware, from transistors to microprocessors.

Consequently, my narratives dwell not only on the twentieth century's iconic inventions, including nuclear reactors, jet engines, plastics, and computers—but also on innovations whose true importance is appreciated only by experts involved in their management and improvement: such critical and often hidden contributions were no less essential for transforming the century's economic fortunes and social organization than the introductions of well-known advances. Materials are a major part of this surprisingly large category, but how many people besides chemical engineers appreciate that no other compound is synthesized (in terms of moles) more abundantly than ammonia, how many know about the quest for continuous casting of sheet steel or what faujasite is good for? And the category of obscure fundamentals also includes machines (planetary grinders, three-cone drilling bits), massive assemblies (continuous casters, tension leg platforms, unit trains), and immaterial electronic codes (programming languages) and management procedures (*kaizen*).

The same dichotomy applies, naturally, to the innovators: their fame is hardly proportional to the fundamental importance of their contributions. While it is indefensible to survey the technical advances of the twentieth century without writing about the famous men who created the nuclear era

(Enrico Fermi, Robert Oppenheimer) or about the pioneers of the electronic age (Walter Brattain, John Bardeen, William Shockley, Jack Kilby, Robert Noyce), I also introduce those innovators whose names are now either entirely forgotten or are familiar only to small groups of experts. Here are six names of inventors whose brilliant ideas had their genesis during the first half of the twentieth century: Joseph Patrick, Waldo Semon, Hans von Ohain, George Devol, Siegfried Junghans, George de Mestral. And here is another set from the second half, names of innovators whose ideas began transforming the world during the century's second half: Taiichi Ōno, Malcolm McLean, Douglas Engelbart, Nils Bohlin, John Backus, Dennis Ritchie.

If you scored 12 out of 12 you are a walking encyclopedia (and a particularly detailed one) of modern technical history. The two lists are carefully constructed. They include the names of people whose innovations were so fundamental that just about everything we touch and use in the modern world is (directly or indirectly) connected to them (Junghans, McLean), those whose ideas keep saving lives (Bohlin) or huge amounts of money (Ōno), and those whose very disparate contributions (made decades apart) helped to create the modern electronic society (Patrick, Ritchie). And they also include de Mestral and Engelbart, two men whose inventions made repetitive tasks easier and whose simple designs, after initially slow starts, came to be replicated in, respectively, many billions and hundreds of millions, of copies. Billions? Indeed!

In 1948, George de Mestral (1909–1990), a Swiss engineer, wondered how cockleburs fastened to his pants and to his dog's coat: under his microscope he saw tiny hooks and spent the next eight years replicating the mechanism in plastic. His solution: a layer of soft loops pressed against a layer of stiffer hooks (made by cutting raised loops) to make *velours croché* that soon began its global conquest as Velcro (de Mestral 1961; Figure II.1.6). This kind of plastic is now not only used every year in hundreds of millions of pieces of clothing and shoes but also sold in sheets, strips, rolls, and coins to make products ranging from display boards to ammunition packs.

Douglas Engelbart's invention did not become as ubiquitous as Velcro's characteristic unfastening sound, but it ranks among the most massproduced electronic devices. He made the first prototype of an "x-y position indicator for a display system" in 1964, at the Stanford Research Institute and filed the patent application in 1967 (Engelbart 1970; Figure II.1.6). The palm size of the device and its wire tail at the end made the nickname

Sept. 13, 1955   G. DE MESTRAL   2,717,437
VELVET TYPE FABRIC AND METHOD OF PRODUCING SAME
Filed Oct. 15, 1952

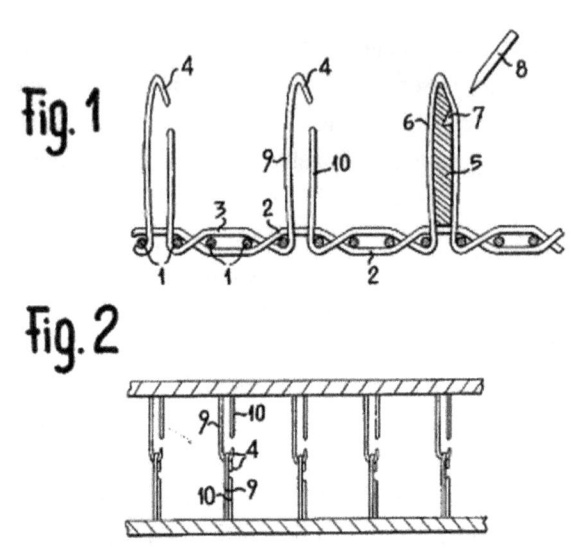

Figure II.1.6 Drawings that accompanied patent applications for two simple but ingenious inventions that have made life easier for hundreds of millions of people around the world: what George de Mestral originally called a "velvet type fabric" and later became universally known as Velcro, and what Douglas Engelbart described as "x-y position indicator for a display system" that later became affectionately called a computer mouse.
Reproduced from de Mestral (1961) and Engelbart (1970).

obvious: when sales of personal computers eventually took off (almost two decades after Engelbart's invention), computer mice were ready to dart around.

You will encounter many other unrecognized innovators as you read the surveys of technical advances that transformed the twentieth century. Once this task is done, I review the key economic, social, and environmental impacts of these contributions in Chapter II.6. Because I have repeatedly demonstrated the futility of long-range forecasts (Smil 2003, 2023a), I do not end the book with any specific predictions regarding the future of modern civilization in Chapter II.7, just with some musings on its durability.

Nov. 17, 1970   D. C. ENGELBART   3,541,541
X-Y POSITION INDICATOR FOR A DISPLAY SYSTEM
Filed June 21, 1967   3 Sheets-Sheet 1

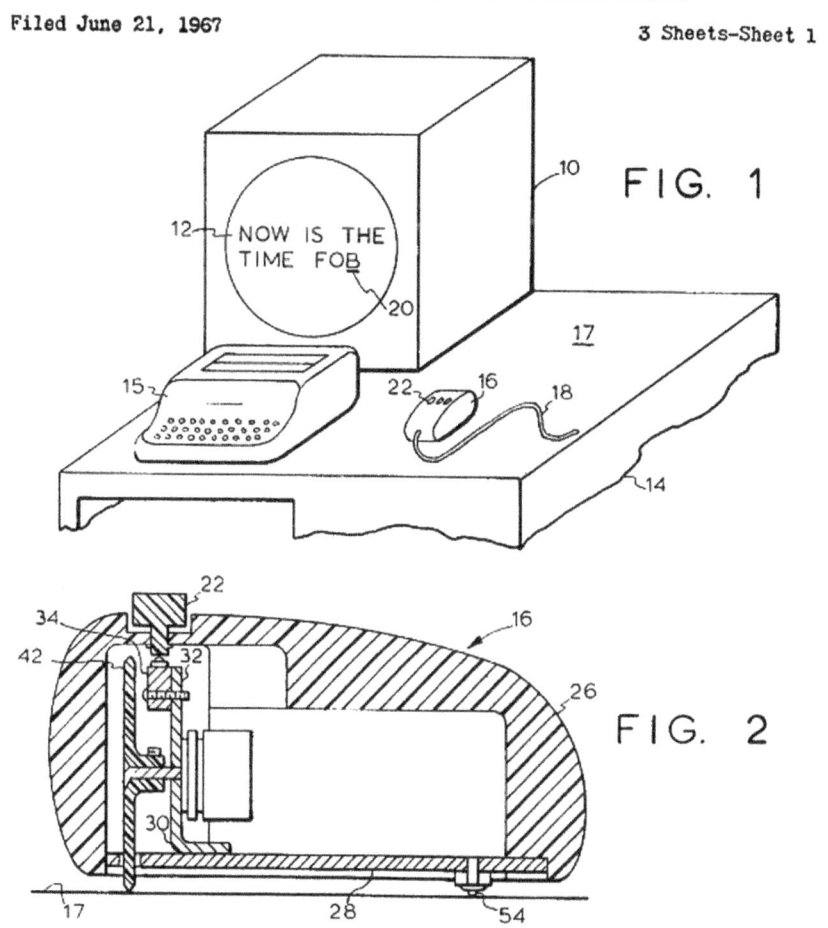

Figure II.1.6 Continued

Frontispiece II.2 High-flow swept fan blades with 3.2-m diameter and a conical spinner of GE90–115B gas turbine aeroengine, the world's most powerful turbofan (top) whose unprecedented size is obvious when mounted on GE's Boeing 747 during the first flight test in the Mojave Desert (bottom). Fiber-and-resin blades are not just elegant and efficient but also quite sturdy, as they can accidentally ingest a bird weighing 1.5 kg and remain fully operational.
Photographs courtesy of GE Aircraft Engines, Cincinnati, Ohio.

2

Energy Conversions

Growth and Innovation

> It is important to realize that in physics today, we have no knowledge of what energy is. We do not have a picture that energy comes in little blobs of a definite amount. It is not that way. However, there are formulas for calculating some numerical quantity.... It is an abstract thing in that it does not tell us the mechanism or the reasons for the various formulas.
> —Richard Feynman, The Feynman Lectures on Physics

Dominant prime movers are the key physical determinants of productive capacities and hence also of the quality and tempo of life in any society. The twentieth century was energized mostly by two prime movers whose genesis during the 1880s and their subsequent evolution have been already detailed: internal combustion engines, whose large numbers (now approaching 2 billion units of many sizes) dominate aggregate installed power capacity of machines, and steam turbines, the most powerful (highest capacities exceeding 1 gigawatt [GW]) energy converters in common use. But one new prime mover rose to great prominence: by the end of the twentieth century, gas turbines became the world's most reliable, as well as highly durable, energy converters. Their two principal uses have been to power both military and commercial flight and to be the most convenient and highly flexible means of electricity generation.

The frontispiece illustration—an oblique view of General Electric's GE90-115B turbofan aeroengine—conveys something of the power and beauty of these extraordinary machines. Their reliability is perhaps best attested by the fact that, during the 1990s, the two-engine Boeing 767 became the dominant trans-Atlantic plane (Figure II.2.1) and that it, with other two-engine planes (Boeing 787, Airbus 340 and 350), eventually replaced three- and

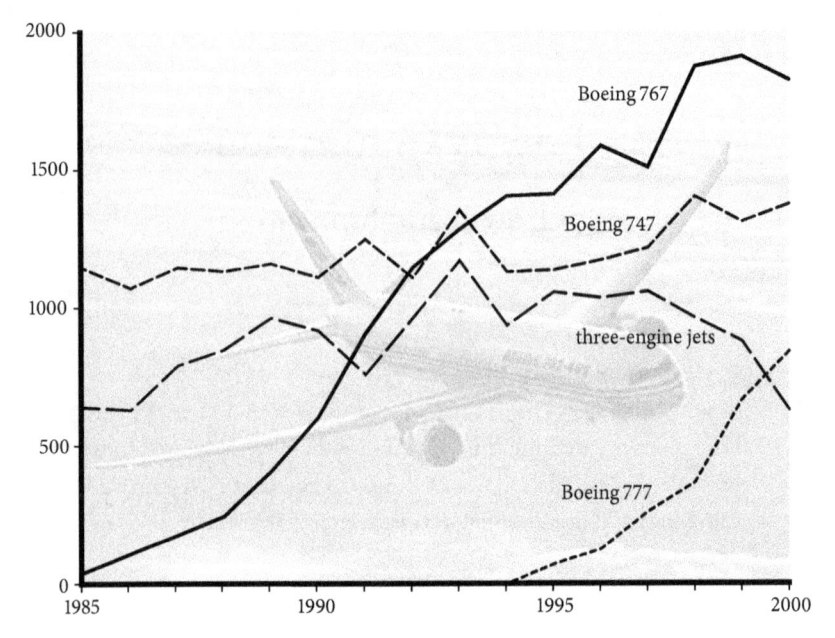

Figure II.2.1 Reliability of modern gas turbines is impressively illustrated by the fact that just six years after its introduction, the two-engine Boeing 767 became the dominant airplane on trans-Atlantic routes.
Based on a graph available from the Boeing corporation.

four-engine aircraft on even longer trans-Pacific routes. During the 1990s, aeroengines could operate, with the requisite maintenance, for up to 20,000 hours—the equivalent of 2.2 years in flight—and stationary gas turbines could work for more than 100,000 hours: at 50% load, this equals more than 22 years of service. Reliability, convenience, and cleanliness made natural-gas–fueled gas turbines the fastest and most economical way of meeting the demand for new electricity generation during the last three decades of the twentieth century.

But this only increased our dependence on fossil fuels and intensified the most fundamental energy trend of the twentieth century, the first period in history when fossil fuels were the globally dominant source of primary energy and when their use created new high-energy societies. Three interrelated trends governed these societies. The first one was the rapid growth and maturation of basic extraction, transportation, conversion, and transmission techniques that resulted in the establishment of new performance plateaus after decades of impressive growth. The second trend was continuous change in the composition of the primary energy supply that brought slow relative decarbonization to the world's primary energy use, and the third was the growing importance of electricity.

Rising demand for electricity was satisfied primarily by the combustion of fossil fuels, but the worldwide construction of large dams and the rapid post-1960 adoption of nuclear generation made important global, and some dominant national, contributions. The history of nuclear generation deserves special attention because this revolutionary innovation is an excellent illustration of exaggerated hopes, unanticipated complications, and failed promises that often accompany technical advances. But the first practical conversion of nuclear energy was to deploy the chain reaction in weapons of unprecedented destructive power: this development, too, deserves closer attention.

Fossil-Fueled Societies

Sometime during the latter half of the 1890s (or, at the latest, before 1905) the aggregate energy content of biomass fuels (wood, charcoal, crop residues, dung) consumed worldwide was surpassed by the energy content of fossil fuels, mostly coal burned by industries, households, and in transport. Pinpointing the year of this transition is impossible because of the absence of reliable data concerning traditional biomass fuel consumption (Smil 2017b). The shift went unnoticed at that time—its occurrence was reconstructed only decades later—and yet this was one of the most important milestones in human history, comparable only to the much more gradual emergence of sedentary farming.

This conclusion defies the common perception of the nineteenth century as the quintessential coal era. That description is correct only as far as England and parts of continental Europe (Belgium, coal-mining regions of France, Germany, Bohemia, and Poland) are concerned, and it does not fit any of the three great economic powers of the twentieth century: Russia, Japan, and the United States. The first two countries became largely fossil-fueled societies only after 1900; the United States began to derive more than half of its primary energy from fossil fuels only during the early 1880s, and, by 1900, it still got a fifth of it from wood (Schurr and Netschert 1960).

The twentieth century was the first period in history when most societies were energized mostly by fossil fuels that had formed in the uppermost strata of the Earth's crust starting mostly about 200 million years ago—while every previous civilization relied on young (mostly just decades-old) carbon in harvested trees and shrubs and, in deforested regions, on even younger (merely months old) carbon stored in crop stalks and cereal straws. My reconstruction of the global supply shows that by 1950 biomass use declined to 28% of the world's energy, and, during the late 1990s, it was just 12% (Smil 2017b). But by the early 1970s, about 90% of all energy in rural China were still derived from biomass (Smil 1988), and similar shares prevailed in the Indian countryside, Southeast Asia, sub-Saharan Africa, and parts of Latin America.

In terms of overall energy use, the global fossil fuel era thus basically coincides with the twentieth century, but more than half of the world's population became dependent on fossil fuels for its everyday energy needs only during the 1970s. As already noted, this global shift was accompanied by three universal transformations: maturation of basic energy techniques, shifting composition of primary energy supply, and the rising importance of electricity that was helped by increasing reliance on two non-fossil sources: hydro energy and nuclear fission. The latter technique was one of the twentieth century's most remarkable innovations, and I will look at both its enormous early promise and at its decades-long stagnation.

Growth and Performance Plateaus

Extraction of fossil fuels is among the most reliably quantified human activities because—in contrast to biomass fuels that are mostly secured by hundreds of millions of individual families for their own use—coal and hydrocarbons have been always produced overwhelmingly by commercial enterprises for sale. During the twentieth century, consumption of fossil fuels rose almost 15-fold, from about 22 exajoules (EJ) per year to 320 EJ/year (UNO 1956; Smil 2017b). Primary electricity generated by water was negligible in 1900, but, by 2000, nuclear, wind, solar, and geothermal electricity added an equivalent of nearly 40 EJ.

An unmistakable formation of S-shaped growth curves was a major recurrent theme in the history of the twentieth-century energy techniques. Some of these curves became first apparent during the 1930s and 1940s, but nearly all these rating and performance plateaus were only temporary, and new records were set after World War II. Fundamentally different plateaus then began to form as maximum specific ratings and conversion efficiencies displayed prolonged spells of minimal or no growth. Moreover, few examples will illustrate how some techniques came close to either thermodynamic or material limits of their performance or how their operation encountered environmental impediments or the lack of social acceptance that proved to be insurmountable.

Crude oil tankers were indispensable for the twentieth-century global energy trade (Ratcliffe 1985; Solly 2022), but between 1884 (when the first vessel of this kind was launched) and 1921, their record capacities increased from slightly more than 2,000 to more than 20,000 dead weight tonnage (DWT); they then stagnated for more than a generation. Capacity growth resumed after World War II as the supertanker designation shifted from 50,000 DWT ships (before the mid-1950s) to 100,000 DWT vessels just a few years later (Figure II.2.2). In 1959, *Universe Apollo* became the first 100,000

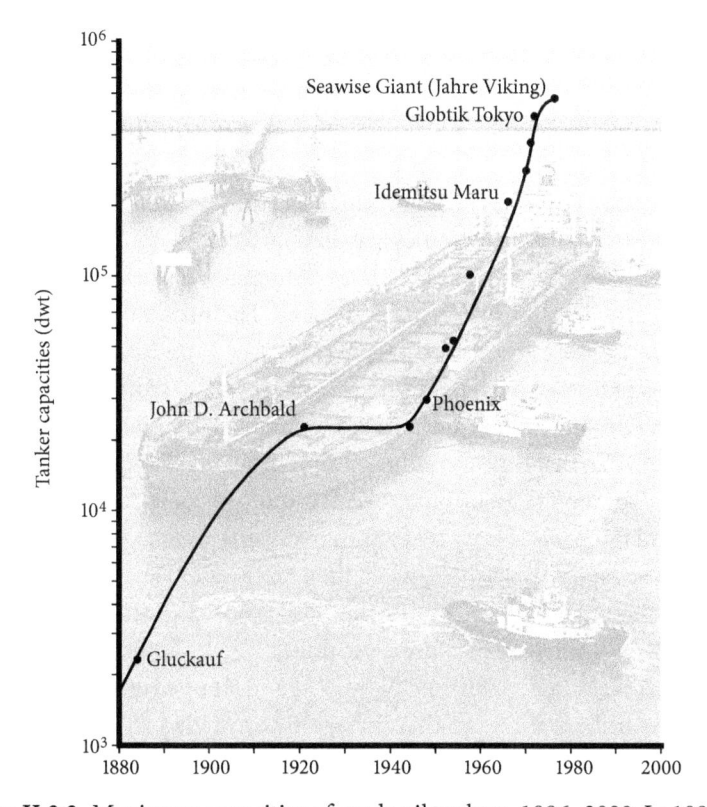

Figure II.2.2 Maximum capacities of crude oil tankers, 1886–2000. In 1886, the *Gluckauf*, a sail-assisted Newcastle-built steamer, became the first bulk carrier with built-in tanks and a capacity of just 2,307 DWT. The *Jahre Viking*, launched as the *Seawise Giant* in 1976, displaced half a million tons.
Based on figure 5.14 in Smil (2017a).

DWT ship; in 1966, Ishikawajima-Harima Heavy Industries completed record-sized 210,000 DWT *Idemitsu Maru*. By 1973, there were 366 very large or ultra-large crude oil carriers, with the largest vessels of more than 300,000 DWT (Kumar 2004).

Both the long-term trend and expert expectations pointed to ships of 1 mega-DWT (MDWT). Instead, the growth peaked with the launching of *Seawise Giant* in 1975, and its enlargement three years later. In 1988, during the Iran-Iraq war, the ship was hit by missiles, but it was subsequently repaired, and the 564,650 DWT (nearly 459-meters [m] long) vessel continued its crude oil deliveries under its new name, *Jahre Viking*; it was then used to store crude oil off Qatar, and, in 2009, it was sold to Indian ship breakers (Konrad 2007). Supertanker growth did not stop because of insurmountable technical problems, but rather because of operational considerations. Very large ships must reckon with the limited depths of many ports and with the long distances needed to stop, and insuring against oil spills and armed attacks became prohibitively expensive. And megaships on the order of 1 MDWT could not pass through the Suez or Panama Canals and could find only a handful of ports to moor, thus losing the flexibility with which multinational oil companies use their tankers.

The history of electricity generation offers many illustrations of temporary pre-World War II performance plateaus. The maximum size of steam turbogenerators rose from 1 megawatt (MW) in 1900 to 200 MW by the early 1930s, and, after a period of stagnation, the post-World War II ascent sent the maximum installed capacities to more than 1,000 MW by 1967 (Figure II.2.3). At that time the industry anticipated machines of more than 2 gigawatts (GW) before 1980, but only a few units of about 1.5 GW went into operation before 2000. Similarly, ratings for transmission voltages in North America rose rapidly until 1923, when they reached 230 kilovolts (kV). This remained the highest voltage until the mid-1950s (the Hoover Dam–Los Angeles 287.5 kV line was the only exception).

The growth resumed with 345 kV in 1954 and stopped at 765 kV by 1965, when the world's first line of that voltage was installed by Hydro-Québec to bring electricity 1,100 kilometers (km) south from Churchill Falls in Labrador to Montréal. A new era of long distance high-voltage direct current (DC) transmission began on June 20, 1972, when Manitoba Hydro's 895-km-long ± 450 kV DC line brought electricity from Kettle Rapids hydro station on the Nelson River to Winnipeg (Smil 1991). The performance plateau that had caused the greatest concern in the electricity-generating industry during

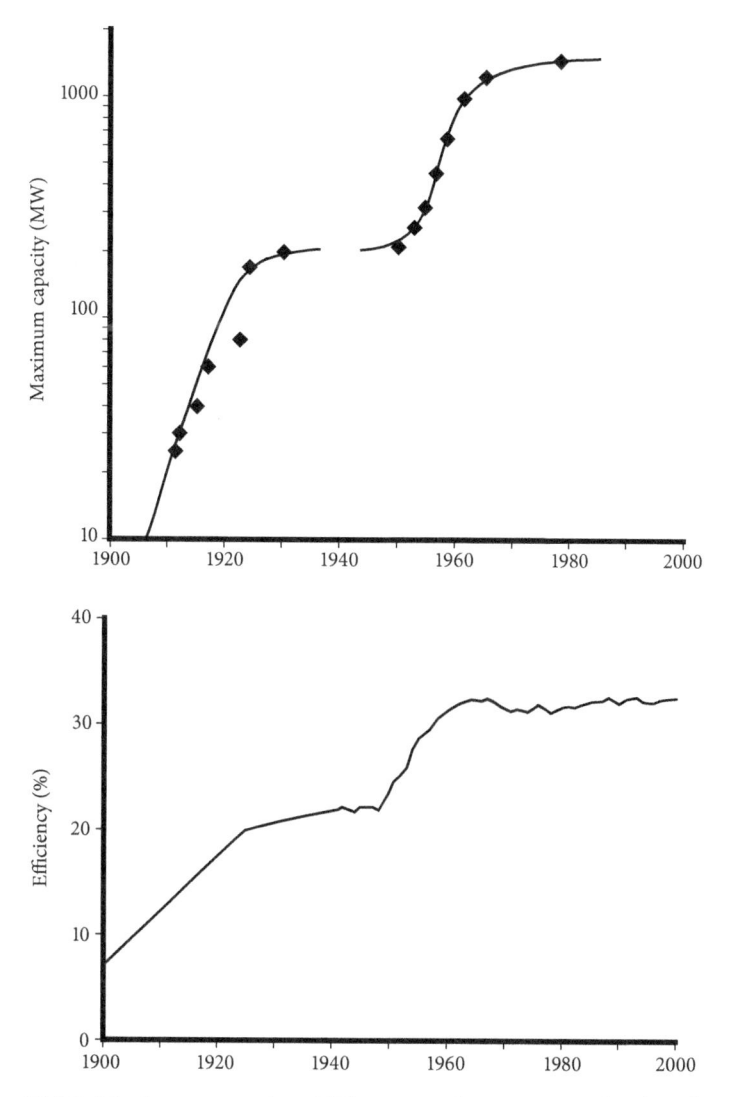

Figure II.2.3 Maximum capacity of US steam turbogenerators (top) and average efficiency of thermal generation of electricity (bottom), 1900–2000. The growth (top) shows two distinct S-curves, the first one with the saturation at 200 MW by 1930, the second one with the saturation at about 1.5 GW by the early 1980s (based on figure 1.9 in Smil 2003). Generation efficiency (bottom) reached its first plateau during the 1930s (slightly more than 20%) and the second one (surpassing 30%) by the early 1960s, and then it stagnated for the rest of the twentieth century.
Based on figure 1.17 in Smil (2003).

the last quarter of the twentieth century was stagnation in the average heat rate of thermal stations. Reliable historical data for the United States show that after this rate rose to just over 30% during the early 1960s, it remained flat (Figure II.2.3).

Energy Transitions and Relative Decarbonization

Energy transitions, whereby the dominant fuel is gradually replaced by a different source or by a mixture of new primary energies, are driven by combinations of environmental, economic, and technical imperatives. When the process is seen from an energetic perspective, there is one overriding explanation: the principal driving force of energy transitions, and of the associated relative decarbonization of the primary supply, has been the rising power density of final energy use (Smil 2015). Modern high-energy societies have benefitted from the high-power densities of fossil fuels to supply their cities (where most people now live), industries (where concentrated mass production is the norm), and transportation (only high-energy-density fuels are easily portable).

Moving from wood (18–20 gigajoule per ton [GJ/t] when dry) to coal (typically between 22 and 26 GJ/t) to refined oil products (40–44 GJ/t) to natural gas (53.6 GJ/t) taps fuels with progressively higher energy density. A home heating example illustrates how the use of these fuels affects storage and supervision requirements: wood requires space-consuming storage and time- and effort-intensive stoking of wood stoves; coal is a better choice on both counts; the same amount of energy in heating oil is easier to deliver, needs even less space to store, and can be fed automatically into a furnace; but natural gas is the best choice: "the strongly preferred configuration for very dense spatial consumption of energy is a grid that can be fed and bled continuously at variable rates" (Ausubel 2003: 2). Natural gas pipelines take care of the feeding; a thermostat regulates the actual consumption rate by a furnace.

The first transition, from wood to coal, had its origins in extensive deforestation (not only because of wood needed for direct combustion and charcoal production but also for wood needed for buildings and ships), the rising costs of wood transported over greater distances to cities, and coal's fundamental advantages. Besides higher energy density, the fuel also has higher flame temperatures and is easier to transport. Origins of the British transition to coal predate the beginning of industrialization, while (as already noted) in other major Western economies, as well as in Japan, this shift began in earnest at different times during the nineteenth century (Smil 2017b). This substitution began the continuing relative decarbonization of the global energy supply.

Wood is made up largely of cellulose, a polymer of glucose, and lignin, and it contains about 50% carbon (C) and 6% hydrogen (H) by weight (Smil 1983). Taking 19 GJ/t as a representative mean, complete oxidation would release about 30 kilograms (kg) C/GJ (or 110 kg carbon dioxide [CO_2]). Ultimate analyses of typical bituminous coals show that they contain between 65% and 70% C and about 5% H. Good bituminous coal will yield about 27 GJ/t, and its combustion will emit roughly 25 kg C/GJ (IPCC 1996; Hong and Slatick 1994). Moving from wood to coal will thus release typically 10–15% less carbon per unit of energy. Coal's primacy was eroded only slowly during the first three decades of the twentieth century: it supplied about 93% of the word's primary commercial energy in 1910, 88% in 1920, and still almost 80% in 1930 (UNO 1956; Smil 2017b).

By 1950, the energy content of global coal consumption was double that in 1900, but coal's share fell to slightly more than 60% as crude oil provided 27% and natural gas 10% of the world's total primary energy. And once post–World War II Europe and Japan began to put in place the foundation of modern consumer societies, coal's retreat and oil's ascent accelerated. By 1962, less than half of the world's primary energy was coming from coal, and even the two rounds of large oil price increases by the Organization of the Petroleum Exporting Countries (OPEC) of the 1970s did not result in coal's comeback. But they did slow down the rate of transition as coal's share declined from 34% of the total in 1970 to 30% by 1990 and 23% by the end of the century (Figure II.2.4). During the entire century, coal supplied about 5,500 EJ compared to about 5,300 EJ for crude oil. This difference is within the margin of statistical and conversion errors: the twentieth century was an energetic draw between coal and oil—but during the century's second half crude oil's total energy surpassed that of coal roughly by a third.

Cost, environmental impact, and flexibility of use were three major reasons for coal's relative retreat (in absolute energy terms, the fuel's output in 2000 was 4.5 times higher than in 1900). Relatively high risks of coal production in underground mines have never limited the use of the fuel. While open-cast extraction greatly reduced the risk of mining accidents, it has only accelerated the trend toward progressively poorer quality of fuel. In 1900, a

372 THE TWENTIETH CENTURY

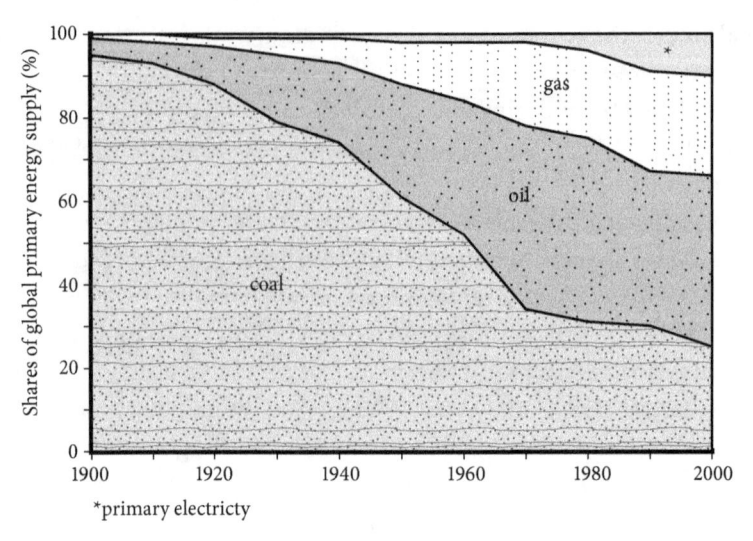

Figure II.2.4 Shares of global primary energy supply, 1900–2000. Coal remained the single most important commercial fuel until 1966, but its subsequent retreat and the loss of major markets make any strong worldwide comeback highly unlikely.
Calculated from data in UNO (1956, 2002) and EI (2023).

ton of mined coal was equivalent to about 0.93 tons of standard fuel (hard coal with 29 GJ/t); by 1950, the global ratio was 0.83, and, by the century's end, it fell to just below 0.7 (UNO 1956, 2002). Declining quality means that larger masses of coal must be extracted and transported and that pollution controls are needed to prevent higher emissions. Beginning in the 1950s, electrostatic precipitators, rated to remove 99.9% of all fly ash, became a standard part of coal-fired stations, and flue-gas desulfurization processes (further increasing both the capital and operation costs) have been commercialized since the late 1960s.

And even the best coal has 30% lower energy density than crude oil; it is much less convenient to store, transport, and distribute; and it cannot directly energize internal combustion engines. Consequently, outside China and India, coal's market shrank to just one large and two secondary sectors: electricity generation (in 2000 almost 40% of the world's electricity was generated in coal-fired plants) and the production of metallurgical coke and cement. In the United Kingdom, the industry's pioneer year and the world's second largest coal producer in 1900, coal extraction declined to less than 20 megaton (Mt) per year by 2000 as the peak labor force of 1.25 million miners (in 1920) was reduced to fewer than 10,000 (Hicks and Allen 1999). And while in 2000 coal still accounted for a quarter of the world's total primary energy output, most of the world's countries never used any of it.

But during the twentieth century every country turned to products refined from crude oil, which became the world's leading source of fossil carbon in 1966; during the 1970s, its share peaked at about 44%, and in 2000 crude oil supplied about 40% of all commercial primary energy (EI 2023). Crude oil has been particularly critical in transportation, where its share was more than 90% of all energy in 2000. Land transport was further enabled by the availability of inexpensive paving materials derived from crude oil, and the fuel is also a valuable feedstock for many chemical syntheses. Modern civilization is thus defined in many ways by its use of liquid fuels, and it has gone to great lengths to ensure their continued supply.

Combustion of liquid fuels liberates much less carbon per unit of energy than does the burning of coal. The ultimate analysis of crude oil shows a carbon content of about 85%, and with 42 GJ/t, their typical emission factor will be about 20 kg C/GJ, 20–25% lower than for coals. Despite some significant differences in density, pour point, and sulfur content, crude oils have very similar energy content (42–44 GJ/t), about 50% higher than standard hard coal. And, unlike coal, they are also conveniently, cheaply, and safely transported by pipelines and by tankers. Global transition to oil was made possible by a combination of rapid technical progress, discoveries of enormous resources in the Middle East, and the extraction of Siberian and American oil.

The fossil fuel with the lowest share of carbon is, of course, methane (CH_4), the principal constituent of natural gases. During the first decade of the twentieth century these gases contributed only about 1% of the world's commercial primary energy consumption, mostly due to slowly expanding US production. Expanded production was predicated on the availability of inexpensive large-diameter seamless pipes and efficient compressors. By 1950, the share of natural gas was still only about 10% of the world's primary commercial energy, but then its cleanliness and convenience made it the preferred fuel for space heating, as well as for electricity generation.

After the OPEC-driven oil price increases of the 1970s, coal's relative retreat proceeded at a slower pace, and crude oil, despite numerous predictions of imminent production decline, was also retreating more slowly. That is why by 2000, natural gas still supplied no more than 25% of the world's commercial primary energy. With 75% of carbon in CH_4 and 35 megajoules per cubic meter (MJ/m^3) (53.6 MJ/kg), its combustion releases only 15.3 kg C/GJ, nearly 20% less than gasoline and 40% less than typical bituminous coal. Global energy supply was further decarbonized by rising shares of primary electricity generation: shares of hydro energy and nuclear fission rose from just 2% of the total in 1950 to about 10% by 2000. The overall carbon intensity of the world's energy supply thus declined from more than 24 kg C/GJ of commercial energy in 1900 to about 19 kg C/GJ in 2000, more than a 20% decrease.

Electrification

The genesis and early achievements of electrification have been already covered in detail. So were the multiple reasons for the superiority of electric lighting and of electric motors in industrial production and in some modes of transportation. Electricity's conversion also delivers high-quality thermal energy for markets ranging from the smelting of aluminum to household space heating. All these advantages were appreciated from the very beginning of the electric era, but technical, economic, and infrastructural imperatives dictated gradual advances of electrification.

Early electricity generation was very inefficient and hence expensive. In 1900, the waste of fuel was astonishingly high, as the average US heat rate of 91.25 MJ per kilowatt hour (kWh) converted less than 4% of coal's chemical energy to electricity. The rate more than tripled by 1925 to nearly 14% and then almost doubled by 1950 to roughly 24% (Schurr and Netschert 1960). The nationwide mean surpassed 30% by 1960, and the best new stations topped the 40% mark. Burning of pulverized coal (for the first time in 1919, at London's Hammersmith power station) and larger turbogenerators (see Figure II.2.3) operating at higher pressures and temperatures (rising, respectively, from less than 1 megapascal [MPa] to more than 20 MPa, and from less than 200°C to more than 600°C) were the principal reasons for the improved performance.

As already noted, these techniques matured by the late 1960s, and average generation efficiency remained stagnant for the remainder of the twentieth century (see Figure II.2.3). This stagnation (during the late 1990s, energy wasted annually in US thermal electricity generation surpassed Japan's total energy consumption!) was one of the greatest weaknesses of the late twentieth-century energy system. Even so, between 1900 and 2000, the average performance of thermal power plants rose by an order of magnitude, and the costs of generation—and electricity prices—declined impressively. The inherently capital-intensive and time-consuming process of building transmission and distribution networks meant that even in the United States fewer than 5% of farms had electric service right after World War I, compared to nearly 50% of all urban dwellings, and the country's rural electrification was rapidly completed only during the early 1950s (USBC 1975).

Electricity's numerous advantages and declining generation and transmission costs resulted in exponential growth of its generation and in a constant rise of the share of the total fossil fuel supply needed to generate it (Figure II.2.5). The US share rose from less than 2% in 1900 to slightly more than 10% by 1950 and to 34% in 2000 (EIA 2004). These shifts proceeded even more rapidly in many modernizing countries. Most notably, China converted only about 10% of its coal to electricity in 1950; by 1980, the share surpassed 20%, and, by 2000, it was about 30%, surprisingly like the US share (Smil 1976; Fridley 2001). The global share of fossil fuels converted to electricity rose from slightly more than 1% in 1900 to 10% by 1950 and to slightly more than 30% by the end of the century.

During the first stage of electrification, which lasted in North America until the early 1930s, there were only two large markets, mechanical drive and lighting, and per capita consumption of electricity remained below 1 MWh/year. The only major appliances that were widely owned in the United States before World War II were vacuum cleaners and refrigerators (by 1940, nearly half of all US households had them) and clothes washers (Burwell and Swezey 1990). In Europe and Japan, these appliances became common only during the 1950s, and successive diffusion waves included frost-free refrigerators and dishwashers, in North America also freezers, clothes dryers, and first room and then central air conditioning. During the same time, markets in affluent countries became saturated with smaller heating appliances ranging from toasters and coffee makers to (beginning in the mid-1970s) microwave ovens.

The variety of these devices meant that no single appliance dominated the household use of electricity but that, during the late 1990s, US households spent more on electricity than on any other source of energy (EIA 1999). After the labor-saving appliances came the acquisition of

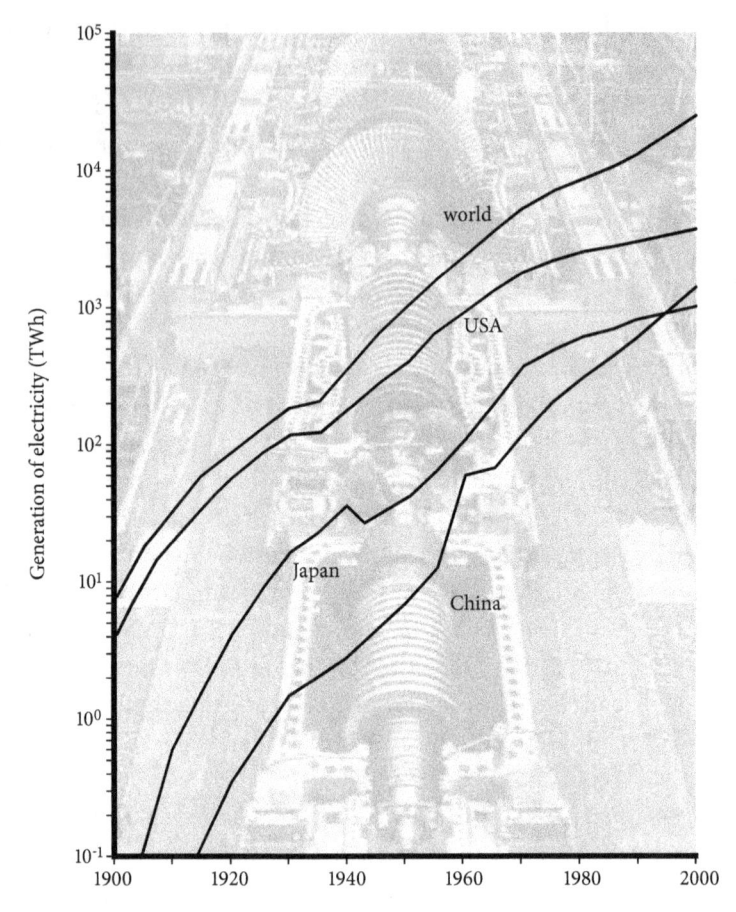

Figure II.2.5 Electricity generation (worldwide and for the three largest producers), 1900–2000.
Based on a graph in Smil (1999a) and data in EIA (2004).

new entertainment devices and personal computers, their attachments (printers, scanners), and other kinds of office equipment such as copiers and fax machines: during the late 1990s, their demand amounted to only about 2% of the total US electricity use (Kawamoto et al. 2000). At the same time, electronic devices and the profusion of remote-ready equipment (TVs, VCRs, and audio) also became the major reason that the average US household leaked constantly about 50 W: these "phantom" loads added up to about 5% of all residential use (Meier and Huber 1997). These demands pushed the US annual household electricity consumption above

4 MWh per capita, and the overall average rate surpassed 12 MWh per capita. In Europe, these rates remained well below 10 MWh, in China still below 1 MWh.

A 1998 US poll showed refrigerators to be the appliances that would be the hardest to live without: 57% of respondents said so, compared to just 12% who chose air conditioning first (Shell Poll 1998). Commercial refrigeration made great advances during the 50 years before 1930, but household refrigeration was slow to take off (Nagengast 2000). Early models were too massive, too expensive, and too unreliable, and they used such flammable or toxic refrigerants as ammonia, isobutane, methyl chloride, or sulfur dioxide. During the 1920s, better electric motors, compressors, and sealants made the machines mechanically adequate, but it was only in 1928, when Thomas Midgley introduced chlorofluorocarbons (CFCs) as coolants, that the ownership of refrigerators took off (Figure II.2.6).

Before World War II nearly half of the US households owned a refrigerator; during the 1950s, the ownership began to spread in Europe and later to richer segments of urban populations in low-income countries, with some markets accomplishing saturation in little more than a single generation. CFCs were readily adopted as ideal refrigerants: they were

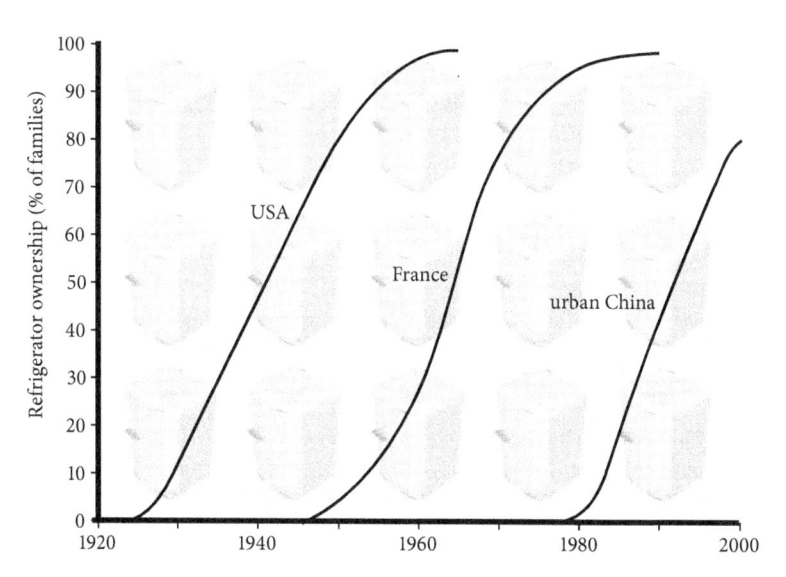

Figure II.2.6 Diffusion of household ownership of refrigerators in the United States and France and in China's urban areas, 1920–2000.
Based on graphs in Burwell and Swezey (1990) and Smil (2003).

stable, noncorrosive, nonflammable, nontoxic, and relatively inexpensive (Cagin and Dray 1993). After World War II their use spread to aerosol propellants, foam blowing, plastics production, cleaning of electronic circuits, extraction of edible and aromatic oils, and air conditioners. Space cooling became common in the United States during the 1960s, and, by the 1990s, it spread among affluent segments of populations living in hot climates.

As microprocessors became a part of myriads of industrial, commercial, transportation, and communication operations, the 99.9% reliability of electricity supply became unacceptably low, and, during the 1990s, the electricity-generating industry aimed at "six nines"—99.9999%—reliability, which would limit outages to just a few seconds a year. Electricity's indispensability now extends to every segment of the national economy and everyday life: automation, robotics, and the entire information age are as unthinkable without it as are mass manufacturing and household chores. These ubiquitous benefits preclude any clear ranking of applications, but there is little doubt that from the existential point of view its most transformative impact was to improve human health (Ausubel and Marchetti 1996; WHO 2023).

Vaccines (requiring refrigeration) eliminated a number of previously widespread infectious diseases (smallpox, diphtheria, whooping cough); incubators (requiring electric heating and constant monitoring of vital functions) provided increased chances for the survival of premature babies; water treatment (requiring electric motors for pumps and mixing) drastically reduced the incidence of gastrointestinal illnesses, as did the refrigeration of food (at the point of production, during transportation, and at home); electric lights made homes (staircase falls are a leading cause of accidents) and streets safer; and a variety of diagnostic, treatment, and surgical equipment (from x-rays to lasers) helped to prolong lives of hundreds of millions of people.

Finally, there is perhaps no better demonstration of electricity's macroeconomic importance than contrasting the intensity of its use in the economy (kWh consumed per unit of gross domestic product) with the long-term direction of overall energy intensity (Figure II.2.7). The latter rate tends to increase during the early stages of industrialization but then it generally declines. The US peak was in the early 1920s; by 1950, the rate declined by a third, and, during the next 50, years it was almost exactly halved. Similarly, the energy intensity of the world economic output was more than halved during the twentieth century (Smil 2003). In contrast, electricity

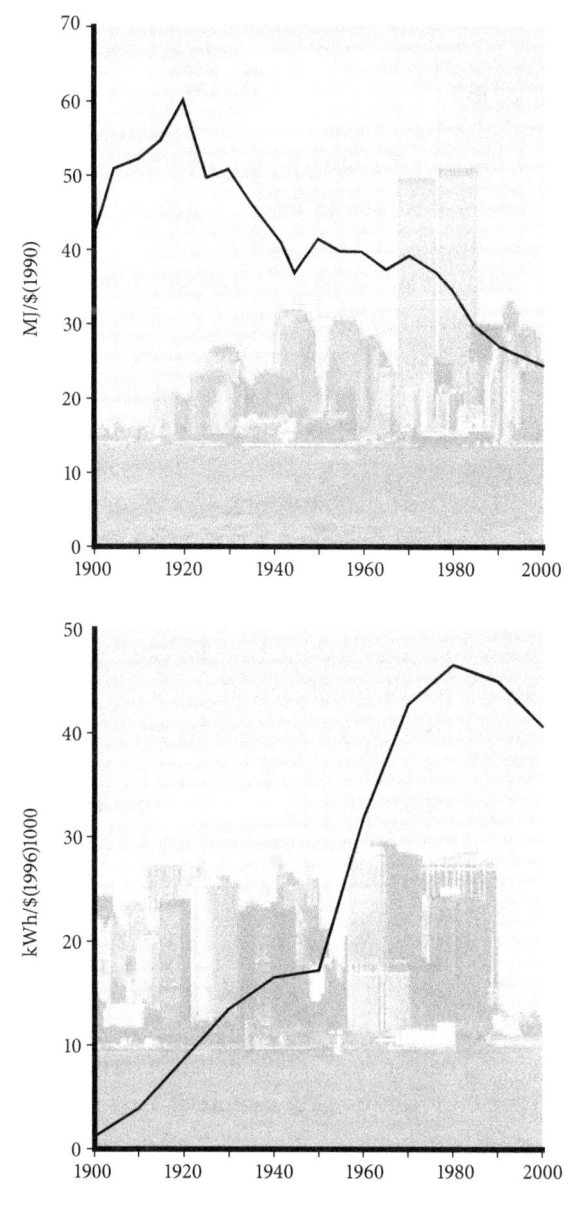

Figure II.2.7 Energy and electricity intensities of the US economy, 1900–2000.
Based on graphs in Smil (2003 and 2017) and data in EIA (2004).

intensity continued to increase even after 1950: electricity intensity of the US economy rose about 2.7 times between 1950 and 1980, then declined a bit, but by 2000 it was still more than twice as high as two generations earlier (Figure II.2.7).

Nuclear Energy

The beginnings of the theoretical foundations of nuclear fission predate World War I. In the spring of 1896, Henri Becquerel (1852–1908) discovered uranium's radioactivity And, not long after the publication of his famous 1905 relativity paper, Albert Einstein (1879–1955) began to develop what he called an "amusing and attractive thought": in 1907, he concluded that "an inertial mass is equivalent with an energy content μc^2" (Einstein 1907: 442). Demonstrating this equivalence with chemical reactions was entirely impractical, but Einstein knew that "for radioactive decay the quantity of free energy becomes enormous." In 1911, Ernest Rutherford's (1871–1937) studies of the penetration of thin layers of gold by alpha particles led him to propose a model of atomic nuclei (Rutherford 1911), and, two years later, Niels Bohr (1885–1962) revealed his simple model of the atomic structure with the nucleus surrounded by orbiting electrons (Bohr 1913).

The two essential advances that opened the road to fission were made only at the beginning of the 1930s, both in Rutherford's famous Cavendish Laboratory in Cambridge. In 1931, John Douglas Cockroft (1897–1967) and Ernest T. S. Walton (1903–1995) achieved the first fission of an element by using high-voltage electricity to accelerate hydrogen protons to disrupt the nucleus of lithium (^{7}Li) and convert it to two alpha (^{4}He) particles. And in February 1932, James Chadwick (1891–1974; Figure II.2.8) provided the correct explanation of some of the earlier experimental results coming from Germany and France that produced "a radiation of great penetrating power."

Chadwick (1932: 312) concluded that

> These results ... are very difficult to explain on the assumption that the radiation from beryllium is quantum radiation. ... These difficulties disappear, however, if it be assumed that the radiation consists of particles of mass 1 and charge 0, or neutrons. The capture of the α-particle by the Be^9 nucleus may be supposed to result in the formation of a C^{12} nucleus and the emission of the neutron. From the energy relations of this process the velocity of the neutron emitted in the forward direction may well be about 3×10^9 cm. per sec.

Neutrons—uncharged elementary particles that are just slightly more massive (about 1.0008 times) than protons and whose fast variety has velocity of about 1.4×10^9 centimeter per second (cm/sec)—do the splitting

Figure II.2.8 In 1932, James Chadwick was the first physicist to posit the existence of the neutron.
Photograph ©The Nobel Foundation.

of nuclei of the heaviest natural elements in chain reactions. In a lecture on September 11, 1933, Rutherford claimed that "anyone who looked for a source of power in the transformation of the atoms was talking moonshine" (Lanouette 1992: 133). One man's irritation with this opinion had enormous consequences for the transformation of the twentieth century. Leo Szilard (1898–1964)—a Hungarian physicist, Einstein's student in Berlin, and, since April 1933, an exile in London—was an intellectual adventurer engaged in several scientific disciplines, from designing (with Einstein) a refrigerator without any moving parts to probing several branches of modern biology.

But Szilard's innate inability to concentrate for long periods on a single research topic made him, using Lanouette's apt label, a genius in the shadows. After reading Rutherford's verdict, Szilard began to think, in his usual way while soaking in a bathtub or walking in a park, about ways to prove that the doyen of nuclear physics was wrong. Szilard's eureka moment came as he stopped for a streetlight on Southampton Row.

As I was waiting for the light to change and as the light changed to green and I crossed the street, it suddenly occurred to me that if we could find an element which is split by neutrons and which would emit two neutrons when it absorbed one neutron, such an element, if assembled in sufficiently large mass, could sustain a nuclear reaction. I didn't see at the moment just how one would go about finding such an element or what experiments would be needed, but the idea never left me. (cited in Lanouette 1992: 133)

And so it was that on September 12, 1932—little more than six months after Chadwick's publication of neutron discovery and more than six years before the first neutron-driven fission took place in a German laboratory—Leo Szilard conceived the basic idea of nuclear chain reaction. This was a fundamental conceptual breakthrough because Szilard's idea encompassed both the nuclear chain reaction and the critical mass of a fissionable element to initiate and sustain it. Unable to get any financial support to investigate the idea, Szilard applied for a British patent on March 12, 1934, identifying (using incorrect data) beryllium as the most likely element to be split by neutrons, but also naming uranium and thorium as other possible candidates.

During the subsequent years Szilard was unable either to get a major research position to pursue the concept or to interest industrial leaders in researching its potential. His only success was that, in 1936, the British authorities agreed to keep his patent secret to prevent its use in developing a nuclear weapon by a hostile power. That possibility appeared all too real as 1938 turned into 1939. In December 1938, Otto Hahn (1879–1968) and Fritz Strassman (1902–1980) irradiated uranium by slow neutrons and produced several new isotopes other than the transuranic elements whose formation was seen in previous experiments.

In their first report of these tests, they noted that this was "against all previous experience in nuclear physics" (Hahn and Strassman 1939a: 15). Unable to explain this outcome, Hahn turned for help to Lise Meitner (1878–1969), Germany's first female (in 1926) professor of physics, with whom he collaborated in research on radioactivity since 1907 (Figure II.2.9). Their association ended with Meitner's forced exile to Sweden in July 1938, after her Austrian passport, which gave her protection against the Nazi anti-Jewish laws, became invalid with Hitler's annexation of Austria. Meitner, and her nephew Otto Frisch (1904–1979), at that time on a visit to Sweden from his exile in Copenhagen, interpreted the result correctly as nuclear fission: Frisch

Figure II.2.9 Otto Hahn (left, in his office during the 1930s) and Lise Meitner (right, photograph from the late 1920s).
Images courtesy of the Hahn-Meitner Institut in Berlin.

chose the term *(Kernspaltung* in German) after William Arnold, a visiting biologist, told him that was the proper way to describe dividing bacteria.

Hahn and Strassman conclusively stated that their experiments produced barium isotopes (atomic number 56) from uranium (atomic number 92) in a paper published on February 10, 1939, in *Naturwissenschaften* (Hahn and Strassman 1939b); a day later, a correct theoretical interpretation by Meitner and Frisch appeared in *Nature*.

> It seems therefore possible that the uranium nucleus has only small stability of form, and may, after neutron capture, divide itself into two nuclei of roughly equal size.... These two nuclei will repel each other and should gain a total kinetic energy of c. 200 Mev, as calculated from nuclear radius and charge. (Meitner and Frisch 1939: 239)

Five years later Hahn was honored for his discovery with a Nobel Prize in Chemistry; the award was revealed only in November 1945, and the award ceremony took place a year later (Hahn 1946). Meitner received neither the prize nor Hahn's generous recognition for her contributions: their long collaboration, in which she was often the intellectual leader, and the letters they exchanged between December 1938 and spring 1939 leave no doubt about Hahn's subsequent biased reinterpretation of actual events (Hahn 1968; Sime 1996; Rife 1999). As pointed out by Meitner and Frisch, nuclear fission could liberate a stunning amount of energy, and physicists in Germany, the United Kingdom, France, the United States, and the USSR immediately concluded that this reality could eventually lead to an entirely new weapon of unprecedented destructive power.

Philip Morrison recalled how, in 1939, physicists in Berkeley sketched "with a crudely correct vision . . . an arrangement we imagined efficacious for a bomb" and how Oppenheimer, in a letter on February 2, 1939, noted "that a ten centimeter cube of uranium deuteride . . . might very well blow itself to hell" (Morrison 1995: 42–43). Soon afterward Szilard, Walter Zinn, and Enrico Fermi (1901–1954), together with Herbert Anderson, observed the release of fast neutrons. This new reality prompted Szilard to draft his famous letter to President Roosevelt that Albert Einstein signed and sent on August 2, 1939. This intervention led (during the summer of 1942) to the establishment of the Manhattan Engineer District (known simply as the Manhattan Project) and to the world's first fission bombs (USDOE 2024).

At that time the world's most expensive research project thus began to develop nuclear weapons before Germany did so. Germans had some of the world's leading nuclear physicists and access to uranium (from mines in the occupied Czechoslovakia), and they began their formal nuclear research program ahead of the United States. The Manhattan Project's success opened the intertwined path of the nuclear arms race and fission-powered electricity generation, both being developments of immense complexity. One retrospective judgment that cannot be challenged is that the extreme expectations were not—fortunately and regrettably, as the case may be—fulfilled. Nuclear weapons did not obliterate, as widely feared, the late twentieth-century civilization, and nuclear fission did not furnish it, as was perhaps equally widely expected, with an unlimited source of inexpensive energy.

Nuclear Weapons

On September 17, 1942, Colonel Leslie Groves (Figure II.2.10)—an Army engineer who supervised the construction of the Pentagon and who was made Brigadier General on September 23, 1942—was named to head the Manhattan Project. Just two days after his appointment he secured the site

Figure II.2.10 The two men oversaw the Manhattan Project: General Leslie Groves (left) and physicist Robert Oppenheimer (right).
Photographs from the Archival Research Catalog of the US National Archives and Records Administration.

of the first uranium isotope separation plant at Oak Ridge, Tennessee. Land for the bomb-making Los Alamos laboratory was purchased in November, and, in December, Robert Oppenheimer (1904–1967) was appointed the scientific director of the project (Figure II.2.10). This was done against the objections of the scientific leaders of the project, who thought Oppenheimer to be an administratively inexperienced theoretician, and even more so against the wishes of the Army Intelligence officers, who doubted his loyalty (Goldberg 1992). The project's challenges and achievements can be followed in fascinating personal reminiscences (Compton 1956; Hawkins, Truslow, and Smith, 1961; Groves 1962), historical accounts (Lamont 1965; Groueff 1967; Rhodes 1986), and extensive files on the World Wide Web (Manhattan Project Heritage Preservation Association 2004; Avalon Project 2003; NWA 2024).

The first sustained nuclear chain reaction was directed by Enrico Fermi at the University of Chicago on December 2, 1942. The first reactor was a large assembly of precisely machined long graphite bricks piled atop one another (hence Fermi's name for it, a "pile") in a squash court underneath the (later demolished) football stadium bleachers at the University of Chicago (Allardice and Trapnell 1946; Wattenberg 1992). Thousands of pseudospheres of metallic uranium and uranium oxide were emplaced in these bricks to form a precise three-dimensional lattice. Graphite acted as a moderator to slow down the released neutrons, and the control rods inserted into slots to stop the reactor from working were strips of wood covered with neutron-absorbing cadmium foil. Fermi calculated that the pile would become critical after emplacing the fifty-sixth layer, and the fifty-seventh layer was laid on December 1, 1942.

The next day's experiments began at 9:45 A.M. with the gradual withdrawal of control rods and proceeded through a series of small adjustments toward the criticality. After lunch and after a further series of adjustments, Fermi gave the order to George Weil to pull out the final control rod by another 30 cm. This was done at 3:25 P.M., and Fermi said to Arthur Compton, who stood at his side, "This is going to do it. Now it will become self-sustaining. The trace will climb and continue to climb, it will not level off" (Allardice and Trapnell 1946: 21). And it did not—as the recording pen began to trace an exponential line and continued to do so for 28 minutes before the world's first nuclear reactor was shut down at 3:53 P.M. Szilard was on Fermi's team of 15 physicists who succeeded in using moderated neutrons to split uranium nuclei, and within days after this achievement he signed over his fission patents to the US government.

Los Alamos scientists concentrated first on a weapon to be exploded by firing one part of a critical mass of ^{235}U into another, a design that did not need full-scale testing. Then they turned to a more complicated design of a plutonium (Pu) bomb whose test became the first fission-powered explosion. Plutonium 239—produced by the irradiation of ^{238}U with neutrons in large, water-cooled graphite-moderated reactors that were designed by Eugene Wigner (1902–1995) and built at Hanford, Washington—was used to make the first nuclear bomb. Its core was a 6-kg sphere of metal that was designed to be imploded, compressed uniformly to supercriticality by explosive lenses. The test took place at Alamogordo, New Mexico, at 5:29:45 A.M. on July 16, 1945, and the explosion was equivalent to about 21 kilotons (kt) of TNT (Figure II.2.11).

Philip Morrison, who was a member of the Los Alamos group that measured the first bomb's critical mass, watched the explosion from the south bank of the base camp. He saw first

> a brilliant violet glow entering my eyes by reflection from the ground and from the surroundings generally.... Immediately after this brilliant violet

Figure II.2.11 The first nuclear explosive device for the test code-named Trinity by Robert Oppenheimer being positioned at Alamogordo, New Mexico, in July 1945 (left). Mushroom cloud rises above Nagasaki on August 9, 1945 (right).
Photographs from the Archival Research Catalog of the US National Archives and Records Administration.

flash, which was somewhat blinding, I observed through the welding glass, centered at the direction of the tower an enormous and brilliant disk of white light.... Beginning at T = +2 to 3 seconds, I observed the somewhat yellowed disk beginning to be eaten into from below by dark obscuring matter.... In a matter of a few seconds more the disk had nearly stopped growing horizontally and was beginning to extend on a vertical direction.... This turbulent red column rose straight up several thousand feet in a few seconds growing a mushroom-like head of the same kind. (Morrison 1945: 1)

In retrospect, the uncharted theoretical and engineering road that led to the first nuclear weapon was traversed in a very short period (in less than 28 months after the project's formal start) thanks both to unprecedented concentrations of intellectual power and to enormous expenditures of energy. The project's participants included eight men who received Nobel prizes before or during World War II (ranging from Niels Bohr in 1922 to

Isidore Rabi in 1944) and 12 others who received them, in physics or chemistry, between 1951 (Glenn Seaborg) and 1989 (Norman Ramsey). The enormous energy needs were due to the separation of the fissile isotope of uranium ^{235}U, which represents only 0.72% of the metal's natural mass, which is dominated by ^{238}U. This was done electromagnetically by a process developed by Ernest O. Lawrence (1901–1958) and later also by thermal diffusion that was developed by Philip Abelson (1913–2004).

Oak Ridge's two uranium enrichment plants furnished the roughly 50 kg of uranium needed for the 12.5 kt bomb that destroyed Hiroshima at 8:15 A.M. on August 6, 1945. The bomb, delivered by a B-29 from Tinian, was dropped by a parachute, and it exploded about 580 m above ground. Its blast wave (maximum speed of 440 m/s) carried about 50% of all energy liberated by the chain reaction and caused massive structural damage over a wide area, destroying all wooden structures within 2 km of the hypocenter and smashing reinforced concrete structures up to about 0.5 km away. The bomb's thermal energy was about a third of the total energy release, and the emitted infrared radiation caused severe burns (Committee for the Compilation 1981).

These two effects, the blast wave and the infrared radiation, were responsible for tens of thousands of virtually instant deaths, while the ionizing radiation caused both instant and delayed casualties. The best available casualty count is 118,661 civilian deaths up to August 10, 1946; military casualties and later deaths attributable to radiation exposure raise the total to about 140,000. A larger (22 kt) plutonium bomb that was dropped at 11:02 A.M. of August 9, 1945, on Nagasaki caused fewer than 70,000 casualties. For comparison, the firebombing of Tokyo on March 9 and 10, 1945, killed 100,000 (Harada, Ito, and Smith 2019). Bombs developed to preempt German nuclear supremacy were thus used for an entirely different, and initially quite unanticipated, purpose. Undoubtedly, they helped to speed up the end of the Pacific war. Arguably, they saved more lives than they took: this view is easily defensible, particularly given the enormous casualties from the conquest of Iwo Jima and Okinawa and given the continuing calls for Japan's resistance by some of its military leaders (Jansen 2000).

These nuclear bombings launched two generations of dangerous superpower conflict. Arguments about the necessity and morality of using nuclear weapons preoccupied many physicists even before the first device was tested. These debates, and efforts to come up with effective international controls of nuclear weapons, have continued in the open ever since August 1945. Isidore

Rabi (1898–1988), one of the key participants in the Manhattan Project, summed up the feeling of most of the bomb's creators on the fortieth anniversary of Los Alamos's founding when he entitled his speech "How Well We Meant" (Sherwin 1985).

Philip Morrison—who helped to assemble the bombs at Tinian, in the Marianas, before they were loaded on planes and who was in a small party of scientists who entered Japan on the first day of US occupation—put the tragedy into a clear historical perspective by contrasting the effect of "old fires" (set by massive and repeated drops of jelly gasoline by huge fleets of highflying B-29 bombers) with the new destruction.

> A single bomber was now able to destroy a good-size city, leaving hundreds of thousands of dead. Yet there on the ground, among all those who cruelly suffered and died, there was not all that much difference between old fire and new. Both ways brought unimagined inferno...the difference between the all-out raids made on the cities of Japan and those two nuclear attacks remains less in the nature of the scale of the human tragedy than in the chilling fact that now it was much easier to destroy the populous cities of humankind. (Morrison 1995: 45)

At least two facts are indisputable: until his death, Harry Truman, who gave the order to drop the bombs, firmly believed that he did so to destroy Japan's capacity to make war and to bring the earliest possible conclusion to the conflict (Messer 1985), and postwar examinations of the German nuclear research showed that Germany's physicists did not make any significant progress toward making a weapon. And we also know that this lack of progress was not because of any deliberate procrastination to prevent these weapons falling under Hitler's command: transcripts of conversations of German nuclear physicists that were secretly recorded during their internment in an English manor in 1945 show that even Werner Heisenberg, the leader of German nuclear effort, had a very poor understanding of how a fission bomb would work (Bernstein 1993; Bethe 1993).

The wartime USSR also had an active nuclear fission program, but its first atomic weapon, which was tested on August 29, 1949, was based on detailed information about the US bomb that the Soviet secret service obtained during the second half of 1945 from Klaus Fuchs, a British physicist who participated in the Manhattan Project (Khariton and Smirnov 1993). The second bomb of Soviet design, smaller yet more powerful, was tested in

1951. The age of thermonuclear weapons began on November 1, 1952, with a 10.4 Mt blast at the Enewetak Atoll when the United States tested its first hydrogen device. Nine months later, on August 12, 1953, the USSR was the first country to test a hydrogen charge that could be used as a bomb (the first US device was an immobile building-sized assembly weighing about 82 t).

The most powerful thermonuclear bomb that was tested by the USSR over the Novaya Zemlya archipelago on October 30, 1961, had an equivalent of 50 Mt of TNT. Its explosion, visible for thousands of kilometers, released 20 times as much energy as all the bombs dropped during World War II, with its mushroom cloud stretching across 20 km and its shock wave circling the planet three times (Stone 1999; Wellerstein 2021). And less than 15 months later Nikita Khrushchev revealed that the country had an untested 100 Mt bomb. Between 1959 and 1961, during the three years of its peak nuclear weapon production, the United States made almost 19,500 new warheads at a rate of 75 a day (Norris and Arkin 1994). In total, the country produced some 70,000 nuclear weapons between 1945 and 1990, the USSR, about 55,000.

The maximum numbers of all kinds of nuclear weapons reached 32,500 in the United States in 1967 and 45,000 in the USSR in 1986; just over a third of this total (more than 13,000 in the United States and more than 11,000 in the USSR) were strategic weapons (mostly between 100 and 550 kt/warhead) that targeted major cities and military installations. These weapons could be delivered by a triad of launchers: bombers, submarine-based missiles, and land-based missiles (NRDC 2002). Long-range bombers were the first means of delivery (B-29s carried the bombs dropped on Japan), and, by the end of the twentieth century, their most advanced varieties include the Soviet Tupolev 95 and the US B-2 stealth planes.

The first nuclear submarine was commissioned on September 30, 1954, and, by the end of the Cold War, submarine-launched missiles (SLBMs) were dominated by powerful Soviet-built SS-N-18 and American Tridents. And beginning in 1959, with the American Atlas 1, and then continuing with Titan 1 and 2 and Minuteman, and in the USSR with SS-24 and SS-25, hundreds of intercontinental ballistic missiles (ICBMs) were emplaced in fortified underground silos in bases largely in the Great Plains and mountain states in the United States and in both European and Asian parts of the USSR (NRDC 2002). Both ICBMs and SLBMs initially carried single warheads, but later they were tipped with multiple independently targeted reentry vehicles (MIRVs); for example, SS-18 carries 10 warheads of 550–750 kt, and, by the late 1990s, Peacekeeper (LGM-118A) launchers have ten 300-kt warheads.

The accuracy of ICBMs (measured in terms of circular error probability [CEP], the diameter of a circle within which half of the missiles would hit) improved thanks to better inertial guidance, midcourse correction, and, eventually, terminal homing using missile-borne radar and satellite images: the CEP of the US ICBMs decreased from around 1,000 m to less than 100 m. Testing of the warheads was an exercise in risky global contamination. During the 1950s, all Soviet and British weapon tests and nearly 90% of the US tests were done in the atmosphere, introducing relatively large amounts of radionuclides that were, inevitably, diffused worldwide and contributed to higher rates of miscarriages, birth defects, and thyroid cancers.

This growing danger was reduced in August 1963, when the United States and the USSR signed a Limited Test Ban Treaty that ended their atmospheric explosions of nuclear weapons (Cousins 1972). Andrei Sakharov (1921–1989), the leading creator of the Soviet hydrogen bomb, contributed greatly to Nikita Khrushchev's acceptance of the ban (Lourie 2002). France and China continued their atmospheric tests until, respectively, 1974 and 1980: by that time 528 weapons had been experimentally exploded in the atmosphere (NRDC 2004). The last underground tests in the United States and the USSR were done, respectively, in 1993 and 1990.

Proliferation of nuclear weapons beyond the two superpowers began with the United Kingdom (fission device in 1952, fusion bomb in 1957). France and China were added, respectively, in 1960 and 1964, and the beginnings of Israel's still unacknowledged arsenal date also to the 1960s. India's first test took place in 1974, South Africa acquired its first (and publicly unacknowledged) nuclear weapons by 1982, and Pakistan announced its first series of underground nuclear tests on May 28, 1998. Other nations, including Iran, Libya, and North Korea, also tried to build nuclear weapons during the last two decades of the twentieth century (and North Korea had eventually succeeded). Four countries—South Africa and the three successor states of the USSR (Ukraine, Belarus, and Kazakhstan)—renounced their nuclear status, but at least three times as many have the intellectual, although not necessarily the technical, capacity to produce them.

The real purpose of the monstrous nuclear arsenal held by the two superpowers was to deter its use by the other side. There is no doubt that the possession of nuclear weapons (the mutually assured destruction [MAD], concept) was the main reason that the United States and the USSR did not fight a thermonuclear war. While the weapons were abominable but effective peacekeepers, the overall level of nuclear stockpiles—eventually nearly 25,000 strategic nuclear warheads with an aggregate capacity of over 5 Gt, equivalent to nearly half a million Hiroshima bombs—went far beyond any rational deterrent level. My very conservative estimate is that at least 5% of all US and Soviet commercial energy consumed between 1950 and 1990 was claimed by developing and amassing these weapons and the means of their delivery.

And the burden of these activities continues with the safeguarding and cleanup of contaminated production sites. The US nuclear weapons complex eventually encompassed about 5,000 facilities at 16 major and more than 100 smaller sites, and estimated costs of its cleanup, maintenance, and surveillance operations have been rising steadily (Crowley and Ahearne 2002). And a much greater investment would be needed to clean up and safeguard the more severely contaminated nuclear weapons assembly and testing sites in Russia and Kazakhstan. The former USSR left behind at least 221 major military nuclear facilities that extensively contaminated surrounding areas (GAO 1995).

Even so, given the potentially horrific toll of a thermonuclear exchange between the United States and the USSR—Natural Resources Defense Council modeling shows that a major US thermonuclear "counterforce" attack on Russian nuclear forces would kill 8–12 million people and a thermonuclear "countervalue" attack on Russian cities would cause 50 million deaths (McKinzie et al. 2001)—one could argue that the overall cost-to-benefit ratio of the nuclear arms race was acceptable. In contrast, there are no positive aspects of unchecked nuclear proliferation, particularly when there are possibilities of fissionable materials, weapon components, or complete weapons coming into the possession of terrorist groups or when a country such as North Korea chooses the threat of nuclear aggression as a tool of its foreign policy. Moreover, research performed since the 1980s demonstrated that effective defense against limited nuclear attacks will not be achieved soon: development of antimissile missiles has yet to produce any deployable, reliable system.

Nuclear weapons obviously transformed the great strategic relationships during the second half of the twentieth century. At the same time, potential for their use by terrorists and blackmail-minded states appeared to be higher than at any time since 1945. At the century's end, we were as much in the need of an effective and truly global nuclear policy as we were half a century earlier, at the very beginning of the nuclear era. Without any doubt, the best outcome would be if Philip Morrison's hopes were to come true.

> The danger remains and will remain until the powers (now almost a dozen of them) realize that practical arms control is the best way out of the dilemma of lawlessness in weapons in the time of modern science. Every day I hope for a world move to end the danger by realizing mutual control. I am an optimist, still, against all my experience, but confirmed by all reason. (P. Morrison, personal communication, May 2004)

Electricity from Fission

Electricity generation powered by nuclear fission owes its genesis entirely to the advances achieved during the unprecedented dash to develop nuclear weapons, and the two endeavors remained closely related during the subsequent decades: countries that pioneered fission-based electricity generation—the United States, the United Kingdom, the USSR, and France—were also those with nuclear arsenals. Concerns and fears surrounding the matters of nuclear armaments have thus cast shadows over the peaceful development of fission for electricity generation. And by the time the demise of superpower confrontation finally weakened that link, commercial nuclear generation was experiencing many difficulties unrelated to its association with weapons.

The conceptual foundations of fission reactors were laid out clearly by Enrico Fermi and Leo Szilard in their patent application for "neutronic reactor," which they filed on December 19, 1944, and which was only granted and published more than 10 years later (Figure II.2.12). The key physical prerequisite of the process was obvious.

> In order to attain such a self-sustaining chain reaction in a system of practical size, the ratio of the number of neutrons produced in one generation by the fissions, to the original number of neutrons initiating the fissions, must be known to be greater than unity after all neutron losses are deducted.... We have discovered certain essential principles required for the successful construction and operation of self-sustaining neutron chain

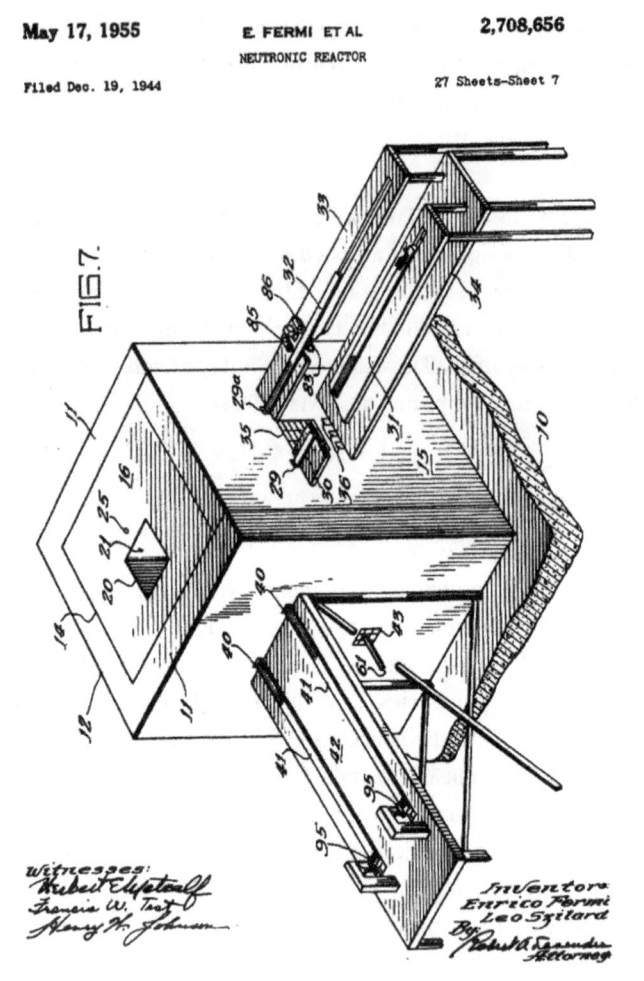

Figure II.2.12 Basic layout of neutronic reactor depicted in the US Patent 2,708,656 filed by Fermi and Szilard in December 1944. The core (*14*) containing uranium and graphite stands on a concrete foundation (*10*) and is encased in concrete walls (*11*). Control rods (*32*) rest on rod platforms (*31*) ready to be inserted through apertures (*29*); in modern reactors, they are inserted from the top.
Image from the US Patent Office.

reacting systems (known as neutronic reactors) with the production of power in the form of heat. (Fermi and Szilard 1955: 1)

The patent anticipated all principal types of future power reactors whose common major components include fissile fuel, moderator, control rods, blanket, and a reactor vessel to hold these parts. The *moderator* is a material of low atomic weight used to slow down the neutrons to increase the probability of their encounters with fissile nuclei. *Coolant* is a fluid used to remove the heat generated by the chain reaction and used for steam production for thermal electricity generation. Water makes both an excellent moderator and an efficient coolant, particularly when under pressure. The *control rods* are made of materials that capture neutrons, and their removal and insertion start and terminate the chain reaction. The *blanket* is a reflector that scatters neutrons back into the reactor core, and reinforced *vessels* that hold the entire assemblies are designed to prevent the leakage of radioactive materials as well as to withstand considerable impacts from the outside.

Fermi and Szilard's patent envisaged a variety of reactors: those moderated by heavy water (D_2O) and using natural uranium (the combination chosen during the late 1950s for CANDU (CANada Deuterium Uranium) reactors of Canada's national nuclear program), those cooled by a gas (a method favored by the British national program), and those using enriched uranium with a light water moderator, the system that, starting during the late 1950s, became the principal choice of the US nuclear industry. While nuclear fission was a new form of energy conversion, heat generated by the splitting of heavy nuclei had to produce steam to be used in the same way as in any thermal station—that is, expanded in turbines to drive large generators. Fermi and Szilard outlined the basic options of recovering the heat released by fission in both low- and high-pressure systems and by using exchangers for indirect steam generation as well producing steam directly in tubes passing through the reactor core in a boiling water reactor.

In 1944, Alvin Weinberg suggested using high-pressure water both as a coolant and moderator for a reactor fueled with slightly enriched in fissionable ^{235}U. This led to practical designs of pressurized water reactors (PWR) that were chosen first for submarine propulsion. A technique of unusual complexity was then transformed into a highly reliable means of providing steam for large-scale commercial thermal electricity generation in just two decades. Britain was the nuclear pioneer, with its 10-year program of nuclear power plant construction announced in 1955 (Bertin 1957). Calder Hall (4 × 23 MW) in West Cumbria—using natural uranium metal fuel rods encased in finned magnesium alloy and pressurized CO_2 for cooling—began generating electricity on October 17, 1956 (Jay 1956). The station was shut down after more than 46 years of operation on March 31, 2003. Ten stations of the Magnox type (total capacity of about 4.8 GW) were commissioned between 1962 and 1971. The United Kingdom's second power program (totaling nearly 9 GW) was based on advanced gas-cooled reactors using slightly enriched UO_2 pellets contained in steel tubes.

The industry's American origins had their genesis in the US Navy's nuclear submarine program. This development succeeded largely thanks to Hyman Rickover's (1900–1986) relentless effort to apply reactor drive to submarine propulsion (Rockwell 1992). The first nuclear-powered submarine, *Nautilus*, was put to sea in January 1955, but already 18 months before that the US Atomic Energy Commission assigned the country's first civilian nuclear power project, the Shippingport station in Pennsylvania, to Rickover. The reactor reached initial criticality on December 2, 1957, more than a year after Calder Hall (Taylor 2016). Real commercial breakthrough came only a decade later: American utilities ordered only 12 reactors before 1965, but 83 between 1965 and 1969, with PWRs as their dominant choice.

In these reactors, water, pressurized to as much as 16 MPa, circulates through the core (packed with fuel in zirconium steel tubes) in a closed loop, where it is heated to between 300°C and 320°C and transfers its energy content in a heat exchanger to a secondary circuit whose steam is used to generate electricity. PWRs use ordinary water as both a coolant and as a moderator. France also chose this design as the basis of its bold effort to produce most of its electricity from fission. In the pressurized heavy water reactors of Canada's national program, D_2O is both a moderator and a coolant. By the time of OPEC's first round of sharp crude oil price increases during the fall of 1973, more than 400 reactors were in operation, under construction, or in various planning stages in 20 countries. OPEC's actions appeared to benefit the industry that would sever the link between fossil fuels and electricity, and, during the 1970s, nuclear enthusiasts expected that the world's electric supply would be soon dominated by increasingly more affordable nuclear energy.

There was also a worldwide consensus that water-cooled reactors were just a temporary choice to be replaced before the end of the century by liquid-metal fast breeder reactors (LMFBR). Their dominant design uses nonmoderated neutrons to convert the much more abundant but non-fissionable ^{238}U to fissile ^{239}Pu. The source of these neutrons is fuel that is enriched to a high degree with ^{235}U (15–30% compared to between 3% and 5% in water-cooled reactors) and is surrounded by a blanket of ^{238}U. Sodium was the liquid metal of choice due to its cheapness and excellent heat transfer properties, and an LMFBR was eventually expected to produce at least 20%

more fuel than it consumed. The LMFBR represents one of the costliest and most spectacular technical failures of the twentieth century. What makes the breeder case so remarkable is an extraordinarily high degree of consensus regarding its desirability and its rapid impact.

The technique's origin coincides with the birth of the nuclear era: Szilard had anticipated breeders (and used the term) already in 1943; Alvin Weinberg and Harry Soodak published the first design of sodium-cooled breeder in 1945, and a small experimental breeder reactor near Arco, Idaho, was the world's first nuclear electricity generator: on December 21, 1951, it powered four 200-W lightbulbs, and the next day it lit the entire building in which it was located (Mitchell and Turner 1971). As the first (PWR) nuclear era was unfolding, Weinberg (1973: 18) concluded that there is not "much doubt that a nuclear breeder will be successful" and that it is "rather likely that breeders will be man's ultimate energy source." During the 1970s, LMFBR projects were under way in the United States, USSR, United Kingdom, France, Germany, Italy, and Japan. General Electric expected that breeders would account for half of all new large thermal generation plants in the United States by the year 2000 (Murphy 1974).

But in the United States these hopes lasted little more than 10 years. The declining cost of uranium and the rising costs of reprocessing facilities needed to separate plutonium from spent nuclear fuel made the option clearly uneconomical (von Hippel and Jones 1997). The US breeder program was abandoned in 1983, but other countries continued to spend billions. France embarked on the construction of a full-scale breeder, the 1,200 MW Superphénix at Creys-Malville, its designers confident that its success would bring a virtually inexhaustible source of energy (Vendryes 1977). But the reactor, completed in 1986, had many accidents and, on February 2, 1998, the French Prime Minister confirmed its final shutdown after spending about US$10 billion and another US$2.5 billion for the decommissioning. Similarly, Japan's breeder commissioned in 1994 was closed after 640 kg of liquid sodium leaked from its secondary coolant loop on December 8, 1995 (JNCDI 2000).

At that time, it became clear that the second nuclear era would not begin in the foreseeable future. During the 1990s, nuclear generating capacities reached their peaks in all Western countries, and their gradual decline appeared inevitable. France was the only affluent economy with a large and successful program. The two keys to this achievement were public acceptance of the technique and the decision by Electricité de France (EdF) to base the entire program on standardized designs of PWRs and entire plants. During the late 1990s, EdF's stations provided nearly 80% of the country's electricity and generated steady export earnings. But even EdF placed its last order for a new nuclear station in 1991, and, during the late 1990s, the only two regions with new reactors under construction were East Asia (PWRs in Japan, China, and South Korea) and India (pressurized heavy water reactors).

Conclusions about the economic benefits of fission-produced electricity were always questionable because they ignore the enormous subsidies spent by many governments on nuclear R&D and a unique treatment of catastrophic risks. The US nuclear industry received more than 96% of all monies that were appropriated by Congress between 1947 and 1998 for energy-related R&D (NIRS 1999). And the US nuclear industry could not have been launched without the Price-Anderson Act, Section 170 of the Atomic Energy Act passed by the Congress in 1954, that reduced private liability by guaranteeing public compensation in the event of a catastrophic accident in commercial nuclear generation (USDOE 2001). And we must also consider the eventual costs of decommissioning the plants (Farber and Weeks 2001) and, perhaps the greatest challenge for the nuclear industry, securing highly radioactive waste for unprecedented spans of time (100–1,000 years).

Alvin Weinberg, one of the architects of the nuclear era, asked a crucial question: "Is mankind prepared to exert the eternal vigilance needed to ensure proper and safe operation of its nuclear energy system?" (Weinberg 1972: 34). In retrospect, Weinberg concluded that "had safety been the primary design criterion, rather than compactness and simplicity that guided the design of submarine PWR, I suspect we might have hit upon what we now call inherently safe reactors at the beginning of the first nuclear era" (Weinberg 1994: 21). Even more important, promoters of nuclear electricity generation did not believe Enrico Fermi's warning that the public may not accept an energy source that produces large amounts of radioactivity as well as fissile materials that might come into the possession of terrorists.

This warning seemed more relevant during the last decade of the twentieth century than it was when Fermi made it before the end of World War II. The plutonium route that requires continuous transporting of the spent fuel to reprocessing plants and the return of mixed oxide fuel to generating stations is even more vulnerable to terrorist action. Public perceptions of unacceptable risk were boosted by an accident at the Three Mile Island plant in Pennsylvania, in 1979 (Denning 1985), and seven years later by an incomparably worse accidental core meltdown and the release of radioactivity during the Chernobyl disaster in Ukraine, in May 1986 (Hohenemser 1988; IAEA 1992).

This accident arose from the combination of a flawed reactor design and unacceptable operating procedures by inadequately trained personnel (WNA 2001). A steam explosion and fire released about 5% of the reactor's radioactive core into the atmosphere, and the drifting plume contaminated large areas of eastern and northern Europe. Actual health consequences were far less tragic than the exaggerated early fears: 31 people were killed almost instantly or shortly after the accident, 134 were treated for the acute radiation syndrome, and 14 years after the accident there were 1,800 cases of additional thyroid cancer, mostly in children—but no evidence of higher overall cancer incidence or mortality (NEA 2002).

Subsequently, neither the rising demand for electricity nor the growing concerns about CO_2 emissions were able to change nuclear power's fortunes: the expansive stage of the first nuclear era in the Western nations ended much sooner than the technique's proponents envisaged (Weinberg 1994). But, as most of the era's stations were still in operation by the century's end, nuclear generation was making a substantial contribution to the world's electricity supply (Beck 1999; IAEA 2001). By the end of 2000, the world had 438 nuclear power plants with a total net installed capacity of 351 GW or about 11% of the global total. Because of their high load factor they generated about 16% of the world's electricity in 2000 (IAEA 2001). But, unlike coal-fired or hydro generation, nuclear generation was highly concentrated: just five countries—United States, France, Japan, Germany, and Russia—produced about 70% of the world total. Nuclear reactors made the greatest difference in France, where they generated 76% of all electricity, with the Japanese and US shares at, respectively, 33% and 20% (IAEA 2001).

In retrospect, nuclear generation could be seen as a significant technical success: after all, by 2000, roughly 700 million people lived in the world's largest affluent economies that derived about 25% of their electricity from splitting uranium, and in 15 European countries the share was higher than 25%. Losing this capacity would have serious economic and social consequences, and even its gradual replacement would call for enormous investment in new generation infrastructure. Many nuclear reactors achieved impressive annual load factors of more than 95%; generation in North America, Western Europe, and Japan had an enviable safety record (in Japan, that changed in 2011 with the Fukushima disaster); and even an alarmist interpretation of low-level radiation's effects resulting in additional mortality was far lower than that attributable to particulate emissions from coal-fired power plants (Levy, Hammitt, and Spengler 2000). Moreover, by the late 1990s, fission generation helped to lower global carbon emissions. A different interpretation sees the technique as a hurried, excessively costly, and unnecessarily risky experiment. That is why my best assessment of the technique combines an approbatory adjective with a condemnatory noun: the twentieth century's use of fission for electricity generation was a successful failure (Smil 2023a).

New Prime Movers

Internal combustion engines suitable for mobile applications, electric motors, and steam turbines, the three kinds of machines that dominated the world's mechanical power capacity during the twentieth century, were the legacy of the 1880s, the most inventive decade in history. All of them were rapidly improved soon after their introduction, and all of them were widely commercialized within a single generation. In contrast, there was only one new prime mover whose practical design was introduced and impressively advanced during the twentieth century: the gas turbine, an internal combustion engine that can burn both gaseous and liquid fuels but that differs from Otto and Diesel engines in three fundamental ways (Bathie 1996; Langston and Opdyke 1997; Saravanamuttoo, Rogers, and Cohen 2001).

In a gas turbine, compression of gas precedes the addition of heat in a combustor, the combustion process is continuous rather than intermittent, and energy from the hot air flow is extracted by a turbine (Figure II.2.13). Compressors are either centrifugal flow, with air aspired near the center of an impeller, or axial, with air flowing along the engine's axis. The gas turbine's power is used in two ways: to drive the compressor and to produce additional useful work. In aviation turbojets, the air is compressed 20–35 times the atmospheric level, and its temperature is raised by more than 500°C. Afterward this air is forced through the combustion chamber and its temperature more than doubles; some of the kinetic energy of this hot gas drives the turbine and the remainder leaves through the exhaust nozzle to generate a powerful forward thrust. In nonaviation gas turbines, this energy

ENERGY CONVERSIONS 401

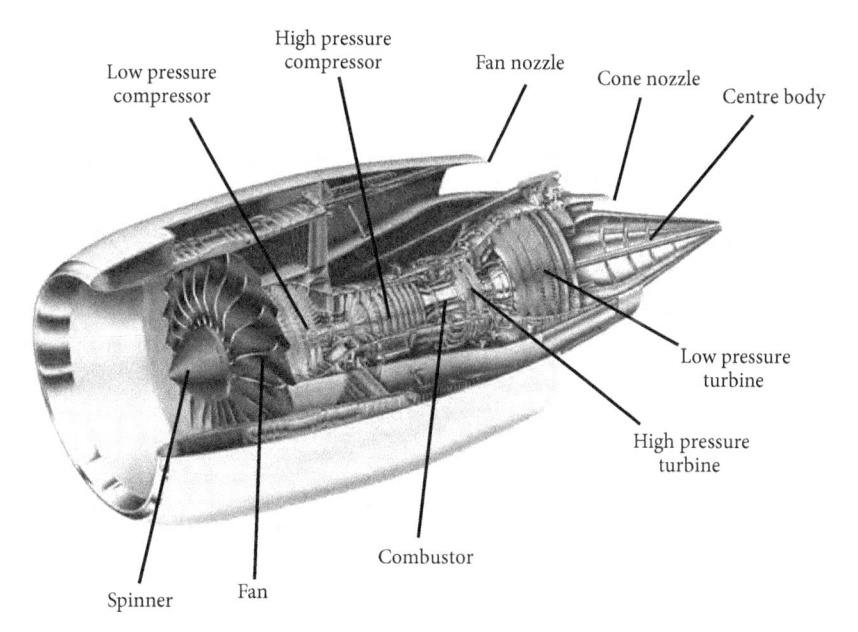

Figure II.2.13 Trimetric view of General Electric's GE90-115B aeroengine showing the principal components of modern gas turbines.
Image courtesy of General Electric Aeroengines, Cincinnati, Ohio.

is transmitted by a rotating shaft to drive electric generators, ship propellers, or compressors.

Serious commercial development of these complex machines began in 1917, when Sanford Moss (1872–1946) set up a turbine research department at General Electric's steam turbine factory. Its first practical product was a turbo supercharger (using hot exhaust gases from a reciprocating engine to drive a centrifugal compressor) that boosted the power of the Liberty piston engine from about 170 to 260 kW (GEAE 2002). Supercharging was practiced more widely after World War I, but no fundamental progress in gas turbine design took place until the 1930s, when advances in aerodynamics and material science made it possible to come up with first commercially viable designs (Constant 1981; Islas 1999). Mass production of aero turbines began during the 1950s, with the advent of jet age, and machines for electricity generation began to find a growing market during the late 1960s, with the need for more peaking power and for more flexible capacities. During the 1980s, the modification of jet engines for stationary applications and their use in marine propulsion began to blur the boundaries between the two turbine categories (Williams and Larson 1988).

The only prime movers with a lower mass-to-power ratio than gas turbines are rocket engines. Their propulsion can develop enormous brief thrusts that can put payloads into terrestrial orbits or even let them escape the planet's gravity to follow trajectories within the solar system. Intensive engineering development of rockets capable of such achievements began only after World War I, and advances that were envisaged by the pioneers of rocket science— Konstantin Eduardovich Tsiolkovsky in Russia (1857–1935) and Hermann Oberth (1894–1989) in Germany—took place during the third quarter of the twentieth century (von Braun and Ordway 1975; Neal, Lewis, and Winter 1995; Furniss 2001).

The most obvious indicator of advances in rocket propulsion is to compare Wernher von Braun's first and last rocket engines. In 1942, the ethanol-powered engine used in the V-2 missile developed a sea level thrust of about 250 kilonewtons (kN). During its 68-second burn it imparted the maximum speed of 1.7 km/s to a 13.8-meter-long missile whose explosive payload was less than 1 t (Encyclopedia Astronautica 2019). In contrast, during their 150-second burn, five bipropellant (kerosene and oxygen) F-1 engines of the first stage developed the sea level thrust of about 35 meganewtons (MN) (140 times that of V-2) that imparted to the nearly 109-m-tall Saturn V rocket a velocity of 9.8 km/h as it sent Apollo 11 on its journey to the Moon (Lea 2022; Figure II.2.14). Post–World War II development of rocket engines was driven primarily by military needs: they propelled ICBMs and SLBMs, whose deployment transformed global strategy and great power politics.

Manned missions to the Moon and later unmanned exploration of Mars attracted a great deal of public attention but had little direct impact on ordinary lives. In contrast, a long array of major socioeconomic transformations could not have taken place without rocket engines used as affordable prime movers in launching increasing numbers of reconnaissance, communication, and Earth observation satellites. The first weather satellites were launched in the early 1960s, and, by the end of the twentieth century, there were hundreds of highly specialized assemblies circling the Earth or matching the rate of its sidereal rotation 35,786 km above the equator and thus appearing to be parked in geostationary mode.

Perhaps the most rewarding use of images transmitted in real time by weather satellites has been to minimize US cyclone (hurricane and typhoon) casualties as early warnings give ample time for orderly evacuation

Figure II.2.14 Kennedy Space Center, Cape Canaveral, Florida, July 16, 1969: fisheye lens view just after the ignition of Saturn V launching the Apollo 11 spacecraft on its journey to the Moon.
NASA image 69PC-0421; available from Great Images in NASA.

of soon-to-be stricken areas, while the Earth observation satellites have been documenting planet-wide changes of land cover and weather patterns. Telecommunication satellites have drastically lowered the cost of long distance telephone calls and mass data transfers, and global positioning systems made it possible to fix one's location on the Earth with a high degree of accuracy, a great advantage for tasks ranging from air and ocean navigation to precise targeting of missiles.

Turbines in Flight

The invention and early development of gas turbines for flight offers another remarkable example of parallel innovations whose most famous pre-1900 examples involving Alexander Graham Bell and Elisha Gray, and Charles Hall and Paul Héroult were described in the first part of this book. The leading inventors of jet engines—Frank Whittle in the United Kingdom and Hans Pabst von Ohain in Germany—started as outsiders and only after years of independent and initially frustratingly slow effort were their designs adopted and improved by major companies that developed them first for military applications. Frank Whittle (1907–1996) joined the UK Royal Air Force (RAF) as a boy apprentice in 1923 and became a pilot officer in 1928 (Golley and Whittle 1987). In the same year, during his last term at the RAF College, he began to think about radically new methods of propulsion, and, in January 1930, he applied for his first gas engine patent (UK Patent 347,206) that outlined a very simple turbojet (Figure II.2.15).

There was no commercial interest in his invention, and when Whittle studied mechanical engineering at Cambridge University (1934–1936), the Air Ministry refused to pay the £5 renewal fee, and the patent lapsed. While still at Cambridge, Whittle was approached by two colleagues who helped to arrange financing for a new company, and Power Jets Ltd. was incorporated in March 1936. The company's first goal was to develop a single-shaft turbojet with single-stage centrifugal compressor. Much like Charles Parsons, the inventor of steam turbines, Whittle approached his design from the first principles. His mastery of thermodynamics and aerodynamics made it possible, in the words of Stanley Hooker at Rolls-Royce, to lay down the engine's performance with the precision of Newton (Fulton 1996). Enormous challenges of this task included the achievement of unprecedented levels of

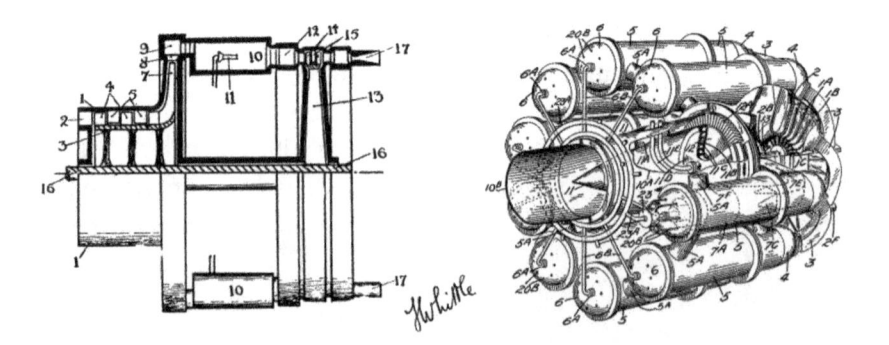

Figure II.2.15 Cross-section of a simple turbojet illustrated in Whittle's patent application (UK Patent 347,206) filed on January 16, 1930 (left), and a drawing of Whittle's first flight engine, W.1 (right), that accompanied the application for US Patent 2,404,334 filed in February 1941 and granted on July 16, 1946. The engine's main components include the impeller with radial vanes (*1A*), 10 combustion chambers (*5*), turbine wheel (*12*), and the nozzle (*10B*).

British patent drawing reproduced from Golley and Whittle (1987). US patent drawing from the US Patent Office.

combustion intensity, pressure ratio in a centrifugal blower, mass-to-power ratio, and material endurance.

British Thomson-Houston Company (BTH) was contracted to build the first experimental (4.4 kN) engine, and its bench tests began in April 1937. The engine's second reconstruction, with 10 separate combustion chambers, assumed the form that was depicted on Whittle's US patent application and that was the basis of W.1, the first flight engine (Figure II.2.15). Meanwhile, an entirely independent effort by Hans-Joachim Pabst von Ohain (1911–1998), who worked since March 1936 at Ernst Heinkel's company, had already produced the first experimental turbine-powered aircraft. Von Ohain began to think about a new mode of propulsion in 1933, as a doctoral student in engineering at Göttingen (Conner 2001).

He patented his design in 1935, and then, in cooperation with Max Hahn and Ernst Heinkel's design team, he tested his first gas turbine able to power an airplane—a centrifugal-flow engine developing nearly 5 kN—in March 1938 (Jones 1989). On August 27, 1939, its slightly more powerful version powered an experimental Heinkel-178 piloted by Erich Warsitz; it made its first test flight at the company airfield. Warsitz carried a hammer as his escape tool to break the cockpit if he got into trouble (Heinkel introduced the first ejection seats before the end of World War II; in the United States, they appeared first on Lockheed's P-80C jet fighter in 1946). But this success did not result in any immediate large-scale commitment to further development of jet aircraft.

Whittle's engine finally took to the air in the experimental Gloster E.28/29 on May 15, 1941. Its smooth performance led to the decision to build the W.2B engine developed by Power Jets for the Meteor fighter. After further delays Rolls-Royce finally replaced Rover as the engine's contractor in December 1942. Whittle recalls that when he stressed the great simplicity of the engine to the company's chairman, he was told that, "We'll bloody soon design the simplicity out of it" (Golley and Whittle 1987: 179). And they did—Rolls-Royce remained in the forefront of jet design for the rest of the century. In October 1941, a Whittle engine was also shipped to General Electric's factory in Massachusetts, and, just 13 months later, on October 2, 1942, the first US jet, XP-59A Airacomet (still inferior to the best propeller-driven planes of the day) made its first flight.

Gloster Meteor's prototype finally flew on March 5, 1943, but the plane entered the service only in July 1944. Its top speed of nearly 770 km/h speed made it possible to fly alongside a German V-1 missile and to flip it on its back with the plane's wing (Walker 1983). As with Whittle's machine, an improved version of von Ohain's turbine was developed by somebody else: in 1942, Anselm Franz, chief engineer at Junkers, designed the axial-flow turbojet Jumo 004, the world's first mass-produced jet engine used to power the Messerschmitt Me-262, the world's first operational jet fighter (Wagner 1998). The plane was tested in March 1942, but it was first encountered in combat only on July 25, 1944. Eventually more than 1,400 Me-262 (Schwalbe) were built before the end of World War II, but only about 300 were deployed, and those in operation had a high failure rate.

This failure was fortunate because the technical performance of Me-262 was superior to that of any Allied fighter even until 1947 (Meher-Homji 1997). A more successful deployment would not have changed the outcome of the war, but it could have made the Allied invasion of Europe, which rested on air supremacy, much more difficult. Jet propulsion entered the war too late to make any difference on either side, but the low weight-to-power ratio of gas turbines and their powerful thrust made it the unrivaled choice for post–World War II fighters and bombers. Germany's defeat spelled the end of turbojets based on von Ohain's design; in contrast, the three gas turbine manufacturers whose aeroengines dominated the global market during the second half of the twentieth century—General Electric, Pratt & Whitney (P&W), and Rolls-Royce—got their start by producing improved designs of Whittle's machine.

Whittle himself was not involved in this post–World War II effort: during the war he suggested that Power Jets should be nationalized (they were), and, in 1948, he retired from the RAF, never to work again directly on new jet engine designs. In 1976, he moved to Maryland with his new American wife. By that time von Ohain had been living in the United States for nearly 30 years, working most of that time at Wright-Patterson Air Force Base Aerospace Research Laboratory in Dayton, Ohio. Both men were eventually much acclaimed and honored for their inventions, but neither had any decisive influence on the post–World War II innovations in aeroengines.

Better gas turbines, together with better aluminum alloys and advances in aerodynamic design, combined first to produce military planes that proceeded to break all kinds of aviation records (Loftin 1985; Younossi et al. 2002). General Electric had a particular success with its J47 turbojet: with 35,000 units by the late 1950s, it was the world's most numerous gas turbine (GEAE 2002). Six J47-GE-11 engines (each rated at 23.1 kN) powered Boeing's B-47 Stratojet, America's first jet bomber (Figure II.2.16), which

ENERGY CONVERSIONS 407

Figure II.2.16 B-47 Stratojet taking off. Because its turbojet engines could not develop enough thrust, the plane had 18 small rocket units in the fuselage for jet-assisted takeoff. Without any thrust reversers and antiskid brakes, the plane had to rely on a ribbon-type drag parachute to reduce its landing distance. The first test flight of the B-47 took place on December 17, 1947.
Image from Edwards Air Force Base.

introduced two lasting design features of all large jet aircraft during the remainder of the twentieth century (Tegler 2000). After World War II George Schairer, Boeing's aerodynamicist, discovered German wind-tunnel data on swept-wing jet airplanes, and this work became the foundation of B-47's slender, 35-degree swept-back wings. The second distinct feature were pod engines hung on struts under wings.

In 1948, Boeing won the competition to build the B-52, which, powered by eight P&W J57 turbojets, flew for the first time in 1955. This design remained in service into the twenty-first century, being the only warplane flown by three generations of crew. The conflicts of the 1990s saw the deployment of both the venerable B-52 and the B-2, a stealth aircraft powered by General Electric's F-118. Any list of great US jet fighters must mention the F-86 Sabre (using one J47); the F-4 Phantom II powered by J79, General Electric's first engine making it possible to fly at Mach 2 or faster; and the F-15 Eagle (Walker 1983; Lombardi 2003). The Sabre broke the sound barrier (at 1,080 km/h) in September 1948. The Phantom (first flown in 1958) set more than a dozen speed and altitude records and served simultaneously with the US Air Force, Navy, and Marine Corps.

The Eagle, introduced in 1979, became the superiority fighter of the last generation of the twentieth century. Its two P&W turbofans can increase their steady thrust for short periods of time by using *afterburners* (i.e., by burning additional fuel in the engine exhaust). This boosts the Eagle's total thrust by 67%, making it possible to accelerate to supersonic speed during a straight-up climb and reach more than Mach 2.5. Soviet fighter jets of the Cold War period came from the two main design bureaus, Mikoyan and Gurevich, and Sukhoi. The MiG-15, powered by a Rolls-Royce engine, resembled the Sabre, which it faced during the Korean War. The MiG-21 eventually became the world's most widely used fighter (Gunston 1995). The delta-wing Su-15, capable of Mach 2.5, was the USSR's leading interceptor, and the Su-27 was the Eagle's counterpart.

The lessons learned from the design and operation of military jets were transferred to commercial aviation (Loftin 1985; Gunston 1997). On May 5, 1952, the de Havilland Comet 1, powered by four de Havilland Ghost engines, became the first passenger jet to enter scheduled service with flights between London and Johannesburg (Cowell 1976). Later in 1952 came the routes to Ceylon and Singapore, and, in April 1953, to Tokyo. The first Comets carried only 36 people, their cabins were pressurized just to the equivalent of about 2.5 km above the sea level, and their maximum range was just 2,800 km. Their engines had a very low thrust-to-weight ratio of 0.17, making the airplanes prone to over-rotation and loss of acceleration during takeoff. The Comet's engines were placed in the wing roots, right next to the fuselage, causing concerns about the consequences of their disintegration. But at 640 km/h the planes were twice as fast as the best commercial propeller airplanes.

The plane's first two accidents during takeoffs were attributed to pilot mistakes, but exactly a year after its introduction one of the airplanes disintegrated in the air shortly after takeoff from Calcutta, and two more catastrophic failures took place above the Mediterranean in January and April 1954 (ASN 2003). More than 20 Comet planes had to be withdrawn, and investigations traced the failures to fatigue and subsequent rupture of the pressurized fuselage: the destructive process began with tiny stress cracks that began forming around the plane's square window frames. A completely redesigned Comet 4 began flying in October 1958, but by that time there were two other turbojets in scheduled operation.

Soviet Tupolev Tu-104 entered regular service on the Moscow-Omsk-Irkutsk route in 1956. Its military pedigree (it shared most of its parts with the Tupolev Tu-16 strategic bomber, code-named Badger by NATO) extended even to two braking parachutes to slow down the plane on landing.

But neither the Tu-104 nor the redesigned Comet proved to be a lasting departure. That came when the Boeing 707—the first US commercial jetliner whose development, based on the B-47, began in 1952 and whose prototype flew in 1954—entered service on October 28, 1958, with a New York–Paris flight (Mellberg 2003; Smil 2018). Its four P&W turbojets had a combined thrust of 240 kN. The Boeing 707 launched a series of commercial jetliners (Figure II.2.17). In order of their introduction, they were the 727, 737 (the best-selling jetliner of the twentieth century with more than 3,800 planes of different versions shipped between 1967 and 2000), 747, 757 and 767 (both introduced in 1982), and 777, the only entirely new Boeing of the 1990s (flown since 1995).

Post-1960 expansion of jet travel called for more power and, after OPEC's oil price rise, also for significantly more efficient engines. This demand was met by the development of large turbofans (Coalson 2003). These machines, anticipated in Whittle's 1940 patents for a thrust augmenter, eliminated the turbojet's low propulsion efficiency at normal cruising speeds (even at 500 km/h, it wasted nearly 80% of its power) by reducing the nozzle exit velocity of combustion gases (Bathie 1996). Turbofans compress more air (to only about twice the inlet pressure) by an additional set of large-diameter fans placed ahead of the compressor; this air then bypasses the combustion chamber, resulting in lower specific fuel consumption. These two streams of exhaust gases (the rapid one from the core and the much slower bypass flow) generate higher thrust.

And, unlike in turbojets, whose peak thrust comes at the very high speeds needed for fighter planes, the peak thrust of turbofans is during low speeds, which makes them particularly suitable for large planes that need the power during takeoff (a fully loaded Boeing 747 weighs nearly 400 t). As a result, takeoffs became much less nerve-wracking: the turbojet Boeing 707 took up to 45 seconds to leave the runway, the first turbofans cut this to 25–30 seconds, and much more powerful engines of the 1990s can make commercial jets airborne in less than 20 seconds. Turbofans are also quieter because the high-speed, high-pitched exhaust from the engine's core is enveloped by slower and more voluminous bypass air.

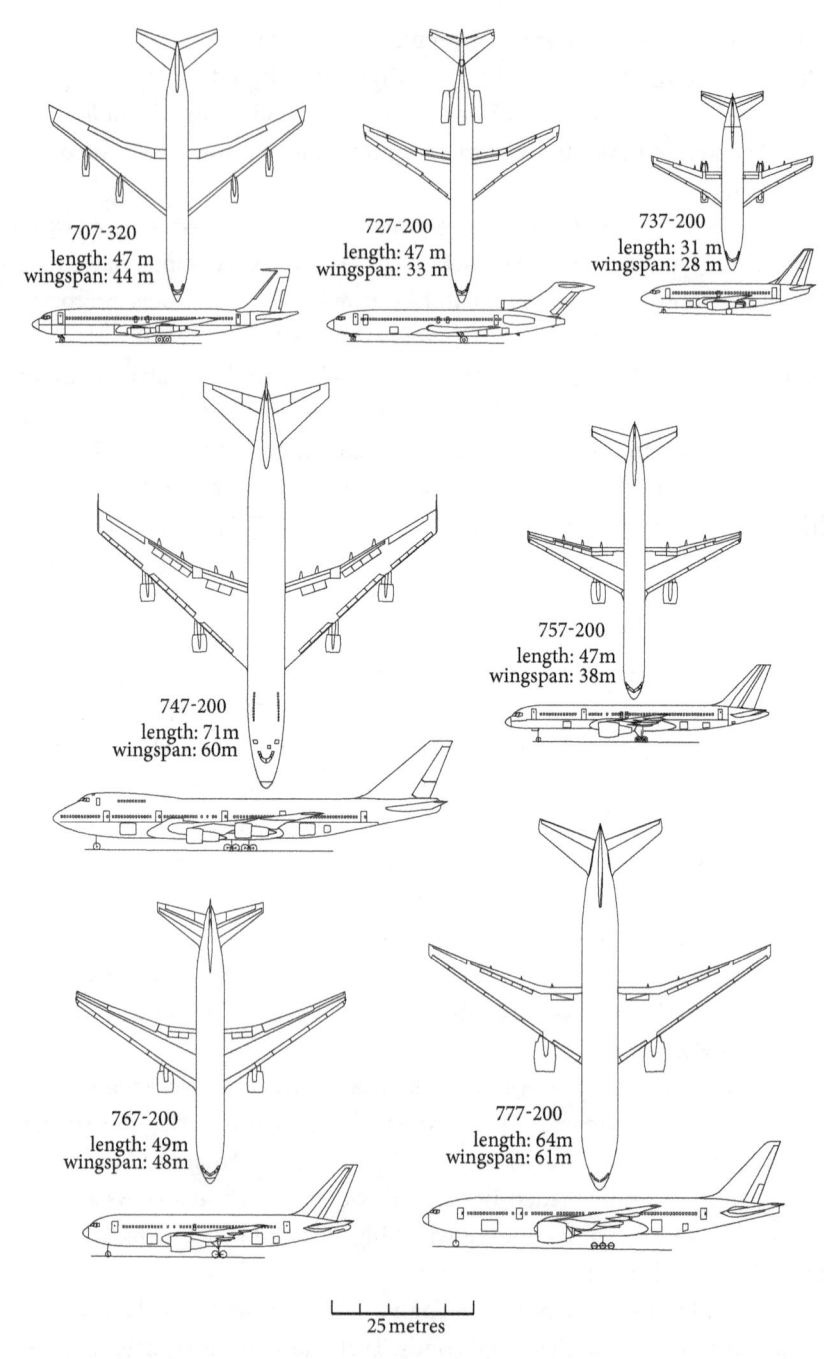

Figure II.2.17 Nearly 70% of roughly 16,000 large commercial jets that were in service in 2000 were Boeings. This drawing shows scaled plans and frontal views of the company's 700 series planes.

Adapted from the aerospace engineering department of California Polytechnic, San Luis Obispo, CA.

In 1955, the Rolls-Royce JT3D Conway became the first commercial turbofan with a bypass ratio (core-to-bypass flow) of just 0.3; seven years later, the Rolls-Royce Spey had a bypass ratio of 1.0, and, in December 1966, P&W's famous JT9D engine, with a peak thrust of about 210 kN and bypass ratio of 4.8, was the first giant turbofan with double-shrouded (narrow and long) blades (Sample and Shank 1985; Pratt & Whitney 1999). This engine was chosen to power the wide-bodied Boeing 747, the plane that revolutionized intercontinental flight: Pan Am's founder and president Juan Trippe ordered it in 1966, and William Allen, Boeing's president, gambled the company's future by investing more than twice its worth in building the world's largest passenger jet (see Figure I.4.13).

Prototype 747 took off on February 9, 1969, and the first scheduled flight was on January 21, 1970 (Smil 2000). A wide body (to allow for the placing of two standard ship containers side by side) and the bubble cockpit (for easy loading through an upturned nose) betray the original design intent: during the late 1960s, it was expected that supersonic jets would soon dominate all long routes and that 747s would become mainly cargo carriers. But supersonic travel remained limited to the expensive, noisy, and uneconomical Concorde—powered by four Rolls-Royce/Snecma turbojets totaling 677 kN with afterburner and capable of Mach 2.04—which operated on a few routes from London and Paris between 1976 and 2003. And the 747, rather than being just another compromise design, turned out to be, in Tennekes's (1997) memorable description, the only plane that obeys ruthless engineering logic.

During the 1990s, Airbus gained a growing share of the large commercial jet market, but many flight experts still believe that the jumbo was the most revolutionary, if not the best, jetliner ever built. The 1,000th 747 was delivered on October 12, 1993, and, by the end of 2000, the count approached 1,300 as different 747 versions had carried some 2 billion passengers, or one third of humanity. The Boeing 747 is an improbably graceful behemoth whose design combined symbolism and function, beauty and economy—a daring revolutionary design embodied in a machine that ushered in the age of mass intercontinental travel and became a powerful symbol of global civilization (Smil 2000).

Turbofans that entered service during the 1980s had bypass ratios between 4.8 and 6.0, and the most powerful engine introduced during the 1990s, General Electric's GE90, has the record high bypass ratio of 9.0. At the time of its introduction in November 1995, the engine was rated for the minimum of 330 kN; it was certified at 512 kN, and, during tests, one of its versions

Figure II.2.18 During the closing years of the twentieth century, General Electric began developing the world's most powerful turbofan aeroengine, the GE90-115B, rated at 511 kN, shown here in the factory setting. See also this chapter's frontispiece for the engine's front view and its first test flight, and Figure II.2.13 for a trimetric illustration.
Photograph courtesy of General Electric Aeroengines, Cincinnati, Ohio.

(GE90-115B) set the world record at 569 kN (GEAE 2003; Figure II.2.18). Its rated thrust is nearly 2.5 times that of the JT9D that was installed on the first Boeings 747. These powerful turbofans operate with temperatures around 1,500°C, hotter than the melting point of rotating blades, which must have internal air cooling. Diameters of their largest fans are slightly more than 3 m (equal to the diameter of the fuselage of Boeing 727); their weight-to-power ratios are less than 0.1 g/W, and their thrust-to-weight ratios have surpassed 6.

Consequently, modern turbofans are not a limiting factor in building larger planes: in 2000, Airbus decided to power the world's biggest commercial jet (the A 380, for up to 840 people) with four GP7200 engines made jointly by General Electric and P&W, each rated at 360 kN. The unmatched reliability of the latest generation of turbofans is the principal reason for the extraordinary safety of modern commercial flying. In the late 1990s, there were fewer than two accidents per million departures compared to more than 50 in 1960—and of these infrequent mishaps fewer than 3% were attributable to engine failure during the 1990s (Boeing 2003a).

And this reliability made it possible to use twin-engine jetliners on routes that take more than one hour of flying time from the nearest airport. This old restriction, dating back to piston engines, was changed in 1985 when the extended range twin engine operations (ETOPS) authority for twin-engine commercial jets was extended to two hours. In 1988, it was raised to three hours, and, in March 2000, some North Pacific flights were given permission to follow trajectories that were up to 207 minutes away from the nearest airport (Kupietzky 2023). By the beginning of the twenty-first century, the record for flying on a single engine, 6 hours and 29 minutes, was set during its ETOPS tests by Boeing 777 powered by two GE900-115B engines (Boeing 2003b).

Industrial Gas Turbines

The world's first industrial gas turbine was built by the Swiss Brown Boveri Corporation for the municipal electricity-generating station in Neuchâtel (ASME 1988). Its specifications show the modest beginnings of the technique, and their comparison with the best year 2000 ratings impressively shows enormous progress made in 60 years of technical development. Brown Boveri's first machine had an inlet temperature of just 550°C, rotated at 3,000 revolutions per minute (rpm), and it developed 15.4 MW, of which 11.4 MW was used by the compressor—hence its efficiency (without any heat recovery) was only 17.4%. Remarkably, the Neuchâtel turbine was still in operation by 2000. In contrast, the most advanced gas turbine of the late 1990s, General Electric's MS9001H, rated 480 MW in 50 Hz combined cycle mode (i.e., with the recovery of exhaust heat) and had an inlet temperature of 1427°C; its rate of 6 MJ/kWh resulted in the unprecedented efficiency of 60% (GEPS 2003).

These impressive advances, charted in Figure II.2.19, resulted in the world's most efficient combustion engines. They were made possible first by using the experience and innovations of jet engine industry and later by deploying new materials and scaling up the established designs. Mass production of gas turbines was stimulated by the November 1965 blackout that left 30 million people in the US Northeast in darkness for up to 13 hours (USFPC 1965). Until that time gas turbines in peaking plants were rare. The first American industrial gas turbine (rated at just 1.35 MW) was made in 1948; by 1965, annual gas turbine shipments reached 840 MW, but three years after the blackout they rose to 4 GW. Total US gas turbine capacity jumped from 1 GW in 1963 to 43.5 GW in 1975, but then the lower growth of electricity demand and large increases in the price of hydrocarbons ended new capacity gains for a decade. Stationary gas turbines became a common choice of utilities only after 1985: by 1990, about half of the 15 GW of new

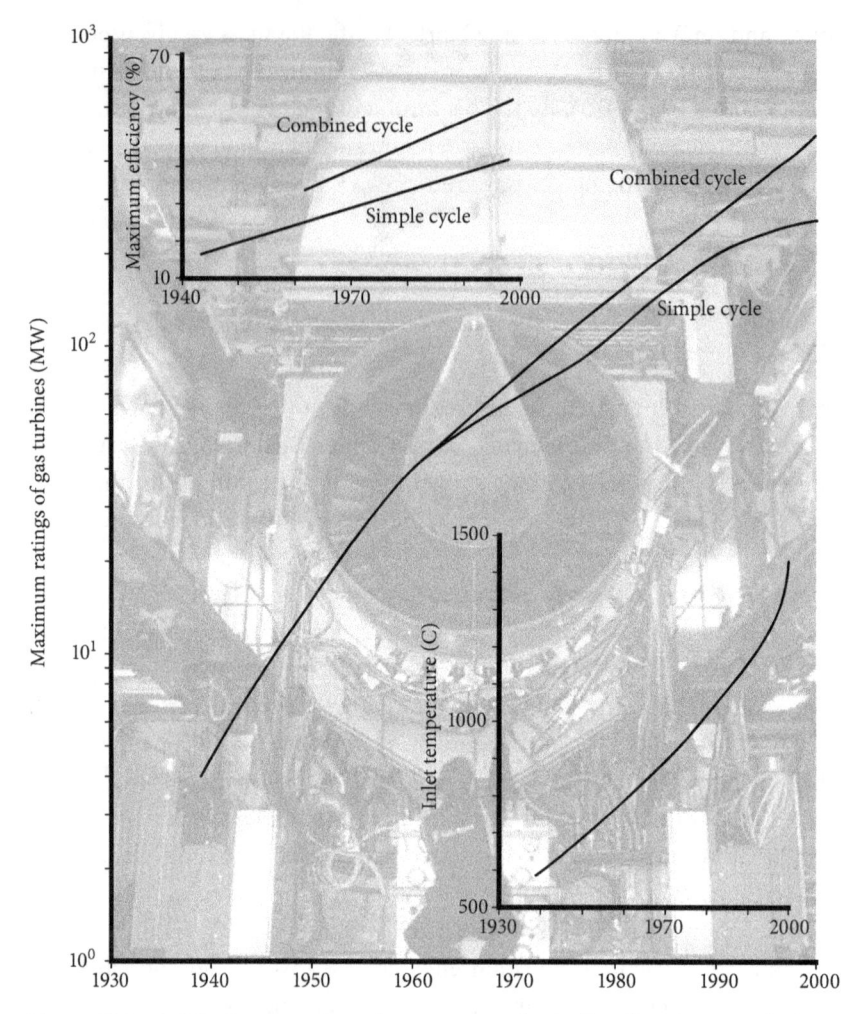

Figure II.2.19 Maximum ratings (in megawatts [MW]), efficiencies, and inlet temperatures of industrial gas turbines, 1938–2000.
Based on figure 1.10 in Smil (2003) and data in Islas (1999).

capacity ordered in the United States employed gas turbines either in simple cycle or combined cycle arrangements (Smock 1991).

The gas turbine's generating potential is best exploited when the machine is paired with a steam turbine (Horlock 2002). Exhaust gases leaving the turbine are hot enough to be used efficiently for downstream production of steam (in a heat recovery steam generator) for a steam turbine. In these combined cycle gas turbines (CCGT), the cycle's efficiency is the sum of individual efficiencies minus their product. This explains why, by the end of the twentieth century, the combination of two state-of-the-art machines (42% efficient gas and 31% efficient steam turbines) made it possible to reach 60% CCGT efficiency (0.42 + 0.31−0.13). Steam produced by the exhaust heat can be also used for industrial processing or to heat buildings in cogeneration (combined heat and electricity) plants.

General Electric, the leader in developing large stationary units, introduced its first 100 MW machine in 1976: its thermal efficiency reached about 32%. By 1990, the worldwide capacity of new gas turbine orders surpassed that of steam turbines (Valenti 1991); by 2000, gas turbines, and steam turbines for CCGT generation, represented 60% of all new capacity orders, and their installed capacity (12% of the total) was just ahead of all fission reactors (Holmes 2001). Gas turbines were also used for mid-range and some base-load generation. By the 1990s, even simple cycle gas turbines were nearly as efficient as the best steam turbines (about 40%), and, in 1990, General Electric's 150 MW turbine was the first machine to reach an efficiency of 50%. High-efficiency gas turbines offered, finally, an economic and flexible opportunity to break the long stagnation of average electricity generation efficiency (see Figure II.2.3), and they were also the major reason for the already-noted decrease in optimum power plant size.

The 1980s were the first decade when the orders for aeroderivative gas turbines (AGTs)—jet engines adapted for stationary application or for marine propulsion—began to rise (Williams and Larson 1988). General Electric began developing AGTs in 1959, and its LM6000 machine, introduced in 1992, was the first design to break the 40% efficiency mark (Valenti 1993). This gas turbine was based on the engine used on Boeing's 747 and 767 as well as on Airbuses, and in 2000, its largest available versions could generate up to 50 MW (GE 2004). On the other end of the ratings spectrum are gas turbines in the 25–260 kW range that were originally developed for ground transportation and that provide inexpensive and reliable electricity in isolated locations or as emergency standby units (Scott 1997).

And during the 1990s, a cooperative effort between General Electric and the US Department of Energy resulted in the development of the most efficient (and the most powerful) series of H System gas turbines rated up to 480 MW whose combined cycle operation resulted in the unprecedented efficiency of 60% (GEPS 2003). One of the greatest engineering challenges of gas turbine design was to develop single-crystal nickel-based superalloys for turbine blades. In addition, these large (up to 18.2 kg) single-crystal alloy blades need a ceramic coating to prevent their melting and provide resistance to oxidation and corrosion. In retrospect, the rise of gas turbines appears inevitable as the machines offer several highly valued advantages. From the basic structural point, the simple rotational mechanics of gas turbines (as opposed to the reciprocating motion of internal combustion engines) makes them relatively compact, extends their useful lifetime, makes them easier to maintain, and simplifies their sealing and lubrication.

Their compactness implies high power density and hence a limited footprint and a higher flexibility of location and, for smaller units, easy transportability when mounted on barges and ships or loaded on trucks. gas turbines must be started externally, but they can reach full load in a matter of minutes (while steam turbines take hours to do so), making them a perfect choice for peaking electricity generation. Gas turbines can use any kind of gas, light distillates, and heavier residual oils, and their unmatched efficiency makes for highly economical operation and translates into a much-reduced specific generation of CO_2. Steam turbine generation fueled by coal produces more than 1 kg CO_2/kg; fuel oil, at least 0.8 kg, while CCGT will emit less than 0.4 kg/kWh (Islas 1999). Gas turbines also do not require any external cooling, and new silencing systems can reduce their noise to below 60 decibels (dBA) at a distance of 100 m. The wide range of gas turbine sizes—from the smallest stationary microunits of just 25 kW to the largest industrial machines whose capacity approached 500 MW by 2000—offers a great deal of flexibility.

Two other important sectors that were transformed after World War II by gas turbines are natural gas transportation and marine propulsion. Modern, large-diameter (up to 142 cm) natural gas pipelines require reliable power for moving the gas, and gas turbines, rated up to more than 30 MW and placed at regular intervals along the pipeline route, do this work by driving large centrifugal compressors. Gas turbines also power pumps that move crude oil through pipelines, maintain pressure in oil wells, generate all required power on offshore hydrocarbon exploration and drilling platforms, and compress gases blown into blast furnaces, and they are used in refineries and chemical syntheses.

Orders for commercial marine AGTs became common during the 1990s: their compactness, low noise, and low emissions make them particularly suitable for passenger ships. During the late 1990s, AGTs became the propulsion of choice for increasingly larger cruise ships, augmenting or replacing cheaper but much noisier and vibrating diesel engines. In contrast, there has been no large-scale commercial deployment of gas turbines in wheeled transportation, and the only notable deployment in land vehicles was in heavy tanks: America's 60-ton M1/A1 Abrams is powered by a 1.1 MW AGT-1500 Honeywell gas turbine (Military Today 2023).

Rocket Engines

Rockets have a long history that began in China soon after the formulation of gunpowder during the eleventh century. The "red glare" of projectiles that William Congreve introduced in 1806 and that were used during the British attack on Maryland's Fort McHenry in 1814 found its way into the US national anthem, but as the accuracy of nineteenth-century artillery improved, the interest in those inaccurate weapons ebbed and rocketry was prominently absent among the innovations of pre-World War I era. Rocket engines had their experimental origins during the 1920s, and, within a single generation, they became the world's most powerful but ephemeral prime movers (Sutton 1992; Sutton and Biblarz 2001).

Large steam turbines rotate for months before they are shut down for scheduled maintenance, gas turbines that cover peak demand are working for hours, and even short car commutes mean that internal combustion engines work for tens of minutes at a time. In contrast, as already noted, Saturn V rocket engines worked for only 150 seconds to put Apollo 11 on its trajectory to the Moon, and the Space Shuttle's rocket engines were designed to fire for a maximum of 8 minutes. And while large steam turbines are as powerful as the rockets (and can work continuously for thousands of hours), no energy converter can even remotely approach the extraordinarily high thrust-to-weight (or very low weight-to-power) ratio of rocket engines.

Serious theoretical proposals for space travel and first rocket engine patents predate World War I. Robert H. Goddard (1882–1945) got his first patent for a two-stage rocket in 1914 (US Patent 1,102,653); in 1915, he began testing rockets with De Laval nozzle (its convergent–divergent profile became eventually a standard feature of large rocket engines) and experimented with solid-propellant rockets in 1918 (Durant 1974). Systematic design and testing of rocket engines began only during the 1920s, using toylike models. Goddard, whose work was supported since 1917 by the Smithsonian Institution, tested the world's first liquid-propelled rocket (liquid oxygen and gasoline) on March 16, 1926 (Figure II.2.20). In his report Goddard (1926: 587), described how during this test of a small rocket

> weighing 5.75 lb empty and 10.25 lb loaded with liquids, the lower part of the nozzle burned through and dropped off, leaving, however, the upper part intact. After about 20 sec the rocket rose without perceptible jar, with no smoke and with no apparent increase in the rather small flame; it increased rapidly in speed and, after describing a semicircle, landed 184 feet from the starting point—the curved path being due to the fact that the nozzle had burned through unevenly, and one side was longer than the other.

And so—with a 2.6-kg projectile traveling at an average speed of a bit less than 100 km/h above a field near Auburn, Massachusetts—began the age of liquid rocket engines; just two generations later their advanced versions carried men to the Moon. Goddard's first version of liquid-fueled rocket (shown in Figure II.2.20) had its small engine at the top, but that was soon changed to the standard configuration of engine surmounted by liquid oxygen and propellant tanks. Many other small models of similar engines were

Figure II.2.20 Robert Goddard testing the world's first liquid-fuel rocket in 1926 (left) and lecturing on space flight (right).
Photographs from Great Images in NASA.

tested subsequently, with different degrees of success, in the Soviet Union, Germany, France, Italy, and Czechoslovakia (Durant and James 1974). As with turbojets, critical advances came only during World War II, but in this case, they were limited to the German development of cruise missiles led by Wernher von Braun (1912–1977).

Walter Dornberger recruited von Braun to work for the Wehrmacht's development of military rockets in 1932, and, 10 years later, they began testing a missile capable of reaching England. Between September 8, 1944, and March 27, 1945, a total of 518 V-2 (*Vergeltungswaffe*) missiles hit London, killed about 2,500 people, and destroyed 20,000 houses but had no effect on the war's outcome (Piszkiewicz 1995). Although the V-1, the first German V weapon, killed almost the same number of people, it did so with nearly five times as many hits as the V-2. And while the V-1 was a cheaply built subsonic missile (maximum speed of 640 km/h) powered by a pulse jet engine, the V-2 was a supersonic missile whose rocket engine was powered by a mixture of liquid oxygen and ethanol–water and whose high impact speed (about 750 m/s) caused considerably greater casualties and material damage (Cooksley 1979). Rather inaccurate (CEP of 17 km), the V-2 carried an explosive payload of 730 kg.

In a well-known case of expertise transfer, virtually the entirety of von Braun's design team was brought to the United States in December 1945, where some its members were soon helping to launch V-2 rockets built from captured parts (Figure II.2.21). Eventually von Braun and his associates formed the core of NASA's Marshall Space Flight Center in Huntsville, Alabama. Their work proceeded slowly until the surprising launch of the Earth's first artificial satellite, the Soviet *Sputnik*, on October 4, 1957, launched by an R-7A rocket designed by a team led by Sergei Korolyov (1907–1966). Its engine, the RD-107 developed by the state Energomash company between 1954 and 1957, had four large main and two small steering combustion chambers, and it was also used to launch Yuri Gagarin's first manned orbital flight less than four years later (April 12, 1951). After several redesigns, it was still in production during the 1990s (Energomash 2003).

Even before John Glenn orbited the Earth on February 20, 1962, US President John Kennedy announced the goal of landing men on Moon within a decade. Well-known milestones of the superpower space race followed (von Braun and Ordway 1975; Heppenheimer 1997). Gemini flights with their spacewalks took place in 1965 and 1966; early Apollo Saturn launches prepared the ground for the first Moon landing on July 20,

Figure II.2.21 Test firing of a V-2 rocket in the United States in 1946 (left; NASA photograph 66P-0631), and Wernher von Braun and Saturn IB (right; NASA photograph 6863092).
Most powerful turbofan aeroengine, the GE90–115B, rated at 511 kN, shown here in the factory setting. See also this chapter's frontispiece for the engine's front view and its first test flight, and Figure II.2.13 for a trimetric illustration.

V-2 and Saturn IB images from Great Images in NASA. Turbofan photograph courtesy of General Electric Aeroengines, Cincinnati, Ohio.

1969; five more landings followed, the last one in 1972. Then, unmanned fly-bys and exploration of Mars (*Viking* landed on July 20, 1976) and Venus (the Soviet *Venera*) informed us about the two nearest planets, and probes were sent to the outer solar system (*Voyager I* launched on September 5, 1977). The United States rapidly gained the upper hand in the space race but then almost as rapidly pulled back to launch just low-orbital Skylab (1973–1980), and Shuttle flights (since April 1981) were used to resupply the International Space Station (it began November 1998, the flights ended in 2011).

The first stages of Saturn rockets of the Apollo program were powered by Wernher von Braun's liquid propellant engines (Bilstein 1996). The single-start H-1 engine, used on the early missions, had 912 kN of thrust (Figure II.2.21). J-2 developed 1.02 MN, and it was used on the second and third stages of the Saturn V. And the F-1, an order of magnitude more powerful than Saturn 1 engines, made the Moon landings possible: five of them (the four outboard ones gimbaled, the central one fixed) powered the first stage of the Saturn V. The Rocketdyne division of Boeing got the contract to build F-1 in 1959, and none of them ever failed in flight. Each liquid oxygen–kerosene engine developed about 6.8 MN (27 GW) at sea level, and, with a mass of 8.391 t its thrust-to-weight ratio was about 83 and its mass-to-power ratio just 0.0003 g/W (Figure II.2.22).

Engines powered by liquid propellants share four common basic features (Huzel and Huang 1992): separate storage tanks for the fuel and oxidizer; powerful turbopumps and requisite pipes and valves to deliver the liquids and to control their flow to a combustion chamber; a combustion chamber to generate high-pressure, high-temperature (2,500–4,000° C) gas; and a convergent–divergent nozzle (bell-shaped ones are most efficient) with a throat to expand the gas to supersonic speeds (2,000–4,000 m/s).

Figure II.2.22 The F-1 was the largest-ever propulsion engine (sea level thrust of 6.8 MN) developed for Saturn rockets of the Apollo program. The first image shows the oxidizer and fuel pumps, valves, and ducts surmounting a conical thrust chamber, enveloped by a circular turbine exhaust manifold. The second image shows the nozzle extension that increased the overall length to 5.7 m and its maximum width to 3.7 m.
Images from NASA.

Low-density propellants require larger tanks while cryogenic storage needs insulation; both options add to the mass of the launcher.

Although solid propellants generate less thrust per unit mass than do liquids, and engines that use them cannot be restarted, they have several obvious advantages: because they combine fuel and the oxidizer they do not need an external source of oxygen; they can be stored for long periods of time (even decades) without leaks or deterioration; engines are simpler because they do not need feed systems or valves; and rocket launches can be done within seconds. Solid propellants are thus an excellent choice for less powerful engines and for secondary boosters, as well as for military missiles kept in place for long periods of time. In 1942, John W. Parsons, an explosives expert at the California Institute of Technology (Caltech), formulated the first castable composite solid propellant by combining asphalt (fuel and binder) with potassium perchlorate (an oxidizer).

Key improvements in the preparation of these propellants were made during the late 1940s and the early 1950s. In 1945, Caltech's Charles Bartley and his coworkers replaced asphalt with Thiokol (synthetic rubber). This polysulfide polymer was synthesized for the first time in 1926 by Joseph C. Patrick, and the eponymous corporation was formed in 1928 (Sutton 1999). The second key innovation was the replacement of potassium perchlorate by ammonium perchlorate (AP) as an oxidizer (Hunley 1999), and Keith Rumbel and Charles B. Henderson found that adding large amounts of aluminum (Al) greatly increased the specific impulse of composite propellants: they used 21% Al, 59% AP, and 20% plasticized polyvinyl chloride (PVC).

Solid propellants were used in America's first SLBMs (Polaris in 1960, with AP/Al/polyurethane propellant) and ICBMs, including the three-stage Minuteman (AP/polybutadiene acrylic acid and Al propellant in 1962) and the first three stages of America's last ICBM deployed during the twentieth century, the four-stage Peacekeeper. And, in 1981, Thiokol introduced the world's largest solid rocket motor ever built for strap-on boosters to supply most of the thrust for the Space Shuttle's liftoff: each booster carried 499 t of propellant and generated as much as 14.68 MN of thrust for around 120 seconds.

Until the mid-1960s, the military importance of these new prime movers, deployed on a large variety of missiles, far surpassed their commercial use. Nonmilitary use of rocket propulsion began in the early 1960s, with the first US meteorological satellites, and, in 1971, came LANDSAT, the first satellite dedicated to monitoring the Earth's land use. By the late 1990s, there were several orbiting astronomical and a growing number of Earth observation satellites providing information ranging from high-resolution ground surveillance to changes in sea level. The first communication satellites launched during the 1960s were one of the key ingredients of the still far from finished revolution in global communication (Chapter II.5).

Frontispiece II.3 The ferrous metallurgy of the second half of the twentieth century was profoundly transformed by continuous casting of steel and the use of basic oxygen and electric arc furnaces. These advances lowered the price and increased the quality of the metal that continues to form the infrastructural foundations of modern civilization. Curved strands of hot metal emerge from a continuous billet caster (top) on their way toward the discharge roller table; in the cooling (spray) chamber area (bottom) the strands exit toward the withdrawal unit.

Photographs courtesy of CONCAST AG, Zürich, Switzerland.

3
Materials

Old Techniques and New Solutions

> More and more one comes to see that it is the everyday things which are interesting, important and intellectually difficult. Furthermore, the materials which we use for everyday purposes influence our whole culture, economy and politics far more deeply than we are inclined to admit.
> —James E. Gordon, *The New Science of Strong Materials* (1976)

The high material intensity of modern societies is inextricably linked with high use of energy. Rapid growth of aggregate material consumption would not have been possible without abundantly available energy in general, and without cheaper electricity in particular. In turn, affordable materials of higher quality opened new opportunities for energy industries thanks to advances ranging from fully mechanized coal mining machines and massive offshore oil drilling rigs to improved efficiencies of energy converters. These gains were made possible not only by better alloys but also by new plastics, ceramics, and composite materials. New materials were also developed to sustain the post-World War II electronic revolution as well as to advance new energy conversions, particularly the photovoltaic generation of electricity.

Dominant twentieth-century trends in the use of materials thus closely resembled those in energy conversion: enormous increase in the overall volume, continued improvements of techniques that were introduced before World War I, and the emergence of entirely new processes and their rapid commercialization after World War II. There is, of course, a fundamental difference in variety. There are only three major classes of fossil fuels (coals, crude oils, and natural gases) and six major renewable energy flows (direct solar, wind, water, geothermal, tides, biomass), and only a handful of prime movers (steam, water, wind, and gas turbines and internal combustion engines) convert these energies into electricity, heat, and motion.

In contrast, modern civilization now extracts huge volumes of inexpensive and abundant basic construction materials (stone, gravel, sand, clay) and scores of metallic ores (from aluminum-yielding bauxite to zinc-yielding sphalerite) and nonmetallic elements (from boron to silicon); produces thousands of alloys; separates huge volumes of industrial gases from the air; synthesizes increasing volumes of raw materials and plastics (overwhelmingly from non-renewable hydrocarbon-based feedstocks); and creates an entirely new class of composite compounds. This chapter traces only the innovations in producing basic, high-volume materials that form the very structure of modern civilization. Steel comes first; then I look at inexplicably neglected advances in producing the most important industrial gases and at the syntheses of a still growing range of plastics. I close the chapter by describing the production of exceedingly pure silicon, the substrate of the late twentieth-century electronics.

But first I must point out the two key material shifts—the overall increase of the annual throughput and a pronounced shift from renewable to non-renewable resources—whose evidence emerged from a detailed US record (Matos and Wagner 1998; Matos 2003). After excluding crushed stone and construction sand and gravel, the total materials used rose from about 100 megatons (Mt) in 1900 to nearly 900 Mt in 2000, and while nearly two-thirds of the 1900 total originated in wood, natural fibers, and agriculture products, a century later that share, including all recycled paper, dropped to less than 25% (Figure II.3.1). Matos and Wagner (1998) also estimated that, in 1995. The global total of consumed materials, after growing nearly twice as fast as did US use during the preceding generation, reached about 9.5 gigatons (Gt). This means that the United States, with about 5% of the world's population, claimed nearly 30% of all materials, a share even higher than its acquisition of fossil fuels (Smil 2003).

Still the Iron (Steel) Age

Despite new alloys, plastics, ceramics, and composite materials, the twentieth century still belonged to the Iron Age, but nearly all the metal was used as steel, an alloy of iron, carbon, and other elements (Smil 2016). In the United States, its absolute consumption peaked in 1973 (146 Mt, nearly

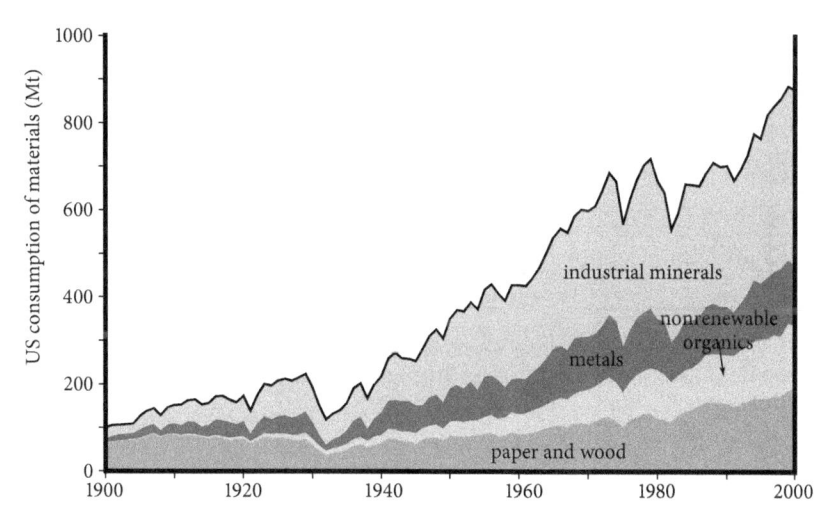

Figure II.3.1 The two most notable conclusions that emerge from tracing the use of materials in the United States during the twentieth century are the nearly ninefold increase in the total annual throughput and the drastic shift from renewable to nonrenewable substances. In 1900, about 65% of all materials (excluding stones, sand, and clay) originated in wood and agricultural products; a century later, that share dropped to slightly more than 20%.
Plotted from data in Matos (2003).

16 times greater than the 1900 level) and declined to as little as 80 Mt by 1982, and it ended the century almost 50% higher. Consequently, the metal's per capita consumption fell by more than a third from the high of about 680 kilograms (kg) in 1973 to just 435 kg by 2000—and the decline per unit of gross domestic product (GDP) was even steeper, from the peak of more than 50 kg/$1,000 in 1950 to just 13 kg by 2000 (Figure II.3.2). Similar shifts in absolute and relative consumption of steel also took place in Japan and in major European Union producers. In contrast, every one of these indicators rose in global terms.

By 2000, the global extraction of iron ore reached 1 Gt a year, a mass surpassed only by the two leading fossil fuels and by the common building materials. Pig (cast) iron output was nearly 580 Mt, and virtually all of it was used to make steel. An additional 270 Mt of steel came from the recycled scrap, and hence the total steel output was almost 20 times as large as the combined total of five other leading metals: aluminum, copper, zinc, lead, and tin (IISI 2002). Steel output rising from less than 30 Mt in 1900 to about

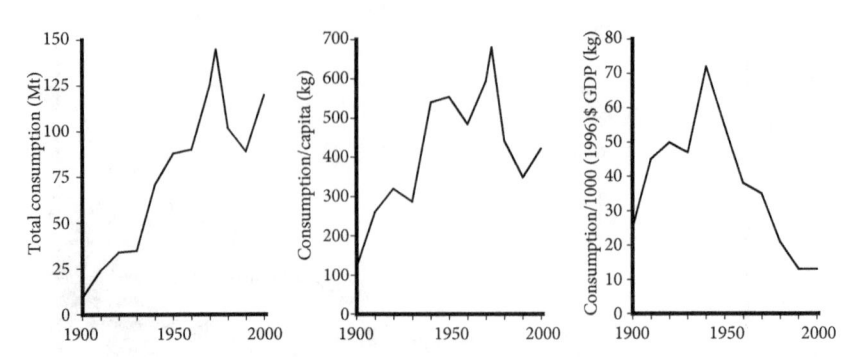

Figure II.3.2 US steel consumption during the twentieth century: annual total and averages per capita and per unit of GDP.
Plotted from data in Kelly and Fenton (2004) and IISI (2002).

850 Mt by 2000 prorated to per capita rates of 18 and about 140 kg/year, and the intensity per $1,000 of the gross world product (GWP; in constant 1996 US$) nearly doubled from 14 to 26 kg. Steel's output reflected the key economic trends: its expansion was checked by the economic downturn brought by crude oil price rises of the 1970s, and this resulted in a spell of 15 years of fluctuating output before global steel production began rising again.

Steel is produced by reducing cast iron's high carbon content (>4%) to mostly between 0.1% and 1% by alloying with other metals (most commonly with chromium [Cr], manganese [Mn], or nickel [Ni]) and by physical treatment to achieve a variety of desired properties. The metal's qualitative importance is easy to see as we are surrounded by products and structures that either are unthinkable without it or could be built otherwise only at a much-increased cost or with compromised performance. As already explained, steel's penetration of manufacturing and construction began before 1900, and it both accelerated and widened during the second half of the twentieth century.

Perhaps the best way to appreciate the importance of steel in modern society is to point out its indispensability (from scalpels to welded tanker hulls, from reinforcement rods in concrete to cars) and to note that just about everything around us is made with it (by tools, machines, and assemblies used in mining, processing, and manufacturing industries as well as in household tasks). Steel's qualities have improved as its uses have spread. High-strength steel, able to withstand stress above 800 megapascals (MPa), is used in car chassis and door guard bars. Steel with 10–12%

chromium and small amounts of other rare metals withstands ultimate stress of more than 900 MPa and is used for gas turbines blades. Steel wires used for suspension cables on the world's longest bridges have a tensile strength as high as 1.8 gigapascals (GPa). High-chromium (17%) steel is suitable for parts subjected to high-temperature and high-pressure steam, while high-carbon (0.8% C, 16.5% Cr) steel is made into cutting tools and ball bearings (Bolton 1989).

An increasing share of the metal has come from recycled material, but, by 2000, two-thirds of the world's steel was still made from pig iron. This means that blast furnaces remained at the core of this vast metal-smelting enterprise. The largest pre-World War I units had an internal volume of slightly more than 500 cubic meters (m^3) and daily capacity of about 500 tons (t). Hot metal from these furnaces was converted into steel predominantly in open hearth furnaces, cast into massive steel ingots, and only then, after reheating in soaking pits, shaped into semifinished products. Blast furnaces grew enormously and so did the performance of open-hearth furnaces, which were eventually displaced by basic oxygen furnaces as the process of ironmaking became integrated with steelmaking. At the same time, small, specialized mills became common, and, by 2000, ironmaking techniques that dispensed with the blast furnace had captured about 7% of the world's iron production.

Blast Furnaces

The blast furnace is an excellent example of a centuries-old technique (its European origins predate 1400) that reached a high level of structural and operational sophistication before 1900 but whose capabilities were transformed by impressive innovations during the twentieth century. Those massive assemblies are undoubtedly among the most remarkable artifacts of industrial civilization (Peacey and Davenport 1979; Walker 1985; Wakelin 1999; Geerdes et al. 2009). Their main sections, from the bottom up, are the circular hearth; the bosh (a short, truncated, outward-sloping cone within which are the furnace's highest temperatures, more than 1,600°C); the stack (the longest and slightly narrowing part where the countercurrent movement of downward-moving ore, coke, and limestone and upward-moving hot carbon monoxide [CO]-rich gases reduce oxides into iron); and the throat surmounted by an apparatus for charging raw materials (Figures II.3.3 and II.3.4).

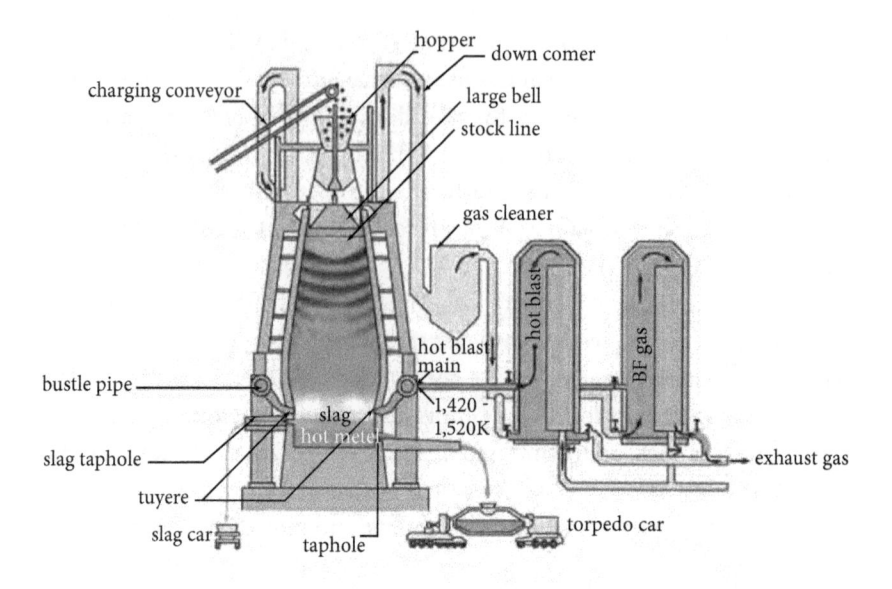

Figure II.3.3 Principal components of a modern blast furnace and its associated equipment.
Modified (with permission) from Morita and Emi (2003).

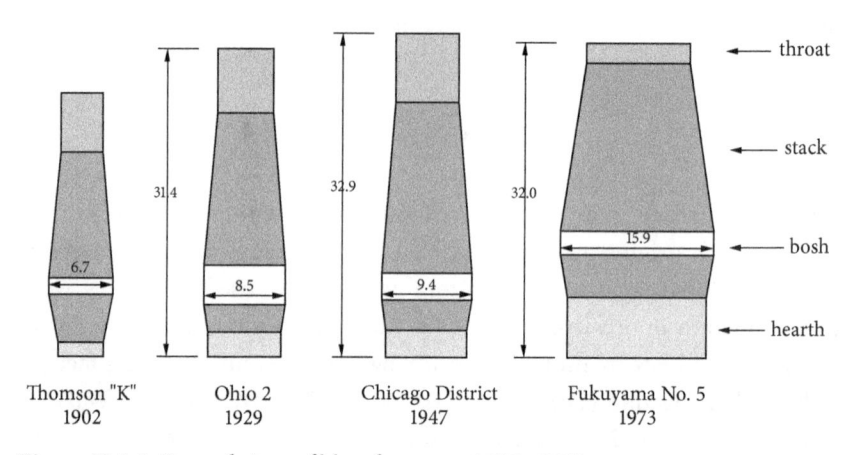

Figure II.3.4 Record sizes of blast furnaces, 1902–1973.
Based on figure 5.11 in Smil (1994).

By 1914, the steady post-1800 growth of blast furnace capacity increased the maximum output of largest units by roughly two orders of magnitude (see Figure I.4.1). The next, US-led period of expansion helped to meet the unprecedented demand for steel during World War II (King 1948). America's ironmaking continued its technical leadership into the 1950s. In 1947, Republic Steel introduced smelting under pressure, which led to considerable savings of coke. Highly beneficiated ores, enrichment of blast air by oxygen, injection of gaseous or liquid fuel through tuyeres, better refractories, carbon hearth lining, automated plug drill, and better operation controls were adopted during the 1950s (Gold et al. 1984). Before the end of that decade, Japan emerged as the most innovative ironmaking power, and for the rest of the century every one of the successive 16 furnaces holding the world record for internal volume was Japanese.

During the 1960s, Japan's pig iron production rose nearly sixfold from 11.9 to 68 Mt, and by 1974 it topped 90 Mt, a total surpassed only by Soviet production that was coming also from large (2,000–3,000 m^3) and highly automated blast furnaces. And then, after centuries of gradual growth, blast furnaces reached a clear growth plateau; by 2000, it appeared unlikely that we would see a significant number of larger units—but (as I will explain in the closing chapter) during the first two decades of the twenty-first century, China's extraordinary expansion of ferrous metallurgy changed that. By 2000, the maximum height of the largest blast furnaces reached about 35 m, internal volumes were up to about 5,200 m^3, with hearth areas up to 150 m^2 (Figure II.3.4). Charged materials were delivered by a conveyor belt from the stock house, and pipes led waste gases to be cleaned and used to preheat the blast and to generate electricity (Figure II.3.3).

These furnaces operated under pressure of up to about 250 kilopascals (kPa), and could produce hot (1,530°C) metal continuously for many years before their refractory brick interior and their carbon hearth were relined. By 2000, the maximum daily output of the largest furnaces was about 10,000 t of hot metal (Figure II.3.5). Mass and energy flows needed to operate large blast furnaces are impressive (Peacey and Davenport 1979; Sugawara et al. 1986; Geerdes et al. 2009). A furnace producing 10,000 t of metal a day needed about 1.6 t of pelletized ore, 400 kg of coke, 100 kg of injected coal (or 60 kg of fuel oil), and 200 kg of limestone for every ton of iron. These charges added up to an annual mass of 8 Mt of raw materials and an equivalent of nearly 2.5 Mt of good steam coal.

By 1980, American ironmaking was in a steep decline: the total number of blast furnaces operating in the United States was down to just 60 (from about 250 during the 1960s and 452 in 1920) and pig iron smelting fell to less than 40 Mt by 1982, from a peak of 91.9 Mt in 1973. European output also decreased and even Japanese ironmaking had its first post-World War II

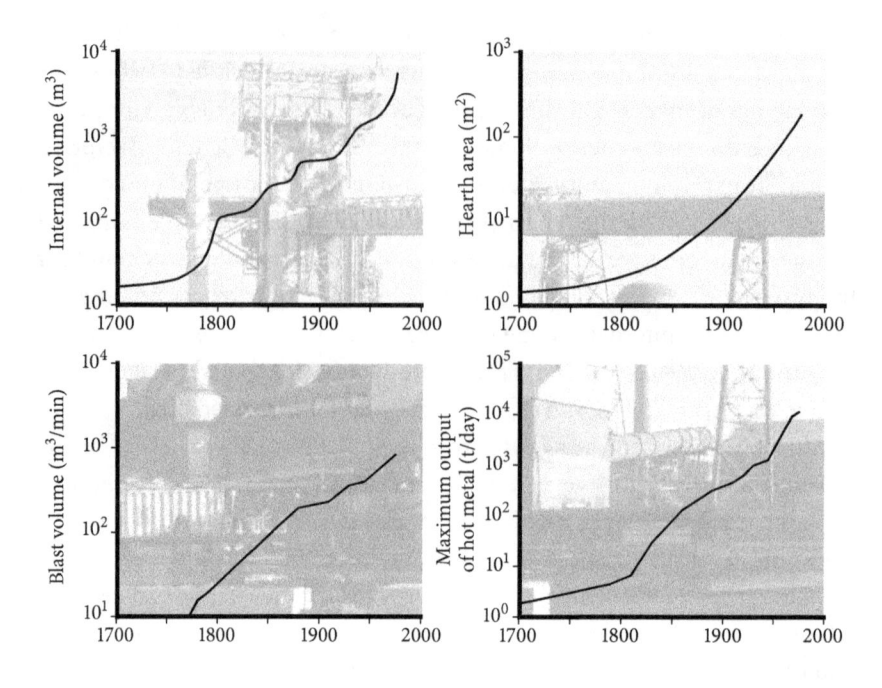

Figure II.3.5 Four semilogarithmic graphs showing the twentieth-century growth of blast furnaces in terms of the internal volume (top left), blast volume (bottom left), (top right) hearth area, and maximum daily output of hot metal (bottom left). Trends assembled from numerous metallurgical publications.

decline. After almost exactly 250 years of growth, it might have appeared that the smelting of iron with coke and limestone had reached its peak. Yet by 1990, it became clear that blast furnaces would produce the bulk of the world's iron needs not only for the remainder of the century, but also for most of the first half of the new one. Why this reprieve? Because blast furnaces are reliable producers of large volumes of pig iron supported by elaborate supply systems of iron ore mines, limestone or dolomite quarries, pelletizing or sintering plants, coal mines, coking batteries, coal trains, ore and coal carriers, and harbors. Such proven, high-volume performers and such expensive infrastructures have considerable operational inertia (McManus 1988; Geerdes, Toxopeus, and van der Vliet 2009). Better linings allowed unprecedented extensions of typical blast furnace campaigns. During the early 1980s, campaigns were usually no longer than 3–5 years; by the late 1990s, 8- to 10-year-long campaigns were not unusual, and the world record-holder,

OneSteel's Whyalla blast furnace in Australia, accomplished a generation-long (1980–2004) campaign (Bagsarian 2001; OneSteel 2005). But the most effective way of improving the blast furnace's longer-term prospects was the reduction of coke consumption thanks to high coal injection rates. This option was neglected during the decades of low-cost hydrocarbons, when increasing amounts of fuel oil and natural gas were used to replace coke.

Kentucky's Armco was the exception, injecting pulverized coal through tuyeres at its Ashland plant since 1963. Rising oil prices led to a rapid Japanese adoption of the technique: Nippon Kokan Kabushiki (NKK) was Armco's first foreign licensee in 1981—and by 1985 oil-less operation was the leading blast furnace technique in Japan. By the late 1980s, the injection's economies (replacing up to 40% of charged coke with coal costing only 35–45% as much per ton) finally brought growing US interest. By 2000, a maximum of up to about 200 kg of coal was injected per ton of hot metal, enough to reduce coke use by about 175 kg/t, and, during the late 1990s, NKK pioneered partial coke replacement by used plastics, charging as much as 120,000 t of the pulverized material per blast furnace in a year, saving coke (1.1 t/t of plastics, an equivalent of 1.5% of annual energy use) and reducing carbon dioxide (CO_2) emissions by about 3.5 kg C per ton of hot metal (NKK 2001).

The world's pig iron smelting grew by about 14% during the twentieth century's last two decades, when China emerged as the metal's leading producer. By 2000, its output of slightly more than 130 Mt accounted for more than 20% of the global output; it surpassed that of the European Union and was three times as large as the US total (see Figure II.3.2 for US figures). By that time East Asia—China, Japan, South Korea, and Taiwan—produced more than two-fifths of the world's pig iron and a third of all crude steel; the latter share is lower because European and North American countries produce relatively more steel from scrap.

Steelmaking: Basic Oxygen Furnace and Electric Arc Furnace

In a 1992 paper, the president of research at Kawasaki Steel argued that the popular conception of steelmaking as the quintessential outmoded kind of industrial production was wrong because the process was transformed by a flood of innovations. He then concluded that "with little fanfare, it has become as impressive as that acme of modern manufacturing practice, integrated-circuit processing" (Ohashi 1992: 540). Modern steelmaking underwent two largely concurrent waves of key technical substitutions: open hearth furnaces (OHFs) were replaced by basic oxygen furnaces (BOFs), and BOFs were replaced by electric arc furnaces (EAFs). OHFs, dominant between the first decade of the twentieth century and the 1960s, eventually were much larger (hearth up to about 85 m^2), and produced up to 200 t per heat (King 1948). But they were inherently inefficient, and once a better alternative became available it quickly prevailed (see Figure I.4.5).

By 2000, a mere 4% of the world's steel came from OHFs, as only the Ukraine and Russia produced major shares of the metal by this quintessential nineteenth-century technique (IISI 2002). The OHF's demise was driven by BOF's rapid ascent, which was possible only once inexpensive pure oxygen became available in large volumes. The BOF process is basically an improved type of Bessemer's converter: Bessemer did take out a patent for oxygen blowing, rather than blowing plain air, to decarburize the metal (UK Patent 2,207 on October 5, 1858), but the gas did not become available at reasonable cost and in large volumes until four generations later. The first adjective in the process's name refers to the use of basic magnesium oxide (MgO) refractory linings that are used to remove trace amounts of phosphorus and sulfur from the molten metal. The story of the BOF is notable for the fact that neither the technique's development nor its commercialization owes anything to the leading steel companies and everything to the perseverance of one man and a vision of a few managers in a small company.

After his graduation in Aachen, in 1915, Swiss metallurgist Robert Durrer (1890–1978) became a professor of *Eisenhüttenkunde* at Berlin's Technische Hochschule, where he began years of experiments with oxygen in steel refining and smelting. After his return to Switzerland in 1943, he became a board member of von Roll AG, the country's largest steel company (Starratt 1960). In the same year C. V. Schwarz was issued a German patent (735,196) for a top-blown oxygen converter, and, in 1946, a Belgian patent (468,316) by John Miles added some refinements to the basic concept (Adams and Dirlam 1966). But neither of these conceptual designs was tested while Durrer continued his experiments at von Roll's plant in Gerlafingen with the help of the German metallurgist Heinrich Hellbrügge.

In 1947, Durrer bought a small (2.5-t) converter in the United States, and with it, as he reported in the plant newspaper in May 1948,

> on the first day of spring, our "oxygen man" Dr. Heinrich Hellbrügge carried out the initial tests and thereby for the first time in Switzerland hot

metal was converted into steel by blowing with pure oxygen.... On Sunday the 3rd of April 1948... results showed that more than half of the hot-metal weight could be added in the form of cold scrap... which is melted through the blast-produced heat. (Durrer 1948: 73)

As in the case of Edison's incandescent light or the Wrights' first engine-powered flight, we thus have the exact date of one of the key (albeit generally unappreciated) modern inventions.

Subsequent developments made the BOF commonly (but inaccurately) known as an Austrian invention. Soon after his success Hellbrügge told Herman Trenkler, the works manager at Vereinigte Österreichische Eisen- und Stahlwerke AG (VÖEST; the largest Austrian steelmaker), that the BOF process was ready for scaling up to commercial production, and, in a matter of weeks, an agreement for a joint study of BOFs was concluded by von Roll, VÖEST, and Alpine Montan AG, another Austrian steelmaker. The first steel was produced on November 27, 1952 (Starratt 1960; VÖEST 2003). Alpine's BOF production began in May 1953.

VÖEST eventually acquired the rights to what became known as the Linz-Donawitz process (LD Verfahren)—although the Bessemer-Durrer process would be a more accurate identification of its intellectual and pilot-plant origins. None of the large US steelmakers showed any interest in the innovation that was perfected by a small European company whose total capacity was less than one-third that of a single plant of US Steel (Adams and Dirlam 1966). McLouth Steel in Trenton, Michigan, with less than 1% of the US ingot capacity, pioneered American BOF steelmaking with three furnaces installed late in 1954 (Hogan 1971; ASME 1985).

Steelmaking experts were convinced that the BOF was an innovation of far-reaching significance (Emerick 1954), but US Steel and Bethlehem Steel installed their first BOFs only in 1964. By 1970, half of the world's steel (and 80% of Japan's production) came from BOFs; 30 years later, as the use of EAFs increased, the US and Japanese BOF shares were, respectively, slightly more than 50% and slightly more than 70%, and globally almost 60% of all steel was made by BOFs in 2000. Oxygen was initially blown in only as a supersonic jet from a water-cooled vertical lance onto molten pig iron, but, starting in the late 1960s, it was also introduced through the tuyeres at the furnace bottom, and combined systems are now common (Barker et al. 1998; Figure II.3.6). Bottom blowing is also used to introduce inert gas (argon) for stirring the charge and lime to remove excess phosphorus. Typically, about

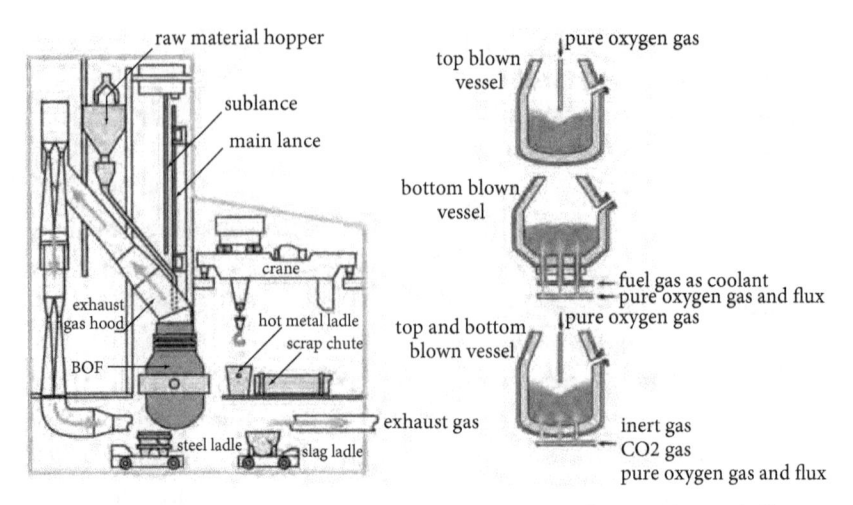

Figure II.3.6 Typical setting of a modern basic oxygen furnace (BOF; left) and the three common ways it operates (right).
Modified (with permission) from Morita and Emi (2003).

50–60 m³ of oxygen is needed per ton of crude steel; the largest BOFs are 10 m high and up to 8 m in diameter, and the capacities range mostly from 150 to 300 t.

The substitution of OHFs by BOFs resulted in enormous productivity gains. BOFs require a much smaller volume of blown-in gas, no nitrogen gets dissolved in the hot metal, and the reaction's surplus heat can be used to melt added cold scrap. And a BOF can decarburize a heat of up to 300 tons of iron from 4.3% to 0.04% C in just 35–45 minutes compared to 9–10 hours in the best OHF, while labor needs dropped from more than 3 worker-hours per ton in 1920 to 0.003 worker-hours by 1999, a 1,000-fold gain. The only disadvantage of the BOF process has been the reduced flexibility of charges: whereas for OHFs scrap can make up as much as 80% of the total, for BOFs it can be only up to 30%.

The EAF was yet another contribution of William Siemens, one of the greatest innovators of the nineteenth century (see Figure I.4.4), to ferrous metallurgy. But because of high cost and limited availability of electricity, his pioneering (1878–1879) experiments did not lead to any immediate commercial applications. Starting in 1888, EAFs made it possible to smelt aluminum and then to produce specialty steels, and the furnace remained in these roles, and with relatively limited unit capacities (first pre-1910 Héroult's EAFs rated just 3–5 t; those of the 1920s, mostly 25 t), until the 1930s. EAF's use for large-scale steelmaking became attractive as electricity prices declined and once large amounts of scrap iron and steel became available.

The US EAF output passed 1 Mt per year only in 1940, and, by the end of World War II, the total was more than 3 Mt. By the early 1970s, the largest capacities were more than 300 tons, and EAFs dominated the smelting of highly alloyed and stainless varieties. The scrap metal to feed EAFs comes from three major sources. The home (circulating) scrap that originates in ironmaking and during the conversion of ingots to finished products represented most of the recycled mass in integrated plants. Prompt industrial scrap comes from a multitude of processes, above all from metalworking and manufacturing. The composition and quality of these two streams is well known. In contrast, obsolete metal, be it in the form of defunct machinery or old vehicles, ships, and appliances, often contains undesirable ingredients either as added metals (e.g., tinned steel cans) or as contaminants.

The shift from BOFs to EAFs severed the link between steelmaking and blast furnaces and coking and made it possible to set up new, smaller mills whose location did not have to consider the supplies of coal, ore, and limestone. This mini-mill option became the most competitive segment of new steelmaking during the last third of the twentieth century. These steelworks— with annual capacities ranging from less than 50,000 to as much as 600,000 t of metal—combine EAFs (charged with cold scrap) with continuous casting and rolling. Initially they made such low-grade products as bars, structural shapes, and wire rods, but during the 1980s they began to produce higher grades of steel (Barnett and Crandall 1986; Szekely 1987; Hall 1997).

Mini-mills, dispensing with blast furnaces, needed much lower capital investment (only 15–20% of the total needed for new blast furnaces) and had considerably lower operating costs because continuous casting eliminated the equipment and energy needed for reheating and shaping the ingots (Jones, Bowman, and Lefrank 1998). By 2000, a third of the world's steel came from EAFs, and the United States, with 47% of its total output, had the highest share among major producers (IISI 2002). Steadily improving performance accompanied the EAF diffusion. In 1965, the best furnaces needed about 630 kilowatt hours per ton (kWh/t) of crude steel; 25 years later the best rate was down to 350 kWh/t (De Beer, Worrell, and Blok 1998).

Concurrently, the worldwide average tap-to-tap times declined from 105 to 70 minutes, while the tap weights rose from 86 to 110 Mt and productivity increased by more than 50% to 94 t per hour. Increased consumption of oxygen, charging of hot and preheated metal, and higher power inputs were

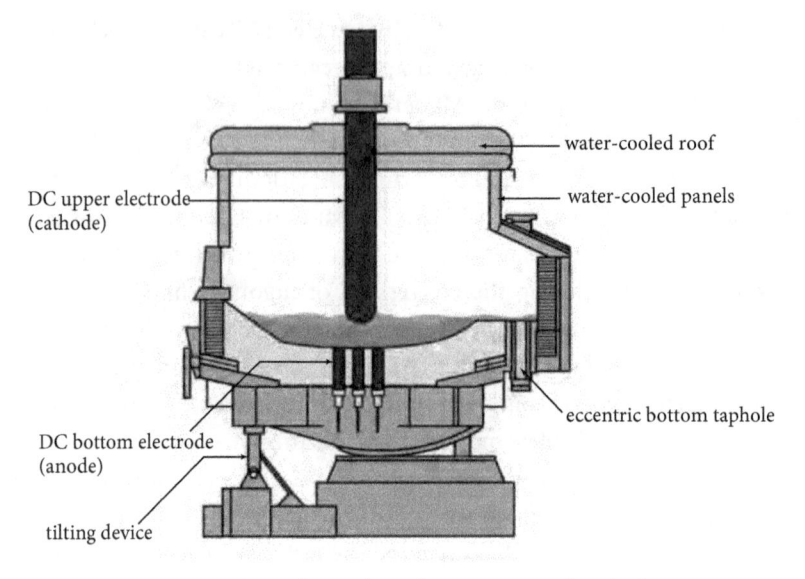

Figure II.3.7 Cross-section of a modern direct-current (DC) electric arc furnace with an eccentric taphole.
Modified (with permission) from Morita and Emi (2003).

the principal reasons for these improvements. With nearly 40 m^3/t, a furnace at the Badische Stahlwerke in Kehl set the world record in April 1999 by producing 72 heats of nearly 80 t each, and the world's largest EAF (375 t) in Sterling, Illinois, consumed oxygen, delivered from eight fuel burners and two lances, at a nearly identical rate (Greissel 2000). Until the 1970s, EAFs were commonly installed at grade level and needed pits for tapping and slag removal, while later furnaces were elevated above the plant floor, and efficient tapping without tilting became possible with the introduction of eccentric bottom taps (Figure II.3.7).

Continuous Casting

The diffusion of BOFs and EAFs was not the only key post-1950 transformation of steelmaking: the subsequent processing of the metal had also undergone a fundamental change. The traditional process involved first the production of steel ingots, oblong pieces weighing 50–100 t that had to be reheated before further processing yielded standard semifinished products: slabs (from just 5 centimeters (cm) to as much as 25 cm thick and up to more than 3 m wide), billets (square profiles with sides of up to 25 cm used mainly to produce bars), and blooms (rectangular profiles wider than 20 cm used to roll I- and H-beams) that were then converted by hot or cold rolling into finished plates and coils (some as thin as 1 millimeter [mm]), structural pieces (bars, beams, rods), rails, and wire. This inefficient sequence, which often consumed as much energy as the steelmaking itself, was eventually replaced by continuous casting (CC) of steel (Schrewe 1991; Fruehan 1998; Tanner 1998; Luiten 2001; Morita and Emi 2003). The sequence blast furnace → OHF → steel ingots, dominant until the 1950s, turned into the blast furnace → BOF → CC and, as more steel scrap was used, increasingly into the EAF → CC process.

The idea of continuous casting was patented by Henry Bessemer in 1865, and was further elaborated by him nearly three decades later (Bessemer 1891), but the process was first successfully deployed only during the 1930s for casting nonferrous metals whose lower melting points made the operation easier. Siegfried Junghans, a German metallurgist, invented a vertically oscillating (reciprocating) mold. This invention eliminated the possibility of solidifying molten metal sticking to the mold and causing shell tearing and costly breakouts of the liquid metal. Junghans designed the first working prototype in 1927 and applied for a patent in October 1933; in 1935, he began a long period of independent development of steel casting.

Irving Rossi (1889–1991), an American engineer who had been involved in several business ventures in pre-World War II Germany, visited Junghans at his workshop in 1936, and, despite witnessing a less than perfect attempt at continuously casting a brass billet, he became convinced of the technique's enormous potential (Tanner 1998). On November 15, 1936, Junghaus gave Rossi exclusive rights to the basic patent and its follow-ups for the United States and England as well as nonexclusive rights for other countries outside of Germany, and, two years later, also a commitment to share all information needed to build CC plants. In return, Rossi was to finance the commercialization of the process outside of Germany and, by 1937, had the first US orders for German-built brass casters. In 1947, Rossi convinced Alleghany Ludlum Steel to use the technique for casting steel in its Watervliet, New York, plant, where the first slabs were produced by an American-made (Koppers) machine in May 1949.

Concurrently, Junghans's post-World War II work led to the first successful production runs at his workshop in Schorndorf in 1949, and, in 1950, he set up an experimental plant in cooperation with Mannesmann AG at the company's works in Huckingen, which began casting in 1952. But Rossi remained the technique's leading pioneer, setting up first the Continuous Metalcast Corporation and then, in October 1954, establishing Concast AG in Zurich (Tanner 1998). The formula of comprehensive licensing, strengthened in 1970 by a patent interchange cooperation agreement with Mannesmann (the company's main competitor), proved so successful that Concast AG eventually controlled more than 60% of the world market for continuous casters, and, in 1981, its increasingly monopolistic behavior led to its forced reorganization ordered by the antitrust authorities in the United States and Europe.

As with the BOF, major US companies largely ignored the technique while Japanese steelmakers embraced it (Figure II.3.8). Japan's first CC line was installed in 1955; two decades later Japan had more than 100 CC machines, and, by the early 1980s more than 90% of Japanese steel came from CC machines, compared to slightly more than 40% in the United States, where the two-thirds mark was passed only by 1990 (Burwell 1990; Okumura 1994). But, by 2000, more than 85% of the world's steel was cast continuously, with the shares (96–97%) basically identical in the European Union, United States, and Japan (Figure II.3.8). CC also accounted for nearly 90% in China, and only Russia (50%) and the Ukraine (20%) were still far behind.

The casting process begins as the molten steel is poured first into a tundish (a large vessel holding enough metal to assure continuous flow), from which it flows steadily through a submerged entry nozzle into a water-cooled copper mold, where it begins to form a solid skin (Schrewe 1991). Vertically oscillating the mold prevents sticking, and, as the metal descends, drive rolls withdraw the strand at speeds of 1.5–2.8 m/min to match the flow of the incoming metal (Figure II.3.9). Support rolls carry and bend the strand, and water sprays cool it until the core becomes solid at the caster's end, 10–40 m from the inlet. At that point strands are cut with oxyacetylene torches.

Vertical casters required either tall buildings or excavations of deep pits; bending-type machines, introduced in 1956, reduced the height requirement by about 30%. Curved mold (bow-type) casters (Figure II.3.9)—designed independently by two Swiss engineers working at von Moos Steelworks and by Otto Schaaber, a metallurgist working for Mannesmann, during the late 1950s and introduced commercially in 1963—reduced the

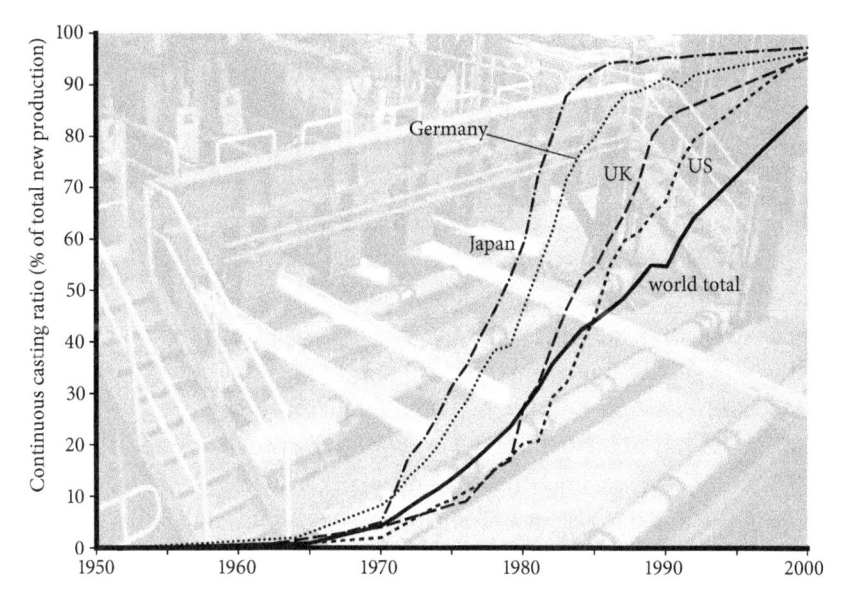

Figure II.3.8 Diffusion of continuous casting, 1952–2000.
Based on data in Okumura (1994) and data in IISI (2002).

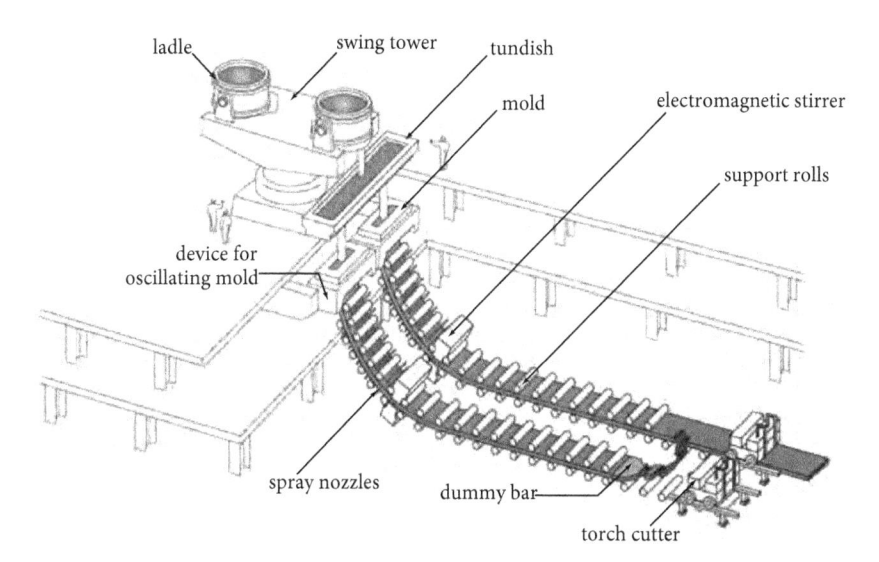

Figure II.3.9 A typical layout of a two-stranded continuous steel slab caster.
Modified (with permission) from Morita and Emi (2003).

height requirement yet again by at least 40%, and these remained the most popular machines until the early 1980s. They bend the metal strand into the horizontal plane as a progressively thicker solidified shell forms on its exterior. Post-1980 introductions included vertical-bending, low-head, and horizontal designs (Okumura 1994).

The resulting semi-finished products depend on the configuration of CC machines. Initially, the process could produce only billets (round or square) of limited cross-section; larger blooms came next, and slabs were introduced during the late 1960s. The advantages of CC compared to ingot casting and subsequent primary rolling are numerous: much faster production (less than one hour vs. 1–2 days), higher yields of metal (up to 99% compared to less than 90% of steel to slab), energy savings on the order of 50–75%, and labor savings of the same magnitude. Going one step further is taking the hot slabs from CC directly to the rolling mill. This method yields the highest energy savings and reduces plant inventories of semi-finished products, but the high capital costs of retrofitting the operations delayed its wider adoption.

Development of thin slab casting began during the mid-1970s. Germany's Schloeman Siemag (SMS) first tested its Compact Strip Production in 1985, and America's Nucor deployed the world's first thin slab caster in Indiana, in 1989. Thin slabs, still retaining a liquid core, are reduced to 50 and 30 mm and are then directly hot-rolled into strips. During the 1980s some steelmakers began working on the production of hot strips (hot rolled coil steel) directly from a caster with no or minimal intermediary rolling. The first commercially successful strip caster designs came from Nippon Steel and from Australia's Broken Hill Proprietary (BHP) Corporation. Nippon Steel started its strip-casting operation at the Hikari plant secretly in October 1997 and made its success and its capacity (producing annually up to 500,000 t of 2–5 mm stainless strip up to 1.33 m wide) public in October 1998.

During the same month BHP revealed its secret Port Kembla (New South Wales) operation: it began casting low-carbon steels from a 60-t ladle in strips of 1.5–2.5 mm, reduced by in-line rolling to 1.1 mm. Eurostrip, a joint undertaking of three major companies, cast its first strips, stainless steel 1.5–4.5 mm thick and 1.1–1.45 m wide (at speeds averaging 60–100 m/min), in Thyssen Krupp's Krefeld plant in December 1999 (Bagsarian 2000). Strip casting does not use an oscillating mold but rather replicates closely Bessemer's original twin-drum design (patented

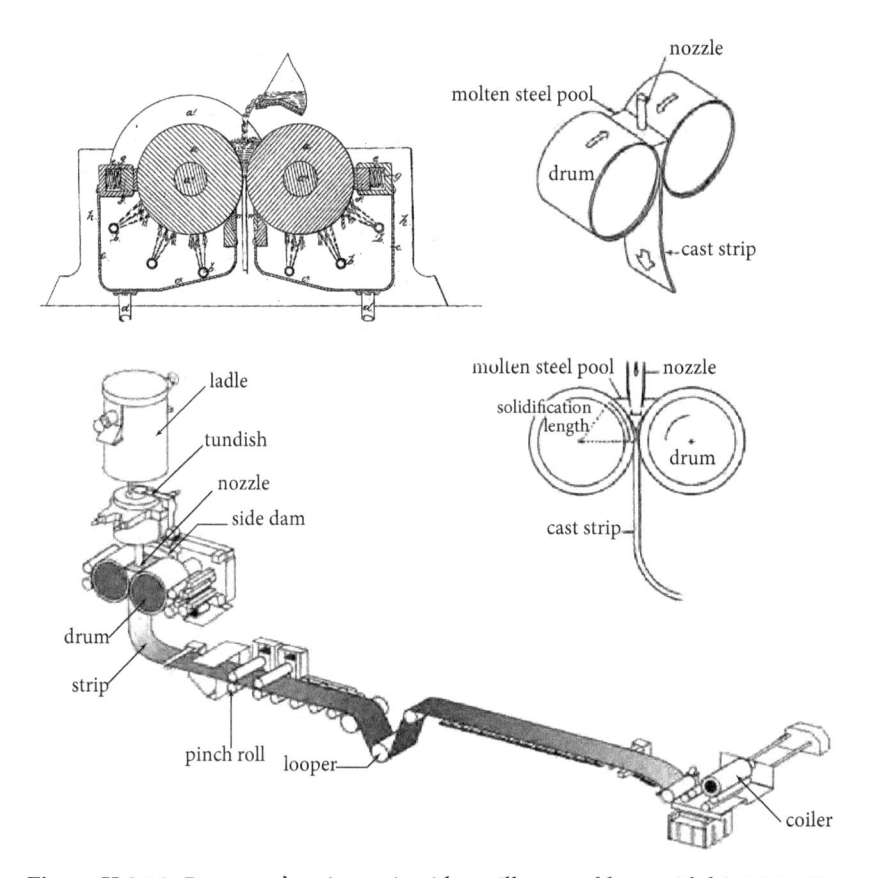

Figure II.3.10 Bessemer's strip casting idea—illustrated here with his 1865 US patent drawing—became commercially successful only during the 1990s (top left). Schematic drawings show a detail of the modern twin-drum caster (right) and of an entire strip casting line (bottom).

Reproduced from Bessemer (1891) and modified (with permission) from Takeuchi et al. (1994) and Morita and Emi (2003).

in 1857) that employed two rotating rolls (Bessemer 1891). Molten steel is poured between these water-cooled drums and cools rapidly, and the two solidifying shells are fused into a single strip as they are compressed between the drums (Takeuchi et al. 1994; Figure II.3.10). Cooling rates are 1,000 times faster than in conventional slab casting, and casting speeds can exceed 100 m/min. The greatest reward is, of course, the elimination of expensive rolling mills, which lowers the overall capital expenditure by 75–90% (Luiten 2001).

New Paths to Iron

The last two decades of the twentieth century also saw the beginning of a radically new way of producing the civilization's dominant metal, the *direct reduction of iron* (DRI) that dispenses with coke and reduces iron ores in their solid state at temperatures well below the melting point of the metal (at only about 1,000°C). Attempts to use DRI on an industrial scale go back to the nineteenth century: most notably, between 1869 and 1877, William Siemens experimented with reducing a mixture of crushed high-quality ore and fuel in rotating cylindrical furnaces (Miller 1976). By 1975, there were nearly 100 different DRI designs but only two approaches emerged as commercially important: processes using reducing gas that was prepared outside the reduction furnace, and techniques whose reducing gas is generated from hydrocarbons introduced into the furnace (Feinman 1999).

Dominant natural-gas–based shaft processes—the pioneering Mexican *Hojalata y Lamina* and the US MIDREX—reduce iron pellets or fines in an ascending flow of reducing gas and crushed limestone (Davis et al. 1982; Tennies et al. 1991; Figure II.3.11). The reducing gas is prepared by catalytic steam reforming of natural gas, which produces a mixture of CO and hydrogen ($CH_4 + H_2O \rightarrow CO + 3H_2$). Coal-based processes use ore with fine fuel, and the reducing gas comes from hydrocarbons concurrently

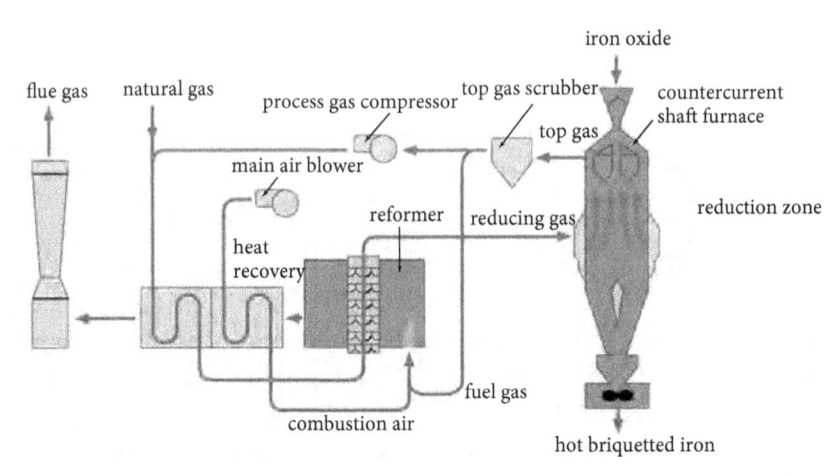

Figure II.3.11 Schematic outline of the MIDREX process, the most successful way to produce iron by direct reduction.
Modified (with permission) from Morita and Emi (2003).

introduced into the furnace. The final product for all processes was initially solid sponge iron whose large specific area makes it prone to reoxidation and difficult to transport. Conversion of the sponge into hot briquetted iron (HBI, first in 1984) makes a product that is easily shipped and that can replace scrap in EAFs. DRI's other advantages are the possibility to produce the metal in countries devoid of coking coal but rich in natural gas, and lower energy cost. And while the blast furnace–BOF route requires an annual capacity of at least 2 Mt, gas-based DRI is feasible with capacity of less than 0.5 Mt/year suitable for mini mills.

There was a great deal of enthusiasm for these processes during the 1970s, but, by 2000, DRI plants produced just 43 Mt, or about 7%, of the world's iron. In 1988, a task force of the American Iron & Steel Institute estimated that it would take a decade to commercialize this new metal-making route, and, five years later, McManus (1993: 20) thought that "the race to find a coal-based replacement for the blast furnace is moving into the home stretch." But that was only wishful thinking because blast furnaces remained the dominant producer of iron until the very end of the twentieth century.

Accomplishments and Consequences

Both the absolute and relative figures of twentieth-century global steelmaking are impressive: a 30-fold increase of annual production (from 28 to 846 Mt) and an eightfold rise in per capita output (from less than 18 to nearly 140 kg/year). So is the cumulative aggregate: during the twentieth century the world's steelmakers produced nearly 31 Gt of the metal, with half of this total made after 1980. What happened to this enormous mass? At least a tenth of all that steel had been oxidized or was destroyed in wars, about a quarter had been recycled, and another 15% was embedded in above-ground structures and underground (or underwater) foundations and hence beyond easy recycling. The remainder represented the accumulated steel stock of about 15 Gt, or about 2.5 t per capita, that could be potentially made into new steel. Close to 10% of the world's accessible steel stock was in motorized vehicles, and about 2% was reused annually as scrap. A detailed Japanese account showed that the country's accumulated steel stock by 2000 was 1.3 Gt, or 10 t per capita (Morita and Emi 2003).

But in high-income economies, these impressive achievements were not matched by the perception of the industry's role in the modern world.

Annual production of raw steel peaked during the early 1970s, and afterward it has been either stagnating or declining, with many producers losing money or surviving on assorted subsidies: an industry with such a dismal record of devouring capital had little chance of being as glamorous as computer chip making, which produced some of the most spectacular investment returns in history. By the end of the twentieth century, North American, European, and Japanese steelmakers were delivering products of unprecedented quality—but they were doing so within the confines of a mature and stagnating industry.

At the beginning of the twentieth century, it was steel that was the driver of America's stock markets (later in this chapter I, describe US Steel's impressive beginnings)—but by its end it was the computer industry in general, and integrated circuits in particular, that was seen as the leading driver of the new economy. A simple what-if experiment allows us to see the relative importance of these sectors. An affluent, advanced, high-energy civilization is perfectly possible without integrated circuits: that is, indeed, what we were until early 1971, when Intel released its first microchip. In contrast, no such society is possible without the large-scale production and use of steel. The fact that, by 2000, the combined market capitalization of the 10 largest US steel companies was just a tenth of the value of Home Depot (which in that year became part of the Dow) says nothing about the relative importance of these two kinds of activities; it merely reveals the choices of investors who, nevertheless, could not make it through a single day without steel.

Both iron- and steelmaking entered the twentieth century as established, mature, mass-producing industries, but, by its end, they were profoundly transformed by technical advances that made the industry vastly more productive, less labor- and energy-intensive, and also environmentally much more acceptable. At the beginning of the twentieth century, typical energy needs in ironmaking (virtually all of it as coke) were around 50 GJ/t of pig iron; by 1950, it was about 30 GJ/t, and, during the late 1990s, the net specific energy consumption of modern blast furnaces was between 12.5 and 15 GJ/t (De Beer et al. 1998), or at roughly twice the theoretically lowest amount of energy (6.6 GJ/t) needed to produce iron from hematite. These efficiency gains came mainly from reduced coke consumption, increased blast temperatures, and larger furnaces.

The combined effects of replacing OHFs with BOFs and rolling from ingots by continuous casting brought energy costs of steelmaking down to 25 GJ/t by 1975 and to below 20 GJ/t during the late 1990s, with nearly two-thirds of that total used by blast furnaces, just a few percent claimed by BOFs, and the rest needed for rolling and shaping (Leckie, Millar, and Medley 1982; De Beer et al. 1998). The US average for the mid-1990s fell to about 19 GJ/t, and similarly, Japan's consumption averaged 19.4 GJ/t of crude steel in 2000 (JISF 2003). In contrast, the most efficient EAF-based producers of the late 1990s needed about 7 GJ/t.

Continuing declines in the energy intensity of ferrous metallurgy meant that, in 2000, the production of roughly 850 Mt of the world's leading alloy needed about 20 exajoules (EJ) or slightly more than 6% of the world's total consumption of fossil fuels and primary electricity (Smil 2003). Had the energy intensity remained at the 1900 level, ferrous metallurgy would have claimed almost 20% of all the world's primary commercial energy in 2000. But despite these impressive efficiency gains, iron- and steelmaking remained the world's largest energy-consuming industries, claiming about 15% of all industrial energy use.

At the beginning of the twentieth century, rails were the single largest category of finished steel products; by its end, sheets and strips, destined for automotive and household goods markets, were dominant: in 2000, they accounted for slightly more than half of US steel output. Another key shift has been toward higher quality and higher performance steels: by 2000, the global production of stainless alloys (containing at least 13% Cr) approached 20 Mt/year (or more than 2% of the total output), and specialty steels are used in applications ranging from deep-water offshore drilling rigs to record-height skyscrapers.

Finally, a few observations concerning the economic and social impacts of the changing pattern and cost of global steel production. Western steelmaking became a mature industry within three generations after the beginning of its substantial expansion during the 1880s. Subsequently, Japan's innovation-driven quality-stressing ascent was much more important than the technically inferior Soviet expansion. Between 1946 and 1974, there were only three years (1954, 1958, and 1971) when the global steel output dipped; in contrast, the last quarter of the twentieth century saw output decline in 10 years (Figure II.3.12). The slow rise of the average price of finished steel that began after the end of the economic crisis of the 1930s changed into a steep ascent after the oil price hike by the Organization of the Petroleum Exporting Countries (OPEC) of the early 1970s; once that rise was broken, steel prices began to fluctuate, sometimes by more than 10% a year. American steelmaking was especially affected by cheaper exports from Russia, Ukraine,

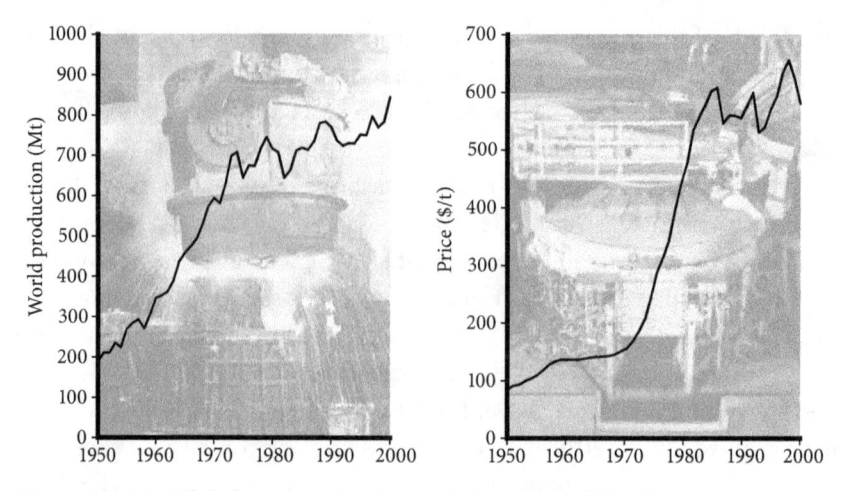

Figure II.3.12 Global steel production and the average US price per ton, 1950–2000.

Plotted from data in Kelly and Fenton (2004) and IISI (2002).

and Asia, and, compared to its global dominance that lasted until the early 1960s, it ended the century as a much-weakened industry.

At the beginning of the twentieth century, US Steel Corporation—formed by Charles M. Schwab, J. Pierpont Morgan, and Andrew Carnegie on April 1, 1901—became the first company capitalized at $1.4 billion, and US steel mills contributed 36% of the global output (Apelt 2001). In 1945, with German and Japanese industries in ruins, the United States produced nearly 80% of the world's steel. Inevitably, that temporarily high share had to drop, but the absolute output kept increasing (with inevitable fluctuations) until 1973. But the post-1973 decline was unexpectedly harsh (Hall 1997). Between 1975 and 1995, 18 out of America's 21 integrated steelmakers merged, were bought, or went bankrupt, and, by the century's end, the United States produced just 12% of the world's steel.

The stagnating and declining stock values of America's steel companies accelerated the loss of prestige that was previously enjoyed by this key industry. During the late 1970s, three out of the world's top seven steelmaking companies were still American (US Steel, Bethlehem, and National Steel, ranking fourth, fifth, and sixth, respectively), but, by 2000, the largest US company (US Steel) placed no higher than fourteenth as surging imports, low prices (they actually declined between 1995 and 2000), and huge foreign excess capacity (at 250 Mt in 2000, almost twice the average US consumption) caused more than a dozen American steelmakers to file for bankruptcy between 1998 and 2000 (AISI 2002). Two other highly symbolic developments signaled US steel's decline: in 1991, USX, the parent company of US Steel, was replaced on the Dow 30 by Disney. Could there be a more incredible mark of decline as Schwab, Morgan, and Carnegie's unmatched original creation was ousted in 1991 by a Mickey Mouse outfit? Six years later Bethlehem Steel was displaced by Johnson & Johnson (Bagsarian 2000).

The socioeconomic dislocations of this decline were substantial. In February 1943, the number of US steelworkers represented by their union peaked at slightly more than 725,000 (Hogan 1971), and, despite the subsequent rise in productivity, the industry still employed half a million workers by 1975 and 425,000 by 1980. But by 1991, the number dropped below 200,000, and the century ended with just 151,000 Americans making iron and steel, only about 20,000 more than were employed by the country's aluminum industry. The consequences of this unraveling of what was once the nation's leading industry can be easily seen by visiting such former strongholds of steelmaking as Pittsburgh, Pennsylvania, or Gary, Indiana.

Industrial Gases

Jumping from the hardest mass-produced solid of modern civilization to the cryogenic distillation of atmospheric gases in air separation plants is not done here for the sake of contrast. Oxygen and nitrogen separated from the air are two inputs that deserve to be ranked, together with steel, among a handful of necessities whose steady and inexpensive supply provides the material foundations of the twentieth century. In 2022, the worldwide industrial gas market reached the total mass of 1.6 Gt worth $100 billion; it was expected to grow by more than 7% a year until 2030, and, according to some estimates, industrial gases are used by producers and services that account for more than half of the global economic outputs (Grand View Research 2024).

During the first decade of the twentieth century, none of the inventors of the separation of gases from liquefied air could have foreseen the decisive role that the two principal elements would have, respectively, in feeding billions of people and in producing the world's most important metal. As already noted, without inexpensive oxygen we would not be producing about three-fifths of the world's steel in basic oxygen furnaces—and without equally inexpensive nitrogen we could not make cheap ammonia, now the precursor of all nearly inorganic nitrogenous fertilizers whose application made it possible to feed a world population of more than 6 billion people (see also Chapter I.4).

I have also already noted the use of oxygen to enrich air and increase combustion temperatures in both blast furnaces and electric arc furnaces, and the gas is used for the same reasons in smelting color metals (copper, lead, zinc) and in the production of glass, mineral wool, lime, and cement. Oxygen and oxygen-acetylene flames are indispensable in welding, particularly of large pieces that had to be previously cast as unwieldy units, and this method is also used for straightening, cutting, and hardening metals. Chemical syntheses, catalytic cracking of crude oil in refineries, coal gasification, wastewater treatment, paper bleaching (to replace chlorine dioxide), and oxygenation of aquacultural ponds (sites of producing animal protein) are other major industrial users, dwarfing the element's well-known therapeutic applications.

Except for the mass flows for ammonia synthesis, nitrogen uses tend to be more low-volume affairs in a wide variety of tasks. Shrink fitting—cooling such inner parts of metal assemblies as liners, pistons, pins or bushings by liquid nitrogen prior to their insertion to create a very tight fit—is an excellent example of a specialized, but now very common, application. Unlike the traditional expansion fitting through the heating of the outer parts, this is a much faster method and one that causes no metallurgical damage (it can be used, naturally, also to disassemble tightly fitted parts). An altogether different class of nitrogen application is to use the gas as an inert blanketing agent to protect flammables and explosives or to prevent degradative reactions (you can use it to protect your opened bottles of wine). Perhaps the best-known use of the gas is for freezing and cooling—be it of the ground (prior to drilling or tunneling) or blood, antiviral vaccines or semen, food, or reaction vessels.

Hydrogen, another major industrial feedstock and the promised energy carrier of the new energy economy—cannot be separated economically from the air where it is present in a merest trace, about 0.00005%. Its first large-scale industrial production (for the Haber-Bosch synthesis of ammonia; see Chapter I.4) was done by the steam reforming of coal, but, since the 1950s, it has been made overwhelmingly by steam reforming of hydrocarbons, above all methane (NH_3). Besides the mass flows for the synthesis of NH_3, hydrogen is also used in such diverse ways as making a multitude of compounds ranging from methanol, polymers, and solvents to vitamins and pharmaceuticals; to produce solid fats through the hydrogenation of unsaturated fatty acids; and to cool large turbogenerators.

Oxygen and Nitrogen

Incremental improvements in the separation techniques introduced by Linde and Claude in the early twentieth century had resulted not only in much higher volumes of gas production but also in lower capital and energy costs. A key innovation that dispensed with the need for any chemicals to remove water (H_2O) and CO_2 from the air and allowed for operations at lower pressure (and hence with reduced energy inputs) was a regenerator designed by German engineer Mathias Fränkl. This simple but highly effective device (initially patented in 1928) was incorporated into the Linde-Fränkl process, whose patents date mostly from the first half of the 1930s and which became commercial in 1934 (Fränkl 1932).

Fränkl's regenerators (heat interchangers) were pairs of cylindrical vessels 4–5 m tall filled with wound corrugated thin metal bands to provide a very large contact surface that received intermittently (for periods of 1–4 min) the cold flow from one of the separated gases and warm flow from the incoming air. Water vapor and CO_2 in the incoming air are deposited as solids on the cooled regenerator packing, to be removed again as vapors by the outgoing gas as it is being warmed. These regenerators are very efficient, and the exchange takes place under less than 600 kPa of pressure. In 1950, Linde AG introduced the first Linde-Fränkl oxygen plant without high-pressure recycling and with stone-filled, rather than metal-filled, regenerators. The Linde-Fränkl process reduced the prices of gaseous oxygen by an order of magnitude, and another key innovation of the 1930s saw expansion turbines replacing piston engines in refrigeration.

An innovation that was pioneered in the 1940s by Leonard Parker Pool of American Air Products brought the production and sales of industrial gases to the places of their use. Instead of continuing the traditional production in central plants and then distributing the gases in cylinders, Air Products began to build on-site cryogenic units, and soon it was scaling them up. Without this combination of low-cost, high-purity oxygen there could have been no BOF revolution in steelmaking: a large BOF needs 20 t of pure oxygen for a single heat. Typical sizes of air separation plants were 50–100 t/day during the 1950s, and, by the end of the 1970s, the ratings reached a plateau as Linde AG was building the world's largest air separation facilities with daily capacities of up to 2,300 t/day of pure O_2 and 800 t/day of N_2 (Hansel 1996; Linde AG 2003). The world's largest air compressors (54 megawatt [MW], built by Siemens) were used in some of these air separation plants, and they could handle volumes of up to 7200,000 m³/h.

Cryogenic separation in these large units begins with the compression of ambient air (normally to about 600 kPa) and with the removal of moisture and carbon dioxide by cooling the gases. The reversing heat exchangers are still used, but most plants designed since the mid-1980s rely instead on molecular (zeolite) sieves (UIG 2003). Pressure swing adsorption for oxygen separation uses synthetic zeolite molecular sieves to selectively adsorb water, CO_2, and nitrogen and to yield 90–95% oxygen (Ruthven, Farooq, and Knaebel 1993). The technique's origins go back to the late 1930s, when Richard M. Barrer began to create new zeolite structures from silicate alumina gels and to study their molecular sieving properties. In 1948, this line of research was taken up by the Linde Division of the Union Carbide Corporation, and, by 1950, its chemists produced two new synthetic zeolites, Linde A and X, and filed a patent in 1953 (Breck 1974).

The final structures of these hydrated aluminosilicates were determined a year later (Figure II.3.13). Linde A, an environmentally friendly compound whose annual production surpassed 700,000 t by the late 1990s (global consumption of all synthetic zeolites reached nearly 1.4 Mt by 2000), proved to be an excellent medium to separate nitrogen and oxygen. Its many other uses range from adding it to detergents to remove calcium and magnesium from water, to binding radioactive cesium and strontium present in nuclear wastes. Activated carbon sieves are used to produce nitrogen with purities between 95% and 99.5%. Vacuum swing absorption produces oxygen of similarly low purity but with lower energy cost, and it has been the fastest growing segment of oxygen production, particularly for plants rated at more than 20–30 t/day.

Pipelines (pressurized to about 3.5 MPa) are the cheapest way to distribute oxygen, and their highest densities are along the US coast of the Gulf of Mexico, and in Germany and Japan. Global production of oxygen was about 75 Gm³ during the late 1990s, with the European Union and the United States each producing more than 15 Gm³ and with five companies—Linde (Germany), Air Liquide (France), BOC (United Kingdom), and American

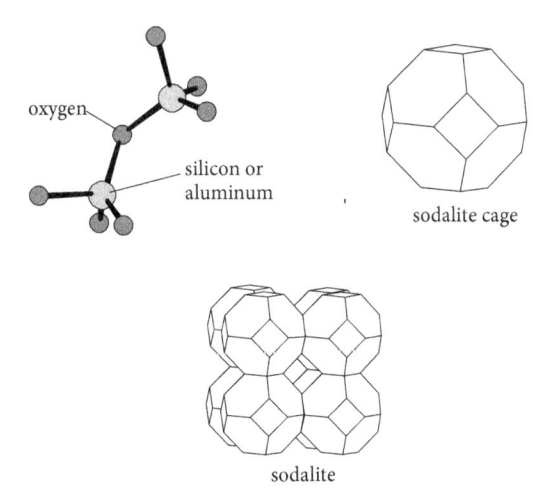

Figure II.3.13 Tetrahedrons, composed of oxygen and aluminum or silicon, are the basic building blocks of zeolites (top left); 24 of them form sodalite cages (top right), which in turn combine to make larger structures such as sodalite (bottom) or synthetic zeolites (Linde A, X, Y). Straight lines in the framework images join the centers of two adjacent tetrahedrons.
Based on illustrations in Kerr (1989).

Praxair and Air Products—accounting for two-thirds of the global output. During the late 1990s, steelmaking accounted for 40–60% of oxygen's use, with the annual US consumption of more than 8 Gm^3, nearly 20 times the early 1950s rate (Hogan 1971; Drnevich, Messina, and Selines 1998).

Oxygen use in blast furnace enrichment and higher specific oxygen use in EAFs resulted in a relative decline of the gas blown into BOFs. In contrast, the market for oxygen in cutting and burning, the first major innovation to use that industrial gas, had been mature for some time. Controlled oxyacetylene flame of unprecedented temperature (3,480°C) that would, as Henri Louis Le Châtelier (1850–1936) demonstrated in 1895, melt every common metal, was made possible by Linde's and Claude's invention and by the accidental discovery of the quenching of calcium carbide in water that was made by James Morehead and Thomas L. Wilson in 1892. Eleven years later, Thomas combined the two gases to create an oxyacetylene torch whose use entails three flame settings: neutral (with equal volumes of the two gases) used for welding, and carburizing (more acetylene) and oxidizing (with higher oxygen flow) used for cutting with high temperatures.

Hydrogen

Badische Anillin- & Soda-Fabrik (BASF)'s first ammonia pilot plant in Ludwigshafen used hydrogen produced from chlorine-alkali electrolysis, a process unsuitable for large-scale operation. Other known methods to produce hydrogen were either too expensive or produced excessively contaminated flow. Given the company's (indeed, the country's) high dependence on coal, Carl Bosch concluded that the best solution was to react the glowing coke with water vapor to produce water gas (about 50% H_2, 40% CO, and 5% each of N_2 and CO_2) and then remove CO cryogenically (Smil 2001). But, in the summer of 1911, Wilhelm Wild (1872–1951) began removing CO by a catalytic process during which the reaction of the gas with water shifted CO to CO_2 and produced additional H_2.

As already noted, this *Wasserstoffkontaktverfahren* dominated hydrogen production for the next four decades. Steam reforming of light hydrocarbons, and particularly pure methane (CH_4), would have been a much better choice, but it was only during the late 1920s that Georg Schiller found the way to reform CH_4 in an externally heated oven in the presence of nickel catalyst. Interwar Germany had no ready access to any hydrocarbons, but Standard Oil of New Jersey licensed the process and began to use it in its Baton Rouge, Louisiana, refinery in 1931. The first ammonia plant based on steam reforming of methane was built by Hercules Powder Company in California in 1939.

After World War II American industry converted rapidly to methane-based hydrogen, while in China coal-based production continued into the twenty-first century. By the late 1990s, roughly half of the world's hydrogen came from the steam reforming of natural gas. The source of carbon is different, but the two steps of this catalytic process are analogous to the pioneering Wild-Bosch method. The first step involves the exothermic reaction of methane and steam to produce H_2 and CO; the subsequent endothermic reaction of CO with water shifts the monoxide to dioxide to form additional hydrogen and, finally, CO_2 and traces of unreacted CO are removed by using standard adsorption processes. Pressure swing adsorption, introduced widely during the 1960s, can produce hydrogen that is 99.9999% pure (Ruthven et al. 1993).

By 2000, about 30% of the world's hydrogen was produced from liquid hydrocarbons (mainly from naphtha) in refineries, where nearly all of it was immediately consumed by two different processes: the gas is catalytically combined with intermediate refined products to convert heavy and unsaturated compounds to lighter and more stable liquids, and it is also used in the desulfurization of crude oil. Rising demand for cleaner burning, low-sulfur gasoline and diesel fuels and the greater use of heavier crudes has been the main reason for rapid increases of hydrogen in refining. In 2000, nearly 20% of hydrogen's global output still came from the steam reforming of coal, but less than 5% of the gas was derived by energy-intensive electrolysis of water that needs 4.8 kWh per cubic meter of hydrogen. In addition, methanol cracking units are used for low-volume and intermittent production of the gas.

By the end of the twentieth century the global demand for hydrogen surpassed 50 Mt a year, with refineries being the second largest consumer after chemical syntheses. Anticipated large-scale use of hydrogen as a key energy carrier in a post-fossil fuel world would make the gas a leading fuel for stationary and some mobile uses (Smil 2003; Scipioni, Manzardo, and Ren 2023). The two best reasons for this anticipated transformation are the high energy density of the gas—at 120.7 MJ/kg, the highest of any fuel, three times as high as that of liquid hydrocarbons and more than five times that of good steam coal—and its pollution-free combustion that generates only water. The well-known disadvantages are the high costs associated with its liquefaction and storage.

Remarkable New Materials

The second adjective in this section's title refers not only to entirely new compounds and combinations that were created or commercialized after World War I but also to many advanced variants of long-established materials. Steel and aluminum are the best examples of these continuing innovations: the hundreds of specialty steels and dozens of aluminum alloys that were used to build cars, appliances, airplanes, and myriads of components during the last decades of the twentieth century did not even exist before World War II. Steelmaking was a mass-scale and mature business even before World War I; in contrast, aluminum's use took off only after 1950, when its alloys became indispensable for the enormous expansion of commercial aviation and when they provided a key substitute for steel wherever the design required a combination of lightness and strength.

The rationale for substitution is always the same: in comparison with specialty steels, aluminum alloys cost more to produce and have a lower modulus of elasticity, but their lower specific density (typical alloys weigh 50% less than steel) and superior strength-to-weight ratio (nearly 50% better than steel) make for lighter yet durable structures. They also resist corrosion (thanks to a very thin layer of hard and transparent aluminum oxide) and can be repeatedly recycled (more than 70% of all automotive aluminum comes from old metal). As a result, these alloys—besides dominating aircraft and spacecraft construction—have made inroads in sectors ranging from road transportation to offshore hydrocarbon production, and from electricity transmission to consumer packaging.

By the late 1990s, an average new American passenger car contained about 120 kg of aluminum, an equivalent of about 7% of the vehicle's total mass, and twice the share during the late 1970s and four times the proportion during the early 1960s (AA 2003). The trend toward substantial replacement of iron and steel by aluminum began during the late 1970s with intake manifolds that were made traditionally of cast iron: by the mid-1980s, nearly all of them were made of aluminum alloy 319 or 356 (Spada 2003). By that time, the two next, and continuing, waves of replacement had begun: cast aluminum wheels instead of fabricated steel rims and cast aluminum engine cylinder heads and blocks. By the end of the 1990s, about 60% of car and light truck wheels, 40% of engine blocks, and 90% of heads, pistons, and transmission cases were aluminum alloys.

In rail transport, aluminum has been used increasingly since the 1970s for large-capacity (more than 50 t) hopper cars as well as for the world's fastest trains. Remarkably, studies show that even after 20 years of service the metal loss on the floors and sidewalls of aluminum hopper cars used to carry different bulk loads (coal, ores, aggregates) was roughly 25% less than for steel cars (AA 2003). The first bridges with aluminum decks or superstructures date from the 1940s, and marine use of aluminum alloys has spread from small boats to cruise ships, yachts, and seawalls. On land, aluminum alloys were increasingly made into irrigation pipes, heat exchangers, and sheet piling.

Since the 1950s, titanium has been replacing aluminum in high-temperature applications, above all in supersonic aircraft. Production of this metal is at least twice as energy intensive as that of aluminum (typically about 400 vs. 200 GJ/t), but it can withstand temperatures nearly three times as high (melting points are, respectively, 660°C and 1,677°C). At the same time aluminum has been losing some of its old markets to a growing array of plastics. Global synthesis of all plastics rose from less than 50,000 t in 1930 to about 2 Mt in 1950 and to slightly more than 6 Mt by 1960. The subsequent exponential growth brought the total to nearly 150 Mt in 2000, a mass surpassing that of aluminum and equal to about 18% of the world's steel production. By the late 1990s, the US plastics industry employed 10 times as many people (1.5 million) as did steelmaking, another 850,000 were in upstream industries that supplied the raw materials, and 50,000 worked in recycling (APC 2003).

Global output of man-made polymers is dominated (nearly 80%) by the synthesis of thermoplastics, materials made up of linear or branched molecules (with no chemical bonds among them) that soften on heating but harden once again when cooled, have relatively high impact strength, and are easy to process. Polyethylene (PE) is the most important thermoplastic: in 2000, it accounted for nearly a third of the world's aggregate production of all polymers; polyvinyl chloride (PVC) accounted for nearly 20%, and polypropylene for roughly 15% (Figure II.3.14). The other major category, thermosets, are compounds whose bonds among molecules resist softening with heating (and hence have a greater dimensional stability than do thermoplastics) but decompose at higher temperatures (Goodman 1999). Production in this category is dominated by polyurethanes and epoxy resins, and other major categories include polyimides, melamine, and urea-formaldehyde (Figure II.3.14). The annual worldwide output of thermoset plastics during the late 1990s was roughly equal to the synthesis of polypropylene.

Steel, aluminum, and plastics were prominent because of their ubiquity, but the twentieth century was transformed with many other new materials whose contribution was due to their unique properties rather than to massive uses. In retrospect, the most important innovation in this category was the discovery of the special conductive properties of certain elements: when extremely pure, they are insulators; when contaminated (doped) with minuscule quantities of other atoms (arsenic, boron, phosphorus), they conduct. This property has been exploited in the design of semiconductors made of germanium and silicon. Silicon's physical properties make it a particularly good choice for transistors that are assembled in increasingly complex integrated circuits at the heart of all modern electronics. Finally, the closing

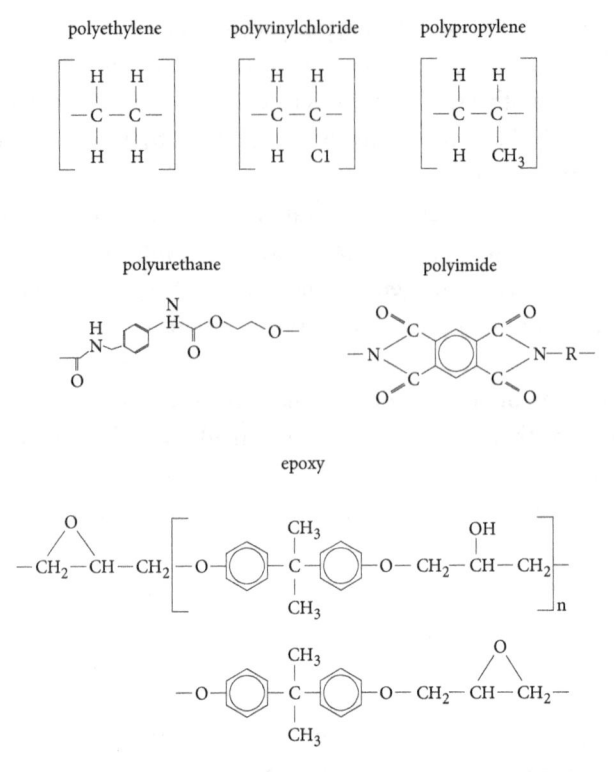

Figure II.3.14 Structures of the three most important thermoplastics (polyethylene [PE], polyvinylchloride [PVC], and polypropylene) and three common thermoset plastics (polyurethane, polyimide, and epoxy).

decades of the twentieth century saw the rise of an entirely new kind of material classed under a broad category of composites, as well as the discovery of nanostructures.

Multitude of Plastics

Plastics are the quintessential materials of the twentieth century. Their well-documented origins, their enormous variety, their rapid conquest of many specialized markets, and their eventual substitution of metals, glass, and wood in countless industrial and household applications attracted a great deal of attention to their discoveries and their commercial success (Brydson 1975; Seymour 1989; Meikle 1995; Fenichell 1996; Marchelli

1996; Mossman 1997; APC 2003; PHS 2003; Vinylfacts 2024). Experimental foundations for their synthesis were laid after 1860, during the pioneering decades of industrial organic chemistry, but several commercially successful compounds that were created during the closing decades of the nineteenth century were based on natural polymers and did not form permanent shapes. The most popular of these was celluloid, made from cellulose nitrate and camphor by the process patented by John Wesley Hyatt in 1870.

The age of durable plastics based on fully synthetic polymers was launched at the very beginning of the twentieth century with the preparation of phenol-aldehyde resins. The basic reaction between phenol and formaldehyde had been studied since the 1870s, but, until the late 1890s, its only practical products were shellac substitutes (Berliner used shellac for his phonograph records). In 1899, Arthur Smith was granted the first patent (UK Patent 16,274) for a phenol-formaldehyde resin that was cast in molds and used for electrical insulation. Leo Hendrik Baekeland (1863–1944), a Belgian professor of chemistry who immigrated to New York in 1898, began to work on reactions between phenol and formaldehyde in his private laboratory in 1902 (Bijker 1995). He could afford to become an independent researcher after he sold his patent of a photographic paper for instantaneous prints (Velox) to Kodak in 1899.

In June 1907, Baekeland succeeded in preparing the world's first thermoset plastic that was molded at temperatures between 150°C and 160°C. The key to his discovery was to use the alkalis or bases "in such relatively small proportions that their presence does not interfere with the desirable qualities of the products.... In fact in most cases the small amount of base persists in the final products and confers upon them new and desirable properties" (Baekeland 1909: 1). Baekeland eventually got more than 100 patents, and his General Bakelite Company, set up in 1910, was the first large-scale producer of a plastic compound. Bakelite would retain its shape without fading or discoloration, would not burn, was impervious to boiling, would not dissolve in any common acid or solvent, and was also electrically resistant. Bakelite's brittleness was solved by the addition of short cellulose (or asbestos) fibers that made the plastic strong enough for many practical uses.

Bakelite enjoyed great popularity between the two world wars, initially as an electric insulator and in manufacturing small household objects, including classic black rotary dial telephones (Figure II.3.15), and it became a common material in lightweight weapons and military machines mass-produced in the United States during World War II. Bakelite was soon joined

Figure II.3.15 Classical rotary black telephone was the most encountered Bakelite object for most of the twentieth century.
Photograph courtesy of Douglas Fast.

by a growing array of new synthetic materials: by the end of the twentieth century, there were more than 50 kinds of plastics. I will outline the main discoveries and then concentrate in some detail on the two leading products whose widespread uses illustrate our dependence on synthetic materials.

A notable commonality of post-Baekeland discoveries was the rising importance of institutionalized research as virtually all major plastic categories were discovered and commercialized thanks to dedicated research that was funded by major American (above all, DuPont) and European (IG Farben in Germany, Imperial Chemical Industries [ICI] in the United Kingdom) corporations. The skills of individual researchers and accidental results played role in several discoveries, but the rise of the modern plastics industry was one of the first instances where systematic corporate quests were decisive in creating new industries.

New discoveries began right after Baekeland's success. Cellophane, the first flexible plastic, was invented by Charles Frederick Cross in 1908, and it was waterproofed by Jacques Edwin Brandenberger in 1912. In 1920, Hermann Staudinger (1881–1965) replaced the long-reigning colloid theory of polymers, which saw the compounds as physical associations of small molecules, by the correct model of small molecular units linked in linear chains to form large macromolecular masses, a paradigm shift that was rewarded in 1953 by a Nobel Prize in Chemistry (Staudinger 1953). The first post-World War I decade saw the spreading uses of Bakelite and cellophane (DuPont began to make it in the United States in 1924) and the introduction of styrene in 1925 (its polymer is now everywhere in insulation and packaging) and cellulose acetate (a nonflammable counterpart of celluloid) in 1927 (Brydson 1975); also, in 1927, DuPont began a systematic program of fundamental research on polymerization.

During the 1930s, this work brought a still unsurpassed concatenation of discoveries and prompt commercialization. In April 1930, a DuPont team led by Wallace Hume Carothers (1896–1937) discovered neoprene (synthetic rubber) and synthesized the first polyester super polymer: an experiment with ethylene glycol and sebacic acid yielded a polyester that could be drawn out as a long fiber both in molten and cold form, but the substance was easily softened by hot water and hence it was of little practical use. Only after Carothers turned to the polyamides, did he find several substances that could produce permanent filaments, and polymer 66, made by the reaction between hexamethylene diamide and adipic acid, was chosen for commercialization under the name Nylon (Hermes 1996). The US patent for linear condensation of polymers was granted in 1937 (Carothers 1937), the pilot plant opened in 1938 (toothbrush bristles were the first application), large-scale synthesis began in December 1939, and stockings were the leading commercial product. DuPont's major rivals had their own successes.

In 1930, IG Farben produced polystyrene, and, in 1932, ICI began to investigate the effects of very high pressure (up to 300 MPa) on organic reactions, an effort that eventually led to PE. During the winter of 1929–1930, William Chalmers, at McGill University in Canada, discovered that methacrylate forms hard transparent polymers; ICI began to commercialize methyl methacrylate in 1933; by 1935, Röhm and Haas in Philadelphia was making sheets of Plexiglas; and, a year later, DuPont was producing Lucite. In 1933, Ralph Wiley at Dow Chemical accidentally discovered polyvinylidene chloride, now commonly known as Saran Wrap, a ubiquitous cling-packaging for food. And in 1936, Pierre Castan (1899–1985) prepared an epoxide resin suitable for dental fixtures; a year later, Otto Bayer (1902–1982), at IG Farben's Leverkusen lab, discovered polyurethanes (first used for bristles).

Polytetrafluoroethylene (branded by DuPont as Teflon) was the last fundamental discovery of the 1930s. In 1938, Roy Plunkett at DuPont accidentally pumped chlorofluorocarbons (CFCs, Freon) into a cylinder that was left overnight in cold storage and found that the gas was converted into a white powder. Teflon's slipperiness, in addition to its resistance to cold, heat, and acids, is perfect for cooking surfaces and for such diverse uses as the edges of windshield wipers, marine coatings, and aircraft toilets (DuPont 2003).

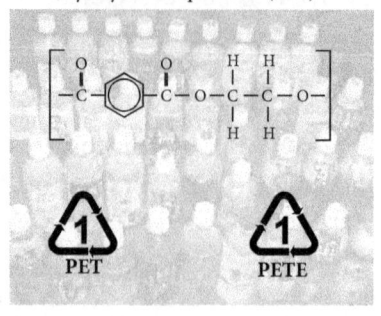

Figure II.3.16 Polyethylene terephthalate (PET)—cheap, clear, lightweight, and shatter and heat resistant—is the dominant worldwide choice of material for water and soft drink bottles. Although the plastic is readily recyclable (identified with number 1 inside the recycling triangle and acronym PET or PETE underneath), tens of millions of such containers are discarded every day.

New plastics discoveries of the 1940s included alkyd polyesters and polyethylene terephthalate (PET) (Figure II.3.16).

PET was initially used as a fiber branded Terylene and Dacron, later as a film best known as DuPont's Mylar, and, since 1973, as the preferred material for soft drink and water bottles as well as for fruit juice, peanut butter, salad dressing, and plant oil containers. Fortunately, this relatively expensive resin is readily reusable, and half of all polyester carpet in the United States is now made from recycled PET bottles; other uses include fiberfill for sleeping bags and coats and car parts such as bumpers and door panels. As noted in Chapter II.1, during the late 1940s, George de Mestral conceived a new plastic fastening system imitating cockleburs. The last year of the 1940s saw the introduction of perhaps the most frivolous plastic as James Wright at General Electric mixed silicone oil with boric acid and created Silly Putty. High-impact polyimide (for bearings and washers) came in 1955, and strong polycarbonate plastics—suited for optical lenses and windows—were introduced commercially in 1958.

But the decade's most significant breakthrough was the synthesis of PE at normal temperatures and low pressures by Karl Ziegler's team in Germany. Soon afterward Giulio Natta (1903–1979) began to use Ziegler's new organometallic catalysts to prepare polypropylene and other stereoregular macromolecules (having spatially uniform structure, with all the side groups pointing in the same direction). These so-called isotactic polymers combine lightness and strength, which are desirable in fabrics or ropes. For their discoveries Ziegler and Natta shared the Nobel Prize in Chemistry in 1963 (Ziegler 1963). During the 1960s came polyimide, polysulfone, and polybutylene; after 1970, thermoplastic polyester; and the 1980s saw the first liquid crystal polymers for high-performance electrical and surgical applications. DuPont's trade-marked successes included Lycra, Kevlar, Nomex, and Tyvek.

PE and PVC

During the late 1990s, PE and PVC accounted for nearly half of the world's synthesis of plastics, and their highly diverse uses continue to expand. Although they are structurally very similar—PE's carbon chain carries only hydrogens, while in PVC every other carbon carries one chlorine atom—and share several commercial applications, they differ greatly where the perception of safe use and environmental impact are concerned. PVC has been seen by many critics as the most environmentally damaging of all common polymers: it requires hazardous chemicals for its production, it releases some of its additives while in use or after it gets landfilled, it emits toxic dioxin when burned, and it is not easily recycled. Consequently, the compound would seem to be a prime candidate for replacement by polymers that can be produced by new methods of green chemistry—and yet its applications, including those in the sensitive medical field, continued to expand (Brookman 1998).

The PE story began with ICI's investigations of high-pressure syntheses in 1930. There were no immediate breakthroughs, but, in March 1933, Eric W. Fawcett and Reginald O. Gibson found a white waxy solid after combining ethylene and benzaldehyde under 200 MPa. This discovery was not followed up, and only two years later, after dismantling a defective apparatus, the white powder was again encountered and recognized as a new remarkable polymer whose synthesis required, as it happened by the accidental loss of pressure, the presence of oxygen (Mossman 1997). By 1936, the company patented a high-volume polymerization process and introduced large compressors for its first PE plant, which began working in September 1939, the day before the United Kingdom's declaration of war on Germany.

The compound's first important use came during World War II as an insulator for underwater cables and radars. The latter use reduced the weight of radar domes by hundreds of kilograms and allowed their placement onboard planes to detect German bombers under all weather conditions. ICI's original process required the unusual and technically demanding combination of high pressures (100–200 MPa) and a temperature of 200°C. In 1953, Karl Ziegler (1898–1973; Figure II.3.17) and his colleagues at the Max Planck Institute for Coal Research, in Mülheim, discovered that ethylene gas could be polymerized very rapidly with some organometallic mixed catalysts that were very easy to prepare and that allowed the synthesis to proceed first at just 20 MPa and eventually at normal pressure (Ziegler 1963).

The catalyst used in the first series of successful experiments was a mixture of aluminum triethyl, or diethyl aluminum chloride, with titanium tetrachloride, and subsequently the team prepared many combinations of aluminum with such heavy metals as cobalt (Co), chromium (Cr), molybdenum (Mo), titanium (Ti), vanadium (V), and zirconium (Zr), mixtures that became known as *Ziegler catalysts*.

Figure II.3.17 Karl Ziegler, whose work on mixed organometallic catalysts led to the discovery of high-density polyethylene (PE), was rewarded in 1963 by the Nobel Prize in Chemistry.
Photograph © The Nobel Foundation.

Our experiment thus destroyed a dogma. It led, in addition, to a PE which differed markedly from the high-pressure product. Low-pressure PE not only has a better resistance to elevated temperatures and a higher density, but is also more rigid.... The difference can be attributed to the fact that in our process molecules of ethylene are joined together linearly, without interruption, whereas in the high-pressure process chain growth is disturbed, so that a strongly branched molecule results. (Ziegler 1963: 8–9)

While the branched low-density PE (LDPE) has specific gravity of 0.92–0.93 g/cm^3, tensile strength as low as 9 MPa, and melting temperature as low as 98°C, the analogous maxima for high-density PE (HDPE) are 0.96 g/cm^3, 32 MPa, and 137°C (EF 2003). Relevant patents to produce HDPE were granted in 1954, and the first industrial production of HDPE began in 1956. The material tended to crack with time, and the solution was to make a compound with a small number of branches in the linear chain. The Phillips Petroleum process—developed during the 1950s by Robert L. Banks and J. Paul Hogan and commercialized in 1961—relies on a chromium catalyst, proceeds at a higher pressure than does Ziegler's synthesis, and produces HDPE with fewer branches and hence with higher specific weight. Finally, by 1980, new catalysts made it possible to make linear LDPE.

Syntheses of PE materials follow the same basic sequence: it starts with the thermal cracking of ethane followed by ethylene purification (a process that requires more than 100 separation stages and results in 99.9% purity), heating and compression, and catalytic conversion to PE. Different PEs now constitute by far the largest category of plastics. Most common are thin but strong LDPE films made into ubiquitous bread, garbage, and grocery bags. Hidden uses range from insulation of electrical cables to artificial hip joints (with ultrahigh-density PE). PE is also spun into fibers and blow-molded into rigid containers (HDPE for milk jugs and detergent and motor oil bottles), gas tanks, pipes, toys, and a myriad of components ranging from delicate pieces to bulky parts. HDPE items are stamped with a number 2 within the recycling triangle and are turned into trash cans, flowerpots, and detergent bottles. LDPE products are stamped with number 4 and reappear as new bags.

In contrast to PE, PVC—the world's second most important plastic (during the late 1990s its global consumption surpassed 20 Mt/year)—has a long history that had already begun in 1835, when Henri Victor Regnault produced the first vinyl chloride monomer. Eugen Baumann prepared its polymer in 1872, but a commercial polymerization that was patented in 1912 by Friedrich Hein-rich Klatte produced a tough, horn-like material that was difficult to fashion into useful articles. In 1926, Waldo L. Semon (1898–1999), a researcher at the blast furnace Goodrich Company in Akron, Ohio, began to experiment with PVC to create a new adhesive for bonding rubber to metal. Instead, he ended up with a new, plasticized vinyl: "This invention, in brief, consists in dissolving a polymerized vinyl halide, at an elevated temperature, in a substantially nonvolatile organic solvent, and allowing the solution to cool, whereupon it sets to a stiff rubbery gel" (Semon 1933: 1).

Although trailing various PE applications in the total global consumption, PVC has acquired an unmatched variety of uses. In the first volume, I used a short fictional narrative to follow a modern office worker in a US city through a part of his morning to highlight the continuing importance of technical inventions made during the 1880s, the most innovative decade of human history; a similar story can feature the ubiquity of today's PVC, be it in North America or East Asia. An alarm clock that wakes a sleeping woman and the lights and kitchen gadgets she uses to make breakfast are all powered by electricity that is delivered by PVC-insulated wires. Water for the shower, tooth brushing, and coffee is led through PVC pipes, as is the outgoing waste. Bathroom and kitchen floors may be made of PVC sheets or tiles, the lunch she packs could be in a PVC wrap, and it may be complemented by a drink in a PVC bottle. The car she drives will have PVC undercoating to delay corrosion, and she will be surrounded by PVC indoors, instrument panel, and upholstery.

But even this level of saturation is modest compared to the degree of PVC dependence that the woman would encounter if a car accident were to send her to a hospital. There she would be enveloped by objects made of different kinds of PVC: disposable and surgical gloves; flexible tubing for feeding, breathing, and pressure monitoring; catheters; blood bags; IV containers; sterile packaging; trays; basins; bed pans and rails; thermal blankets; and labware. In aggregate, by 2000, PVC was the primary component in at least a quarter of all medical products. And, contrary to earlier expectations, PVC uses were still rising (Brookman 1998). In construction, it is now found in siding, roofing membranes, window frames, Venetian blinds, and shower curtains; it is used for blister packaging, hoses, garden furniture, greenhouse sheeting, electronic components, office supplies, toys, cable insulation, and credit cards.

Besides its obvious versatility, PVC's two great advantages are its water- and flame-resistance. Its main drawbacks are emissions of dioxin during its incineration and the use of possibly carcinogenic plasticizers used to soften the compound. Consequently, Greenpeace (2003) campaigned for phasing out all PVC uses. In contrast, the polymer's defenders point to its low cost, its contributions to the quality and safety of products (particularly of cars, where the undercoating extends vehicle life, carpets and coatings reduce noise, and the lowered weight saves fuel), minuscule volumes of released dioxins (of which not all are potentially dangerous), and the biodegradability of leached plasticizers (Vinylfacts 2003).

Silicon

I have taken another jump in my survey of key materials that transformed the twentieth century, and hence this section is not a continuation of the preceding look at polymers, of which silicones (plural) are a smaller but important category. Scientific foundations to produce these oxide polymers were laid between 1899 and 1944, by F. S. Kipping's research at Nottingham University; their commercial production began during the 1940s, and their relative constancy of properties across a large range of temperatures makes them the best possible lubricants and resins in environments with wide temperature variations as well as good insulators and excellent water repellents.

My concern here is not with these silicon-oxygen polymers that have attached methyl or phenyl groups (i.e., silicones), but with the element itself, indeed, with its purest crystalline form without whose large-scale production there would have been no affordable microprocessors and hence no modern inexpensive consumer electronics, as well as no photovoltaic cells and no optical fibers (Seitz and Einspruch 1998). Silicon is, after oxygen, the second most abundant mineral in the Earth's crust. These two elements make up, respectively, nearly 28% and 49% of the crustal mass—but silicon never occurs in a free state, being always bound in silica (SiO_2) that is commonly encountered as quartz, sand, and sandstone or in silicates (numerous compounds including feldspar and kaolinite). The quartz (or quartzite) is the best starting material for producing commercial silicon compounds as well as the pure element. Ferrosilicon's use began in 1902, as a deoxidant in steelmaking, and, after World War I, polycrystalline silicon became important in aluminum alloying.

Reduction of SiO_2 with carbon (coal or charcoal: $SiO_2 + 2C \rightarrow Si + 2CO$) uses graphite electrodes in large electric furnaces, and it produces silicon with a purity (99%) that is quite unacceptable for electronic components. Nearly 60% of this metallurgical-grade silicon (whose output was just more than 1 Mt during the late 1990s) is still used for deoxidation and alloying, a third goes for the synthesis of silicones and other chemicals, and the small remainder (roughly 60,000 t) was the raw material to produce semiconductors, solar cells, and optical fibers. The semiconductor silicon is about 1 billion times purer than the metallurgical material (Föll 2000). Its production starts with trichlorosilane ($SiHCl_3$) that is purified by distillation, and high-purity silicon is produced from it by the process of chemical vapor deposition.

At this stage dopants, precisely calibrated amounts of impurities, are added to increase the conductivity. Pentavalent impurities (antimony [Sb], arsenic [As], phosphorus [P]) produce n-type semiconductors that conduct by additional free electrons; trivalent impurities (aluminum [Al], boron [B], gallium [Ga]) produce p-type semiconductors, which conduct through electron deficiencies ("holes"). This production of doped polysilicon, pioneered by Siemens during the 1960s, must reckon with the extraordinary toxicity of the leading dopants (AsH_3 and PH_3), high combustibility of both H_2 and Si HCl_3, and extreme corrosivity of gaseous HCl. At the same time, extremely precise input controls are needed to produce homogeneous material at a slow (about 1 kg/hr) rate.

In 2000, about 20,000 t of semiconductor-grade polycrystalline silicon were produced by this exacting technique. At roughly $50/kg, this output was worth about $1 billion. Converting this extremely pure polycrystalline material into a single crystal is done overwhelmingly (about 99% of the element for electronic applications is produced this way) by a method that was discovered, accidentally, by a Polish metallurgist working in Berlin during World War I. When Jan Czochralski (1885–1953) dipped his pen into a crucible of molten tin rather than into the inkpot, he drew a thin thread of solidified metal that proved to be a single crystal. Once a small seed crystal replaced the nib, its slow pulling from the melt, accompanied by slow rotation and gradual lowering of the temperature, made it possible to produce crystals of metals (tin [Sn], lead [Pb], zinc [Zn]) a few millimeters across and up to 150 mm long (Czochralski 1918). For 30 years this method was used just to study the growth and properties of metallic crystals.

Then came the invention of transistors (see Chapter II.4), and, in 1948, Gordon K. Teal and J. B. Little, working at Bell Laboratories, used Czochralski's pulling technique to grow the first germanium crystals. The first silicon-based transistor followed in 1949, and, in 1951, Teal and Ernest

Buehler were growing the element's crystals and had patented improved crystal-pulling (as well as doping) designs (Buehler and Teal 1956; Figure II.3.18). By 1956, the largest crystal diameter doubled to 2.5 cm, and crystal weight increased eightfold compared to 1950. The subsequent performance growth was steady but slow until the mid-1970s (7.5 cm and 12 kg by 1973). The first 10-cm, 14-kg crystals came in 1980, and, by the century's end, the largest 200-kg crystals were 30 cm across (Zulehner 2003). While the amorphous silicon is a dark brown powder, the crystals have a lustrous black to gray appearance, and no material made by modern civilization is as perfect as they are: their bulk microdefect density is below the detection limit of the best measuring tools we have, and impurities are present in concentrations below parts per billion, even parts per trillion.

Turning these perfect objects into wafers requires no less admirable fabrication techniques. As diamond-edged circular saws slice wafers (typically 0.1 mm thick, with ultrathin wafers of 25 μm), they also turn nearly half of the expensive crystals into dust, but the final polished product deviates no more than 1 μm from an ideally flat plane. This extraordinary accuracy had to be maintained for each of the roughly 100 million wafers that were produced in 2000 to serve as platforms for integrated circuits. The entire process of purifying silicon, growing its perfect crystals, and fabricating identical wafers became a key paragon of modern technical progress and a perfect example of a steeply rising value-added sequence. During the late 1990s, pure quartz was sold for about $0.017/kg; metallurgical silicon cost $1.10/kg; trichlorosilane, $3/kg; purified polycrystalline silicon, $50–100/kg; monocrystalline rod, at least $500/kg; polished silicon slices, at least $1,500 (and as much as $4,000)/kg; and epitaxial slices, as much as $14,000/kg (Jackson 1996; Williams 2003).

Another critical use for crystalline silicon emerged during the 1950s, almost concurrently with its electronic applications when the material was used to produce the first practical photovoltaic (PV) solar cells. A team at Bell Laboratories led the way in 1954, and, by March 1958, Vanguard I became the first PV-powered satellite, drawing a mere 0.1 W from about 100 cm^2 of cells to power a 5-mW transmitter that worked for eight years. Just four years later Telstar, the first commercial telecommunications satellite, had 14 W of PV power (Figure II.3.19) and, in 1964, Nimbus, a weather satellite, rated 470 W. During the remainder of the twentieth century the world had acquired entire fleets of satellites whose operation is predicated on silicon PV cells.

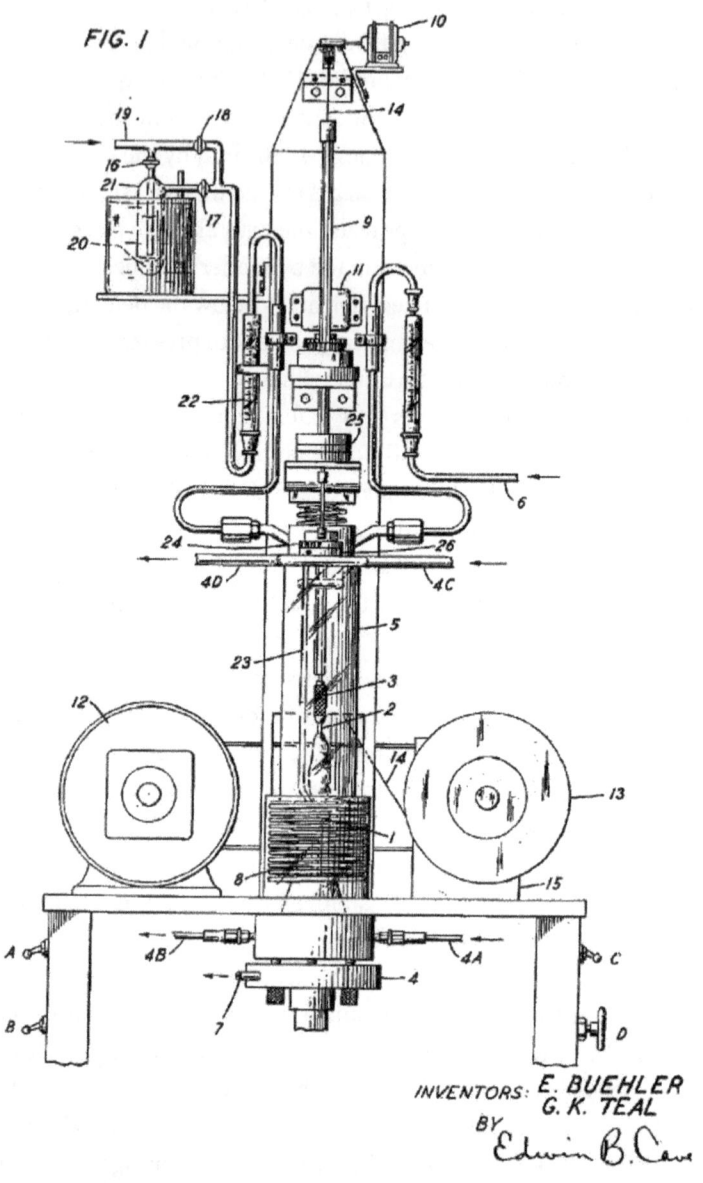

Figure II.3.18 Cross-section of a device for making semiconductive crystals of uniform resistivity designed by Ernest Buehler and Gordon K. Teal in 1951. After the ingot (Ge in the original application) placed in the crucible (*1*) is completely molten, the spindle (*9*) with attached seed (*20*) is lowered to touch the melt. The vibrator (*10*) and rotator (*11*) are turned on, and then the motor (*12*) drives disks (*13*) so that the cable (*14*) draws up the spindle (*9*). Dopants are added from liquid (*20*) and gaseous (*21*) reservoirs.

Illustration from Buehler and Teal (1956). Images from the US Patent Office.

MATERIALS 471

Figure II.3.19 Telstar I, designed and launched by Bell Laboratories in 1962, was the world's first active telecommunication satellite powered by photovoltaic (PV) cells. The spherical satellite was nearly 1 m in diameter and weighed 80 kg.
Photograph used with permission of Lucent Technologies.

PV cells energize the increasingly complex geosynchronous assemblies that carry voice, now with such clarity that intercontinental conversations sound as if spoken in the same room, and data streams that include everything from billions of transactions that converge at the great bourses of the world every day to processed credit card records. Weather satellites in high orbit improve forecasting and warn about incoming cyclones, and Earth observation satellites of many kinds that constantly monitor the oceans and the planet's surfaces allow us to follow the changing qualities of the biosphere. And, of course, there are also scores of low-orbiting spy satellites, some delivering images with resolutions as low as 5–10 cm.

In contrast to rapidly diffusing space applications, land-based uses of PV cells remained rare even after David Carlson and Christopher Wronski fabricated the first amorphous silicon cells (cheaper than the crystalline ones) at RCA Laboratories in 1976 (Carlson and Wronski 1976). A quarter century later steady improvements pushed the efficiencies of new single-crystal modules to as much as 12–14%; thin-film cells convert 11–17% in laboratory settings, but their field performance was as low as 3–7% after a few months, while multijunction amorphous silicon cells did better at 8–11%. Global annual PV production for land-based electricity generation began advancing rapidly only during the 1990s, and it reached 200 MW$_p$ (peak capacity) by 2000.

With about 85% of total PV production based on crystalline silicon (the rest using cheaper amorphous silicon) and with typical rates of 100 peak W/m^2, PV cell thickness of 300 μm, and silicon density of 2.33 g/cm^3, this means that the annual requirement for crystalline PV cells amounted to almost 900 t in 2000, less than 5% of the total production (Föll 2000). This relatively small need was easy to satisfy by the material that failed to meet specifications for wafer production (10–12% of the total output), and it was available at a considerable discount. The total installed PV capacity reached only about 1 GW (about half of it in the United States), still only a tiny fraction of more than 2.1 terawatt (TW) installed in fossil-fueled generators (Smil 2003). The largest PV producers were BP Solarex in the United States, Japan's Kyocera and Sharp, and Germany's Siemens (Maycock 1999).

Silicon's extraordinary purification also became essential for producing optical fibers, whose installation has revolutionized long-distance telecommunication and mass data transfer. Their principal raw material, as in all glass, is silica (about 75%); the remainder is a mixture of Na$_2$O, CaO, and Al$_2$O$_3$. The best lenses produced around 1900 were about 10,000 times more transparent than the first Egyptian glass made more than 5,000 years ago. After the 1960s, production of pure silicon dioxide improved that performance another 10,000 times, and even greater transparencies were possible with new non-silica glass.

Frontispiece II.4 Robots are the most advanced form of mechanization in industrial manufacturing. The top image shows a common application in the automobile industry, in this case Japanese FANUC robots working on a Volkswagen car. The bottom photograph shows coordinated action of multi-material handling FANUC robots.
Photographs courtesy of FANUC Ltd.

4
Rationalized Production
Mechanization, Automation, Robotization

> The world is nothing but raw material. The world is no more than unexploited matter.... Gentlemen, the task of industry is to exploit the entire world.... Everything must be speeded up. The worker's question is holding us back.... The worker must become a machine, so that he can simply rotate like a wheel. Every thought is insubordination! ... A worker's soul is not a machine, therefore it must be removed. This is my system.
> —Factory owner John Andrew Ripraton revealing his vision for organizing mass production in *The System*, a story by Karel and Josef Čapek published in 1908

One of the most remarkable, and the most fundamental, attributes of modern civilization is that we are surrounded by an unprecedented variety of products whose abundance and affordability stem from their mass production that has been not only mechanized but often fully automated and robotized. These tasks are overwhelmingly powered by electricity, and the final products also embody variable amounts of energy that was expended previously on raw materials or feedstocks. Yet by the end of the twentieth century, the twin realities of complete mechanization of many productive tasks and astounding economies of scale were taken so much for granted that most people would see them as unremarkable realities and would not single them out as a key characteristic of modern civilization.

But it was only during the twentieth century that the increasing availability of fossil fuels and electricity provided inexpensive means to energize en masse such affordable and highly versatile energy converters as internal combustion engines and electric motors whose power was used to mechanize just about every conceivable productive task and to take advantage of many obvious economies of scale. This trend was pervasive in both its categorical extent and its spatial reach: no kind of economic activity remained unaffected by it as ingenious machines had mechanized tasks that range (to give two mundane examples) from cutting and debarking of trees to filling and sealing drug containers or (to offer two more spectacular cases) from micro slicing ultrapure silicon crystals to emplacing entire genome arrays on microchips.

One of the best-known images of this reality is shown as this chapter frontispiece: a view of a hall in a car-making factory where robots move in programmed steps to reach out and spot weld car bodies that move along on a belt. This is a mesmerizing choreography of metallic arms, welding sparks, and precise motion—with not a human in sight. This activity is a perfect example of the twentieth century's productive transformations: energetic and material fundamentals (electric motors, high-quality steels) have their origins during the remarkable era of pre-World War I innovations, but the complexity of the entire assembly and its electronic controls are quintessential late-twentieth-century advances.

And the absence of people in the hall is a perfect illustration of the attendant labor shift from physical exertion and dexterity to the intellectual effort of designing, maintaining, and improving the mechanized processes as we increasingly transformed ourselves from low-power, low-efficiency prime movers to highly versatile controllers of high-power flows. Perhaps the best-known illustration of this shift is the decline in agricultural employment that has been reduced to the barest functional minimum in all high-income societies. The United States was a leader in this process: total farm employment was below 50% already by 1880; by the mid-1950s, it dropped below 10%; and, during the late 1990s, it was just 2% (USDA 2003).

And during the twentieth century this trend was evident all around the world. Moreover, in some countries it progressed at an unusually rapid pace, a fact that becomes obvious when one compares, for example, the post-1900 rates of decline in the US and French agricultural labor force with those in post-World War II Japan or post-1980 China (Figure II.4.1). In this chapter, I concentrate on tracing the underlying technical advances that were first evident in crop production and later affected every aspect of biomass harvesting, from field crops to high-sea fishing, from giant cattle feedlots to intensive aquaculture, and from the destruction of tropical forests to short-rotation tree plantations. In most of these cases, there were also major productive gains due to the inputs of chemicals (fertilizers, pesticides, herbicides, hormones,

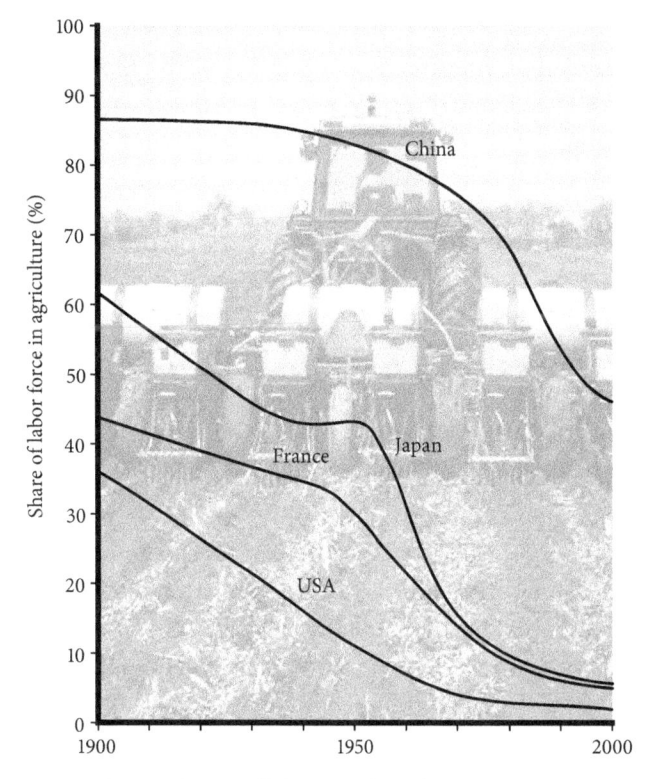

Figure II.4.1 Declining shares of labor employed in US, Japanese, French, and Chinese agriculture, 1900–2000.

Plotted from data in Ogura (1963), USBC (1975), Mitchell (2003), and FAO (2024).

antibiotics) and the modification of the genetic makeup of domesticated land animals, aquatic species, and planted trees.

During the twentieth century the extraction of sand, gravel, and stone, the inexpensive materials used mainly in construction, expanded much faster than the production of fossil fuels. But because of its inherently distributed nature, this extraction was not affected by the economies of scale to such a high degree as the mining of coal and the recovery of hydrocarbons that are reviewed in this chapter, whose last section deals with the diverse category of advances in industrial production, the activity that is still commonly called "manufacturing" even though fewer and fewer hands touch the products. This activity offers many spectacular examples of mechanization ranging from giant hydraulic presses that monotonically stamp out large steel parts for making cars or white goods to the extremely complicated robotization of such rapidly evolving processes as the fabrication of microprocessors. And the mechanization of industrial manufacturing is also the best example of the most dizzying economies of scale and the most impressive rise in the productivity of labor.

Agricultural Harvests

From the fundamental physical point of view, the great agricultural transformation of the twentieth century can be best defined as the change of energy foundation (Smil 2022). All traditional agricultures, regardless of their outward features or the levels of their productivity, were powered solely by photosynthetic conversion of solar radiation, which produced food for people, feed for animals, recyclable organic wastes for the replenishment of soil fertility, and household fuel. Fuel needed for small-scale manufactures and for the smelting of metals that were used to make simple farm tools was also derived from the photosynthetic conversion by cutting firewood and producing charcoal. Consequently, if properly practiced, traditional agricultures were energetically fully renewable as they did not require any depletion of accumulated energy stocks and did not use any non-solar energy subsidies.

But this renewability did not guarantee either the sustainability of the traditional practices or reliable supply of food. Poor agronomic practices and overgrazing reduced soil fertility, and they also commonly resulted in excessive erosion and in the abandonment of arable or pasture lands. Low yields, often hardly changed over centuries, typically provided only marginal food surpluses in good years and resulted in recurrent food shortages, widespread malnutrition, and recurring famines. Intensification of traditional farming through fertilization with organic matter, irrigation, multicropping, plant rotations, and mechanization of some basic tasks that was accompanied by more efficient use of draft animals could eventually support higher population densities, but it demanded higher inputs of human and animal labor (Smil 2017a).

Four universal measures that transformed traditional agriculture into a modern economic activity include the mechanization of field and crop processing tasks; use of inorganic fertilizers, and particularly of synthetic nitrogen compounds; applications of agrochemicals to combat pests and weeds; and development of new high-yielding cultivars. None of these advances could have happened without the input of fossil fuels and electricity or without the introduction of new prime movers. Consequently, modern farming is not just a skillful manipulation of solar energy flows, but it has also become an activity unthinkable without massive fossil fuel energy subsidies that are channeled directly through fuels and electricity used to power farm machinery and indirectly as energy embedded in numerous industrial products and used to support extensive agricultural research. Some of these innovations began during the nineteenth century, but most of them saw widespread adoption only after World War II.

Similarly, it was also only during the 1950s that mechanization and a variety of other energy subsidies began to transform both the scale of typical operations and the level of productivity in feed-grain–dependent animal husbandry in North America and Europe, and a generation later this new mode of production began to diffuse throughout Asia. This transformation resulted in animal concentrations up to five orders of magnitude larger than was usual in traditional settings. At the same time, wood harvesting and commercial fisheries also became highly mechanized and more productive. Logging with chainsaws and trucks (even with helicopters) opened previously untouched areas of boreal and tropical forest for cutting and colonization.

And, by the late 1990s, there were only three major fishing areas in the world ocean that had a history of increasing catches during the previous three decades. Saturation and decline of ocean fish catches stimulated the post-1980 expansion of aquaculture, another intensive method of food production that is highly dependent on energy subsidies. By the end of the twentieth century all biomass harvests, with the sole exception of grazing on unimproved pastures, became parts of the fossil-fuel–subsidized economy—with all attendant benefits (high productivity, low-cost food, fiber, and wood) and concerns (above all, many forms of environmental degradation).

The transformation of cropping and animal husbandry was by far the most remarkable accomplishment because it made possible the near quadrupling of global population during the twentieth century. The resulting productivity gains were impressive. In 1900, US wheat averaged about 40 hours of labor per hectare, but, during the late 1990s, it was no more than 100 minutes. And because of increased yields (from just 0.8 tons per hectare [t/ha] in 1900 to 2.8 t/ha in 2000) the time needed to produce a kilogram (kg) of wheat was reduced from about three minutes in 1900 to only two seconds in the year 2000, and the best producers did it in less than two seconds! (Smil

2022). Similarly, the best broiler operations of the late 1990s could produce a marketable bird in less than 6 weeks with as little as 2 kg of feed per kilogram of live weight—compared to about three months and about 5 kg of feed in the early 1930s (Smil 2022). But these achievements came at a considerable price.

Machines and Chemicals

At the beginning of the twentieth century productivity of traditional agriculture in North America and in parts of Europe was already benefiting from innovations in materials and machinery design. Inexpensive steel brought curved mold-board plows into general use, mechanical grain harvesters became common, horse-drawn combines were used on some large US farms—by 1915 they harvested more than 90% of California's small grains (Olmstead and Rhode 1988)—and threshing machines became widespread. But the field draft was still dominated by animals, above all by heavier and more powerful horses: at the beginning of the twentieth century a standard California combine with a cutting bar more than 6 m wide was pulled by a team of 25–35 horses.

But there were clear logistic limits in using more horses to pull a single machine, and when the total number of horses and mules working in the US agriculture peaked during the second decade of the twentieth century, their feeding required about a quarter of the country's extensive farmland (Smil 2017a). Consequently, horse-drawn mechanization could progress only in countries with large amounts of arable land while early steam engines were simply too heavy for any fieldwork. Eventually, the machines became light enough to be used for stationary threshing, often directly in fields, and to pull gang plows. But steam plows saw only limited use, particularly in the heavy clay soils of the Northern Plains, and the mechanization of field tasks began in earnest only during the first decade of the twentieth century once internal combustion engines had emerged as reliable and versatile prime movers. The Dakotas and California pioneered the trend, with about 20% of all farms reporting tractors by 1925, but it was only by the end of the 1920s that power capacity of gasoline tractors surpassed that of American horses and mules.

Basic configuration of modern tractors, including Harry Ferguson's hydraulic draft control system and low-pressure rubber tires, came together by the mid-1930s (Dieffenbach and Gray 1960). Afterward there were two trends that dominated the evolution of North American tractors: one toward lighter engines and frames, the other one toward more power. The latter trend included the shift to four-wheel drive and diesel engines that began during the 1950s. The mass-to-power ratio of early tractors was as high as 450–500 grams per watt (g/W), while modern tractors rated below 50 kilowatts (kW) weigh 85–95 g/W and machines more than 100 kW as little as 70 g/W. The largest machines that worked on the US Great Plains and Canadian prairies during the 1990s rated around 300 kW (Figure II.4.2). Rollover bars, first available in 1956, reduced the number of fatalities from tipping, and later they became incorporated into a closed cabin design that protects against noise and allows for air conditioning.

By 1950, the aggregate power capacity of tractors surpassed the power of all US draft animals by nearly an order of magnitude, and, by 1963, the US Department of Agriculture stopped counting horses and mules that still worked on the farms. The USSR also produced more powerful tractors (including heavy caterpillar models), but in Europe, Asia, and Latin America the sizes remain much smaller, and the adoption of tractors proceeded much

Figure II.4.2 One of five models in the 2000 series four-wheel-drive tractors that were designed by Winnipeg's Bühler Versatile for large prairie fields. Their engines were rated between 216 and 317 kW (290–425 hp), and they could use implements up to 21 m wide.
Photograph from the Buehler Corporation.

more slowly. Even in the richest European countries they did not displace horses until after 1950, and in the eastern part of the continent many working animals remained even after 1980. Post-World War II Asian mechanization relied on hand-guided two-wheel tractors suited for small rice fields: Japan pioneered the trend, and, during the late 1980s, some 40% of Chinese tractor power was in such units rated at about 3.5 kW. By the 1990s, sub-Saharan Africa remained the only major region largely untouched by widespread mechanization of fieldwork.

With more powerful tractors came specialized implements: unit frame plows in 1915, power take-off–operated corn pickers in 1928, potato harvesters in 1940. After World War II many of them were replaced by specialized self-propelled machines: corn-silage harvesters (invented in 1915), self-tying hay and straw balers (since 1940), a mechanical spindle cotton picker that was developed by John and Mack Rust in 1949, and, in 1976, rotary and tine separator combines (Holbrook 1976; NAE 2024). Chemicals, seeds, and fuel were brought to fields and harvests taken away by trucks. Internal combustion engines and electric motors also changed crop irrigation by tapping deeper wells and by greatly increasing the total volume of delivered water. Most field crops need mostly between 500 and 800 millimeters (mm) of water during their growing period: this was traditionally delivered by gravity through open ditches or lifted by a variety of simple irrigation machines powered by animals and people.

The first step in modernizing irrigation was to use mechanical pumps driven by internal combustion engines: when well-maintained they can work for more than 20 years. Better methods that did away with furrow-and-ridge irrigation spread only after 1950. The most efficient sprinkler technique, the center pivot type, is able to deliver at least 65% and even more than 90% of available water to plant roots in fields of 60–240 ha; it was patented by Frank Zybach in 1952 (Splinter 1976). This technique distributes water (and dissolved fertilizers or pesticides) from a central well through a long rotating arm carried by a series of wheeled supports.

By the early 1970s, this became the leading method of crop irrigation in the United States: patterns of circles dot the Great Plains and arid Western states. By the early 1990s, the country had about 110,000 center pivots irrigating nearly 6 megahectares (Mha). And, after 1980, center pivots appeared in deserts of Libya, Egypt, and Saudi Arabia. Bu the only rational use of water in such environments is drip irrigation: plastic pipes to individual plants or trees to deliver slowly the needed amounts of water to vegetables, fruits, and flowers. Pumping water from deep wells can be the most energy-intensive part of modern cropping. Despite these costs, irrigation expanded rapidly during the twentieth century. In the United States, it rose nearly fourfold, and globally it went from less than 50 Mha in 1900 to about 270 Mha by 2000 (or from about 5–19% of cultivated area), with half of the latter total irrigated by mechanical means (FAO 2024).

Production of inorganic fertilizers dates to 1841, when James Murray (1788–1871) began to sell liquid superphosphate made by dissolving bones in H_2SO_4. Soon the extraction and treatment of phosphates created a new large-scale industry. Relatively small English phosphate production was surpassed, beginning in the late 1860s, by phosphate mining in North Carolina, and, since 1883, in Florida. The industry's principal product, the ordinary superphosphate, contained between 7% and 10% of the nutrient, an order of magnitude more than most recycled organic wastes. Potassium is mostly retained in crop residues, and their recycling usually provides enough of the nutrient, but a new large-scale industry began in 1861 with the exploitation of the Stassfurt potash deposit in Saxony (Kleine-Kleffmann 2023).

Securing enough nitrogen, usually the most important macronutrient to limit crop growth, was not that easy. During the latter half of the nineteenth century there were three possibilities: imports of guano and Chilean nitrates and the recovery of ammonia from coke ovens. But by 1900, these fertilizers delivered only about 340,000 tons (t) of nitrogen or no more than about 2% of the nutrient that was removed by that year's global harvest. Recycling of crop residues and manures and intercropping of cereals with legumes remained the dominant source of the nutrient; the nitrogen barrier to higher yields was finally broken by Fritz Haber's invention of ammonia synthesis in 1909 and by Carl Bosch's leadership that turned a small bench demonstration into a new industrial process in less than four years (see Part I).

Rapid post-1950 expansion peaked during the late 1980s at about 85 megatons (Mt) of N (Figure II.4.3). Demand for nitrogen rose with new crop varieties whose high productivity required intensive nutrient applications. Hybrid corn was introduced in the 1930s, raising average yields from less than 2 t/ha to 8.6 t/ha by 2000 (USDA 2003). High-yielding cultivars of wheat and rice were developed during the 1960s at, respectively, the International Maize and Wheat Improvement Center (CIMMYT) in Mexico and the International Rice Research Institute (IRRI) in the Philippines. Their widespread adoption doubled the world's average rice yields and led to a 2.5-fold increase in the wheat yield (FAO 2004). Urea, a solid fertilizer

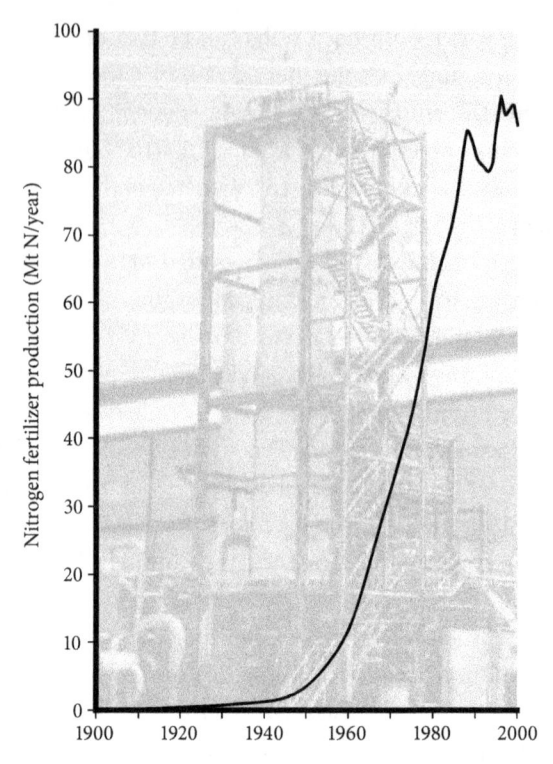

Figure II.4.3 Global consumption of inorganic nitrogenous fertilizers rose from less than 400,000 t of nitrogen in 1900 to slightly more than 85 Mt in 2000. In 1900, most of this nutrient originated in Chilean $NaNO_3$ and in $(NH_4)_2SO_4$ recovered from coke ovens. Haber-Bosch synthesis of ammonia, first commercialized in 1913, became globally dominant by the early 1930s, and, in 2000, more than 99% of nitrogenous fertilizers were compounds derived from improved versions of the process.
Plotted from data in Smil (2001).

with the highest nitrogen content, became the world's leading nitrogenous compound, and, by 2000, about half of the nutrient used annually by the world's crops (and about 40% of the element available in the world's food proteins) originated in the Haber-Bosch process (Smil 2001). At that time the world's crops also received 16 Mt of phosphorus and about 23 Mt of potassium (FAO 2024).

Post-1950 growth of fertilizer applications was accompanied by the widespread adoption of synthetic pesticides to control invertebrates (mainly insects) and herbicides to destroy weeds. Both insect and weed controls were commercially introduced during the 1940s. Dichlorodiphenyl trichloroethane (DDT) was the first and for many years by far the most important insecticide (Figure II.4.4). Its first synthesis was made by Othmar Zeidler in 1874, but for more than half a century nobody recognized the compound's insecticidal properties (Mellanby 1992). In 1935, Swiss chemist Paul Hermann Müller (1899–1965) began his search for an ideal insecticide (effective against many species, stable, having minimal impact on plants and higher animals) and four years later his tests singled out the previously known DDT (Smil 2023a). Post-World War II commercial sales, driven first by the insecticide's use against typhus-carrying lice and malarial mosquitoes, grew rapidly, and Müller was awarded a Nobel Prize in Physiology (Müller 1948).

DDT applications were initially quite effective, and they proved to have low toxicity in mammals. But the compound is highly toxic in fish, and its persistence in the environment (half-life of about 8 years) led to its long-distance transport: it was found in Antarctic penguins and seals (Roberts 1981). Its storage in body lipids eventually resulted in relatively high DDT levels in some human populations and in dangerously high concentrations in wild species, particularly in raptors, where it caused thinning of eggshells and hence the failure to reproduce (Smil 2023a). In addition, many insect species gradually developed resistance against DDT. General use of the compound was banned in the United States as of December 31, 1972, after some 6,000 t were applied during the preceding three decades. Less toxic and more rapidly degradable insecticides took DDT's place.

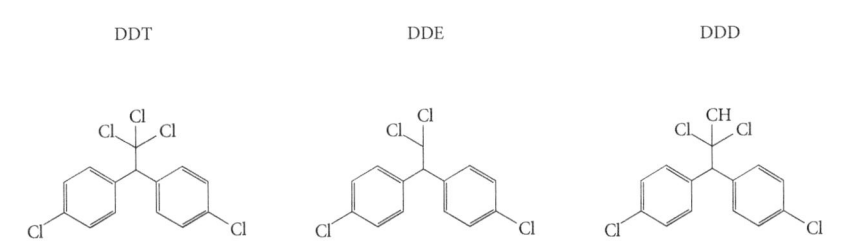

Figure II.4.4 Structural formula of DDT (dichlorodiphenyl trichloroethane), the first commercial synthetic insecticide whose applications were eventually banned in affluent countries, and of its two breakdown products, DDE (dichlorodiphenyl dichloroethylene) and DDD (dichlorodiphenyl dichloroethane), which are among the most persistent environmental contaminants.

486   THE TWENTIETH CENTURY

Discovery of the first, and still very commonly used hormone herbicides—2,4-D, 2,4,5-T, and methylphenoxyacetic acid—has a remarkable history: it was made independently by not just two but four groups of researchers—two in the United Kingdom and two in the United States during World War II between 1940 and 1943—causing a great deal of lasting confusion regarding both the timing and contributions of individual teams (Troyer 2001). The compounds are chlorophenoxyacetic acids, growth-regulating substances whose small quantities kill many broad-leaved weeds without any serious injury to crops (Figure II.4.5). Their first application as defoliants was by the British in Malaya between 1951 and 1960, and the much better known and massive American use of Agent Orange in Vietnam began in 1962 (Stellman et al. 2003). Agricultural uses started immediately after World War II.

Many broad-spectrum and highly specific herbicides have been introduced since 1945, but 2,4-D remained the world's leading defense against field weeds for the rest of the century, and, in 2000, it was still the third widely

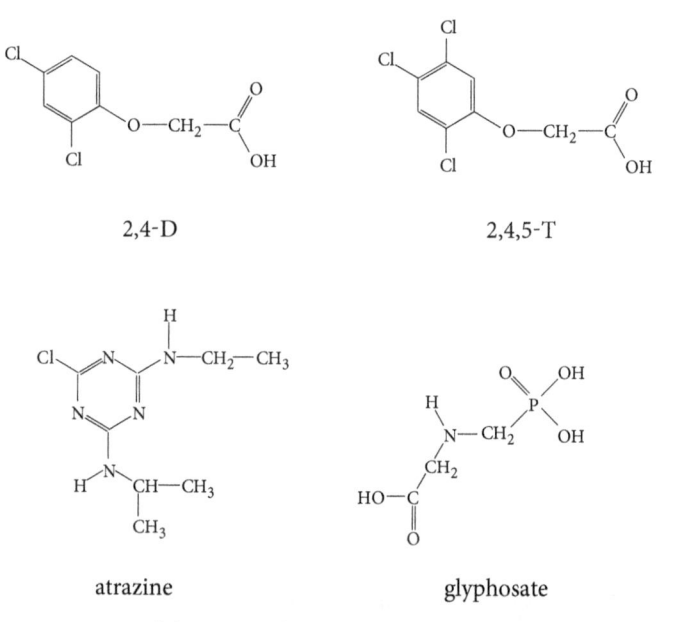

Figure II.4.5 Structural formulas of four widely applied herbicides: 2,4-D and 2,4,5-T, two chlorophenoxyacetic acids that were first synthesized during World War II; atrazine, introduced in the early 1960s and a leading herbicide of the 1990s; and glyphosate (N-phosphonomethylglycine), patented by Monsanto in 1971.

used antiweed compound in North America. The most innovative addition was glyphosate, formulated by John E. Franz, at the Monsanto Company, in 1971 (Franz 1974). This broad-spectrum post-emergence herbicide, the active ingredient of Monsanto's Roundup, shuts down plant metabolism; it acts only upon contact with green leaves, and it does not move through soil to affect other plants or to contaminate water. (Figure II.4.5). During the 1990s, Monsanto developed transgenic varieties of glyphosate-resistant crops that were adopted rapidly in the United States and Canada but met with a great deal of resistance in Europe (Ruse and Castle 2002).

Animal Foods

Production of animal foods was transformed no less radically than the cultivation of crops. At the beginning of the twentieth century only some pastoralist groups had large herds of cattle or sheep reared in a purely extensive manner by grazing. In contrast, animal husbandry integrated with crop farming relied on small groups of animals whose ownership filled several important roles. They were the source of meat, eggs, milk, leather, and wool; several large species were indispensable draft animals; and all species produced wastes whose recycling was essential for replenishing the fertility of nearby fields. In North America, abundant land made it possible to practice both extensive grazing and feeding with grains, while the limited amount of farmland in monsoonal Asia made grain feeding an exceptional practice. Transformation of these traditional practices took place in North America and Western Europe between 1920 and 1960; in East Asia and parts of Latin America, it began only after World War II.

Our species is naturally omnivorous, and eating animal foods in general and meat in particular is clearly a part of our evolutionary heritage (Smil 2024a). Limited meat consumption in traditional societies reflected inadequate supply rather than dietary choices, and rising incomes created higher demand for meat. As internal combustion engines displaced draft animals, more land became available to produce high-quality feed for meat and dairy animals, and even more land could be planted to feed grains because of the rising yields of food grains and their declining direct consumption. Animal husbandry changed from the traditional pattern of small-scale operations integrated with cropping to increasingly concentrated large-scale animal feeding operations located far from major feed-producing regions, many even importing most of their feed from different continents.

These operations benefited from economies of scale in housing and breeding the animals, mixing their specific feeds, and processing their products. During the last quarter of the twentieth century this shift toward larger units spread beyond North America and Europe, and two sets of examples illustrate its magnitude. Around 1900, livestock farms in Illinois, on some of the world's richest farmland, had 100–200 hogs, which were fed grain but kept on pasture as much as possible (Spillman 1903). In 2000, all feeding in the United States was in confinement; operations with more than 2,000 animals accounted for two-thirds of all hogs, and piggeries with fewer than 100 animals had less than 2% of all animals in the inventory (USDA 2003).

In China, the world's largest pork producer, extensive rural surveys done between 1929 and 1933 showed that the usual number of livestock per farm included just two hogs and three chickens (Buck 1937). A 1996 survey found that half a million Chinese farms had more than 50 pigs, and there were more than 600 operations with 10,000–50,000 animals (Simpson et al. 1999). In the Netherlands during the late 1990s, broiler production was shifting from operations with flocks of up to 25,000 birds to farms with more than 50,000 birds (van Middelkoop 1996), and China had 250,000 farms with more than 1,000 laying hens and two with more than 10 million birds (Simpson et al. 1999). Such gigantic operations can be supplied only from large feed mills that process raw materials from a wide variety of sources into balanced mixtures designed to maximize productivity.

Animal foodstuffs are consumed above all for their high-quality protein, whose production entails inevitable metabolic losses. To produce 1 kg of protein, animals must consume anywhere between 2.5 kg (dairy cows) and more than 20 kg (cattle) of feed (Smil 2000a and 2024). Milk protein can be produced most efficiently, and aquacultures of herbivorous fish come in a close second, followed by eggs and chicken meat. Pork production converts feed protein to lean meat only about half as efficiently as broilers do, and beef production is inherently the least efficient way of supplying meat protein: large bodies, long gestation and lactation periods (requiring large amount of feed for breeding females), and relatively high basal metabolic rates guarantee high feeding requirements.

These realities did not matter when the animals were totally grass-fed or fed only crop and food processing residues that are indigestible (or unpalatable) by non-ruminant species. But with rising supplies of corn and soybeans, most US beef was converted to a combination of grass and grain feeding, and, during the late 1990s, the average amount of concentrate feed needed to produce 1 kg of live weight for all beef cattle was 5.5 times higher than for broilers (USDA 2003). Since a larger share of a broiler's weight is edible, the multiple rises to about 7.5 times in terms of lean meat (Smil 2000a and 2024).

The scale of modern meat, egg, and dairy operations is unthinkable without mechanized feeding, watering, and waste removal, and a great deal of energy must be spent on lighting and air conditioning of the structures housing often high densities of animals. This crowding led to the use as prophylactic antibiotics, other chemicals are used to control infestations, and growth hormones have been used to stimulate meat and milk production. Prophylactics are also used in aquaculture, with various species of carp and shrimp leading the production in Asia, and with salmon being by far the most important cultured fish species in Europe and North America (Lee and Newman 1997).

Potatoes Partly Made of Oil

This is how Howard Odum explained one of the fundamental misconceptions regarding the success of modern agriculture:

> The great conceit of industrial man imagined that his progress in agricultural yields was due to new know-how in the use of the sun ... and that higher efficiencies in using the energy of the sun had arrived. This is a sad hoax, for industrial man no longer eats potatoes made from solar energy; now he eats potatoes partly made of oil. (Odum 1971: 115–116)

This perfect summary of modern agricultural practices is also valid for all animal foods, as we are now also eating meat, eggs, and salmon partly made of fossil fuels and electricity.

The overall magnitude of these inputs (external energy subsidies) is insignificant in comparison with the input of solar energy that powers photosynthesis, the water cycle, and atmospheric circulation without which there could be no biomass production. While an intensively cultivated cornfield in Iowa may receive annually 30 gigajoules (GJ)/ha in direct and indirect energy subsidies, solar energy that will reach it during 150 days between planting and harvesting of the crop will amount to 30 terajoules (TJ)/ha, a 1,000-fold difference (Smil 2024a). Solar energy is essential and free, but without fossil fuels and electricity—whose acquisition carries significant energy costs—the field's productivity would be a fraction of the possible harvests.

Quantifying how much energy is needed is complicated not only because of the problems that beset energy accounting in general but also because of some unavoidably arbitrary decisions (Fluck 1992). Moreover, there are wide performance ranges in every category of major inputs. By 2000, the most efficiently produced fertilizers—in terms of pure nutrients, representative means they were 60 GJ/t nitrogen, 10 GJ/t phosphorus, and just 5 GJ/t potassium—required 25% or even 50% less energy than those coming from aging enterprises or from less productive extraction. Synthesis of pesticides is a highly energy-intensive process that required 100-200 megajoules (MJ)/kg of active ingredients (Helsel 1992).

Pumping water from deep aquifers and distributing it by center pivots may need more than 10 GJ/ha. The energy cost of producing field machinery may prorate to anywhere between 80 and 160 GJ/t, and fueling it has a similarly wide range of needs. Those rain-fed crops that receive only a limited applications of fertilizers and a minimum of cultivation—such as North American wheat—were grown with energy subsidies of no more than about 15 GJ/ha, while intensive corn production and irrigated high-yielding rice required between 40 and 60 GJ/ha. The year 2000 comparisons per kilogram of the final product converted to around 5 MJ/kg of Manitoba spring wheat, 3 MJ/kg of Iowa corn, and up to 10 MJ/kg of Jiangsu rice. Vegetable fields, orchards, and vineyards needed much higher energy subsidies per hectare (>200 GJ/ha), but the rates per kilogram are like those for cereals. Large modern greenhouses were the highest energy consumers, with as much as 40 MJ/kg of peppers or tomatoes (Dutilh and Linnemann 2004).

Published nationwide evaluations based on the practices that prevailed during the last 30 years of the twentieth century show means ranging from just 15 GJ/ha in US agriculture to 78 GJ/ha in Israeli farming (Smil 1992). Typical Western European averages were 25–35 GJ/ha (but the Dutch mean was at least 60 GJ/ha), while the Chinese rate was above 30 GJ/ha. My conservative global account of all major direct and indirect energy inputs showed that, during the late 1990s, food production received at least 15 EJ of subsidies, an equivalent of about 360 Mt of crude oil, or just over 10 GJ/ha. Slightly more than 40% of it was used to produce inorganic fertilizers, and a similar share went to field machinery and its fuel. Giampietro (2002), whose global calculation also included the cost of preparing animal feeds and energy invested in the cultivation of forages, came up with a total of slightly more than 21 EJ in 1997.

These annual energy subsidies meant that, during the late 1990s, the world's agriculture was consuming about 5–6% of the world's primary commercial energy. In affluent countries, with high energy use in households and services, the rate could be as low as 3%, while in some large populous nations it was closer to 10% (Smil 1992; FAO 2000). Several detailed accounts of energy use in US agriculture ended up with very similar shares in 1970 (3%), 1974 (2.9%), and 1981 (3.4%) (Steinhart and Steinhart 1974; USDA 1980; Stout, Butler, and Garett 1984). Shares of 2.5–5% appear to be very modest amounts given the quantity and variety of the produced food—but they cover only a part of energy needed by the modern food system.

Traditional growers produced most of their own food, but a high degree of specialization, and hence of spatial concentration, has led to increasing distances between growers and consumers and hence to more energy-intensive food transport (often including cooling and refrigeration). In North America, fruits and vegetables commonly travel 2,500–4,000 kilometers (km) from fields to homes, and the global total of international food shipments surpassed 800 Mt in 2000, four times the mass in 1960 (Halweil 2002). To these high shipping costs, energies required for food processing, packaging, storing, retailing, cooking, washing dishes, shopping trips, and waste disposal must be added.

None of these diverse activities can be quantified by a single mean, but rates of 50–100 MJ/kg of product are common in processing and packaging, 5–7 MJ/kg of food are usual for home cooking, and 2–4 MJ/kg are typically needed for a dishwashing event (Dutilh and Linnemann 2004). As a result, most food systems claim at least twice as much energy as is used in food production, and, in North America, the combination of excessive processing and packaging, long distribution trucking to supermarkets, ubiquitous refrigeration, automatic dishwashing, and electricity-powered cooking appliances raises that multiple even higher. Even after excluding the energy cost of food-related car trips, Steinhart and Steinhart (1974) calculated that, in 1970, energy use in the US food system added up to almost 13% of the country's primary energy consumption.

Mineral Extraction and Processing

Extraction, transportation, and processing of all minerals (aggregates, ores, and fuels) had become highly mechanized in all affluent countries, but, given the great variety of raw materials, the progress and the scale of extractive mechanization as well as the degree to which the production became concentrated, were far from uniform. As already noted, only the extraction of bulky construction materials—clays, sand, gravel, and crushed stone—remained highly decentralized to serve local markets: during the late1990s, more than 90% of US counties produced one or more types of these minerals. Fairly reliable US historical statistics show that the consumption of construction materials, which accounted for more than a third of the 160 Mt of all materials used in 1900, expanded to constitute nearly 75% of 3.4 Gt of nonfuel minerals and biomass used by the country's economy by 2000 (Matos 2003).

This more than 40-fold increase in absolute terms was nearly a 12-fold rise in per capita terms (from less than 0.8 to slightly more than 9 t/year). Most of this expansion took place only after 1950, with more than half of the total mass consumed after 1975 (Matos and Wagner 1998; Figure II.4.6). Extraction of billions of tons of natural aggregates receives little attention, but no other modern material is used as much as concrete, which combines cement, aggregates (sand and gravel), and water. Similarly, more than 90% of asphalt pavements consist of aggregates, as do all bricks, wallboard, and glass and roofing tiles, and finely ground aggregates go into products ranging from medicines to paints, from plastics to paper, and help to protect the environment as they purify water or remove sulfur dioxide from flue gases.

New mining techniques were needed to exploit progressively poorer ore deposits at competitive cost. Iron ores provide a revealing example. Some of the best nineteenth-century ironmaking was based on ores that contained more than 65% iron (pure magnetite, Fe_3O_4, has 72% Fe). During World War II the United States relied overwhelmingly on Lake Superior's hematite that averaged slightly more than 50% iron (Kakela 1981). By the early 1950s, US Steel had developed a process to crush and grind hard, fine siliceous Mesabi Range taconite that contains no more than 20–30% iron and to concentrate the metal with magnetic separation to the range of 60–64%. In the case of many other ore deposits, the main challenge was simply to reach them with the most economical opencast mining techniques, and this quest resulted in larger and deeper surface mines.

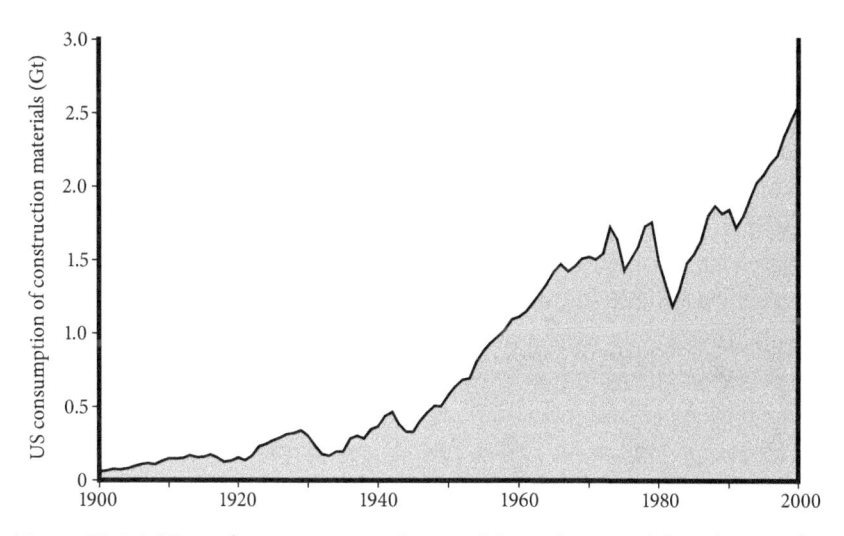

Figure II.4.6 Natural aggregates are the uncelebrated material foundation of modern societies. No other minerals are extracted in such quantities as sand, gravel, and stone: by the end of 1990s the US production surpassed 2.5 Gt, the total that amounted to more than a 40-fold increase since 1900.
Plotted from data in Matos (2003).

Larger operations were also used to recover nonmetallic industrial minerals whose consumption growth was led by increased mining of phosphates and potash for fertilizer. But the most conspicuous case of pervasive mechanization was the extraction of fossil fuels. During the first six decades of the twentieth century this effort was dominated by coal production, and, once oil and gas production became dominant, it was revolutionized by shifts from land to offshore deposits (made possible by exploration drilling rigs and platforms able to work in greater depths) and by deeper and more productive drilling.

As a result, by 2000, the extraction of fossil fuels ranked among the least labor-intensive (and the least often seen) tasks in modern economies. Moreover, during the last generation of the twentieth century the media focus on technical gigantism was displaced by the infatuation with the smallness of microchips, genomic arrays, or putative nanomachines. And so even the images of extractive giants became rare in mainstream media, and most people are entirely unaware of the existence of these remarkable machines and structures; the largest moving or fixed assemblies ever built on land or offshore do their work in obscurity.

Coal Mining Transformed

Traditional underground coal mining was one of the most taxing and most hazardous occupations, killing the miners even long after they stopped working because of fatal lung diseases that resulted from long-term exposures to coal dust. Mechanization of underground mining began at the very outset of industrialization with steam-powered water pumping and hoisting, but actual extraction still used muscle power until the very end of the nineteenth century. Advances in mechanization of extraction and loading followed in successive waves, beginning in the United States. Pickaxes and shovels were replaced by hand-held pneumatic hammers and mechanized cutters and loaders. Joseph Francis Joy (1883–1957)—the inventor and manufacturer of efficient loading machines (the first US patent obtained in 1919)—was perhaps the most important contributor to this shift.

Continuous miners, machines that grind coal from the face and dump it on an adjoining belt, spread rapidly, but they were eventually superseded by the most economical and safest method of underground extraction, long-wall mining (Ward 1984; Barczak 1992). The technique uses large drum-shaped shearers that move down the length of a coal face, cutting out panels up to 2.5 meters (m) high, 80–200 m wide, and as much as 1.5 km long that are delimited by two side tunnels (Figure II.4.7). Coal ground from the mine face is dumped onto a conveyor, but, unlike in continuous mining, the workers remain under a protective canopy of hydraulic steel supports that are, once a complete pass along the mining face is completed, lowered, moved forward, and reset, leaving the unsupported roof behind to cave in. This became possible only with new, strong self-advancing steel supports introduced in the early 1960s. The technique can recover more than 90% of all coal, compared to just 50% in the traditional room-and-pillar operations.

But the largest coal production increases during the last third of the twentieth century came from new large-scale opencast operations in the United States, Russia, Australia, and China. During the late 1990s, the largest of these mines produced annually more coal that did the entire formerly prominent coalmining nations from their underground seams. With more than 60 Mt a year mined in 2000, the Powder River Coal Company's North Antelope/Rochelle mine in Wyoming extracted twice as much as energy in solid fuel as did all of the United Kingdom's remaining deep mines—and even America's

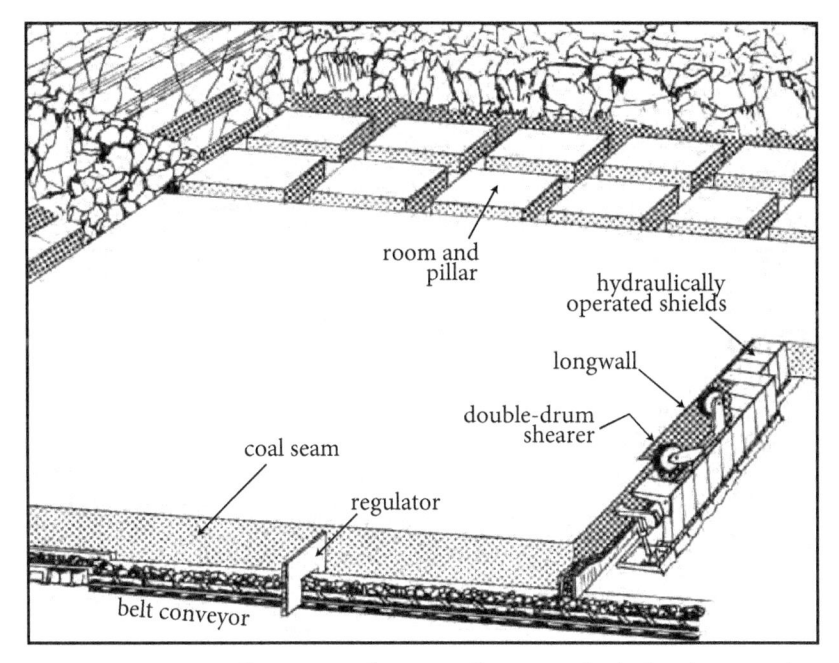

Figure II.4.7 Longwall mining is the most efficient method of underground coal extraction, compared with room-and-pillar operations.

tenth largest surface mine (Belle Ayr, also in Wyoming) still outproduced, with about 15 Mt/year, about half of the world's 20 coal-mining nations (EIA 2003).

This was made possible through the deployment of increasingly larger excavation machines (shovels and walking draglines) to remove thick overburden layers (Hollingsworth 1966). In the United States, Bucyrus-Erie and Marion were the dominant makers of these machines (Anderson 1980). Capacities of the largest shovel dippers rose from 2 m^3 in 1904 to 138 m^3 in 1965; similarly, bucket volumes of the largest walking draglines increased from less than 3 m^3 before World War I to 168 m^3 before they stopped growing. These machines had to be erected on site from components shipped by several hundred railroad cars, and, because they were also too large to be powered by diesel engines, they had to rely on dozens of electric motors (the world's largest dragline used more electricity than a US city of 100,000 people). The Marion 6360 shovel, built in 1965 for stripping work at Captain Mine in Illinois, was the world's heaviest excavating machine: it weighed 12,700 t (or as much as 32 fully loaded Boeing 747s), was 21 stories tall, and its bucket was at the end of 65-m boom.

Mechanization resulted in the consolidation of coal mining into a smaller number of larger operations, falling labor force totals, and plummeting occupational deaths and injuries. In Germany's Ruhr region, one of Europe's oldest coal-mining areas, the share of mechanical coal extraction rose from less than 30% in the early 1950s to more than 95% by 1975, while the total number of operations fell by about 90% (Erasmus 1975). Average productivity of US underground mining rose from less than 1 t per worker per shift in 1900 to more than 3 t per hour per worker by 2000; in surface operations, it surpassed 7.5 t/worker-hour (Darmstadter 1997). But productivity was as high as 6 t/miner-hour in longwalls and up to 300 t/worker-hour in the largest surface coal mines in Wyoming.

By the end of the twentieth century, US coal mining produced almost exactly four times more fuel than it did at its beginning, with less than 20% of labor force it had in 1900 (Figure II.4.8). At the same time, accidental deaths declined by 90% since the early 1930s (MSHA 2004). Among the world's remaining large coal producers, only China followed a peculiar path, as a much slower rate of mechanization in its large, state-owned enterprise was accompanied by indiscriminate openings of unmechanized small mines. By 1997, half of China's 1.3 Gt of raw coal originated in some 82,000 small mines that were distinguished by low productivity and dangerous working conditions. Their number was cut to 36,000 by 2000.

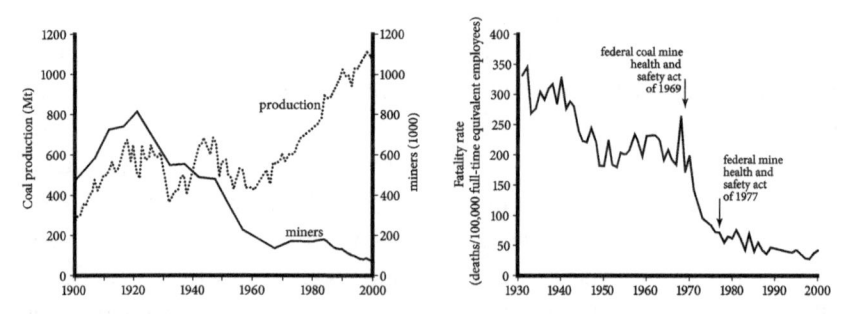

Figure II.4.8 Production and work force in US coal mines, 1900–2000 (left), and accidental death rate in coal mining, 1931–2000 (right).
Plotted from data in Schurr and Netscherrt (1960), Darmstadter (1997), and MSHA (2004).

Oil and Gas Industries

Until the end of the nineteenth century, hydrocarbon production relied on percussion drilling whose earliest version was introduced during the early Han dynasty more than 2,000 years ago (see Figure I.1.2). The deepest well completed by this process in China, in 1835, was 1 km deep (Vogel 1993). The first US percussion-drilled oil well, completed on August 27, 1859, at Oil Creek, Pennsylvania, struck oil at a depth of 21 m after penetrating 10 m of rock (Brantly 1971). During the next 50 years, steam engines did the raising and dropping of heavy metal bits hung from manila ropes or wires to fracture and pulverize the drilled substrate, and the bailing out of the cuttings from the holes remained laborious. Even so, the record depth reached with this technique, at a gas well in Pennsylvania in 1925, was more than 2,300 m.

The first rotary rig was introduced in the Corsicana field in Texas in 1895, but the technique began to spread only after World War I, and, even in the US, cable tool rigs outnumbered rotaries until 1951 (Brantly 1971). The main components of a rotary rig are a heavy rotating circular table, drill pipes inserted in its middle, and the hole-bottom assembly at the drill string's end (Devereux 1999). Tables were driven by gearing, originally powered by steam engines and later by diesel engines or diesel-powered electric motors. As the drilling progresses, sections of threaded drill pipes are added, but as a drilling bit wears out the pipes must be withdrawn from the well, stacked, and reattached after a new bit is mounted (the process is called *tripping*). The weight of the drill string is increased by heavy drill collars placed just above the bit.

Drilling mud (water-, oil-, or synthetics-based fluid) pumped at high pressure down the drill string and through the bit cools the rotating bit, makes the removal of cuttings a continuous and easy operation, and exerts pressure on the well sides to help prevent the hole from caving in (Van Dyke 2000). Casings of varying diameters are then installed and cemented in place to stabilize the well. The performance of rotary drilling was much improved with the invention of a superior bit by Howard Robard Hughes (1869–1924), who invented the tool after he failed to drill through extremely hard rock while wildcatting in Texas in 1907.

Hughes's expertise consisted only of unfinished law studies and seven years of experience in oil drilling, but, in November 1908, after he spent just two weeks on a new design (while visiting his parents in Iowa), he filed his application for a rotary cone drill provided with two cutting members 4 which preferably consist of frustoconical-shaped rollers having longitudinally extending chisel teeth 5 that disintegrate or pulverize the material with which they come in contact.... These cutting members 4 are arranged at an angle to each other... and are rotatably mounted on stationary spindles 6. (Hughes 1909: 1).

The numbers refer to the figures that accompanied the patent (Figure II.4.9), and this simple design made it possible to drill wells 10 times faster than with fishtail bits.

In 1909, Hughes formed a partnership with Walter B. Sharp (1870–1912), and the Sharp-Hughes Tool Company (since 1918 just Hughes Tool) began to make the bits and lease them at $30,000 per well (Brantly 1971). For more than two decades it was considered impossible for the teeth on the cutters of a three-cone bit to fit, but the problem was solved by Floyd L. Scott and Lewis E. Garfield by a design patented in 1934 (Figure II.4.9). Compared to standard two-cutter bits, this provided much better support on the well bottom, reduced vibration, and allowed faster yet smoother drilling. Key improvements included the first diamond drill (in 1919), heavy drill collars for added weight and rigidity, and various well control devices to cope with high pressures in the well and to prevent catastrophic blowouts. Modern compact bits are covered with a layer of fine-grained synthetic diamonds, and they may last long enough to complete a well 2 km deep. Leading companies in efficient cementing of wells and automatic well logging bear the names of two individuals who pioneered the innovations.

In 1912, Conrad Schlumberger (1878–1936), a physics lecturer at École des Mines, proposed to use electrical measurements to map subsurface rock bodies, and, in 1919, Erle P. Halliburton (1899–1957) started his oil well cementing company in Oklahoma (Haley 1959; Allaud and Martin 1976). Halliburton patented his cement jet mixer in 1922 (US Patent 1,486,883), and Henri Doll, Schlumberger's son-in-law, produced the first electrical resistivity well log in Pechelbronn field in 1927. In 1931, the company introduced electrical *well logging*, the simultaneous recording of resistivity and spontaneous potential produced between the drilling mud and water present in permeable beds. A multivalve stack to control well flow ("Christmas tree") was introduced in 1922, and, by the 1990s, the best control-flow devices could resist up to 103 megapascals (MPa). Procedures were developed to control well blowouts, extinguish fires, and cap the wells: Myron Kinley and Red Adair were pioneers of these often-risky operations (Singerman 1991).

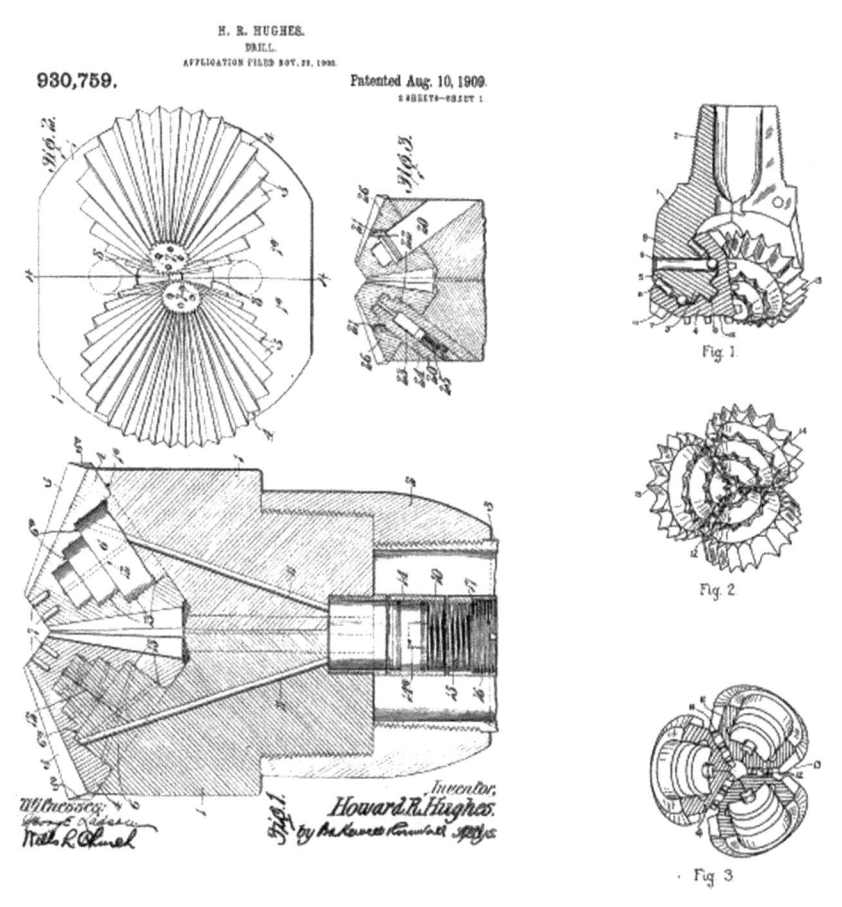

Figure II.4.9 The first three figures that accompanied Howard Hughes's patent application for a conical drilling bit filed in November 1908 (US Patent 930,759), and the figures from the patent application for a three-cone bit submitted by Floyd L. Scott and Lewis E. Garfield of the Hughes Tool Company in April 1933.
Images from the US Patent Office.

Faster operation and increased drilling depths—records progressed from less than 2,000 m before World War I to 4,500 m by the late 1930s and 6,000 m a decade later—led to a spate of new discoveries. In the United States, 64 giant oilfields were discovered between 1900 and 1924, but 147 were added during the subsequent 25 years (Brantly 1971). The largest pre-1950 finds included the first giant oilfields in Iraq (Kirkuk in 1927), Saudi Arabia (Abqaiq in 1940; Ghawar, the world's largest oilfield, in 1948), Kuwait (Burgan, the second largest, in 1938), and Venezuela (Nehring 1978). By the 1970s, production from wells deeper than 5,000 m became common in some hydrocarbon basins, particularly in Oklahoma's Anadarko Basin. With rapid post-World War II expansion of oil industry—from less than 500 Mt in 1950 to about 3.5 Gt by 2000, when there were almost 1 million wells in more than 100 countries—every one of its infrastructural elements had to get bigger to meet rising demand.

These challenges were particularly demanding in offshore drilling. Drilling from wharves was done in California as early as 1897, the first platforms were extended into Venezuela's Lake Maracaibo in 1924, and the first well drilled out of the sight of land (nearly 70 km from the Louisiana shore) was completed by the Kerr-McGee Corporation in the Gulf of Mexico, in just 6 m of water, only in 1947 (Brantly 1971). Offshore drilling then progressed from small jack-up rigs for shallow near-shore waters to drill ships able to work in up to 3,000 m of water and semisubmersible rigs for year-round drilling even in such stormy waters as the Barents and North Seas. The Offshore Company, which launched the first jack-up, eventually combined with three other pioneers of marine drilling to form Transocean, Inc., whose long list of firsts includes self-propelled jack-ups, dynamically positioned semisubmersibles, and rigs capable of year-round drilling in extreme environments (Transocean 2003).

During the 1970s, the company introduced the Discoverer-class drill ships that repeatedly set new drilling records: by the end of the twentieth century, their fifth generation could work in waters up to 3,000 m deep. But Shell Oil was the first company to deploy a semisubmersible rig, the Bluewater I, in 1961, in the Gulf of Mexico. By the end of the century there were nearly 380 jack-ups (mostly in waters between 30 and 90 m), about 170 semisubmersibles (for depths up to 300 m), and about 80 drill ships and barges for the total nearly 640 marine drilling rigs (*World Oil* 2000). The concurrent increase in the size of offshore production platforms produced some of the most massive structures ever built (Figure II.4.10). In November 1982, the Statfjord B platform began its operation in the Norwegian sector of the North Sea: its four massive concrete columns and storage tanks at its base made it, at slightly more than 800,000 t, the heaviest object ever moved by man.

By 1989, Shell's Bullwinkle, sited in 406 m of water and weighing about 70,000 t, became the world's tallest pile-supported fixed steel platform (SEPCo 2004). In 1983, the first tension-leg platform (TLP) in the Hutton

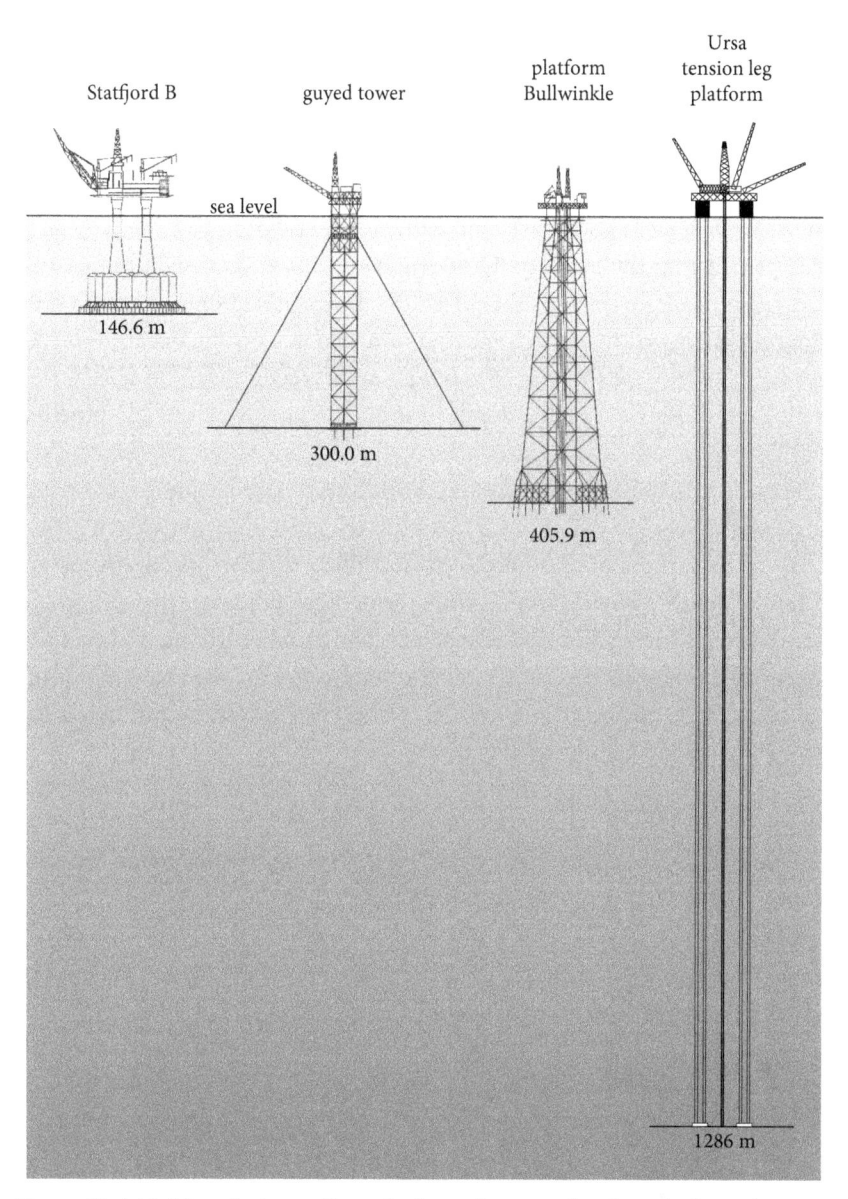

Figure II.4.10 New designs of large hydrocarbon production platforms have resulted in more massive structures anchored in increasing depths of water. Simplified drawings based on a variety of graphs and photographs in trade publications.

field in the UK sector of the North Sea was anchored by slender steel tubes to the seafloor 146 m below the surface. By 1999, the Ursa TLP, a joint project of a group of companies led by the Shell Exploration and Production Company, was the largest structure of its kind (SEPCo 2004). Its total displacement of about 88,000 t surpasses that of a Nimitz-class nuclear aircraft carrier; it rises 146 m above water and is anchored with 16 steel tendons to massive (340 t) piles placed into the seafloor 1,140 m below the surface (Figure II.4.11). Drilling in deeper waters also required the development of new subsea production systems with wells connected to manifolds on the ocean floor. By the century's end, the world record holder in this respect was the Mensa area in the Gulf of Mexico, with a depth of about 1.6 km and a 100-km pipeline (SEPCo 2004).

By the late 1990s, nearly 30% of the world's crude oil was coming from underneath the sea, and two innovative trends changed the actual extent of fuel recovery. As exploratory and production drilling reached into deeper strata and as wells could deviate farther from the vertical, it became possible to tap more deposits through extended reach and horizontal drilling. In addition, improved methods of secondary oil recovery increased the share of liquids that could be extracted from parental rocks. The problem of unintentional

Figure II.4.11 The Ursa tension leg platform (TLP) is operated by SEPCo about 210 km southeast of New Orleans. Its height from the seafloor to the top of the drilling rig is 1,286 m. Its fabrication began in July 1996; oil and gas production began in March 1999. The project cost $1.45 billion.
Photograph courtesy of SEPCo, Houston, Texas.

deviation from the vertical was recognized for the first time only during the mid-1920s, in Oklahoma, and its eventual studies by Arthur Lubinski in the 1950s resulted in the development of directional drilling, which made it possible to complete several wells from a single location. This became a common practice, and even more complicated extended-reach drilling at more extreme angles and horizontal drilling began to be commercialized during the 1980s (Society of Petroleum Engineers 1991; Cooper 1994).

Directional drilling uses steerable down-hole motors that are powered by the pressurized mud flowing in cavities between a spiral-fluted steel rotor and a rubber-lined stator. A bend is used to point the bit away from the vertical axis when the drill string is not rotating and then a mud motor powers the drill in the pointed direction. This new course can be then maintained by rotating the entire drill string, including the bent section. Steerable downhole motors are also used by a new drilling technique that was introduced during the late 1990s that uses narrow (5–7 centimeters [cm] in diameter) steel tubing wrapped on a large drum mounted on a heavy trailer (Williams et al. 2001).

The tubing is unreeled into a well, where the bit is rotated by a down-hole motor powered by mud pressure. This slim-hole drilling eliminates the tripping (removing or replacing the drilling pipe to change the bit or the drill string). Benefits of directional drilling are obvious; for example, if the oil-bearing strata are 3 km below the surface, then drilling at 70 degrees rather than at a 60 degree angle off the vertical will make it possible to develop an area 55% larger from a single site. Pressure that is naturally present in crude oil reservoirs rarely suffices to release more than 30% of the oil that is originally in place. Secondary recovery techniques use water flooding or gas injection to force more of the fuel from the parental rock. These interventions increased the typical rate of recovery close to, or even above, 40% (Schumacher 1978).

Crude oils undergo a complex process of refining, a combination of physical and chemical treatments. Refining separates the complex mixture of hydrocarbons into more homogeneous categories and adds a great deal of value to final products. In terms of specific weight and progressively higher boiling points, these categories range from petroleum gases (short-chain alkanes, C_nH_{2n+2}, that boil away at temperatures lower than 40°C and are used as fuel and petrochemical feedstocks) and naphtha through gasoline (mixtures of alkanes and cycloalkanes, C_nH_{2n}, boiling at 40–200°C), kerosene (jet fuel), and diesel fuel to long-chained (more than 20 atoms) lubricating and heavy oils. Residual solids—coke, asphalt, tar, waxes—boil only at temperatures higher than 600°C.

Early refining used high-pressure steam (at 600°C) to separate these complex mixtures into their constituent fractions: if a particular crude oil contains only a small share of highly volatile hydrocarbons, its refining produced largely medium and heavy products. This was a highly unsatisfactory outcome once growing car ownership began to increase the demand for gasoline. Most of the world's crude oils are not rich in light fractions: Saudi exports contain 48–55% heavy gas oil and residual oil, some West Texas crudes had less than 15% of the lightest fractions. Without an effective technical solution, the extent of driving and flying would have remained a direct function of crude oil quality. The first relief came in 1913, when William Burton (1865–1954) obtained his patent (US Patent 1,049,667) for thermal cracking of crude oil that relied simply on the combination of heat and high pressure to break heavier hydrocarbons into lighter mixtures (Burton 1913).

A year after Burton was granted his patent, Almer M. McAfee patented the first catalytic cracking process (US Patent 1,127,465): it relied on heating crude oil in the presence of aluminum chloride, a compound able to break long hydrocarbon molecules into shorter, more volatile chains. But because the relatively expensive catalyst could not be recovered, thermal cracking remained dominant until 1936, when Sun Oil's Pennsylvania refinery in Marcus Hook put on line the first catalytic cracking unit designed by Eugène Houdry (1892–1962) to produce a larger share of high-octane gasoline. His invention, developed with the help of Socony Vacuum and Sun Oil, was based on a patent originally filed in France, in 1927 (Houdry 1931).

Warren K. Lewis and Edwin R. Gililand of the Massachusetts Institute of Technology (MIT) replaced Houdry's fixed catalyst with a more efficient moving-bed arrangement, with the catalyst circulating between the reaction and regeneration vessels. This innovation spread rapidly, and, by 1942, 90% of all aviation fuel produced in the United States was obtained through catalytic cracking, providing us with yet another instance of an invention that was ready at the time of the greatest need—to defeat the Axis in World War II (Houdry's second invention that made a key contribution to the war effort was his butane dehydrogenation process to make synthetic rubber).

And even greater yields of high-octane gasoline were achieved in 1942, with the commercialization of airborne powdered catalyst that behaves like a fluid in "a process and apparatus in which solid material in finely divided form is intermingled in a gaseous medium and the resulting mixture passed through a treating zone" (Campbell et al. 1948). This fluid catalytic cracking, invented by a group of four Standard Oil chemists in 1940, was further improved in 1960, with the addition of a synthetic zeolite, a crystalline aluminosilicate with a structure of uniform pores (see Figure II.3.14) that provided an exceptional active and stable catalyst composite to facilitate the cracking of heavy hydrocarbons (Plank and Rosinski 1964:1). Zeolite Y improved the gasoline yield by as much as 15%.

For the remainder of the twentieth century fluid catalytic cracking remained the leading process in all modern refineries that used a variety of crudes to produce large quantities of high-octane gasoline and other light fractions. During the 1950s, Union Oil Company developed the process of *hydrocracking*, which combines catalysis at temperatures higher than 350°C with hydrogenation at relatively high pressures, typically 10–17 MPa. Large-pore zeolites loaded with a heavy metal (platinum [Pt], tungsten [W], or nickel [Ni]) are used as dual-function (cracking and hydrogenation) catalysts, and the main advantage is that high yields of gasoline are accompanied by low yields of the lightest and less desirable alkanes (methane and ethane).

But gasoline would have been used inefficiently without solving a problem that is inherent in Otto's combustion cycle: violent engine knocking caused by spontaneous ignition of the fuel–air mixture that produces a pressure wave traveling in the direction opposite to that of the spreading flame. To minimize this destructive knocking, the compression ratios of early internal combustion engines were held below 4.3:1, limiting engine efficiency and leading to concerns about the adequacy of crude oil supplies in America's expanding post-World War I gasoline market. General Motor's research, led by Charles F. Kettering (1876–1958), identified the addition of ethanol as an effective solution, but ethanol from crops was expensive, and there was no commercial way to make it from cellulosic biomass. As a result, General Motors began a systematic research program to explore beyond such known but expensive additives as bromine and iodine.

This search was led by Thomas Midgley (1889–1944; Figure II.4.12). After promising trials with tetraethyl tin, the group tried tetraethyl lead—$(C_2H_5)_4Pb$—on December 9, 1921, and found it highly effective even when added as a mere 1/1,000th of the fuel's volume (Wescott 1936). The first leaded gasoline was marketed in February 1923, and the compression ratio of engines eventually rose to the range of 8:1–10.5:1. Besides saving energy in driving, leaded aviation fuel made it possible to develop more powerful, faster, and

Figure II.4.12 In December 1921, Thomas Midgley, Jr., found that the troublesome knocking of Otto-cycle engines can be eliminated by the addition of small amounts of tetraethyl lead to gasoline.

Midgley's photograph reproduced here with the permission of Thomas Midgley IV.

more reliable aeroengines. Midgley's solution was inherently risky, but leaded gasoline became a prominent environmental concern only when deteriorating air quality forced the adoption of catalytic converters (Smil 2023a). These devices, based on Eugène Houdry's third fundamental invention patented in 1962, drastically cut the emissions of carbon monoxide, nitrogen oxides, and unburned hydrocarbons, but to prevent the poisoning of the platinum catalyst, the vehicles must use unleaded gasoline.

Advances in Factory Production

The transformation label is particularly apposite for factory manufacturing because, by 1900, its mechanization was well established and highly diversified. Many ingenious machines were designed even during the pre-industrial era, including such curiosities as intriguing anthropomorphic or zoomorphic automata (Wood 2002). Some relatively large factories were in place even before the commonly accepted beginning of the Industrial Revolution, and the rise of the factory, led by textile manufactures, was obviously one of the signature trends of Western industrialization. But Mokyr (2002: 150) reminds us that the transition to large-scale production was "more gradual and nuanced than mass-production enthusiasts have allowed for" and that very-small-scale business was still very much in place until the very beginning of the twentieth century: he cites the French census of 1906 that showed a third of the country's manufacturing labor force working at home.

Early industrial mechanization was limited by the low power ratings of available prime movers and by the properties of materials used in machine building. The steam engine was far from being an ideal prime mover able to provide the reliable and flexible drive needed for many productive tasks, early steels did not have qualities that were required by many industrial processes, and machine-tool makers could not meet many high-performance demands. Only the combination of electric motors, specialty steels, and better machine tools provided the energetic and material foundations for efficient and ubiquitous mechanization.

The complex history of machine tools is among the most neglected topics in the study of modern technical advances, a reality that is particularly puzzling given their key position in the creation and transformation of modern societies. Pre-1900 advances left machine-makers with a wealth of carbon steel tools whose maximum hardness is obtained only after tempering (heating and quenching). Even then, their high carbon content (0.7–1.2%) makes them brittle (Bolton 1989). Better tools were introduced at the very beginning of the twentieth century thanks to the determination of Frederick Winslow Taylor (1856–1915), a rich Philadelphia Quaker whose career progressed from an apprentice patternmaker, engineer, and machinist to a manager and the world's first efficiency expert.

While working as a senior manager at Bethlehem Steel in 1898, Taylor decided to investigate the reasons for repeated failure of tempered cutting tools. Together with a metallurgist, J. Maunsel White, he set out to determine how heating affects the cutting speed. They found a marked improvement with temperatures higher than 1,010°C, and the greatest increase in performance after heating the tools to the maximum temperature to which it was possible to raise the carbon-tungsten steel without destroying it (Taylor

1907). Bethlehem Steel's lathe using Taylor's tool was a surprising magnet for visitors at the Paris Exposition in the summer of 1900. As noted by Henri Le Châtelier (quoted in Copley 1923: 116),

> nobody quite believed at first in the prodigious result which was claimed by the Bethlehem Works, but we had to accept the evidence of our eyes when we saw enormous chips of steel cut off from a forging by a tool at such high speed that its nose was heated by the friction to a dull red color.

High-speed steel came along just in time to be incorporated into more productive machines to supply the growing automotive market. In 1903, Charles Norton (1851–1942) introduced a crankshaft-grinding machine whose wide wheel could reduce a steel journal to the desired diameter in a single cut in just 15 minutes—compared to the 5 hours previously needed for slow lathe turning, filing, and polishing. Just two years later, James Heald (1864–1931) launched his planetary grinding machine whose unprecedented accuracy turned out cylinders with deviations of less than 0.0062 mm (Heald Machine Company 2024). This design was retained by all similar machines during the twentieth century.

Well-known designers of automobiles and airplanes could not have turned their ideas into mass-produced realities without Taylor's tools and Norton's and Heald's machines. But, as Rolt (1965: 220) noted, "one may search the pages of popular motoring histories in vain for the names of Heald or Norton. Like all great toolmakers, their fame has never penetrated far beyond the four walls of the machine shop." Further development of original high-speed steels led to Stellite, a cobalt–chromium–tungsten alloy introduced in 1917 whose performance did not deteriorate even at red heat stage. The next qualitative jump took place in 1928, when Krupp unveiled at the Leipzig Fair its tungsten-carbide tools made by mixing pulverized tungsten carbide with cobalt, pressing the mixture into a required shape, and sintering it. These small cutting tips could go through 120 m of cast iron in 1 minute compared to 45 m for Stellite and 23 m for the Taylor-White high-speed alloy (Rolt 1965). Further speed was gained by the cam-type regulating wheel for precision grinding machines, an innovation developed by the Cincinnati Milling Machine Company between 1932 and 1935, and by internal centerless grinders, introduced by the Heald Machine Company in 1933.

Mass Production

Traditional artisanal production could not be rapidly dislodged by the impressive technical innovations of the two pre-World War I generations. Mechanization of textile mills was a major exception, as it had been advancing for more than a century, but the first light bulbs of the early 1880s had to be mouth-blown, the first electric motors of the late 1880s had to be hand-wound, and the first automobiles of the 1890s had to be built by skilled machinists and mechanics who transferred their know-how from carriages and bicycles to car chassis and Otto engines. And the idea of assembly production also had a long gestation period. Its obvious precursor (in reverse, disassembling) was the work of Cincinnati's meat packers of the 1830s, who pioneered the stepwise butchering of hogs. By 1890, Chicago's Union stockyards fine-tuned these disassembly lines to process 14 million animals a year (Wade 1987). A more kindred experience came from large-scale bicycle assembly during 1890s, the decade of runaway demand for those simple machines.

Ford's system (see Chapter I.3) was fully deployed at the Highland Park factory in 1913 on a chassis assembly line that accommodated 140 workers along its 45 m length (Lewchuk 1989). In 1913, Ford made nearly 203,000 cars, more than five times as many as the second-ranked Willys-Overland, about 40% of the US production, and roughly 70 times the output of the largest British automakers. By 1915, Ford was also making, rather than buying, all major car components, and, in 1917, he began to build a new River Rouge plant in Dearborn, near Detroit, Michigan (Figure II.4.13). By 1928, the plant became the world's largest manufacturing complex and the paragon of vertical integration (HFMGV 2004). Within its "ore to assembly" confines, Ford not only made all car components (from engines to upholstery) but also produced electricity, coke, pig iron, rolled steel, and glass, with raw materials brought in by Ford's fleet of freighters and by his own railroad company.

At its peak operation during the 1930s, the plant covered 800 ha, employed more than 100,000 people, and produced a car every 49 seconds. And Ford carried his pursuit of high-volume, low-cost production of one standardized item not only to kindred manufacturing of tractors and airplanes but also to soybean extraction (providing feedstock for plastics). Under Alfred Sloan's (1875–1966) leadership, General Motors, Ford's greatest rival, pursued a

Figure II.4.13 Ford's mammoth River Rouge plant in 1952, during the peak years of its production. The complex, built between 1917 and 1928 at the confluence of the Rouge and Detroit Rivers, was the ultimate embodiment of vertical integration: its facilities included wharves and marshaling yards; an electricity-generating plant; blast furnaces; steel, tire, glass, and paper production; large assembly halls; and a peak labor force of more than 100,000 people.
Photograph courtesy of Ford Motor Company.

different strategy by operating separate car divisions and specialized component makers. The divisions offered an enduring range of choices—from affordable Chevrolets to luxurious Cadillacs—to capture all segments of the market, and the company's annual change of its cars' exterior appearance was hiding a high degree of standardized parts that were mass produced for long periods.

Ford's innovations were complemented by Taylor's scientific management, with its attention to individual labor tasks. Taylor believed that human work is essentially a set of techniques that should not just be improved but possibly optimized to maximize productivity, lower prices, and bestow benefits on all (Taylor 1911). Taylor's quest arose from personal observations of ubiquitous inefficiency in America's pre-1900 factories, particularly from the disdain for what he called "systematic soldiering" of deliberately underperforming workers. His remedy was systematic management of labor based on principles that he believed to be applicable to all social activities, but, after three years of trying to introduce scientific management at Bethlehem Steel, he was fired by the company's president in May 1901.

But eventually both his grand ideas and specific methods of detailed task-timing studies attracted an enormous amount of attention that ranged from condemnation through enthusiastic but often misinformed acceptance to careful implementation (Copley 1923; Kanigel 1997). As a quote from his 1911 summations shows, Taylor could sound callous: "[T]he first requirement for a man who is fit to handle pig iron … is that he be so stupid and so phlegmatic that he more nearly resembles in his mental make-up the ox." But he also believed that the combined knowledge and skills of managers "fall far short of the combined knowledge and dexterity of workmen under them" and that using his method to set quotas to overtire a worker "is as far as possible from the object of scientific management."

He had also little use for the antagonistic postures of workers and management and argued for "the intimate, hearty cooperation" between the two "in doing every piece of work" (quoted in Copley 1923: 159). He was an excellent researcher: a century later his exercise of evaluating 12 variables that affect the performance of metal cutting remains a tour de force of engineering and managerial inquiry. And any admirers of the great originality of modern Japanese factory management (more on this below) who may have never heard about the rich Philadelphia Quaker who worked for years as an ordinary machinist should read Taylor's brief 1911 magnum opus. They would find that the three great alliterative tenets of Toyota's famed production system—*muda, mura*, and *muri*, reductions of, respectively, non–value-adding activities, uneven pace of production, and excessive workload—are nothing but pure Taylor.

Ford's classical process of mass production was widely copied. Ford's factories in Dagenham in Essex (since 1931) and in Germany introduced the production model directly to pre-World War II Europe. Among the many future top car company executives who made the pilgrimage to pre-World War II Detroit were Giovanni Agnelli (later the CEO of Fiat) and Kiichiro Toyoda (of what was at that time the fledgling Toyota Motor Company), who came in 1937 and spent a year acquiring an intimate understanding of Ford's system. World War II provided great stimulus for the diffusion of mass production patterned on Ford's great example. Sudden demands for unprecedented numbers of machines, weapons, and specialized components led to extensive copying of mass production procedures. For example, during the last quarter of 1940, the US military received just 514 aircraft; four years later, the country's aircraft production capacity was 20 times larger, and actual deliveries in 1944 amounted to 96,000 airplanes (Holley 1964).

To meet this need, companies had to build plants of unprecedented size and capacity, including Ford's Willow Run bomber plant near Detroit; Douglas Aircraft's plant in Long Beach, California (Figure II.4.14); Dodge's Chicago aircraft engine plant; and the Bell Bomber (B-29) plant in Marietta, Georgia. And, after the war, the methods of mass production were applied to the manufacturing of passenger cars and extended to the making of consumer items for a growing population. Some wartime advertisements explicitly promised this shift as they portrayed new conveniences and appliances filling the American dream (Albrecht 1995), and many of these promises became realities for the country's large middle class within the first two post–World War II decades. For some 60 years after its introduction, modified and improved versions of Ford-type assembly (repetitive, indeed numbing, tasks performed by expendable workers along moving assembly lines) remained the dominant method of mass manufacturing.

But the growing complexity of modern automobiles—Ford's Model T could be assembled in just 84 steps, while assembling the cars of the 1950s required commonly more than 5,000 parts—began to force the refinement of the traditional system, and then, during the 1970s, two different kinds of innovations began its dismantling. The first one was overwhelmingly nontechnical, relying on a superior form of manufacturing organization, fine-tuned coordination of supplies, and ongoing quality control. Its birthplace was not the giant assembly factories or the leading business schools in the

Figure II.4.14 Mass production for war: women workers grooming lines of transparent noses for A-20 attack bombers. The original 1942 caption read, "Stars over Berlin and Tokyo will soon replace these factory lights reflected in the noses of planes at Douglas Aircraft's Long Beach, Calif., plant."
National Archives photograph 208-AA-352QQ-5 by Alfred Palmer.

United States but the floors of Japanese automobile plants whose managers were determined to produce the world's highest-quality cars (Womack, Jones, and Roos 1991; Fujimoto 1999). Genesis of this approach goes back to the 1930s, and, in 1970, this new way of mass production became formally known as the Toyota Production System (TPS).

TPS's common informal generic name is "lean production," and its origins were laid by the efforts of Kiichiro Toyoda (1894–1952)—the son of Sakichi Toyoda (1867–1930), the inventor of an automatic loom (in 1924)—who established a new car company, Toyota Motor, in 1937, and adapted Ford's system for smaller production volume. Toyoda also tried to reduce the component inventories by developing a subcontractor network that was able to supply the parts in a more timely manner. Major management changes were under way only after Eiji Toyoda (1913–2013), Sakichi's nephew, became the company's managing director in 1950, and after the Japanese car industry surpassed its pre–World War II production peak (in 1953) and began its climb to global leadership. By 1960, Japan produced nearly half a million cars (more than 10 times the 1941 total); by 1970, the total surpassed 5 million, and, in 1990, the annual output peaked at nearly 13.5 million units.

Toyota's new-style manufacturing combined elements of the Ford system and other American approaches with Japanese practices and original ideas (Fujimoto 1999). Ford Motor Company's training within the industry and William Deming's (1900–1993) emphasis on statistical quality control evolved into two key conjoined ingredients of TPS: *kaizen*, the process of continuous improvement; and the dedication to total quality control, whose key components are not just automatic detection of defects and shutdown of machines (*jidoka*) but also foolproof prevention of errors (*poka-yoke*). Traditional mass production facilities had a great deal of space devoted to postproduction quality control, and extensive reworking areas were required to fix numerous problems. In contrast, TPS aims to prevent big problems by doing things right the first time, by building in quality (*tsukurikomi*).

This goal is achieved by constantly detecting and immediately fixing any defects. Any worker who notices a problem can pull the overhead line-stop cord, and the team leader is called in to resolve the matter; in addition, some lines are designed to stop automatically when a serious problem is encountered. The emphasis on *muda*, *mura*, and *muri*; quality circles, small groups of workers aiming to solve arising problems; workers' involvement in preventive maintenance; and general adherence to cleanliness and neatness of the workplace are among other key ingredients of the constant quest for quality. Taiichi Ōno (1912–1990), an engineer who joined the company in 1932 and who became an assembly shop manager at the Nagoya plant in 1950, was the chief proponent of some of these innovative solutions, which he began to put in place during the 1950s (Ōno 1988).

Early innovations that sought to overcome the company's precarious financial situation by raising productivity without much capital investment included multitask job assignment (*takotei-mochi*), a major departure from traditional Fordist practice of single repeated motions; and levelization of production pace (*heijunka*), which mixes the making of different kinds of products on the same line and which required detailed planning of the assembly sequence. But Ōno's main contribution was to perfect the *just-in-time* system of parts delivery organized by using the *kanban* (signboard) technique aimed at producing a smooth, continuous workflow. Printed cards attached to parts or assemblies carry the relevant information on these inputs, with the production *kanban* circulating within each assembly process and withdrawal *kanban* used to transfer parts from the preceding process to replace the used parts (Toyota Motor Company 2004). Fine-tuning of this practice cut inventories as subcontractors delivered the needed parts as close to the time of their actual use as was logistically possible.

Reduced inventories minimize the space needed inside factories and limit the amount of capital that is tied up in accumulated parts. These are significant considerations given the fact that the late-1990s car models were assembled from about 30,000 parts. In 1986, the inventory in Toyota's Takaoka plant averaged two hours, compared to 2 weeks in a General Motors plant in Massachusetts. Most of the TPS features were embraced and further developed by other Japanese carmakers. By the century's end Japanese companies were making nearly 30% of all passenger cars made in the United States, while American and European automakers not only adopted many elements of the Japanese system but also purified some of them into even more extreme practices (Fujimoto 1999). And during the 1980s, the just-in-time approach became perhaps the most widely accepted Japanese innovation to improve productivity in many other branches of manufacturing (Voss 1987).

What emerged before the end of the twentieth century was yet another management hybrid. Taylor's concern about waste remains universally valid, but other assumptions that underpinned its operation proved to be counterproductive. While recognizing the value of workers' collective expertise, he failed to take advantage of it because he reduced their participation to the efficient execution of single tasks and relied on a one-way, downward flow of information. And post-1950 linear programming models demonstrated that the optimization of individual tasks does not necessarily optimize the performance of the entire process. Japanese innovations proved immensely influential as the quest for lean production was adopted by Western manufacturing. Its benefits were also extended to multiproject management, also pioneered by Toyota Motor, which aims to introduce products that share key components (e.g., engines or chassis in cars) but that are developed by separate teams for specific markets. Cusumano and Nobeoka (1999) called this "thinking beyond lean."

Automation and Robotization

Unlike the primarily organizational approach of lean production, the second category of innovations that transformed the classical industrial factory combined innovative machine design with increasing computing power. "Automatization" would be a more correctly derived noun—from the Greek term for statues that divine Hephaistos imbued with life, αύτόματοι—but the shorter "automation" prevailed after John Diebold (1952) used it in the title of his pioneering book. In turn, robotization, which made its first commercial inroads in the early 1960s, is qualitatively superior to mere automation.

By the beginning of the twentieth century, the mechanization of tasks and processes produced some amazingly ingenious machines capable of fully automatic operation. Among my favorites are the first fully automatic bottle-making machines that were designed by the Ohio glassblower Michael Joseph Owens (1859–1923). With no formal education and lacking the skills to build the machines himself, he described and sketched his ideas so that his engineering friends could convert them into pioneering machines (Owens 1904). The first model, offered in 1905, could make 17,280 bottles a day at 10 cents per gross, compared to 2,880 bottles made in a shop employing six workers at $1.80 per gross (ASME 1983). An entirely new design introduced in 1912 increased the capacity to an average of 50,400 bottles a day, and it could make them in sizes ranging from prescription ware to gallon containers (Figure II.4.15). By 2000, Owens-Illinois and its affiliates shipped some 100 million bottles every day.

But complete automation, the ultimate stage of mechanization, must go beyond an ingenious machine: it implies controlled deployment of many

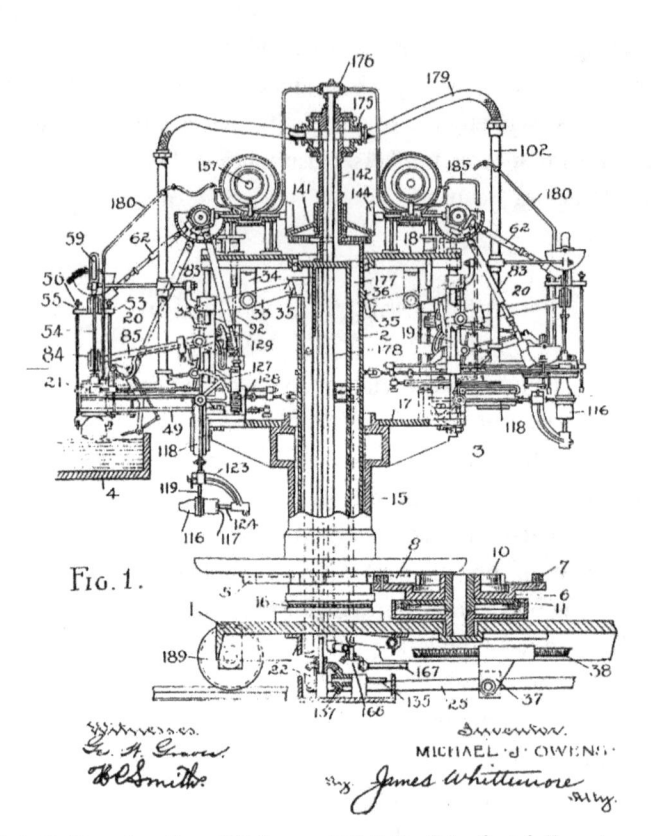

Figure II.4.15 Drawing from US Patent 766,768 of the first fully automatic glass bottle machine designed by Michael J. Owens in 1903.
Available from the US Patent Office.

mechanical devices to perform manufacturing tasks in a timed sequence without (or with minimal) human intervention. Its success rests not only on integrated deployment of machines but also on rationalizing complete production sequences, and its prerequisites include optimized space allocation and coordinated and streamlined production flows. Not surprisingly, the automobile industry, the most complex manufacturing activity before World War II, was the birthplace of many key procedures. But the most important post-1945 advances came in response to the unprecedented requirements of military aircraft construction.

Machining large series of the complex surfaces of wings for fighter planes or rotors for helicopters and doing so within extremely narrow tolerances could not be done in traditional ways by highly skilled operators using accurate jigs and templates. Electromechanical tracer machines, introduced in the early 1920s to duplicate carefully made models, were an unacceptable solution: too many models were needed, and copied surfaces deteriorated with time. By 1947—when Ford set up a new automation department—it became clear that a very different approach relying on numerically controlled (NC) machine tools held the best promise. But NC tools became a commercial reality only after more than a decade of complicated development (Noble 1984). In 1947, John T. Parsons, president of a company making helicopter rotor blades, and Frank L. Stulen, one of his engineers, conceived the first design of a milling machine whose automatic contour cutting was to be controlled by using IBM punched cards (Figure II.4.16). In June 1949, Parsons signed a contract with the US Air Force to develop a prototype for machining wing surfaces.

No commercial model resulted from this effort. In 1985, when Parsons received a National Medal of Technology, he remarked that "the impact of this invention is little understood, yet its applications range from computer chips to jet aircraft machine tools, even to the production of clothing" (White House Millenium Council 2000). Parsons's initial concept of a relatively simple electromechanical machine was transformed once MIT's Servomechanisms Laboratory was brought into the Air Force project. The first NC machine was publicly displayed in September 1952, but the early models were clumsy (250 electronic tubes and 175 relays) and for many years their economics remained questionable. The prospects improved with MIT's development of the Automatically Programmed Tools (APT) computer language and, in the spring of 1957, with the establishment of a consortium of aircraft companies dedicated to the advancement of NC machining (Ross 1982).

By the early 1960s, computer-assisted manufacturing (CAM) was a reality, and it received an enormous boost from integrated circuits and increases in computing power. Eventually the assembly of many identical products became completely automated by single-purpose NC machines (Reintjes 1991). While the early NC machines were recognizable by bulky control boxes and panels, with instructions stored on cards, tapes, or

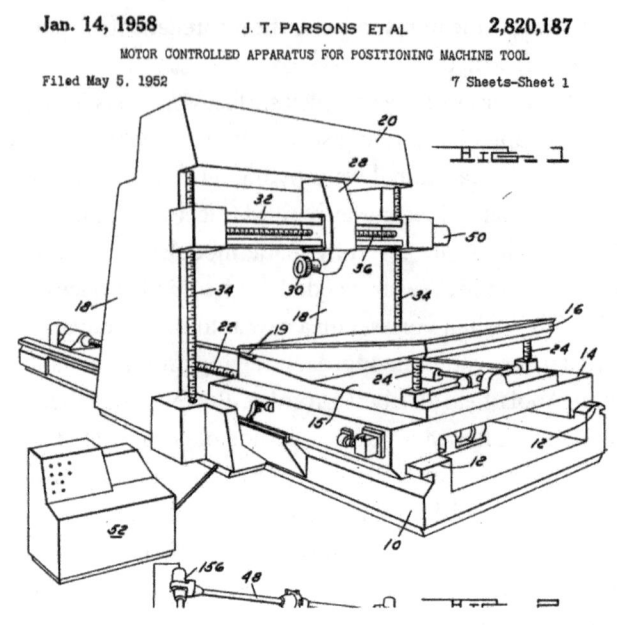

Figure II.4.16 Parsons and Stulen's pioneering patent for a "motor-controlled apparatus for positioning machine tool." The drawing, from US Patent 2,820,187, shows a perspective view of a bridge-type planer mill. A traveling table (*14*) supports a tiltable sine plate (*16*), the cutter head (*30*) moves horizontally by means of a crosshead (*32*) that can also move vertically (*34*), and the control device (*52*) delivers commands to various electric motors (e.g., *28* and *50*).
Available from the US Patent Office.

magnetic drums, post-1980 machines were run by progressively smaller microcomputers. The principal disadvantage of fixed tooling was a lack of flexibility, a drawback that was effectively addressed by the development of programmable industrial robots. Much as with NC tools, their success was a direct function of improved computing capabilities. A small factual correction first: robot histories credit Karel Čapek (1890–1938), a Czech writer whose plays, novels, and short stories were known worldwide during the 1920s and 1930s, with coming up with the name for an obedient mechanical automaton.

The noun is derived from *robota*, the Czech word for forced labor, and it was used for the first time in 1921, in Čapek's play *R.U.R.: Rossum's Universal Robots*, his "collective drama" about artificial people (Čapek 1921). But it was Karel's brother Josef, a painter and playwright, who suggested the term (Klíma 2001). And the concept itself was clearly formulated by the Čapek brothers more than a decade before R.U.R., in their story "The System," which was published for the first time in 1908 (see this chapter's epigraph). For decades, the heirs of Čapeks's literary creation could come alive only in other works of fiction, and industrial robots could be deployed only after 1960, when there was a sufficient control capacity to program and operate them.

Accuracy, repeatability, speed, and quality are the most obvious advantages of industrial robots. Accuracy begins with precise positioning, and robots with only Cartesian or cylindrical configuration have limited flexibility. To perform such common tasks as spot welding a car body, a robot must have three degrees of freedom in its arm to position the welding gun close to the spot to be welded and three degrees of freedom for its wrist movement to make the weld. With these six degrees of freedom, a robotic arm can work with an error of less than 1 mm to make thousands spot welds on a car body much faster than people ever could. And no group of skilled workers could perform around 4,000 spot welds per car and go for more than 60,000 hours (about 7 years) between failures.

The economic benefits of robotization also include major improvements in worker safety and flexibility that far surpass those of any previous industrial tools. Gains in occupational safety are due to the elimination of such risky tasks as welding, spray painting, and manipulation of hazardous materials. Flexibility came to mean not just machines that could be easily reprogrammed to take on a different set of tasks but also systems that could be preprogrammed for a wide array of missions. Robotization is inextricably bound with the advances in computer science and engineering. Intellectual milestones along this difficult path included George C. Devol's patent for controlling machines by magnetically stored instructions, Alan Turing's paper on computable numbers (Turing 1936), and Norbert Wiener's pioneering writings on cybernetics (Wiener 1948).

But the real intellectual breakthrough came only in 1954, when George Devol designed the first machine for "programmed article transfer" in December of that year (Figure II.4.17). His patent application spelled out both the novelty of the invention and the critical difference between automation and robotization.

> The present innovation makes available for the first time a more or less general-purpose machine that has universal application to a vast diversity

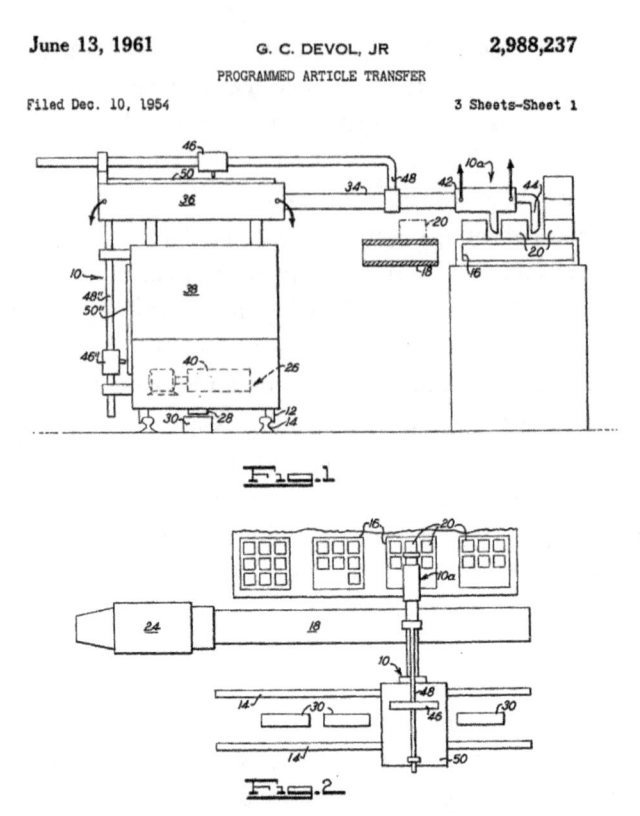

Figure II.4.17 The first two drawings of George Devol's patent (US Patent 2,988,237 granted in 1961, filed in 1954) for programmed article transfer, the automatic operation of machinery for materials handling. Control unit (*26*) contains the program drum (*40*) to direct the transfer apparatus (*10*), whose head (*10a*) contains a hydraulic actuator (*42*) and a jaw (*44*) to handle articles (*20*) on pallets (*16*).
Available from the US Patent Office.

of applications where cyclic control is to be desired; and in this aspect, the invention accomplishes many important results. It eliminates the high cost of specially designed cam controlled machines; it makes an automatically operating machine available where previously it may not have been economical to make such a machine with cam-controlled, specially designed parts; it makes possible the volume manufacture of universal automatic machines that are readily adaptable to a wide range of diversified applications; it makes possible the quick change-over of a machine adapted to any particular assignment so that it will perform new assignments, as required from time to time. (Devol 1961: 1)

By the time the patent was granted, in June 1961, the invention moved from a concept to a working machine. In 1956, Devol met at a party Joseph F. Engelberger, an aerospace engineer, and they decided to set up the world's first robot company, Unimation (contracted from "universal automation"). They built the first prototype transfer machine in 1958, and, in 1961, their first industrial robot was sent from their Danbury, Connecticut, factory to a General Motors plant in Trenton, New Jersey, to be used for lifting and stacking hot metal parts made by a die-casting machine. Commands for the operation of its 1.8 t motorized arm were stored on a magnetic drum. Unimation did not make any profit until 1975, but both men continued to advance the process of industrial automation: Devol by working on machine vision (including the now ubiquitous bar coding) and Engelberger by designing automatic assembly systems and promoting robots for service tasks (Engelberger 1989).

While the first Unimation robot was readied for industrial deployment, the American Machine and Foundry offered its first Versatran, a cylindrical robot designed by Harry Johnson and Veljko Milenkovic, in 1960. In 1969, the Stanford Arm, designed by Victor Scheinman, was the first computer-controlled and electrically powered kinematic configuration. In 1973, Cincinnati's Milacron offered the world's first minicomputer-operated industrial robot designed by Richard Hohn (Scott 1984). In May of 1978, Unimation brought out its Programmable Universal Machine for Assembly (PUMA) designed to duplicate the functions of the human arm and hand. But despite these American advances, Japanese companies became the world's leaders in robotics. In 1967, the first Versatran was imported to Japan, and, a year later, Kawasaki Heavy Industries began to produce licensed Unimation machines; by 1980, Japan had 47,000 robots, compared to fewer than 6,000 in Germany and 3,200 in the United States.

But Japanese statistics also included data on simple manipulators that are controlled by mechanical stops, and these machines would not pass a stricter definition of industrial robots used in the United States and the European Union. Comparison of more narrowly defined robots shows 3,000 units in Japan in 1980, compared to 2,145 in the United States (Engelberger 1980). Even so, there is little doubt that Japan's industries were able to introduce robots at a much faster rate, and, after they gained the global primacy during the late 1970s, they held it for the rest of the century. By 2000, the country had 389,400 robots, one-third more than the combined total in the United States and the European Union (UNECE 2001). The country's many makers of robots included such leading producers as FANUC, Fujitsu, Kawasaki, Mitsubishi, Muratec, Panasonic, and Yaskawa (Motoman). American car makers embraced robotization only after the post-1974 success of Japanese cars, but the extent of US robotization remained far behind the Japanese level, and, in relative terms, this was also true for Germany and Italy (see Figure II.4.18 for numbers of industrial robots in use 1960–2000).

During the 1990s, the price of industrial robots in the United States fell by 63% per unit and by 82% when adjusted for quality changes (UNECE 2001). Improvements included increased speed and reach and a more than doubling of the mean time between failures. As the number of product variants quadrupled, surveys showed that the actual length of service may be as high as 15 years and that would translate to 975,000 robots at work in 2000. Japan had 52% of the world's total (but keep in mind the definitional disparity), followed by Germany and the United States (each about 12%). Japan also led in terms of robots per 10,000 manufacturing workers (nearly

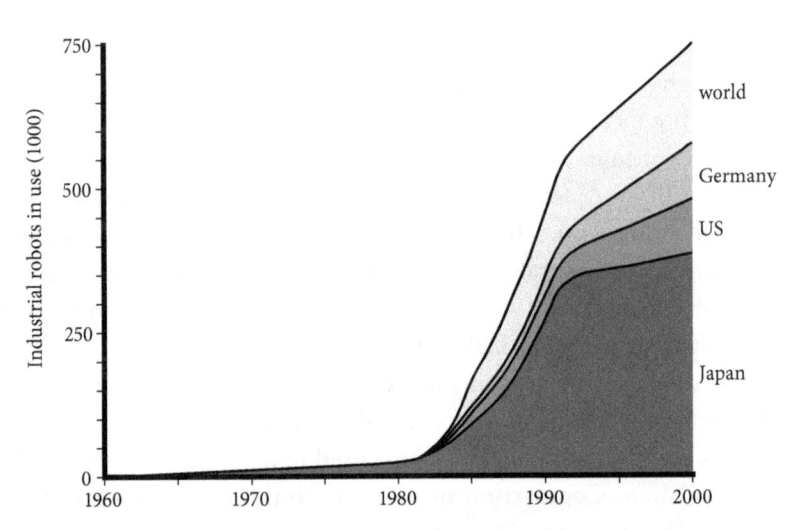

Figure II.4.18 Industrial robots in use, 1960–2000, worldwide and in Germany, Japan (using a broader definition), and the United States.

300); South Korean and German rates surpassed 100, the United States was just short of 50. The high German rate was due to their robotized car industry: more than 800 per 10,000 production workers, compared to fewer than 600 in the United States.

Besides the well-established tasks that marked the beginning of their use in the 1960s and 1970s—loading and unloading, spot and arc welding, and paint spraying—by the end of the twentieth century industrial robots were widely used for such diverse operations as palletized handling in warehouses and high-speed precision machining. They were also increasingly used by lasers cutting shapes in metal, composites, plastics, and cloth; for the application of adhesives and sealants (to glass, furniture, vehicles); precision wire cutting (needed to make electronic devices); plastic injection molding; and for failure-proof quality control (inspecting with visible, ultraviolet, and infrared light or with x-rays or lasers). Multipurpose robots can also use sensors to locate, distinguish, select, and grasp parts of many sizes and perform increasingly complicated assembly tasks (see the bottom image of this chapter's frontispiece).

Engelberger's (1989) vision of household service robots helping the elderly and infirm did not become a reality during the 1990s, but many hospitals came to rely not just on trackless couriers but also on robotic operating systems. Autonomous robotic couriers could transport supplies, meals, medical records, laboratory specimens, or drugs around a hospital, getting on and off elevators without any assistance and navigating through cluttered hospital corridors with interruptions only to recharge their batteries (Pyxis 2004). Development of surgical robots began during the late 1980s; the first robot that helped to perform an operation on a human was a Robodoc machine that assisted in a hip replacement surgery in November 1992, and the first robotic knee replacement was done in February 2000 (Integrated Surgical Systems 2004).

Since 1995, Intuitive Surgical, set up by Fred Moll, Rob Younge, and John Freud, has led in the design and marketing of machines that integrate surgical skills with the ability to operate through tiny ports (Intuitive Surgical 2004). By the late 1990s, computerized machines also provided many highly visible, as well as largely unseen, tasks in information, communication, and transportation sectors, which are the topic of Chapter II.5. But before turning to those advances, I should describe the key features of the most innovative late twentieth-century factories.

Flexibility and Variety

No other two nouns sum up better the goal of advanced manufacturing: large volumes of production coming from flexible setups capable of delivering an unprecedented choice of products. This approach is completely antithetical to the Ford-Taylor pattern that was created at the beginning of the century. Instead of mass production of identical items made by single-purpose machines, programmable, flexible machining and assembly is capable of rapid switches and adjustments and can yield a mix of low-volume products at competitive costs (Ayres and Haywood 1991). This transformation came about as the markets in high-income countries became saturated with many consumer goods, be it refrigerators (see Figure II.2.6) or TVs. Greater rewards were clearly waiting in numerous specialized markets where smaller lots could be profitably sold to different age, education, lifestyle, and income groups.

This diversification provided effective solutions to the productivity paradox, also known as *Abernathy's dilemma*: increasing standardization and product specialization drives down unit costs but the growing investment in inflexible automation discourages product innovation (Abernathy 1978). But the shift from mass to batch production (there is no binding definition of the divide, but 50 units may do) and eventually even to a single bespoke product could begin in earnest only when increasing computing capabilities opened several new paths to more flexible manufacturing and when digital controls became indispensable for all major industrial activities (Ouelette et al. 1983; Noble 1984; Cortada 2004).

NC machines introduced the age of digital automation. Their diffusion, led by the aircraft industry, was slow as they accounted for less than 5% of all US manufacturing machines in 1964 and for about 15% by 1969 (Cortada 2004). The rise of microprocessors did away with punched media and led to direct computer numerical control (CNC) that allowed for easy modifications of stored programs and for making a range of products with a single tool. One of the most remarkable examples of the CNC machine innovations introduced during the 1990s is Ingersoll's Octahedral Hexapod, a flexible platform that is five times as stiff as conventional machining centers while having only a fifth of their mass (Lindem and Charles 1995). The machine's framework of eight triangular faces joined at six corner junctions is self-contained and hence it needs no foundation for stability. Its cutting edge can be positioned anywhere within the struts' reach and pointed in any direction (Figure II.4.19). This makes the hexapod an ideal choice for multiple-axis contour precision-machining (with tolerances of less than 1 μm) of such large and heavy objects as turbine blades, stamping dies, or compressor bodies.

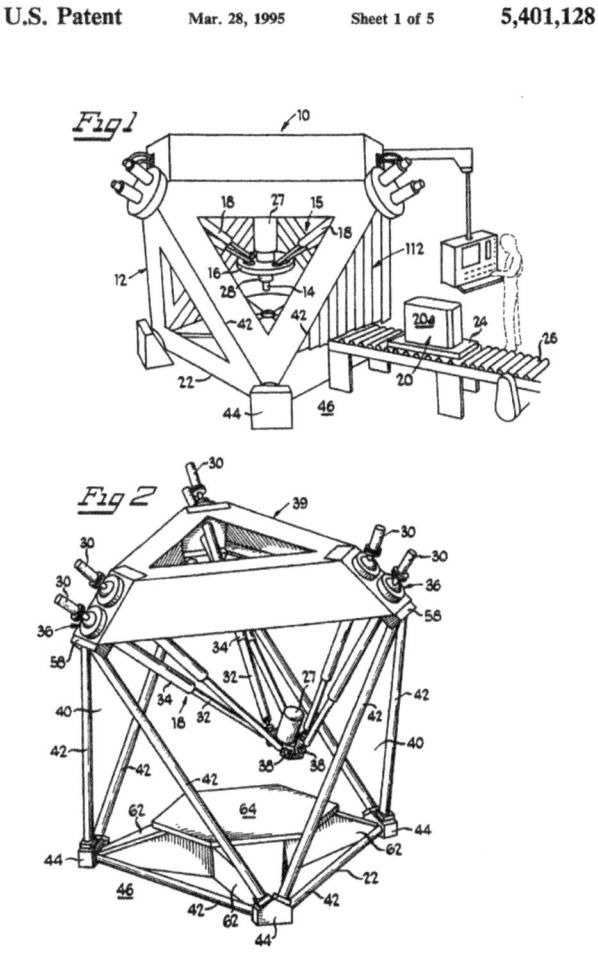

Figure II.4.19 Two perspective views of Ingersoll's Octahedral Hexapod® CNC machining platform (US patent 5,401,128). Its cutting tool (14) is carried by a servostrut support (15) mounted on extendable and retractable struts (18). The tool engages a workpiece (20) mounted on a table (64). The hexapod can be used either as a horizontal or vertical machining center or as a vertical turret lathe.

Automation was also extended to any imaginable control function. Monitoring production processes became commonplace as computers were installed to maintain optimal conditions (temperature, pressure, flow rates, concentrations) in chemical syntheses, crude oil refining, metallurgy, glass, pulp and paper, pharmaceutical, and food and beverage industries. The diffusion of CNC machines paralleled and stimulated the development of computer-assisted design (CAD) and the closely allied computer-assisted manufacturing (CAM). CAD applications, beginning with two-dimensional designs and progressing to complex three-dimensional capabilities, spread much faster than CAM. Military projects came first, but General Motors began its CAD development during the late 1950s, and commercial aircraft makers joined during the 1960s. CAD did away with the drafting boards that used to fill large rooms and with masses of drawings that were needed for major engineering products. Bulky blueprints stored in steel cabinets were replaced by electronic graphics, and painstaking redrafting gave way to a few additional mouse clicks.

This substitution made the greatest difference in aerospace designs: a World War II bomber required 8,000 drawings, but the Boeing 747 needed about 75,000. That is why, during the 1970s, the company developed a mainframe-based CAD system that eventually contained more than 1 million drawings and that was replaced, after more than 25 years of service, by a server-based Sun Solaris operating system (Nielsen 2003). During the 1990s, commercial jetliners became the most impressive materialization of pure CAD. Even the earliest CAD systems often halved the time spent on a particular task, and later advances reduced labor requirements to between 10% and 20% of the original need. CAD also allows one to try out many variants and adjustments: the software follows the changes through the entire design and amends everything.

And CAD made it possible to dispense with building expensive models, mockups, and prototypes to be subjected to actual physical testing and substitute them with exhaustive virtual reality tests. With rising computing power, even such complex phenomena as turbulent airflow or flight emergencies could be simulated with a high degree of accuracy. This led to much shortened lead times from product to market, and this time was further compressed by directly linking CAD with CAM. CAD/CAM combinations, pioneered by the aerospace industry, spread rapidly to semiconductor fabrication. Once the design is finalized, it can be transmitted from an office into CNC machines in the same building or across an ocean. CAD/CAM also benefited producers of smaller items who may need to judge an object's appearance during the development stage.

CAD can feed alternative forms directly into rapid prototyping machines that can build three-dimensional objects. This can be done either by subtractive or additive autofabrication (Burns 1993). Subtractive prototyping is a special application of the just-described CNC tools whose deployment became common by the late 1980s: the final shape will emerge from unattended shaping through cutting, drilling, milling, and polishing or, when working with such hard materials as specialty steels and ceramics, by electrical spark erosion. In contrast, additive fabrication (3D printing), an innovation of the early 1990s, forms products by aggrading thin cross sections that are outlined by lasers and built with powder (sintered after a layer is deposited) or photosensitive polymers hardened by exposure.

CAD/CAM became the foundation of flexible manufacturing system (FMS), whereby CNC machines are linked together to form increasingly more productive and versatile production sequences. By the mid-1980s, FMS could be found in companies ranging from aerospace (General Dynamics) to farm machinery (John Deere), and, during the 1990s, these systems made major inroads particularly in Europe where producers facing high labor costs used them to remain competitive (Figure II.4.20). The goal of digitally controlled industrial automation is computer-integrated manufacturing (CIM) that is capable of an almost infinite flexibility at a highly competitive cost: producing small batches tailored to specific consumer demands at prices that beat the formerly limited choice of mass-produced objects (Ayres and Butcher 1993). CIM goes beyond production to embrace the entire information system (ordering, accounting, billing, payroll, employee records) needed to guide the development of new products and to market them in the most profitable manner.

The cost of many products eventually plummeted even as their quality and reliability improved. Studies showed an average 500% gain in a draftsman's output with CAD, and 44–77% reductions of average unit labor costs with FMS (Ayres and Haywood 1991). But the most obvious consequence of automation was the multiplication of product variety, with new models introduced at what just two decades ago would seem to be impossibly short intervals (Patterson 1993). Stalk and Hout (1990) concluded that reducing the time elapsed between an idea and a product determined success in late twentieth-century markets. They believe that it was a stubborn refusal of many American companies to face the problem of long product development

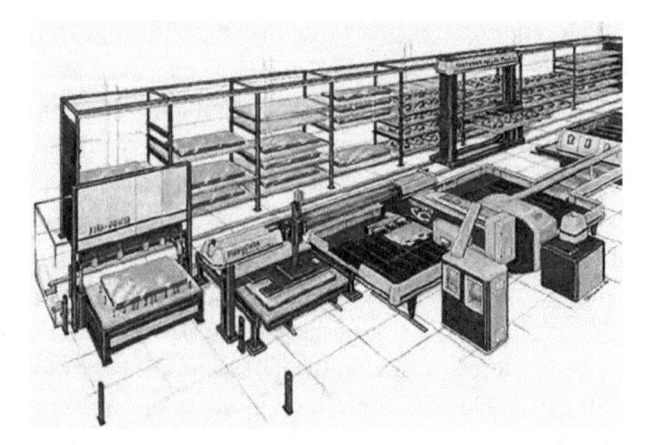

Figure II.4.20 Night Train, designed by Finn-Power, is a leading flexible manufacturing system with automatic loading and unloading of shelved sheet metal.
Digital image courtesy of Finn-Power.

cycles that was mainly responsible for the decline of the country's competitive advantage.

Gerber Scientific's (2003) surveys found that since the late 1980s the number of automobile models increased from 140 to 260. The real choice was much wider, as most models came in several configurations, some models allowed for further in-plant customization before the car's shipment to a dealer, and the workshops doing post-purchase alterations were doing record business. As market diversity approached 1,000 more or less distinct vehicles, the size of production series declined: in 1979, GM offered 28 models in its five divisions, with the annual output averaging almost exactly 1 million vehicles per division; 20 years later the company was selling 36 models in seven divisions, averaging fewer than 400,000 vehicles per division (Ward's Communications 2000). And while during the 1970s it took more than 5 years to introduce a new car model, lean production techniques reduced the lead time by more than a third (Cusumano and Nobeoka 1999). And before getting into their cars for their daily commute, US customers could choose from 340 kinds of breakfast cereals and 70 styles of Levi jeans. And while driving, too many of them kept on talking on their cell phones, devices that offer one of the best illustrations of a transformed mass market. The basic hardwired black Bakelite telephone was a perfect expression of long-reigning mass-production–based monopolies: the first one-piece set incorporating transmitter and receiver in the same unit (see Figure II.3.16) was introduced during the late 1920s, and it endured until the 1960s, when pushbutton dialing (first in 1963) began to take over and the devices started to change color and shape. But only electronic telephony (starting in the late 1970s) and the breakup of telephone monopolies during the 1980s ushered in the age of unbridled design that only intensified with the diffusion of cellular phones.

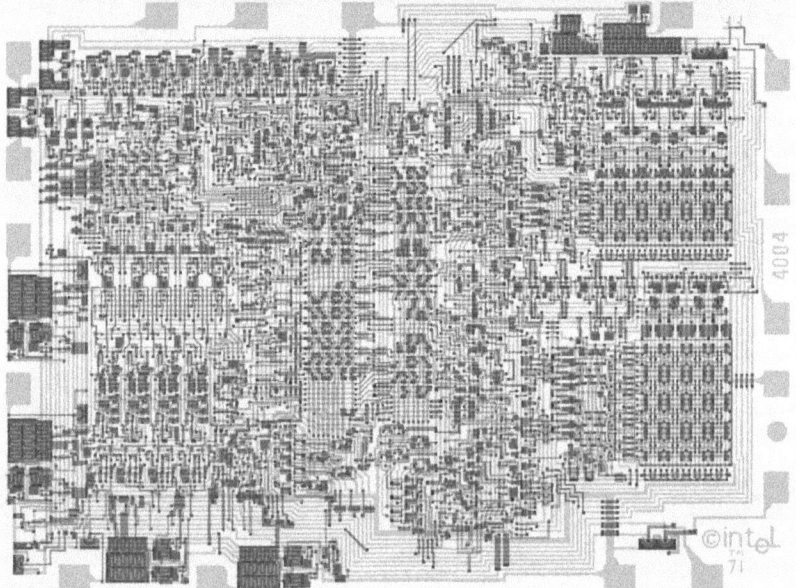

Frontispiece II.5 Two images, only apparently unconnected, exemplify the late twentieth-century innovations in transportation, information, and communication. The top photograph shows the container ship Cornelius Maersk (347 m long vessel with bow width of almost 43 m and able to carry 6,600 standard 20-ft containers in its hold and in up to seven layers on its deck). The bottom image depicts the layout of the Intel 4004, the world's first microprocessor. Like most modern industrial and service activities, container shipping, now the dominant worldwide means of cargo transport, could not be efficiently managed and continuously monitored without microprocessors in computers, machines, and devices.

Cornelius Maersk photograph is from the Maersk Corporation; microprocessor image from Intel.

5

Transportation, Communication, Information

Mass and Speed

> "Oh, I don't know that. The world is big enough."
> "It was once," said Phileas Fogg, in a low tone.
> —Jules Verne, *Around the World in Eighty Days*

Many images could exemplify the theme of this chapter, and shipping steel containers stacked on the deck of a large cargo ship on a long intercontinental voyage is both an impressive and perhaps an unexpected choice—but one that is easily justifiable. Just about every consumer item that was purchased during the last decade of the twentieth century spent some time in a freight container—and it did so most often by traveling via a combination of two or three transport modes, including trucks, railroad flat cars, ships, and the cargo holds of big jetliners. Some of these containers came from domestic factories or from neighboring states, others had traveled across a continent, and hundreds of millions of them arrived after intercontinental journeys.

Shipping containers are also perfect embodiments of this book's key message about transformative technical advances: their deployment and widespread use required ingenious design and radical organizational changes—but there was no need to invent any new materials or prime movers. Containers are just big steel boxes with reinforced edges, trucks and railway cars take them to ports, large cranes handle them, and diesel engines propel the fast ships. The enormous transformation of freight transport was thus achieved with techniques that originated before World War I and whose genesis and rapid evolution are traced in Part I of this book.

But the full scale and impact of this transformation would have been impossible without combining these techniques with devices and processes based on solid-state electronics. What underpins the entire system of global cargo movement are the unseen complex electronic networks that allow a container to arrive just when it is needed after being carried halfway around the world by a succession of transport modes and that make it possible to check its status at any point on this long journey. Microprocessors are at the heart of these electronic networks; Intel's pioneering Model 4004, released in 1971, is shown in the lower part of the frontispiece to this chapter.

And containerized shipping is also a perfect illustration of this chapter's subtitle and common denominator, the twin advances of mass and speed. At the beginning of the twentieth century, *break-bulk* (general purpose cargo) ships powered by steam engines had typical capacities of 5,000–6,000 deadweight tonnage (DWT), and they traveled at about 15 kilometers per hour (km/h). After World War II, surplus military ships of the *Liberty* class became the dominant carriers: their capacity was 10,500 DWT, and they sustained nearly 20 km/h. In 2000, the *Cornelius Maersk* (launched by Lindø Shipyard and shown in the frontispiece) belonged to the largest class of container vessels whose capacity of slightly more than 100,000 DWT allowed them to carry 6,600 standard steel containers and whose 55-megawatt (MW) diesel engines propelled them at up to 46 km/h.

But before dealing with cargo, I look at advances in transporting people, focusing first on the automobile, the key technical marker of the twentieth century, the most cherished mode of personal mobility and a source of many intractable problems. Despite the car's global appeal, railway transport persisted, and it was transformed by the introduction of high-speed trains, while air travel evolved from a rare experience to just another way of mundane mass transport. The second part of this chapter follows the greatest technical transformation of the last two generations of the twentieth century, the rise and rapid advances of solid-state electronics.

The invention and evolution of transistors, integrated circuits, and microprocessors had first transformed computers from expensive, bulky, and relatively low-performance rarities to affordable, portable devices whose gains in processing powers rose by many orders of magnitude. And soon microprocessors penetrated just about every niche of modern life: they are at the heart of cellular phones and programmable thermostats, they guarantee perfectly cooked rice as well as ultra-low emissions in car engines. But why not an entire chapter on microelectronics? Because of temporal and functional proportionality. Commercial computers became available only in 1951; they were not widely embraced until integrated circuits made them affordable, and they became ubiquitous only during the 1980s. And, of course, microelectronics could not exist without the first-order innovations (electricity generation, electric motors, transformers, high-voltage [HV] transmission) that originated during the two pre-World War I generations.

Transport on Mass Scales

The speed of the journeys in pre-industrial societies was inherently limited by the metabolic capacities of human and animal bodies or by the inefficient and often frustrating conversion of wind into sailing motion. After 1830, steam engines were put on rails, and ocean-going vessels, trains, and ships could accommodate increasing numbers of travelers at higher speeds and lower cost. The twentieth century was fundamentally different because of the unprecedented combination of the total number of people who were taking both relatively short everyday trips as well as journeys not just to distant cities but to other continents and doing so not only in the quest for livelihood but increasingly as a pastime. The speed and cost of new modes of travel made this both convenient (or at least bearable) and widely affordable.

The new reality of travel as a mundane personal choice was expressed most obviously by the rising ownership of passenger cars. As a result, in many countries, passenger railways lost their former preeminence, but Japan and France demonstrated the possibilities of modern train travel as they reinvented wheeled machines on steel rails in the form of high-speed trains. Ships as the means of mass intercontinental travel did not make it beyond the 1960s (once jetliners took over), and, during the late 1990s, nearly all passengers boarding ocean and river vessels were tourists as the cruise industry created demand for travelling not only to popular destinations but also to nowhere in particular (Figure II.5.1).

The Automobile Century

The car's allure could not be transformed into a mass-ownership phenomenon without a complex concatenation of technical and organizational innovations that helped to turn automobiles from rare and expensive possessions in the century's beginning to ubiquitous complex machines that were regularly used by billions of people at its end. Rapid pre-1914

Figure II.5.1 Bow and stern views of *Crystal Harmony*, the highest ranked cruise ship in the medium size category (49,400 t, 940 passengers), exemplifies the 1990s vogue of luxury cruising with overfurnished staterooms, private verandahs, pools, whirlpools, gyms, saunas, courts, movie theaters, shopping centers, casinos, and restaurants. The cruising industry offered everything from short-budget runs leaving from Miami to months-long circumnavigations of the world.
Photographs courtesy of John Patrick.

maturation of the Otto cycle engines (see Chapter I.3) left little room for any fundamental changes, but incremental improvements of every one of its parts added up to major performance gains—as illustrated by contrasting the Mercedes 35 from 1901 with Mercedes S500 from 2000 in the first chapter's frontispiece. Foremost among them were higher compression ratios (commonly 8–10), lean combustion (excess air-to-fuel ratio of around 22, well above the stoichiometric ratio of 15), electronic fuel injection, higher-octane fuel, and turbocharging (Flink 1988; McShane 1997).

Abernathy, Clark, and Kantrow (1984) identified slightly more than 600 innovations in the US automobile industry between 1900 and 1981, with only two World War II years (1943 and 1944) not making any contributions. During the 1920s, open car bodies were replaced by closed structures and wood frames by sheet steel; solid metal wheels, four-wheel brakes, and balloon tires became standard, and new lacquers brought a choice of body colors. The 1930s saw independent front-wheel suspension (Mercedes in 1931), power brakes, front-wheel drive (Cord in 1930), better transmissions, and air conditioning (Packard in 1938). Radial tires were introduced in 1948, and automatic transmission using a fluid torque converter came with the 1949 Buick. Power steering, patented by Francis Wright Davis in

1926, became common only during the 1950s. And in 1954, Mercedes-Benz introduced the first passenger car with fuel injection, which was previously used only on racing cars. A transistorized electronic ignition appeared in 1963, the collapsible steering column in 1967, and energy-absorbing bumpers a year later.

Ever since Ford's Model T, automakers have tried to come up with basic, affordable cars for first-time buyers of limited means. Many of these attempts failed, some, because of their innovative design or funky appeal, became classics (Siuru 1989). No such vehicle emerged in the United States after the Model T, and, in terms of aggregate production, size, and longevity (albeit with updates), no car comes close to the one that Adolf Hitler decreed as the most suitable for his people (Nelson 1998; Patton 2003). In autumn 1933, Hitler set down the car's specifications—top speed of 100 km/h, 7 L/km, two adults, three children, air cooling, and the cost below 1,000 reichsmarks—and Ferdinand Porsche (1875–1951) had it ready for production (rather ugly and looking, at Hitler's insistence, like a beetle) in 1938.

Its serial assembly began only in 1945, under British Army command. During the early 1960s, Volkswagen became the most popular imported car in the United States. A redesigned (less ugly) version became available in 1994 (Figure II.5.2), production in Germany stopped in 1977, but it continued in Brazil until 1996 and in Puebla, Mexico, until 2003: the last car

Figure II.5.2 Front views of the end of the twentieth-century versions of two small passenger cars whose predecessors changed the modern automotive history: the New Beetle, redesigned by J. Mays and Freeman Thomas in California and produced since 1998, and the Honda Civic, first introduced in 1972.

Photographs courtesy of Douglas Fast.

produced at the Puebla plant was number 21,529,464. France's answer to the Beetle was the Renault 4CV, with more than 1 million units made between 1945 and 1961. But the country's most famous basic car was the Citroen 2CV, nicknamed "Deux Cheveaux": the numeral stood just for cylinders, the 0.6-liter (L) engine rated 29 horsepower (hp), and the car could reach about 65 km/h.

Many scorned it, and Siuru (1989: 45) found it "hard to believe that people would ever have bought such an unattractive car." Others made it a cult object, and its production continued until 1988. Italy's most famous basic car—Fiat's little mouse, Topolino, a two-seater with a wheelbase just short of 2 meters (m) and an engine of 0.569 L—was made between 1936 and 1955. British contributions included the Austin Seven, produced for 16 years during the 1920s and 1930s, and the Morris Minor, whose total production between 1948 and 1971 was about 2 million units. Morris had a conventional design, but the British Motor Company's Mini, introduced in 1959, was more than a cute small vehicle. Its transverse-mounted four-cylinder engine (initially just 0.91 L) driving the front wheels was a layout that was subsequently much copied.

The interwar years saw an inevitable consolidation in the industry (in the United States, there were more than 40 automakers in 1925) and the demise of such once-famous marques as Marmon, Nash, Pierce-Arrow, Stutz, and Willys. Yet technical innovations and business consolidations did not result in better mass-produced American cars. Automatic transmissions, power steering, power brakes, air conditioning, and other add-ons needed more powerful engines, and the typical mass and power of bestselling vehicles increased to more than 1.5 tons (t) and 110 kilowatt hours (kWh) by the late 1960s. Despite the introduction of compacts—offered for the first time in the 1960 model year to compete with European imports—the average performance of new American passenger cars, which was about 15 L/100 km (16 miles per gallon [mpg]) during the early 1930s, kept slowly deteriorating, with the worst rate at 17.7 L/100 km (13.4 mpg) in 1973 (EIA 2004).

Moreover, frivolous styling was considered more important than safety and reliability. By the mid-1960s "overpowered, under-tired, and under-braked cars were traveling at faster speeds on the highways" (Flink 1985: 158). And defective: the average American car was sold with 24 defects, and, as Ralph Nader so memorably documented, these vehicles were unsafe at almost any speed (Nader 1965). The first post-World War II generation of US car making was little more than a quest for quantity and profit based on built-in obsolescence and on the complete neglect of fuel economy, resulting in waste of energy and high levels of air pollution. This was mass manufacturing reduced to its basest denominator of inelegant quantity and inefficiency.

All that changed due to three profound transformations driven by photochemical smog, rising exports of Japanese cars, and large increases in the price of crude oil. How long would have the inefficient, dangerous shoddiness of American cars continued if it were not for smog, Honda, and Organization of Petroleum Exporting Countries (OPEC)? As the number of American vehicles rose and their fuel efficiency deteriorated, photochemical smog—first noted in Los Angeles of the 1940s and traced by Arie Haagen-Smit (1900–1977) primarily to automotive emissions—began to affect other large metropolitan regions (Colbeck and MacKenzie 1994). Finally, in 1970, the US Clean Air Act called for major reduction of automotive emissions.

The more desirable technical fix was to prevent or at least lower the formation of undesirable gases; the other way was to take care of them just before they were emitted to the atmosphere. The second option became the North American norm in 1974, with the introduction of catalytic converters. This had to be preceded (in 1971) by the formulation of unleaded gas to avoid poisoning the platinum catalyst. Improved catalytic converters made an enormous difference, eventually reducing hydrocarbon and carbon monoxide (CO) emissions by 96% and nitrogen oxides (NO_x) emissions by 90%. The first path was followed most innovatively by Honda Motor, the company established by Soichirō Honda (1906–1991) in 1946, to make light motorcycles (Sakiya 1982).

In February 1971, Honda revealed its Compound Vortex Controlled Combustion (CVCC, and hence Civic) engine with a small auxiliary combustion chamber that was supplied through a separate carburetor with a rich air-to-fuel mixture (ratio no higher than 5:1); it was attached to the main chamber supplied with a lean mixture (18:1 or 20:1). This arrangement kept the main combustion chamber to about 1,000°C, low enough to reduce the formation of NO_x as well as the emissions of CO. Honda also took care of hydrocarbon emissions through oxidation within an improved manifold system. In 1972, Honda's CVCC engines were thus the first design in the world to meet the US Clean Air Act standard prescribed for 1975 and did so without resorting to heavy metal catalysts.

The most remarkable transformative factor in the history of post-World War II car making was the surprising rise of the Japanese auto industry, which combined rapid growth with unprecedented efficiency of production and, for many years, unmatched quality. The country's annual car production rose from slightly more than 30,000 units made in 1950 strictly for domestic sales to nearly half a million a decade later. By 1970, with the first exports to North America under way, the output of 5.3 million vehicles made the country the world's second largest car maker, well ahead of Germany; in 1980, Japan's production surpassed the US total for the first time, and, by the century's end, Japanese companies at home and abroad were making nearly 30% of the world's passenger and commercial vehicles while remaking the entire manufacturing process (see Chapter II.4).

The first popular postwar models, such as the Toyopet Crown and Datsun 110 (both introduced in 1955), reflected their respective American and English pedigrees. But subsequent new designs were unlike either American or European models: they had a much lower number of defects yet making them required less time than in North America and much less time than in Europe (Womack, Jones, and Roos 1991). European import barriers kept them away from the continent, but they were welcome in the United States. Honda Motor Company was an unlikely pioneer of aggressive exports: it made its first car, the small sporty S500, in 1963; its first export to the United States, the tiny two-door sedan N600 in March 1970, was followed in 1972 by the Civic, an entry-level car that combined economy and innovative engineering (overhead-cam 1.2-L CVVC engine and four-wheel independent suspension); 4 years later came the first Honda Accord (Figure II.5.2).

During the early 1970s nobody—and certainly least of all the (at that time) four large American automakers, which looked at these upstarts with condescension—could imagine that by the century's end the car model made in the greatest quantity in North America would be the Honda Civic, followed by the Ford Taurus, followed by the Honda Accord (Ward's Communications 2000). Both the Civic and Accord were soon recognized by automotive experts as the benchmark cars in their class, and Honda kept improving their reliability, performance, and fuel economy through subsequent redesigns. By the end of the twentieth century both the Civic and the Accord were in their sixth generation, and the cumulative production of Hondas reached nearly 43 million vehicles.

After the surge of exports during the 1970s came North American–based Japanese manufacturing. Honda was again the pioneer with its Accord plant in Marysville, Ohio, in 1982, and, by the century's end, Japanese car makers made about 30% of all cars produced in the United States and Canada, and three of the five bestselling models were Japanese (Toyota Camry, Honda Accord, and Honda Civic). These developments were unfolding against the background of dramatically changing crude oil prices: they quintupled between October 1973 and May 1974, and then they rose by 2.6 times between January 1979 and January 1981. Smaller Japanese and European models were always more fuel efficient, and US car makers were forced to approach their performance by complying with Corporate Average Fuel Economy (CAFE) standards and double the fleet's performance to 27.5 mpg (8.6 L/100 km) between 1973 and 1987 (EIA 2004).

These gains were achieved by engine and transmission improvements, lighter bodies, and reduced aerodynamic drag. Material substitutions (high-strength steel for ordinary steel, aluminum and plastics for steel) reduced the mass of average American car by nearly 30% between 1975 and 1985. The subsequent fall in crude oil prices removed the immediate incentive to continue with these improvements, which would have put the average performance at about 40 mpg by 2000. Instead, and despite major post-1985 oil price fluctuations, the US market regressed in a way that would have been unpredictable at the beginning of the 1980s. During the late 1980s, vans and pickup trucks began to be increasingly purchased instead of sedans, and they were joined by a new category of oversize and ridiculously named sport utility vehicles (SUVs). By 1999, these machines had gained almost exactly half of new US vehicle sales (Ward's Communications 2000; Bradsher 2002).

Few things can illustrate better the US SUV addiction than the fact that it became quite normal to use vehicles with a power of more than 100 kW and with a mass that approached the maximum of 4 t (the Chevrolet Suburban weighed 3.9 t) to carry a single person on a short trip to buy a small carton of milk from a nearby convenience store. Because they were classified as light trucks, these vehicles were exempt from CAFE standards, and the fuel efficiency of many models was well below 20 mpg, some even less than 15 mpg. But these wasteful machines would have had an even more devastating impact without the benefits of electronic controls introduced during the early 1960s: 1964 was the last year that Ford's cars were made without any electronic components (Horton and Compton 1984).

These applications increased steadily to control not just panel displays and audio systems but also such key operations as the air-to-fuel ratio, spark timing, recirculation of exhaust gases, and antilock braking (to prevent an uncontrolled slide), which appeared for the first time during the 1980s. By the late 1990s, passenger cars contained commonly between 50 and 80 microprocessors: cars thus became complex mechatronic machines combining innovative mechanics with advanced electronics. They also became safer to drive thanks to stronger structural materials, safety glass, and better brakes. But the first mandatory seatbelt laws came into effect in Europe only during the 1960s, and in the United States (New York State) in 1984.

The three-point automotive seat belt (combining a lap belt running across the pelvis and a diagonal belt running across the rib cage; Figure II.5.3) was designed only in 1959, by Nils Ivar Bohlin for Volvo (Bohlin 1962). Ohlin's last project at Saab, before moving to Volvo in 1958, was developing the ejector seat for the supersonic fighter J35 Draken. In 1985, the German Patent Office ranked the seat belt among the eight most important innovations of the preceding 100 years: undoubtedly, during the past two generations it has saved the lives of several hundreds of thousands of people and prevented serious injuries for millions of drivers and passengers. Few innovations can compete with the seat-belt's cost-benefit ratio: saving a life with an investment of several hundred dollars. Airbags, first just on the driver's side, came during the late 1980s.

The greatest accomplishment of twentieth-century car making was the transformation from artisanal manufacture of small series of simple but expensive machines to the world's leading industrial enterprise that turned out annually more than 50 million increasingly complex mechatronic vehicles in

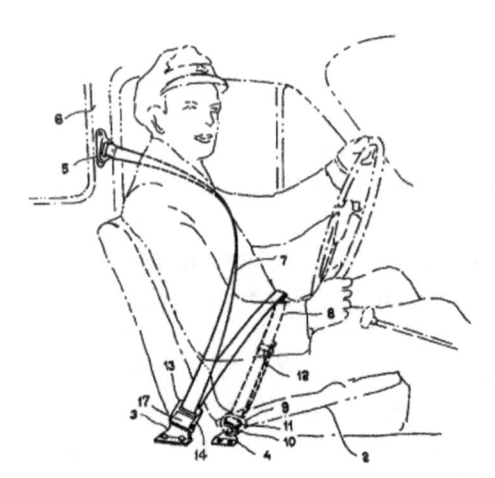

Figure II.5.3 The first figure in Bohlin's patent (US Patent 3,043,625) for the three-point safety belt, one of the most beneficial inventions of the twentieth century.
Available from the US Patent Office.

highly automated plants at rates of hundreds of thousands of units per year and sold most of them at prices that were well within the reach of hundreds of millions of people. The car's twentieth-century history is thus a conjoined sequence of transformations from simplicity to complexity, from expensive oddity to affordable ubiquity. Centennial statistics span orders of magnitude. In 1900, artisanal producers were putting annually fewer than 2,000 cars on the market; during the last 5 years of the twentieth century that rate, including both passenger cars and commercial vehicles, was above 50 million units.

In 1900, the world's passenger car fleet was less than 10,000 vehicles, and most of them were electric or steam-powered. The count surpassed 1 million before World War I, 10 million in 1920, 100 million in 1960, and, by the century's end, it passed half a billion, with the addition of commercial vehicles raising the total to nearly 700 million (Figure II.5.4).

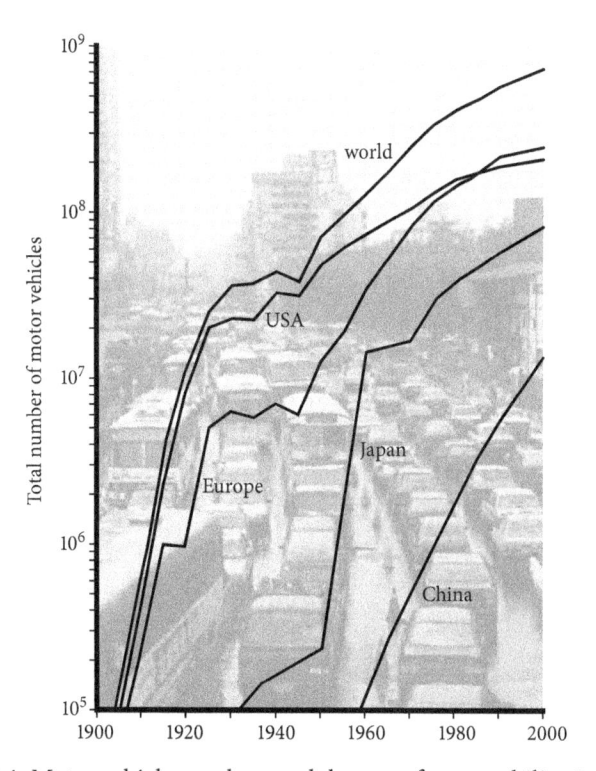

Figure II.5.4 Motor vehicle numbers and the rate of auto mobilization, 1900–2000.

Based on figure 5.19 in Smil (1994) and on data in Ward's Communications (2000).

In terms of people per passenger car, the global ratio dropped to below 50 by 1950 and to 12 by 1999, the mean hiding very large regional and national disparities. By the late 1990s, Africa's mean was about 110; Western European, 2.2. The United States had the lowest ratio for more than seven decades after Ford's Model T went on sale in 1908: down to 3.3 already by 1950 and 2.1 in 1999. European pre-World War II progress was slow, but the post-1960 spurt brought many national ratios to the US level, or even marginally lower, by 1999 (Germany at 2.0, Italy at 2.1, France at 2.2), while Japan's ratio fell to 2.5.

New High-Speed Trains

Steam-driven railways were technically mature well before 1900, and the speed of a mile a minute (96 km/h) was first reached, briefly, on a scheduled English run already by 1847 (O'Brien 1983). By the beginning of the twentieth century speeds on many sections in Europe and United States exceeded 100 km/h, and, during the 1930s, engineers designed powerful streamlined locomotives. In 1934, a Union Pacific train averaged nearly 150 km/h between North Platte and Alda, Nebraska, with a brief maximum of 180 km/h, and, in 1936, the German *Borsig* reached the world record speed of 200.4 km/h (Ellis 1977). After World War II many engineers favored gas turbines, and, in 1971, France's first experimental turbotrain, TGV 001 powered by Sud Aviation's Super Frelon helicopter turbines, was tested at speeds exceeding 300 km/h.

But that route was abandoned when OPEC quintupled crude oil prices, and the French efforts turned, following the Japanese example, to electric traction. Direct current (DC)-driven trains had been commercialized already during the 1890s, but the tests of single-phase alternating current (AC) traction began only in 1936, in Germany, and were eventually abandoned (Bouley 1994). After World War II three international conferences examined this technique, and the last one, in Lille in 1955, was attended by a Japanese delegation. Soon afterward, four major Japanese companies, Hitachi, Toshiba, Fuji, and Mitsubishi Electric, produced suitable AC-commutator motors and rectifier-type locomotives (Sone 2001). Instead of using locomotives, Hideo Shima (1901–1998), at that time the vice president for engineering of Japanese National Railways (JNR), advocated the use of multiple electric motors in individual carriages as the best way to limit axle loads, reduce stress on tracks and structures, and use the motors for regenerative braking (Shima 1994).

Japan's pioneering role in rapid rail transport is remarkable for several reasons. First is its boldness of vision: the development of *shinkansen* (a new trunk line, distinct from the old narrow gauge) began when the country just emerged from the enormous destruction of World War II. In May 1956, when JNR launched its feasibility study, it was just 2 years since the country's gross national product finally surpassed its peak (1939) pre-World War II level and when the average disposable income was only a small fraction of the US rate. Financing the project was a challenge. The World Bank's loan of $80 million was less than 15% of the original budget that JNR set on the low side, at ¥194.8 billion (or US$541 billion in 1958 monies), to ensure the government's funding approval. As the cost overruns continued (the final total was ¥380 billion), Shinji Sogo (1884–1981), the president of JNR who promoted the project, resigned in 1963, and Shima joined him (Suga 1998).

Second, the commitment to fast trains ran against the prevailing consensus that saw the ascendant turbojet-powered flight as the unbeatable choice for rapid travel. Third, although the project's initial phase, the Tokaido shinkansen between Tokyo and Osaka, had to solve several unprecedented technical problems, it was completed successfully on time (just before the Olympics), 5 years and 5 months after its construction began in April 1959. The scheduled service was launched on October 1, 1964, and it was improved and expanded with new lines, new trains, and gradual increases in top operating speeds and train frequency (Figure II.5.5). By 1974, the Sanyo line running to the west of Osaka was extended to Hakata in northern Kyushu. The Tohoku line to Morioka in northwestern Honshu, and the Joetsu line to Niigata on the Sea of Japan began service in 1982. Tokaido trains ran at 210 km/h (for faster *hikari*) until November 1986, when their speed was increased to 220 km/h.

In 1987 JNR, deeply in debt, was privatized and divided into six regional companies. In 1990, Japan's bubble economy burst—but *shinkansen*'s high standards were maintained, and its progress continued (Fossett 2004). In March 1992, the new (300 series) Nozomi trains reached speeds of 270 km/h. But since March 1997, the record belongs to the Sanyo line, where the 500 series Nozomi, easily distinguished by their 15-m-long aerodynamic nose and wing-shaped pantograph, have speeds up to 300 km/h (Figure II.5.5). In October 1997, the Hokuriku line connected Nagano in the Japanese Alps for the 1998 winter Olympics, and, by 2000, extensions were under

Figure II.5.5 Shinkansen Hikari 100 series (left; now retired) at Kyoto station and 500 series Nozomi (right), the last new type of train introduced before 2000. Photographs from author's collection.

way in Kyushu and in northernmost Honshu. The frequency of trains on the Tokaido line increased from 60 in 1964 to 285 by 2000, and, during the busiest period of the day, they followed in intervals of just 3.5 minutes with an average delay of just 24 seconds (JR Central 2004).

The Tokaido line alone carried annually about 132 million people, slightly more than Japan's total population, and all lines were boarded by nearly 300 million people a year. Remarkably, none of the *shinkansen* trains had derailed or collided, so, by the end of 2000, the system carried 4.5 billion people without a single accidental fatality. All *shinkansen* lines have standard gauge (1,435 mm), continuously welded rails (60 kg/m) and prestressed concrete sleepers, but ballasted track gradually gave way to reinforced concrete slab tracks (Miura et al. 1998). Electricity is supplied by catenary wires (copper or copper-clad steel) with spans of 50 m as 25 kilovolt (kV) AC at 60 hertz (Hz). Sixteen-car trains, seating up to 1,323 people and lasting for 15–20 years, soon became standard; the 700 series has 64 AC motors of 275 kW each for a total of 13.2 MW. This makes frequent accelerations and decelerations, needed for relatively short interstation runs, easier, and the motors also function as dynamic brakes once they become generators driven by the wheels and exert drag on the train. Pneumatic brakes are used for speeds below 30 km/h and as a backup.

Original track specifications were for a minimum of 2.5 km for the radius of curves and the steepest gradient of 2%; maximum for faster trains is 4 km and 1.67%. There are no level crossings, multipurpose inspection trains run between midnight and 6 A.M., and everything is subject to automatic controls, including the Urgent Earthquake Detection and Alarm System that picks up the very first seismic waves reaching the Earth's surface and can halt trains, or at least slow them down, before the main shock arrives (Noguchi and Fujii 2000).

France's state railway corporation (SNCF) began testing its experimental high-speed electric locomotive, Zébulon, in 1974, and the country's first highspeed service, *train à grand vitesse* (TGV), between Paris and Lyon, began in September 1981 (Soulié and Tricoire 2002). After TGV Sud-Est came TGV Atlantique (western branch to Brittany in 1989, southwest line to Bordeaux a year later), TGV Nord (to Calais in 1993, connected to the Channel Tunnel in 1994), the second international link with Belgium (1997), and the first exported TGV to Spain for the Madrid and Seville (AVE) line in 1991. The more powerful TGV Atlantique (8.8 megawatts [MW] compared to 6.45 MW for the Sud-Est) was the first train to travel at up to 300 km/h in regular commercial service, and, in May 1990, the world speed record of 515.3 km/h was set on this line (Figure II.5.6).To accommodate growing demand on the Paris–Lyon line, two trains were coupled together, and then a double-decker TGV was introduced in 1996.

TGV, unlike *shinkansen*, has two locomotives in every train, each weighing 68 t and capable of 4.4 MW (TGVweb 2000). Electric supply and subsequent conversions are as in Japan: the pantograph picks up AC at 25 kV and 50 Hz, the rectifier converts it to 1,500 V DC, and traction inverters convert DC to

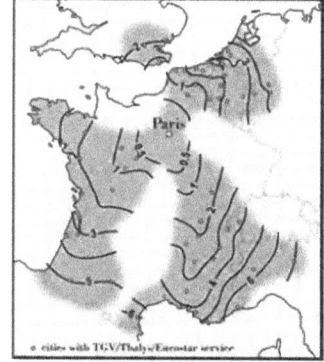

Figure II.5.6 TGV Sud-Est (left)—200 m long, 2.81 m tall, powered by motors capable of 6.45 MW—carries 414 passengers at speeds of up to 300 km/h, and isolines (right) indicating travel time from Paris to cities with high-speed train service.
Photograph courtesy of SNCF. Isolines based on a detailed map from Railfan Europe.

variable frequency AC that is fed to synchronous motors that are also used for dynamic braking at high speeds. During the late 1990s, SNCF—despite the transfer of its huge debts to Réseau Ferré de France, the company that owns the track and signals—continued to operate with large annual losses.

By 2000, the Sanyo Nozomi between Hiroshima and Kokura and TGV Sud-Est between Valence and Avignon had almost identical speeds as the world's two fastest scheduled trains at, respectively, 261.8 and 259.4 km/h (Taylor 2001). Besides the TGV exported to Spain, several European countries gradually acquired their own high-speed trains (European Commission 1995). In contrast, more than a dozen North American high-speed rail projects remained on paper (GAO 1993). The Boston-Washington Acela train was upgraded to go as fast as 240 km/h but its average speed was a mere 116 km/hour, hardly better than the best steam locomotives could manage a century ago. North America's low population density and highly competitive road and air links were at least as important a reason for the continued decline of rail travel as was its infatuation with cars (Thompson 1994). In contrast, railway travel in Japan more than tripled and in France more than doubled during those same 50 years.

Daily passenger density per kilometer of rail route is nearly six times higher in Japan than it is in France, and in the region served by Tokaido shinkansen it is about five times Japan's national mean and not too far below the density experienced by Tokyo's subways (Smith 2001). New trains also demonstrate the limits of wheeled transportation. Indisputable advantage of energy efficiency—per passenger-kilometer, fast trains use 20–25% of the energy needed by airplanes and only about half as much as cars—began to erode with high speeds (Thompson 1986). Higher speeds yield diminishing returns in time saved per unit of distance, centrifugal forces that people can comfortably tolerate limit the radius of curves, permissible height discrepancies between the levels of the rails decline, and energy consumption rises exponentially (Figure II.5.7).

These problems have only partial technical solutions. Tilting trains, whose development was pioneered in Italy beginning in the late 1960s, allow for higher speeds in sharper curves, and massive concrete beds can provide desirable rail-level tolerances—but both costs more. And while aerodynamic design can reduce high-speed drag, noise energy becomes much more difficult to control at more than 300 km/h as it goes up in proportion to the 6th power of the speed. And in Japan, where residential and commercial areas amount to 86% of the 513 km between Tōkyō and Ōsaka, the noise must be

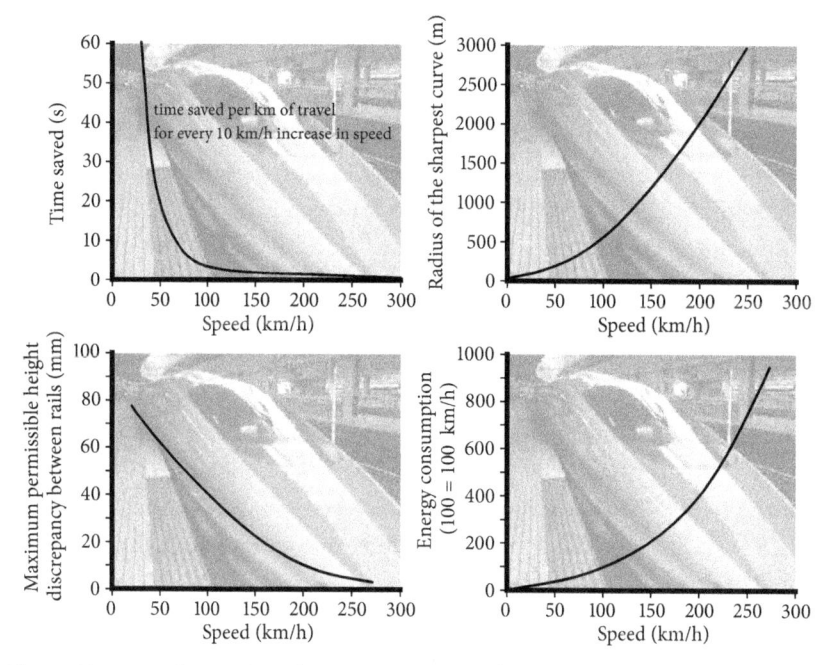

Figure II.5.7 Relationships between train speed and time saved (top left), radius of the sharpest curve (top right), maximum height discrepancy between rails (bottom left), and relative energy consumption (bottom right).
Based on data from Thompson (1986).

limited below 70–75 dB, and hence the *shinkansen* trains must slow down far below their capability.

Commercial Aviation

Flying during the pre-jet era was an uncommon, expensive, and trying experience. The best planes were well appointed—in 1939, Boeing's 314 trans-Pacific Clipper had a stateroom, dressing rooms, and a dining room, and its seats could convert into bunks (Bauer 2000)—but the travelers had to endure inadequately pressurized cabins, propeller noise, vibration induced by reciprocating engines, and frequent turbulence encountered at cruising altitudes that put the aircraft in the midst of tropospheric flows (also the tedium of long hours in the air without any audio-visual diversions). The first airplane to make commercial aviation profitable—the DC-3 by Douglas, originally a sleeper with 14 berths for transcontinental flights of American Airlines—had a maximum cruising speed of 274 km/h, just above the landing speed of modern jetliners (Davies, Thompson, and Veronica 1995).

A few more comparisons highlight the advances achieved by the end of the twentieth century. Maximum cruising speeds on long-distance flights tripled: in 1938, Boeing's 314 Clipper could reach 320 km/h; 20 years later, the Boeing 707 could go 920 km/h. Before World War II, a New York to Los Angeles flight took, with the three necessary stops, 15.5 hours; two decades later, it lasted just 6 hours. During the propeller age the flying time from London to Tokyo was 85 hours; Comet 1 cut it to 36 hours, and nonstop flights of the 1990s took 11 hours. Passenger capacities increased from the maximum of 74 people in the pre–World War II Boeing 314 Clipper to 120–160 for short-hauled flights with such planes as the Boeing 737 or Airbus 320, and to 300–500 in various versions of the Boeing 747. And, of course, jetliners cruising near the upper edge of the troposphere can experience protracted periods devoid of any turbulence.

Gas turbines were just one key component behind the success of the late twentieth-century commercial flying. New materials—aluminum alloys, titanium, plastics, and composite materials—were indispensable to remove the previous structural constraints on engines, fuselages, and wings. Microelectronics brought unprecedented levels of automation to both in-flight monitoring and control and ground operations. But arguably the most important technical innovation that transformed commercial aviation was the rapid development of radar: without it there could be no accurate aircraft traffic control and no all-weather flying.

Standard accounts of the history of radar (abbreviated from "radio detection and ranging," the name chosen by the US Navy) start with British concern about German bomber attacks and the Air Ministry's call for effective defensive measures. A memorandum by Robert Watson-Watt (1892–1973) of the Radio Research Station at Slough held out, based on calculations done by Arnold F. Wilkins, the possibility of using the reflected waves for long-distance aircraft detection (Rowe 1948; Watson-Watt 1957). But detection of vessels with continuous-wave shipborne broadcasts had already been demonstrated by Christian Hülsmeyer in 1904, and rudimentary radar demonstrations had taken place in Germany, the USSR, Italy, and France. In December 1934, Albert H. Taylor and Leo C. Young of the US Naval Research Laboratory used pulsed signals to detect a small airplane flying above the Potomac (Buderi 1996; Brown 1999; Seitz and Einspruch 1998).

The first British demonstration was on February 26, 1935, when continuous waves from the BBC's 10 kW Daventry transmitter were bounced off a Royal Air Force (RAF) bomber and received from a more than 10 km. What set the subsequent British developments apart is that they quickly improved the technique (by shortening the wavelength to avoid interference with commercial radio signals and designing better metal transmitters and wooden receivers) and, by the outset of World War II, the country had a chain of 19 radar stations along its southern and eastern coasts. Moreover, massive ground units were transformed into the first airborne radar, and, late in 1939, the first identification friend or foe (IFF) transponder made it possible to control an aerial battlefield. Radar's first remarkably successful use came during the Battle of Britain, and, most fortuitously, in 1940, two Birmingham University physicists, Harry A. Boot and John T. Randall, built a cavity magnetron that made it possible to deploy high-power microwaves, starting with a 9-centimeter (cm) device in 1941.

If modern air warfare is unimaginable without radar, the case is even stronger for commercial aviation: How else could we coordinate and control millions of takeoffs that were made every year during the late 1990s and keep track of more than 16,000 airplanes that spent nearly 100,000 h airborne every day during the late 1990s (Figure II.5.8)? Modern air traffic control began in the United States in 1936 (Gilbert 1973). After World War II controls changed radically, with constant and multilayered radar

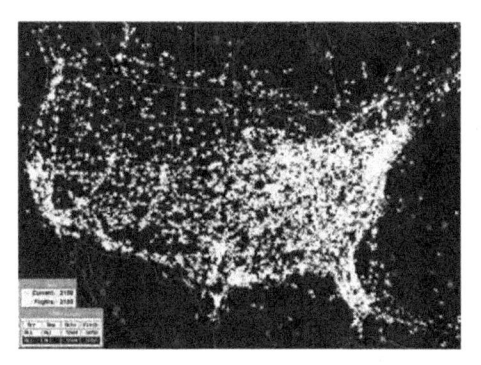

Figure II.5.8 This image of the master screen of the US Air Traffic Control System Command Center from the late 1990s showed more than 5,000 active flights above the United States and in the nearby airspace. The Northeast, Midwest, Florida, Texas, and California had the highest density of flights.
Available from the US Federal Aviation Administration.

detection—operating near the bottom of the ultrahigh-frequency range, between 1 m and 10 cm (300 megahertz [MHz] to 3 gigahertz [GHz])—and they further improved during the 1960s with the introduction of computers.

Primary radars receive echoes from all aircraft flying within the control area, while secondary radars interrogate aircraft transponders to identify a particular flight, its altitude, and call sign. Pilots received warnings about possible conflicting flight paths, insufficient altitude, or intrusion into a restricted air space; queuing and scheduling programs sequence the flights with minimum delays in the regions of high air traffic frequency and hand them over to a neighboring flight control center. Improvements in radar join more reliable jet engines as one of the key reasons behind the extraordinary safety of modern aviation. As noted in Chapter II.2, worldwide accidents per million departures declined by more than 96% between 1960 and 2000, and analysis shows that human error was the primary factor in 67% of all cases (Boeing 2003b). Airplane problems caused only 12% of all accidents, and the latest generation of jets has a safety record about 18 times better than the first one.

Flying with certified air carriers remained the safest mode of transportation in terms of the risk of fatality per person per hour of exposure: during the late 1990s, the risk was about 10^{-7}, an order of magnitude lower than the risk of dying of disease and about 20 times safer than driving a car. While the risk of driving depends clearly on the distance traveled, the risk of flying is primarily affected by the number of takeoffs (including climbs to cruising altitude) and landings (including descent), the two operations that put the greatest demand on engines and stress on wings and fuselage, but US statistics show that flying is the safest mode of transportation for any distance for which using a scheduled airline is an option (Sivak and Flannagan 2003). And when accidents happen, flight recorders (two black boxes for cockpit voice and flight data), first introduced during the late 1950s, make it much easier to find their cause.

Size and organization of airports had to change to cope with the growing traffic (Horonjeff and McKelvey 1983). Many major airports rank among the largest civil engineering projects of the post-World War II era. Notable jet-age expansions took place at what were the world's two busiest airports of the late 1990s, Chicago's O'Hare and Atlanta's Hartsfield, the largest regional hubs in the world's largest commercial aviation market. During the late 1990s, each handled annually more than 70 million passengers. On its busiest day ever, after Thanksgiving in 1998, O'Hare handled 252,000 passengers as nearly 3,000 aircraft took off and landed.

Aprons, approaches, and runways of modern airports must be built on excellent foundations to avoid any shifting or waterlogging of the ground. Reinforced concrete used for these surfaces is up to 1.5 m thick, able to support planes whose mass during the late 1990s ranged mostly between 150 t (Boeing 737) and almost 400 t (fully loaded Boeing 747), and on the runways it must take the impact of planes landing at very short intervals. Large airports also need multiple runways, and, to accommodate the largest wide-body jets, their length must be at least 3,000 m: Los Angeles airport has two parallel runways of 3,630 m, and, in Japan, Narita's principal runway is 4,000 m.

The organization and management of the airline industry had to undergo many adjustments as the total number of passengers soared. US airlines led this expansion since the beginnings of the industry during the 1920s (Solberg 1979; Heppenheimer 1995). They carried 3 million people in 1940, more than 50 million in 1955, and 665 million (i.e., roughly 2.4 flights a year for every inhabitant) in 2000. By the century's end, when the country had 14 of the world's 20 busiest airports, Dallas–Fort Worth (America's fourth busiest) handled more traffic than did Frankfurt or Roissy-Charles de Gaulle, continental Europe's two busiest airports. Increased traffic also required international rules and institutional commitments (ranging from strict engine licensing to cooperative aircraft control) to achieve high levels of reliability and safety. And bookings, many including multiple connections, would be a logistic challenge without computerized reservation and baggage-tagging systems.

Competition resulted in the demise of some iconic companies—notably PanAm, the pioneer of intercontinental flight and the jumbo jet, in 1991—and in a changing balance between the two remaining major aircraft producers. Until the 1970s, the large jetliner industry was dominated by three US companies that were established during the early years of aircraft production (Pattillo 1998). Allan and Malcolm Lougheed opened their San Francisco seaplane factory already in 1912, and Lockheed Company (spelled the way Allan pronounced his name) began in 1926. William Boeing (1881–1956) set up his Seattle company in 1916, and Donald Douglas (1892–1981) began his four years later in California (Figure II.5.9). Lockheed's last jetliner

Figure II.5.9 William Boeing (left) and Donald Douglas (right), two leading pioneers of innovative aircraft design and construction.
From Aiolos Production.

(between 1968 and 1984) was the Tristar; afterward, the company (since 1995 Lockheed Martin) concentrated on military and aerospace designs. McDonnell Douglas—maker of the DC-3, the most durable propeller aircraft in history—merged with Boeing in 1997.

During the 1990s, Boeing itself began to lose its global market share to Airbus. This European consortium flew its first plane, the wide-bodied twinjet A300 designed for 226 people, in 1974. Afterward the company introduced a range of planes designed to compete with Boeing's offers. In 2000, Airbus sold for the first time more planes than Boeing (300 vs. 285), and, in December 2000, it announced that it would proceed with building a double-decker super-jumbo A380 to carry 555 passengers between large international hubs (Airbus 2004). Competition among airlines, particularly after their deregulation (in the United States starting in 1978, in the European Union since 1997), increased the number of flights and destinations and lowered the average cost: by 2000, it was down by more than two-thirds in inflation-adjusted terms compared to the late 1970s. Commercial carriers posted annual net operating loss only four times during the third quarter of the twentieth century, but between 1975 and 1999 they had, despite direct and indirect subsidies by many governments, 11 loss years.

Moving Freight

Many technical advances also made it possible to move unprecedented volumes of goods over longer distances, and all the forms of freight movement gained from computerization that improved the speed of cargo handling, optimized scheduling and routing, and shipment tracking. The earliest comparative US statistics are available for 1939 when (leaving oil pipelines aside) railways accounted for slightly more than 70% of total freight movements measured by ton-kilometers; waterborne transport was second with almost 20%, and trucking came third with 10% (USBC 1975). These shares began to shift immediately after World War II, and, by the early 1970s, the railway share fell just below 50%.

And there it stayed for the remainder of the century: in 1999, nearly 49% of all US freight ton-kilometers were on rails, 35% on roads, and a mere 0.5% went by domestic airlines (USCB 2004). Railroads retained their freight traffic share because they aggressively participated in the most important advance in the twentieth-century transportation: the adoption of intermodal containers that can be carried by various means from the place of a product's origins all the way to a retail store, a storage depot, or a construction site. Many simple eurekas had profound economic effects, but it is hard to think about a simpler idea with a greater global impact.

Intermodal containers are just rectangular steel prisms with reinforced edges and corners. Their most common dimensions are 20 or 30 ft (6 or 9 m) long, 8 ft (2.5 m) wide, and 8.5–9.5 ft (2.5–3 m) high. Only the width remains fixed (to fit on trucks), the weight ranges between 1.8 and 4.2 t, and the payload is between 15 and 30 t (Figure II.5.10). Small containers of various sizes had already been used by some traditional break-bulk shippers at the beginning of the twentieth century and during World War II, but containerization on industrial scale began only during the late 1950s with two innovators on America's opposite coasts. Malcolm Purcell McLean (1914–2001), a North Carolina trucker, must be credited with the most radical transformation of global freight delivery. He had to wait nearly two decades between his eureka moment and his first steps toward commercializing his idea.

As he later recalled, the idea came naturally one day in 1937. because of frustration with the traditional freight-handling practices.

> I had to wait most of the day to deliver the bales, sitting there in my truck, watching stevedores load other cargo. It struck me that I was looking at a

Figure II.5.10 A Maersk Sealand container being unloaded by a large gantry crane that can lift one 45 t unit or two 30 t units.
Image available from the A. P. Moller-Maersk Group.

lot of wasted time and money.... The thought occurred to me, as I waited around that day, that it would be easier to lift my trailer up and, without any of its contents being touched, put it on the ship. (quoted in The Economist 1991, p. 91)

Only in 1955 was he ready to act: he sold his share in the family's trucking business and bought seven ships of the Pan-Atlantic Steamship Corporation that operated small tankers between the US Northeast and the Gulf of Mexico (Kendall and Buckley 2001). The next step was a new truck trailer that was reinforced vertically for stacking and horizontally for lifting and crane loading. Then a steel skeleton deck built above the main tanker deck was equipped with sockets to accept stubby container feet for fastening. On April 26, 1956, the first adapted World War II-era T-2 tanker, *Ideal X*, set out from Port Newark, New Jersey, to Houston carrying 58 large (11-m [35-ft]) containers (altogether about 1,000 t) on its strengthened deck.

With the concept proven, McLean converted other ships, and, on October 4, 1957, *Gateway City*, a C-2 freighter fitted with cellular compartments and able to transport up to 226 large containers stacked within a steel framework erected in the hold became the first true container ship to operate between New York and Texas. Twist-locks were used to fasten the containers into huge blocks, and spreader machines lifted the containers by using hooks inserted into corner holes. In 1960, McLean renamed the company Sea-Land. In 1956, just as McLean began his East Coast operations, the Matson

Navigation Company, which had served Hawaii from its California ports since 1882, opted for containerization (Matson 2024). In August 1958, the *Hawaiian Merchant*, a converted freighter, shipped 20 containers on its deck from Los Angeles, and, in April 1960. The *Hawaiian Citizen* became the company's first all-container vessel carrying 436 24-ft (7-m) units.

In 1962, container operations began at the port of Oakland and at the Port Elizabeth Marine Terminal in Newark. Shipments to Europe started in April 1966, when the SS *Fairland* carried 226 containers to Rotterdam. Containerization received a major boost because of the Vietnam War: by 1968, the US Department of Defense used more than 100,000 containers to speed up the delivery of military supplies. In that year McLean also launched the first fully containerized service to Japan. A year later, when he sold Sea-Land to R. J. Reynolds, the company had 27,000 containers. By 1980, there were more fully cellular ships than semi-containerized vessels (Pearson and Fossey 1983). In 1986, Sea-Land was sold to CSX, and, in 1999, it was bought by the Danish Maersk to form Maersk Sealand, the world's largest container shipper.

Railroad companies first used flat cars for intermodal, piggyback shipments of truck trailers, the rectangular boxes with permanently attached wheels. Before the end of the 1960s, this became a common practice with increasingly longer flatcars (up to 25 m). But just a decade later the railroad companies, which invested heavily into trailer on flat car (TOFC) shipping, had to switch to container on flat car (COFC). US statistics show the rapid switch between the two modes: in 1985, nearly 80% of all intermodal moves on railways involved trailers; by 2000, 75% of all carried units were containers (IANA 2004).

Southern Pacific designed the first double-stack car in 1977, APL Stacktrain pioneered the new practice with large 48-ft (14.6 m) containers in 1986, and that configuration dominates modern railroad freight transport. Stacktrains must minimize longitudinal and lateral movements through special connections and reduce vertical motion by setting wheel assemblies closer than in other trains. Initially Sealand used 35-ft (10.5-m) containers, Matson preferred a smaller (24 ft [7.2 m]) size, and subsequently different international shippers introduced sizes ranging from 20 to 48 ft (6 to 14.5 m). But there are also smaller containers (10 ft [3 m]), primarily used in Europe), and, in 1989, Stacktrain introduced 53-ft (16 m) units. During the 1990s, marine transport was dominated by two common ISO sizes, 20 and 40 ft (6 and 12 m), which accounted, respectively, for nearly 60% and 40% of all shipborne units.

In 1969, faced with this lack of standardization when compiling international vessel statistics, the British journalist Richard F. Gibney chose the 20-foot equivalent unit (TEU) as the common denominator, and the acronym became a standard term in the industry. The first container ships of the late 1950s carried fewer than 300 TEU, a decade later up to 1,300 TEU, and, during the 1970s, as many as 3,000 TEU. The largest ships able to pass through the Panama Canal carried 4,500 TEU—and in 1996 *Regina Maersk* was the first ship with 6,000 TEU. By the late 1990s, Maersk Sealand had S-class ships that carried 6,600 TEU. All other major shipping lines (Hanjin, MSC, P&ONL) had vessels approaching or exceeding the 6,000 TEU capable of speeds up to about 45 km/h. These ships, double-hulled and double-bottomed, typically had eight layers of containers stacked below the deck line and up to seven above it (see the top image of this chapter's frontispiece).

By the late 1990s, nearly 60% of the total value of international trade moved in containers. Leaving aside commodities that are obviously shipped better in bulk—pumped on board (crude oil, refined products, many liquid chemicals) or dumped into a hold (metallic ores, coal, grains)—the dominance of containerized transport was complete. Besides manufactures, containers also carry perishable food in refrigerated units. Worldwide there were about 6.7 million TEU in use by the year 2000, and the total throughput of about 200 million units meant that a unit was handled nearly 30 times a year, the total that includes the repositioning of empty containers and trans-shipments using smaller feeder vessels. Empty containers handled at ports reached 41 million TEU by 1999, and, because of the large trade imbalance between North America and East Asia, the Asian ports had to handle increasing shares of empties (ESCAP 2000).

On the high seas these containers were carried by more than 7,000 vessels, but the traffic was dominated by roughly 3,000 fully cellular ships that plied regularly the busiest fixed routes between East Asia and North America (two-way shipments of almost 13 million TEU in 2000) and Asia and Europe with almost 7 million TEU (ESCAP 2000). Hong Kong port retained its long-held global primacy as it handled slightly more than 18 million TEU, just ahead of Singapore, while Rotterdam, Los Angeles, Long Beach, and Hamburg were the busiest receiver ports. With larger and faster ships, the speed of offloading and loading had to increase.

Early gantry cranes could not handle more than 20 containers per hour, and, by the 1980s, the rate was up to 30 (Gubbins 1986). Two cranes could thus unload a 3,000-TEU ship in two days. By the late 1990s, a single yard crane spreader could lift two 20-ft containers at one time (see Figure II.5.10), and cranes could clear 75–100 TEUs per hour. Designs were under way to raise this productivity to 200 moves so that even the largest container vessels could be turned around in a single day. Every container is identified by the Bureau des Containers code, a combination of four letters (for the operator) and seven digits (inscribed on both outer and inner walls) that is used for computerized routing and tracking.

Benefits of containerization are obvious. The technique created the first truly global, unbroken chain of freight delivery with sealed containers protected against easy pilfering and shipping damage and hence eligible for reduced insurance rates. The time needed to load and unload the containers was reduced close to the physical minimum dictated by the power of the dedicated machines and safety of operations: traditional break-bulk loading of a 10,000-t ship would take 7–10 days, the same mass of containerized cargo can be handled in a day, and the entire shipping sequence from producer to buyer had been cut by more than 95%.

Other important innovations that made mass-scale long-distance trade faster and cheaper included liquefied natural gas (LNG) vessels that carry methane at −162°C and roll-on/roll-off (RO/RO) ships to transport passenger cars and trucks. LNG tankers were used for the first time in 1958, to ship Algerian gas to the United Kingdom (*Methane Pioneer*) and France and, in 1960 (*Bridgestone Maru*), to take Indonesian gas to Japan (Corkhill 1975). By the late 1990s, more than 100 LNG vessels, most with capacities of more than 100,000 m^3, carried about a quarter of global natural gas exports, with shipments from Indonesia, Algeria, Brunei, and Malaysia to Japan, South Korea, and the United States dominating the trade. Before the port of Dover introduced its first two drive-on berths in 1953, it handled only 10,000 cars loaded laboriously by cranes; by the late 1990s, the total number of vehicles driving on RO/RO ferries was approaching 5 million. By that time the worldwide total of RO/RO ships was close to 5,000, with Europe having their largest concentration.

And as airplanes increased their payloads, they began carrying more than just mail, their original cargo assignment. Air transport was developed for the first time on a mass scale during World War II when more than 9,500 C-47 Skytrains, a version of the DC-3, were produced for the Allied forces. Worldwide air cargo was just 200 million ton-kilometers (t-km) in 1950, but then it grew faster (more than 500-fold rise of freight t-km between 1950 and 1999) than passenger traffic: by the end of the twentieth century, it reached 108 billion t-km, with 85% logged on international routes (ICAO 2001). Every passenger jet also carries cargo (manufactures, flowers, or live animals): in the United States, slightly more than 20% of all air freight was moved on passenger planes, earning the airlines 10–15% of total revenues. And most large jetliners have been available in cargo versions that carry 50–100 t and that operate on regular daily schedules (the fleets of FedEx or UPS planes) or are leased or chartered as needed.

During the late 1990s, cargo Boeing 747s accounted for half of the world's international air freight: the 747-400 version had a cargo capacity of 105 t loaded on pallets up to 6 m long. But it was not the largest cargo carrier: that primacy belonged to the originally Soviet-designed (in service since 1986) and then Ukrainian-produced Antonov AN-124 (Ruslan), whose capacity far surpassed that of the previous record holder, the US military transport C-5 Galaxy that can carry almost 120 t. The AN-124 can load up to 150 t, and their worldwide heavy-lift services provided much-needed foreign earnings for the Ukraine economy during the 1990s (Figure II.5.11).

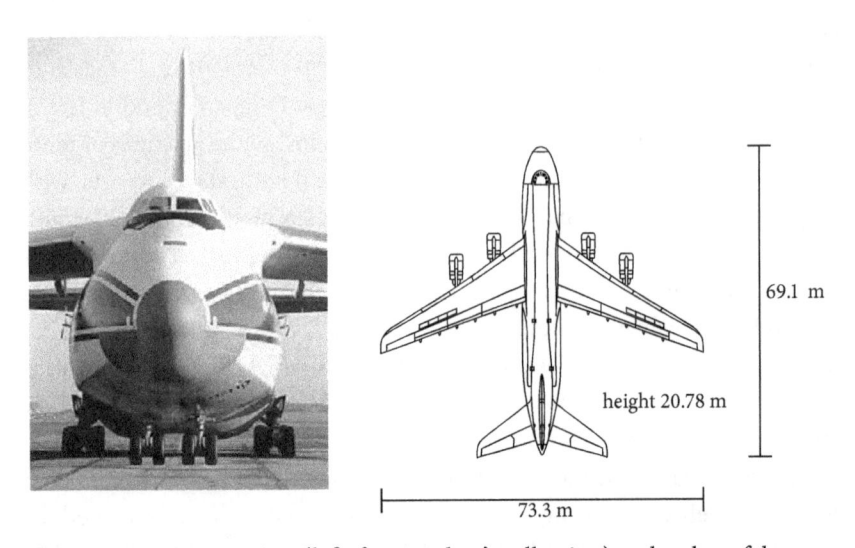

Figure II.5.11 A nose view (left; from author's collection) and a plan of the Antonov AN–124 (right; from Aerospace.org) convey the enormous size of the world's largest commercial cargo airplane.

Computing Revolutions

Electronic computing, a new way of information handling—the term understood here in the broadest sense of ordering, processing, and transferring any data, text, or images—was invented during the 1940s. During the next two decades these new powers were used mostly for advancing true computing, with tasks ranging from complex scientific and engineering calculations and statistical evaluations to basic business inventory and payroll operations. Afterward, increasingly more powerful microprocessors made it possible to computerize almost any conceivable scientific, industrial, or managerial task—and many creative and leisure processes as well. By the century's end, there were programs with 10^7 lines of code; microprocessors inside supercomputers could do virtual tests of thermonuclear weapons, inside fabrication units they could run complex machining processes (see Chapter II.4), and inside laptops they could be used to write books or play fast games.

There was no such thing as the first computer, a fully electronic machine capable of rapid binary-based operations that use stored instructions (Rojas and Hashagen 2000). Instead, between the late 1930s and the mid-1940s, the most advanced electromechanical devices became the first electronically controlled externally programmed computing machines. Their history is a classic case of independent inventions and incremental improvements. Until the early 1950s, British scientists and engineers were ahead in several important respects of what soon became the dominant US computer establishment. There were many firsts and no Edisonian figure to pull the known strands together, to invent some missing parts, and launch a new fully operational and commercially viable system (Moreau 1984; Campbell-Kelly and Aspray 2004). What is most remarkable about the history of electronic computing is its rapid sequence of transformations that magnified the performance of dominant designs to an extent that has no match in any other human endeavor.

In 1951, UNIVAC, America's first commercial mainframe, cost $930,000, could execute about 60,000 instructions per second, and needed 160 kW of electricity; in 2000, Intel's Pentium III processor, used in personal computers and servers, cost $990 (in large lots), could handle 1 billion instructions per second, and, in maximum performance mode, drew 20.5 W (Figure II.5.12). Computing revolutions of the second half of the twentieth century thus raised the calculating speed (even when leaving supercomputers

Figure II.5.12 The room-size UNIVAC 1, the first US commercial computer, offered by Remington Rand in 1951 (left; from the National Energy Research Scientific Computing Center), weighed more than 13 t, but its computing power was only 0.006% of the Pentium III microprocessor (with 9.5 million transistors) released by Intel in 1999 (right; from Intel), designed to enhance Internet browsing and download high-quality video.

aside) by four orders of magnitude, while the cost of computing and energy requirements fell by eight orders of magnitude. If a passenger car of the early 1950s would have undergone similar transformations, the late 1990s models would cruise at speeds about 40 times higher than the velocity needed to escape Earth's gravity, 1 L of gasoline would be enough to go 20,000 times around the equator, and even the smallest parking fine would be orders of magnitude larger than the car's utterly negligible cost.

The First Electronic Computers

Theoretical foundations came from two mathematicians, Alan Turing (1912–1954) at Cambridge and Claude Shannon (1916–2001) at the Massachusetts Institute of Technology (MIT). Shannon, in his master's thesis in 1937, pioneered the use of Boolean algebra for a working model of digital circuits, and a decade later he presented the fundamental mathematical theory of communication (Shannon 1948). Turing (1936) introduced the concept of a universal computing machine that could be programmed to do any calculation a human is capable of as long as it receives clear instructions. The most advanced mechanical precursor of electronic computers was Vannevar

Bush's 1932 differential analyzer, an analog mechanical device that did calculus by rotating gears and shafts.

The first electromechanical machines, whose relays represented digits in the binary system with on and off positions, were built by Georg Stibitz (1904–1995) at Bell Laboratories in 1936 and by Konrad Zuse (1910–1995) in Germany between 1936 and 1941, while Howard Aiken's (1900–1973) massive Harvard Mark I (about 5 t, 760,000 parts, 17,000 vacuum tubes) used decimal numbers (Cohen 1999). The first small electronic computer that combined digital arithmetic with vacuum tubes was conceived by John Atanasoff (1903–1995) of Iowa State College late in 1937 (Burks and Burks 1988). Atanasoff's great pioneering departures were to design a parallel machine (capable of up to 30 simultaneous operations) that was fully electronic, used memory capacitors to store binary numbers, and separated memory from processing (Gustafson 2000). Electronic switching and the use of capacitors (now a part of microprocessors) remain among the critical ingredients of every modern computer, as does the separation of memory from processing.

The first prototype was built, with the assistance of Atanasoff's graduate student, Clifford Berry (1918–1963), by October 1939, and a larger machine designed to solve simultaneous linear equations was completed by 1942 (Figure II.5.13). Before war commitments interrupted further development

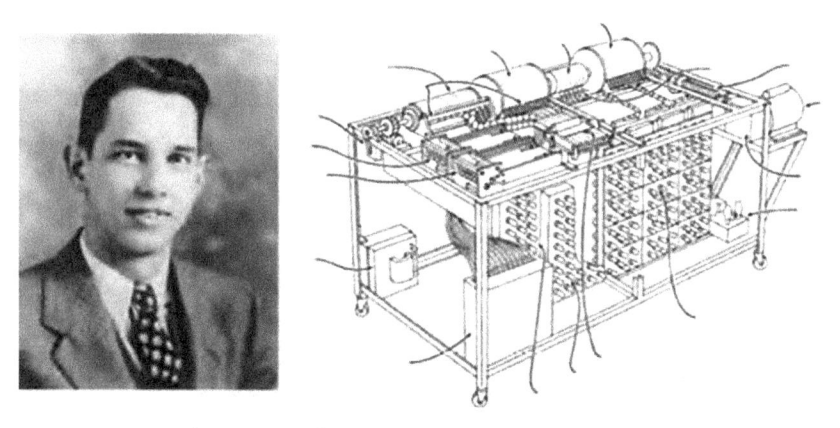

Figure II.5.13 John Atanasoff in 1938, when he and Clifford Berry were building the first prototype of their electronic computer; major components of a larger machine, completed in 1942, are shown in the drawing.
From the Iowa State University Department of Computer Science.

of the machine, Atanasoff showed it to and discussed it in detail with John Mauchly (1907–1980), who visited him in Iowa in June 1941. Between July 1943 and November 1945, Mauchly and John Presper Eckert (1919–1995) built the Electronic Numerical Integrator and Computer (ENIAC) at the University of Pennsylvania. The machine, programmed by manually interconnecting its units with tangles of wires and patented in 1947, was generally recognized as the first American electronic computer. But Mauchly's 1941 Iowa visit played a key role in the decision of the US District Court in Minneapolis in October 1973 to assign, as a part of the lawsuit between Sperry Rand (owners of the ENIAC patent) and Honeywell, the invention priority to Atanasoff, who never patented his design (Burks 2002).

The first operational externally programmable vacuum-tube logic calculator was the British Colossus, built in 1943 by Tommy Flowers (1905–1998) at the Post Office Research Laboratories in London. Bletchley Research Establishment eventually deployed 10 of them to decipher the German code generated by the Lorenz SZ42 cipher machines (Hinsley and Stripp 1993; Sale 2000). The existence of Colossus was kept secret until 1974, opening the way for ENIAC's priority claim. In 1945, John von Neumann (1903–1957) outlined the principles of a stored program, and its first installation, at the University of Manchester between 1948 and 1951, was based on F. C. Williams's observation of the stored patterns of electrostatic charge inside a cathode-ray tube (Lavington 2001). The first program run by the university's Small Scale Experimental Machine (SSEM) on June 21, 1948, was written by Tom Kilburn (1921–2001) to find the highest factor of x.

The SSEM served as the prototype of the world's first commercially available electronic computer, the Ferranti Mark I, delivered in February 1951, four months ahead of Eckert and Mauchly's ENIAC-based UNIVAC (Universal Automatic Computer), the first device with magnetic tape storage rather than with punched cards: its first owner was the US Bureau of Census (see Figure II.5.12). And the British also had the first business computer: in November 1951, LEO (Lyons Electronic Office), a copy of the EDSAC (the first stored-program computer in regular service completed at Cambridge University in 1949) was delivered to J. Lyons & Company operating a chain of British teashops. In 1954, J. Lyons began building and selling its more advanced versions, with deliveries ending in 1966, when the business was bought by English Electric (Ferry 2003).

Between 1951 and 1958, came the first generation of commercial computers, still very expensive, massive, and of limited capacity. Between 1959 and 1964, they were followed by new models equipped with transistors, but their high cost (in 1962, the IBM 7090 sold for US$15 million) began to decline only with the introduction of the third generation of machines based on integrated circuits. This trend was carried to its logical conclusions with the emergence of microprocessors that combined all functions in a single microchip. Development of the fourth generation of computers proceeded along two paths: the first one producing smaller but more powerful mainframes, the other leading to personal computers. Yet another fundamental shift came during the 1990s, when much of the mass-produced hardware became portable (laptops, hand-held units).

A software-guided classification, focusing on the way computers are instructed to perform their operations, is much simpler, falling into just three longer categories. Until 1957, the only way to instruct computers was in tediously produced numerical machine code. Finally, in 1957, IBM released FORTRAN (FORmula TRANslation), developed by a group of nine experts led by John Backus (Backus 1979). FORTRAN was the first programming language that made it possible to instruct machines in terms akin to simple mathematical formulas that were translated by a compiler into machine code. ALGOL and business-oriented COBOL followed shortly afterward, and, in 1964, came the easier-to-use BASIC (Beginners All-Purpose Symbolic Instruction Code) designed by John Kemeny (1926–1992) and Thomas Kurtz at Dartmouth University.

Development of the most popular programming language of the last three decades of the twentieth century began at Bell Labs in 1971, when Dennis Ritchie began to write C, based on Ken Thompson's B (Kernighan and Ritchie 1978). Ritchie and Thompson used the language for the UNIX operating system that eventually spread to operate millions of laptops as well as some of the most complex mainframe systems, including airline reservations. Until the late 1970s, the only way to interact with a computer was to learn one of the programming languages. The great divide was the introduction of personal desktop computers, above all, the Apple II in 1977 (the first machine with color graphics) and IBM's PC in 1981 (and its many clones)—which could be used by anybody able to peck at a keyboard and master a modicum of instructions.

Transistors and Integrated Circuits

The miniaturization trend began with the invention of transistors (devices transferring current across a resistor) in 1947 and 1948, but, until the late 1950s, electronic switching in all computers depended on vacuum tubes.

These were either advanced versions of two pre-World War I devices, Fleming's diode and De Forest's triode (see Chapter I.5) or their subsequent elaborations (tetrodes, pentodes). These devices control the flow of electrons in a vacuum and can amplify the current and switch it on and off rapidly, but they were inherently bulky, failure-prone, and inefficient. When deployed in large numbers, they added up to sizable masses of hot glass: a single tube weighed mostly between 50 and 100 g and so even in a smaller computer with just 5,000 tubes they alone would have added up to close half a ton, and overheating shortened their durability, which was just 50–100 hours for some models.

ENIAC parameters illustrate these challenges (Kempf 1961; Van der Spiegel et al. 2000). The machine had 17,648 vacuum tubes of 16 different kinds, as well as 70,000 resistors, 10,000 capacitors, 6,000 manual switches, and 1,500 relays; its 30 separate units occupied about 80 m^3, covered an area of 167 m^2 (roughly the size of two badminton courts), and, including its power supply and forced air cooling, weighed about 30 t—but it ran at a mere 100 kHz and its random-access memory (RAM) was only about 1 kilobytes (kB). About 90% of all interruptions were due to tube failures. And although the computer needed 174 kW to operate (a bit more than the power of a 1965 Fastback Mustang), air conditioning installed to take care of the large heat dissipation load proved inadequate, and high temperatures increased the frequency of failures.

Half a century after the machine's completion, a team at the University of Pennsylvania reconstructed ENIAC on a microchip, on a 7.4 × 5.3 mm sliver of silicon that contained 174,569 transistors (Van der Spiegel et al. 2000). The order of magnitude difference between the number of vacuum tubes and transistors was because transistors also replaced all resistors, capacitors, and switches. Differences between ENIAC and ENIAC-on-a-Chip are staggering: the original machine was more than 5 million times heavier, required 78 different DC voltage levels (compared to just one 5-V input), and consumed about 40,000 times more electricity, yet its speed was at best 0.002% that of the reconstructed processor (100 kHz vs. at least 50 MHz): it needed 24 hours of operations to do what the chip does in less than 3 minutes. This astonishing advance had its origins in the discovery of semiconductivity in germanium and silicon that are insulators when extremely pure but conductors when doped with minuscule quantities of other atoms (see Chapter II.3).

The first rectifier using metallurgical-grade crystalline silicon for reception had already been patented (US Patent 836,531) in 1906, by Greenleaf W. Pickard (1877–1956). That was the silicon point-contact diode—commonly known as "cat's whisker" or the crystal detector—that was used to receive radio signals before it was displaced by Forest's triode. By the 1930s, silicon–tungsten whisker diodes were already being used as crystal detectors in the microwave range (Seitz and Einspruch 1998), and, after he became the head of Bell Laboratories in 1936, Mervin Kelly (1894–1971), impressed by the simplicity, reliability, and low cost of copper-oxide rectifiers used to convert low-frequency AC to DC, initiated a research program to develop a similar semiconductor triode. The first success came only a decade later when Walter Brattain (1902–1987) and John Bardeen (1908–1991) tried to understand the reasons for failure of the experiments done by their group leader, William Shockley (1910–1989), as he tried to make what later became known as a *field-effect transistor* (Riordan and Hoddeson 1997).

They found that when injecting a current to flow between two close point-contact electrodes, they could amplify power and, by December 16, 1947, both power and voltage, and do so at audio frequencies. Their bipolar point-contact transistor (Figure II.5.14) was demonstrated to the management on December 23, 1947, and publicly unveiled at a press conference on January 27, 1947. The discovery led to a split between the two inventors and their group leader, who tried to take credit for their invention. But eventually Brattain and Bardeen's names were the only ones on the patent for a "three-electrode circuit element utilizing semiconductive materials" (Bardeen and Brattain 1950). Shockley intensified his efforts to come up with a more practical solution. The path to it opened once John Shive discovered that, contrary to Bardeen's conclusion, electron flow is not confined to the surface and can travel through the body of a material.

By January 23, 1948, Shockley proposed a major modification of Brattain–Bardeen design. Rather than using point contacts to inject positive charges, he used larger junctions and higher currents. Shockley's three-layer p-n-p design—known as the *bipolar junction transistor* (with the two outer layers doped to supply electrons and the middle layer to create "holes")—thus functions much like a triode with the n-layer analogous to a vacuum tube grid (Shockley 1964). A small current in the center (base) region controls a larger current between the emitter and collector and amplifies that flow (Figure II.5.15). The three physicists were awarded a joint Nobel Prize in

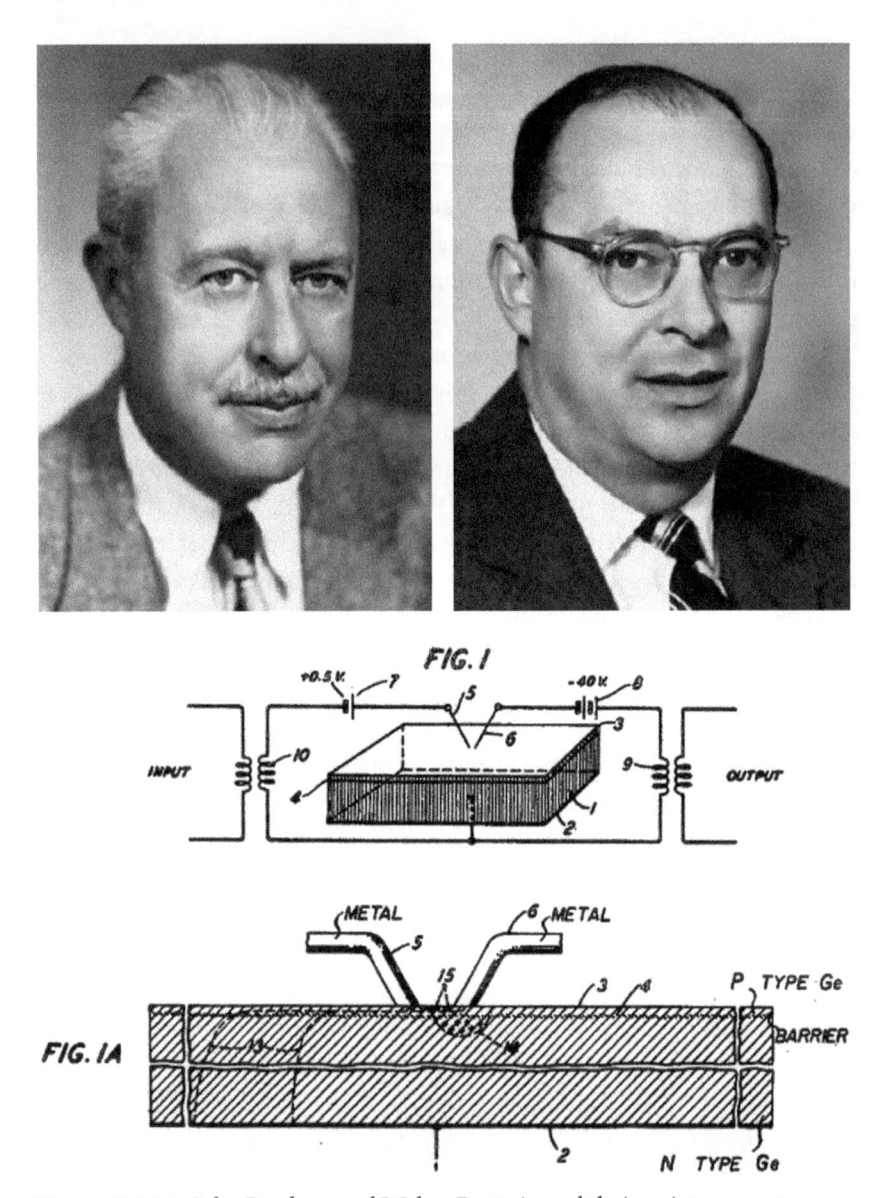

Figure II.5.14 John Bardeen and Walter Brattain and their point-contact transistor (US Patent 2,524,035).

Top image, © The Nobel Foundation; bottom image, from Bardeen and Brattain 1950).

1956, and soon afterward, with the advent of integrated circuits, the importance of their inventions became universally obvious: progressively smaller metal-oxide-silicon field-effect transistors (MOSFETs) carried microelectronic circuitry to astonishing densities and amazingly low costs.

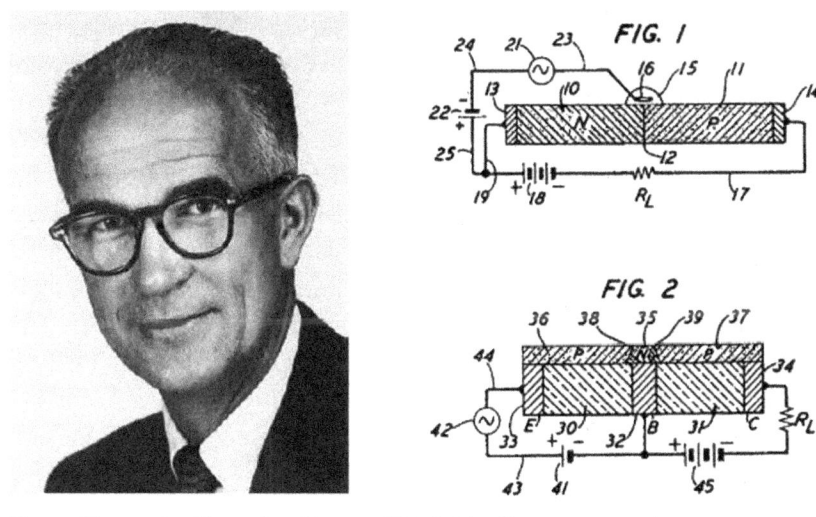

Figure II.5.15 William Shockley and his field-effect transistor (US Patent 2,569,347).
Shockley image, © The Nobel Foundation.

The advantages of transistors were obvious even with the earliest designs: the first prototype occupied only 1/200th of the space of an equally effective vacuum tube while its solid state assured higher reliability and lower energy use without any warmup time. Bell Labs began licensing transistor production in 1951 (for a $25,000 fee): hearing aids were among the first transistorized devices, and the first mass-produced consumer item was a transistor radio in 1954. The first use of transistors in computers came only in 1956, and, in 1959, IBM's Stretch and Sperry Rand's LARC were the first large, transistorized machines. Wider use of transistors came only when improved silicon fabrication methods (see Chapter II.4) replaced germanium with a cheaper material.

The conversion was further accelerated by another important discovery made at Bell Labs. In 1955, Carl Frosch and Link Derick found that heating a silicon wafer to about 1,200°C in the presence of water or oxygen creates a thin SiO_2 film on its surface. Selective etching of this film can be used to make p-n junctions with doped silicon. In an all-diffused silicon transistor, impurities permeate the entire wafer while the active parts are protected by the oxide layer. But even this switch would not have made semiconductors as inexpensive as the next radical innovation that led to their miniaturization and mass production. This step was imperative to produce more complex circuits. But the traditional fabrication of miniaturized circuit components (resistors, capacitors, transistors) required a variety of substrates, materials, heat treatments, and protection measures, and some of these steps were incompatible.

Making the components separately can cause damage to sensitive parts during the assembly: all of them must have individual terminations, and hand-soldered connections between them must be faultless or the circuit fails; moreover, in very complex computing circuits where speed matters, long connections would necessarily slow down the signals. A complex, highly miniaturized circuit would be thus extremely difficult to build. The ingenious solution, the monolithic idea, was first proposed in July 1958, by Jack S. Kilby, who came to Texas Instruments (TI) from a Milwaukee-based electronics company in May of that year. Kilby proposed to use only one material for all circuit elements and a limited number of compatible process steps to produce

> a novel miniaturized electronic circuit fabricated from a body of semiconductor material containing a diffused p-n junction wherein all components of the electronic circuit are completely integrated into the body of semiconductor material. (Kilby 1964: 1)

Transistors and diodes had already been made of semiconducting materials, but to make resistors and capacitors from expensive silicon wafers seemed counterproductive—but it could be done. Kilby detailed two such possibilities in his patent application but noted that "there is no limit upon the complexity or configuration of circuits which can be made in this manner" (Kilby 1964: 7). Simple integrated circuits demonstrating the principle were completed early in 1959. By the time the company's lawyers were rushing to patent the invention, the only model they could depict was Kilby's simple demonstration chip with gold wires arching above the chip's surface (Figure II.5.16). Kilby knew that such a design was antithetical to the idea of an integrated circuit but because he did not have a better solution ready, he only added a paragraph to the application (submitted on February 6, 1959) noting that the connections may be provided in other ways and gave an example of gold deposited on silicon oxide. An effective solution was devised, unknown to Kilby, just before he filed his patent application.

In late January 1959, in yet another case of parallel invention, Robert Noyce (1927–1990), director of research at Fairchild Semiconductors (FS), wrote in his lab notebook (cited in Reid 2001: 13) that

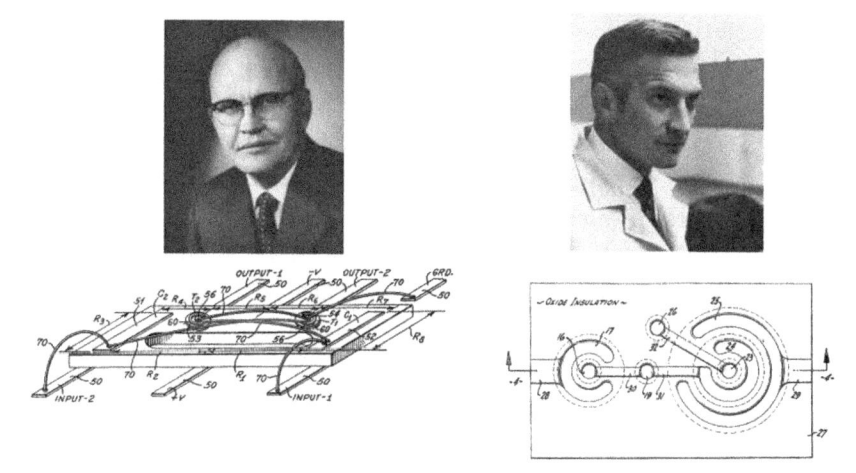

TRANSPORTATION, COMMUNICATION, INFORMATION   569

Figure II.5.16 Inventors of integrated circuit—Jack Kilby (left) and Robert Noyce (right)—and key figures from their patents (US Patent 3,138,743, from Kilby 1964 and US Patent 2,981,877; Noyce 1961). Kilby's figure shows the impractical flying wire connections; Noyce's cross-section clearly outlines neighboring n-p-n junctions on a single silicon wafer.

Kilby image reproduced courtesy of Texas Instruments; Noyce image reproduced courtesy of Intel.

it would be desirable to make multiple devices on a single piece of silicon, in order to be able to make interconnections between devices as part of the manufacturing process, and thus reduce size, weight, etc. as well as cost per active element.

The key idea was identical to Kilby's proposal, but Noyce's design of a planar transistor also solved the problem of miniaturized connections by taking advantage of the SiO_2 layer that Jean Horni (1924–1997) at FS proposed, in 1958, to place on top of the n-p-n sandwich to protect it from contamination.

Noyce's patent application of July 30, 1959, "Semiconductor device-and-lead structure," depicted a design that looked unmistakably like an integrated circuit (Figure II.5.16) and specified

> dished junctions extending to the surface of a body of extrinsic semiconductor, an insulating surface layer consisting essentially of oxide of the same semiconductor extending across the junctions, and leads in the form of vacuum-deposited or otherwise formed metal strips extending over and adherent to the insulating oxide layer for making electrical connections to and between various regions. (Noyce 1961: 1)

While Noyce's patent (US Patent 2,981,877) was granted in April 1961, Kilby's application (US Patent 3,138,743) was approved only in July 1964. The ensuing lengthy litigation was settled only in 1971, when the Supreme Court granted Noyce (i.e., FS) priority in the matter. This made no practical difference: already by the summer of 1966, the two companies granted each other production licenses and agreed that outside fabricators would have to arrange separate licenses with both. Kilby and Noyce were awarded the National Medal of Science and inducted to Inventor's Hall of Fame, but only Kilby lived long enough to share a Nobel Prize in Physics in 2000. As with the first transistors, there was no rush to commercial applications of what TI's first advertisements in 1961 called Micrological Elements. They were expensive and untried, their individual components could not be tested for reliability, and their design could not be changed by buyers. The prospects changed once the Apollo missions got under way and once the designers of the first US ICBMs decided to control them with integrated circuits.

Microprocessors

In 1965, when integrated circuits contained up to 50 elements, Gordon Moore, at that time the director of research at FS, wrote a paper in which he not only unequivocally stated that "the future of integrated electronics is the future of electronics itself," but also predicted that the trend of complexity for minimum cost, which had increased roughly by a factor of two per year since 1959, would continue over the short term and, although he naturally considered a longer term forecast to be more uncertain, put the minimum number of components per integrated circuit at 65,000 by 1975 (Moore 1965). In 1968, when Burroughs released the first computers made with integrated circuits, Moore and Noyce co-founded Intel (Integrated Electronics), and the company set out to fulfill Moore's forecast.

Also in 1968, another fundamental semiconductor discovery was made at Bell Labs by a group led by Alfred Y. Cho that developed the process of molecular beam epitaxy, whereby single-crystal structures are deposited one atomic layer at a time (Cho 1994). This method, unmatchable in its precision, became the mainstay for making microchips. Intel's first profitable product was the static random access memory chip, and, in 1969, the company contracted to design 12 custom large-scale integration (LSI, with more than 1,000 components) chips for a programmable calculator to be made by Busicom, a Japanese company that funded the development. Intel's Marcian Hoff opted for a programmed computer solution composed of a single central processing unit (CPU) that retrieved its instructions from a read-only memory (ROM). Hoff and Stanley Mazor made the original proposal for the MCS-4 chip, Federico Faggin did the logic and circuit design, and Mazor wrote the necessary software (Mazor 1995). The resulting four-chip set could be used for other applications besides running calculators.

Interestingly, as Mazor noted, the Intel patent of the design (US Patent 3,821,715) did not list the single-chip processor among its 17 original claims (Hoff, Mazor, and Faggin 1974). Busicom made only a few large calculator models using the MCS-4 chip set before it went bankrupt. Fortuitously, Intel bought back the rights for the processor for $60,000 before that, and the world's first universal microprocessor was formally released in November 1971. The $200 chip was a tiny rectangle of 3 × 4 mm, processed four bits of information at a time at the speed of 400 kHz, and, with 60,000 operations per second, was the functional equal of the ENIAC (Intel 2024). Its success was its simplicity: by the early 1970s, it was clear that a computer on a chip would be eventually built, but it was difficult to see how it could be done with fewer than 20,000 transistors. The Intel 4004 had only 2,250 metal oxide semiconductor transistors, but it succeeded, as did other early Intel designs, "because they were scaled down computers. Like a golf cart, they were very limited, 'but got across the green'" (Mazor 1995: 1068).

One of the first uses of the Intel 4004 was on the Pioneer 10 spacecraft launched on March 2, 1972. But none of the established makers of integrated circuit-based computers rushed to use the processor for designing the first personal microcomputer: that revolution came from a Cupertino, California, garage where Steven Jobs and Steven Wozniak designed the first Apple machines between 1975 and 1977 (Moritz 1984). And just 7 months before Intel released its first microprocessor, in April 1971, TI introduced the first mass-produced consumer item based on integrated circuits: the Pocketronic, the world's first pocket calculator. Its development was launched in 1965, by Patrick Haggerty (1914–1980), the company's CEO, and Kilby was the engineer in charge. The basic architecture of the machine proved very durable. I bought TI's TI-35 Galaxy Solar when it came out in 1984, and, until 2017, nearly all calculations for my books and papers were done with it: the first patent listed on its back (US Patent 3,819,921) was the original Kilby-led design (Kilby, Merryman, and Van Tassel 1974).

Moore's original forecast of the doubling time for transistor density turned out to be too optimistic: in 1974, Intel's 8080 microprocessor had 5,000 transistors; in 1979, the 8088 reached 29,000. But the slower rate of component doubling—18, rather than just 12, months starting with Intel's 4004 in 1971—held for the remainder of the twentieth century, defying many predictions of its impending end, and, by 2000, there were no reasons for its imminent demise (Figure II.5.17). IBM's choice of the 8088 as the CPU in its first PC pushed the demand to more than 10 million microprocessors a year. By 2000, Intel's Pentium 4 64-bit microprocessor had 42 million transistors, and looking back, Moore, the company's Chairman Emeritus, was

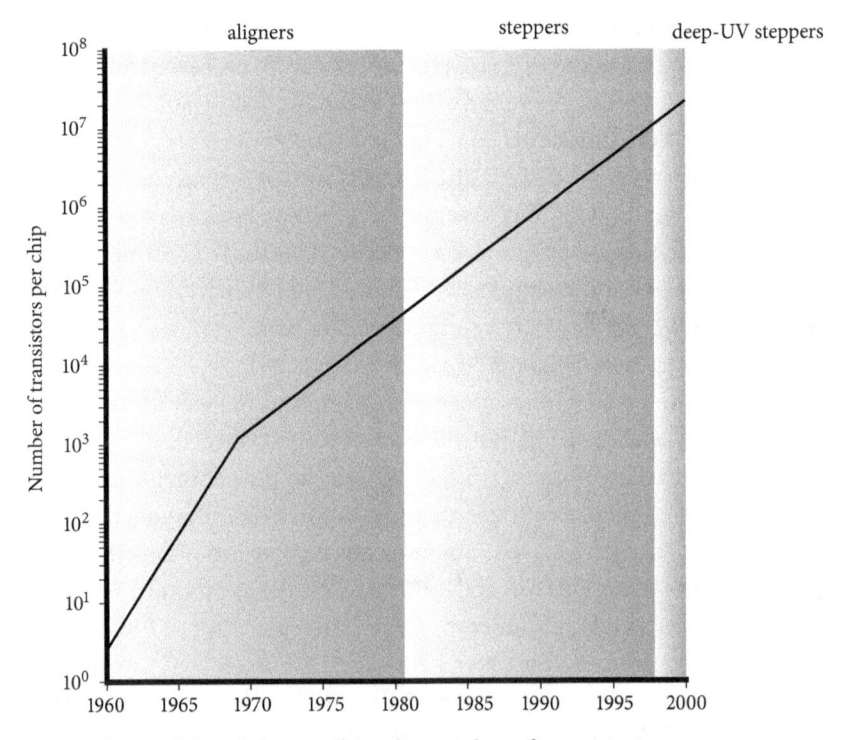

Figure II.5.17 Moore's law predicts the number of transistors per microchip. Between 1959 (Noyce's first planar transistor) and 1972, the density of components doubled annually. Afterward, with successive Intel microprocessors, it doubled every 18 months. Vertical divides show the time frames for higher-resolution lithography systems.

Based on graphs in Moore (1965), Hutcheson and Hutcheson (1996), and Intel (2024).

amazed we can design and manufacture the products in common use today. It is a classic case of lifting ourselves up by our bootstraps—only with today's increasingly powerful computers can we design tomorrow's chips. (Moore 2003:57)

Maintaining the trend required an unending stream of innovations needed to place more circuits on a silicon wafer. After LSI (up to 100,000 transistors on a chip) came very large-scale integration (VLSI, up to 10 million transistors) and ultra large-scale integration (ULSI). VLSI design, first proposed in 1969, was developed, optimized, and integrated by Carver Mead of Caltech and Lynn Conway of Xerox: thanks to their work, custom design of VLSI circuits became universally possible (Mead and Conway 1980). Density of data stored on high-capacity magnetic hard-drive disks experienced a similar rate of growth: between 1957, when IBM introduced the first commercial disk drive, and the late 1990s, the gain was 1.3 million-fold. Increasingly more powerful microprocessors made it possible to make themselves rapidly obsolete while they helped to design even more powerful, as well as more flexible, versions whose sales financed the next round of innovation.

But the fabrication of these microchips also required ever more expensive facilities: by the late 1990s, the cost was $3 billion/plant, after doubling every 4 years (an expensive corollary of Moore's law?). Fabrication requires extraordinarily clean rooms, and it minimizes human handling and exposure to airborne particles of even the smallest size. The process uses chemicals, gases, and light to produce complicated sequences of layers on a silicon wafer (Van Zant 2004). A layer of SiO_2 is deposited first, and then it is coated with a light-sensitive material (photoresist). Photolithography is used to project ultraviolet light through a patterned mask (stencil) and a lens onto a coated wafer. After removing the exposed photoresist (unprotected by the mask), the revealed SiO_2 is etched away and the remaining photoresist is removed, leaving behind ridges of the oxide on a wafer. A thinner deposit of SiO_2 is then overlaid by polysilicon and photoresist and exposed through a second mask. Dissolved photoresist exposes the areas of polysilicon and SiO_2 that are to be etched away.

Repetition of this process produces up to 20 layers in a micro-3D structure with windows for interconnections between the layers. Deposition of a metal layer and another round of masking and etching produce desired connections. The simplest MOSFET based on silicon substrate has just two islands (source and drain) of the element doped with either n-type or p-type channels overlaid with a SiO$_2$ insulator topped with a metal electrode (gate). Their mode of operation could not be simpler: low-voltage gated electrons cannot flow from source to drain, while a high-voltage gate lowers the energy barrier, sending a signal to the next circuit stage (Figure II.5.18). By the late 1990s, it became possible to fashion gate lengths shorter than 100 nanometers (nm) as photolithography techniques advanced from contact aligners of the early 1970s to deep-ultraviolet steppers after 1998 (Figure II.5.17). Yet even this incredible feat was still far below the fundamental physical limits. At the same time, the thickness of deposited oxide was just 25 silicon atoms (<3 nm), and, again, further reductions were under way.

The rising density of microchips had the already-noted obverse trend in falling costs. Writing in 1993, Robert W. Keyes highlighted the trend by comparing transistors with staples, the simplest of tiny steel products. He noted that the computer he used to type the article had some 10 million transistors, "an astounding number of manufactured items for one person to own. Yet they cost less than the hard disk, the keyboard, the display and the cabinet. Ten million staples, in contrast, would cost about as much as the entire computer" (Keyes 1993: 70). Or, as Moore recalled, "initially we looked forward to the time when an individual transistor might sell for a dollar. Today that dollar can buy tens of millions of transistors as a part of a complex circuit" (Moore 2003: 57).

In less than two decades, microprocessors were transformed from rare, proprietary devices to a mass-produced commodity that was globally traded in bulk. Strong post-1975 expansion of the Japanese semiconductor

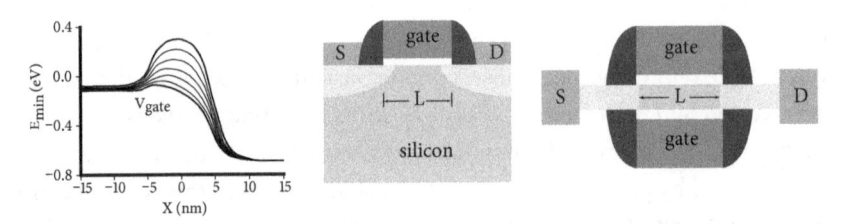

Figure II.5.18 The principle of the metal-oxide field-effect transistor (MOSFET). Minimum electron energy (left) is plotted against position from source to drain (X) under high drain voltage; sections through a planar MOSFET (middle) and a double-gate MOSFET (right) show source (S) and drain (D). L denotes channel length.
Based on graphs in Lundstrom (2003).

industry erased the pioneering US lead, and, by the mid-1980s, there were doubts about its very survival. In 1987, a government–industry partnership, SEMA-TECH, headed by Robert Noyce, was set up to help US semiconductor companies. Nevertheless, Intel remained the industry's global leader for the remainder of the century, but by 2000, only two other US companies, TI and Motorola, ranked among the world's 10 largest chipmakers. South Korean Samsung and Japan's Hitachi, Mitsubishi, and Toshiba were the Asian leaders.

Electronic Control and Communication

No other fabricated product in history had ever experienced such a cost reduction as did microprocessors, and hence, to quote Moore (2003: 57) once more, "nearly any application that processes information today can be done most economically electronically." During the last four decades of the twentieth century, entire industries, individual processes and products, and everyday actions were sequentially transformed by the introduction of transistors, integrated circuits, and microprocessors: very few complex items now remain outside their orbit. Much of this transformation was visible as inexpensive computers diffused throughout the industrial and service sectors and, after 1980, became common household items in most affluent nations.

But the hidden impact of microprocessors was no less important. Outwardly, countless processes and functions—be it drilling for oil, driving a car, or preparing a dinner—remained the same but most of their controls were transferred from direct human action to automated guidance through microprocessors (Cortada 2004). Every aspect of scientific instrumentation was transformed by reliance on computers, and many methods, relying on massive data processing, are simply unthinkable without it. Several of these methods, including magnetic resonance imaging (the first US patent by Raymond Damadian filed in 1972), became indispensable diagnostic tools of modern medicine, whose principal mode of intervention has been also greatly aided by computers. Modern drug discovery relies on computer-controlled analytical machinery, clinical trials require complex statistical analyses, and safe dispensation of prescriptions is best monitored by computerized databases.

The combination of inexpensive computing and instant telecommunications has transformed nearly every aspect of the energy business. Modern remote sensing depends on advanced electronics to store and process enormous volumes of data acquired by geophysical prospecting. By the mid-1990s, traditional two-dimensional seismic data used in oil and gas exploration were almost completely replaced by 3-D images, and 4-D monitoring (time-lapse 3D) of reservoirs made it possible to trace and simulate the actual flow of oil in hydrocarbon-bearing formations and to double the oil recovery rates (Morgan 1995; Lamont Doherty Earth Observatory 2001). Computers also control refinery operations and maintain (still imperfectly, as recurrent blackouts make clear) the stability of high-voltage transmission networks. And I have already given some examples of how modern industrial production and transportation were transformed by microprocessors.

From Mainframes to the World Wide Web

The shrinking size of dominant machines was the most obvious physical proof of successive computer revolutions. The earliest room-filling commercial mainframes, vacuum tube computers used by large research organizations and major businesses, weighed more than 10 t, but their transistorized progeny dropped below 5 t: UNIVAC I totaled 13.5 t; UNIVAC Solid State 90, released in 1958, weighed 3.28 t. A decade later the CPU of the IBM 360/67, the machine that claimed a major part of my days after we arrived in the United States in 1969, was 1.26 t, but the entire system, including core and disk storage, control units, punches, and card readers, added up to 12 t. Contrast this with the late 1990s laptops, most of which weighed less than 5 kg, and personal digital assistants brought the weight reduction (albeit with reduced power) to mere hundreds of grams. But the key objective of computer design was higher speed to execute increasingly complicated tasks in less time.

The fastest supercomputers of the late 1990s were more than 10 trillion times speedier than the best electromechanical calculators. As a result, complex global atmospheric circulation models were used for 5-day weather forecasts that would have taken at least half a year to process with UNIVAC. At the other end of the celerity spectrum were the increasingly more powerful microcomputers that introduced an entirely new combination of capabilities beginning with word processing and simple games and progressing to stunning complex graphics. Nearly all other key components that ingeniously combined hardware and software and made user-friendly desktop computing possible—stand-alone personal computer; mouse; WYSIWYG editing; graphical user interface (GUI) with icons and pop-up menus; laser printing; spell checkers and thesaurus; the combination of computing, text editing, and graphics; and access to file servers and printers with point-and-click actions—were created during the 1970s by Xerox's Palo Alto Research Center (PARC) that was set up in 1970 (PARC 2003).

A generation later—after Xerox Corporation ignored the invention (Smith and Alexander 1988) and had nearly disintegrated—how many people now appreciate the enormous role that PARC had in creating the PC era? Without PARC's array of techniques, Apple's Steven Jobs and Steven Wozniak could not have launched the first successful commercial PC, the Apple II with color graphics, in 1977 (Moritz 1984). Apple was also the first company that brought the graphical user interface (GUI) to the mass market with its Macintosh in 1984, and soon afterward Microsoft's PARC-derived Windows 1.0 was released for IBM PCs (Stross 1996). Microsoft's rise to megacompany status epitomized the emergence of software, an intangible commodity that is easy to steal but whose economic contributions are difficult to quantify, as one of the leading products of the 1990s.

Early mainframes had software as part of the machine; early programming was done by switches (in 1975, the Altair 8800, the first hobby kit microcomputer, had them), and programming languages, no matter how much closer they were to human language than was machine code, could never enter universal use. Readily available software, for purchase or free to download, greatly reduced the need for individual programming, and the largest programs adding up to tens of millions of lines of code became indispensable drivers of tasks ranging from word processing to flexible manufacturing, and from mindless and violent games to ingenious virtual reality tests.

The utility of PCs was greatly potentiated by the launching of the World Wide Web in 1990, and its rapid adoption deserves to be singled out as the century's last great transformative computing trend. The Internet's origins go to 1962, when Joseph Licklider (1915–1990) became the first director of the Pentagon's Advanced Research Project Agency's (ARPA) Command and Control Research and soon proposed a nationwide "Intergalactic Computer Network" (Waldrop 2001). His successor, Robert W. Taylor, decided to start its construction, and ARPANET began its operation, initially with just four sites at the Stanford Research Institute, University of California at Los Angeles (UCLA), University of California at Santa Barbara, and the University of Utah, in 1969. By 1971, there were nearly two dozen sites using protocols for remote terminal access (Telnet) and file transfer (ftp).

A year later Ray Tomlinson of BBN Technologies designed two programs for sending messages to other computers: "It just seemed like a neat idea. There was no directive to 'go forth and invent e-mail'" (Tomlinson 2002: 1). He also picked the "@" sign as the locator symbol for e-mail addresses.

In 1974, Vinton Cerf and Robert Kahn released a protocol that enabled computer communication across a system of networks (TCP/IP). By 1980, ARPANET connected more than 400 computers that could be accessed by about 10,000 users. By 1983, ARPANET officially converted to TCP/IP, and the Internet was in place with fewer than 600 hosts; six years later, when ARPANET ended its existence, there were more than 100,000 hosts. In 1990, Tim Berners-Lee created the hypertext-based World Wide Web at Geneva's CERN to organize online scientific information (Abbate 1999). But the World Wide Web would have been restricted to its initially specialized existence without an easy means to move around in it.

This was first supplied in 1993, by Mosaic, designed by Mark Andreessen and his colleagues at the University of Illinois, and, a year later, this software morphed into Netscape's Navigator, and the publicly accessible electronic superhighway was born. Domain surveys regularly undertaken by the Internet Systems Consortium showed the host total surpassing 5 million during the first half of 1995 and 100 million by the end of 2000 (ISC 2004), and worldwide PC sales nearly doubled between 1996 and 2000, to 130 million units a year. Text, still images, video, movies, and sound recordings became accessible to anybody able to click and follow simple instructions.

The Internet brought a new vocabulary of inevitable acronyms (ftp, http, URL) and new means of looking for information by using browsers and search engines, with Google rising to the leading spot. Soon the ease with which new websites could be created led to the emergence of an unprecedented depository of information that spans a complete range from archival repositories of books and scientific periodicals and fascinating real-time, or near-real-time, sites fed by satellites and remote cameras, to a flood of misleading, offensive, and outright deranged and criminal postings. New ways of doing business emerged with online retail and company-to-company sales and auctions and were almost immediately followed by new ways of distributing offensive advertising (spam), breaching privacy, and facilitating criminal activities ranging from credit card and identity theft to mass disruption of traffic by hacker-generated viruses, which led to rising outlays on data security, encryption, firewalls, and virus identification.

Mass Media and Personal Choices

The pre-World War I advances gave only an inkling of possibilities that were transformed into commercial realities of the twentieth century when such disparate information and communication means as publishing, telephony, sound recordings, movies, radio, and television benefited from replacement of mechanical, electromechanical, and vacuum tube devices by solid-state electronics. Typesetting, dominated by linotypes, did not see any fundamental innovations until the 1950s. In September 1946, Louis M. Moyroud and René Alphonse Higonnet revealed their photo composing machine that replaced cast metal with the strobe light projection of characters on a spinning disk onto photographic paper. The machine made "it possible to obtain directly composed text of justified lines" and to change "the length of lines, sizes of characters and fonts ... by simple means" (Higonnet and Moyroud 1957: 1).

The Graphic Arts Research Foundation of New York improved the design, and the first book composed by the Photon machine came out in 1953. Eventually all mechanical parts were eliminated as solid-state components took over after 1960. But linotypes dominated composition well into the 1960s. John Updike's most famous fictional character, Rabbit Armstrong, became one of the late victims of a switch ("So no Linotypers, huh?") in 1969, when his boss tells him about the new technique: "Offset, you operate all from film, bypass hot metal entirely. Go to a cathode ray tube, Christ, it delivers two thousand lines a minute" (Updike 1971: 296). During the 1980s microcomputers, inexpensive and easy-to-use typographic software, and laser printers ushered in the era of desktop publishing, and direct computer-to-plate techniques did away with resin-coated paper or film.

Unlike typesetting, the classical pre-World War I mode of movie making was transformed twice before the middle of the century by the introduction of color (first Technicolor in 1922, and an improved version during the 1930s) and electronically recorded sound after 1926 (Lloyd 1988; Friedman 1991). But, except for wide screens (Cinerama starting in 1952) and surround sound, there were no fundamental changes to the movie-going experience. The first Kodachrome (color) film was released in 1935, but color film began to be used more widely in cameras only after 1950, when the traditional dominance of German optics began to be challenged by Japanese companies. By the early 1970s, Nikon, Pentax, Minolta, and Fujica dominated the world single-lens reflex market, while Kodak kept a substantial share only in the low-end amateur niche. Transistors and then microchips changed cameras from mechanical to electronic devices, and the first digital cameras for consumer market were introduced in 1994 and 1995: the end of film was near.

Much like typesetting, telephony remained relatively static until the early 1960s. For decades telephone ownership was limited to hardwired rotary dialing models (see Figure II.3.16), long-distance calls remained expensive, and intercontinental traffic was rare: the first trans-Atlantic telephone cable was laid from Scotland to Nova Scotia only in 1956, and it could carry just 36 simultaneous calls. But soon afterward the cost of long-distance calls plummeted as satellites opened the path toward inexpensive intercontinental calling. The first overt transformation (in 1963) was to replace the rotary dial with the touch-tone telephone, but beneath the buttons there was still the decades-old electromechanical assembly. Still, touch-tone dialing made possible new telephone services and uses (voice mail, call centers), including exasperating automated recordings with numbered menus.

Development of the electronic telephone began during the 1970s (Luff 1978). The bell ringer, the tone oscillator, and the speech network were replaced by integrated circuits, and two dynamic transducers acted as transmitter and receiver. But the most important post-1960 advances were not in the handheld devices but in techniques not seen by the user that increased the throughput of the system while cutting its cost and improving its quality. Bell Laboratories in Murray Hill, New Jersey, was at the forefront of many of these innovations (Bernstein 2011). As already noted, the first telecommunication satellite was launched in 1962 (see Figure II.3.20). In the same year Bell Labs introduced the first digitally multiplexed transmission of signals that created a more economical yet more flexible and more robust way for voice traffic and led to services ranging from 911 to 800 calls to call waiting.

In 1979 came the first single-chip digital signal processor, the forerunner of cellular phones and modems of the 1990s, and, in 1988, the first fiber-optic trans-Atlantic cable could handle 40,000 calls. A much larger and much faster—and largely unanticipated—wave of cellular telephony began to roll during the early 1990s. Bell Labs had already made the first demonstration of cellular telephony in 1962, and it introduced digital cellular transmission, with higher channel capacity and lower cost, in 1980. But despite the pioneering work at Bell Labs, the first markets that were nearly saturated with cellular telephones were in Western Europe and in Japan and Hong Kong.

World War I delayed radio's commercial expansion, and limited ownership of receivers marked radio's first public decade. But by the mid-1930s the medium was well established throughout the Western world. The most important technical innovation of the 1930s was Edwin Armstrong's (1890–1954) development of frequency modulation (FM) broadcasting that eliminated the notorious static interference in amplitude modulation (AM). But the broadcasting industry with its heavy investment in AM resisted the innovation, and FM became common only after 1950. Radio's capability to captivate instantly an unprecedented number of people was famously demonstrated by Orson Welles's 1938 *War of the Worlds* broadcast. The first portable radio went on sale in October 1954: the TR-1 by Regency Electronics had four of TI's germanium transistors, was available in six colors, and cost $49.95 (Lane and Lane 1994).

While TI did not pursue the further development of the portable radio, a small Japanese company, Tōkyō Tsushin Kōgyo, set up in 1946 by Akio Morita and Masaru Ibuka and renamed Sony to have an internationally pronounceable name, followed its release of the TR-55 radio in August 1955 by the world's first transistorized television (in 1960) and compact videotape recorder (in 1963). And although its Betamax video cassette recorder, introduced in May 1975, lost to Matsushita's VHS design, its stereo Walkman (1979), Discman (1984), Watchman (handheld micro-TV, 1988), camcorder (1989), digital camcorder (1995), and Memory Stick Walkman (1999) continued the tradition of small portable solid-state devices that found a ready global market and made the company the world's second largest (after Matsushita) maker of electronic goods (Sony 2004). The century's last major radio innovation was the rapid expansion of World Wide Web stations, with listeners using their PCs. In 1995, Radio Hong Kong became the first 24-hour Internet-only station.

By the time radio's audience was peaking, television was making rapid inroads. Its commercial beginnings in the United States and United Kingdom during the 1930s were so tentative that the true TV era began only after 1945. As noted in Chapter II.1, television could have been born before World War I. Its early development is akin to the evolution of movies, as many innovations led to a commercial product. Its American origins center on the patent dispute between Philo T. Farnsworth (1906–1971) and Vladimir K. Zworykin (1889–1982). Farnsworth, a self-educated Mormon, conceived the idea of an all-electronic TV at age 14 and transmitted the first picture (of a piece of glass painted black with a scratched center line) using a cathode ray tube (CRT) camera in 1927, when he patented a complete television system (Godfrey 2001; Figure II.5.19).

Figure II.5.19 Philo Farnsworth in 1926 (left), just before he filed his patent for television system, and inspecting (right), with Mable Bernstein, one of the first portable TV cameras built in 1934.
From Donald G. Godfrey, Arizona State University.

Zworykin, a Russian engineer who immigrated to the United States in 1919, claimed that he designed a similar CRT camera (iconoscope) in 1923 (Friedman 1991). But he did not build any working device at that time; his first patent for "an apparatus for producing images of objects" was granted only in 1935, and his 1923 application was approved, after an extraordinarily long period of delays and revisions, only in 1938. But in 1935, the US Patent Office ruled that Zworykin's original 1923 design would not work and awarded the invention priority to Farnsworth. It was RCA, Zworykin's corporate sponsor, whose engineers developed the system for commercial use and began broadcasting in 1939. Meanwhile, in the United Kingdom, John L. Baird demonstrated his low-resolution (30 lines per picture) electromechanical design in 1926, and, by 1930, he also used it to test color broadcasts. The BBC began regular broadcasts based on Baird's invention in 1932 and replaced them with the world's first high-resolution (405 lines per picture) service in 1936 (Smith and Paterson 1998).

Transistorizing and later using microchips to shrink the size cut the cost and boosted TV's capabilities. Bulky black-and-white TVs of the late 1940s cost $500, an equivalent of nearly a fifth of an average annual salary; half a century later, a standard CRT color TV could be bought for less than a tenth of an average monthly salary. Global TV count rose from 100 million units in 1960 to 300 million by 1979 and to about 1.2 billion by 2000, with saturation complete in affluent countries and in many urban areas in low-income nations. Introduction of cable TV (for the first time in the United States in 1948, but widely available only since the 1970s) led to a proliferation of channels and to a predictable correlation between quantity and quality. Finally, before the end of the century, a superior alternative to bulky and energy-inefficient CRTs became available as liquid crystal display panels allowed perfectly flat and much wider screens, and high-definition TV (first demonstrated by Phillips in 1988) began its commercialization process.

Frontispiece II.6 Two images from 1950 illustrate the progress of the twentieth century's technical transformation. TV's key components (top) were available before World War I, regular broadcasts began before World War II, but rapid diffusion of the medium took place only after World War II. This console receiver with a 30-cm cathode ray tube screen was made by the long-defunct Capehart-Farnsworth Corporation. Flying underwent a fundamental technical transformation (bottom) as gas turbines entered commercial service after World War II: a young American boy contemplates a ("someday") trip to far-distant lands by a machine that became a commercial reality 8 years later.

Cathode ray screen image from author's collection; drawing reproduced courtesy of INCO.

6
New Realities and Counterintuitive Worlds
Accomplishments and Concerns

> The changes have come so fast that man's customs, mores, ethics, and religious patterns may not have adapted to them.... Man's role in the environment is becoming so enormous that his energetic capacity to hurt himself by upsetting the environmental system is increasing.... We may wonder whether the individual human being understands the real source of the bounty to him of the new energy support.
>
> —Howard Odum, *Environment, Power, and Society*

The great technical saltation that took place during the two pre-World War I generations introduced new forms of energy; new prime movers; new materials; new industrial, agricultural, transportation, and household machines; and new ways of communication and information processing. Advances of the twentieth century improved all these techniques and brought major innovations in energy conversion, materials, agricultural mechanization, industrial manufacturing, data processing, and telecommunication, many of them possible only thanks to solid-state microelectronics that had rapidly penetrated every segment of modern economies. These technical transformations created a civilization that differs in several fundamental ways from all previous societies.

Our lives now depend on using a still growing array of artifacts and on following a still expanding set of procedures that are required to accomplish countless productive activities and that gives us more leisure. In return for affluence, comfort, convenience, speed, accuracy, and reliability, we must conform to the requirements of these complex systems. Even those individuals or small groups of people who claim that they have opted out of these confines could do so in only partial ways—and as large collectivities (towns, cities, nations) we have no choice but to participate. As modern economies became more dependent on concatenated technical advances, they passed the point of no return: this dependence now assures not only our affluence but also our very survival. An inkling of this new reality came in parts of the Western world before World War I, but most pre-1914 societies still had a long way to go before technical transformations made our dependence on complex techniques so pervasive and so inescapable

Two mid-century images used as this chapter's frontispiece exemplify this process of gradual technical transformation. TV's key components were available before World War I, but practical service was developed only during the 1930s and began to diffuse rapidly only after World War II. Flying underwent a fundamental transformation as gas turbines entered commercial service after World War II. In the United States, these new realities were largely in place soon after World War II, in Europe they became common only during the 1960s, and the pervasiveness and complexity of artificial environments reached new heights everywhere with post-1970 automation and computerization. As if this were not enough, the 1990s brought a new trend that has been detaching individuals and societies even further from the physical world and submerging them in "virtual" reality.

Higher energy use is a key prerequisite of all technical transformations, and its most obvious payoffs have been higher physical quality of life and vastly expanded individual and collective horizons as a surfeit of information, affordable travel, and instant means of communication changed the patterns of work, education, leisure, and social interaction. Displacement of animate energies by fossil fuels and primary electricity, the rise of mass production, and new means of control—progressing from ingenious mechanical devices to electromechanical setups to solid state electronics—transformed the structure, efficiency, and labor intensity of all economic activities. The quest for higher productivity, improved quality, minimized labor force, and perpetual expansion of output resulted in unprecedented rates of economic growth.

Although the world's population nearly quadrupled during the twentieth century, both the average per capita economic product and typical disposable income grew much faster. At the same time, a highly skewed distribution of these accomplishments left humanity with greater relative welfare disparities. The highly uneven distribution of natural resources needed to generate these advances, as well as many instances of comparative advantage in producing goods or offering specific services, led to unprecedented levels of interdependence. This trend had gone beyond the expansion of international trade as it embraced leisure travel (the world's single largest economic activity), migration, flows of information (patents, international consulting, education, scientific collaboration), and personal styles and tastes. These shifts have been closely connected with urbanization. At the same time, precipitous urbanization was also a major reason for persistent, even deepening, social and economic inequalities.

Complexities of global interdependence are just one of many new concerns that arose from the success of the twentieth century's technical advances. New risks were created by new techniques, and it has not been easy to subject them to dispassionate and convincing cost-benefit assessments. And no century had seen so many deaths in violent conflicts as aggression was made easier thanks to the vastly increased destructiveness of weapons. And while the concerns about the unthinkable—the thermonuclear exchange between the two superpowers—receded with the peaceful disintegration of the USSR in 1991, the 1990s were marked by a heightened awareness of a multitude of environmental changes. These ranged from destruction and degradation of ecosystems and biodiversity on local and regional scales to concerns about global effects, above all about the consequences of rapid climate change whose genesis is directly linked to the key driver of modern civilization, the combustion of fossil fuels.

Gains and Losses

On a personal level, modernity's most obvious manifestation is the increasing accumulation of possessions. On the collective level, more complex and more durable artifacts combine to produce artificial environments of concrete, metals, glass, and asphalt, where trees are planted according to design, beaches are trucked in, streams are trained, air is polluted, and invisible greenhouse gases keep changing the Earth's radiation balance. None of this has diminished the drive to amass material possessions. In some places, the first steps toward this new reality began many generations ago. During the fifteenth and sixteenth centuries, homes of wealthy Dutch merchants displayed paintings, maps, rugs, tea services, musical instruments, and upholstered furniture; soon afterward, incipient mass consumption of "frivolities" spread to population segments with lower incomes as modern materialism was born (Mukherji 1983). Similar acquisition of luxuries was taking place at the same time in Muromachi and Tokugawa Japan, and in Ming and early Qing China (Pomeranz 2000).

Industrialization enriched this trend by adding affordable mechanical devices (clocks, watches, sewing machines, cameras), increasingly more complex machines (bicycles, automobiles), and ingenious converters of electricity (lights, motors). The unprecedented technical saltation of the two pre-World War I generations began to extend new infrastructures as households acquired indoor plumbing, electricity, and telephones. Concurrently, new factories, transportation networks, and rapidly expanding industries began to transform previously tightly confined settlements—maximum diameters of pre-industrial walled cities were on the order of 5 kilometers (km)—as they encroached on the surrounding countryside. Eventually this anthropogenic world began

> to impose itself everywhere, to change everything, to take over all social activities and forms, and to become a true environment.... We can survive neither in the natural environment nor in a social environment without our technical instruments. Our gadgets are as necessary to us as food. (Ellul 1989: 133)

From the fundamental physical point of view, the most obvious outcome of the twentieth century's technical transformation was the creation of high-energy societies where even ordinary citizens could command, affordably and also effortlessly, power that was just a few generations ago beyond anybody's reach (or even imagination). Consequences of this achievement ranged from the welcome increases in longevity and reductions in morbidity to ostentatious and wasteful overconsumption. Perhaps the most fascinating truth about this situation is that, as discussed further below, no important indicators of a rationally defined quality of life require the tremendous levels of per capita energy use that were taken for granted in the world's most affluent countries during the 1990s. Higher energy use also transformed our lives by expanding both physical and mental horizons thanks to affordable and speedy transportation and communication. But these advances also offer perfect examples of serious, unintended, and often unanticipated side effects and the worrisome consequences of technical innovations.

Expanded transportation became a leading contributor to environmental degradation, mainly because of automotive and airplane emissions and the effects of transportation infrastructure on the reordering of inhabited space;

it also remained a major cause of accidental death and injuries, and, as its intensity grew, even its most cherished benefits, speed and saving of time, declined. New means of communication and new ways of disseminating and processing information—from inexpensive printed matter to the multifaceted capabilities of computers—can obviously be tools of both formal and informal education and expanded knowledge and more subtle understanding of complex realities. They can be also valuable repositories of historical records, collective memories, and cultural accomplishments, as well as the foundations needed for informed decision-making. But during the twentieth century these powerful technical means were often used as the tools of propaganda and misinformation, were deployed as agents of deliberate forgetting and mass control and were found to be ideal choices to debase cultural standards and weaken social structures.

High-Energy Societies and the Quality of Life

An improved quality of life was bought with higher flows of useful energy. Transformations of pre-World War I prime movers and the post-World War II deployment of gas turbines made it possible to produce, move, process, and convert fossil fuels with throughputs and efficiencies that were unprecedented in history. A new high-energy civilization was born as global consumption of fossil fuels and primary electricity expanded 16-fold between 1900 and 2000 and as the share of biomass energies fell from nearly 50% to just 10% of the total (Figure II.6.1). This large expansion meant that even with the near quadrupling of the global population (from 1.6 to 6.1 billion), the average annual per capita supply of primary energies still more than quadrupled. The century's global cumulative production of coal and crude oil was roughly equal in energy terms, but oil's post-1960 global dominance makes it a better choice for equivalent measures.

Using kilograms of oil equivalent (kgoe) as the common denominator, average per capita consumption of fossil energies rose from about 300 kgoe in 1900 to more than 1,250 kgoe by 2000: that is roughly nine barrels using the oil industry units or less than 13 gigajoules (GJ) using the scientific unit (UNO 2002). Addition of primary electricity (since the 1880s hydrogeneration, since the mid-1950s also nuclear, and small amounts of geothermal, wind, and solar generation) raises the 1900 total only marginally, but it adds roughly 11% to the 2000 rate. But because of the enormous inequalities of

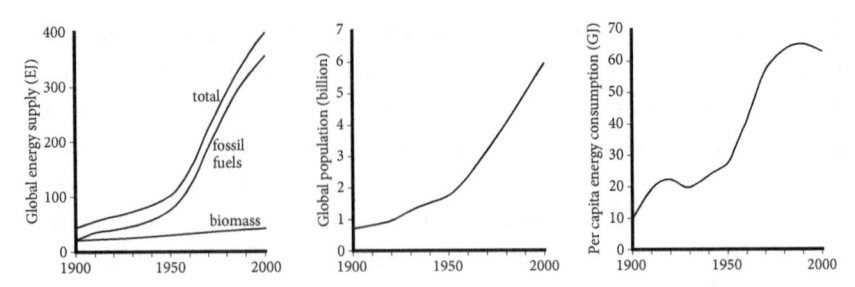

Figure II.6.1 Global energy supply (left), population (middle), and per capita energy consumption (right), 1900–2000. Although the worldwide biomass consumption doubled during the twentieth century, by its end, it provided only about 10% of the global primary energy supply dominated by fossil fuels and primary (hydro and nuclear) electricity.
Based on figures in Smil (2017a).

commercial energy use, it is much more important to know the modal rate and the global range of values. In 2000, the modal rate (for more than 50 countries) was below 250 kgoe; the global ranges go, on the national level, from less than 100 kgoe in sub-Saharan Africa to about 8,000 kgoe for the United States and Canada, and from nothing (for the poorest populations relying still solely on biomass fuels) to about 12,000 kgoe for the most affluent populations living in the richest US counties.

In the United States and in many European countries, coal consumption was already high in 1900 and hence the per capita supply of fossil fuels "just" tripled or quadrupled by 2000. But Japan's use of fossil fuels grew nearly 30-fold, and because in China or India consumption of fossil fuels in 1900 was limited to small amounts of coal, the centennial multiples were extremely large. But all these comparisons are flawed because they are based on the gross energy content of fossil fuels and primary electricity and do not reflect higher conversion efficiencies that delivered much more useful energy (heat, motion, light) from the same inputs in 2000 than they did in 1900. Space heating illustrates these gains. Wood stoves were usually less than 20% efficient; coal stoves doubled that rate, and fuel oil furnaces brought it to nearly 50%. Natural gas furnaces initially rated up to 65%, but, by the 1990s, many of them delivered more than 90%. Consequently, every unit of gross heating energy now provides roughly five times more useful service than it did in 1900.

In affluent countries, the overall efficiencies of primary energy use at least doubled and even tripled during the twentieth century. When coupled with the absolute rise in energy use, this means that these countries had experienced 8- to 12-fold increases in per capita supply of energy services. But even this adjustment fails to capture fully the actual gain because the new converters and new ways of consumption also brought improvements in comfort, safety, and reliability, all of them most welcome but none of them easy (and some impossible) to quantify. Another way to illustrate increased energy use in modern societies is to contrast the flows controlled directly by individuals during their daily activities. Peak unit capacities of prime movers that can deliver sustained power rose about 15 million times in 10,000 years—from 100 watts (W) from sustained human labor to 1.5 gigawatts (GW) for the largest steam turbogenerators—with more than 99% of the rise taking place during the twentieth century (Figure II.6.2).

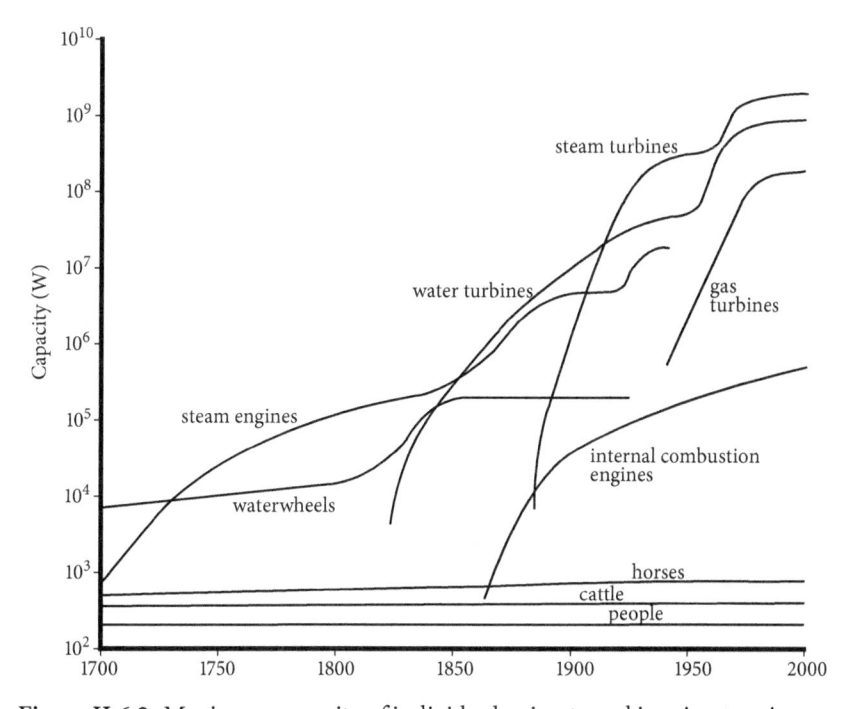

Figure II.6.2 Maximum capacity of individual animate and inanimate prime movers, 1700–2000. The background shows an internal combustion engine (an early Rolls-Royce model) because no other prime mover has reached such a large aggregate installed capacity: in 2000, the engines (mostly in passenger vehicles and trucks) had more than 20 times as much installed power as did all the world's steam turbogenerators.

Based on a graph in Smil (2017a).

Although the citizens of affluent societies take these levels of individually and collectively commanded energies for granted, their magnitudes still astonish (Smil 2003). In 1900, even a prosperous Great Plains farmer controlled no more than 5 kW of sustained animate power as he held the reins of six large horses when plowing his fields—and he had to do so sitting in the open on a steel seat; a century later, his great-grandson performed the same task from air-conditioned comfort of an insulated cabin of a huge tractor capable of 300 kW. In 1900, engineers of the most powerful locomotives that could pull transcontinental trains at about 100 kilometers per hour (km/h) commanded about 1 megawatt (MW) of steam power while exposed to blasts of heat and rushing air; in 1999, a captain of Boeing 747-400 could let onboard microprocessors control four jet engines whose aggregate cruise power of some 45 MW could retrace the train's route 11 km above the Earth's surface at an average speed of 900 km/h.

In 1900, even well-off US urban households had only half a dozen low-power lightbulbs and one or two small appliances whose combined power rating was far below 1 kW; during the late 1990s, an all-electric, air-conditioned exurban house had more than 80 switches and outlets ready to power lights and appliances that could draw more than 30 kW, and adding the family's three vehicles would raise the total power under the household's control close to 500 kW! In the past this power—although nowhere near a comparable level of convenience, versatility, flexibility, and reliability—could be commanded only by an owner of a Roman *latifundia* with about 6,000 strong slaves, or during the 1890s by a landlord with 3,000 workers and 400 big draft horses.

Fossil fuels and electricity were indispensable to increase farm productivity and hence drastically reduce the agricultural population, to mechanize industrial production and release the labor force to move into the service sector, to make megacities and conurbations a reality, and to globalize trade and culture. And the energy industries also had to energize themselves as they imposed many structural uniformities onto the diverse world. Fortunately, given the high energy densities (J/g) of fossil fuels and high-power densities (W/m^2) of their deposits, net energy return in coal and hydrocarbon industries has been very high. This is particularly impressive because these industries require many expensive infrastructures—coal mines, oil and gas fields, pipelines, tankers, refineries, power stations, transformers, high-voltage transmission, and overhead and underground distribution lines—and a great deal of energy to operate them.

While higher energy flows correlate highly with greater economic outputs, all of the quality of life variables relate to average per capita energy use in a nonlinear manner as there are some remarkably uniform saturation levels beyond which further increases of fuel and electricity consumption produce hardly any additional gains. No matter which quality of life indicator one graphs against the average per capita energy use (Figure II.6.3), clear inflections become evident at annual consumption levels of between 1 and 1.6 tons of oil equivalent (toe) per capita, with obviously diminishing returns afterward and with basically *no* additional gains accompanying primary energy consumption above 2.6 toe per capita. These realities led me to conclude that a society concerned about equity and willing to channel its resources into the provision of adequate diets, healthcare, and basic schooling could guarantee decent physical well-being with an annual per capita use of as little as 1–1.2 toe (Smil 2003).

A more desirable performance, with infant mortalities below 20, female life expectancies above 75, and the United Nations Development Programme (UNDP) Human Development Index (HDI) above 0.8, appears to require annually at least 1.4–1.5 toe per capita, while currently the best global rates (infant mortalities below 10, female life expectancies above 80, high rates of house ownership, good access to postsecondary education, HDI above 0.9) need about 2.6 toe per capita. And the level of political freedom has little to do with any increases in energy use above the existential minimum (Figure II.6.3). By the end of the twentieth century US per capita energy consumption was more than three times the desirable minimum of 2.6 toe, and almost exactly twice as much as in Japan or the richest countries of the European Union (8 vs. 4 toe/year)— yet it would be ludicrous to suggest that American quality of life was twice as high. In fact, the United States ranked behind Europe and Japan in several important indicators because of its higher infant mortality, more homicides, lower levels of scientific literacy and numeracy, and less leisure time.

Pushing beyond 2.6 toe per capita has brought no fundamental quality of life gains, only further environmental degradation. And given still very large potential for higher energy conversion efficiencies, it is realistic to conclude that lifestyles that were supported by 2.6 toe during the late 1990s could have been achieved by using less than 2 toe per capita, close to the actual average global consumption of about 1.5 toe per capita per year. This fact puts one of the greatest concerns of modern affluent civilization in a different perspective: equitable distribution of available energy supplies and their rational use alone could have guaranteed high worldwide quality of life!

594 THE TWENTIETH CENTURY

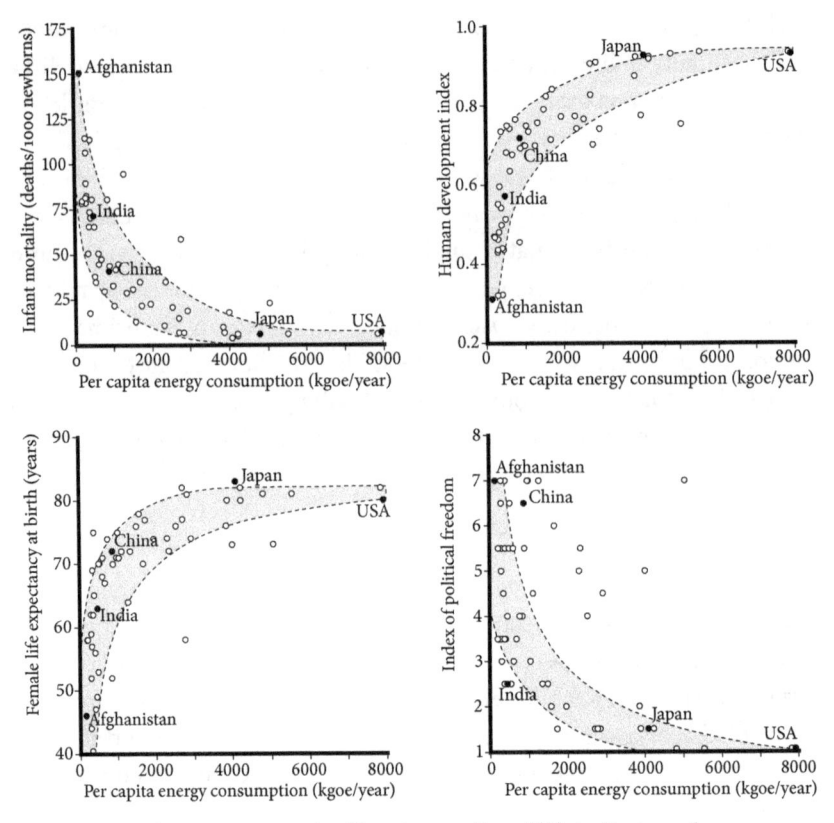

Figure II.6.3 Improvements in all major quality of life indicators show diminishing returns with increasing per capita energy consumption and virtually no gains above the average level of 2.6 toe per capita.
Based on graphs in Smil (2003).

Mobility and Its Price

New means of transportation have expanded physical horizons to the extent that is quite magical when seen from the pre-1900 perspective. Coaches, averaging about 10 km/h, were the fastest means of land transport until the 1830s; by 1900, steam-powered trains averaged 60–80 km/h; reciprocating aeroengines pushed cruising speeds above 200 km/h by the early 1930s and to more than 600 km/h by the early 1950s. Then travel speeds rose and leveled off abruptly with the introduction of gas turbines as jetliners began to cruise at the edge of the stratosphere at subsonic speeds of up to about 950 km/h or Mach 0.9 (Figure II.6.4). In cities, the introduction of streetcars

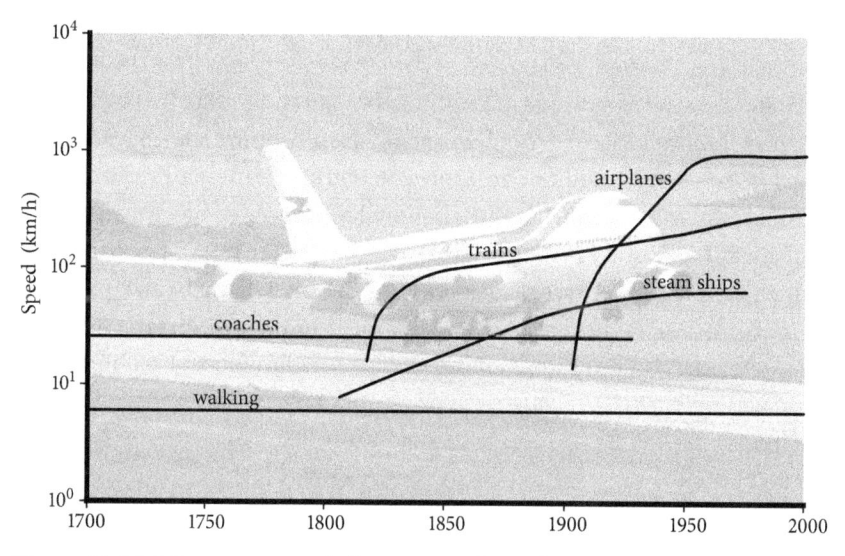

Figure II.6.4 Maximum speed of commercial travel, 1800–2000. The supersonic Concorde cruised at Mach 2, but it operated only on limited routes between 1976 and 2003, when the fleet was retired, and it was an uneconomical oddity.
Based on Shevell (1980) and Smil (2017a).

and buses, averaging about 20 km/h (but only 10 km/h in congested downtowns), quadrupled walking speed, and passenger cars (and subways in cities with high train frequency) traveling at 40–50 km/h at least doubled the speed of surface public transport (Ausubel, Marchetti, and Meyer 1998).

Time allocation studies show that people, regardless of the transportation mode, spend about 1 hour a day traveling; during that time, an urban driver, going about eight times faster than a pedestrian, can make return trips inside a roughly 60 times larger area. A driver traveling on an interstate highway at 120 km/h acquires access to an area nearly 580 times larger than available to a pedestrian, and, because cars eliminate waiting time for departure and transfers (often between two transport modes), they also offer a considerable advantage compared with trains. The price paid for these magically expanded horizons included not only the construction and maintenance of the requisite infrastructures, inevitable accidents, and environmental degradation but also some less noted negatives that include a counterintuitive decline of typical car travel speeds and the geochemical supremacy of automobiles over people.

Road construction was one of the twentieth century's most remarkable, most massive, and hence also most destructive enterprises. The US example illustrates best the magnitude of this transformation: in 1900, there were just a couple hundred kilometers of paved roads; a century later paved roads added up to nearly 4 million km. Only 256,000 km of these belonged to the National Highway System, but these national roads carried more than 40% of all highway traffic and 75% of heavy truck traffic (Slater 1996). The backbone of this network is the nearly 75,000 km of the Interstate Highway System, whose construction, begun in 1956, amounted to the world's largest public works construction program. But this enviable achievement also poses a forbidding maintenance challenge: repairing and replacing the enormous mass of reinforced concrete adds up to expenditures that vastly surpass the original investment.

Absolute death rates in car accidents declined or stabilized in all high-income countries where stricter safety standards, better roads, and better enforcement of seat belt legislation not only stopped but reversed the fatality trend. The US peak in 1972 was 54,589 deaths, but during the 1990s, the annual total never surpassed 44,000 even though the total distance driven by vehicles grew by almost 25%. But in low-income countries, total death rates continued to rise (WHO 2002). In 2000, traffic deaths amounted to 1.26 million people and 25% of all fatal injuries, roughly equal to the combined total of all homicides and suicides and four times as many as casualties of wars—and 90% of these fatalities occurred in low- and middle-income countries.

Among populous countries the highest road traffic mortality was in China (50 deaths/100,000 people) and India (30 deaths/100,000 people), while the North American and Western European rates were between 10 and 15/100,000. These differences are not surprising given that the total number of traffic deaths increases rapidly during earlier stages of automobilization while the rate of fatalities per 1,000 vehicles declines (Evans 2002; Figure II.6.5). Because of the disproportionately large share of young victims (particularly males in their late teens and early 20s), every traffic fatality also represented roughly 33 disability-adjusted life years (defined as years of healthy life lost either due to premature death or disability).

Numerous improvements of the car's overall design, better fuel, greatly improved driving conditions, and, after 1970, the effort to lower its polluting emissions combined to make the passenger automobile a considerably more efficient, more reliable, safer, and less offensive machine. But—notwithstanding the growing sales of "sport" cars, the myth propagated by

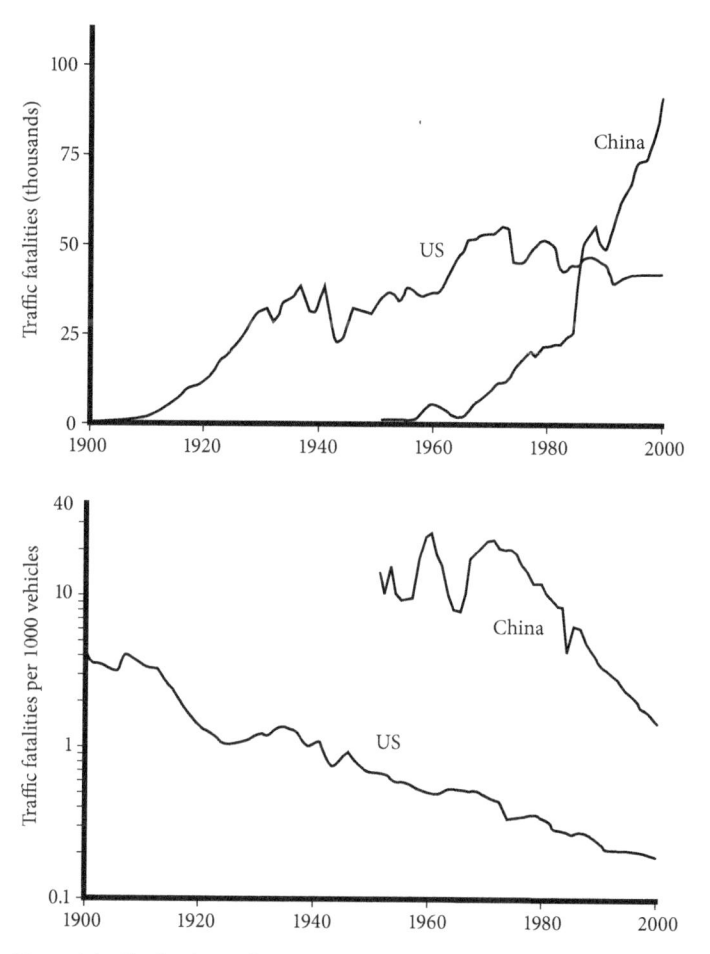

Figure II.6.5 Traffic fatalities (annual totals, top, and deaths per 1,000 vehicles, bottom) in the United States and China, 1900–2000. In spite of the much steeper post-1970 decline of relative fatality rates in China, driving there during the late 1990s was still about 10 times riskier than in the United States.
Based on graphs in Evans (2002).

their makers, and the impression that immature buyers of these vehicles wish to convey—there were no corresponding increases in speed for most car trips (i.e., daily commutes and errands). Road congestion that limits average speed has been the norm in every large city, no matter whether its traffic was dominated by horse-drawn vehicles (see Figure I.1.7), low-power automobiles (in the 1920s), or a mix of increasingly more powerful vehicles in the 1990s. But the rate of automobilization presented an unprecedented escalation of transport density.

For example, in 1939, the Paris region had about half a million cars; by 1965, the total was 2 million, but the street surface in the city had increased by just 10% since 1900 (Ross 1996). Consequently, parking has become as much of, if not more of, a problem as driving. Many studies demonstrated that new roads brought only temporary time savings before they, too, became clogged or encouraged longer drives for the same pattern of activities. And while the average distance driven per year had stabilized in Europe and Japan, in the United States it began to rise (after 50 years of remarkable constancy at around 15,000 km) during the 1980s, and, by 2000, it was about 19,000 km/car.

More fundamentally, hundreds of millions of machines that were driven every year during the late 1990s over the cumulative distance of more than 1 trillion km became not only the embodiments of speed and providers of choice, but they also demanded significant investment of time and capital. But a pioneering attempt to quantify these burdens got it wrong. Ivan Illich, an Austrian Catholic priest and a famous social critic, made an appealing (and seemingly convincing) case for walking (rather than driving a car) in 1974, in his book on energy and equity: "People on their feet are more or less equal" but putting "more than a certain horsepower behind any one passenger . . . has reduced equality among men, restricted their mobility to a system of industrially defined routes and created time scarcity of unprecedented severity" (Illich 1974: 16).

And after he accounted for the time a typical American needed to earn money for buying (or leasing) the vehicle, insuring, fueling, and repairing it, Illich concluded that "the model American puts in 1,600 hours to get 7,500 miles: less than five miles per hour. In countries deprived of a transportation industry, people manage to do the same, walking wherever they want to go" (Illich 1974: 18–19). Illich's estimate of 8 km/h for the early 1970s is equivalent to a fast walk, and it might be easily assumed that, by 2025, the higher cost of cars, insurance premiums, and repair bills and growing traffic congestion reduced that de facto speed even further, to perhaps as low as 5 km/h, a pace that millions of fit retirees could match.

Attempts at total cost assessments should always be welcome as they offer a more realistic appraisal of overall burdens—but Illich's conclusion was wrong. He did not explain how he arrived at his totals, and, regrettably, I did not try to reconstruct the entire set of calculations. When I did so—using the best available 2022–2023 US statistics (for averages of driving time, distance driven, after-tax income, vehicle, gasoline, insurance, and maintenance costs)—I ended up with average driving time of 293 hours prorated to 74 km/h, and with 333 hours needed to earn monies for annual vehicle use, resulting in the adjusted speed of about 35 km/h rather than just 7.5 km per hour (Smil 2025). And the National Highway Traffic Safety Administration put the cost of car accidents at $340 billion in 2019, the last year unaffected by COVID restrictions (NHTSA 2023). With cars involved in just over half of all accidents, that works out to be about $700 per licensed driver, and it adds about 25 hours of additional earnings, lowering the average cost-adjusted speed to 33.3 km/h. This must be compared with a similarly adjusted speed of walking whose energy cost, obviously, is not free. An hour of moderately fast walk claims about 300 kcal of energy and in pre-industrial societies where walking was the only locomotion mode for most people, it would have required at least 1 hour of agricultural labor to secure that food. That would halve the adjusted walking speed to less than 4 km/h, making driving nearly 10 times faster than walking.

Obviously, such a gain (widely intuited and attested by experience) has a strong appeal attested by the universal embrace of cars—as soon as incomes allow that. China is a perfect illustration of this reality. Maoist China had no private cars; when Deng Xiaoping's economic reforms began in the early 1980s, there were just 1 million government and army vehicles; the annual output of cars surpassed 20 million units in 2015; and, in 2023, China also became the world's largest car exporter (Li 2021). So much for the legacy of decades-long Maoist exhortations of frugality and for Illich's arguments about the superiority of walking!

But even with this correction the car's negative impacts on the environment and the quality of life have been substantial. Recall how I opened the first volume of this book by imagining an exceedingly sapient civilization that periodically explores our star system for signs of life. Then let us assume that it defines life not quite like we do, but as a combination of carbon-based metabolism, large size, and mobility. Its recurrent monitoring of the Earth would have shown slow evolution and diversification of large quadrupeds and then the appearance and relatively rapid diffusion of large-brained bipeds whose numbers would rise to more than 1 billion by the beginning of what was our nineteenth century.

Then, a century later, the composition of the dominant mobile mass began to change with the appearance of motorized wheeled organisms that also released carbon dioxide (CO_2) but metabolized carbon in a different way. While the bipeds subsist largely on carbohydrates that they converted (digested) at ambient temperature, the four-wheeled creatures required large volumes of liquid hydrocarbons that they metabolized (by combustion) at high temperatures and at elevated pressures. Just before the end of the twentieth century the planet's anthropomass (dry matter of humans) was about 100 megatons (Mt), while the mass of motorized wheeled organisms was an order of magnitude larger, more than 1 billion t, and they became metabolically dominant.

Every year during the 1990s, the bipeds harvested about 1.3 billion t of carbon for their food while they were busily engaged, in servant- or slave-like capacities, in securing almost 1 billion t of fossil carbon (coke and hydrocarbon feedstocks) to produce the wheeled creatures (initially almost solely with iron, later with increasing use of aluminum, glass, rubber, and plastics) and about 1.5 billion t of gasoline and diesel fuel to energize them. Further studies would show that the wheeled organisms, besides killing annually more than 1 million bipeds and maiming more than 10 million of them, were also responsible for much of the accelerating increase of atmospheric CO_2 that is making life for the bipeds more precarious.

Despite all of this, the bipeds were spending increasingly more time inside these organisms (in some countries averaging up to 90 minutes every day), and they worshipped the metallic creatures through many peculiar rituals that range from elaborate grooming and safeguarding to fanciful displays and expensive accessories and decorations. Kenneth Boulding's insight explains best this extraordinary human devotion to cars. The car, he concluded,

> turns its driver into a knight with the mobility of the aristocrat.... The pedestrian and the person who rides public transportation are, by comparison, peasants looking up with almost inevitable envy at the knights riding by in their mechanical steeds. Once having tasted the delights of a society in which almost everyone can be a knight, it is hard to go back to being peasants. (Boulding 1974: 255)

Automobilization is a perfect example of the difficulties faced by any attempt at comprehensive cost-benefit analyses of the twentieth century's technical advances: it is often impossible even to come up with any conclusive verdict. Cars, besides being efficient killing machines, have been also immensely beneficial tools for saving lives. As Ling (1990: 27) concluded, "no amount of nostalgia can obscure the needless suffering and fatalities in rural America prior to the motor car." By the time a horse was harnessed and a victim of a farm accident was transported to the nearest town, it was too often too late. And a fast ambulance ride may make the difference between life and death for heart attack victims or, indeed, for people severely injured in traffic accidents.

Cars that isolate drivers in private cocoons have also increased social interactions and personal choices. In rural areas, they turned an uncomfortable daylong drive in a horse-pulled buggy to visit friends or relatives into an easy 1-hour trip and opened the choice of shops and schools and entertainment. Being great social equalizers—the road belongs to everybody, and once cars became broadly affordable, previously hard-to-reach destinations and uncommon experiences were within everybody's reach—cars were the key tool of *embourgeoisement* that had sequentially swept industrializing countries during the twentieth century. They were also the keystone machines of twentieth-century modernity, the leading prime movers of modern economies, and critical indicators of their performance. By the mid-1920s, the US car industry led US manufacturing in terms of value, and this accomplishment was later repeated in every nation that concentrated on automobile design and production: after Europe, the last wave of the twentieth century was the rise of Japanese and then South Korean car making and their determined pursuit of the global market.

Cars, perhaps more than any other machine in history, also became cultural icons, a phenomenon as obvious in the young and expanding United States as it was in old and introspective France (Ross 1996). Their presence permeates modern literature, popular music, and, to a particularly large extent, movies, where cars play roles in both romance and thrills. Their allure combines industrial design, mechanical ingenuity, and perceptions of control, power, daring, and adventure. But these perceptions may be also seen as immature illusions, even delusions, and their combination with alcohol is repeatedly deadly: although down from a peak of about 60%, during the late 1990s, some 40% of all US traffic deaths were alcohol-related.

Cars also became prime—both economically and environmentally—and very expensive examples of planned obsolescence, embodiments of an enormous waste of resources, and ephemeral showcases of poor taste and ostentatious display. Average longevity of cars had increased with improved quality: in most Western countries, it was between 10 and 15 years during the 1990s, but cars can be built to last 20 (Swedish Volvos averaged that much during the 1990s) to 25 years (Nieuwenhuis and Wells 1994). Proper engineering should have also facilitated car dismantling, and sensible regulations should have mandated their complete recycling. Moves in both directions came after 1970, but millions of abandoned cars still litter the environment, many in ugly car cemeteries.

Flying, too, has its many yin-yang qualities. No other means of transportation has expanded physical horizons and created previously unthinkable personal experiences and business opportunities. The annual total of passenger-kilometers flown globally by scheduled airlines rose from less than 100 million during the late 1920s to nearly 1.5 billion in 1939; it surpassed 40 billion in the early 1950s, and then its doubling time averaged less than 6 years for roughly a 75-fold increase to nearly 3 trillion passenger-kilometers in 2000 (Figure II.6.6; ICAO 1971, 2001). With the introduction of more affordable European flights, many offered by smaller discount airlines, and with the economic rise of East Asia, the US share of this global aggregate declined from 50% during the late 1960s to about 38% in 2000.

Intercontinental flight also became a routine experience, with steadily declining real costs, as tens of millions of people travel every year to family

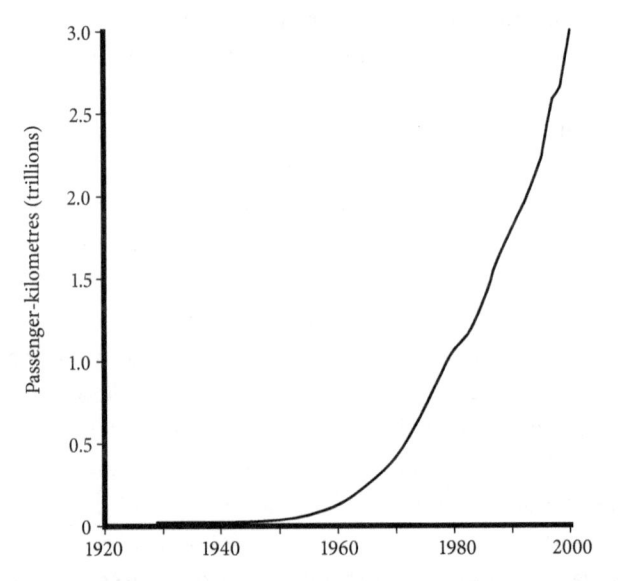

Figure II.6.6 Historical graph shows sustained exponential growth of passenger-kilometers flown worldwide by commercial airlines between 1920 and 2000.

Plotted from data in ICAO (2001) and previous annual ICAO reports.

reunions, business meetings, or beaches. Although travel on the first generation of jetliners was relatively expensive, it marginalized ship travel on its most lucrative crossing in a matter of years: when North Atlantic shipborne passenger traffic peaked in 1957, airlines already carried the same number of people (slightly more than 1 million), and, 10 years later, the ratio was better than 10:1 in favor of jet airplanes, about 5.5 million versus 500,000 (ICAO 1971). By 2000, discounters offered last-minute trans-Atlantic trips for $100 and trans-Pacific flights from the West Coast for less than $1,000. In contrast, in 1939, one trans-Pacific ticket on the Clipper cost $10,000 (in 2000 US$). As soon as an "exotic" place got a runway that accommodated wide-bodied jets, travel agencies found ways to offer affordable package tours: that is how Bali (Denpasar airport), Mauritius (Seewoosagur Ramgoolam airport), and Fiji (Nadi airport) changed from rarely visited islands to tourist destinations.

Negative aspects of this unprecedented mobility include the threat of aircraft hijacking, widespread drug smuggling, and virtually instantaneous diffusion of many infectious diseases. The first two waves of aircraft hijacking were from Castro's Cuba beginning in the early 1960s and the Palestinian attacks of the late 1960s and the early 1970s. Of course, the deadliest of all hijackings—the one that killed nearly 3,000 people, destroyed New York's World Trade Center, damaged the Pentagon, and caused enormous losses to the US economy—took place on September 11, 2001. By the 1970s, airborne drug smuggling evolved into a truly global business, and, before the century's end, its most profitable routes were Colombian shipments of cocaine and Asian exports of heroine. And rapid airborne diffusion of infectious diseases is particularly worrisome because it reintroduces previously eradicated diseases from those regions where they are still endemic: tuberculosis is the most common example of this risk.

The Electronic World

Four waves of technical innovations that transformed the world of electronics came in progressively shorter intervals: the first wave of electronic devices used vacuum tubes (invented before World War I), and, in the United States, it crested with mass ownership of radios and with the first generation of television and computers between the late 1930s and the early 1950s. Then, during the mid-1950s, came the first transistorized versions of these devices. During the 1960s, the new integrated circuits were not yet used widely in consumer products, but since the early 1970s declining costs and rising speeds of microprocessors allowed for previously unthinkable applications that had transformed, subtly or radically, nearly every aspect of economic activities and many habits and routines of everyday life.

The pervasive (albeit usually well hidden) presence of microprocessors affected not only every branch of traditional electronics (from radios and TVs to recorders and phones) but also those industrial and transportation techniques and processes that used to be purely mechanical. Microprocessors transformed computers from unwieldy, costly, energy-intensive, and rather inflexible tools of scientific inquiry and business management to common, inexpensive, and continuously evolving personal possessions. During the 1980s, their deployment also transformed older electronic designs into affordable, mass-produced personal possessions.

The underlying impulse, the unwavering increase in computing speed, had roughly followed Moore's law, and comparisons are done commonly in terms of *flops* (floating point operations per second), that is, basic arithmetic procedures done with numbers whose decimal point is not fixed. Depending on the complexity of a task, the head-hand-paper combinations will take mostly between 1 and 100 seconds (1–0.01 flop). The best electromechanical machines of the early 1940s could not average more than 1 flop; a decade later, UNIVAC topped 1,000 flops, and by the early 1960s, the transistorized LARC reached 1 megaflop (Figure II.6.7). Seymour Cray's (1926–1996) first supercomputer, the CRAY-1 completed in 1976, reached 160 megaflops, and 3 years later, the CRAY Y-MP was the first machine to break through 1 gigaflop. IBM's ASCI Red, the first teraflop (1 trillion operations per second) supercomputer, was in operation by 1996, at Sandia Laboratory in New Mexico.

By the century's end, the fastest machine, IBM's ASCI White, surpassed 12 teraflops: it was installed at the Lawrence Livermore National Laboratory to do virtual testing of thermonuclear weapons and other complex computations. But at that time much more powerful supercomputers, including Japan's NEC Earth Simulator with 35.86 teraflops completed in 2002, were in design stage, all intended for extremely demanding problems in such fields as hydrodynamics, nuclear fusion, and global atmospheric circulation (Top500 2024). But in personal terms nothing came close to the transformation of telephones.

The oldest of all personal electrical communication techniques was made much cheaper and much faster thanks to solid-state electronics. During the

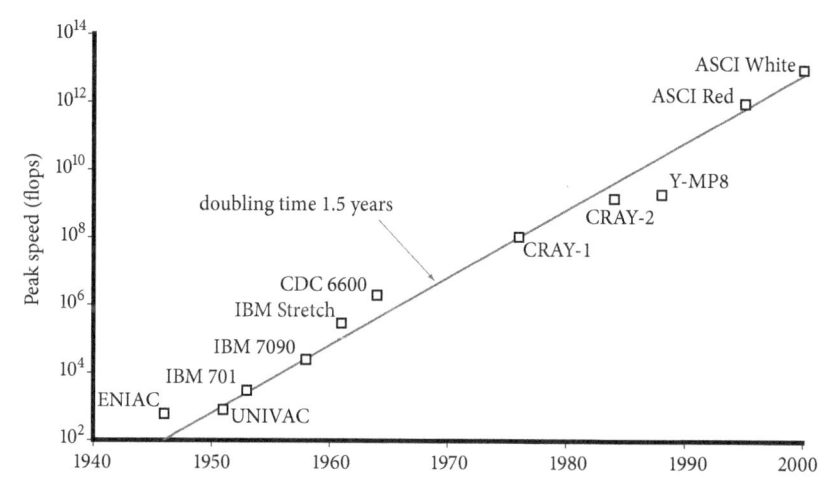

Figure II.6.7 Peak speed of digital computing increased more than 10 billion times between 1944 and 2000, from less than 1,000 flops for ENIAC to 12 teraflops to IBM's ASCI White.
Based on graph in Pulleyblank (2003).

late 1990s, any one of the world's more than 1 billion telephones was almost instantly accessible (and most of the time with clarity comparable to face-to-face conversation) from any other machine just after punching about a dozen digits, forming the world's most impressive anthropogenic network (the area code scheme for the entire world was put in place in 1969). Time–space convergence in the electronic universe, the rate at which places approach one another, was complete as there was no significant difference between the time required to get a local connection and to complete the process of automatic switching for a long-distance call—whereas in 1915, the first US transcontinental call needed five operators and 23 minutes to arrange.

And although cost–space differentials were not yet eliminated on the global scale, they were gone, or reduced to minima, on many national levels as it was possible to make calls for the same rate regardless of the distance between the parties even within such large countries as United States and Canada. Conversation via Hertzian waves is telephone's dominant role—yet this was not always so as telephones were initially seen only as replacements for telegraph messages, and an analysis of AT&T's advertisements showed that, before World War I, only a few percent of them featured social uses of the apparatus (Fischer 1997). By the 1930s, close to or more than half of advertisements promoted sociability; between 1979 and 1984, AT&T's biggest advertising hits were its commercials based on the "Reach out and touch someone" tag line (Bell System 2024), and contentless yakking on mobile phones was emblematic of the 1990s.

Turning to television, I do not intend to recount the charges against this indubitably addictive medium: "[T]hese spurts of light are drunk in by my brain, which sickens quickly, till it thirst again" (Updike 2003: 104). Since the 1950s, too much has been written about biased and manufactured newscasts (who decides what makes news?), repeated use of lacerating images, and the instantaneous global coverage of violence and catastrophes that creates the impression of unceasing crises. And so perhaps just a single noteworthy addition to that long list of charges: after controlling for other effects, Hu et al. (2003) found that each 2-hour increase in daily TV watching was associated with a 23% increase in obesity and a 7% increase in the risk of diabetes.

While TV's cost-benefit appraisals remain highly controversial, there is no doubt that only a few electronic innovations can ever equal the payoff from satellites for communication and Earth observation. Affordable intercontinental calls and (beginning with TIROS 1 in April 1960) more reliable weather forecasting had the most widespread impact, and real-time monitoring of tropical cyclones has prevented many thousands of deaths because early warnings made it possible to safely evacuate low-lying coastal areas subject to sea surges. Satellites also made it possible to monitor the Earth's land-use changes, vegetation, and oceans and to pinpoint locations through the global positioning system (Hofmann-Wellenhof et al. 1997).

During the last two decades of the twentieth century, the multiple benefits of personal computers have been uncritically extolled by too many irrepressible promoters of high-tech nirvanas and the New Economy. True, by the late 1990s, their capabilities were quasi-magical. Any person with a PC (and a credit card) could book a flight without a travel agent's help, send e-mails around the world, read instantly hundreds of newspapers in scores of languages, find the level of air humidity in Tonga and the cost of a ferry from Positano, download one of the classical books in Greek or Latin, or search enormous databases. Or spend all their free time in obscure chatrooms or on shopping websites.

I will address the link between computers and economic growth and labor productivity later in this chapter; here I confine myself to a brief annotated listing of things that computers cannot do and to some undesirable consequences and negative side effects of their use. First, a note about processing speed: it is obvious that the PCs that were in widespread use during the late 1990s were not 10,000 times faster than the pioneering machines of the late 1970s. Most of the newly available cycles went into delivering lots of features quickly, into integration (universal character sets, drag-and-drop functions, etc.) and compatibility with older hardware and software—and only a factor of 10 went into faster responses (Lampson 2003).

Perhaps the least arguable statement regarding what computers cannot do is that they cannot provide a superior teaching and learning experience (Oppenheimer 1997). This delusion, widely held since the 1980s by many unimaginative experts and by too many American school boards, is just the latest transmutation of the overrated promise of image-driven education: Edison believed that movies would become a superior teaching tool (in 1922, he said that the "motion picture is destined to revolutionize our educational system"); during the early 1960s, the famous behaviorist B. F. Skinner believed that teaching machines would accelerate learning; and, during the 1990s, Clinton and Gore believed that computers should be as much a part of every classroom as blackboards. Their goal was largely accomplished—but functional illiteracy and stunning innumeracy in American schools has only deepened.

Jumping from K-12 education to graduate studies and scientific research, it might seem gratuitous to stress that while many computer simulations of physical and chemical processes are very faithful replications of complex realities, simulations of social and economic affairs are only as good as their inevitably incomplete and biased inputs. Unfortunately, too many researchers, enamored of their increasingly more complex models, have been putting too much faith in their creations that will always be subject to the most inexorable of all computer simulation laws—garbage in/garbage out. And when IBM's Deep Blue II beat Garry Kasparov in 1998, many artificial intelligence enthusiasts saw this as a critical milestone in the coming triumph of machines—but, as John Casti (2004: 680) noted, this "taught us nothing about human thought processes, other than that world-class chess-playing can be done in ways completely alien to the way in which human grandmasters do it."

The environmental impact of computers surpasses that of any manufactured device of comparable mass (Kuehr and Williams 2003). While the production of cars or refrigerators needs fossil fuel that is roughly equal to their mass, a PC needs about 10 times its weight and as much as 1.5 t of water. This is not surprising given a large share of such energy-intensive inputs as microchips, rare metals, and plastics. An even greater impact arises from the failure to recycle these machines. Discarded PCs contain relatively large amounts of lead (in cathode ray tubes, circuit-board solder), mercury (in backlighting, boards, switches), beryllium, cadmium (in batteries), hexavalent chromium, and several rare elements (europium [Eu], niobium [Nb], yttrium [Y]), and, during the late 1990s, they were the fastest growing solid waste stream in the United States. This led to their exports to low-income countries and to the first regulations regarding better design, proper disposal, and mandatory recycling.

Finally, a few observations that embrace the entire universe of electronic communication and information. One thing that its expansion did not do was to make any real dent in the frequency of transportation: in fact, the intensity of travel increased concurrently with the pervasiveness of affordable communication. No less remarkably, the great flood of words and images did not turn out to be a zero-sum affair. In 1950, it would have been logical to expect that radio broadcasters, publishers, and movie producers would struggle as television ownership spread, and that conclusion would only have been reinforced with the introduction of VCRs, CDs, PCs, and the World Wide Web. Yet none of this turned out to be true, at least not in the simplest terms of the overall output. During the 1990s, movie attendance increased, newspapers got thicker, and the variety of periodicals and annual totals of new book titles and reprints reached unprecedented numbers.

Unfortunately, as Aldridge (1997: 389) noted, this flood led some theorists to leap "from quantitative measurement of the volume of information and the velocity of its circulation to sweeping conclusions about the qualitative changes in culture and society." This was an unfortunate delusion. To begin with, it is unclear how to identify and measure the overall flux of information, and, given the quantity of text, data, and images that is now generated, it is unavoidable that the quality and significance of what is communicated must be rapidly declining. This conclusion is supported by the best available estimates of aggregate global information recordings and flows in 1999–2000 (Lyman and Varian 2003). The grand total of recorded information amounted to at least 2 exabytes per year, with more than 97% of this put into magnetic storage media: videotapes, digital tapes, computer hard drives, audiotapes, and floppy disks. Most of the rest were photographs and x-rays, while all new printed matter added up to a mere 0.01% (and no more than 0.04%) of the total.

This is not surprising given the high byte density of images—this book's text adds up to less than 3 megabytes (MB), which is less than a single high-resolution photograph or 10 seconds of high-fidelity sound—all it tells us is that we live in a world suffused with images and sound, but even a perfunctory perusal of the bestselling varieties of the late 1990s (*National Enquirer*, World Wrestling Federation TV, rap music) made the inverse relationship between information volume and quality painfully evident. As for the annual information flows, they were at least three times as large as all new information stored on all media, but about 97% of this huge total was due to voice telephone traffic: again, only a delusionary mind could see the largely contentless yammering on cell phones as a sign of enhanced quality.

Person's (1930) observation, made long before it became fashionable to analyze the power of media, provides an apposite ending to these musings. In 1930, he noted that technical advances that "multiplied the number and the variety of stimuli with which the individual may pay" created

> a culture without a unifying dominant interest and without stabilizing patterns of conduct.... It is causing social forces to become centrifugal instead of centripetal, and the individual's interests and standards of conduct to be conceived in terms of self-satisfaction without a stabilizing sense of group responsibility. (Person 1930: 91)

And that was generations before the access to hundreds of cable channels, thousands of rental movies, tens of thousands of chatrooms and blogs, and billions of websites.

Changing Economies

Technical advances have transformed modern economies in many fundamental ways. The three most conspicuous consequences of technical transformations are their indispensable role in raising the quality of life through mass production, their radical restructuring of the labor force, and their critical role in achieving high rates of economic growth. I pay particular attention to several factors that make standard quantifications of such essential concepts as manufacturing, services, gross domestic products, and their growth surprisingly questionable. And I also stress the persistence of major inequalities on both national and global scales: more worrisome than their presence is the fact that, in too many cases, the gap had widened during the twentieth century.

And yet there were also matters that were not affected at all by technical advances or that were changed only marginally. Technical transformations did little to eliminate the traditional segregation of occupations by sex: in affluent Western countries, secretarial, nursing, and retail jobs remained overwhelmingly female throughout the century. Nor did they make the fate of businesses less risky and firm management more rational and more predictable, something one might have expected given the vastly increased flow of information and many unprecedented opportunities for its analysis and evaluation. Yet this was not enough to protect even some of the companies that had pioneered entirely new industrial fields or that transformed major segments of modern economies. Unexpected hardships and stunning declines experienced by these firms proved that inventing new keys (or perhaps especially by doing so?) is not enough to open the doors to long-term prosperity.

Mass Production and Innovation

By the twentieth century's end, we lived in a world that produced annually more than 2 quintillion bytes of new information and that had more than 2 billion radios, more than 1 billion phones and TVs, more than 700 million motor vehicles, and about 500 million personal computers (UNESCO 1999; Lyman and Varian 2003). And in a world of nearly 30 million km of roads, almost 1 million km of high-voltage transmission lines, and more than 10,000 jetliners aloft at any given time. Producing these prodigious inventories and emplacing and operating these complex and far-flung infrastructures could not have been done without resorting to mass production, be it of microprocessors or concrete—and the requisite economies of scale would have been impossible without mechanization that culminated in automation and robotization (see Chapter II.4). Mechanized production clearly favored larger enterprises not only in the manufacturing/fabrication sector but also all along the productive chain, starting with agriculture, forestry, and extraction of natural resources and including transportation and distribution of retail items.

Some notable examples of this universal concentration trend are cited throughout this book, so here is just a handful of additional illustrations.

A commonly used measure of this process is the *four-firm concentration ratio* (percentage of the market held by four leading companies); in the United States of the late 1990s, its highs ranged from 100% for commercial airplanes to about 80% for beef packing and soybean crushing (Heffernan and Hendrickson 2002). Concentration at the global level has been equally high. By the late 1990s, a duopoly of Airbus and Boeing made all large jetliners (and another duopoly, Bombardier and Embraer, made all commuter jets) and just four companies (General Electric, CFM International, Pratt & Whitney, Rolls-Royce) supplied their engines. Four chipmakers (Intel, Advanced Micro Devices, NEC, and Motorola) made about 95% of all microprocessors, three companies (Bridgestone, Goodyear, and Michelin) sold some 60% of all tires, two producers (Owens-Illinois and Saint Gobain) pressed two-thirds of the world's glass bottles, and four carmakers (General Motors, Ford, Toyota-Daihatsu, and DaimlerChrysler) assembled nearly 50% of the world's automobiles.

This concentration trend did not bypass highly specialized industries or the makers of very small series of products or even unique items, as just one or two—and commonly no more than half a dozen—companies serve the entire global market with their unique assemblies and machines. Japan's Jamco, which dominates the global market for lavatories and custom-built galleys and inserts in commercial jetliners, and the world's largest construction equipment made by Caterpillar (United States), Komatsu (Japan), and Liebherr (Germany) are excellent examples in this category. And the companies that specialized in the worldwide fighting of oil and gas well blowouts (Red Adair of Houston and Safety Boss of Calgary) exemplify the concentration of unusual expertise in the field of uncommon services. Another way to demonstrate high economic concentration is to compare revenues of the world's largest companies with national gross domestic products (GDPs). By 2000, General Motor's annual revenues surpassed $170 billion, and those of General Electric reached $130 billion: even the lower figure was higher than the aggregate GDP of about 120 of the world's 180 countries!

All innovative industries had to go through distinct stages of market expansion, saturation, maturity, and retreat as their aggregate production rates first tended to rise exponentially and then relatively brief peaks were followed by often rapid declines. These declines were caused by the inevitable saturation of respective markets or, even before that could take place, by the introduction of new replacement products. The first universal trend can be illustrated by US tractor production, which rose from a few hundred machines a year before 1910 to nearly 600,000 units by 1951 and then declined to fewer than 200,000 since 1969; the second process is exemplified by successive waves of new electronic devices (gramophones, tape decks, CD players) used to play recorded music.

Inevitably, these production waves were reflected in often rapid changes in total employment in the affected manufacturing sectors and in surprisingly high rate of creative destruction: a US study found that between 1987 and 1997, 27% of all plants with more than 10 employees shut down during any 5-year period (Bernard and Jensen 2004). Even some of the largest companies found it difficult to maintain their place in the hierarchy of mass production. As noted in Chapter II.4, the practice of simple mass production of identical copies of items began to unravel more than a generation ago as the overall rise in production volumes had to reckon with the tendency toward shorter product cycles (under the mantra of "new and improved") that affected everything from small household items to expensive electronic devices ("nth generation") and passenger cars (increasingly "loaded"). A single miss of an unpredictable shift in consumer tastes could result in punishing revenue losses—no matter how large or how well established a leading maker may be.

And it is remarkable how many individual inventors or early corporate pioneers that developed major innovations turned out to be either outright losers or at best also-rans while nimble marketing made fortunes for latecomers. The lives of quite a few innovators ended in bitter disappointment or marginal existence, some even in suicide. William C. Durant (1861–1947), the founder of General Motors in 1908, lost, regained, and lost again control of the company; he died in 1947 as a manager of a bowling alley in Flint, Michigan. Philo Farnsworth's Radio and Television company could not compete with RCA in developing commercial TV, and, after years of patent disputes and business problems, he finally sold out in 1949, depressed, alcoholic, and addicted to painkillers. Edwin H. Armstrong did not get any royalties for his FM radio and ended his impoverished existence in suicide. Many inventors are remembered only by a small number of experts because their ideas became associated with the names of those individuals or companies that succeeded in turning the inventions into commercial realities: discoveries are only as successful as their applications!

Nothing illustrates this better than the genesis of DOS, Microsoft's disk operating system. Gary Kildall (1942–1994; Figure II.6.8) wrote Programming Language/Microprocessor for Intel's 4004, the first microchip released in

NEW REALITIES AND COUNTERINTUITIVE WORLDS 613

Figure II.6.8 Gary Kildall wrote the first control program for microcomputers, and, by 1980, its commercial version, CP/M, was the world's most successful operating system. But he did not jump on IBM's offer to supply an operating system for their new PC, a fateful decision that led to Microsoft DOS, its global monopoly, and Gates's wealth.
Kildall's photograph courtesy of Joe Wein.

1971, and for its successors (8008 and 8080); in 1973, he and his students at the Navy's Postgraduate School in Monterey, California, created a short Control Program for Microcomputers (CP/M) that made it possible to read and write files from the newly developed 8-inch floppy disk. Later he made it independently installed on the computer hardware by adding a Basic Input/Output System (BIOS) module, and, by 1980, more than 600,000 copies of enhanced CP/M were sold by Kildall's Digital Research Inc. (DRI) to operate early PCs by Apple, Commodore, Radio Shack, Sharp, and Zenith.

But Kildall, preoccupied by other projects, was slow in upgrading the system, first for hard disks and then for new 16-bit Intel microprocessors. That is why Tim Paterson of Seattle's Computer Products, a maker of memory cards, decided to write his own CP/M-based operating system for Intel's 16-bit 8086 and spent about two man-months doing so (Conner 1998). Paterson's Quick and Dirty OS (QDOS) was released in August 1980, and an enhanced 86-DOS version 1.00 came out in April 1981. Meanwhile, DRI released its CP/M-86 and IBM began to cooperate with Microsoft on its belated launch of a PC. Microsoft recommended CP/M as the obvious choice, but when IBM approached DRI to get a license, it could not make a quick deal.

As Kildall's former colleague recalls, "Gary was very laid back. He didn't care that much. Dorothy [his former wife] ran the business and he ran the technical side and they did not get on" (Eubanks 2004). Besides, IBM was at that time just one of many companies in the new microcomputer business, and, even more important, it was impossible to foresee that it would let its PC design be freely copied, a decision that eventually made the clones—and not the IBM machines—globally dominant. IBM then contracted Microsoft to provide a new operating system, and Microsoft delivered by simply buying SPC's 86-DOS in July 1981 in exchange for $50,000 and a license, and then renaming it MS-DOS (Paterson Technology 2004). Five years later, after a lawsuit regarding the specifics of that license, Microsoft bought it back for $975,000. The success of IBM's PC and its clones made the upgraded versions of 86-DOS the world's best-selling software, and DRI's last attempt at an operating system comeback, DR-DOS, released in 1989, could not survive Microsoft's aggressive marketing.

Paterson eventually rejoined Microsoft as a programmer, but he turned his main attention to off-road racing. His 1997 *Forbes* profile made him sound egomaniacal—"I was 24 when I wrote DOS. It's an accomplishment that probably can't be repeated by anyone ever"—but a year later, talking about the QDOS origins, he was realistic: "I prefer 'original author.' I do not like the word 'inventor' because it implies a certain level of creativity that wasn't really the case" (Conner 1998). Quite: he bought the CP/M manual before he started to write QDOS! Kildall died in 1994, and his former partner's words could be his epitaph: "Gary could have owned this business if he had made the right strategic decisions" (Eubanks 2004). Bill Gates made them instead, and outside of a narrowing circle of people who witnessed or studied the genesis of modern software, nobody now recognizes the name of the man who wrote the first desktop operating system and hence opened the way to the PC era.

The closing decades of the twentieth century also had no shortage of failed or struggling corporate innovators, and, again, the history of computing offers many examples of the fate common in the early phases of technical revolutions, the demise of pioneers. How many people remember, a generation later, such pioneering microcomputer models as Radio Shack's TRS-80 (with Zilog's Z-80 microprocessor), Sinclair Spectrum, and Commodore

PET? More surprising was the retreat or demise of companies that emerged a long time ago as the winners of innovative contests but, despite their size and experience, could not maintain that leading status.

The two most prominent generic examples are the US automobile and aircraft industries. In 1974, when the first oil crisis struck, US automakers had 85% of the North American market. Afterward they could never quite meet the challenge posed by the Japanese companies that mastered and transformed the lessons of US mass production (see Chapter II.4). Chrysler survived only thanks to the 1979–1980 federal bailout, and at that time Ford also narrowly escaped bankruptcy. Chrysler was soon again in trouble, and it merged with Daimler-Benz or, more accurately, was taken over by it (Vlasic and Stertz 2000). During the boom years of the 1990s, US car makers came closer to Japan in average productivity and made some big profits on the newly fashionable SUVs, but they continued to lose the market share to Japanese vehicles made in the United States and to Asian imports. The market total for the big three fell from 73% in 1996 to 63% by 2000, and the prospects were dim enough to contemplate the end of Detroit (Maynard 2003).

And the US aerospace companies were in even greater disarray. European aircraft makers were the dominant innovators until the early 1930s, but then the US companies took over with their iconic commercial designs (exemplified by DC-3 and Boeing 314 Clipper). This strong leadership continued for three decades following World War I, and, by the mid-1970s, the US aviation industry dominated the global jetliner market, and its military planes were at least a generation ahead of foreign designs (Hallion 2004). But the remainder of the twentieth century saw such an appalling decline in the US aerospace industry that a commission convened to deliberate on its future concluded that the country had come dangerously close to squandering the unique advantage bequeathed to it by prior generations (Walker and Peters 2002). By 2000, only five major airplane makers remained (from the high of 47 on VJ-Day), the labor force had declined by nearly half during the 1990s, Boeing was steadily yielding to Airbus in the global commercial market, and most top-performing military planes (F-16, F/A-18, A-10) had been flying for more than a quarter-century.

Two high-profile instances that illustrate the impact of rapid technical advances on the fortunes of communication and information business are the cases of AT&T and Xerox. American Telephone & Telegraph was formed in 1885, as a subsidiary of American Bell Telephone, to develop and operate long-distance telephony, but in 1899, it became the parent of the Bell System.

An abbreviated list of its remarkable firsts must include transcontinental calls (1915), dial phones (1919), commercial radio stations (1922), demonstration of TV (1927), concept of cellular telephony (1947), direct long-distance dialing (1951), trans-Atlantic (1956) and trans-Pacific (1964) telephone cables, communication satellites (1962), the touchtone keypad (1963), electronic call switching (1965), and 911 (1968). Moreover, its famous Bell Labs in Murray Hill, New Jersey, produced seven Nobel Prize winners and many fundamental technical innovations. Without exaggeration, the progress of telephone techniques is the history of AT&T.

The company became a government-sanctioned monopoly in 1913, and it divested itself of all local telephone services only on January 1, 1984, because of an antitrust suit launched in 1974. Its post-breakup future seemed assured, and the outlook appeared even better for Lucent Technologies, a spin-off company created in 1996 that included Bell Labs. Values of both companies rose to record highs during the economic bubble years of the late 1990s, to nearly $100 per share for AT&T and to more than $60 per share for Lucent (which was worth at one time more than a quarter trillion dollars) before their precipitous fall reduced them by, respectively, more than 85% and 95%.

Xerox's prehistory goes back to 1938, when Chester Carlson (1906–1968) began to offer his patented process of xerography to more than 20 companies, including the electronic giants General Electric, IBM, and RCA: they all turned him down (Alexander and Smith 1988). Finally, in 1944, Batelle Memorial Institute of Columbus, Ohio, agreed to develop the machine, and, in 1946, it was joined by Haloid Company from Rochester, New York; the company, renamed Xerox, marketed the first xerographic machine in 1949 but did not produce the first plain-paper copier until 1959. During the 1960s, the company's pioneering machines provided a convenient and affordable way for storing and spreading information. The company's profits allowed it to set up a large research establishment for electronic information and communication techniques: as noted in Chapter II.5, PARC became a leading innovator in interactive computing, including Robert Metcalfe's design of Ethernet, the first local area network (PARC 2024).

The first system that combined the key components of the modern graphical user interface (GUI)—a mouse, bit-mapped display, and graphical windows—was PARC's Alto completed in April 1973, and the first Alto-based commercial model was the Xerox Star in 1981, which also featured

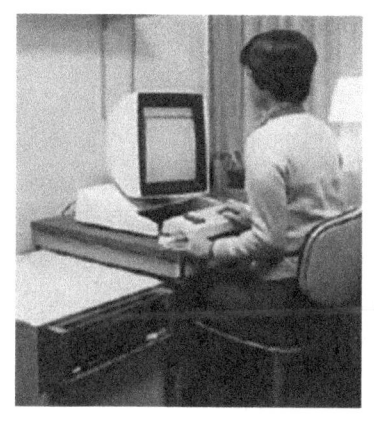

Figure II.6.9 Alto (left) and Star (right), the two pioneering desktop computers designed by the Xerox PARC in, respectively, 1973 and 1981.
Courtesy of Palo Alto Research Center, Palo Alto, California.

double-clickable icons, overlapping windows, and dialog boxes (Figure II.6.9). But Xerox did not pursue this innovative advantage, and, in 1983, as the late-starting IBM and Apple were claiming the PC business, it diversified instead into financial services (Alexander and Smith 1988). First the Japanese copier makers and then cheap desktop printers cut its share of the global market for its core products. The company's insurance group was finally sold at a huge loss; there were repeated layoffs of thousands of employees, and, by 2000, the company's stock fell to $7 from the peak of $64 in 1999.

Transformation of Labor

Technical advances of the twentieth century brought several universal transformations of the labor force: their onset and rates of change had differed, but no modernizing economy could follow a different path. I illustrate these trends largely with reliable historical US statistics (Fisk 2001). Perhaps the most important change is that child labor, common before World War I, had almost disappeared in all affluent countries. US federal law prohibits any full-time work for teenagers under the age of 16, whereas in 1900, 6% of the labor force, or 1.75 million workers, were children 10–15 years of age (USBC 1975). For example, before Michael Owens invented automatic bottle-making machines (see Chapter II.4), boys worked commonly in glass-blowing workshop. Owens himself began shoveling coal at age 10, carried the hot glassware at 11, and worked with adult glass blowers since he was 15.

Although he later recalled these early years with fondness, his inventions made such experiences, thankfully, unrepeatable: in 1913, the national Child Labor Committee credited the machines' rapid commercial acceptance with doing more to eliminate child labor in glass-container plants than it was able to accomplish through legislation (ASME 1983)—and similar effects were seen thanks to mechanization in other industries. While children left the workplaces, women entered the labor market in unprecedented numbers: in the United States, only 19% of females of working age were employed in 1900, compared with 60% in 1999. The dominant nature of their jobs shifted from domestic to public services, while shares in the best-paying professions rose substantially, from 1% to 29% for lawyers and from 6% to 24% for physicians.

As already noted, continuing mechanization and chemicalization of food production reduced the rural labor force to less than 5% of the total in nearly all Western countries (see Chapter II.4, and Figure II.4.1). Shares of workers employed in mining, construction, and manufacturing decreased less precipitously in the United States from the post-World War I peak of about 47% to about 19% in 2000. The decline in manufacturing employment is seen to be a much more consequential development than job losses in agriculture or transportation but advancing automation and robotization transformed the entire sector to such an extent that the very term is misleading—many products are not touched by a human hand before they are integrated into finished items or shipped to consumers.

More important, if we attribute job losses in manufacturing to automation that is impossible without special microchips and complex software, should not we extend the sector's new definition to embrace custom chip design and code writing? After all, during the last quarter of the twentieth century these activities became as indispensable for industrial production as did dexterous manual control of new electricity-driven machines a century earlier. Long-term declines in manufacturing employment had differed substantially among the leading economies: the US rate peaked at above 39% right after World War I, its decline was briefly reversed during World War II, and, by the century's end, the share was just 14%. But in absolute terms US manufacturing employed as many people in 2000 as it did in 1965, and more than it did in the early 1940s, when 32% of the workforce were in wartime industries.

Moreover, while the annual increase of the overall nonfarm productivity averaged about 2% between 1950 and 2000, manufacturing productivity advanced at 2.9% per year, which means that an hour of factory labor generated about 4.2 times as much value in 2000 as in 1950 (CEA 2000; Cobet and Wilson 2002). This, in turn, increased real wages and improved the country's competitiveness. By these standards even US manufacturing did rather well, but many people still see the sector's relative decline in employment and increased manufactured imports as the more important trends. These are emotional matters where economics clashes with national pride and strategy: it is one thing when the world's largest economy loses nearly all of its domestic capacity to make plastic toys or children's shoes and another thing entirely when it sheds large shares of the labor force in industries that were seen for most of the twentieth century as the keystones of economic progress, including steelmaking, automobiles, machine tools, and electronics.

Technical advances dictated this general outcome: they raised industrial productivity to such an extent that declining employment in manufacturing was inevitable and that increased production (the total value of the US industrial output rose more than 11-fold between 1940 and 2000) could be accommodated without hiring more workers. And higher returns on investment in more capital-intensive sectors mean that affluent countries will inevitably tend to lose many kinds of basic manufacturing to economies at lower levels of overall development. The same trend was evident in Japan, where manufacturing employment was down to about 20% from its peak in 2000, and in the European Union, where it slipped during the century's last two decades from 37% to less than 25% in Germany and from 27% to less than 14% in the United Kingdom. During the same time, vigorous growth pushed manufacturing employment up in China, South Korea, and Malaysia.

With every other sector in decline, services absorbed the growing labor force: in the United States, their share rose from 31% in 1900 to slightly more than 80% by 2000, the European Union's overall rate in 2000 was 68%, and in Japan it was nearly 60%. Expanded governments were an important part of this trend: in the United States, their share of the labor force doubled between 1900 and 2000. Service jobs also exemplified the trend toward a more educated labor force and higher investment per worker. In 1900, only one in seven Americans graduated from high school, and only about two out of 100 had postsecondary education; a century later, these rates were, respectively, 83% and 25%. Technical transformations have also redefined many job categories. Drafting and graphic design are excellent examples of this transformation: pencils, pens, and hand-drafting were replaced by mice and electrons on screens.

These universal labor transformations brought three fundamental personal benefits to every segment of the modern workforce: real wages rose, average work hours declined, and workplace safety improved. In the United States, average per capita income (in 1999 US$) was $2,400 in 1900 and $33,700 a century later, and manufacturing wages rose from about $3.50 to $13.90 per hour (Fisk 2001). Pay-offs for higher qualifications were substantial: hourly wages of university graduates were 60–65% above those of people with only a high school diploma. The average workweek in American manufacturing was cut from 53 hours in 1900 to 39–41 hours after World War II, and in retail to only 29 hours (more part-time workers); the French working week was legally limited to 35 hours. Improved workplace safety had already been illustrated by the example of US coal-mining accidents (see Figure II.4.8), and that improvement is indicative of the general trend toward safer workplaces.

But technical advances did little to eliminate or even to attenuate business cycles: in the United States, 19 of them occurred during the twentieth century (USCB 2004). They brought periods of unemployment ranging from very low (between 1900 and 1908 less than 3%, between 1995 and 2000 less than 5%) to very high (more than 8% during the recessions that peaked in 1921, 1975, and 1982) and to the record level of about 25% in 1933, the worst year of the Great Depression. What technical advances did, particularly because of instantaneous transfers of complex designs overseas and because of cheap intercontinental transportation, was to extend the potential pool of the labor force worldwide. With the right conditions (political stability and access to containerized shipping) a multinational corporation could use labor forces in new plants on any continent more profitably than at its home locations.

Growth, Interdependence, and Inequalities

High rates of economic growth became one of the least questioned components of good governance. Of course, even slightly higher average rates make a huge long-term difference: a 2.4% rate will raise total economic product 11-fold in a century; 3% will result in a 20-fold increase. As "healthy" economic growth became a key indicator of national fortunes, even an economically illiterate public learned that the GDP measures the aggregate output of goods and services of an economy. Unfortunately, GDP does not make any distinction as to the quality and desirability of its contents. The aggregate GDP rises as money is invested in more efficient and more profitable methods of industrial fabrication or in measures to reduce soil erosion and hence improve crop yields or the purity of local water supplies—but it also grows as more asthma inhalers are bought and more emergency hospital visits are made in cities with high levels of photochemical smog or as more policemen are hired to control violence in ghettoized cities. And there is no consideration of quality and desirability when the GDP grows with goods made by child labor or based on theft of intellectual property, or with sales of repulsively violent computer games or as psychotic crowds fight over a piece of sale-priced ephemeral junk.

Despite these shortcomings, long-term GDP trends clearly indicate the overall performance of economies. That is why many economists devoted their time and ingenuity to the reconstructions of past GDPs (or GNPs) for individual nations, why others prepared extensive international comparisons (Summers and Heston 1991; Prados de la Escosura 2000), and why Angus Maddison (1995, 2001) had boldly undertaken the integration of GDP estimates at the global level. These reconstructions (expressed in constant monies) leave no doubt that the twentieth century experienced economic growth that was unmatched in history. Maddison's (2001) approximations show the compound growth rates of gross world product (GWP) averaging only about 0.2%/year (0.05% in per capita terms) for the first eight centuries of the second millennium, rising to 0.93% for the 50 years between 1820 and 1870, and to 2.1% between 1870 and 1913. Despite the two world wars and the Great Depression of the 1930s, GWP grew an average of 1.85% between 1913 and 1950, or 0.91% in per capita terms (Figure II.6.10).

The years between 1950 and 1973 saw unprecedented rates of annual per capita GDP growth on all continents, resulting in a nearly 5% annual rise of the GWP (almost 3% in per capita terms). Rates were generally lower afterward, and average per capita GWP grew by about 1.35% between 1973 and 2000. For the world's largest economies, the record during the twentieth century was as follows: American per capita GDP increased about eightfold (it doubled between 1966 and 2000), and the average Western European economic product rose about sixfold; Japanese per capita GDP grew about 17 times, with the fastest rate (about 8%) between 1956 (when it surpassed the 1941 peak) and 1973 (when it was checked, as was the rest of the world, by the

622   THE TWENTIETH CENTURY

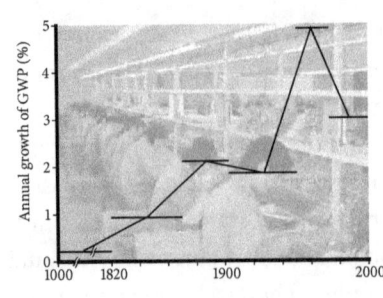

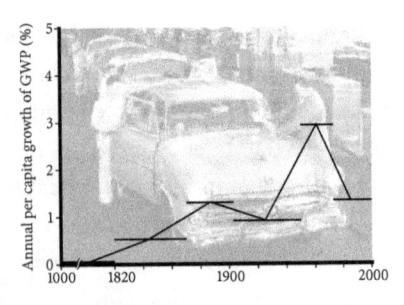

Figure II.6.10 Average compound rates of growth of world economic product (gross world product, GWP): annual growth (left) and annual per capita growth (right).
Plotted from data in Maddison (2001).

sudden oil price rise by the Organization of Petroleum Exporting Countries [OPEC]). By far the most impressive achievement during the century's last quarter was post-Mao China's rise, with real per capita GDP quadrupling in a single generation!

But all these multiples showing the secular rise of per capita GDP or average family income are substantial underestimates of real gains. In 1900, an income eight times larger than the average could not buy the American standard of living in 2000 because so many goods and services taken for granted in 2000 did not even exist in 1900. And, as we have already seen with the example of US electricity, its real prices fell by about 98% (factor of 50), but because of huge increases in efficiency, the same monies bought a century later more than 4,000 times more illumination (see Chapter II.1 and Figure II.1.5). Even greater multiples would be obtained for electricity that is used to do calculations (large and heavy electromechanical machines of 1900 vs. microprocessors of 2000). Consequently, in many ways the real gains have been practically infinite.

Two serious weaknesses of standard economic accounts make all GDP accounts questionable: their exclusion of underground economic activities (undervaluing the actual GDP) and the fact that they ignore cumulative changes in natural assets and environmental services that are indispensable for any economy (this omission greatly overvalues the real achievements). The black (shadow, underground) economy was growing during the last decades of the twentieth century, and even conservative estimates put its size at about 15% of officially reported GDP in high-income nations and at one-third in low-income countries, with the highest shares more than 40% in such countries as Mexico, the Philippines, and Nigeria (Lippert and Walker 1997; Schneider and Ente 2003). The global shadow economy added up to about $9 trillion in 1998 compared to $39 trillion of the world economic product (at purchasing power parity) or nearly a quarter of all activities!

More fundamentally, standard economic accounts ignore the degradation of the environment caused by human activities and treat the depletion of irreplaceable or difficult-to-regenerate natural assets as current income that boosts GDP totals—although it is obvious that no society can keep running down its fundamental natural capital, treat this process as income, and hope to survive (Smil 1993). Because macroeconomic growth encroaches on a larger, and obviously limited, whole of the Earth's biosphere, it incurs an opportunity cost that does eventually surpass its marginal benefit (Daly 2001). There is no consensus about quantifying these impacts, but some studies suggested their cost reduces standard GDP by significant margins (Daly and Cobb 1989; Repetto et al. 1991; Smil 1996).

Finding the real rates of growth is thus a surprisingly elusive task. But no matter if GDP growth rates are exaggerated or overcorrected, a large fraction of the overall increase must be attributed to technical advances. Remarkably, this seemingly obvious conclusion began to be quantified only during the 1950s. Contrary to the common belief that saw the combined input of labor and capital as the key factor in explaining economic growth (Rostow 1990), Abramovitz (1956) demonstrated that, since 1870, this combination accounted for just 10% of the growth of per capita output in the US economy. His result for labor productivity growth was similar, with capital per worker-hour explaining just 20% of the total. Technical advances had to be a major part of the residual (known as the *total factor productivity*), and Solow (1957) came out with a startling attribution: between 1909 and 1949, the US economy doubled its gross output per hour of work, and some 88% of that increase could be attributed to technical changes in the broadest sense, with the higher capital intensity accounting for the small remainder.

The following decades saw many studies elaborating and extending these basic conclusions. Edward Denison found that, between 1929 and 1982, 55% of US economic growth was due to advances in knowledge, 16% to improved resource allocation (labor shift from farming to industry), and 18% to economies of scale (Denison 1985). As the last two factors depend on technical advances, Denison's disaggregation implied that at least three-quarters of all growth was attributable to technical innovation. In his Nobel lecture, Solow concluded that "the permanent rate of growth of output per unit of labor input ... depends entirely on the rate of technological progress in the broadest sense" (Solow 1987: xii). Solow, Denison, and others considered technical change as an exogenous variable: innovators make fundamental discoveries or come up with new ideas that are eventually adopted by enterprises and converted into commercial products or practices.

But such a unidirectional interpretation ignores the learning process and numerous feedbacks that exist in any economy. Arrow (1962) was the first economist to link the learning function with already accumulated knowledge, and, during the 1960s, others related technical advances to labor resources devoted to R&D. Endogenous models of economic growth, the idea that technical change is induced by previous actions within an economy and not the result of autonomous forces that impinge from outside, became very popular during the 1980s and 1990s (Romer 1990; Grossman and Helpman 1991). These models saw a continuous increase in the level of resources devoted to innovation as the key factor in continued increase in economic growth.

Solow (2001: 14) did not find their key claim—"that the rate of growth of this is a function of the level of that, where 'that' is some fairly simple and accessible variable that can be maneuvered by policy"—persuasive. Others pointed out that the claim is inconsistent with historical evidence. Jones (1995) found that despite a substantial increase in the number of US scientists engaged in R&D, there was no comparable rise in economic growth during the previous 40 years. De Loo and Soete (1999) concluded that the lack of correlation between rising R&D and productivity growth during the post-World War II period was mainly because those efforts became increasingly devoted to product differentiation rather than to product or process innovation: this improves consumer's welfare but has little effect on economic growth. Moreover, R&D expenditures were highly skewed toward a handful of sectors (precision instruments, electrical machinery, pharmaceuticals, chemicals).

David (1990) explained the productivity paradox of the 1980s—the failure of the wave of microprocessor and memory chip innovations to bring about a surge in US productivity—by a historical analogy with the dynamo: electricity generation had little impact on manufacturing productivity before the early 1920s, four decades after the commissioning of the first power plants. The slowdown in labor productivity growth (to less than 1.5%/year between 1973 and 1995) was eventually reversed, and, during the late 1990s, it was almost as high (2.5%) between 1960 and 1973 (USDL 2002). And Landau

(2003: 219) turned the basic premise on its head, arguing that "what needs to be explained is not growth but the failure to grow, the failure to engage in the natural process of investment and innovation." I find this search for disaggregating and quantifying the drivers of economic growth and productivity fascinating, but I am not surprised that it has brought so many mixed answers.

Decompositions of the famous Solow residual performed by Denison and others may be inappropriate for drawing any inferences about the underlying causes of economic growth (Grossman and Helpman 1991). Identified factors are not independent entities but rather dynamically linked variables of a complex system, and the exogenous–endogenous dichotomy is to a large extent artificial because all of these impulses are at play. Perhaps the best illustration of these preconditions, interactions, and synergies is the genesis of the proliferating Silicon Valley microelectronic companies that began in September 1957, when eight young scientists and engineers left Shockley Semiconductor (Moore 1994). They pooled $4,000 of their own money, but Sherman Fairchild (1896–1971), inventor of an aerial camera, advanced $1.3 million to establish Fairchild Semiconductors (FS). Six months later the new subsidiary made a profit by selling custom transistors to IBM; in 1958 came its first planar transistor, and, in 1959, Robert Noyce designed the first integrated circuit (see Chapter II.5).

Key factors that determine the growth of productivity clicked in this case: top education (six of the eight men had a PhD), existence of adequate venture capital and relatively open access to it, innovative drive, willingness to take risks, ability to tap into established commerce and opportunities to create new demand, a legal framework that made it comparatively easy to set up new ventures, and effective protection of intellectual property. And FS was just the beginning. In 1968, Robert Noyce and Gordon Moore left it to establish Intel Corporation. In 1972, Eugene Kleiner (1923–2003), who helped to arrange the initial FS financing, set up Kleiner Perkins (later Kleiner Perkins Caufield & Byers), the country's leading venture capital firm that helped to create new companies with combined assets in the tens of billions of dollars. Eventually at least 200 companies besides Intel (including Advanced Micro Devices and National Semiconductor) were founded by former employees of FS.

Qualitative conclusions regarding the sources of productivity growth remain preferable, even more so when one considers the already outlined inherent flaws of aggregate measures of economic performance. Significant shares of modern economic growth must be attributed to narrowly targeted R&D and to endogenous technical innovation in general, and even larger shares are due to broadly conceived technical advances (regardless of their origins) and to such first-order preconditions as the need for a more educated labor force and adequate investment in ancillary techniques (in turn, these factors are stimulated by successful advances). What is much more important is to know what economic growth, regardless of its respective drivers, has done in the long run: Has it reduced the gap between rich and poor countries, has it helped to eliminate some of the worst income inequalities?

There is no clear-cut answer to these critical questions. Neoclassical growth theory predicts convergence of average per capita incomes around the world, and an unmistakable post-World War II trend toward increasing interdependence of national economies should have accelerated this process. I prefer the term "interdependence" to "globalization" because it describes more accurately the realities of modern economies that came to rely on more distant and more diverse sources of raw materials, food, and manufactured products and on increasingly universal systems of communication and information processing. The genesis of this reality goes back to the beginning of the early modern world, and the process was accelerated with the expansion and maturation of the British Empire and the US economy during the nineteenth century (O'Rourke and Williamson 2000; Cain and Hopkins 2002).

During the last generation of the nineteenth century, food grains (from the United States and Canada) and frozen meat (from the United States, Australia, and Argentina) joined textiles as major consumer items of new intercontinental trade that was made possible by inexpensive steam-driven shipping. By the middle of the twentieth century, the combination of progressing conversion from coal to liquid fuels and of the highly uneven distribution of crude oil resources elevated tanker shipments to perhaps the most global of all commercial exchanges. But only after 1950 did the trend embrace all economic activities. Inexpensive manufacturing was the first productive segment affected by it, beginning with the stitching of shirts and stuffing of toy animals and culminating, during the century's last quarter, with the assembly of intricate electronic devices.

During the 1990s, the level of attendant specialization and product concentration reached unprecedented levels as increasing shares of consumer products originated in highly mechanized assembly plants or in both highly automated and highly labor-intensive fabrication facilities located increasingly in just a few countries in East Asia. Microprocessors are the best example of fully automated industrial production, while cellular phones are among the most prominent illustrations of labor-intensive assembly. And the 1990s was also the first decade when globalization extended into many services. The trend, both extolled and reviled by clashing constituencies of free market promoters and anti-globalization activists, had not run its course. But has it helped to narrow huge income gaps?

Reconstructions of historic datasets and more reliable macroeconomic statistics of the second half of the twentieth century made it possible to quantify inequality in global terms, between countries and regions and within a particular country. The simplest approach is to use national averages of GDP or GNP per capita (exchange rated or in terms of purchasing power parity, or PPP) derived from standard national accounts; a more realistic choice is to weigh the national averages by population totals, and the most revealing approach is to use average disposable incomes (from household surveys) and their distribution within a country in order to assign a level of income to every person in the world. Comparisons of the best available studies (Figure II.6.11) show that unadjusted global inequality changed little between 1950 and 1975 but that it subsequently increased, and not only because Africa has been falling behind (Milanovic 2002).

In contrast, population-weighted calculations show a significant convergence of incomes across countries since the late 1960s—but a closer look reveals that this desirable shift has been entirely due to China's post-1980 gains: weighted analysis for world income without China shows little change between 1950 and 2000. And comparisons of post-1970 studies that use household incomes and within-country distributions indicate only a minor improvement (Sala-i-Martin 2002), slight deterioration, or basically no change (Figure II.6.11). These disparities are attributable to inaccuracies of different household surveys and to the use of different inequality indexes. Still, a few conclusions are clear. Except for the World War II years, inequality of per capita GDPs among the Western nations has been declining throughout the entire twentieth century. As this included the interwar period of reduced internationalization and post-World War II decades of vigorous globalization, the convergence does not appear to be related to openness but rather to rapid application of technical advances that has characterized all Western economies.

As a result, North America and Western Europe pulled ahead of the rest of the world. The US GDP per capita was 3.5 times the global mean in 1913, 4.5 times in 1950, and nearly 5 times in 2000 (Maddison 2001). Even

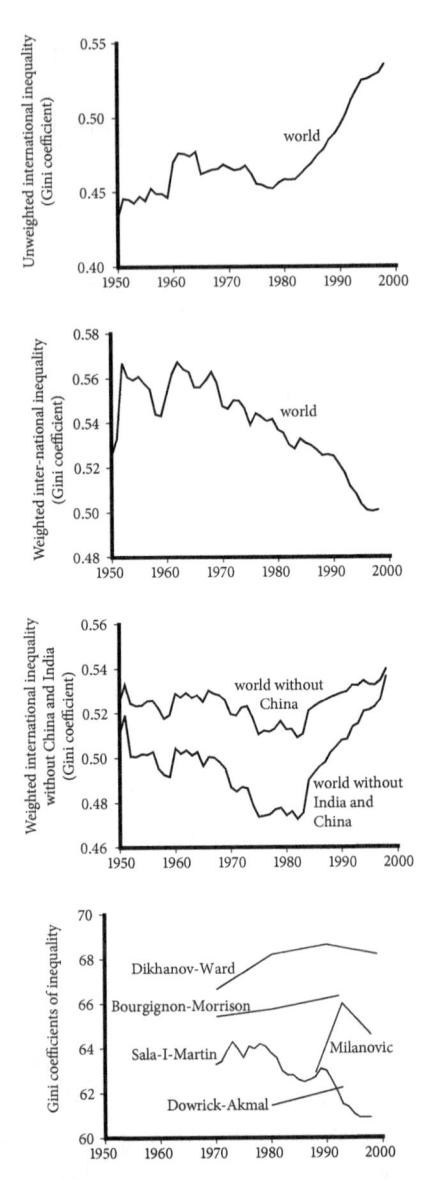

Figure II.6.11 Four views of global inequality of per capita GDPs and incomes, 1950–2000. Unweighted international inequality (top left) shows considerable divergence; population-weighted assessment (top right) indicates an encouraging convergence, but that trend was almost solely due to China's post-1980 progress (bottom left); and Gini coefficients for evaluations based on income surveys show no clear trend (bottom right).
Based on Milanovic (2002).

China's spectacular post-1980 growth did not narrow the income gap with the Western world very much (but China pulled ahead of most of the low-income countries). The obverse of this trend has been the growing number of downwardly mobile countries: in 1960, there were 25 countries whose GDP per capita was less than a third of the poorest Western nation; 40 years later there were nearly 80. Africa's post-1960 economic (and social) decline accounted for most of this negative shift (by the late 1990s, more than 80% of the continent's countries were in the poorest category), but, China's and India's progress aside, the Asian situation also deteriorated. Income-based calculations confirm this trend as most global disparities can be accounted for by cross-country and not within-country disparities (Sala-i-Martin 2002).

Perhaps the most stunning consequence of these inequalities is "the emptiness in the middle" (Milanovic 2002: 92), the emergence of a world without a middle class. By 1998, fewer than 4% of world population lived in countries with GDPs between $(PPP)8,000 and $(PPP)20,000 while 80% were below the threshold. Everything we know about social stability tells us that this is a most undesirable trend. And yet our technical capabilities are more than adequate to secure a decent quality of life for the entirety of humanity; our inability to do so is one of the greatest failures of rational governance. In November 1946, George Orwell wondered about this disparity between what was and what could be in one of his incisive "As I Please" *Tribune* essays.

> [N]ow we have it in our power to be comfortable, as our ancestors had not.... And yet exactly at the moment when there is, or could be, plenty of everything for everybody, nearly our whole energies have to be taken up in trying to grab territories, markets and raw materials from one another. (Orwell 1946: 1137)

During the last decades of the twentieth century there were still many violent (and some seemingly interminable) conflicts in Africa, Asia, Latin America, and even Europe, but most of our inability to use technical advances for securing a decent quality of life on the global scale had to be ascribed to a multitude of causes. On the one hand is the heedlessly wasteful Western lifestyle of excessive consumption; on the other are horrendous corruption and greed of governing elites in most low-income countries and the readiness to resort to violence, often ethnically motivated, in many of them. There is also general unwillingness to compromise, cooperate, and (notwithstanding all the talk about sustainability) work within new economic paradigms. But the great and persistent global inequality gap is just one of many risks that modern civilization faces and that raise questions about its longevity.

Levels of Risk

The great pre-World War I technical saltation followed by the advances of the twentieth century added entire new categories of risk to the unavoidable threat of natural catastrophes and to historically ever-present dangers of food shortages and famines, tainted water, pandemics, and violent conflicts. These new risks included many forms of environmental pollution (ranging from exposures to gases and particulates released by combustion of fossil fuels to contaminated water and long-lasting pesticide residues), the already-described transportation accidents (motor vehicles, airplanes, crude oil tankers), and the chance of major malfunctions of complex technical systems (radioactivity from nuclear electricity-generating plants, failed high-voltage transmission networks instantly depriving tens of millions of people of electricity).

At the same time, the overall level of everyday risks faced by an average citizen of an affluent country had substantially decreased. Famines were eliminated (in fact, overeating became a major risk); safe water became available by turning on a tap (as a result, US cases of typhoid fever fell from nearly 100/100,000 in 1900 to 0.1 in 2000); the incidence of infectious diseases plummeted thanks to the combination of vaccination and antibiotics, better materials, and better emergency responses; and stricter safety rules reduced the frequency and consequences of home and industrial accidents. Yet paradoxically, the emergence of many new risks arising from technical advances and often exaggerated reporting of their possible effects created a widespread impression of having to face a more perilous world. Statistical evidence firmly contradicts this perception. No matter what has been alleged about the perils of persistent pesticides, particulate air pollution, dental amalgam, or low-intensity electromagnetic fields, none of their effects was reflected deleteriously in basic demographic statistics.

Life expectancies at birth kept rising to unprecedented levels (see Figure II.1.2), overall morbidity rates kept on declining, and when mortalities due to cardiovascular diseases and various forms of cancer are corrected for aging populations, there was no evidence of runaway increases in their frequency. Just the opposite: US heart disease mortality was cut in half between

1950 and 2000, and, between 1975 and 2000, primary cancers declined in all age groups younger than 54 years (Ries et al. 2003). Moreover, epidemiological evidence shows that a significant number of these diseases are linked to lifestyle rather than to unavoidable environmental factors caused by technical advances.

Mokdad et al. (2004) estimated that tobacco use, poor diet, inadequate physical activity, and alcohol consumption caused about 900,000 deaths in the United States in 2000 compared to 127,000 for all toxic agents, motor vehicle accidents, and firearm use. Consequently, I do not offer quantifications of major risks arising from technical advances: whatever their undeniable effects may be, they have been more than compensated for by numerous benefits brought by other, or in many cases the very same, modern techniques. I do concentrate on two classes of risks that really matter: violent conflicts and their enormous death toll, and global biospheric change.

Although many accounts must remain imperfect, statistics confirm that in absolute terms the twentieth century was by far the deadliest in history, and technical advances greatly contributed to this toll because of the greatly increased destructiveness of weapons. Development of nuclear weapons, enmity between the two superpowers, and tens of thousands of accumulated warheads made the possibility of a massive thermonuclear exchange the most worrisome of all risks the modern world had faced between the late 1950s and the late 1980s. Easing of superpower tensions with Gorbachev's post-1985 reformist policies was followed by the unexpected nonviolent disintegration of the USSR and the end of the Cold War in 1990. But just as the risk of a catastrophic superpower confrontation was receding, new concerns about the global consequences of environmental change began to emerge during the late 1980s.

None of these concerns was as acute as the danger of global thermonuclear war, but all of them could bring unprecedented changes whose eventual consequences were difficult to predict. Resolute attention to one of those changes—loss of stratospheric ozone—and the fortuitous availability of an effective technical solution combined to produce a very rapid response to avert its further destruction by chlorofluorocarbons (CFCs). But this success cannot serve as a model for dealing with the immensely more complex challenge of greenhouse gas emissions whose rising atmospheric concentrations increase the risks of relatively rapid and rather pronounced global warming. There were other major environmental problems—including the loss of biodiversity in general and tropical deforestation in particular, the biosphere's enrichment with reactive nitrogen, and elevated concentrations of tropospheric ozone—but none of them posed as broad and as radical ecosystemic, economic, and social consequences as does the prospect of living on a planet whose average surface temperature could be several degrees warmer in less than 100 years.

Weapons and Conflicts

Explosives of unprecedented power and new weapons to deliver them were first introduced during the two pre-World War I generations (see Chapter I.4). Pursuit of these murderous innovations continued throughout the twentieth century when the more advanced designs were augmented by three techniques that further raised the destructiveness of modern warfare: nuclear weapons, and jet and rocket engines (see Chapter II.2). Consequently, the century saw gains in destructive power even more stunning than the already-cited gains in useful energy controlled by individuals. The kinetic energy of a World War I shrapnel shell was about 50,000 times that of a prehistoric hunter's stone-tipped arrow, but the kinetic energy of the largest tested thermonuclear weapon (Soviet 100 Mt bomb in 1961) was 240 billion times that of the gun shrapnel (Smil 2017a). For the first time in history, it became possible to envisage the almost instantaneous death of tens of millions of people. Even a limited thermonuclear exchange (targeting only strategic facilities and no cities) between the two superpowers would have caused at least 27 million and up to 59 million deaths (von Hippel et al. 1988).

This horrifying trend began with Nobel's dynamite and other high-power explosives of the late nineteenth century. Pre-World War I and World War I innovations included machine guns, tanks, dreadnoughts, submarines, gas warfare, and the first military planes, including light bombers (Waitt 1942; O'Connell 1989). The two interwar decades saw the rapid development of weapons that defined World War II: battle tanks, fighter planes, long-range bombers, and aircraft carriers. The war's closing months brought the deployment of two new prime movers (gas turbines and rockets) and the destruction of Hiroshima on August 6, 1945, and of Nagasaki three days later. These destructive innovations were further developed during the decades of superpower confrontation.

Increased destructiveness of weapons led to much higher combat casualties. When expressed as fatalities per 1,000 men fielded at the start of a conflict, these were below 200 during the Crimean War (1853–1856) and the Franco-Prussian War (1870–1871), but they surpassed 2,000 for both the United Kingdom and Germany during World War I, and during World War II they were about 3,500 for Germany and almost 4,000 for Russia (Richardson 1960). Civilian casualties grew even faster, and during World War II they reached at least 30 million people or 60% of the 50 million total (White 2003). The most devastating prenuclear techniques of mass destruction could kill as many people as the first atomic bombs. Allied bombing of Germany killed nearly 600,000, and about 100,000 people were killed by nighttime raids by large fleets of B-29 bombers as incendiary bombs were used to level about 83 km^2 of Japan's four principal cities between March 10 and 20, 1945 (Odgers 1957). Five months later, two nuclear bombs killed close to 200,000 people in Hiroshima and Nagasaki. Modern wars also caused an enormous economic drain on the warring parties (Smil 2004b).

Total US expenditures (all in 2000 US$) were about $250 billion for World War I, $2.75 trillion for World War II, and $450 billion for the Vietnam War (1964–1972). Peak World War II spending, expressed as a share of national economic product, ranged from 54% in the United States in 1944 to 76% in the USSR in 1942 and in Germany in 1943 (Harrison 1988). And wartime economic losses took a long time to repair; for example, Japan's per capita GDP surpassed its pre-World War II peak only in 1956. Peacetime expenditures on military forces were also large. The highly militarized Soviet economy was spending for decades on the order of 15% of its GDP on the development of weapons and on its large standing army, border guard, and paramilitary forces. In contrast, US defense spending reached nearly 10% of GDP during the peak year of the Vietnam War (Smil 2003).

The two world wars stand out because of their combat casualties (nearly 9 and 20 million) and total deaths (approximately 15 and 50 million) and because they involved all major military and economic powers in conflicts whose outcomes shaped so much in modern history—but the century's most regrettable statistic is the never-ending list of other armed conflicts. Interstate and civil conflicts that caused the death of more than about 30 people had roughly the same frequency (20–40 per decade) during the first half of the twentieth century and the frequency of interstate conflicts fell to about 10 per decade by the 1990s, while the civil conflicts peaked at nearly 80 during the 1960s and then leveled off at about 50 per decade (Brecke 1999). The duration of major conflicts ranged from days (Six Day War of June 1967) to decades (Sudan's civil war started in 1955 and was not over by 2000), and the deadliest of them claimed millions of lives: Vietnam (Second Indochina War, 1964–1973) about 3.5 million, and Korea (1950–1953) about 2.8 million (White 2003).

Technical advances also made it much easier and cheaper to use small arms—assault rifles (Russian AK-47, US M-16, Israeli Uzi), hand grenades, mortars, and land mines—to wage new total wars where gratuitously destruction, looting, burning, absence of any sanctuaries, and indiscriminate killings are the norm (Figure II.6.12). During the 1990s, nearly 100 countries were producing small arms, and the decade saw more than 100 largely small-weapons conflicts, from Afghanistan and Angola to Sudan, and Uganda (Boutwell and Klare 2000).

And even after the century's killing sprees were over, millions of mines remained in place in some 40 countries, many of them hard-to-detect plastic varieties made in colors and shapes that help them to blend with the background, continuing to kill and maim at a worldwide rate of more than 15,000 a year during the late 1990s (Strada 1996). Technical advances also made it much more deadly for terrorists to strike at their targets. A suicide bomber whose belt contains a few kilograms of high explosives (RDX, often spiked

Figure II.6.12 These mass-produced assault rifles—Russian AK-47 (top), US M-16 (middle), and Israeli Uzi (bottom)—became the weapons of choice in scores of the conflicts during the last quarter of the twentieth century, when they killed more people than did the two nuclear bombs dropped in Japan in August 1945.

Reproduced from weapon catalogues.

with metal bits) unleashes about 50 times more explosive energy than a TNT-filled grenade thrown into a crowd. And car bombs can be 10–20 times more devastating than a suicide bomber. These devices—able to destroy massive buildings and kill hundreds of people (as they did at the US Marine base in Lebanon in October 1983 or at the Murrah federal building in Oklahoma in May 1996)—are (as already noted in Volume I) mixtures of two readily available materials: ammonium nitrate, a common fertilizer, and fuel oil.

Technical advances provided new, more affordable, and readily accessible weapons with which to carry on old dreams of aggression, revenge, or aimless brutality. At the same time, nuclear weapons clearly acted as an effective war deterrent, and a major, but hardly noticed, landmark was passed in 1984, as the major powers of the affluent world had managed to remain at peace with each other for the longest continuous stretch since the emergence of nation states in Europe more than a millennium ago (Mueller 1989). And, by the century's end, the probability of a major war between any two (or more) leading economic powers—United States, European Union, Japan, China, and Russia—became comfortably low, and affluent countries became more concerned about very different risks that could undermine their future: about the changing biosphere.

Global Environmental Change

Human actions were shaping the Earth's envelope of life long before the beginning of recorded history, and centuries of pre-industrial agriculture, slow urban expansion, expanding metallurgy, shipbuilding, and artisanal manufactures had left many locales and even extensive regions scarred with excessive soil erosion, deforestation, landscape degradation, and air and water pollution (Krech, McNeill, and Merchant 2004). The great technical saltation of the two pre-World War I generations began to change the magnitude of these impacts as previously localized effects began to spread to large urban and industrial areas—and even more worrisome was the emergence of two kinds of global environmental changes.

The first category includes problems that affect nearly every major region of the inhabited world—but whose progression in one region has little or no direct impact on a situation elsewhere: excessive soil erosion, depletion of aquifers, deforestation, and irrigation-induced salinization are among old environmental problems that were elevated by technical advances into this new category. The worldwide presence of countless bits of degradation-resistant fragmented plastic found in the oceans and marine sediments (Thompson et al. 2004) and contamination of biota with leaded gasoline exemplify new twentieth-century additions. By the early 1970s, every large urban area with high vehicular traffic had elevated lead levels as well as recurrent periods of photochemical smog, high levels of sulfur dioxide, acid precipitation, and most likely also inadequately treated (or untreated) wastewater and problems with the disposal of solid waste that included higher shares of plastics.

Many of these global problems were effectively eliminated or reduced to tolerable levels by technical fixes. In the United States, the phasing out of leaded gasoline began in 1975; catalytic converters cut dramatically the precursor emissions of photochemical smog; the switch to low-sulfur fuels and flue gas desulfurization of large power plants cut sulfur dioxide levels and began to lower precipitation acidity; better water treatment techniques and recycling of many solid waste products became compulsory in many places. But the closing decades of the twentieth century brought the second category of global environmental problems, whereby anybody's actions anywhere are eventually translated into biosphere-wide impacts that could alter the fundamental biophysical parameters within which any civilization has to operate.

By far the most worrisome challenges in this new category were the destruction of stratospheric ozone and human interference in global cycles of carbon and nitrogen (Smil 2002). The ozone saga began, again, with Thomas Midgley. Unlike the choice of a lead compound for gasoline (its toxicity was well known, and Midgley himself had lead poisoning in 1923), Midgley's selection of CFCs as perfect refrigerants made impeccable sense (Midgley and Heene 1930). These compounds were not only inexpensive to synthesize but also completely inert, noncorrosive, nonflammable, and nontoxic, a perfect choice for household refrigerators and air conditioners, and later also for aerosol propellants, as foam-blowing agents, and for cleaning electronic circuits and extracting plant oils (Cagin and Dray 1993; Figure II.6.13).

What was totally unsuspected for some four decades after their introduction is that, once these gases get mixed into the stratosphere, they can be dissociated there by light wavelengths (between 290 and 320 nanometers [nm], the ultraviolet B [UVB] part of the spectrum) that are entirely screened from the troposphere by the stratospheric ozone. This breakdown releases free chlorine atoms, which then break down ozone (O_3) molecules and form chlorine oxide (ClO); in turn, ClO reacts with oxygen to produce O_2 and

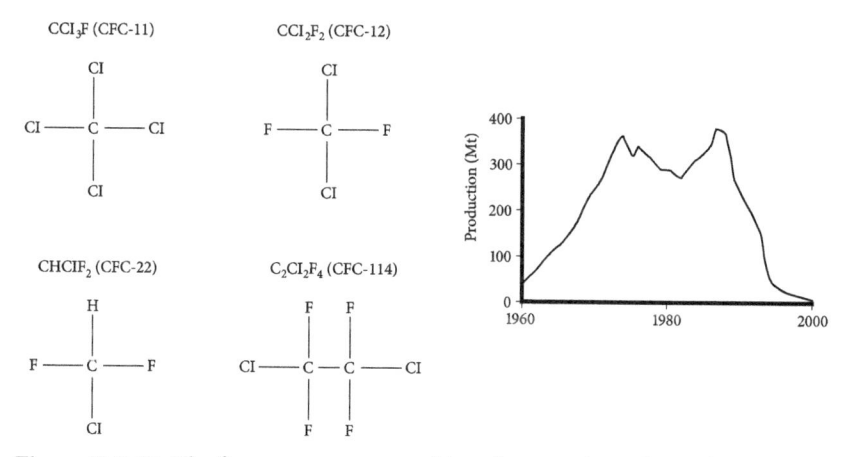

Figure II.6.13 The four most common chlorofluorocarbons (CFCs) and their annual global production, 1960–2000.
Plotted from data from the Alternative Fluorocarbons Environmental Acceptability Study.

frees the chlorine atom. Before its eventual removal, a single chlorine atom can destroy on the order of 10^5 O_3 molecules. This destructive sequence was outlined first by Molina and Rowland (1974), and its confirmation came in 1985, from the British Antarctic Survey base at Halley Bay, Antarctica, where measurements showed that the lowest spring (October) O_3 levels had declined by about a third (Farman, Gardiner, and Shanklin 1985).

The continuing decline of stratospheric ozone and its possible extension beyond Antarctica posed very serious concerns. Complex terrestrial life developed only after the oxygenated atmosphere gave rise to a sufficient concentration of stratospheric ozone that prevented all but a tiny fraction of UVB radiation from reaching the biosphere. Without this molecular shield, productivity of phytoplankton, the base of complex marine food webs, would decline, crop yields would be reduced, and both humans and animals would experience higher incidences of basal and squamous cell carcinomas, eye cataracts, conjunctivitis, photokeratitis of the cornea, and blepharospasm as well as effects on the immune system (Tevini 1993). That is why the annual, and more pronounced, recurrence of what became known as the "Antarctic ozone hole" led to rapidly concluded treaties aimed at cutting and then eliminating all CFC emissions (Figure II.6.13) and replacing them by less dangerous hydrochlorofluorocarbons (UNEP 1995).

Because of their long atmospheric lifetimes, CFCs' stratospheric effects will be felt for decades to come, but atmospheric concentrations of these compounds have been falling since 1994, and the stratosphere was expected to return to its pre-CFC composition before 2050. Just as this concern eased, the Earth's radiation balance caused by rising concentrations of anthropogenic greenhouse gases began to be widely recognized as the most worrisome environmental challenge in modern history. The greenhouse effect is indispensable for life: if several atmospheric gases (mainly water vapor, CO_2, and methane [CH_4]) were not selectively absorbing some of the outgoing radiation (before reradiating it both down- and upward), the Earth's surface temperature would be 33°C colder, and there would be no liquid water and no life.

Long-term feedback between CO_2 levels, surface temperature, and the weathering of silicate minerals explains a relative stability of tropospheric temperatures (Berner 1998). Atmospheric CO_2 levels are now known accurately for the past 740,000 years thanks to the analyses of air bubbles from ice cores retrieved in Antarctica (Petit et al. 1999; EPICA 2004). This record shows that during all recorded history, before the beginning of fossil-fueled era, CO_2 levels had fluctuated within a very narrow range of 250–290 ppm (Figure II.6.14).

Accelerating combustion of fossil fuels pushed the concentration to about 320 ppm by 1958, the year when continuous CO_2 measurements began in Hawaii and at the South Pole (Keeling 1998). By 2000, Mauna Loa's mean CO_2 levels surpassed 370 ppm, a more than 30% increase in 150 years (Figure II.6.14). The combined effect of all anthropogenic greenhouse gases had already burdened the troposphere with an additional 2.8 W/m² by the late

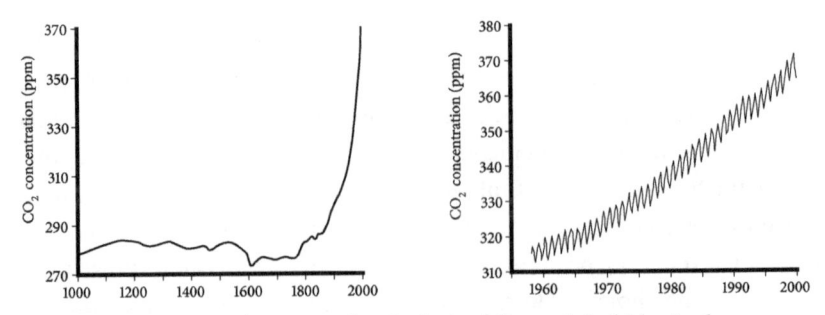

Figure II.6.14 Atmospheric CO_2 levels derived from air bubbles in the Antarctic ice (1000–1958) and concentrations of the gas from continuous monitoring at Mauna Loa (1958–2000).

Plotted from data from the Carbon Dioxide Information Analysis Center at Oak Ridge National Laboratory.

1990s (Hansen et al. 2000). This is equivalent to little more than 1% of solar radiation reaching the ground. Further increases would eventually double the pre-industrial greenhouse gas levels and raise the average tropospheric temperatures by several degrees Celsius above the 2000 mean.

At the end of the twentieth century the broad scientific consensus on all of these matters was best summarized in the reports of the Intergovernmental Panel on Climatic Change (Houghton et al. 2001). The warming would be more pronounced on land and during nights, with winter increases about two to three times the global mean in higher latitudes than in the tropics, and greater in the Arctic than in the Antarctic. Its major impacts would include intensification of the global water cycle, changed precipitation, higher circumpolar runoffs, later snowfalls and earlier snowmelts, thermal expansion of seawater and gradual melting of mountain glaciers leading to appreciable (up to 1 m) sea level rise, changed photosynthesis and shifts of ecosystemic boundaries, and poleward extension of some tropical diseases.

These possible threats generated an enormous amount of research and a great deal of policy debates, but the challenge cannot be tackled without limiting, and eventually replacing, the energetic foundation of modern civilization. Because of the rising combustion of fossil fuels, annual CO_2 emissions rose from less than half a gigatonne of carbon (0.5 Gt C) in 1900 to 1.5 Gt C in 1950, and, by 2000, it surpassed 6.5 Gt C, with about 35% originating from coal and 60% from hydrocarbons (Marland et al. 2000). There was also a net release of 1–2 Gt C from the conversion of natural ecosystems, mainly from tropical deforestation. In addition, methane levels have roughly doubled since 1860, with the gas coming from anaerobic fermentation in landfills and rice paddies, ruminant livestock, and direct emissions from coal mines, natural gas wells, and pipelines (Warneck 2000).

Human interference is relatively much more intensive in the global nitrogen cycle than in carbon's flows. During the 1990s, losses of nitrogen from synthetic fertilizers and manures, nitrogen added through biofixation by leguminous crops, and nitrogen oxides released from combustion of fossil fuels added up to about as much reactive nitrogen as the element's total mass fixed by terrestrial and marine ecosystems and by lightning (Smil 2002). In contrast, carbon from fossil fuel combustion and land-use changes was equal to less than 10% of the element's annual photosynthetic fixation. But various impacts of this interference—nitrogen enrichment of aquatic and terrestrial ecosystems, water pollution, contributions to photochemical smog and acid precipitation—do not pose such a large global risk as does relatively rapid climate change.

Frontispiece II.7. We have no clearer idea of the world in the year 2100 than we had of the world of 2000 in 1913, when *Scientific American* (in its July 26 issue) published this cover illustration of elevated sidewalks that "will solve city transportation problems" as each kind of transport "will be then free to develop itself along its own lines." But during the twentieth century, rather than elevating sidewalks, we had elevated highways and railways, and in many cities the situation is exactly the *reverse* of this image.

7

A New Era or an Ephemeral Phenomenon?

Outlook for Technical Civilization

> "Where are you riding to, master?" "I don't know," I said, "only away from here, away from here. Always away from here, only by doing so can I reach my destination." "And so you know your destination?" he asked. "Yes," I answered, "didn't I says so? Away-From-Here, that is my destination.
>
> —Franz Kafka, *Das Ziel*

Unavoidable natural catastrophes—such as prolonged mega volcanic eruptions or a collision of the planet with a large asteroid that could imperil or even destroy any civilization—are not, fortunately, common. Although there is nothing regular about those phenomena, their recurrence is mostly in the order of many thousands to many millions of years, compared to the less than 10,000 years that have elapsed since the beginning of settled human cultures. Consequently, there is a very high probability that we could double, triple, or even quintuple the age of our civilization without encountering any extraordinarily destructive natural events.

Yet this is not as reassuring as it might seem because there are other perils that pose increasing risks to the long-term survival of a civilization that is based on high energy flows and mass consumption. We were able to identify and rapidly reduce one of those dangers (reduction of stratospheric ozone by CFCs), but so far, we have adopted only an inadequate, piecemeal (and, most definitely, not truly global) set of actions to deal with the potentially enormously disruptive risk of rapid global warming. As expected, some global concerns recede with time, others persist (or can even intensify), and new worries emerge. At the very beginning of the twenty-first century there were many non-environmental concerns besides Islamic terrorism and running out of cheap crude oil that preoccupied the affluent world: poor governance, the persistence of economic inequality, violent intra-national conflicts, illegal migration.

In the original (2006) version of the second part of this book, I noted the virtual inexorability of another major pandemic, and in *Global Catastrophes and Trends* (Smil 2008) I concluded that we were, probabilistically speaking, already very much inside a high-risk zone and, with another event being highly likely before 2021. The COVID-19 pandemic officially began in March 2020. But even after setting these major uncertainties aside, I singled out the energy transition from fossil fuels to renewable (or other non-fossil carbon) energies as an epochal event of the twenty-first century because it will be protracted and complex and because it will affect every infrastructure, process, and habit that was created by the enormous fossil-fuel flows of the twentieth century.

The concatenation of these concerns leads to some very worrisome scenarios. Think of increases of political instability, unmanageable migrations, intra- and international conflicts, economic crises, pandemics, nuclear proliferation taking place alongside a rapid and very costly energy transition accompanied by climate change whose progress will be faster than at any time in history. Consequently, an unavoidable question to be asked about humanity's prospects is simply this: Have the last six generations of great technical innovations and transformations been merely the beginning of a new extended era of unprecedented accomplishments and sustained affluence—or have they been a historically ephemeral aberration that does not have any realistic chance of continuing along the same, or a similar trajectory, for much longer?

Contrasting Views

Some techno-optimists have no doubt about prospects for the future. Saying that "We are convinced that the progress of the twentieth century is not a mere historical blip but rather the start of a long-term trend of improved life on earth" (Moore and Simon 1999: 7) sounds all too modest besides the predictions of imminent (2035–2045) "Singularity," when machine intelligence will have surpassed human intelligence and will spread into the universe at infinite speed (Kurzweil 2005). Extreme singularity claims aside, several facts bolster the claims of continuing advances. To begin with, correction or elimination of enormous inefficiencies and irrational policies could extend the benefits of technical advances to billions of people without any new, untried techniques. More important is the well-documented recurrent and profound underestimation of human inventiveness and, even more so, of our adaptability. New anthropological evidence indicates that, in contrast to all other organisms, including our hominin ancestors, humans have evolved not to adapt to specific conditions but to cope with change (Potts 2001).

In contrast to optimistic views of a future that is elevated by technical advances, there is a long and continuously flourishing tradition of highlighting the negatives and unforeseen consequences of technical change, which is never seen as real progress. These writings range from Jacques Ellul's eloquent condemnations of *la technique* (Ellul 1964, 1990) and enumerations of the deleterious side effects and costly surprises (Tenner 1996) to arguments about the limits of technical efficiency and diminishing returns of technical innovation (Giarini and Loubergé 1978). In Ellul's bleak view, technical advances bring consequences that are independent of our desires, leaving no room for moral considerations, no choice but "all or nothing": "If we make use of technique, we must accept the specificity and autonomy of its ends, the totality of its rules. Our own desires and aspirations can change nothing" (Ellul 1964: 141).

And the environmental costs of creating fabulous new opportunities and rising incomes have been high not only in the case of many heavy industries but also with the latest high-tech endeavors. No other place exemplifies this better than Santa Clara County, south of San Francisco. Until the early 1950s, this was one of the country's richest agricultural areas, with highly fertile alluvial soil and a climate ideal for fruit: it produced nearly half of the world's prunes, apricots, and cherries, and people came to admire some 8 million blooming trees in the world's largest near-continuous orchard spread over more than 50,000 hectares (ha) (Sachs 1999). This Valley of the Heart's Delight was transformed into Silicon Valley, the paragon of the late twentieth century's technical progress and an area with one of the highest per capita incomes in the world. But the price of this achievement was destroyed orchards, the densest concentration of hazardous waste dumps in the United States, polluted aquifers, shellfish contaminated by heavy metals, the near-continuous ugliness of thousands of nondescript buildings housing electronics companies, exorbitantly priced inadequate housing, and chronic traffic congestion.

As different as they are, these opposite reactions to technical advances share two traits with the tenets of bizarre Pacific cargo cults that expect goods to descend from the heavens and to change everything. These beliefs—in autonomous technical advances that exert either uplifting (*pace* Simon or Kurzweil) or downgrading (*pace* Ellul) powers—err by exaggerating the impacts of technical innovations. Techno-optimists and techno-utopians see them as solutions for even the most intractable problems (Simon and Kahn 1984; Segal 1985; Kurzweil 2005). During the late 1990s, the Internet was added to this miracles-generating list: previously both telegraph and radio were initially expected to reduce or eliminate conflicts as they promoted better communication, and Michael Dertouzos (1997) succumbed to similar wishful thinking when he saw the Internet as a tool that may bring "computer-aided peace." At the same time, critics saw technical advances as a source of unpredictable problems, as destroyers of the best human qualities and despoilers of natural environments (Lightman, Sarewitz, and Desser 2003; McKibben 2003).

Realities are much more complex. Many innovations arise organically and seemingly inevitably from previous accomplishments, but generally there is nothing automatic about the process, and there are many examples of unfulfilled expectations and failed cargo deliveries. After half a century of promises, by 2000 we were still waiting for true artificial intelligence, for nuclear electricity that is too cheap to meter, for energy from fusion, and for victory over viruses and cures for common cancers. On the other hand, there have been many technical advances whose truly miraculous nature, from the late nineteenth-century point of view, have vastly outweighed any undesirable side effects. Electricity did not do away with economic cycles, and electronic communication did not eliminate conflicts, but how many of us—even when mindful of air pollutants from fuel combustion, electromagnetic fields along high-voltage transmission lines, or risks of electrocutions—would volunteer for a nonelectric economy and pre-electronic society?

And as much as I like some of Ellul's penetrating criticisms of *la technique*, I think he is wrong when he concludes that human desires and aspirations are powerless against the inexorable, self-augmenting progression of technical advances. Sensible public policies (still too rare, admittedly) have stopped several advances that 20–30 years ago were seen by many experts as near-certainties by the century's end and that received temporarily large government funding. The most prominent items on this list are intensive manned exploration of Mars, large-scale supersonic air travel (the Concorde remained an expensive oddity with a limited service), mass extraction of crude oil from shales, and accelerated development of nuclear breeder reactors (Smil 2003).

And even the most destructive technical advances were not responsible for the worst death tolls of the twentieth century. Horrific as their contribution was, technical innovations were not the leading cause of what Rhodes (1988) labeled "public man-made death" and what he called perhaps the most overlooked cause of twentieth-century mortality. Most of the 160–180 million people who died in those ways were not killed by nuclear bombs, napalm, or air raids but by the ageless disdain for life, by hatred, cold, disease, or lack of food or medical care. History's largest famine did not take place in antiquity or in the Middle Ages, it was a horrific man-made event caused by Mao's economic delusions that killed at least 30 million Chinese between 1959 and 1961 (Smil 1999b; Yang 2008); during the 1970s, the deaths due to Stalin's long reign of terror (1924–1953) were put at no less than 20 million (Medvedev 1972; Solzhenitsyn 1973–1975).

During the late 1970s, some 20% of Cambodia's people (about 1.7 million) died because of Pol Pot's deranged creation of a new society; between 1915 and 1923, some 1.5 million Armenians died in Turkey just because they were Armenians, and, for the same reason, 1.2 million Tutsis and Hutus were hacked to death or shot in Rwanda and Burundi between 1959 and 1995 (White 2003). And Mesquida and Wiener (1999) demonstrated that a key human factor that is responsible for violence and that is highly correlated with a society's chance to be involved in a conflict is simply the ratio of young (15–29) men in a country's population. Evolutionary psychology thus provides a better explanation for the genesis of killings than does the availability and destructiveness of weapons: as the share of young men in a population declines, weapons of any kind become less used.

Human evolution has always been contingent on the complex unfolding and synergy of many factors, and although undeniably the progress of technical advances has its special dynamism, it is not exempt from this general rule. In 1987, when I was writing a book on the complex interplays of modern realities, myths, and options, I chose to end it by a back-casting exercise, by describing the world of the early 1980s through the eyes of the early 1930s. I concluded that "there is no reason to believe that the next fifty years will see changes less profound and less far-reaching," that we cannot foresee the totality of these changes, that the effects of the unpredictable will remain strong, and that "this uncertainty cannot be shed: it is the quintessence of the human condition" (Smil 1987: 348).

These lasting realities mean that tomorrow's technical advances may lead us into a civilizational cul-de-sac, or, after a period of stagnation, their new wave can be generated by a non-Western society. Jacob Bronowski (1973: 437) closed his sweeping examination of human history by concluding that humanity's ascent will continue.

> But do not assume that it will go on carried by Western civilisation as we know it. We are being weighed in the balance at this moment. If we give up, the next step will be taken—but not by us. We have not been given any guarantee that Assyria and Egypt and Rome were not given. We are waiting to be somebody's past too, and not necessarily that of our future.

Siding with Bronowski, I would just note that six generations is not enough time to embed any long-lasting automatism that will self-perpetuate for centuries. The era of astonishing technical innovations energized by fossil fuels and electricity has been a very short one when compared to the tens of thousands of years that our species had spent in paleolithic foraging (circumscribed by the deployment of nothing but somatic energy and fire for warmth and primitive cooking), or the thousands of years it lived in traditional societies based on biomass fuels, animate energies, and a limited array of simple mechanical devices. The era's brevity is even greater when one dates it not from its genesis during the two pre-World War I generations but from the time when the great innovations had spread sufficiently to benefit majorities of industries and households in the innovating nations.

Many statistics illuminate the recent nature of affluent societies that were created by technical advances. For example, in 1900, only 8% of American households were connected to electricity or telephone networks, fewer than 15% had a bathtub and an indoor toilet, but 40% of them had incomes below the poverty line (USBC 1975). The European situation was in some ways distinctly premodern even after World War II. In 1954, a French census showed only a quarter of homes with an indoor toilet and just 10% with a bathtub, shower, or central heating (Prost 1991). At that time even a lengthy interruption of electricity supply in most European cities would have made no impact on cooking or heating (done by coal or wood stoves) or access to money or shopping (all transactions were in cash, and there were no electronic registers).

As for transportation, the entirely car-dependent structure of US cities is largely a post-1960 phenomenon, while in most European countries more distance was covered by bicycles than by cars until the 1950s; for example, the break-even point in the Netherlands was reached only in 1962 (Ploeger 1994). In the same year, still fewer than 10% of French homes had a telephone (and new subscribers had to wait one and a half years to get it), and, on July 10, 1962, the first intercontinental TV transmission took place: Telstar I sent a black-and-white picture of the US flag flapping in a Maine breeze; two weeks later it carried the first live images (including the Statue of Liberty and the Sistine Chapel) between the United States and Europe. And, of course, the world of complicated electronics could come about only after transistors were miniaturized and crowded as microprocessors onto slivers of silicon—that is, after the early 1970s.

Consequently, by the beginning of the twenty-first century, even in the world's most affluent countries, most people have enjoyed a large range of these advances, now taken so much for granted, only for two generations, a historically ephemeral span. And if we were to specify the beginning of technically advanced civilization only from the time when its benefits embraced more than half of the humanity, then—even if our criteria include just such basic quality-of-life indicators as nationwide longevity in excess of 70 years, adequate food supply, and universal school attendance until the age of 14—the new age would have begun, largely thanks to the accomplishments of China's post-Mao economic growth, only during the late 1990s!

Given these realities, any forecasts of the long-term course of a high-energy mechatronic civilization might be as informed, and as reliable, as projections about the long-term prospects of steam engines made during the 1830s, six generations after Newcomen's first clumsy designs and just as the machine began to conquer both land and waterborne transportation. Who would have predicted that during the following six generations even the most advanced steam engines would become rare museum pieces? The same may happen within a century to our turbines and engines. But, unless one believes that technical progress is an inexorably autonomous process, this ignorance cuts also the other way: we may, as Bronowski put it, give up and find ourselves on a regressive course. What would matter most then would be the rate of decline, but chances are that it would not be a slow unraveling akin to the centuries-long demise of the Western Roman Empire.

The *magnum mysterium* of unknown tomorrows remains as deep as ever: despite our armies of scientists and engineers, our observational and analytical capabilities, complex computer models, think tanks, and expert meetings, we have no clearer idea of the world at the end of the twenty-first century than we had of the world of the 1990s in 1900. We can only ask all those fascinating questions. Has the era of ingenious machines, ever more accomplished techniques, and complex interconnections been nothing but a historically ephemeral phenomenon—or have we lived in a mere prelude to even more impressive accomplishments? Did its genesis during the two pre-1914 generations see the peak of hopes that were subsequently crushed by world wars and then, after a short post-World War II interlude, darkened by the growing realization of the possibly irreparable harm that technical advances inflict on the biosphere? Will this complex, high-energy, machine-dependent civilization survive at least until the end of the twenty-first century?

We cannot know. What we do know is that the six generations preceding the end of the twentieth century saw the most rapid and the most profound change our species has experienced in its long evolution. During the two pre-1914 generations we laid the foundations for an expansive civilization based on the synergy of fossil fuels, science, and technical innovation. The twentieth century had followed along the same path—and hence the profound indebtedness and the inevitable continuities. But we had traveled along that path at a very different pace—and hence many deep disparities and many more disquieting thoughts accompany the process of technical transformation. There is much in the record to be proud of, and much that is disappointing and abhorrent. And the record contains the promise of both yet greater ascent and of unprecedented failure. Neither of these courses is preordained, and one thing that no technical transformation could have done is to eliminate the unpredictable: the future is as open as ever—and as uncertain.

PART III
LOOKING BACK, LOOKING AHEAD
Two Decades Later

When I suggested reissuing the two-book set narrating the story of the pre-World War I fundamental technical inventions and their transformations during the twentieth century in a single volume, I did not have any extensive changes in mind. I wanted to keep the original approach: factual, with accurate technical details and with interdisciplinary perspectives. And because the two books under one cover make a substantial volume, I did not envisage any major extensions. Just the opposite: I made some cuts, streamlined some sections, and left out some unnecessary comments and asides (topical at the time of writing, superfluous more than two decades later) to make space for this brief coda: it offers no new grand summaries or verdict reversals, just some musings—while looking back and ahead—falling into five categories.

The first one includes just two additional contrasts that vividly capture those remarkable accomplishments brought by the Age of Synergy and that were noted by two famous contemporary witnesses. The second one provides a brief reminder of broader historical settings and of the coexistence of rational and creative progress with irrational and destructive claims and ideologies because both eras, distinguished by their technical ingenuity and inventive rationality, had their dark obverse. The third one points out the continued (in many instances increasing) existential need for many fundamental advances introduced during the five decades before World War I and transformed by twentieth century improvements: by far the best illustration of this dependence during the first quarter of the twenty-first century has been the economic rise of China.

The fourth one addresses the matter of innovative acceleration, the supposed early arrival of the "Singularity," when the unstoppable expansion of artificial intelligence (AI) will take over. I see few signs of such an early takeover but there is a great deal of evidence indicating that modern civilization's grandest attempt at engineering a fundamental system transition is not proceeding as fast as many activists, politicians, and commentators claimed (and still claim) it would. That is why the last section of this coda looks at the actual extent of new, post-2000 advances whose goal is to accomplish a fundamental transformation of the global energy system by replacing its dominant component—the extraction, processing, distribution, and combustion of fossil fuels—by energy conversions devoid of any fossil carbon. Such options are restricted: they mean either electricity generated by wind, photovoltaics, or by advanced designs of nuclear reactors, or by fuels derived from biomass or synthesized by using carbon taken from carbon dioxide (CO_2) in the atmosphere.

Great Contrasts

Fittingly, one of the great, memorable contrasts between the traditional and new, electrified, and mechanized arrangements was pointed out by Thomas Edison when he recalled his 1882 census of buildings in the part of Manhattan that he intended to electrify.

> I had a great idea of the sale of electric power . . . and I got all the insurance maps in New York City, and located all the hoists, printing presses, and other places where they used power. . . . There were, I remember, 554 hoists in that district. In some places, a horse would be taken upstairs to run a hoist and would be kept there until he died. (quoted in Martin 1922:120)

There are countless examples of such contrasts arising from the juxtaposing of traditional realities (working animals, toiling people) and modern norms (electric motors and engines started by turning a knob or pushing a button), but this image of aging harnessed horses going nowhere at the top of buildings strikes me as particularly poignant, an unforgettable example of old practices ended by a radical change. No less poignant is to think about the common nineteenth-century underground counterpart of the practice, of horses employed in coal mining (pit ponies) who were bundled and lowered into shafts, to live, strain, and die without ever again seeing daylight (Figure III.C.1; Evans 2023).

And then, as Thomas Martin (who spent 40 years in Edison's service) noted in his memoirs, elevator (and pit) horses ceased to exist, along with myriads of other uses relying on animate energies: because of electricity they "utterly vanished from the scene as the dinosaur and the dodo" (Martin 1922:8). A bizarre thought: teams of harnessed horses at the top of Burj Khalifa, Dubai's—and the world's—tallest skyscraper! Of course, no practically possible concentration of draft animals housed at the top of modern skyscrapers could

Figure III.C.1 Before engines and motors changed everything: A bundled horse is lowered into a coal-mine shaft to spend the rest of its life underground. French engraving from the 1870s.

furnish the energy needed to operate rapid lifts, rising scores of stories! And compared to this admirable—efficient, reliable, and painless—substitution of muscles by electric motors, today's quotidian, instant, and inexpensive global communication is truly miraculous—but its functional origins are the same, set in the generation, transformation, transmission, and conversion of electricity, that peerless set of 1880s advances crowning the Age of Synergy.

And a very different observation, published in 1919, by John Maynard Keynes (who subsequently became one of the world's most influential economists) reminds us that today's easy access to everything via wireless electronics (think of next-day Amazon deliveries) is not so unprecedented. As Keynes recalled, in London before World War I, anybody whose income exceeded the average could afford, at

> low cost and with the least trouble, conveniences, comforts, and amenities beyond the compass of the richest and most powerful monarchs of other ages. The inhabitant of London could order by telephone, sipping his morning tea in bed, the various products of the whole earth, in such quantity as he may see fit and reasonably expect their early delivery upon his doorstep. (Keynes 1919: 9)

Of course, today's orders are done on wireless phones (or online from laptops)—but (at least as far as the higher-income Western urban populations were concerned), we (almost) were here more than a century ago!

Missing Contexts

The second point is to provide a wider context, to recall the historical settings of post-1860s technical advances. The original text was kept from excessive length by concentrating (as the title promised) on the history of these advances and their impacts. Undoubtedly, most of the readers were aware of many less admirable historical contexts, but perhaps I should have included a reminder that both the years between 1867–1914 (the Age of Synergy) and the post-World War I decades of transformative technical advances were not only periods of invention, discovery, progress, critical thinking, and enlightenment but also eras of pseudoscience, cults, manias, clairvoyants, mediums (producing countless "unexplainable" tappings, voices, moving tables, and messages from afterlife), reincarnations (of self-proclaimed gurus), dubious prophets of doom, and, worst of all, also the periods when murderous ideologies were formulated and then deployed with terrible outcomes.

In 1928, Gilbert Seldes published *The Stammering Century*, a lengthy book about the cults and manias of nineteenth-century America, abounding (even after leaving out the coverage of exaggerated claims, outrageous advertising, and scandalous public stunts connected with P. T. Barnum's name) with fanatics, radicals, and mountebanks (Seldes 1928). These realities are even more remarkable because the acceptance of such lies and claims took place amid an unprecedented surge of facts and verifiable information, and because even some famous scientists, inventors, and observant thinkers joined the ranks of misguided true believers.

The case of phrenology (linking personality traits to the skull's shape) is particularly troubling yet instructive (Poskett 2019). Although beliefs in its efficacy were largely discredited before the middle of the nineteenth century, decades afterward, this racist pseudoscientific construct was embraced by Thomas Edison ("I never knew I had an inventive talent until Phrenology told me so. I was a stranger to myself until then!"), it was believed by Francis Dalton (Darwin's half-cousin and polymath, also a proponent of eugenics, another of the era's misleading advocacies), and by Arthur Conan Doyle (Figure III.C.2). But perhaps most surprisingly, Alfred Russell Wallace, one of the century's greatest biologists, who independently formulated the theory of evolution and who deserves equal billing with Charles Darwin (Lloyd et al. 2010), was phrenology's strong advocate.

In 1899, when writing about "the wonderful" nineteenth century, he asserted that in the coming century that phrenology,

> a science of whose substantial truth and vast importance I have no more doubt than I have of the value and importance of any of the great intellectual advances already recorded ... will assuredly attain general acceptance. ... Its practical uses ... will give it one of the highest places in the hierarchy of the sciences; and its persistent neglect and obloquy during the last sixty years, will be referred to as an example of the almost incredible narrowness and prejudice which prevailed among men of science at the very time they were making such splendid advances in other fields of thought and discovery. (Wallace 1899: 160, 193)

And Wallace's peculiar advocacies did not stop there. In the book's longest chapter, Wallace gathered arguments against vaccination and concluded that

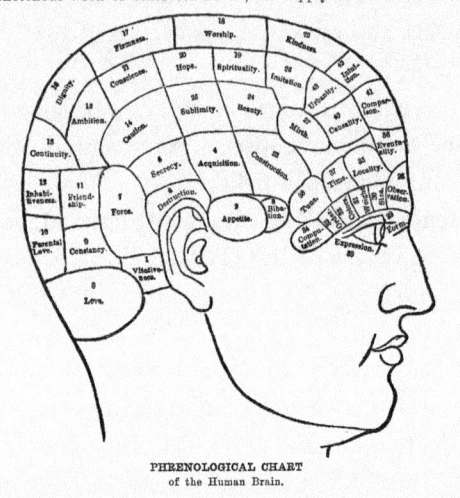

Figure III.C.2 A guide to a nineteenth-century pseudoscience: phrenological chart showing the skull regions supposedly associated with specific mental attributes (Hadden et al. 1914).

the practice of inoculating people "is utterly opposed to the whole teaching of sanitary science, and is one of those terrible blunders which, in their far-reaching evil consequences, are worse than the greatest of crimes" (Wallace 1899: 315). His false claims (unsafe, unnecessary, "gross interference with personal liberty," and even claiming "infants being killed") could be heard, verbatim, 120 years later as campaigns of misinformation and delusions tried to stop or at least disrupt vaccination against the pandemic COVID-19: *nihil novi sub sole* (Skafle et al. 2022).

At the same time, it is obvious that this coexistence of technical and scientific advances and deluded and pseudoscientific beliefs is not a unique feature of the late nineteenth century. The combination was also much in evidence during the eighteenth century, now known as the era of "enlightenment" and superstition, and Internet-enabled deluded claims are thriving in the early twenty-first century. This persistence of incompatible approaches to reality shows how profound contradictions and inexplicable complexities and biases coexist among human beliefs, the combination whose existence appears to be immune to anything labeled as cognitive progress and that indicates the limits of reason.

But the consequences of widespread belief in media, occult practices, powers of phrenology, and false arguments against vaccination were mild in comparison with eventual effects of new ideologies that took preliminary hold (often justified by pseudoscientific claims) during the latter half of the nineteenth century and that rose to murderous heights during the twentieth century. Explicitly defined racism and antisemitism of the nineteenth century led to mass-scale genocides in the twentieth century, and even greater numbers of people perished as new ideologies promising a thousand-year rule by a superior race or the creation of utopian societies under the leadership of parties led by ruthless individuals killed tens of millions (Arendt 1951; Medvedev 1972; Payne 1996; Lovell 2019).

The histories of these prolonged periods of madness, terror, and deaths are well documented, and I have already noted some estimates of their aggregate tolls in the closing chapter of this book's second part. Here I just want to stress that these quantifications depend on which categories of direct and indirect victims are included. For example, the best surveys after the fall of the USSR showed that close to 2 million people died in the Stalinist Gulag, but twice, and perhaps three times, as many died in the regime-induced Ukrainian and Kazakh famines of the early 1930s, and Stalin's pre-war military purges and the country's lack of preparedness for war resulted in even more deaths in 1941 and 1942; hence the total toll attributable to Stalinism (1924–1953) may be anywhere between 7 and 20 million (Snyder 2011). Hitler's decisions led to the deaths of at least 17 million people (Holocaust Memorial Museum 2024), and Mao's purges of 1950s, the Cultural Revolution (1966–1976), and, above all, the regime-induced famine of 1959–1961 cost as many as 40 million lives (Yang 2008).

This coexistence of reason and murderous violence is as readily explained as the coexistence of technical and scientific advances and pseudoscientific or outright deluded notions. The voices of restraint, nonviolence, and moderation have been with us for as long as the calls for conquest, territorial expansion, domination, and the killing of "others." And while it is true that technical advances (explosives, more lethal individual and heavy weapons, aerial bombing, gas chambers) contributed to the unprecedented death toll of the twentieth century, perhaps as many people died in ways that had no link to technical progress, above all in famines and due to abandonment and destitution.

Lasting Need for Age of Synergy Advances

The third point, the enduring importance of pre-World War I technical advances, is easy to demonstrate. Realities of energy and material supply, food and industrial production, construction and transportation that prevailed during the first quarter of the twenty-first century only strengthen the conclusion that many epoch-making innovations that were introduced before World War I have a much greater staying power than may be commonly believed. To be sure, there were a few notable partings, perhaps none more important—yet accomplished with relatively little notice—than the demise of the light bulb. As already noted, this quintessential invention of the Age of Synergy retained its strong market presence until the 1990s, but then came the inevitable retreat. In 2005, Brazil was the first country to ban the sales of incandescent lights; the European Union (EU) followed in 2009, Canada in 2014, and the United States in 2023 (Koretsky 2021; USDOE 2022; Figure III.C.3). After nearly 150 years, the age of incandescent lighting was over, but the achievement—the world's first practical, affordable, safe, and long-lasting means of abolishing natural darkness—remains.

Other consumer devices (and establishments associated with their use) that became rarities by the 2020s belong overwhelmingly to a large class of information and image recording, processing, and retrieval practices that were displaced by the rapid adoption of wireless electronics. They range from fountain pens; notebooks; inexpensive cameras; films; slides; overhead projectors; and photo labs, booths, and albums to VCRs, CD players, Rolodexes, road maps, atlases, and encyclopedias. Telephone landlines are still with us, but in the United States only a quarter of adults have them, only some 5% adults (mostly rural and older than 65 years) depend on them, and more than 70% use only mobiles (Kelly 2024). In several years, the inevitable will happen and companies will cut the service whose unique advantage is that it works even during electric blackouts.

Figure III.C.3 Doing away with incandescent lights after 140 years of service: smart Vgogfly 7 W LED lights (equivalent to 60 W incandescent bulb) have a plastic mask rather than glass and are turned on and off automatically.

The disappearance and declining use of these formerly ubiquitous devices and services have created biased perceptions of rapid change: in most instances, the fundamental technical capabilities of the global economy either trace their origins to those two incomparable pre-World War I generations or are the result of transformations introduced during the twentieth century, not of recent innovations. This is true, above all, about the material foundations of our civilization. Cast iron continues to flow from large blast furnaces (as it has done for some 150 years) and it is converted to steel in basic oxygen furnaces (introduced in the 1950s) or made from scrap metal in electric arc furnaces (whose development began with William Siemens's patents of the late 1870s). Ammonia comes from the Haber-Bosch synthesis first commercialized in 1913 (now, as expected, more efficiently, requiring less of the processing energy), Hall and Héroult would readily recognize the essence of their late 1880s aluminum-making method (except for much higher volumes of daily output), and cement is produced in large rotary kilns introduced during the 1890s.

And most of the world's electricity continues to be generated by steam turbines invented by Charles Parsons in the 1880s, even as their capacities (typically less than 1,000 megawatts [MW], although the largest one rates 1.9 gigawatts [GW]) and efficiencies (up to 47% with reheat, compared to just 4% in 1890) have approached the limits of performance (GE 2024; Siemens 2024). And fossil-fueled central power plants, transformers, electric motors, and transmission lines are conceptually identical to their pioneering designs of the 1880s, while some of their top performances still showed some growth during the first quarter of the twenty-first century.

More than ever before, the world is now dependent on electricity, be it converted into heat and kinetic energy (by electric motors) or enabling the functioning of modern electronics. Without nitrogen fertilizers (all derived from ammonia, with solid urea a favorite) the global population would be halved, and there is no early prospect for having cereals or oil with nitrogen-fixing capabilities akin to those of leguminous plants (Smil 2024a). Critical mass-produced materials (steel, aluminum, cement) are needed in rising quantities, and internal combustion engines (Otto and Diesel) continue to dominate all modes of land and ocean transportation, with the world's largest diesel engines powering bulk carriers (ores, cement, grain) and container ships, the prime movers of globalization.

No other development demonstrated our lasting dependence on pre-1914 technical advances and their subsequent transformations than the economic rise of China. This enormous transformation began four years after Mao Zedong's death in 1976, unfolded relatively slowly during the 1980s, accelerated during the 1990s, and reached unprecedented levels of performance during the first quarter of the twenty-first century. Simply put, most of the road that many Western countries took more than a century to traverse was covered by China in less than 50 years, and the country's broad and rapid advance during the first quarter of the twenty-first century is best summarized by a brief comparison of its global rankings and actual output levels.

To energize this economic expansion China had more than tripled its primary energy consumption between 2000 and 2025 as it became by far the largest user (more than a quarter of the global total) of fossil fuels and electricity, with heavy emphasis on domestic coal extraction (more than 4 Gt mined every year, or very close to half of the world's output) and rising hydrocarbon imports. Electricity generation rose more than sevenfold in 25 years, accounting for half of the world's total, with about three-fifths coming from large coal-fired central stations (with steam turbine capacities of mostly 200–1,000 MW) and one-seventh from water, including four of the world's ten largest hydro stations. To tie large coal-fired plants and hydro sites with the most populated provinces, China built the world's most extensive ultra-high-voltage direct current (DC) grid, breaking distance (>3,300 kilometers [km]) and voltage and transformer (1,100 kV) records (Fairley 2019).

Advances in energy conversions were paralleled by the rise of new industrial capacities (by 2025, China accounted for nearly 30% of the world's manufacturing) and by mass-scale construction of housing and new transportation infrastructures resulting in soaring material demands. China has emerged as by far the world's largest producer of cement (by 2025, more than half of the global output), crude steel (about 60%), aluminum (nearly 60%), ammonia (almost 30%), and plastics (a third of the total). Unlike in the affluent countries, where increasing shares of steel come from scrap processed in electric arc furnaces, China's steel output comes overwhelmingly (90%) from coke-fueled blast furnaces: no country built as many of them as rapidly as did China since the 1980s (Global Energy Monitor 2024). These enormous material flows were embedded in new industrial capacities as well as in new mass-produced consumer items and transportation infrastructures, all based on US, European, and Japanese designs.

Companies ranging from Nippon Steel to Volkswagen, from Bosch to Unilever, and from Boeing to Apple set up production and joint ventures in China (Hurun 2021), and, since 1980, foreign investors transferred into the country more than $4 trillion (Macrotrends 2024). In 1980, China produced 220,000 cars; since 2009, it has been the world's largest motor vehicle producer (by 2025, more than 30 million a year). The country now also has six of the world's ten busiest container ports (led by Shanghai, Ningbo, and Shenzhen), and, since 2007, it built the world's most expansive network of rapid trains totaling more than 45,000 km with speeds up to 350 km/h. The country has about 500 general-purpose airports, most of them built after 1990, and, in 2019, it completed the world's largest airport (Beijing Daxing) in less than 5 years. Given these infrastructural feats it is not surprising that, since 2008, China has been emplacing every 3 years more concrete than the United States did during the entire twentieth century (Smil 2023b).

Compared to all previous industrialization waves, this post-1980 Chinese wave had the well-known advantages of a late starter (able to rely on the adoption of highly perfected techniques) as well as the benefits of readily available advanced electronics deployed in design, production, and distribution. And India, since 2023 the world's most populous nation, is now trying to realize its own version of a spectacular rise (if not to replicate China's post-1980 ascent) as it modernizes its economy, and builds new industries, coal-fired electricity generating stations, rapid rail links, and concrete residential high-rises, all predicated on expanded production of steel, cement, ammonia, motor vehicles, and of a wide array of consumer items.

And ranking below India on the overall development scale and waiting to improve its quality of life by expanding its use of energy materials and industrial processes is nearly all of sub-Saharan Africa. The size of that need is easily illustrated by comparing prevailing per capita levels of primary energy use or gross domestic product (GDP). On the first score, sub-Saharan Africa (leaving the RSA aside) averages less than 15 gigajoules (GJ) per year compared to India's 25, China's 110, and the EU's 130 GJ/year (EI 2024). On the second one, the respective rounded values are $1,700, $2,500, $13,000, and $37,500 (World Bank 2024): to equal recent Chinese gains, sub-Saharan Africa needs to boost its annual per capita energy and material flows eightfold!

This means that the world will continue to rely on the fundamental pre-1914 advances transformed by post-World War I improvements for a large part of the twenty-first century and that, in some cases, the need is not only for marginal increments of global production but for further mass-scale expansion of annual output. These realities make it even more desirable, as well as more rewarding, to understand the genesis and evolution of these innovations as well as their energy and material requirements and their technical capabilities and performance limits.

Accelerating Innovations?

In the Part II of this book, I brought the story of great technical advances right to the end of the twentieth century. An inevitable question is how much has changed during the first quarter of the twenty-first century, the time too often touted as the era of unprecedented and accelerating innovation, where every improvement (or just a distant promise of it) is seen as "transformative" and "revolutionary." Critical examinations can separate the now ubiquitous media-generated hype from indisputable quotidian realities: I have shown how many mature production, conversion, and assembly techniques have either already reached or are closely approaching their maximum sizes (be it blast furnaces or steam turbogenerators) or performance ratings (efficiencies of aluminum smelting, ammonia synthesis, or gas turbine-driven electricity generation).

And even people unfamiliar with various technical specifications are aware of performance plateaus in transportation. Typical speeds of all major transportation modes whose beginnings date to 1880s and 1890s (internal combustion engines and electric trains) and to 1960s (jetliners and container ships) have been at, or near, their affordable limits for decades. Jetliners do not fly faster than a lifetime ago when gas turbines began to power long-distance flying, most rapid trains go as fast as they did 20–40 years ago (although some scheduled runs show marginal speed gains), and to save fuel and reduce emissions, many large container ships now practice slow steaming, limiting their speed to 18 knots (33 km/h) or even lower (Sanguri 2012). But, in one of the most remarkable post-2000 developments, maximum capacities of container ships have increased nearly fourfold, from 6,600 20-foot equivalents (TEUs) in 2000 to 24,346 TEU (China-built *Irina* and *Loreto* for MSC) in 2023 (Marine Traffic 2024).

But unlike in the case of many mature production and transportation techniques, the performance of microprocessors, the foundations of modern computation, data processing, and telecommunication, has continued to advance while the cost of computation continues to decline rapidly (Coyle and Hampton 2024). But a closer look reveals that the performance gains of early twenty-first century electronics have not been due to any fundamentally new technical departures: they have come from pursuing well-established (albeit increasingly more challenging) expected trajectories that began during the early 1970s as they followed expected (and sometime accurately predicted) course. To a historian of technical advances, the admirable miniaturized electronics of the early twenty-first century appears as the fifth-order elaboration of two fundamental advances (Smil 2017c). First is James Clerk Maxwell's identification of electricity, magnetism, and light as different manifestations of electromagnetic radiation (Maxwell 1865). The second one is miniaturized fabrication whose development followed Richard Feynman's much-quoted 1959 dictum that "there's plenty of room at the bottom" (Feynman 1960). By identifying electricity, magnetism, and light as different aspects of electromagnetic waves, Maxwell had, in Max Planck words, "achieved greatness unequalled" and, in Feynman's judgment,

> From a long view of the history of mankind—seen from, say, ten thousand years from now—there can be little doubt that the most significant event of the nineteenth century will be judged as Maxwell's discovery of the laws of electrodynamics. (Nature Physics 2011: 441)

The second-order elaboration (previously detailed in this volume) came between 1886 and 1888, when Heinrich Hertz generated and received electromagnetic waves (just across his laboratory) whose frequencies he placed between the acoustic and the light oscillations (Hertz 1887). The third-order elaboration began with the first radio broadcasts (Oliver J. Lodge and Alexander S. Popov, in 1894 and 1895), and it culminated in the invention of vacuum tubes (John A. Fleming, Greenleaf W. Pickard, and Lee de Forest's, between 1904 and 1907). The fourth-order elaboration was made possible by solid-state devices, progressing from transistors to integrated circuits and microprocessors.

The fifth-order elaboration was the next expected step as microprocessors enabled portable wireless devices and made them cheap enough to become ubiquitous. As I wrote in 2017,

> As I pass the zombielike figures on the street, oblivious to anything but their cell phone screens, I wonder how many of them know that the most fundamental advances enabling their addictions came not from Nokia, Apple, Google, Samsung, or LG. These companies' innovations are certainly admirable, but they amount only to adding a few fancy upper floors to a magnificent edifice whose foundations were laid by Maxwell . . . and whose structure depends on decades-old advances that made it possible to build electronics devices ever smaller. (Smil, 2017: 24).

How much smaller? Visible light (435 nanometers [nm]) was used to produce the first microprocessors by photolithography in the early 1970s; by 1999, leading semiconductor companies were using the 180 nm process; and, in 2023, ASML Holding was the sole producer operating with 5 nm and 3nm (extreme ultraviolet) lithography nodes (ASML Holding 2024). As a result, the total number of components on a single microchip rose from 42 million in 2000 (Intel's Pentium 4) to more than 100 billion for the largest

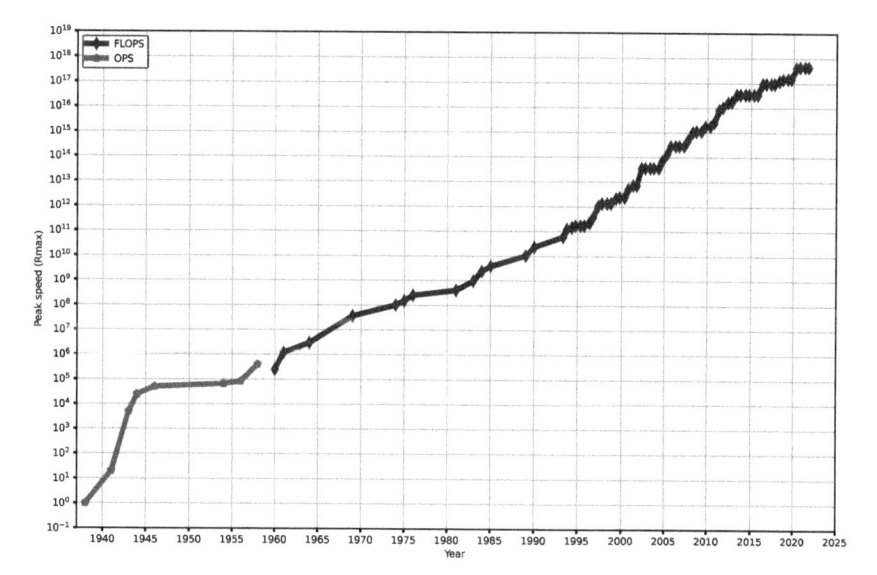

Figure III.C.4 Peak speed of supercomputers measured in floating operations per second keeps on increasing at the exponential rate, creating a straight line on a semilogarithmic graph.
Plotted from data in annual Top500 reports.

consumer microprocessor, to more than 200 billion in for Nvidia's 2024 AI chip, while commonly used storage capabilities went from floppy disks of the 1990s (1.44 megabytes [Mb]) to flash memories of the 2020s (128–256 gigabytes [Gb]). And the speed of supercomputers used for the most complex scientific calculations and simulations rose from 7.226 teraflops in 2000 to 1.102 exaflops, a 150,000-fold gain (Top500 2024; Figure III.C.4).

Not surprisingly, many people impressed by these continuing gains in computing performance have claimed that we are already experiencing a profoundly new technical revolution that is extending and multiplying our mental abilities and that it is only a matter of time before the thinking machines (AI) cut loose, develop hyperintelligence, and make us superfluous or engineer our outright demise. No other enthusiast detailed this promise as clearly, and as overwhelmingly, as Ray Kurzweil, and an extended quote is needed to convey his astonishing claim.

> An analysis of the history of technology shows that technological change is exponential, contrary to the common-sense "intuitive linear" view. So we won't experience 100 years of progress in the twenty-first century—it will

be more like 20,000 years of progress (at today's rate).... There's even exponential growth in the rate of exponential growth. Within a few decades, machine intelligence will surpass human intelligence, leading to The Singularity—technological change so rapid and profound it represents a rupture in the fabric of human history. The implications include the merger of biological and nonbiological intelligence, immortal software-based humans, and ultra-high levels of intelligence that expand outward in the universe at the speed of light. (Kurzweil 2001: 1)

In later writings, Kurzweil put the Singularity's arrival as early as 2040 (Kurzweil 2005), and similar bold claims (albeit without the "speed-of-light" expansion) became common in 2023 and 2024 with the release of new chatbots (based on large language models and producing humanlike conversations) and with promises of astounding progress resulting from the deployment of generative AI solving hitherto insurmountable challenges in fields ranging from drug discovery to company management and from new energy conversions to genetically engineered crops. This is not a place to adjudge the plausibility of these new, sometimes truly apocalyptic claims regarding Singularity's arrival or AI's eventual impact. Younger readers of this book will have a chance to verify the fate of this supposedly omnipotent development, but my concern is with the past—and here the verdict is clear.

Claims of accelerating innovation have been ahistorical, proffered by the zealots of the electronic faith, by the true believers and overenthusiastic advocates of AI, e-life forms, and spiritual machines. If these claims are based on incontrovertible foundations, then the first quarter of the twenty-first century should have already seen many clear signs signaling this exponentially advancing transformation of our civilization. But there is no clear evidence of this progress when trying to assess it on two fundamental grounds: the claimed advances have not improved labor productivity (a key indicator of progress driven by new knowledge), and they have not changed the physical foundations of modern economies. So far, just the opposite has been true as the gains in labor productivity are now running well below the long-term historical average and as the advances in computing performance have led to higher demand for energy and materials and hence a greater impact on the biosphere.

Historical studies of productivity growth show some notable patterns. While during the first half of the nineteenth century, capital and labor explained all but a small fraction of the US annual economic growth, between 1870 and 1900, the contribution of technical advances (subsumed under the label of *total factor productivity*, the share of the growth in output that is not explained by changing the three standard inputs) rose to the levels surpassing those of the last three decades of the twentieth century, proving the extraordinary nature of this period with its knowledge-based progress (Kendrick 1961; Field 2009). Solow (1957) found that between 1909 and 1949 technical advances were responsible for nearly 90% of the doubling of the US gross output per hour of work, while Denison (1985) attributed at least 75% of the US economic growth to technical innovation.

But the US data show that a new period of lower and fluctuating labor productivity growth began in the early 1970s, precisely when the first microprocessors made appearance. The 5-year rolling mean of labor productivity gains was 2.8% for the period of 1948–1971, 1.8% for 1971–1989, then came a brief temporary rise peaking in 2003 at above 3%, followed by a decline to less than 1% by 2015, and then a weak gain to 1.4% in 2023 (Sprague 2021; USBLS 2024). This has been a worrisome trend for all affluent economies facing the gradual contraction of the labor supply during the coming generations. How to explain this post-1971 historically below-average rate?

De Loo and Soete (1999) ascribed this lack of correlation between productivity growth and the rise of electronic performance to R&D efforts increasingly concerned with product differentiation rather than with product innovation: that was good for consumers, but it did little for productivity growth. Nobody devoted as much attention to this counterintuitive reality as Robert Gordon, who saw the Second Industrial Revolution—the term he used for the period of 1870–1900 that introduced electricity, internal combustion engines, running water, indoor toilets, wired communications, new forms of entertainment, oil extraction, and chemical industries—being far more important for economic growth than both the First Revolution (his term for 1750–1830, the age of steam and railroads) and the Third, starting in 1960 and proceeding with the successive waves of mainframe computing, personal computers, the Internet, and mobile devices (Gordon 2012, 2016).

Gordon saw the Second Revolution as the principal reason for some 80 years (1890–1972) of comparatively rapid productivity growth between 1890 and 1972 and concluded that, after 1973—that is, after the economy adapted the post-1945 spin-offs derived from those basic advances (including air conditioning, jetliners, and interstate highways)—US productivity growth (except for a temporary uptick between 1996 and 2004, entered a period of decline. As Gordon (2016) noted, growth in the early twenty-first century was centered on personal communication and entertainment devices whose mass-scale ownership and incessant use do not raise labor productivity in ways comparable to the adoption of such fundamental advances as electricity, internal combustion engines, or modern healthcare.

That, of course, is the key message of *Creating the Twentieth Century*—except that it is supported by detailing the period's (slightly differently timed) technical advances and synergies, rather than by focusing on the growth of economic productivity. The two pre-World War I generations lived in a unique period that may never have a comparable follow-up: coming generations may never again witness such an astonishing concatenation of technical progress subsequently reflected in labor productivity gains. That is why Gordon and John Fernald (2014) suggest that the ongoing productivity slowdown should be seen as an expected reversion to lower gains that prevailed before the unique impulse provided by what I have called the Age of Synergy and its aftermath.

Other reasons for the slowdown of productivity growth have been suggested: from diminishing returns on R&D to the cyclical nature of innovations, and from the difficulties in measuring productivity gains modern service-dominated economies to an inevitable lag between the adoption of new advances and their productivity benefits (Manyika et al. 2017). And Kelly (2021) argued that not only is it too early to see the full benefits of the latest electronic advances (whose transformative phase, he says, began only in the early twenty-first century "when computers married the telephone" and "everything went online") but that, in this new shift to a networked world, the real wealth "will not be found merely in greater productivity, but in greater degrees of playing, creating, and exploring. We don't have good metrics for new possibilities."

Still, the basic fact is indisputable: continued and impressive (and exponential) gains in computing and in mass ownership of portable electronics have not brought any unprecedented gains in labor productivity—and without such gains it is hard to imagine an early arrival of anything approaching the Singularity. And there is yet another way to contrast the unequaled, fundamental importance of pre-World War I advances with the latest wave of supposedly unprecedented, accelerating, epoch-making electronic gains, one that does not depend on constructing arguable economic measures but on observing the basic material endowments of modern civilization. Unlike the innovations adopted between 1867 and 1914 (and transformed and augmented during the twentieth century), the latest wave of computational and data-processing advances has not fundamentally changed the physical foundations of modern high-energy, food-rich, material-intensive civilization.

Moreover, the post-2000 expansion of high-speed data processing, instant telecommunication and data storage, and mass-scale ownership of portable device has combined to add a major source of demand for more energy, materials, and cooling water, adding further environmental burdens.

I will note just a few prominent components of this new demand. Nearly all intercontinental Internet traffic is carried by undersea fiber-optic cables (by 2023, totaling nearly 1.5 million km, about half of that laid since 2010) whose capacity increased from 70 to 500 terabytes per second (Tb/s) between 2000 and 2023 and whose product of capacity and length is expected to increase about ninefold between 2023 and 2030 (Papapavlou et al. 2022).

And despite their improving efficiency, data centers consumed 460 terawatts per hour (TWh) of electricity in 2023 (i.e., 2% of the global total), and the IEA expects that to rise to as much as 1.05 petawatts per hour (PWh), or an increase as large as Germany's total electricity use, while the US data centers will claim 6% of the country's generation by 2026 (IEA 2023). The largest one, in northern Virginia, consumed 3.5 GW in 2023, or as much as a modern city of 4 million people (Newmark 2024). And with the need for 1.8 liters (L) of water per kWh for evaporative cooling, a large date center requires as much water as a city of 50,000 people, putting additional stress on water supply in arid regions.

And, obviously, the centers demand a highly reliable electricity supply that could not be guaranteed by intermittent wind and solar conversion without additional large-scale addition of electricity storage or high-voltage links. To this must be added electricity required by personal computing (desktops and laptops) and by the mass ownership of portable electronic devices. US data show that, in 2013, home electronics (some 3.8 billion of various devices) consumed 12% of total residential demand (Frauenhofer USA 2023). As for microprocessors and electronic devices, their production demands expensive ultra-pure silicon, common metals (steel, copper, aluminum), special glass, plastics, and more than dozen rare and heavy metals (arsenic [As], cadmium [Cd], chromium [Cr], cobalt [Co], mercury [Hg], lead [Pb], antimony [Sb], and selenium [Se]). They are needed in minute quantities, but, given the large production runs of mobile devices, they now represent unequaled concentration of rare elements: a million discarded mobile phones contain

16,000 kg of copper, 350 kg of silver, 24 kg of gold, and 14 kg of palladium (Nicolai and Lana 2019).

All of this means that, so far—despite higher efficiencies, better controls, optimized production, and impressively declining cost of computing—those more powerful, more reliable, and more affordable microprocessors and the global ubiquity of mobile devices have not helped either to boost labor productivity to levels that prevailed for decades before 1970 or to moderate (forget about substantially reducing) either global energy or material demands. This has been true even in those cases where recent technical advances (undoubtedly aided by rising electronic capabilities) did result in significant savings of materials, gains in energy efficiency, and overall improvements in operation.

Commercial flying is a perfect illustration of these realities. The latest generation of large airplanes (Boeing 787, Airbus 350) is designed *in silico* (no laborious drafting and paper blueprints); the system is operated (from ticketing to flying-by-wire and air traffic controls) electronically; the planes use lighter yet stronger materials (carbon fiber rather than aluminum alloys); and their operation now needs 50–60% less fuel per passenger-kilometer than that of the first wide-body predecessors half a century ago. But all these efficiency and performance gains have been either marginalized or erased by enormous growth in air travel: revenue passenger-kilometers were 28 billion in 1950, about 9 trillion in 2024, more than a 300-fold rise (IATA 2024).

This means that the consequences of rising microprocessor performance have been most evident when executing demanding computational and data-processing tasks ranging from genome sequencing to the generation of large language models, most of whose eventual social and economic benefits are (as their proponents argue) yet to come. Undoubtedly (and reinforcing Gordon's entertainment argument), the broadest socioeconomic impact of microprocessors has been to affect human behavior and transform quotidian time allocations thanks to ubiquitous ownership of portable electronic devices (laptops, tablets, mobile phones) used to search, communicate, take, send, and watch images and videos; listen to music; and navigate. These applications also began before the end of the twentieth century, but only the first quarter of the twenty-first century witnessed their mass-global, and truly global, impacts.

Information became organized and readily accessible thanks to global aggregation provided by the World Wide Web and convenient browsers, with Netscape and Yahoo rapidly surpassed by Google (the company went public in 2004). In 2001, the company launched Google Earth, and then came the ascent of what became known as *social media*: in 2004, MySpace had fewer than 1 million visitors; Facebook (launched in 2004) surpassed 1 billion users in 2012 and had 3.3 billion users in 2023. YouTube (since May 2005, 2.49 billion users in 2023) became by far the largest video-sharing service, with its content now ranging from millions of useful how-to tutorials to displays of inanities and immature behavior. Altogether nearly 5 billion people (that is 61% of the world's population accessing, on average, nearly seven different platforms) were on social media in 2023 (Dean 2024).

Most importantly, access to all electronic media became easier by the diffusion of touch-screen mobile phones (the first iPhone was released in June 2007), now ubiquitously hand-held devices that are used to take pictures, listen to music, watch videos, navigate, order food and goods, send and receive messages, and talk. Global count of mobile phones rose from less than 500 million in 1999 to 8.5 billion in 2023 (Ericsson 2023). Inevitable obverses of this mass adoption have included mass addiction to mobile phones use, the rising spread of faulty information ranging from dubious claims (by individuals and advertising businesses) to deliberate government-sponsored disinformation, and increased vulnerability to hacking, service denial, and extortion.

Data for 2023 show that the average global user spent 2 hours and 24 minutes every day on social media (but the Nigerian mean was nearly 4 hours) and that people check their phones between 150 and 250 times a day—and that all of that comes *on top of* watching TV for nearly 3 hours (Dean 2024). Even if these numbers are off by 50%, those average time commitments would present a serious diversion of time from pre-mobile phone existence. On March 18, 2024, a *New Yorker* cartoon had facetiously captured the eventual outcome of these trends, showing a tombstone with a mobile phone image on the top and two inscriptions below: 50% looking at phone, 50% looking for phone. Are these realities and trends among the demonstrable onsets of an imminent, epoch-making AI takeover and approaching Singularity?

Global Energy Transition

While the first quarter of the twenty-first century may not bring us closer to an AI-dominated world and a Singularity, it did see the beginnings of a fundamental system change as global energy production began its gradual shift away from fossil carbon, the leading form of global energy use since the beginning of the twenty-first century. This transition aims to limit the degree of global warming, a gradual anthropogenic process generated primarily by emissions of CO_2 from the combustion of fossil fuels and by releases of methane (CH_4) and N_2O. So far, the global mean of that increase has amounted to a little more than 1°C, and it would be preferable to keep it if not below 1.5°C then certainly below 2°C. A series of international meetings and national commitments eventually arrived at 2050 as the date for achieving net zero fossil carbon: "net" meaning that a small amount of CO_2 would be still generated by fossil fuel combustion but that it would be captured and sequestered before reaching the atmosphere.

During the first two decades of the twenty-first century the unfolding energy transition was driven by a rapid rise in wind and solar electricity generation accompanied by rising capacities of wind turbines and photovoltaic (PV) cell efficiencies and by falling prices making both sources competitive with, and even cheaper than, fossil carbon alternatives. I have already described the basic timeline and the state of the art by the end of the twentieth century: the first expensive PV cell in space in 1958, the first (much cheaper) amorphous silicon cells in 1976, followed by a slow commercial adoption with the best year 2000 field efficiencies at 8–11%, a global installed capacity of only about 1 GW, and an annual generation of 1.1 TWh, with BP Solarex (US), Kyocera and Sharp (Japan), and Siemens (Germany) being the largest panel producers. The first quarter of the twenty-first century saw enormous PV progress. Best research efficiencies rose to 26% for the single-crystal cells and to more than 30% for silicon-perovskite cells (NREL 2024).

As the price of PV panels declined (by more than 90% since 2000), their installed capacities surpassed 500 GW by the end of 2023, generation reached more than 1,600 TWh, and five of the world's top solar panel manufacturers were in China (Lebreton 2024). Even so, in 2023, PV panels contributed only about 5% of the world's electricity generation, and their load factors outside the sunniest places remained low (Germany 11%, global mean 17%, US Southwest >24%). Modern wind turbines (three blades on tall steel towers) were introduced in Europe and the United States during the 1970s and 1980s and diffused slowly; the capacities of the largest units rose from 1 MW in 2000 to 16 MW offshore and 7.5 MW onshore in 2024. During the same period, worldwide generation rose 80-fold from 31 TWh to about 2,500 TWh equal to 8% of the world's total (EI 2024). Global weighted average capacity factors, now mostly between 30% and 45%, are expected to increase to more than 40% for onshore and more than 50% for offshore wind farms (IRENA 2019).

Soon these two renewable conversions will generate 15% of the world's electricity, more than the water turbines whose deployment in commercial electricity generation began in 1882: truly, there is something new under the sun! But advances in renewable electricity generation should not be mistaken (as, unfortunately, they sometimes are) for overall progress in the global transition measured in terms of primary energy, that is, all fuels (fossil and biomass) and primary electricity (hydro, nuclear, solar, wind, geothermal). In those terms, there has been no decarbonization at the global level, the only one that matters for generating further increases in tropospheric temperature: in 2023, halfway between Kyoto and 2050, the world burned 55% more fossil carbon than in 1997, and annual emissions from the energy sector had surpassed 37 Gt CO_2 (EI 2024).

This is not surprising given the size of the global energy demand, the inertia of established conversion techniques, and problems with large-scale integration of intermittent electricity generation along with the need for an unprecedented scale of electricity storage and for new high-voltage transmission lines. Decarbonization of energy supply that has taken place on the global level by 2025 has been only relative: when counting only modern energies (leaving out the burning of fuelwood, crop residue, and dung by poor households in Africa and Asia), fossil fuels supplied 98% of all primary energy in 1900 and 1950, 90% in the year 2000, and 81% in 2024.

The scale of the needed transformation and the pace of the primary energy transition during the first quarter of the twenty-first century make it highly unlikely that the global energy supply will be decarbonized (fossil fuels at zero) by 2050 (Smil 2024b; Figure III.C.5). Most people do not realize that even the first global energy transition is not yet complete: more than two centuries after we began shifting our primary energy supply from biomass fuels (wood, charcoal, crop residues, dried dung) to fossil fuels there are still more than 2.5 billion people (mostly low-income families in Africa, Asia, and Latin America, but also wood-burning Swedes and Finns) who rely on these traditional fuels (whose harvesting often causes environmental degradation) for cooking and heating.

And the unfolding transition calls for energy replacements on much larger scales: just two prominent examples indicate the magnitude of this new challenge. After years of substantial annual increases in the sale of electric vehicles (EVs; averaging more than 30% per year since 2020) some 40 million EVs were operating by the end of 2023, less than 3% of more than 1.47 billion motor

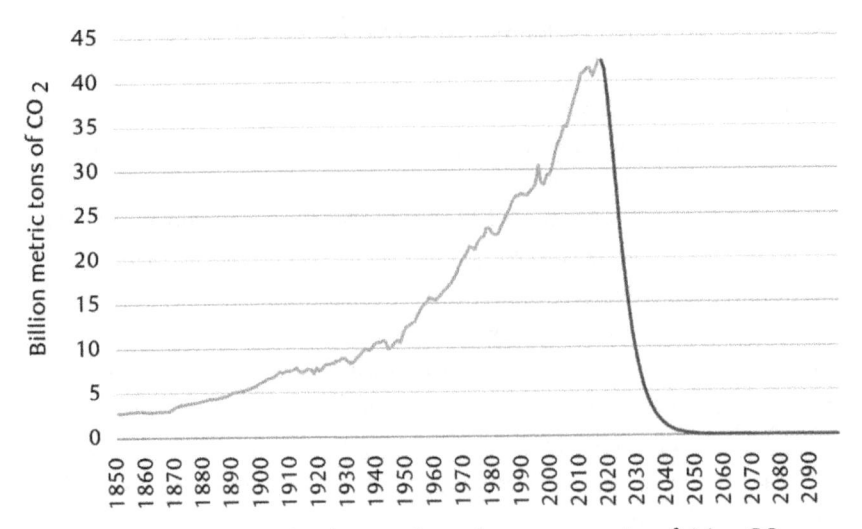

Figure III.C.5 What are the chances that, after two centuries of rising CO_2 emissions from the combustion of fossil fuels, we will decarbonize so rapidly that we get from about 40 billion t emitted in 2025 to zero by 2050?

vehicles on the road (Hedges & Company 2024). By 2040, there may be 2 billion vehicles on the road and simple algebra shows how unlikely it is to have all of them (including heavy trucks) powered by electric batteries. And smelting all the primary iron demand expected by 2050 with hydrogen rather than with coke would require (starting essentially from zero) putting in place electrolyzers and a reliable electricity supply capable of turning out annually (even with perfect efficiency) no less than 90 million tons (t) of green hydrogen.

These examples alone make it clear that fossil fuels converted into useful energies by designs introduced before 1914 (furnaces, boilers, steam turbines, internal combustion engines) and during the 1930s (gas turbines) will remain with us for decades to come, that the world's most complex and most extensive technical system, now supplying some 500 exajoules (EJ) of fossil energy, will not be gone in just 25 year. The global economy will not be carbon-free by 2050: that goal, however desirable, remains unrealistic. The International Energy Agency *World Outlook* confirms that: it sees the energy-related CO_2 emissions peaking in 2025, and the demand for all fossil fuels peaking by 2030, it projects a significant fall in coal consumption (although the total would be still about half of the 2023 level), but it sees only marginal declines in the demand for crude oil and natural gas by 2050 (IEA 2023).

And while replacing some fossil fuel conversions with electricity (most notably, car engines by electric motors) will result in substantially higher conversion efficiencies, the unfolding electrification of transportation and household energy uses (electric road vehicles, heat pumps) and adoption of green hydrogen (made by electrolysis of water) by many industrial processes (including primary iron smelting and ammonia synthesis) will result in substantially higher global demand for electricity: forecasts are for as much as a doubling the 2025 demand to some 70 PWh a year (Enerdata 2024) and that means sustained strong demand for such quintessentially 1880s techniques as transformers, transmission lines, and electric motors.

And there is yet another still insufficiently appreciated dimension to moving away from fossil carbon: some new energy techniques require much more material per unit of capacity than the fossil fuel converters they displace. Most notably, this is true about EVs (with their heavy battery packs and heavier tires that wear faster) and about wind turbines, now the world's leading producers of renewable electricity. While the complete installation of a small gas turbine (including foundations, enclosing structure, connections to a pipeline, and venting) will require some 30 t of metals (steel, copper, aluminum) and concrete per MW, the rate for an onshore wind turbine is commonly 15 times higher, and even more concrete and steel is needed for machines anchored far offshore in the ocean. Again, the green transition will be impossible without relying on a massively expanded supply of materials whose large-scale production was commercialized during the late nineteenth century.

Envoi

Claims of special or historically unique periods are common but singling out the period encompassing two pre-World War I generations as the most consequential time for the development of technical advances that laid the lasting foundations of modern societies is just a factual observation whose strength becomes incontrovertible once a full inventory of those synergetic innovations is presented. That was the intent of *Creating the Twentieth Century*, and the follow-up volume explained the transformations and augmentations of those epoch-making advances whose commercialization created the modern world. A quarter-century after I began writing the first book, and nearly 20 years after publishing the second one, there is no need to change these basic conclusions, and brief reviews of the early twenty-first century's developments further confirm the enormous debt we owe to the late nineteenth-century Age of Synergy and to subsequent developments based on those admirable foundations.

Units and Abbreviations

Units

A	ampere (unit of current)
C	degree of Celsius (unit of temperature)
G	gram (unit of mass)
H	Hour
Hp	horsepower (traditional unit of power = 745.7 W)
Hz	hertz (unit of frequency)
J	joule (unit of energy)
K	degree of Kelvin (unit of temperature)
Lm	lumen (unit of luminosity)
M	Meter
Pa	pascal (unit of pressure)
S	Second
T	metric ton (= 1,000 kg)
V	volt (unit of voltage)
W	watt (unit of power)

Prefixes

N	nano 10^{-9}
μ	micro 10^{-6}
M	milli 10^{-3}
C	centi 10^{-2}
H	hecto 10^{2}
K	kilo 10^{3}
M	mega 10^{6}
G	giga 10^{9}
T	tera 10^{12}
P	peta 10^{15}
E	exa 10^{18}

References

AA (The Aluminum Association). 2003. *The Industry*. Washington, DC: AA.
ABB (ASEA Brown Boveri). 2012. *ABB Develops World's Most Powerful High-Voltage Direct Current Converter Transformer*. ABB develops world's most powerful high-voltage direct current converter transformer
Abbate, Janet. 1999. *Inventing the Internet*. Cambridge, MA: MIT Press.
Abbott, Charles G. 1934. *Great Inventions*. New York: Smithsonian Institution.
Abeles, Paul W. 1949. *The Principles and Practice of Prestressed Concrete*. London: Charles Lockwood.
Abernathy, William J. 1978. *The Productivity Dilemma*. Baltimore, MD: Johns Hopkins University Press.
Abernathy, William J., Kim B. Clark, and Alan M. Kantrow. 1984. *Industrial Renaissance: Producing a Competitive Future for America*. New York: Basic Books.
Abramovitz, Moses. 1956. Resource and output trends in the United States since 1870. *American Economics Review* 46:5–23.
Adams, Walter, and Joel B. Dirlam. 1966. Big steel, invention, and innovation. *Quarterly Journal of Economics* 80:167–189.
Adler, Dennis. 2006. *Daimler & Benz: The Complete History: The Birth and Evolution of the Mercedes-Benz*. New York: William Morrow.
Adriaanse, Albert, et al. 1997. *Resource Flows: The Material Basis of Industrial*. Washington, DC: World Resources Institute.
AISI. 2002. *Perspective: American Steel & Domestic Manufacturing*. Washington, DC: AISI.
Aitken, Hugh G. J. 1976. *Syntony and Spark: The Origins of Radio*. New York: Wiley.
Albrecht, Donald, ed. 1995. *World War II and the American Dream*. Cambridge, MA: MIT Press.
Aldred, James. 2010. Burj Khalifa: A new high for high-performance concrete. *Proceedings of the Institution of Civil Engineers - Civil Engineering* 163(2):66–73.
Aldridge, A. 1997. Engaging with promotional culture. *Sociology* 37(3):389–408.
Alexander, George E. 1974. *Lawn Tennis: Its Founders and Its Early Years*. Lynn, MA: H. O. Zimman.
Alexander, Robert C., and Douglas K. Smith. 1988. *Fumbling the Future: How Xerox Invented, Then Ignored, the First Personal Computer*. New York: William Morrow & Company.
Allardice, Corbin, and Edward R. Trapnell. 1946. *The First Reactor*. Oak Ridge, TN: US Army Environmental Center.
Allaud, Louis, and Maurice Martin. 1976. *Schlumberger: Histoire d'une Technique*. Paris: Berger-Levrault.
Allen, Robert C. 1981. Entrepreneurship and technical progress in the northeast coast pig iron industry: 1850–1913. *Research in Economic History* 6:35–71.
Allen, Robert C. 2017. *The Industrial Revolution: A Very Short Introduction*. Oxford: Oxford Academic.
Allen, Roy. 2000. *The Pan Am Clipper: The History of Pan American's Flying Boats 1931 to 1946*. New York: Barnes & Noble.
Almond, J. K. 1981. A century of basic steel: Cleveland's place in successful removal of phosphorus from liquid iron in 1879, and development of basic converting in ensuing 100 years. *Ironmaking and Steelmaking* 8:1–10.
Almqvist, Ebbe. 2003. *History of Industrial Gases*. New York: Kluwer Academic.
Anderson, Edwin P., and Rex Miller. 1983. *Electric Motors*. Indianapolis, IN: T. Audel.

Anderson, George B. 1980. *One Hundred Booming Years: A History of Bucyrus-Erie Company, 1880–1980*. South Milwaukee, WI: Bucyrus-Erie.
Andreas, John C. 1992. *Energy-Efficient Electric Motors: Selection and Application*. New York: M. Dekker.
Anonymous. 1877. The telephone. *Scientific American* 36:1.
Anonymous. 1883. We have some very interesting figures. *Nature* 28:281.
Anonymous. 1889. The electric lighting of London. *Illustrated London News* 95(2636):526–527.
Anonymous. 1901. The telephone auto-commutator. *Scientific American* 84:85.
Anonymous. 1906. Mr. Punch's great offers. *Punch* 131:380.
Anonymous. 1913a. The first Diesel locomotive. *Scientific American* 109:225–226.
Anonymous. 1913b. The motor-driven commercial vehicle. *Scientific American* 109:168,170.
Anonymous. 1913c. The electric production of steel. *Scientific American* 109:88–89.
Anonymous. 1914. The defense of the "America's" Cup. *Scientific American* 110:865–866.
APC (American Plastics Council). 2003. *About Plastics*. Arlington, VA: APC, 86.
Apelt, Brian. 2001. *The Corporation: A Centennial Biography of United States Steel Corporation, 1901–2001*. Pittsburgh, PA: Cathedral Publishing.
Arendt, H. 1951. *The Origins of Totalitarianism*. New York: Schocken Books.
Arrillaga, Jos. 1983. *High Voltage Direct Current Transmission*. London: Institution of Electrical Engineers.
Arrow, Kenneth. 1962. The economic implications of learning by doing. *Review of Economic Studies* 29:155–173.
ASHRAE (American Society of Heating, Refrigerating and Airconditioning Engineers). 2001. *Handbook of Fundamentals*. Atlanta, GA: ASHRAE.
Ashton, Thomas S. 1948. *The Industrial Revolution, 1760–1830*. Oxford: Oxford University Press.
ASME (American Society of Mechanical Engineers). 1983. *Owens "AR" Bottle Machine (1912)*. New York: ASME.
ASME (American Society of Mechanical Engineers). 1985. *Oxygen Process Steel-Making Vessel*. New York: ASME.
ASME (American Society of Mechanical Engineers). 1988. *The World's First Industrial Gas Turbine Set at Neuchâtel (1939)*. New York: ASME.
ASME (American Society of Mechanical Engineers). 2024. *Edison "Jumbo" Engine-driver Dynamo (1882)*. New York: ASME.
ASML Holding. 2024. Changing the world, one nanometer at a time. ASML_TECHNOLOGY_TAIWAN_LTD_DM_1691466267.pdf
ASN (Aviation Safety Network). 2003. *De Havilland DH-106 Comet*. Amsterdam: ASN.
Atlas Obscura. 2024. Edison's cast-in-place concrete houses. Thomas Edison's Concrete Houses – Montclair, New Jersey. Thomas Edison's Concrete Houses – Montclair, New Jersey - Atlas Obscura
Ausubel, Jesse H. 2003. *Decarbonization: The Next 100 Years*. New York: Rockefeller University Press.
Ausubel, Jesse H., and Cesare Marchetti. 1996. Elektron: Electrical systems in retrospect and prospect. *Daedalus* 125:139–169.
Ausubel, Jesse H., Cesare Marchetti, and Perrin S. Meyer. 1998. Toward green mobility: The evolution of transport. *European Review* 6:137–156.
Avalon Project. 2003. *The Atomic Bombings of Hiroshima and Nagasaki*. New Haven, CT: Yale Law School.
Ayres, Robert U. 1969. *Technological Forecasting and Long-Range Planning*. New York: McGraw-Hill.
Ayres, Robert U., and Duane C. Butcher. 1993. The flexible factory revisited. *American Scientist* 81:448–459.
Ayres, Robert U., and William Haywood. 1991. *The Diffusion of Computer Integrated Manufacturing Technologies: Models, Case Studies and Forecasts*. New York: Chapman & Hall.

Backus, J. 1979. The history of FORTRAN 1, II, and II. *Annals of the History of Computing* 1:21-37.
Baekeland, Leo H. 1909. *Condensation Product and Method of Making Same*. Specification of Letters Patent 942,809, December 7, 1909. Washington, DC: USPTO. http://www.uspto.gov
Bagsarian, Tom. 2000. Strip casting getting serious. *Iron Age New Steel* 16(12):18-22.
Bagsarian, Tom. 2001. Blast furnaces' next frontier: 20-year campaigns. *Iron Age New Steel* July:1-6.
Bailey, Brian J. 1998. *The Luddite Rebellion*. New York: New York University Press.
Bailey, B.F. 1911. *The Induction Motor*. New York: McGraw-Hill.
Bairoch, Paul. 1991. The city and technological innovation. In: Paul Higonnet, ed., *Favorites of Fortune*. Cambridge, MA: Harvard University Press, pp. 159-176.
Baldwin, Neil. 2002. *Henry Ford and the Jews: The Mass Production of Hate*. New York: Public Affairs.
Bancroft, Hubert H. 1893. *The Book of the Fair*. Chicago: Bancroft Co.
Bannard, Walter. 1914. Modern electric engine starters and electric illumination systems. *Scientific American* 110:18-19.
Bannister, R. L., and G. J. Silvestri. 1989. The evolution of central station steam turbines. *Mechanical Engineering* 3(2):70-78.
Barczak, T. M. 1992. *The History and Future of Longwall Mining in the United States*. Washington, DC: US Bureau of Mines.
Bardeen, John, and Walter H. Brattain. 1950. Three-Electron Circuit Element Utilizing Semiconductive Materials. US Patent 2,524,035, October 3, 1950. Washington, DC: USPTO. http://www.uspto.gov
Barker, K. J., et al. 1998. Oxygen steelmaking furnace mechanical description and maintenance considerations. In: Richard Fruehan, ed., *The Making, Shaping and Treating of Steel, Steelmaking and Refining Volume*. Pittsburgh, PA: AISE Steel Foundation, pp. 431-474.
Barnett, F. D., and R. W. Crandall. 1986. *Up from Ashes: The Rise of the Steel Minimill in the United States*. Washington, DC: Brookings Institution.
Basalla, George. 1988. *The Evolution of Technology*. Cambridge: Cambridge University Press.
BASF. 1911. *Patentschrift Nr 235421 Verfahren zur synthetischen Darstellung von Ammoniak aus den Elementen*. Berlin: Kaiserliches Patentamt.
Batchelor, Ray. 1994. *Henry Ford, Mass Production, Modernism and Design*. Manchester: Manchester University Press.
Bathie, William W. 1996. *Fundamentals of Gas Turbines*. New York: Wiley.
Bauer, Eugene E. 2000. *Boeing: The First Century*. Enumclaw, WA: TABA Publishers.
Beales, Hugh L. 1928. *The Industrial Revolution, 1750-1850: An Introductory Essay*. London: Longmans, Green & Co.
Beauchamp, Kenneth G. 1997. *Exhibiting Electricity*. London: Institution of Electrical Engineers.
Beaumont, W. Worby. 1902/1906. *Motor Vehicles and Motors: Their Design Construction and Working by Steam Oil and Electricity*. 2 volumes. Westminster: Archibald Constable & Co.
Beck, P. W. 1999. Nuclear energy in the twenty-first century: Examination of a contentious subject. *Annual Review of Energy and the Environment* 24:113-137.
Beck, Theodore R. 2001. Electrolytic production of aluminum. In: Zoltan Nagy, ed., *Electrochemistry Encyclopedia*. Cleveland, OH: Case Western Reserve University. Electrochemistry Encyclopedia -- Aluminum production
Behrend, Bernard A. 1901. *The Induction Motor: A Short Treatise on Its Theory and Design, with Numerous Experimental Data and Diagrams*. New York: Electrical World and Engineer.
Bell, Alexander G. 1876a. Letter from Alexander Graham Bell to Alexander Melville Bell, March 10, 1876. Family Papers, Folder: Alexander Melville Bell, Family Correspondence, Alexander Graham Bell, 1876. Letter from Alexander Graham Bell to Alexander Melville Bell, March 10, 1876. Library of Congress (loc.gov).

Bell, Alexander Graham. 1876b. Improvement in Telegraphy. Specification Forming Part of Letters Patent No. 174,465, Dated March 7, 1876. Washington, DC: US Patent Office. http://www.uspto.gov

Bell, I. Lothian. 1884. *Principles of the Manufacture of Iron and Steel.* London: George Routledge & Sons.

Bell System. 2024. Bell System (1876–1983). Bell System Memorial Home Page. Bell System Memorial Home Page

Belrose, John S. 1995. Fessenden and Marconi: Their differing technologies and transatlantic experiments during the first decade of this century. In: Institute of Electrical Engineers, *100 Years of Radio.* London: Institute of Electrical Engineers, pp. 32–43.

Belrose, John S. 2001. A radioscientist's reaction to Marconi's first transatlantic wireless experiment—revisited. In: IEEE Antennas & Propagation Society, *2001 Digest Volume One.* Ann Arbor, MI: IEEE, pp. 22–25.

Berlanstein, Lenard R. 1964. *The Working People of Paris, 1871–1914.* Baltimore, MD: Johns Hopkins University Press.

Berliner, Emile. 1888. *The Gramophone.* Montreal: Berliner Gramophone Co. (paper read before the Franklin Institute, May 16, 1888, reprinted in 1909). Emile Berliner | Articles and Essays | Emile Berliner and the Birth of the Recording Industry | Digital Collections | Library of Congress

Bernard, Andrew B., and J. Bradford Jensen. 2004. *The Deaths of Manufacturing Plants.* Cambridge, MA: National Bureau of Economic Research.

Berner, Robert A. 1998. The carbon cycle and CO_2 over Phanerozoic time: The role of land plants. *Philosophical Transaction of the Royal Society London B* 353:75–82.

Bernstein, Jeremy. 1993. Revelations from Farm Hall. *Science* 259:1923–1926.

Bernstein, Peter. 2011. The top Bell Labs innovations. Part I: The Game-Changers. tmcnet.com

Bertin, Leonard. 1957. *Atom Harvest: A British View of Atomic Energy.* San Francisco: W. H. Freeman.

Bessemer, Henry. 1891. On the manufacture of continuous sheets of malleable iron and steel, direct from the fluid metal. *Journal of the Iron and Steel Institute* 6(10): 23–41.

Bessemer, Henry. 1905. *Autobiography.* London: Offices of Engineering.

Bethe, Hans. 1993. Bethe on the German bomb program. *Bulletin of the Atomic Scientists* 49(1):53–54.

BIA (Buro-und Industrie-Anrüstungen Vertriebs). 2001. *Wiring Devices.* Weiterstadt: BIA.

Biermann, Christopher J. 1996. *Handbook of Pulping and Papermaking.* San Diego: Academic Press.

Bijker, Wiebe E. 1995. *Of Bicycles, Bakelites, and Bulbs: Toward a Theory of Sociotechnical Change.* Cambridge, MA: MIT Press.

Billington, David P. 1989. *Robert Maillart's Bridges: The Art of Engineering.* Princeton, NJ: Princeton University Press.

Bilstein, Roger. 1996. *Stages to Saturn: A Technological History of the Apollo/Saturn Launch Vehicles.* Washington, DC: NASA.

Binczewski, George J. 1995. The point of a monument: A history of the aluminum cap of the Washington Monument. *Journal of Metals* 47(ii):20–25.

Black, Edwin. 2001. *IBM and the Holocaust.* New York: Crown Publishers.

Black, Joseph. 1803. *Lectures on the Elements of Chemistry Delivered in the University of Edinburgh by the Late Joseph Black.* London: Longman & Rees.

BLS (Bureau of Labor Statistics). 1934. *History of Wages in the United States from Colonial Times to 1928.* Washington, DC: USGPO.

Boddy, William. 1977. *The History of Motor Racing.* London: Orbis.

Boeing. 2003a. *Statistical Summary of Commercial Jet Airplane Accidents.* Seattle, WA: Boeing.

Boeing. 2003b. *Boeing 777-300ER Completes Extended Operations Flight Testing.* Seattle. WA: Boeing.

Boeing. 2024. Boeing 787 Dreamliner. www.boeing.com

Boggess, Trent. 2000. *All Model T's Were Black*. Centerville, IN: Model T Ford Club of America. http://www.mtfca.com/encyclo/P-R.htmpaint4

Bohlin, Nils I. 1962. Safety Belt. US Patent 3,043,625, July 10, 1962. Washington, DC: USPTO. http://www.uspto.gov

Bohr, Niels. 1913. On the constitution of atoms and molecules. *Philosophical Magazine* 26(series 6):1—25.

Bolton, William. 1989. *Engineering Materials Pocket Book*. Boca Raton, FL: CRC Press.

Bond, A. Russell. 1914. Going through the shops—I. *Scientific American* 110:8–10.

Bondyopadhyay, Probir K. 1993. Investigations on the correct wavelength of transmission of Marconi's December 1901 transatlantic wireless signal. In: *IEEE Antennas and Propagation Society International Symposium*, vol. 1. Ann Arbor, MI: IEEE, pp. 72–75.

Borchers, Wilhelm. 1904. *Electric Smelting and Refining: The Extraction and Treatment of Metals by Means of the Electric Current*. London: Charles Griffin & Co.

Borlaug, Norman. 1970. *The Green Revolution: Peace and Humanity*. A speech on the occasion of the awarding of the 1970 Nobel Peace Prize in Oslo, Norway, December 11, 1970. Stockholm: Nobel e-Museum. www.NobelPrize.org

Bosch Global. 2024. Bosch Global. Invented for life | Bosch Global

Boselli, Primo. 1999. *Una Vita per la Radio: Guglielmo Marconi, Cronologia Storica, 1874–1937*. Firenze: Medicea.

Boulding, K. E. 1974. The social system and the energy crisis. *Science* 184:255–257.

Bouley, Jean. 1994. A short history of "high-speed" railway in France before the TGV. *Japan Railway & Transport Review* 3:49–51.

Boutwell, Jeffrey, and Michael T. Klare. 2000. A scourge of small arms. *Scientific American* 282(6):48–53.

Bouwsma, William J. 1979. The Renaissance and the drama of Western history. *American Historical Review* 84:1–15.

Bowers, Brian. 1998. *Lengthening the Day: A History of Lighting Technology*. Oxford: Oxford University Press.

Bowley, Arthur L. 1937. *Wages and Income in the United Kingdom since 1860*. Cambridge: Cambridge University Press.

Boylston, H. M. 1936. *An Introduction to the Metallurgy of Iron and Steel*. New York: Wiley.

Brachner, Alto. 1995. *Röntgenstrahlen: Entdeckung, Wirkung, Anwendung*. München: Deutsches Museum.

Bradsher, Keith. 2002. *High and Mighty SUVs: The World's Most Dangerous Vehicles and How They Got That Way*. New York: Public Affairs.

Brandt, Leo, and Carlo Schmid. 1956. *The Second Industrial Revolution*. Bonn: Social Democratic Party of Germany.

Brantly, J. E. 1971. *History of Oil Well Drilling*. Houston, TX: Gulf Publishing.

Braudel, Fernand. 1950. Pour une économie historique. *Revue Economique* 1:37–44.

Brayer, Elizabeth. 1996. *George Eastman: A Biography*. Baltimore, MD: Johns Hopkins University Press.

Breck, Donald D. 1974. *Zeolite Molecular Sieves: Structure, Chemistry and Use*. New York: Wiley.

Brecke, Peter. 1999. *Violent Conflicts 1400 A. D. to the Present in Different Regions of the World*. Atlanta, GA: Sam Nunn School of International Affairs. http://www.inta.gatech.edu/peter/taxonomy.html

Brenner, Joel G. 1999. *The Emperors of Chocolate: Inside the Secret World of Hershey and Mars*. New York: Random House.

Bright, Arthur A., Jr. 1949. *The Electric Lamp Industry: Technological Change and Economic Development from 1800 to 1947*. New York: Macmillan.

Brinkley, Douglas. 2003. *Wheels for the World: Henry Ford, His Company, and a Century of Progress*. New York: Viking.

Broadberry, Steven N. 1992. *The Productivity Race: British Manufacturing in International Perspective, 1850–1990*. Cambridge: Cambridge University Press.

Brock, William H. 1992. *The Norton History of Chemistry.* New York: W. W. Norton.
Bronowski, Jacob. 1973. *The Ascent of Man.* Boston: Little, Brown & Company.
Brookman, Robert S. 1998, Vinyl usage in medical plastics: New technologies. Vinyl Usage in Medical Plastics: New Technologies
Brown, Alford E., and H. A. Jeffcott. 1932. *Beware of Imitations!* New York: Viking Press.
Brown, Barbara. 2002. The first photograph. www.culturalheritage.org
Brown, Louis. 1999. *A Radar History of World War II: Technical and Military Imperatives.* Bristol, UK: Institute of Physics Publishing
Buck, John L. 1937. *Land Utilization in China.* Nanjing: Nanjing University Press.
Buderi, Robert. 1996. *The Invention That Changed the World: How a Small Group of Radar Pioneers Won the Second World War and Launched a Technical Revolution.* New York: Simon & Schuster.
Buell, William C. 1936. *The Open-Hearth Furnace: Its Design, Construction and Practice.* Cleveland, OH: Penton Publishing.
Bruce, Robert V. 1973. *Bell: Alexander Graham Bell and the Conquest of Solitude.* Boston: Little, Brown.
Brydson, J. A. 1975. *Plastic Materials.* London: Newnes-Butterworth.
Buehler, Ernest, and Gordon K. Teal. 1956. *Process for Producing Semiconductive Crystals of Uniform Resistivity.* US Patent 2,768,914, October 30, 1956. Washington, DC: USPTO. http://www.uspto.gov
Bureau of Engraving and Printing. 2024. The buck starts here: How money is made. https://www.bep.gov/currency/how-money-is-made:~:text=Since%201862%2C%20BEP%20been%20entrusted,of%20US%20government%20security
Burj Khalifa. 2024. Burj Khalifa. Buy Online & Book Now to Visit the Burj Khalifa | Burj Khalifa
Burks, Alice R. 2002. *Who Invented the Computer? The Legal Battle That Changed Computing History.* New York: Prometheus Books.
Burks, Alice R., and Arthur W. Burks. 1988. *The First Electronic Computer: The Atanasoff Story.* Ann Arbor: University of Michigan Press.
Burns, Marshall. 1993. *Automated Fabrication: Improving Productivity in Manufacturing.* Engelwood Cliffs, NJ: PTR Prentice-Hall.
Burton, William M. 1913. Manufacture of Gasolene. US. Patent 1,049,667, January 7, 1913. Washington, DC: USPTO. http://www.uspto.gov
Burwell, Calvin C. 1990. High-temperature electroprocessing: Steel and glass. In: Sam H. Schurr, et al., eds., *Electricity in the American Economy.* New York: Greenwood Press, pp. 109–129.
Burwell, Calvin C., and Blair G. Swezey. 1990. The home: Evolving technologies for satisfying human wants. In: Sam H. Schurr, et al., eds., *Electricity in the American Economy.* New York: Greenwood Press, pp. 249–275.
Byington, Margaret F. 1910. *Homestead: The Households of a Mill Town.* New York: Charities Publication Committee.
Byrn, Edward W. 1896. The progress of invention during the past fifty years. *Scientific American* 75:82–83.
Byrn, Edward W. 1900. *The Progress of Invention in the Nineteenth Century.* New York: Munn & Co.
Cagin, S., and P. Dray. 1993. *Between Earth and Sky: How CFCs Changed Our World and Endangered the Ozone Layer.* New York: Pantheon.
Cain, P. J., and A. G. Hopkins. 2002. *British Imperialism 1688–2000.* Harlow: Longman.
Calder, Nigel. 2006. *Marine Diesel Engines: Maintenance, Troubleshooting, and Repair.* Camden, MN: International Marine.
Cameron, Rondo. 1982. The Industrial Revolution: A misnomer. *History Teacher* 15: 377–384.
Cameron, Rondo. 1985. A new view of European industrialization. *Economic History Review* 38:1–23.

Campbell, Harry R. 1907. *The Manufacture and Properties of Iron and Steel*. New York: McGraw-Hill Publishing.
Campbell, Donald L., et al. 1948. Method of and Apparatus for Contacting Solids and Gases. US. Patent 2,451,804, October 19, 1948. Washington, DC: USPTO. http://www.uspto.gov
Campbell-Kelly, Martin, and William Aspray. 2004. *Computer: A History of the Information Machine*. Boulder, CO: Westview Press.
Capek, Karel. 1921. *R. U. R.: Rossum's Universal Robots*, Prague: Aventinum.
Carlson, D.E. and C.R. Wronski, C.R. 1976. Amorphous silicon solar cell. *Applied Physics Letters* 28: 671. https://doi.org/10.1063/1.88617
Carlson, W. Bernard. 2015. *Tesla: Inventor of the Electrical Age*. Princeton, NJ: Princeton University Press.
Carnegie, Andrew. 1886. *Triumphant Democracy*. New York: Scribner.
Carnot, Sadi. 1824. *Reflexions sur la Puissance Motrice du Feu et Sur les Machines Propres a Developper Cette Puissance*. Paris: Librairie Bachelier.
Carothers, Wallace H. 1937. Linear Condensation Polymers. US. Patent 2,071,250, February 16, 1937. Washington, DC: USPTO. http://www.uspto.gov
Carr, Charles C. 1952. *ALCOA: An American Enterprise*. New York: Rinehart & Co.
Casson, Herbert N. 1910. *The History of the Telephone*. Chicago: A. C. McClurg & Co.
Casti, John L. 2004. Synthetic thought. *Nature* 427:680.
Caterpillar. 2024. About Caterpillar. Caterpillar | Company | About Caterpillar
CEA (Council of Economic Advisors). 2000. *Economic Report to the President, 2000*. Washington, DC: US Government Printing Office.
Chadwick, James. 1932. Possible existence of a neutron. *Nature* 129:312.
Chanute, Octave. 1894. *Progress in Flying Machines*. New York: American Engineer and Railroad Journal.
Chase, George C. 1980. History of mechanical computing machinery. *Annals of the History of Computing* 2:198–226.
Chauvois, L. 1967. *Histoire merveilleuse de Zenobe Gramme*. Paris: Albert Blanchard.
Cheney, Margaret. 1981. *Tesla: Man out of Time*. New York: Dorsett Press.
Cho, Alfred, ed. 1994. *Molecular Beam Epitaxy*. Woodbury, NY: American Institute of Physics.
Cipolla, Carlo M. 1966. *Guns, Sails and Empires*. New York: Pantheon Books.
Clark, George B. 1981. Basic properties of ammonium nitrate fuel oil explosives (ANFO). *Colorado School of Mines Quarterly* 76:1–32.
Clarke, I. F. 1985. American anticipations. *Futures* 17:390–402.
Classic Typewriter Page, The. 2024. Writers and their typewriters. www.xavier.edu
Clausius, Rudolf. 1867. *Abhandlungen über die mechanische Wärmetheorie*. [*The Mechanical Theory of Heat, with Its Applications to the Steam-Engine and to the Physical Properties of Bodies*]. London: John van Voorst.
Clerk, Dugald. 1909. *The Gas, Petrol, and Oil Engine*. London: Longmans, Green.
Clerk, Dugald. 1911. Oil engine. In: *Encyclopaedia Britannica*, 11th ed., H. Chisholm, ed.,vol. 20. Cambridge: Cambridge University Press, pp. 25–43.
Coalson, Michale. 2003. Jets fans. *Mechanical Engineering* (Suppl.), 125(12):16–17, 38.
Cobet, Aaron E., and Gregory A. Wilson. 2002. Comparing 50 years of labor productivity in US, and foreign manufacturing. *Monthly Labor Review*, June, 51–65.
Coe, Brian. 1977. *The Birth of Photography*. New York: Taplinger Publishing.
Coe, Brian. 1988. *Kodak Cameras: The First Hundred Years*. Hove: Hove Foto Books.
Cohen, Bernard. 1999. *Howard Aiken: Portrait of a Computer Pioneer*. Cambridge, MA: MIT Press.
Colbeck, Ian, and A. R. MacKenzie. 1994. *Air Pollution by Photochemical Oxidants*. Amsterdam: Elsevier.
Collins, James H. 1914. The electric industry and the young man. *Scientific American* 110:419.
Coltman, John W. 1988. The transformer. *Scientific American* 258(1):86–95.
Committee for the Compilation of Materials on Damage Caused by the Atomic Bombs in Hiroshima and Nagasaki. 1981. *Hiroshima and Nagasaki: The Physical, Medical, and Social Effects of the Atomic Bombings*. New York: Basic Books.

Compton, Arthur H. 1956. *Atomic Quest.* New York: Oxford University Press.
Condit, Carl W. 1968. The first reinforced-concrete skyscraper: The Ingalls building in Cincinnati and its place in structural history. *Technology and Culture* 9:1–33.
Conner, Doug. 1998. Father of DOS having fun at Microsoft. *MicroNew.* Issaquah, WA: Paterson Technology. www.landley.net
Conner, Margaret. 2001. *Hans von Ohain: Elegance in Flight.* Washington, DC: American Institute of Aeronautics and Astronautics.
Conreur, Gérard. 1995. *Les annees Lumière: De l'age de pierre a l'age d'or du cinéma en France.* Paris: Editions France-Empire.
Constable, A. R. 1995. The birth pains of radio. In: Institute of Electrical Engineers, *100 Years of Radio.* London: Institute of Electrical Engineers, pp. 14–19.
Constant, E. W. 1981. *The Origins of Turbojet Revolution.* Baltimore, MD: Johns Hopkins University Press
Cooksley, Peter. 1979. *Flying Bomb, The Story of Hitler's V-Weapons in World War II.* New York: Scribner.
Cooper, G. A. 1994. Directional drilling. *Scientific American* 270(5):82–87.
Copley, Frank B. 1923. *Frederick W. Taylor: Father of Scientific Management.* New York: Harper & Brothers.
Coraglia, Giorgio. 2024. Circa Giorgio Coraglia. Giorgio Coraglia, Autore presso MetaPrintArt.
Coren, Richard. 1998. *The Evolutionary Trajectory: The Growth of Information in the History and Future of Earth.* Boca Raton, FL: CRC Press.
Corkhill, Michael. 1975. *LNG Carriers: The Ships and Their Market.* London: Fairplay Publications.
Cornejo, Leonardo M. 1984. *Fundamental Aspects of the Gold Cyanidation Process: A Review.* Golden, CO: Colorado School of Mines.
Cortada, James W. 1993. *Before the Computer: IBM, NCR, Burroughs and Remington Rand and the Industry They Created.* Princeton, NJ: Princeton University Press.
Cortada, James W. 2004. *The Digital Hand: How Computers Changed the Work of American Manufacturing, Transportation, and Retail Industries.* New York: Oxford University Press.
Cotter, Arundel. 1916. *The Story of Bethlehem Steel.* New York: Moody.
Courland, Robert. 2011. *Concrete Planet.* Amherst, NY: Prometheus Books.
Cousins, Norman 1972. *The Improbable Triumvirate: John F. Kennedy, Pope John, Nikita Khrushchev.* New York: W. W. Norton.
Covington, Edward J. 2002. *Early Incandescent Lamps.* Millfield, OH.
Cowell, Graham. 1976. *D. H. Comet: The World's First Jet Airliner.* Hounslow, UK: Airline Publications and Sales.
Crafts, N. F. R., and C. K. Harley. 1992. Output growth and the British Industrial Revolution. *Economic History Review* 45:703–730.
Craig, Norman C. 1986. Charles Martin Hall: The young man, his mentor, and his metal. *Journal of Chemical Education* 63:557–559.
Crookes, William. 1892. Some possibilities of electricity. *Fortnightly Review* 102:173–181.
Crookes, William. 1899. *The Wheat Problem.* London: Chemical News Office.
Crowley, Kevin D., and John F. Ahearne. 2002. Managing the environmental legacy of US nuclear-weapons production. *American Scientist* 90:514–523.
Culick, F.E.C. 1979. The origins of the first powered, man-carrying airplane. *Scientific American* 241(1):86–100.
Cummins, C. Lyle. 1989. *Internal Fire.* Warrendale, PA: Society of Automotive Engineers.
Cusumano, Michael M., and Kentaro Nobeoka. 1999. *Thinking Beyond Lean: How Multiproject Management Is Transforming Product Development at Toyota and Other Companies.* New York: Free Press.
Czochralski, Jan. 1918. Ein neues Verfahren zur Messung des Kristallisationsgeschwindigkeit der Metalle. *Zeitschrift der Physikalische Chemie* 92:219–221.
Dalby, William E. 1920. *Steam Power.* London: Edward Arnold.

Daly, H. E., and J. B. Cobb, Jr. 1989. *For the Common Good*. Boston: Beacon Press.
Daly, Herman. 2001. Beyond growth: Avoiding uneconomic growth. In: Mohan Munasinghe, Osvaldo Sunkel, and Carlos de Miguel, eds., *The Sustainability of Long-Term Growth: Socioeconomic and Ecological Perspectives*. Cheltenham, UK: Edward Elgar, pp. 153–161.
Damer, B. 2006. *Broke Boering Demonstrates Comptometers*. DigiBarn TV: Brooke Boering demonstrates comptometers
Daniel, Eric D., C. Denis Mee, and Mark H. Clark, eds. 1999. *Magnetic Recording: The First 100 Years*. Ann Arbor, MI: IEEE Press.
Darmstadter, J. 1997. *Productivity Change in US Coal Mining*. Washington, DC: Resources for the Future.
Davey, Owen A. 1971. The origins of the Legion des Volontaires Francais contre le Bolchevisme. *Journal of Contemporary History* 6:29–45.
David, Paul A. 1990. The dynamo and the computer: An historical perspective on the modern productivity paradox. *American Economic Review* 80:355–361.
David, P. A. 1991. The hero and the herd in technological history: Reflections on Thomas Edison and the Battle of the Systems. In: P. Higonett, D. S. Landes, and H. Rosovsky, eds., *Favorites of Fortune: Technology, Growth and Economic Development since the Industrial Revolution*. Cambridge, MA: Harvard University Press, pp. 72–119.
Davies, Ed, and Nicholas A. Veronico, Scott A. Thompson. 1995. *Douglas DC-3: 60 Years and Counting*. Elk Grove, CA: Aero Vintage Books.
Davies, R. E. G. 1987. *Pan Am: An Airline and Its Aircraft*. New York: Orion Books.
Davis, A. B. 1981. *Medicine and Technology: An Introduction to the History of Medical Instrumentation*. Westport, CT: Greenwood Press.
Davis, C. G., et al. 1982. Direct-reduction technology and economics. *Ironmaking and Steelmaking* 9(3):93–129.
Davis, Jim. 2011. Mercedes-Benz History: Diesel Passenger Car Premiered 75 Years Ago. https://www.mbusa.com/en/home
Davy, Humphry. 1840. *The Collected Works of Sir Humphry Davy*. London: Smith, Elder & Co. Cornhill.
DCS Documentation. 2024. The Nimitz Class Aircraft Carrier. DCS Documentation. www.lordvesel.win
Dean, B. 2024. Social Network Usage & Growth Statistics: How Many People Use Social Media in 2024? www.backlinko.com
Dearborn Independent. 1922. *The international Jew, the world's foremost problem [Series]*. Dearborn, MI: Dearborn Publishing.
De Beer, Jeroen, Ernst Worrell, and Kornelis Blok. 1998. Future technologies for energy-efficient iron and steel making. *Annual Review of Energy and the Environment* 23: 123–205.
De Loo, Ivo, and Luc Soete. 1999. *The Impact of Technology on Economic Growth: Some New Ideas and Empirical Considerations*. Maastricht, NE: Maastricht Economic Research Institute on Innovation and Technology.
De Vries, Jan. 1994. The Industrial Revolution and the industrious revolution. *Journal of Economic History* 54:249–270.
De Vries, Tjitte. 2017. The 'Cinématographe Lumière' a Myth? www.xs4all.nl
Delhumeau, Gwenael. 1999. *L'invention du béton armeé—Hennebique 1890-1914*. Paris: Norma.
Del Vecchio, Robert et al. 2017. *Transformer Design Principles*. Boca Raton, FL: CRC Press.
de Mestral, George. 1961. Separable Fastening Device. https://www.deutzusa.com/home Washington, DC: USPTO.
Denison, Edward F. 1985. *Trends in American Economic Growth, 1929–1982*. Washington, DC: Brookings Institution.
Denning, R. S. 1985. The Three Mile Island unit's core: A post-mortem examination. *Annual Review of Energy* 10:35–52.

Dertouzos, Michael L. 1997. *What Will Be: How the New World of Information Will Change Our Lives*. San Francisco: HarperEdge.

Devereux, Steve. 1999. *Drilling Technology in Nontechnical Language*. Tulsa, OK: PennWell Publications.

Deutz. 2024. Deutz: The Engine Company. https://www.deutzusa.com/home

Devine, Warren D. 1990a. Early developments in electroprocessing: New products, new industries. In: Sam H. Schurr, et al., eds., *Electricity in the American Economy: Agent of Technological Progress*. New York: Greenwood Press, pp. 77–98.

Devine, Warren D. 1990b. Coal mining: Underground and surface mechanization. In: Sam H. Schurr et al., eds., *Electricity in the American Economy*. New York: Greenwood Press, pp. 181–208.

Devol, George C. 1961. Programmed Article Transfer. US Patent 2,988,237, June 13, 1961. Washington, DC: USPTO. http://www.uspto.gov

Diebold, John. 1952. *Automation: The Advent of the Automatic Factory*. New York: Van Nostrand.

Dichman, Carl, and Alleyne Reynolds. 1911. *The Basic Open-Hearth Steel Process*. New York: D. Van Nostrand.

Dickinson, H. W. 1939. *A Short History of the Steam Engine*. Cambridge: Cambridge University Press.

Diderot, Denis, and Jean Le Rond D'Alembert. 1751–1777. *L Encyclopedie ou Dictionnaire Raisonne des Sciences des Arts et des Metiers*. Paris: Ave Approbation and Privilege du Roy.

Dieffenbach, E. M., and R. B. Gray. 1960. The development of the tractor. In: A. Stefferud, ed., *Power to Produce: 1960 Yearbook of Agriculture*, Washington, DC: US Department of Agriculture, pp. 24–45.

Diesel, Eugen. 1937. *Diesel: Der Mensch, das Werk, das Schicksal*. Hamburg: Hanseatische Verlagsanstalt.

Diesel, Rudolf. 1893. *Theorie und Konstruktion eines rationellen Waörmemotors zum Ersatz der Dampfmaschinen und der heute bekannten Verbrennungsmotoren* [Theory for the Construction of a Rational Thermal Engine to Replace the Steam Engine and the Currently Known Internal Combustion Engines]. Berlin: Springer Verlag.

Diesel, Rudolf. 1913. *Die Entstehung des Dieselmotors* [The Origin of the Diesel Engine]. Berlin: Verlag von Julius Springer.

Donkin, Bryan. 1896. *A Text-book on the Gas, Oil, and Air Engines*. London: C. Griffin & Co.

Donovan, John J. 1997. *The Second Industrial Revolution: Business Strategy and Internet Technology*. Upper Saddle River, NJ: Prentice Hall.

Dr Pepper. 2024. Dr Pepper Musym. https://drpeppermuseum.com

Drnevich, R. F., C. J. Messina, and R. J. Selines. 1998. Production and use of industrial gases for iron and steelmaking. In: Richard Fruehan, ed., *The Making, Shaping and Treating of Steel, Steelmaking and Refining Volume*. Pittsburgh, PA: AISE Steel Foundation, pp. 291–310.

Dunsheath, Percy. 1962. *A History of Electrical Industry*. London: Faber & Faber.

DuPont. 2003. *Teflon*. Wilmington, DE: DuPont.

Dupuy, T. N. 1977. *A Genius for War: The German Army and General Staff, 1807–1945*. Englewood Cliffs, NJ: Prentice-Hall.

Durant, Frederick C. 1974. Robert H. Goddard and the Smithsonian Institution. In: Frederick C. Durant and George S. James, eds., *First Steps Toward Space*. Washington, DC: Smithsonian Institution, pp. 57–69.

Durant, Frederick C. and George S. James, eds. 1974. *First Steps Toward Space*. Washington, DC: Smithsonian Institution.

Durrer, Robert. 1948. Sauerstoff-Frischen in Gerlafingen. *Von Roll Werkzeitung* 19(5):73–74.

Dutilh, Chris E., and Anita R. Linnemann. 2004. Modern agriculture, energy use in. In: Cutler J. Cleveland, ed., *Encyclopedia of Energy*, vol. 2. Amsterdam: Elsevier, pp. 719–726.

Dvorak, August, et al. 1936. *Typewriting Behavior*. New York: American Book Co.

Dyer, Frank L., and Thomas C. Martin. 1929. *Edison: His Life and Inventions.* New York: Harper & Brothers.

Dyson, James. 1998. *Against the Odds: An Autobiography.* London: Orion Books.

The Economist. 2001, June 2. Obituary: Malcolm McLean. *The Economist,* 91. Malcolm McLean

Edge, Stephen R. F. 1949. Papermaking. In: Jocelyn Thope and M. A. Whiteley, eds., *Thorpes's Dictionary of Applied Chemistry,* vol. 9. London: Longmans, Green & Co., pp. 215–225.

Edison, Thomas A. 1876. Improvement in Autographic Printing. Specification Forming Part of Letters Patent No. 180,857, Dated August 8, 1876. Washington, DC: USPTO. http://www.uspto.gov

Edison, Thomas A. 1880a. Electric Lamp. Specification Forming Part of Letters Patent No. 223,898, Dated January 27, 1880. Washington, DC: US Patent Office. http://www.uspto.gov

Edison, Thomas A. 1880b. Electric Light. Specification Forming Part of Letters Patent No. 227,229, Dated May 4, 1880. Washington, DC: US Patent Office. http://www.uspto.gov

Edison, Thomas A. 1881. System of Electric Lighting. Specification Forming Part of Letters Patent No. 239,147, Dated March 22, 1881. Washington, DC: US Patent Office. http://www.uspto.gov

Edison, Thomas A. 1883. System of Electrical Distribution. Specification Forming Part of Letters Patent No. 274,290, Dated March 20, 1883. Washington, DC: US Patent Office. http://www.uspto.gov

Edison, Thomas A. 1884. Electrical Indicator. Specification Forming Part of Letters Patent No. 307,031, Dated October 21, 1884. Washington, DC: US Patent Office. http://www.uspto.gov

Edison, Thomas A. 1889. The dangers of electric lighting. *North American Review* 149: 625–634.

Edwards, Junius D. 1955. *The Immortal Woodshed.* New York: Dodd, Mead, & Co.

EF (Engineering Fundamentals). 2003. *Polymer Material Properties.* Sunnyvale, CA: EF.

EI (Energy Institute). 2023. *Statistical Review of World Energy 2023.* London: Energy Institute. www.energyinst.org

EI (Energy Institute). 2024. *Statistical Review of World Energy.* London: EI.

EIA (Energy Information Administration). 1996. *Residential Lighting Use and Potential Savings.* Washington, DC: EIA.

EIA (Energy Information Administration). 1999. *A Look at Residential Energy Consumption in 1997.* Washington, DC: EIA.

EIA (Energy Information Administration). 2003. *Annual Coal Report 2002.* Washington, DC: EIA.

EIA (Energy Information Administration). 2004. *Historical Data.* Washington, DC: EIA.

Einstein, Albert. 1905. Zur Elektrodynamik bewegter Körper. *Annalen der Physik* 17: 891–921.

Einstein, Albert. 1907. Über das Relativitätsprinzip und die aus demselben gezogenen Folgerungen. *Jahrbuch der Radioaktivität und Elektronik* 4:411–462.

Electricity Council. 1973. *Electricity Supply in Great Britain: A Chronology—From the Beginnings of the Industry to 31 December 1972.* London: Electricity Council.

Ellis, C. Hamilton. 1977. *The Lore of the Train.* New York: Crescent Books.

Ellis, L. W. 1912. The International Motor Contest at Winnipeg. *Scientific American* 107:113, 124.

Ellul, Jacques. 1964. *The Technological Society.* New York: A. A. Knopf.

Ellul, Jacques. 1989. *What I Believe.* Grand Rapids, MI: W. B. Eerdmans.

Ellul, Jacques. 1990. *The Technological Bluff.* Grand Rapids, MI: W. B. Eerdmans.

e-Manufacturing Networks. 2001. John T. Parsons, "Father of the Second Industrial Revolution," Joins e-Manufacturing Networks Inc. Board of Advisors. www.mcadcafe.com

Emerick, H. B. 1954. European oxygen steelmaking is of far-reaching significance. *Journal of Metals* 6(7):803–805.

Encyclopedia Astronautica. 2019. http://www.astronautix.com/

Enerdata. 2024. Total electricity generation. Global 2050 Projections for Total Electricity Generation | Enerdata

Energomash. 2003. *NPO Energomash Engines History.* Moscow: Energomash.

Energy Institute. 2024. *Statistical Review of World Energy.* London: EI. https://www.energyinst.org/statistical-review

Engelbart, Douglas C. 1970. X-Y Position Indicator for a Display System. US Patent 3,541,541, November 17, 1970. Washington, DC: USPTO. http://www.uspto.gov

Engelberger, Joseph F. 1980. *Robotics in Practice: Management and Applications of Industrial Robots.* New York: AMACOM.

Engelberger, Joseph F. 1989. *Robotics in Service.* Cambridge, MA: MIT Press.

EPICA (European Project for Ice Coring in Antarctica). 2004. Eight glacial cycles from an Antarctic ice core. *Nature* 429:623–628.

Erasmus, Friedrich C. 1975. Die Entwicklung des Steinkohlenbergbaus im Ruhrrevier in den siebziger Jahren. *Glückauf* 11:311–318.

Ericsson. 2023. Ericsson Mobility Report November 2023. Ericsson Mobility Report | Read the latest edition

Escales, Richard. 1908. *Nitroglycerin und Dynamit.* Leipzig: Veit & Co.

ESCAP (Economic and Social Commission for Asia and the Pacific). 2000. *Major Issues in Transport, Communications, Tourism and Infrastructure Development: Regional Shipping and Port Development Strategies Under a Changing Maritime Environment.* Bangkok: ESCAP.

Essig, Mark. 2003. *Edison & the Electric Chair: A Story of Light and Death.* New York: Walker & Company.

Eubanks, Gordon. 2004. *Recollections of Gary Kildall.* Digital Research. Digital Research - Recollections of Gary Kildall by Gordon Eubanks

European Commission. 1995. *L'Europe à Grande Vitesse.* Luxembourg: European Commission.

Evans, Leonard. 2002. Traffic crashes. *American Scientist* 90:244–253.

Evans, M. 2023. Pit ponies: Ghosts of the coal mines. Horse Journals. Pit Ponies - Ghosts of the Coal Mines | Horse Journals

Ewing, James A. 1911. Steam engine. In: *Encyclopaedia Britannica,* 11th Ed., H. Chisholm, ed., vol. 25. Cambridge: Cambridge University Press, pp. 818–850.

Fahie, J. J. 1899. *A History of Wireless Telegraphy.* London: Blackwood.

Fairley, P. 2019. China's Ambitious Plan to Build the World's Biggest Supergrid. *IEEE Spectrum.* (PDF) China's Ambitious Plan to Build the World's Biggest Supergrid A massive expansion leads to the first ultrahigh-voltage AC-DC power grid

Fant, Kenne. 1993. *Alfred Nobel: A Biography* (trans. M. Ruuth). New York: Arcade Publishing.

FAO (Food and Agriculture Organization). 2000. Energy and Agriculture Nexus. The Energy and Agriculture Nexus Table of Contents. www.fao.org

FAO (Food and Agriculture Organization). 2024. FAOSTAT. https://www.fao.org/faostat/en/

Faraday, Michael. 1839. *Experimental Researches in Electricity.* London: Richard & John Edward Taylor.

Farber, Darryl, and Jennifer Weeks. 2001. A graceful exit? Decommissioning nuclear power reactors. *Environment* 43(6):8–21.

Farman J. C., B. G. Gardiner, and J. D. Shanklin. 1985. Large losses of total ozone in Antarctica reveal seasonal ClOx/NOx interaction. *Nature* 315:207–210.

Feinman, J. 1999. Direct reduction and smelting processes. In: David H. Wakelin, ed., *The Making, Shaping and Treating of Steel, Ironmaking Volume.* Pittsburgh, PA: AISE Steel Foundation, pp. 763–795.

Feldenkirchen, Wilfried. 1994. *Werner von Siemens.* Columbus: Ohio State University Press.

Fenichell, Stephen. 1996. *Plastic: The Making of a Synthetic Century.* New York: HarperBusiness.

Fermi, Enrico, and Leo Szilard. 1955. Neutronic Reactor. US Patent 2,708,656, May 17, 1955. Washington, DC: USPTO. http://www.uspto.gov

Fernald, J. G. 2014. Productivity and potential output before, during, and after the Great Recession. *NBER Macroeconomics* 29(12014):1–456.

Ferry, Georgina. 2003. *A Computer Called Leo: Lyons Teashops and the World's First Office Computer.* London: Fourth Estate.
Fessenden, Helen M. 1940. *Fessenden, Builder of Tomorrows.* New York: Coward-McCann.
Feynman, R. P. 1960. There's plenty of room at the bottom. *Engineering and Science*, 23 (5):22–36. https://resolver.caltech.edu/CaltechES:23.5.1960Bottom
Ffrench, Yvonne. 1934. *News from the Past: The Autobiography of the Nineteenth Century.* New York: Viking Press.
Field, Alexander J. 2009. The Procyclical Behavior of Total Factor Productivity: 1890–2004. SSRN: https://ssrn.com/abstract=1095891 or http://dx.doi.org/10.2139/ssrn.1095891
Figuier, Louis. 1888. *Les Nouvelles Conquetes de la Science: L'Electricité.* Paris: Manpir Flammarion.
Fischer, Claude S. 1997. "Touch someone": The telephone industry discovers sociability. In: Stephen H. Cutcliffe and Terry S. Reynolds, eds., *Technology and American History.* Chicago: University of Chicago Press, pp. 271–300.
Fisher, A. Hugh. 1903. Steam turbines for Channel vessels: The system explained. *Illustrated London News* 122:898.
Fisk, Donald M. 2001, Fall. American labor in the 20th century. *Compensation and Working Conditions*, 3–8. http://www.bls.gov/opub/cwc/cm20030124ar02p1.htm
Flink, James J. 1975. *The Car Culture.* Cambridge, MA: The MIT Press.
Fleck, Alexander. 1958. Technology and its social consequences. In: Charles Singer, ed., *A History of Technology*, vol. 5. Oxford: Clarendon Press, pp. 814–841.
Fleming, John A. 1901. *The Alternate Current Transformer.* London: The Electrician Printing & Publishing.
Fleming, John A. 1911. Electricity. In: *Encyclopaedia Britannica*, 11th ed., H. Chisholm, ed., vol. 9. Cambridge: Cambridge University Press, pp. 179–193.
Fleming, John A. 1934. *Memories of a Scientific Life: An Autobiography.* London: Marshall, Morgan & Scott.
Flink, James J. 1988. *The Automobile Age.* Cambridge, MA: MIT Press.
Flood, Raymond, Mark McCartney and Andrew Whitaker, eds. 2008. *Kelvin: Life, Labours and Legacy.* Oxford: Oxford University Press.
Flower, Raymond, and Michael W. Jones. 1981. *100 Years of Motoring: An RAC Social History of Car.* Maidenhead: McGraw-Hill.
Fluck, Richard C., ed. 1992. *Energy in Farm Production.* Amsterdam: Elsevier.
Föll, Helmut. 2000. *Electronic Materials.* Kiel: University of Kiel.
Ford, Henry. 1922. *My Life and Work.* New York: Doubleday.
Fores, Michael. 1981. The myth of a British Industrial Revolution. *History* 66:181–198.
Fossett, David A. J. 2004. *Byun2 Shinkansen.* Tōkyō: David A. J. Fossett. http://www.h2.dion.ne.jp/dajf/byunbyun/
Fouquin, Michel, and Jules Hugot. 2016. Two Centuries of Bilateral Trade and Gravity Data: 1827-2014. CEPII Working Paper 2016- 4, May 2016, CEPII. cepii.fr/pdf_pub/wp/2016/wp2016-14.pdf
Fox, Nicols. 2002. *Against the Machine: The Hidden Luddite Tradition in Literature, Art, and Individual Lives.* Washington, DC: Island Press.
Franz, John E. 1974. N-Phosphonomethyl-glycine Phytotoxicant Compositions. US Patent 3,799,758, March 26, 1974. Washington, DC: USPTO. http://www.uspto.govFMC (Ford Motor Company). 1908. *Ford Motor Cars.* Detroit, MI: FMC. http://www.mtfca.com
FMC. 1909. *Ford Motor Cars.* Detroit, MI: FMC. http://www.mtfca.com
Frauenhofer USA. 2023. Energy Consumption of Consumer Electronics. www.fraunhofer.org
Frazier, Ian. 1997. Typewriter man. *Atlantic Monthly* 280(ii):81–92.
Freeberg, Ernest. 2013. *The Age of Edison: Electric Light and the Invention of Modern America.* London: Penguin.
Freeman, Christopher, ed. 1983. *Long Waves in the World Economy.* London: Frances Pinter.
Fridley, David, ed. 2001. *China Energy Databook.* Berkeley, CA: Lawrence Berkeley Laboratory.

Friedel, Robert, and Paul Israel. 1986. *Edison s Electric Light.* New Brunswick, NJ: Rutgers University Press.
Friedman, Jeffrey, ed. 1991. *Milestones in Motion Picture and Television Technology: The SMPTE 75th Anniversary Collection.* White Plains, NY: Society of Motion Picture and Television Engineers.
Frizot, Michel. 1998. *The New History of Photography.* Köln: Kunemann.
Fruehan, Richard, ed. 1998. *The Making, Shaping and Treating of Steel Steelmaking and Refining Volume.* Pittsburgh, PA: The AISE Steel Foundation.
Fujimoto, Takahiro. 1999. *The Evolution of a Manufacturing System at Toyota.* New York: Oxford University Press.
Fulton, Kenneth. 1996. Frank Whittle (1907–96). *Nature* 383:27.
Furniss, Tim. 2001. *The History of Space Vehicles.* San Diego, CA: Thunder Bay Press.
GAO (General Accounting Office). 1993. *High-Speed Ground Transportation.* Washington, DC: GAO.
GAO (General Accounting Office). 1995. *Nuclear Safety: Concerns with Nuclear Facilities and Other Sources of Radiation in the Former Soviet Union.* Washington, DC: GAO.
Garcke, Emil. 1911a. Electric lighting. In: H. Chisholm, ed. *Encyclopae dia Britannica,* 11th ed., vol. 9. Cambridge: Cambridge University Press, pp. 651–673.
Garcke, Emil. 1911b. Telephone. In: H. Chisholm, ed. *Encyclopaedia Britannica,* 11th ed., vol. 26. Cambridge: Cambridge University Press, pp. 547–557.
Garratt, Gerald R. M. 1994. *The Early History of Radio from Faraday to Marconi.* London: Institution of Electrical Engineers.
Garwin, Richard L., and Georges Charpak. 2001. *Megawatts and Megatons: A Turning Point in the Nuclear Age.* Chicago: Universiy of Chicago Press.
Gay, Albert, and C. H. Yeaman. 1906. *Central Station Electricity Supply.* London: Whittaker & Co.
GE (General Elecric). 2004. *LM6000 Sprint Aeroderivative Gas Turbine.* New York: GE.
GE (General Elecric). 2018. *GE Power Conversion.* GE Successfully Completed the No-Load Testing of One of the World's Largest 80-Megawatt Induction Motors for the LNG Industry | Power Conversion
GE (General Elecric) 2024. Steam turbines for power generation. Steam Turbines for Power Generation. GE Steam Power. Steam Turbine Power Plant Solutions | GE Vernova
GEAE (General Electric Aircraft Engines). 2002. *GEAE History.* Cincinnati, OH: GEAE.
GEAE (General Electric Aircraft Engines). 2003. *The GE90 Engine Family.* Cincinnati, OH: GEAE.
Geerdes, M., H. Toxopeus, and Cor van der Vliet 2009. *Modern Blast Furnace Ironmaking: An Introduction.* Amsterdam: IOS Press.
GEPS (General Electric Power Systems). 2003. *Gas Turbine and Combined Cycle Products.* Atlanta, GA: GEPS.
Gerber Scientific. 2003. *What Is Mass Customization?* South Windsor, CT: Gerber Scientific.
Giampietro, Mario. 2002 Fossil energy in world agriculture. In: *Encyclopedia of Life Sciences.* London: ELS. eLS | Major Reference Works
Giarini, O., and H. Loubergé. 1978. *The Diminishing Returns of Technology.* Oxford: Pergamon.
Giedion, Siegfried. 1948. *Mechanization Takes Command.* New York: Oxford University Press.
Gilbert, Glen A. 1973. *Air Traffic Control: The Uncrowded Sky.* Washington, DC: Smithsonian Institution Press.
Ginzberg, Eli. 1982. The mechanization of work. *Scientific American* 247(3):67–75.
Global Energy Monitor. 2024. Global blast furnace tracker. Global Blast Furnace Tracker - Global Energy Monitor
Goddard, Robert H. 1926. Goddard to Abbot, 29 June 1926. In: Esther C. Goddard and G. Edward Pendray, eds., *The Papers of Robert H. Goddard,* vol. 2. New York: McGraw-Hill, pp. 587–590, 1970.
Godfrey, Donald G. 2001. *Philo T. Farnsworth: The Inventor of Television.* Salt Lake City, UT: University of Utah Press.

Gold, Bela, et al. 1984. *Technological Progress and Industrial Leadership: The Growth of the US Steel Industry, 1900-1970*. Lexington, MA: Lexington Books.
Goldberg, Stanley. 1992. Groves takes the reins. *Bulletin of the Atomic Scientists* 48(10): 32-39.
Goldblith, Samuel A. 1972. Controversy over the autoclave. *Food Technology* 26(12):62-65.
Golley, John, and Frank Whittle. 1987. *Whittle, the True Story*. Washington, DC: Smithsonian Institution Press.
Gomery, Douglas. 1996. The rise of Hollywood. In: Geoffrey Nowell-Smith, ed., *The Oxford History of World Cinema*. Oxford: Oxford University Press, pp. 43-53.
Goodman, Sidney H. 1999. *Handbook of Thermoset Plastics*. Norwich, NY: Noyes Publications.
Gordon, Robert J. 2000. Does the "New Economy" measure up to the great inventions of the past? *Journal of Economic Perspectives* 14:49-74.
Gordon, R. J. 2012. *Is US Economic Growth Over? Faltering Innovations Confront the Six Headwinds*. Cambridge, MA: NBER.
Gordon, Robert J. 2016. *The Rise and Fall of American Growth*. Princeton, NJ: Princeton University Press.
Gould, Stephen J. 1989. *Wonderful Life: The Burgess Shale and the Nature of History*. New York: Norton.
Görlitz, Walter. 1953. *History of the German General Staff*. New York: Praeger.
Graham, L.R., ed. 1990 *Science and the Soviet Social Order*. Cambridge, MA: Harvard University Press.
Grahame, Kenneth. 1908. *The Wind in the Willows*. New York: Scribner.
Grand View Research. 2024. Industrial gases market size, share & trends analysis report by product, 2030. www.grandviewresearch.com
Gray, Thomas. 1890. Telephony. In: *Encyclopaedia Britannica*, 9th ed., T.S. Baynes, ed., vol. 23. Chicago: R. S. Peale & Co., pp. 127-135.
Greenpeace. 2003. *Polyvinyl Chloride*. Amsterdam: Greenpeace.
Greissel, Michael. 2000. The power of oxygen. *Iron Age New Steel* 16(4):24-30.
Grossman, Gene M., and Elhanan Helpman. 1991. *Innovation and Growth in the Global Economy*. Cambridge, MA: MIT Press.
Grotte, Jupp, and Bernard Marrey. 2000. *Freyssinet: La Précontrainte et l'Europe, 1930—1945*. Paris: Editions du Linteau.
Groueff, Stephane. 1967. *Manhattan Project*. Boston, MA: Little, Brown and Company.
Groves, Leslie R. 1962. *Now It Can Be Told: The Story of the Manhattan Project*. New York: Harper.
Grünewald, Martin. 2013. *Carl Benz: A Life Dedicated to Cars*. Offenburg: Sadifa-Media.
Gubbins, Edmund J. 1986. *The Shipping Industry: The Technology and Economics of Specialisation*. New York: Gordon & Breach.
Gunston, Bill. 1986. *World Encyclopaedia of Aeroengines*. Wellingborough: Patrick Stephens.
Gunston, Bill. 1995. *The Encyclopedia of Russian Aircraft, 1875-1995*. Osceola, WI: Motorbooks International.
Gunston, Bill. 1997. *The Development of Jet and Turbine Aero Engines*. Sparkford, UK: Patrick Stephens.
Gunston, Bill. 1999. *The Development of Piston Aero Engines*. Sparkford: Patrick Stephens.
Gunston, Bill. 2002. *Aviation: The First 100 Years*. Hauppauge, NY: Barron's Educational Series.
Gustafson, John. 2000. Reconstruction of the Atanasoff-Berry computer. In: Raúl Rojas and Ulf Hashagen, eds., *The First Computers: History and Architectures*. Cambridge, MA: MIT Press, pp. 91-105.
Haber, Fritz. 1909. Letter to the Directors of the BASF, 3 July 1909. Fritz Haber, Allgemeine Correspondenz II, no. 92. Ludwigshafen: BASF Unternehmensarchiv.
Haber, Fritz. 1911. *Patentschrift Nr 238450 Verfahren zur Darstellung von Ammoniak aus den Elementen durch Katalyses unter Druck beerhöhter Temperatur*. Berlin: Kaiserliche Patentamt.

Haber, Fritz. 1920. *The Synthesis of Ammonia from Its Elements.* Nobel Lecture, June 3, 1920. Stockholm: Nobel e-Museum. http://www.nobel.se/chemistry/laureates/1918/haber-lecture.html

Hadden, W., et al. 1914. *The Science of Eugenics and Sex Life.* Philadelphia: National Publishing Co.

Hahn, Otto. 1946. *From the Natural Transmutations of Uranium to Its Artificial Fission.* Nobel Lecture, December 13, 1946. Stockholm: Nobel Foundation. www.NobelPrize.org http://www.nobel.se/chemistry/laureates/1944/

Hahn, Otto. 1968. *Mein Leben.* Munich: Verag F. Bruckmann KG.

Hahn, Otto, and Fritz Strassman. 1939a. Über den Nachweis und das Verhalten der bei der Bestrahlung des Urans mittles Neutronen entstehenden Erdalkalimetalle. *Naturwissenschaften* 27:11–15.

Hahn, Otto, and Fritz Strassman. 1939b. Nachweis der Entstehung activer Bariumisotope aus Uran und Thorium durch Neutronenbestrahlung. *Naturwissenschaften* 27:89–95.

Haley, J. Evetts. 1959. *Erle P. Halliburton, Genius with Cement.* Duncan, OK: Halliburton Oil Well Cementing Company.

Hall, Christopher G. L. 1997. *Steel Phoenix: The Fall and Rise of the US Steel Industry.* New York: St. Martin's Press.

Hallion, Richard P. 2004. Remembering the legacy: Highlights of the first 100 years of aviation. *The Bridge* 34:5–11.

Halweil, Brian. 2002. *Home Grown.* Washington, DC: Worldwatch Institute.

Hamilton, Maurice. 2022. *Formula 1: The Official History.* London: Wellback Publishing.

Hammer, William J. 1913. The William J. Hammer Historical Collection of Incandescent Electric Lamps. Reprinted from *The Transactions of the New York Electrical Society*, No. 4. The William J. Hammer Historical Collection of Incandescent Electric Lamps | Smithsonian Institution

Hammond, Rolt. 1964. *The Forth Bridge and Its Builders.* London: Eyre & Spottiswoode.

Hansen, J., et al. 2000. Global warming in the twenty-first century: An alternative scenario. *Proceedings of the National Academy of Sciences of the USA* 97:9875–9880.

Harrison, M. 1988. Resource mobilization for World War II: The USA., U. K., USS. R., and Germany, 1938–1945. *Economic History Review* 41:171–192.

Hands, R. 1996. Ammonium nitrate on trial. *Nitrogen* 219:15–18.

Hansel, James G. 1996. Oxygen. In: *Kirk-Othmer Encyclopedia of Chemical Technology*, vol. 17. New York: Wiley, pp. 919–940.

Hawkins, David, Edith C. Truslow, and Ralph C. Smith 1961. *Manhattan District History: Project Y, the Los Alamos Project.* Los Alamos, NM: Los Alamos Scientific Laboratory.

Hawkins, Laurence A. 1951. *William Stanley (1858–1916)—His Life and Work.* New York: Newcomen Society of America.

Haynes, William. 1954. *American Chemical Industry: Background and Beginnings.* New York: D. Van Nostrand.

Heald Machine Company. 2024. Heald Machine Co. History. www.VintageMachinery.org

Hedges & Company. 2024. How many cars are there in the world? Statistics by country. www.hedgescompany.com

Heffernan, William D., and Mary K. Hendrickson. 2002. Multinational Concentrated Food Processing and Marketing Systems and the Farm Crisis. Paper presented at Science and Sustainability, a AAAS Symposium, Boston, MA, February 14–19, 2002.

Heilbroner, Robert L. 1967. Do machines make history? *Technology & Culture* 8:335–345.

Helsel, Zane R. 1992. Energy and alternatives for fertilizer and pesticide use. In: Richard C. Fluck, ed., *Energy in Farm Production.* Amsterdam: Elsevier, pp. 177–201.

Hendricks, Gordon. 2001. *Eadweard Muybridge: The Father of the Motion Picture.* Mineola, NY: Dover.

Heppenheimer, T. A. 1995. *Turbulent Skies: The History of Commercial Aviation.* New York: Wiley.

Heppenheimer, T. A. 1997. *Countdown: A History of Space Flight.* New York: Wiley.
Herlihy, David V. 2006. *Bicycle: The History.* New Haven, CT: Yale University Press.
Hermes, Matthew E. 1996. *Enough for One Lifetime: Wallace Carothers, Inventor of Nylon.* Washington, DC: American Chemical Society and Chemical Heritage Foundation.
Hertz, Heinrich. 1887. Uber sehr schnell elektrischen Schwingungen. *Annalen der Physik* 21:421–448.
Hertz, Heinrich. 1893. *Electric Waves; Being Researches on the Propagation of Electric Action with Finite Velocity through Space* (English trans. by D. E. Jones). London: Macmillan.
Hertz, Mathilde, and Charles Susskind, eds. 1977. *Heinrich Hertz: Erinnerungen, Briefe, Tagebücher.* Weinheim: Physik-Verlag.
Heywood, John B. 1988. *Internal Combustion Engine Fundamentals.* New York: McGraw-Hill.
HFMGV (Henry Ford Museum and Greenfield Village). 2004. *The Great American Production.* Greenfield, MI: HFMGV.
Hicks. J., and G. Allen. 1999. *A Century of Change: Trends in UK Statistics since 1900.* London: House of Commons Library
High Speed Rail Alliance. 2024. What are the world's fastest trains? High Speed Rail Alliance. www.hsrail.org
Higonnet, René, and Louis Moyroud. 1957. Photo Composing Machine. US Patent 2,790,362, April 30, 1957. Washington, DC: USPTO. http://www.uspto.gov
Hill, Martin. 2018. Modern Electric Motor Technology and Applications. Technical Articles. www.eepower.com
Hiltpold, Gustav F. 1934. *Erzeugung und Verwendung motorischer Kraft.* Zürich: Girsberger.
Hinsley, F. Harry, and Alan Stripp, eds. 1993. *Codebreakers: The Inside Story of Bletchley Park.* Oxford: Oxford University Press.
History-Computer. 2023. The complete history of the Burroughs adding machine. The Complete History of the Revolutionary Burroughs Adding Machine - History Tools
History-Computer. 2024. Mechanical Calculators. History Computer.
Hofer, Margaret K., and Kenneth T. Jackson. 2003. *The Games We Played.* Princeton, NJ: Princeton Architectural Press.
Hoff, Marcian E., Stanley Mazor, and Federico Faggin. 1974. Memory System for a Multi-chip Digital Computer. US Patent 3,821,715, June 28, 1974. Washington, DC: USPTO. http://www.uspto.gov
Hoff, Nicholas J. 1946. A short history of the development of airplane structures. *American Scientist* 1946:212–225, 370–388.
Hofmann-Wellenhof, B., et al. 1997. *Global Positioning System: Theory and Practice.* New York: Springer-Verlag.
Hogan, William T. 1971. *Economic History of the Iron and Steel Industry in the United States.* Lexington, MA: Lexington Books.
Hohenemser, Christopher. 1988. The accident at Chernobyl: Health and environmental consequences and the implications form risk management. *Annual Review of Energy* 13:383–428.
Holbrook, Stewart H. 1976. *Machines of Plenty: Chronicle of an Innovator in Construction and Agricultural Equipment.* New York: Macmillan.
Holdermann, Karl. 1954. *Im Banne der Chemie: Carl Bosch Leben und Werk.* Düsseldorf: Econ-Verlag.
Hollerith, Herman. 1894. Machine for tabulating statistics. 1498395268949793004-00526130. www.storage.googleapis.com
Holley, I. B. 1964. *Buying Aircraft: Material Procurement for the Army Air Forces.* Washington, DC: Department of the Army.
Hollingsworth, John A. 1966. *History of Development of Strip Mining Machines.* Milwaukee, WI: Bucyrus-Erie Company.
Holmes, Harry N. 1914. Powdered coal for fuel. *Scientific American* 110:330–331.

Holmes, Peter. 2001. *WP 6—RTD Strategy*. Plenary address presented at the CAME—GT Workshop, October 2001. Brussels: Thematic Network for Cleaner and More Efficient Gas Turbines. http://www.came-gt.com/second-workshop/2wStrategy.pdf

Holocaust Memorial Museum. 2024. Documenting numbers of victims of the Holocaust and Nazi persecution. Holocaust Encyclopedia. www.ushmm.org

Holzmann, Gerard J., and Bjorn Pehrson. 1994. The first data networks. *Scientific American* 270(1):124–129.

Honda Engines. 2024. GX Series. Honda Engines. GX Commercial Series Engines. Honda Engines | GX Commercial Series Engines

Hong, B. D., and E. R. Slatick. 1994. Carbon emission factors for coal. *Quarterly Coal Report* 1994(1):1–8.

Hopkins, Eric. 2000. *Industrialisation and Society: A Social History, 1830–1951*. London: Routledge.

Horlock, J. H. 2002. *Combined Power Plants Including Combined Cycle Gas Turbine Plants*. Malabar, FL: Krieger Publishers.

Horonjeff, Robert, and Francis McKelvey. 1983. *Planning and Design of Airports*. New York: McGraw-Hill.

Horton, Emmett J., and W. Dale Compton. 1984. Technological trends in automobiles. *Science* 225:587–592.

Hoshide, R. K. 1994. Electric motor do's and don'ts. *Energy Engineering* 91:6–24.

Hossli, Walter. 1969. Steam turbines. *Scientific American* 220(4):100–110.

Houdry, Eugène. 1931. Process for the Manufacture of Liquid Fuels. US Patent 1,837,963, December 22, 1931. Washington, DC: USPTO. http://www.uspto.gov

Houghton, John T., et al., eds. 2001. *Climate Change 2001: The Scientific Basis*. New York: Cambridge University Press.

Hounshell, David A. 1981. Two paths to the telephone. *Scientific American* 244(i):157–163.

Houston, R. E. 1927. *Model T Ford Production*. Centerville, IN: Model T Ford Club of America.

Howell, J. W., and H. Schroeder. 1927. *The History of the Incandescent Lamp*. Schenectady, NY: Maqua Co.

Hoy, Anne H. 1986. *Coca-Cola: The First Hundred Years*. Atlanta, GA: Coca-Cola Co.

Hu, Frank, et al. 2003. Television watching and other sedentary behaviors in relation to risk of obesity. *JAMA* 289:1785–1791.

Hughes, A., and B. Drury. 2019. *Electric Motors and Drives: Fundamentals, Types and Applications*. Amsterdam: Elsevier.

Hughes, David E. 1899. Researches of Professor D. E. Hughes, F. R. S. In: Electric Waves and Their Application to Wireless Telegraphy, 1879–1886. Appendix D of Fahie, J. J., *A History of Wireless Telegraphy*. London: Blackwood, pp. 305–316.

Hughes, Howard R. 1909. Drill. US Patent 930,759, August 10, 1909. Washington, DC: USPTO. http://www.uspto.gov

Hughes, Thomas P. 1962. British electrical industry lag: 1882–1888, *Technology and Culture* 3:27–44.

Hughes, Thomas P. 1983. *Networks of Power*. Baltimore, MD: Johns Hopkins University Press.

Humphrey, William S., and Joe Stanislaw. 1979. Economic growth and energy consumption in the UK, 1700–1975. *Energy Policy* 7:29–42.

Humphreys, Richard. 1999. *Futurism*. Cambridge: Cambridge University Press.

Hunley, J. D. 1999. *The His*tory of Solid-Propellant Rocketry: What We Do and Do Not Know*. Paper presented at American Institute of Aeronautics and Astronautics/ASME/SAE/ASEE (American Society of Mechanical Engineers/Society of Automotive Engineers/Society for Engineering Education) Joint Propulsion Conference and Exhibit, Los Angeles, CA, June 20–24, 1999.

Hunter, Louis C. 1979. *A History of Industrial Power in the United States, 1780–1930*. vol. 1: *Waterpower in the Century of the Steam Engine*. Charlottesville, VA: University Press of Virginia.

Hunter, Louis C., and Lynwood Bryant. 1991. *A History of Industrial Power in the United States, 1780–1930*. vol. 3: *The Transmission of Power*. Cambridge, MA: MIT Press.

Hurun. 2021. Hurun Largest Foreign Companies in China 2021. Hurun Report - List

Hutcheson, G. Dan, and Jerry D. Hutcheson. 1996. Technology and economics in the semiconductor industry. *Scientific American* 274(1):54–62.

Huzel, Dieter K., and David H. Huang. 1992. *Modern Engineering for Design of Liquid-Propellant Rocket Engines*. Washington, DC: American Institute of Aeronautics and Astronautics.

IAI (International Aluminium Institute). 2024. Primary aluminium smelting energy intensity. www.international-aluminium.org

IAEA (International Atomic Energy Agency). 2001. *Status of Nuclear Power Plants Worldwide in 2000*. Vienna: IAEA.

IANA (Intermodal Association of North America). 2004. *Industry Statistics*. Calverton, MD: IANA.

IATA (International Air Transport Association). 2024. *World Air Transport Statistics*. IATA - World Air Transport Statistics (WATS)

ICAO (International Civil Aviation Organization). 1971. *The ICAO Annual Report*. Montreal: ICAO.

ICAO (International Civil Aviation Organization). 2001. *The ICAO Annual Report*. Montreal: ICAO.

ICOLD (International Commission on Large Dams). 2024. *World Register of Dams*. Paris: ICOLD. https://www.icold-cigb.org/GB/world_register/world_register_of_dams.asp

IEA (International Energy Agency). 2023. *Data Centers and Data Transmission Networks*. Data centres & networks - IEA

IISI (International Iron and Stee Institute). 2002. *World Steel in Figures*. Brussels: IISI.

Illich, Ivan D. 1974. *Energy and Equity*. New York: Harper & Row.

Iles, George. 1906. *Inventors at Work*. New York: Doubleday, Page & Co.

Ingels, M. 1952. *Willis Haviland Carrier, Father of Air Conditioning*. New York: Arno Press.

Institute Lumiére. 2024. Le Cinématographe Lumière. www.institut-lumiere.org

Intel. 2024. Moore's law. www.intel.com

Integrated Surgical Systems. 2004. *Redefining Surgery . . . Around the World*. Davis, CA: Integrated Surgical Systems.

Intuitive Surgical. 2004. *Da Vinci Surgical System*. Sunnyvale, CA: Intuitive Surgical.

IPCC (Intergovernmental Panel for Climatic Change). 1996. *Revised 1996 IPCC Guidelines for National Greenhouse Gas Inventories*. Geneva: IPCC.

IRENA (International Renewable Energy Agency). 2019. Future of wind. https://www.irena.org/media/Files/IRENA/Agency/Publication/2019/Oct/IRENA_Future_of_wind_2019.pdf

ISC (Internet Systems Consortium). 2004. *ISC Domain Survey*. Redwood City, CA: ISC.

Islas, Jorge. 1999. The gas turbine: A new technological paradigm in electricity generation. *Technological Forecasting and Social Change* 60:129–148.

Israel, Paul. 1998. *Edison: A Life of Invention*. New York: Wiley.

ITA (International Trade Administration). 2002. Electric current abroad. Washington, DC: US Department of Commerce. Electric current worldwide (ECW). www.trade.gov

Ives, Frederic E. 1928. *The Autobiography of an Amateur Inventor*. Philadelphia: Private print.

Jackson, Kenneth A., ed. 1996. *Materials Science and Technology: A Comprehensive Treatment*. New York: Wiley.

Jacot, Bernard L., and D. M. B. Collier. 1935. *Marconi—Master of Space: An Authorized Biography of the Marchese Marconi*. London: Hutchinson.

Jansen, Marius B. 2000. *The Making of Modern Japan*. Cambridge, MA: Belknap Press.

Jay, Kenneth E. B. 1956. *Calder Hall: The Story of Britain's First Atomic Power Station*. London: Methuen.

Jeffries, Zay. 1960. Charles Franklin Kettering. kettering-charles.pdf, www.nasonline.org

Jehl, Francis. 1937. *Menlo Park Reminiscences*. Dearborn, MI: Edison Institute.

Jewett, Frank B. 1944. *100 Years of Electrical Communication in the United States.* New York: American Telephone & Telegraph.

JISF (Japan Iron and Steel Federation). 2003. *Energy Consumption by the Steel Industry at 2,013 PJ.* Tokyo: JISF.

JNCDI (Japan Nuclear Cycle Development Institute). 2000. *The Monju Sodium Leak.* Tokyo: JNCDI.

Johansson, Carl H., and P. A. Persson. 1970. *Detonics of High Explosives.* London: Academic Press.

Johnson, Benjamin. 2022. *Making Ammonia: Fritz Haber, Walther Nernst, and the Nature of Scientific Discovery.* Cham: Springer Nature.

Johnson, Elmer D. 1973. *Communication: An Introduction to the History of Writing, Printing, Books and Libraries.* Metuchen, NJ: Scarecrow Press.

Jolly, W. P. 1974. *Sir Oliver Lodge.* Rutherford, NJ: Fairleigh Dickinson University Press.

Jones, Glyn. 1989. *The Jet Pioneers.* London: Methuen.

Jones, Howard M. 1971. *The Age of Energy: Varieties of American Experience 1865–1915.* New York: Viking Press.

Jones, Jeremy A. T., B. Bowman, and P. A. Lefrank. 1998. Electric furnace steelmaking. In: Richard Fruehan, ed., *The Making, Shaping and Treating of Steel, Steelmaking and Refining Volume.* Pittsburgh, PA: AISE Steel Foundation, pp. 525–660.

Josephson, Matthew. 1959. *Edison: A Biography.* New York: McGraw Hill.

Joule, James P. 1850. *On Mechanical Equivalent of Heat.* London: R. & J. E. Taylor.

Kahan, Basil. 2000. *Ottmar Mergenthaler: The Man and the Machine.* New Castle, DE: Oak Knoll.

Kakela, Peter J. 1981. Iron ore: From depletion to abundance. *Science* 212:132–136.

Kander, Astrid, Malanima, P., and P. Warde. 2013. *Power to the People.* Princeton, NJ: Princeton University Press.

Kanemitsu, Nishio. 2001. *The Spark Plug.* Lausanne: FontisMedia.

Kanigel, Robert. 1997. *The One Best Way: Frederick Winslow Taylor and the Enigma of Efficiency.* New York: Viking.

Kawamoto, Kaoru, et al. 2000. *Electricity Used by Office Equipment and Network Equipment in the US* Berkeley, CA: Lawrence Berkeley National Laboratory.

Keeling, C. D. 1998. Reward and penalties of monitoring the Earth. *Annual Review of Energy and the Environment* 23:25–82.

Keenan, Thomas J. 1913. How trees are converted into paper. *Scientific American* 109: 256–258.

Keklik, Mümtaz. 2003. *Schumpeter, Innovation and Growth: Long-cycle Dynamics in the post-WWII American Manufacturing Industries.* Ashgate: Ashgate Publishing.

Kelly, H. 2024. *Why people are holding onto landline phones in rural areas.* The Washington Post. Why people are holding onto landline phones in rural areas - The Washington Post

Kelly, K. 2021. *Benefits of the Second Industrial Revolution vs the Benefits of the Third Industrial Revolution.* Online: P2P Foundation. Benefits of the Second Industrial Revolution vs the Benefits of the Third Industrial Revolution - P2P Foundation

Kelly, Thomas D., and Michael D. Fenton. 2003. *Iron and Steel Statistics.* Washington, DC: USGS.

Kempf, Karl. 1961. *Electronic Computers Within the Ordnance Corps.* Aberdeen Proving Ground, MD: US Army Ordnance Corps.

Kendall, Lane C., and James J. Buckley. 2001. *The Business of Shipping.* Centreville, MD: Cornell Maritime Press.

Kendrick, John W. 1961. *Productivity Trends in the United States.* Princeton, NJ: Princeton University Press.

Kennedy, Edward D. 1941. *The Automobile Industry: The Coming of Age of Capitalism's Favorite Child.* New York: Reynal & Hitchcock.

Kernighan, Brian, and Dennis Ritchie. 1978. *The C Programming Language.* Englewood Cliffs, NJ: Prentice-Hall.

Kerr, George T. 1989. Synthetic zeolites. *Scientific American* 261(1):100–105.

Keyes, Robert W. 1993. The future of the transistor. *Scientific American* 268(6):70–78.
Keynes, John M. 1919. *The Economic Consequences of the Peace*. London: Macmillan.
Khariton, Yuli, and Yuri Smirnov. 1993. The Khariton version. *Bulletin of the Atomic Scientists* 49(3):20–31.
Kilby, Jack S. 1964. Miniaturized Electronic Circuits. US Patent 3,138,743, June 23, 1964. Washington, DC: USPTO.
Kilby, Jack S., Jerry D. Merryman, and James H. Van Tassel. 1974. Miniature Electronic Calculator. US Patent 3,819,921, June 25, 1974. Washington, DC: USPTO. http://www.uspto.gov
King, Clarence D. 1948. *Seventy-five Years of Progress in Iron and Steel*. New York: American Institute of Mining and Metallurgical Engineers.
King, W. I. 1930. The effects of the new industrial revolution upon our economic welfare. *Annals of the American Academy of Political and Social Science* 149(1):165–172.
Kiple, Kenneth F., and Kriemhild C. Ornelas, eds. 2000. *The Cambridge World History of Food*. Cambridge: Cambridge University Press.
Kirsch, David A. 2000. *The Electric Vehicle and the Burden of History*. New Brunswick, NJ: Rutgers University Press.
Kleine-Kleffmann, Ulrich. 2023. The discovery of the first potash mine and the development of the potash industry since 1861. *Journal of Plant Nutrition and Soil Science*. https://doi.org/10.1002/jpln.202300382
Klíma, Ivan. 2001. *Karel Čapek: Life and Work*. North Haven, CT: Catbird Press.
Kline, Ronald R. 1992. *Steinmetz: Engineer and Socialist*. Baltimore, MD: Johns Hopkins University Press.
Kodak. 2024. *History of Kodak*. Rochester, NY: Kodak. History | Kodak
Kondratiev, Nikolai D. 1935. The long waves in economic life. *Review of Economic Statistics* 17:105–115.
Konrad, J. 2007. *Knock Nevis*. Knock Nevis - Video Of The World's Largest Ship
Koretsky, Z. 2021. Phasing out an embedded technology: Insights from banning the incandescent light bulb in Europe. *Energy Research & Social Science* 82:102310.
Kramer, Deborah A. 1997. *Explosives*. Washington, DC: USGS.
Kranzberg, Melvin. 1982. The industrialization of Western society, 1860–1914. In: Bernhard, Carl G., et al., eds., *Science and Technology and Society in the Time of Alfred Nobel*. Oxford: Pergamon Press, pp. 209–230.
Krause, Paul. 1992. *The Battle for Homestead, 1880–1892: Politics, Culture, and Steel*. Pittsburgh, PA: University of Pittsburgh Press.
Krech, Shepard, John R. McNeill, and Carolyn Merchant, eds. 2004. *Encyclopedia of Environmental History*. New York: Routledge.
Kuehr, Ruediger, and Eric Williams, eds. 2003. *Computers and the Environment: Understanding and Managing Their Impacts*. Dordrecht: Kluwer Academic.
Kumar, Shashi N. 2004. Tanker transportation. In: Cutler J. Cleveland, ed., *Encyclopedia of Energy*, vol. 6. Amsterdam: Elsevier, pp. 1–12.
Kupietzky, R. 2023. *ETOPS: history, evolution, current applications*. ETOPS: History, Evolution, Current Applications
Kurzweil, R. 2001. The law of accelerating returns. http://www.kurzweilai.net/the-law-of-accelerating-returns.
Kurzweil, R. 2005. *The Singularity Is Near*. New York: Penguin.
Lamont Doherty Earth Observatory. 2001. *Lamont 4D Technology*. New York: Lamont Doherty Earth Observatory.
Lamont, Lansing. 1965. *Day of Trinity*. New York: Atheneum.
Lamson, Alexander. 1890. Press the button. *Beacon* 2(19):154.
Lampson, Butler. 2003. Computing meets the physical world. *The Bridge*, Spring, 1–7.
Landau, Daniel. 2003. A simple theory of economic growth. *Economic Development and Cultural Change* 52:217–235.

Landau, Sarah B., and Carl W. Condit. 1996. *Rise of the New York Skyscraper, 1865—1913*. New Haven, CT: Yale University Press.

Landes, David. 1969. *The Unbound Prometheus: Technological Change and Industrial Development in Western Europe from 1750 to the Present*. Cambridge: Cambridge University Press.

Lane, David R., and Robert A. Lane. 1994. *Transistor Radios: A Collector's Encyclopedia and Price Guide*. Radnor, PA.: Wallace-Homestead Books.

Langen, Arnold. 1919. *Nicolaus August Otto, der Schöpfer des Verbrennungsmotors*. Stuttgart: Franck.

Langston, Lee S., and George Opdyke. 1997. *Introduction to Gas Turbines for Nonengineers*. Atlanta, GA: ASME International Gas Turbine Institute.

Lanouette, William. 1992. *Genius in Shadows: A Biography of Leo Szilard*. New York: Scribner.

Lavington, Simon. 2001. Tom Kilburn (1921–2001). *Nature* 409:996.

Law, Edward F. 1914. *Alloys and Their Industrial Applications*. London: Charles Griffin & Co.

Leach, Barry. 1973. *German General Staff*. New York: Ballantine Books.

Lebrun, Maurice. 1961. *La Soudure, le Brasage et l Oxycoupage des Métaux; 3500 ans d Histoire*. Paris: Académie de la Marine.

Leckie, A. H., A. Millar, and J. E. Medley. 1982. Short- and long-term prospects for energy economy in steelmaking. *Ironmaking and Steelmaking* 9:222–235.

Lee, Jasper S., and Michael E. Newman. 1997. *Aquaculture: An Introduction*. Danville, IL: Interstate Publishers.

Lebreton, T. 2024. The 7 largest solar panel manufacturers in the world (2024). www.theecoexperts.co.uk

Lemelson-MIT Program. 1996. *1996 Invention Index*. Cambridge, MA: MIT.

Levy, Jonathan I., James K. Hammitt, and John D. Spengler. 2000. Estimating the mortality impacts of particulate matter: What can be learned from between-study variability? *Environmental Health Perspectives* 108:109–117.

Lewchuk, Wayne. 1989. Fordism and the moving assembly line: The British and American experience, 1895–1930. In: Nelson Lichtenstein and Stephen Meyer, eds., *On the Line: Essays in the History of Auto Work*. Urbana: University of Illinois Press, pp. 17–41.

Lewis, David L. 1986. The automobile in America: The industry. *Wilson Quarterly* Winter, 47–63.

Li, A. 2021. *Decoding China's Car Industry: 40 Years*. Singapore: World Scientific.

Liebig, Justus von. 1840. *Die chemie in ihrer Anwendung auf Agricultur und Physiologie*. Braunschweig: Verlag von F. Vieweg und Sohn.

Lightman, Alan, Daniel Sarewitz, and Christina Desser, eds. 2003. *Living with the Genie: Essays on Technology and the Quest for Human Mastery*. Washington, DC: Island Press.

Lilley, Samuel. 1966. *Men, Machines and History*. New York: International Publishers.

Linde, Carl P. 1916. *Aus meinem Leben und von meiner Arbeit*. München: R. Oldenbourg.

Linde AG. 2003. *Linde and the History of Air Separation*. Höllriegelskreuth, Germany: Linde AG.

Lindem, Thomas J., and Paul A. S. Charles. 1995. Octahedral Machine with a Hexapodal Triangular Servostrut Section. US Patent 5,401,128, March 28, 1995. Washington, DC: USPTO. http://www.uspto.gov

Lindren, Michael. 1990. *Glory and Failure: The Difference Engines of Johann Möller, Charles Babbage and Georg and Edvard Scheutz*. Cambridge, MA: MIT Press.

Ling, Peter J. 1990. *America and the Automobile: Technology, Reform and Social Change*. Manchester: Manchester University Press.

Lippert, Owen, and Michael Walker. 1997. *The Underground Economy: Global Evidence of Its Size and Impact*. Vancouver, BC: Fraser Institute.

Lloyd, Ann, ed. 1988. *History of the Movies*. London: Macdonald Orbis.

Lloyd, D., et al. 2010 Alfred Russel Wallace deserves better. *Journal of Biosciences* 35 339–349. doi:10.1007/s12038-010-0039-x

LOC (Library of Congress). 2024. Inventing entertainment: The motion pictures and sound recordings of the Edison Companies. Washington, DC: LOC. www.loc.gov
Lodge, Oliver. 1894. The work of Hertz and some of his successors. *Electrician* 27:221–222, 448–449.
Lodge, Oliver. 1908. *Signalling through Space without Wires.* London: Electrician Printing & Publishing.
Lodge, Oliver. 1928. *Why I Believe in Personal Immortality.* London: Cassell.
Loftin, Laurence K. 1985. *Quest for Performance: The Evolution of Modern Aircraft.* Washington, DC: NASA.
Lombardi, Michael. 2003. *Century of Technology.* Seattle, WA: Boeing.
Lord, Barry. 2014 *Art & Energy.* Washington, DC: American Alliance of Museums.
Lourie, Richard. 2002. *Sakharov: A Biography.* Waltham, MA: Brandeis University Press.
Lovell, J. 2019. *Maoism: A Global History.* London: The Bodley Head.
Lucchini, Flaminio. 1966. *Pantheon: Monumenti dell Architettura.* Roma: Nuova Italia Scientifica.
Lüders, Johannes. 1913. *Dieselmythus: Quellenmassige Geschichte der Entstehung des heutingen Ölmotors.* Berlin: M. Krayn.
Luff, Peter P. 1978. The electronic telephone. *Scientific American* 238(3):58–64.
Luiten, Ester E. M. 2001. *Beyond Energy Efficiency: Actors, Networks and Government Intervention in the Development of Industrial Process Technologies.* Utrecht: Universiteit Utrecht.
Lumley, Roger, ed. 2010. *Fundamentals of Aluminium Metallurgy.* Amsterdam: Elsevier.
Lundstrom, Mark. 2003. Moore's law forever? *Science* 299:210–211.
Lyman, Peter, and Hal R. Varian. 2003. *How Much Information?* Berkeley: University of California.
MacKeand, J. C. B., and M. A. Cross. 1995. Wide-band high frequency signals from Poldhu? In: Institute of Electrical Engineers, *100 Years of Radio.* London: Institute of Electrical Engineers, pp. 26–31.
MacLaren, Malcolm. 1943. *The Rise of the Electrical Industry During the Nineteenth Century.* Princeton, NJ: Princeton University Press.
Macrotrends. 2024. *China Foreign Direct Investment 1979–2024.| MacroTrends.* China Foreign Direct Investment 1979-2024 | MacroTrends
Maddison, Angus. 1991. *Dynamic Forces in Capitalist Development.* Oxford: Oxford University Press.
Maddison, Angus. 1995. *Monitoring the World Economy 1820–1992.* Paris: Organization for Economic Cooperation and Development.
Maddison, Angus. 2001. *The World Economy: A Millennial Perspective.* Paris: OECD.
MAN (Maschinenfabrik Augsburg Nürnberg). 2024. MAN Trucks, Buses, Vans and Services. MAN Trucks, Buses, Vans and Services | MAN Global.
Manhattan Project Heritage Preservation Association. 2004. *Manhattan Project History.* Mountour Fall, NY: MPHPA.
Mannesmannröhren-Werk GmbH. 2024. Home page. www.mrw.de
Manyika, J., et al. 2017. *A Future That Works: Automation, Employment and Productivity.* New York: McKinsey & Company.
Marchelli, Renzo. 1996. *The Civilization of Plastics: Evolution of an Industry Which Has Changed the World.* Pont Canavese, Italy: Sandretto Museum.
Marconi, Guglielmo. 1909. *Wireless Telegraphic Communication.* Nobel Lecture December ii, 1909. Stockholm: Nobel Foundation. www.NobelPrize.org
Marder, Tod A., and M. W. Jones, eds. 2018. *The Pantheon: From Antiquity to the Present.* Cambridge: Cambridge University Press.
Marine Traffic. 2024. Vessel characteristics: Ship MSC LORETO (container ship) registered in Liberia - Vessel details, current position and voyage information. IMO 9934735MMSI 9934735Call Sign 5LJP2. AIS Marine Traffic.

Marland, G., et al. 2000. *Global, Regional, and National CO$_2$ Emission Estimates for Fossil Fuel Burning, Cement Production and Gas Flaring*. Oak Ridge, TN: Oak Ridge National Laboratory.

Marrey, Bernard. 2013. *Joseph Monier et la naissance du ciment armé*. Paris: Editions Du Linteau.

Marrus, M. R. 1974. *The Emergence of Leisure*. New York: Harper & Row.

Marshall, Arthur. 1917. *Explosives*. London: J. & A. Churchill.

Martin, Thomas C. 1894. *The Inventions Researches and Writings of Nikola Tesla*. New York: Electrical Engineer.

Martin, Thomas C. 1922. *Forty Years of Edison Service, 1882–1922: Outlining the Growth and Development of the Edison System in New York City*. New York: New York Edison Co.

Matos, Grecia. 2003. *Materials 2000*. Microsoft Excel File. Washington, DC: US Geological Survey.

Matos, Grecia, and Lorie Wagner. 1998. Consumption of materials in the United States, 1900–1995. *Annual Review of Energy and the Environment* 23:107–122.

Matson. 2024. Matson company history. Founding, timeline, and milestones. History - MATSON

Maxwell, James C. 1865. A dynamical theory of the electromagnetic field. *Philosophical Transactions of the Royal Society London* 155:495–512.

Maxwell, James C. 1873. *A Treatise on Electricity and Magnetism*. Oxford.

May, George S. 1975. *A Most Unique Machine: The Michigan Origins of the American Automobile Industry*. Grand Rapids, MI: William B. Eerdmans Publishing.

Maycock, P. D. 1999. *PV Technology, Performance, Cost: 1975–2010*. Warrenton, VA: Photovoltaic Energy Systems.

Mayer, Julius R. 1851. *Bemerkungen über das mechanische Aequivalent der Wärme*. Heilbronn: J. V. Landherr.

Maynard, Micheline. 2003. *The End of Detroit: How the Big Three Lost Their Grip on the American Car Market*. New York: Doubleday/Currency.

Mazor, Stanley. 1995. The history of microcomputer: Invention and evolution. *Proceedings of the IEEE* 83:1600–1608.

McClure, Henry H. 1902. Messages to mid-ocean: Marconi's own story of his latest triump. *McClure's Magazine*, April, 525–527.

McCullough, David. 2016. *The Wright Brothers*. New York: Simon and Schuster.

McDonald, Ronald. 1997. *The Complete Hamburger: The History of America's Favorite Sandwich*. New York: Birch Lane Press.

McElroy, Robert C., Reuben W. Hecht, and Earle E. Gavett. 1964. *Labor Used to Produce Field Crops*. Washington, DC: USDA.

McKibben, Bill. 2003. *Enough: Staying Human in an Engineered Age*. New York: Times Books.

McKinzie, Matthew G., et al. 2001. *The US Nuclear War Plan: A Time for Change*. Washington, DC: Natural Resources Defence Council.

McManus, George J. 1988. Blast furnaces: More heat from the hot end. *Iron Age* 4(8):15–20.

McManus, George J. 1993. The direct approach to making iron. *Iron Age* 9(7):20–24.

McMenamin, M. A. S., and D. L. S. McMenamin. 1990. *The Emergence of Animals: The Cambrian Breakthrough*. New York: Columbia University Press.

McNeill, William H. 1989. *The Age of Gunpowder Empires, 1450–1800*. Washington, DC: American Historical Association.

McShane, Clay. 1997. *The Automobile: A Chronology*. New York: Greenwood Press.

Mead, Carver, and Lynn Conway. 1980. *Introduction to VLSI systems*. Reading, MA: Addison-Wesley.

Medvedev, Roy A. 1972. *Let History Judge: The Origins and Consequences of Stalinism*. London: Macmillan.

Meher-Homji, Cyrus B. 1997. Anselm Franz and the Jumo 004. *Mechanical Engineering* 119(9):88–91.

Meier, A., and W. Huber. 1997. *Results from the Investigations of Leaking Electricity in the USA.* Berkeley, CA: Lawrence Berkeley Laboratory.
Meikle, Jeffrey L. 1995. *American Plastic: A Cultural History.* New Brunswick, NJ: Rutgers University Press.
Meitner, Lise, and Otto R. Frisch. 1939. Disintegration of uranium by neutrons: A new type of nuclear reaction. *Nature* 143:239–240.
Mellanby, Kenneth. 1992. *The DDT Story.* Farnham, UK: British Crop Protection Council.
Mellberg, William F. 2003. Transportation revolution. *Mechanical Engineering*, Suppl., 100 Years of Flight, 125(12):22–25.
Melville, George W. 1901. The engineer and the problem of aerial navigation. *North American Review*, December, 825.
Mendeleev, Dimitrii I. 1891. *The Principles of Chemistry.* London: Longmans, Green.
Mengel, Willi. 1954. *Ottmar Mergenthaler and the Printing Revolution.* Brooklyn, NY: Mergenthaler Linotype Co.
Mensch, Gerhard. 1979. *Stalemate in Technology.* Cambridge, MA: Ballinger.
Menzel, Peter. 1994. *Material World: A Global Family Portrait.* New York: Sierra Club Books.
Mercedes-Benz. 2024. 1901: The first Mercedes. Mercedes-Benz Corporate history. Mercedes-Benz Corporate history.
Mercedes-Benz Group. 2024. Bertha Benz. Bertha Benz | Mercedes-Benz Group > Company > Tradition > Founders & Pioneers
Messer, Robert L. 1985. New evidence on Truman's decision. *Bulletin of the Atomic Scientists* 41(4):50–56.
Mesquida, Christian G., and Neil I. Wiener. 1999. Male age composition and severity of conflicts. *Politics and the Life Sciences* 18:181–189.
Michelin. 2024. Over 130 years of adventures. Michelin Heritage | Michelin
Midgley, Thomas, and Albert L. Heene. 1930. Organic fluorides as refrigerants. *Industrial and Engineering Chemistry* 22:542–545.
Milanovic, Branko. 2002. *Worlds Apart: International and World Inequality, 1950–2000.* Washington, DC: World Bank.
Military Today. 2023. M1A2 Abrams Main Batlle Tank. www.Military.com
Miller, J. R. 1976. The direct reduction of iron ore. *Scientific American* 235(1):68–80.
Mitchell, Brian R. 1998. *International Historical Statistics Europe 1750–1993.* London: Macmillan.
Mitchell, Brian R., ed. 2003. *International Historical Statistics.* London: Palgrave.
Mitchell, Sally. 1996. *Daily Life in Victorian England.* Westport, CT: Greenwood Press.
Mitchell, Walter, and Stanley E. Turner. 1971. *Breeder Reactors.* Washington, DC: US Atomic Energy Commission.
Mitry, Jean. 1967. *Histoire du Cinema I (1895–1914).* Paris: Editions universitaires.
Mitsubishi. 2003. *The Man Who Started It All.* Tokyo: Mitsubishi.
Mittasch, Alvin. 1951. *Geschichte der Ammoniaksynthese.* Weinheim: Verlag Chemie.
Miura, Shigeru, et al. 1998. The mechanism of railway tracks. *Japan Railway & Transport Review* 15:38–45.
Mokdad, Ali H., et al. 2004. Actual causes of death in the United States, 2000. *Journal of American Medical Association* 291:1238–1245.
Mokyr, Joel. 1990. *The Lever of Riches: Technological Creativity and Economic Progress.* New York: Oxford University Press.
Mokyr, Joel. 1999. The Second Industrial Revolution, 1870–1914. In: V. Castronovo, ed., *Storia dell Economia Mondiale.* Rome: Latreza, pp. 219–245.
Mokyr, Joel. 2002. *The Gifts of Athena: Historical Origins of Knowledge Economy.* Princeton, NJ: Princeton University Press.
Molina, Mario J., and F. Sherwood Rowland. 1974. Stratospheric sink for chlorofluoromethanes: Chlorine atom catalyzed destruction of ozone. *Nature* 249:810–812.

Montgomery Ward. 1895. *Montgomery Ward & Co. Catalogue and Buyer's Guide No. 57, Spring and Summer 1895*. Chicago: Montgomery Ward & Co.
Moore, Dylan. 2011. Rotary kilns. Cement Kilns: Early rotary kilns. Cement Kilns: Rotary kilns
Moore, Gordon. 1965. Cramming more components onto integrated circuits. *Electronics* 38(8): 114–117.
Moore, Gordon. 1994. The accidental entrepreneur. *Engineering & Science* 57(4).
Moore, Gordon. 2003. *No Exponential Is Forever... But We Can Delay Forever*. Paper presented at International Solid State Circuits Conference, February 10, 2003. Santa Clara, CA: Intel.
Moore, Stephen, and Julian L. Simon. 1999. *The Greatest Century That Ever Was: 25 Miraculous Trends of the Past 100 Years*. Washington, DC: Cato Institute.
Moravec, Hans P. 1999. *Robot: Mere Machine to Transcend Mind*. New York: Oxford University Press.
Moreau, R. 1984. *The Computer Comes of Age: The People, the Hardware, and the Software*. Cambridge, MA: MIT Press.
Morgan, N. 1995. 3D popularity leads to 4D vision. *Petroleum Economist* 62(2):8–9.
Morita, Zen-ichiro, and Toshihiko Emi, eds. 2003. *An Introduction to Iron and Steel Processing*. Tokyo: Kawasaki Steel 21st Century Foundation.
Moritz, Michael. 1984. *The Little Kingdom: The Private Story of Apple Computer*. New York: W. Morrow.
Morris, Edmund. 2020. *Edison*. New York: Random House.
Morrison, Philip. 1945. Observations of the Trinity Shot, July 16, 1945. Record Group 227, OSRD-S1 Committee, Box 82 folder 6, "Trinity." Washington, DC: National Archives.
Morrison, Philip. 1995. Recollections of a nuclear war. *Scientific American* 273(2):42–46.
Mossman, Susan, ed. 1997. *Early Plastics: Perspectives, 1850–1950*. London: Science Museum.
Mott, Frank L. 1957. *A History of American Magazines, 1885–1905*. Cambridge, MA: Harvard University Press.
MSHA (Mining Safety and Health Administration). 2004. *Coal Fatalities by State*. Washington, DC: MSHA. http://www.msha.gov/stats/charts/coalbystate.asp
Mueller, John. 1989. *Retreat from Doomsday: The Obsolescence of Major War*. New York: Basic Books.
Mukherji, Chandra. 1983. *From Graven Images: Patterns of Modern Materialism*. New York: Columbia University Press.
Müller, Paul H. 1948. Dichloro-diphenyl-trichloroethane and newer insecticides. Nobel Lecture, December 11, 1948. Stockholm, Sweden: Nobel Foundation. www.NobelPrize.org
MTFCA (Model T Ford Club of America). 2024. Model T Ford Club of America. Largest Model T Club in the world. www.mtfca.com.
MTS (Midwest Tungsten). 2024. Tungsten Detailed History. Tungsten Detailed History | Midwest Tungsten
Murphy, P. M. 1974. *Incentives for the Development of the Fast Breeder Reactor*. Stamford, CT: General Electric.
Musser, Charles. 1990. *The Emergence of Cinema: The American Screen to 1907*. New York: Scribner.
Musson, Albert E. 1978. *The Growth of British Industry*. New York: Holmes & Meier.
Muybridge, Eadweard. 1887. *Animal Locomotion: An Electro-photographic Investigation of Consecutive Phases of Animal Movements*. Philadelphia, PA: University of Pennsylvania.
Myhrvold, Nathan, and F. Migoya. 2021. *Modernist Pizza*. Bellevue, WA: The Cooking Lab.
Nader, Ralph. 1965. *Unsafe at Any Speed: The Designed-in Dangers of the American Automobile*. New York: Grossman.
NAE (National Academy of Engineering). 2024. Greatest engineering achievements of the twentieth century. Washington, DC: NAE. www.greatachievements.org
Nagengast, Bernard. 2000. It's a cool story. *Mechanical Engineering* 122(5): 56–63.
NASA. 2021. Sunglider builds on legacy of solar aircraft. NASA. Sunglider Builds on Legacy of Solar Aircraft - NASA

NASM (National Air and Space Museum). 2024a. *Langley Aerodrome A*. Washington, DC: NASM. http://www.nasm.si.edu/nasm/aero/aircraft/langleyA.htm

NASM (National Air and Space Museum). 2024b. https://airandspace.si.edu/stories/editorial/eugene-ely-and-birth-naval-aviation-january-18-1911

NHTSA (National Highway Safety Administration). 2023. The economic and societal impact of motor vehicle crashes, 2019. Traffic crashes cost America billions in 2019. NHTSA.

Nature Physics. 2011. Editorial: Keep it simple? *Nature Physics* 7:441. Keep it simple? | Nature Physics

NBER (National Bureau of Economic Research). 2003. *Index of the General Price Level*. Cambridge, MA: NBER.

NEA (Nuclear Energy Agency). 2002. *Chernobyl: Assessment of Radiological and Health Impacts*. Paris: NEA.

Neal, Valerie, Cathleen S. Lewis, and Frank H. Winter. 1995. *Spaceflight: A Smithsonian Guide*. New York: Macmillan USA.

Needham, Joseph, et al. 1965. *Science and Civilisation in China*, vol. 4, Pt. II: Physics and Physical Technology. Cambridge: Cambridge University Press.

Needham, Joseph, et al. 1971. *Science and Civilisation in China*, vol. 4, Pt. III: Civil Engineering and Nautics. Cambridge: Cambridge University Press.

Needham, Joseph, et al. 1986. *Science and Civilisation in China*, vol. 5, Pt. VII: Military Technology: The Gunpowder Epic. Cambridge: Cambridge University Press.

Nehring, Richard. 1978. *Giant Oil Fields and World Oil Resources*. Santa Monica, CA: Rand Corporation.

Nelson, Walter H. 1998. *Small Wonder: The Amazing Story of the Volkswagen Beetle*. Cambridge, MA: Robert Bentley.

Németh, J. 1996. *Landmarks in the History of Hungarian Engineering*. Budapest: Technical University of Budapest.

New Steam Age. 1942. First practical gas turbines. *New Steam Age* 1(1):9–10, 20.

Newby, Frank, ed. 2001. *Early Reinforced Concrete*. Burlington, VT: Ashgate.

Newmark. 2024. 2023 US data center market overview & market clusters. Newmark. www.nmrk.com

Nielsen, Stephanie. 2003. Boeing turns to Russian programming talent in massive database project. *Serverworld* 2003(1):1–4.

Nieuwenhuis, Paul, and Peter Wells, eds. 1994. *Motor Vehicles in the Environment: Principles and Practice*. Chichester: Wiley.

Nilsson, Bengt V. 1987. Ernst Frederik Werner Alexanderson. www.philpem.me.uk

NIRS (Nuclear Information and Resource Service). 1999. *Background on Nuclear Power and Kyoto Protocol*. Washington, DC: NIRS.

NKK (Nippon Kokan Kabushiki). 2001. *Development and Application of Technology to Utilize Used Plastics for Blast Furnace*. NKK: Tokyo.

Nobel, Alfred. 1868. Improved Explosive Compound. US Patent 78,317. Washington, DC: USPTO. http://www.uspto.gov

Nobel, Alfred. 1895. *Alfred Nobel's Will*. Stockholm: Nobel Foundation. www.NobelPrize.org

Noble, David F. 1984. *Forces of Production: A Social History of Industrial Automation*. New York: A. A. Knopf.

Noguchi, Tatsuo, and Toshishige Fujii. 2000. Minimizing the effect of natural disasters. *Japan Railway & Transport Review* 23:52–59.

Norris, Robert S., and William M. Arkin. 1994. Estimated US, and Soviet/Russian nuclear stockpiles, 1945–94. *Bulletin of the Atomic Scientists* 50(6):58–60.

Nowell-Smith, Geoffrey, ed. 1996. *The Oxford History of World Cinema*. Oxford: Oxford University Press.

Noyce, Robert N. 1961. Semiconductor Device-and-Lead Structure. US Patent 2,981,877, April 25, 1961. Washington, DC: USPTO. http://www.uspto.gov

NRDC (Natural Resources Defence Council). 2002. *US, and Russian Strategic Nuclear Forces.* Washington, DC: NRDC.
NRDC (Natural Resources Defence Council). 2004. *Table of Known Nuclear Test Worldwide: 1945–1996.* Washington, DC: NRDC.
NREL (National Laboratory for Renewable Energy). 2024. Best research-cell efficiency chart. Photovoltaic Research. NREL. Best Research-Cell Efficiency Chart | Photovoltaic Research | NREL
NWA (Nuclear Weapons Archive). 2024. A guide to nuclear weapons. NWA. http://nuclearweaponarchive.org/
Nye, David E. 1990. *Electrifying America: Social Meaning of a New Technology.* Cambridge, MA: MIT Press.
Nye, David E. 2013. *America's Assembly Line.* Cambridge, MA: The MIT Press.
O'Brien, Geoffrey. 1995. First takes. *New Republic* 212:33–37.
O'Brien, P., ed. 1983. *Railways and the Economic Development of Western Europe, 1830–1914.* New York: St. Martin's Press.
O'Connell, Frobert L. 1989. *Of Arms and Men: A History of War, Weapons, and Aggression.* New York: Oxford University Press.
O'Connor, J. J. 1913. The genealogy of the motorcycle. *Scientific American* 109:12–13.
Odgers, George. 1957. *Air War Against Japan, 1943–1945.* Canberra: Australian War Memorial.
Odhner, Clas T. 1901. Award ceremony speech. Nobel Prize in Physics 1901. Nobel Foundation. www.NobelPrize.org
Odum, Howard T. 1971. *Environment, Power, and Society.* New York: Wiley-Interscience.
Ogura, Takekazu, ed. 1963. *Agricultural Development in Modern Japan.* Tokyo: Japan FAO Association.
Ohashi, Nobuo. 1992. Modern steelmaking. *American Scientists* 80:540–555.
Ohm, Georg S. 1827. *Die galvanische Kette, mathematisch bearbeitet.* Berlin: T. H. Rieman.
Okumura, Hirohiko. 1994. Recent trends and future prospects of continuous casting technology. *Nippon Steel Technical Report* 61:9–14.
Olivetti, Claudia. 2014. *The Female Labor Force and Long-Run Development: The American Experience in Comparative Perspective.* Cambridge, MA: National Bureau of Economic Research.
Olmstead, Alan L., and Paul Rhode. 1988. An overview of California agricultural mechanization, 1870–1930. *Agricultural History* 62:86–112.
OneSteel. 2005. *Interim Report.* Whyalla, South Australia, Australia: OneSteel.
Ōno, Taiichi. 1988. *Toyota Production System: Beyond Large-Scale Production.* Cambridge, MA: Productivity Press.
Oppenheimer, Todd. 1997. The computer delusion. *Atlantic Monthly* 280(1):45–62.
Ordóñez, José A. F. 1979. *Eugène Freyssinet.* Barcelona: Coop Industrial Trabajo.
O'Rourke, Kevin H., et al. 1996. Factor price convergence in the late nineteenth century. *International Economic Review* 37:499–530.
O'Rourke, K. H., and J. G. Williamson. 2000. *When Did Globalization Begin?* Cambridge, MA: National Bureau of Economic Research.
Orwell, George. 1942. The re-discovery of Europe. In: George Orwell, *Essays.* New York: Alfred A. Knopf (1969) p. 411.
Orwell, George, 1946. As I Please 63. In: George Orwell, *Essays.* New York: Alfred A. Knopf (1969) p. 1137.
Osborn, Fred M. 1952. *The Story of the Mushets.* London: T. Nelson.
Osler, A. 1981. *Turbinia.* Newcastle-upon-Tyne: Tyne and Wear County Council Museums. Turbinia | What's On | Discovery Museum
OTO (Ordo Templi Orientis). 2024. Carl Kellner. US Grand Lodge. www.oto-usa.org
Otto, Nicolaus A. 1877. Improvement in Gas-Motor Engines. Specification Forming Part of Letters Patent No. 194,047, Dated August 14, 1877. Washington, DC: USPTO. http://www.uspto.gov

Ouelette, Robert P., et al. 1983. *Automation Impacts on Industry*. Ann Arbor, MI: Ann Arbor Science.
Ovington, Earle L. 1912. The Gnome rotary engine. *Scientific American* 107:218–219,230.
Owens, Michael J. 1904. Glass-Shaping Machine. US Patent 766,768, August 2, Washington, DC: USPTO. http://www.uspto.gov
Painter, William. 1892. Bottle-Sealing Device. Specification Forming Part of Letters Patent No. 468,226, Dated February 2, 1892. Washington, DC: US Patent Office.
Papapavlou, C., et al. 2022. Toward SDM-based submarine optical networks: A review of their evolution and upcoming trends. *Telecom* 2022, 3, 234–280. https://doi.org/ 10.3390/telecom3020015
PARC (Palo Alto Research Center). 2024. *PARC History: Innovation and Inventing the Future*. Palo Alto, CA: PARC.
Park, Benjamin. 1898. *A History of Electricity, The Intellectual Rise in Electricity, from Antiquity to the Days of Benjamin Franklin*. New York: Wiley.
Park, James. 1900. *The Cyanide Process of Gold Extraction*. London: Charles Griffin & Co.
Parshall, Jonathan, and Anthony Tully. 2006. *Shattered Sword: The Untold Story of the Battle of Midway*. Sterling, VA: Potomac Books
Parsons, Charles A. 1911. *The Steam Turbine*. Cambridge: Cambridge University Press.
Parsons, John T., and Frank L. Stulen. 1958. Motor Controlled Apparatus for Positioning Machine Tool. US Patent 2,820,187, January 14, 1958. Washington, DC: USPTO. http://www.uspto.gov
Parsons, Robert H. 1936. *The Development of Parsons Steam Turbine*. London: Constable & Co.
Passer, Harold C. 1953. *The Electrical Manufacturers, 1875–1900: A Study in Competition, Entrepreneurship, Technical Change and Economic Growth*. Cambridge, MA: Harvard University Press.
Paton, J. 1890. Gas and gas-lighting. In: *Encyclopaedia Britannica*, 9th ed., T.S. Baynes, ed., vol. 10. Chicago: R. S. Peale & Co., pp. 87–102.
Patterson, Marvin L. 1993. *Accelerating Innovation: Improving the Process of Product Development*. New York: Van Nostrand Reinhold.
Pattillo, Donald M. 1998. *Pushing the Envelope: The American Aircraft Industry*. Ann Arbor: University of Michigan Press.
Patton, Phil. 2003. *Bug: The Strange Mutations of the World's Most Famous Automobile*. New York: Simon & Schuster.
Payen, Jacques. 1993. *Beau de Rochas: Sa Vie, son Oeuvre*. Digne-les-Bains: Editions de Haute-Provence.
Payne, S. G. 1996. *A History of Fascism, 1914–1945*. London: Routledge.
Peacey, J. G., and W. G. Davenport. 1979. *The Iron Blast Furnace*. Oxford: Pergamon Press.
Pearson, Roy, and John Fossey. 1983. *World Deep-Sea Container Shipping*. Aldershot, UK: Gower Publishing.
Perez, Carlota. 2002. *Technological Revolutions and Financial Capital*. Cheltenham: Edward Elgar.
Perkins, Frank C. 1902. High-speed German railway at Zossen. *Scientific American* 86: 91–92.
Perrodon, Alain. 1981. *Histoire des Grandes Découvertes Pètrolieres*. Paris: Elf Aquitaine.
Perry, H. W. 1913. Teams and motor trucks compared. *Scientific American* 108:66.
Person, H. S. 1930. Man and the machine: The engineer's point of view. *Annals of the American Academy of Political and Social Science* 149:88–93.
Pessaroff, Nicky. 2002. An electric idea Edison's electric pen. *Pen World International* 15(5):1–4.
Peters, Tom F. 1996. *Building the Nineteenth Century*. Cambridge, MA: MIT Press.
Petit, J. R., et al. 1999. Climate and atmospheric history of the past 420,000 years from the Vostok ice core, Antarctica. *Nature* 399:429–426.
Petroski, Henry. 1993. On dating inventions. *American Scientist* 81:314–318.
Phillips, David C. 1996. *Art for Industry s Sake: Halftone Technology, Mass Photography and the Social Transformation of American Print Culture, 1880–1920*. New Haven, CT: Yale University.

PHS (Plastics Historical Society). 2003. *The Plastics Museum*. London: PHS.
Piszkiewicz, Dennis. 1995. *The Nazi Rocketeers: Dreams of Space and Crimes of War*. Westport, CT: Praeger.
Plank, Charles J., and Edward J. Rosinski. 1964. Catalytic Cracking of Hydrocarbons with a Crystalline Zeolite Catalyst Composite. US Patent 3,140,249, July 7, 1964. Washington, DC: USPTO. http://www.uspto.gov
Platt, Harold L. 1991. *The Electric City: Energy and the Growth of the Chicago Area, 1880–1930*. Chicago: University of Chicago Press.
Ploeger, Jan. 1994. The bicycle as part of a green integrated traffic system. In: Paul Nieuwenhuis and Peter Wells, eds., *Motor Vehicles in the Environment: Principles and Practice*. Chichester: Wiley, pp. 47–62.
Pocock, Rowland F. 1995. Improved communications at sea: A need and a new technology. In: Institute of Electrical Engineers, *100 Years of Radio*. London: Institute of Electrical Engineers, pp. 57–61.
Pohl, E., and R. Müller, eds. 1984. *150 Jahre Elektromotor, 1834–1984*. Würzburg: Vogel-Verlag.
Pomeranz, Kenneth. 2000. *The Great Divergence: China, Europe, and the Making of the Modern World*. Cambridge: Cambridge University Press.
Pope, Franklin L. 1894. *Evolution of the Electric Incandescent Lamp*. New York: Boschen & Wefer.
Popović, Vojin, R. Horvat, and N. Nikić, eds. 1956. *Nikola Tesla: Lectures, Patents, Articles*. Beograd: Nikola Tesla Museum.
Porter, Horace C. 1924. *Coal Carbonization*. New York: Chemical Catalog Co.
Poskett, J. 2019. *Materials of the Mind: Phrenology, Race, and the Global History of Science, 1815–1920*. Chicago: University of Chicago Press.
Potts, R. 2001. *Complexity and Adaptability in Human Evolution*. Paper presented at Development of the Human Species and Its Adaptation to the Environment, a AAAS conference. Cambridge, MA, July 7–8, 2001. http://www.uchicago.edu/aff/mwc-amacad/biocomplexity/conference_papers/PottsComplexity.pdf
Powell, Horace B. 1956. *The Original Has This Signature: W. K. Kellogg*. Englewood Cliffs, NJ: Prentice-Hall.
Power-Technology. 2024. Three Gorges Dam hydro electric power plant, China. www.power-technology.com
Prados de la Escosura, L. 2000. International comparisons of real product, 1820–1890: An alternative data set. *Explorations in Economic History* 37:1–41.
Pratt & Whitney. 1999. *JT9D Engine*. East Hartford, CT: Pratt & Whitney.
Press, Frank, and Raymond Siever. 1986. *Earth*. New York: W. H. Freeman.
Priestley, Joseph. 1768. *The History and Present State of Electricity with Original Experiments*. London: J. Dodsley.
Prost, Antoine. 1991. Public and private spheres in France. In: Antoine Prost and Gérard Vincent, eds., *A History of Private Life*, vol. 5. Cambridge, MA: Belknap Press, pp. 1–103.
Prout, Henry G. 1921. *A Life of George Westinghouse*. New York: American Society of Mechanical Engineers.
Pulleyblank, William R. 2003. *Application Driven Supercomputing: An IBM Perspective*. Yorktown Heights, NY: IBM.
Pursell, C. 2007. *Technology in Postwar America: A History*. New York: Columbia University Press.
Pyxis. 2004. *Pyxis HelpMate®, the Trackless Robotic Courier*. San Diego, CA: Cardinal Health.
Raby, Ormond. 1970. *Radio's First Wave: The Story of Reginald Fessenden*. Toronto: Macmillan.
Radovsky, M. 1957. *Alexander Popov: Inventor of Radio*. Moscow: Foreign Languages Publishing House.
Rae, John B. 1971. *The Road and the Car in American Life*. Cambridge, MA: MIT Press.
Ratcliffe, K. 1985. *Liquid Gold Ships: History of the Tanker (1859–1984)*. London: Lloyds.

Rathke, Kurt. 1953. *Wilhelm Maybach. Anbruch eines neuen Zeitalters. (Von der Pferdekutsche zum Mercedes-Wagen)*. Friedrichshafen: Verlag Robert Gessler.
Ratzlaff, John T., and Leland I. Anderson. 1979. *Dr. Nikola Tesla Bibliography*. San Carlos, CA: Ragusan Press.
Reid, T. R. 2001. *The Chip: How Two Americans Invented the Microchip and Launched a Revolution*. New York: Simon & Schuster.
Reintjes, J. Francis. 1991. *Numerical Control: Making a New Technology*. New York: Oxford University Press.
Repetto, Richard, et al. 1991. *Accounts Overdue: Natural Resource Depletion in Costa Rica*. Washington, DC: World Resources Institute.
Rhodes, Frederick L. 1929. *Beginnings of Telephony*. New York: Harper.
Rhodes, Richard. 1986. *The Making of the Atomic Bomb*. New York: Simon & Schuster.
Rhodes, Richard. 1988. Man-made death: A neglected mortality. *JAMA* 260:686–687.
Richardson, Kenneth. 1977. *The British Motor Industry 1896–1939*. London: Macmillan.
Richardson, Lewis F. 1960. *Statistics of Deadly Quarrels*. Pacific Grove, CA: Boxwood Press.
Riedel, Gerhard. 1994. *Der Siemens-Martin-Ofen: Rückblick auf eine Stahlepoche*. Düsseldorf: Stahleisen.
Ries, L. A. G., et al., eds. 2003. *SEER Cancer Statistics Review, 1975–2000*. Bethesda, MD: National Cancer Institute.
Rife, Patricia. 1999. *Lise Meitner and Dawn of the Nuclear Age*. Boston: Birkhäuser.
Riis, Jacob. 1890. *How the Other Half Lives*. New York: Scribner.
Rinderknecht, Peter, ed. 1966. *75 Years Brown Boveri, 1891–1966*. Baden: Brown, Boveri & Co.
Riordan, Michael, and Lillian Hoddeson. 1997. *Crystal Fire: The Birth of the Information Age*. New York: Norton.
Ristori, Emanuel J. 1911. Aluminium. In: *Encyclopaedia Britannica*, 11th ed., H. Chisholm, ed., vol. I. Cambridge: Cambridge University Press, pp. 767–772.
Rittaud-Hutinet, Jacques. 1995. *Les frères Lumière: L invention du cinéema*. Paris: Flammarion.
Robinson, Eric, and Albert E. Musson. 1969. *James Watt and the Steam Revolution*. New York: Augustus M. Kelley.
Roberts, John. 1981. *The Circulation of D. D. T. Throughout the World Biosphere*. Newcastle, NSW: University of Newcastle.
Robson, Graham. 1983. *Magnificent Mercedes: The Complete History of the Marque*. New York: Bonanza Books.
Rockwell, Theodore. 1992. *The Rickover Effect: How One Man Made a Difference*. Annapolis, MD: Naval Institute Press.
Rogge, Michael. 2022. More than one hundred years of film sizes. Almost one hundred film widths and perforations were experimented with. www.xs4all.nl
Rogin, L. 1931. *The Introduction of Farm Machinery*. Berkeley: University of California Press.
Rojas, Raúl, and Ulf Hashagen, eds. 2000. *The First Computers: History and Architectures*. Cambridge, MA: MIT Press.
Rolls, Charles S. 1911. Motor vehicles. In: *Encyclopaedia Britannica*, 11th ed., H. Chisholm, ed. vol. 18. Cambridge: Cambridge University Press, pp. 914–921.
Rolt, L. T. C. 1965. *Tools for the Job: Short History of Machine Tools*. London: B. T. Bats-ford.
Romer, Paul. 1990. Endogenous technological change. *Journal of Political Economy* 98: 71–102.
Rose, Philip S. 1913. Economics of the farm tractor. *Scientific American* 109:114–115.
Rosen, William. 2012. *The Most Powerful Idea in the World: A Story of Steam, Industry, and Invention*. Chicago: University of Chicago Press.
Ross, Douglas T. 1982. *Origins of the APT Language for Automatically Programmed Tools*. New York: ACM Press.
Ross, Kristin. 1996. *Fast Cars, Clean Bodies: Decolonization and the Reordering of French Culture*. Cambridge, MA: MIT Press.
Rostow, Walt W. 1971. *The Stages of Economic Growth: A Non-Communist Manifesto*. Cambridge: Cambridge University Press.

Rostow, Walt W. 1990. *Theories of Economic Growth from David Hume to the Present.* New York: Oxford University Press.

Rott, N. 1990. Note on the history of the Reynolds number. *Annual Review of Fluid Mechanics* 22:1–11.

Rowe, Albert P. 1948. *One Story of Radar.* Cambridge: Cambridge University Press.

Ruse, Michael, and David Castle, eds. 2002. *Genetically Modified Foods: Debating Biotechnology.* Amherst, NY: Prometheus Books.

Rushmore, David B. 1912. The electrification of the Panama Canal. *Scientific American* 107:397, 405–406.

Rutherford, Ernest. 1911. The scattering of α and β particles by matter and the structure of the atom. *Philosophical Magazine*, Series 6, 21:669–688.

Ruthven, Douglas M., S. Farooq, and K. S. Knaebel. 1993. *Pressure Swing Adsorption.* New York: Wiley.

Sachs, Aaron. 1999. Virtual ecology: A brief environmental history of Silicon Valley. *WorldWatch*, January/February, 12–21.

Sakiya, Tetsuo. 1982. *Honda Motor: The Men, the Management, the Machines.* New York: Kodansha International.

Sala-i-Martin, Xavier. 2002. *The Disturbing "Rise" of Global Income Inequality.* Cambridge, MA: National Bureau of Economic Research.

Sale, Anthony E. 2000. The colossus of Bletchley Park: The German cipher system. In: Raúl Rojas and Ulf Hashagen, eds., *The First Computers: History and Architectures.* Cambridge, MA: MIT Press, pp. 351–364.

Sample, Frank R., and Maurice E. Shank. 1985. Aircraft turbofans: New economic and environmental benefits. *Mechanical Engineering* 107(9):47–53.

Samuelson, James. 1893. *Labour Saving Machinery.* London: K. Paul, Trench, Trüber & Co.

Samuelson, James, ed. 1896. *The Civilisation of Our Day.* London: Sampson, Low, Marston & Co.

Sanguri, M. 2012. The guide to slow steaming on ships. Slide 1. www.marineinsight.com

Saravanamuttoo, H. I. H., G. F. C. Rogers, and H. Cohen. 2001. *Gas Turbine Theory.* New York: Prentice Hall.

Sauvage, Léo. 1985. *L'Affaire Lumiere Du Mythe a l'Histoire: Enquête sur les Origines du Cinéma.* Paris: L'Herminier.

Schaaf, Larry J. 1992. *Out of the Shadows: Herschel, Talbot, and the Invention of Photography.* New Haven, CT: Yale University Press.

Schellen, Heinrich, and N. S. Keith. 1884. *Magneto-Electric and Dynamo-Electric Machines,* vol. 1. New York: Van Nostrand.

Schlebecker, John T. 1975. *Whereby We Thrive: A History of American Farming, 1607–1971.* Ames: Iowa State University Press.

Schneider, Friedrich, and Domink H. Ente. 2003. *The Shadow Economy: An International Survey.* Cambridge: Cambridge University Press.

Schrewe, H. F. 1991. *Continuous Casting of Steel: Fundamental Principles and Practice.* Düsseldorf: Stahl und Eisen.

Schumacher, M. M., ed. 1978. *Enhanced Oil Recovery: Secondary and Tertiary Methods.* Park Ridge, NJ: Noyes Data.

Schumpeter, Joseph A. 1939. *Business Cycles: A Theoretical, Historical, and Statistical Analysis of the Capitalist Process.* New York: McGraw-Hill.

Schurr, Sam H., and Bruce C. Netschert. 1960. *Energy in the American Economy 1850–1975.* Baltimore, MD: Johns Hopkins University Press.

Schurr, Sam H., et al. 1990. *Electricity in the American Economy: Agent of Technological Progress.* New York: Greenwood Press.

Scipioni, A., A. Manzardo and J. Ren. 2023. *Hydrogen Economy: Processes, Supply Chain, Life Cycle Analysis and Energy Transition for Sustainability.* Amsterdam: Elsevier.

Scott, Peter B. 1984. *The Robotics Revolution.* Oxford: Basil Blackwell.

Scott, R. H. 1951. *Mechanism and Operation of Modern Linotypes*. London: Linotype and Machinery Ltd.
Scott, Walter G. 1997. Micro-size gas turbines create market opportunities. *Power Engineering* 101(9):46–50.
Seeger, Hendrik. 1999. The history of German wastewater treatment. *European Water Management* 2:51–56.
Segal, Howard P. 1985. *Technological Utopianism in American Culture*. Chicago: University of Chicago Press.
Seitz, Frederick, and Norman G. Einspruch. 1998. *Electronic Genie: The Tangled History of Silicon*. Urbana: University of Illinois Press.
Selden, George B. 1895. Road-Engine. Specification Forming Part of Letters Patent No. 549,160, Dated November 5, 1895. Washington, DC: US Patent Office. http://www.uspto.gov
Seldes, G. 1928. *The Stammering Century*. New York: John Day.
Sellen, Abigail, and Richard Harper. 2002. *The Myth of the Paperless Office*. Cambridge, MA: MIT Press.
Semon, Waldo L. 1933. Synthetic Rubber-like Composition and Method of Making Same. US. Patent 1,929,453, October 10, 1933. Washington, DC: USPTO. http://www.uspto.gov
SEPCo (Shell Exploration and Production Company). 2004. *Shell in the Gulf of Mexico*. Robert, LA: SEPCo.
Seymour, Raymond B., ed. 1989. *Pioneers in Polymer Science*. Boston: Kluwer Academic Publishers.
Shaeffer, R. E. 1992. *Reinforced Concrete: Preliminary Design for Architects and Builders*. New York: McGraw-Hill.
Shankster, H. 1940. Dynamite. In: Jocelyn Thope and M. A. Whiteley, eds., *Thorpes's Dictionary of Applied Chemistry*, vol. 4. London: Longmans, pp. 239–245.
Shannon, C. E. 1948. A mathematical theory of communication. *Bell System Technical Journal* 27:379–423, 623–656.
Sharlin, Harold I. 1967. Applications of electricity. In: Melvin Kranzberg and Carroll W. Pursell, Jr., eds., *Technology in Western Civilization*, vol. 1. New York: Oxford University Press, pp. 563–578.
Shevell, R.S. 1980. Technological development of transport aircraft – past and future. *Journal of Aircraft* 17:67–80.
Shell Poll. 1998. The millennium. *Shell Poll* 1(2).
Sheppard, D. 2017. Robert le Rossignol, 1884-1976: Engineer of the "Haber" process. *Notes and Records* 71:263–296.
Sherwin, Martin. 1985. How well they meant. *Bulletin of the Atomic Scientists* 41(7):9–15.
Shima, Hideo. 1994. Birth of the Shinkansen: A memoir. *Japan Railway & Transport Review* 3:45–48.
Shockley, William. 1964. Transistor technology evokes new physics. In: *Nobel Lectures: Physics 1942–1962*. Amsterdam: Elsevier, pp. 344–374.
Siemens, C. William. 1882. Electric lighting, the transmission of force by electricity. *Nature* 27:67–71.
Siemens, Werner von. 1893. *Personal Recollections of Werner von Siemens* (trans. W. C. Coupland). New York: D. Appleton & Co.
Siemens. 2024. Reliable Gas Turbines. www.siemens-energy.com
Sime, Ruth Lewin. 1996. *Lise Meitner: A Life in Physics*. Berkeley: University of California Press.
Simpson, James R., et al. 1999. *Pig, Broiler and Laying Hen Structure in China, 1996*. Davis, CA: International Agricultural Trade Research Consortium.
Simon, J.L. and Kahn, H. eds. 1984: *The Resourceful Earth: A Response to Global 2000*. Oxford and New York: Basil Blackwell.
Sinnott, Terri, and John Bowditch. 1980. Edison "Jumbo" engine-driven dynamo. 48.pdf. www.asme.org
Sittauer, Hans L. 1972. *Geba'ndigte Explosionen*. Berlin: Transpress Verlag für Verkehrswesen.

Siuru, B. 1989. Horsepower to the people. *Mechanical Engineering* 111(2):42–46.
Sivak, Michael, and Michael J. Flannagan. 2003. Flying and driving after the September 11 attacks. *American Scientist* 91:6–8.
Skafle, I., et al. 2022. Misinformation about COVID-19 vaccines on social media: Rapid review. *Journal of Medical Internet Research* 2022 Aug 4; 24(8):e37367. doi: 10.2196/37367. PMID: 35816685; PMCID: PMC9359307
Slater, Rodney E. 1996. The National Highway System: A commitment to America's future. *Public Roads* 59(4):1–3.
Smil, Vaclav. 1976. *China's Energy.* New York: Praeger.
Smil, Vaclav. 1983. *Biomass Energies.* New York: Plenum.
Smil, Vaclav. 1987. *Energy Food Environment: Realities, Myths, Options.* Oxford: Oxford University Press.
Smil, Vaclav. 1988. *Energy in China's Modernization.* Armonk, NY: M. E. Sharpe.
Smil, Vaclav. 1991. *General Energetics.* New York: Wiley.
Smil, Vaclav. 1992. Agricultural energy cost: National analysis. In: Richard C. Fluck, ed., *Energy in Farm Production.* Amsterdam: Elsevier, pp. 85–100.
Smil, Vaclav. 1993. *Global Ecology.* London: Routledge.
Smil, Vaclav. 1996. *Environmental Problems in China: Estimates of Economic Costs.* Honolulu, HI: East-West Center.
Smil, Vaclav. 1999a. Nitrogen in crop production: An account of global flows. *Global Biogeochemical Cycles* 13:647–662.
Smil, Vaclav. 1999b. China's great famine: 40 years later. *British Medical Journal* 7225: 1619–1621.
Smil, Vaclav. 2001. *Enriching the Earth: Fritz Haber, Carl Bosch and the Transformation of World Food Production.* Cambridge, MA: MIT Press.
Smil, Vaclav. 2002. *The Earth s Biosphere: Evolution, Dynamics and Change.* Cambridge, MA: MIT Press.
Smil, Vaclav. 2003. *Energy at the Crossroads: Global Perspectives and Uncertainties.* Cambridge, MA: MIT Press.
Smil, Vaclav. 2008. *Energy in Nature and Society.* Cambridge, MA: MIT Press.
Smil, Vaclav. 2009. *Prime Movers of Globalization.* Cambridge, MA: MIT Press.
Smil, Vaclav. 2015. *Power Density.* Cambridge, MA: MIT Press.
Smil, Vaclav. 2016. *Still the Iron Age.* Oxford: Elsevier.
Smil, Vaclav. 2017a. *Energy and Civilization: A History.* Cambridge, MA: MIT Press.
Smil, Vaclav. 2017b. *Energy Transitions.* Santa Monica, CA: Praeger.
Smil, Vaclav. 2017c. Thank Maxwell for cellphones. *Spectrum* March 2017:24.
Smil, Vaclav. 2018. October 1958: First Boeing 707 to Paris. *IEEE Spectrum* October 2018:23.
Smil, Vaclav. 2019. *Growth: From Microorganisms to Megacities.* Cambridge, MA: MIT Press.
Smil, Vaclav. 2022. *How the World Really Works.* London: Viking.
Smil, Vaclav. 2023a. *Invention and Innovation.* Cambridge, MA: MIT Press.
Smil, Vaclav. 2023b. *Materials and Dematerialization.* Chichester: Wiley.
Smil, Vaclav. 2024a. *Food.* London: Viking.
Smil, Vaclav. 2024b. *Halfway Between Kyoto and 2050: Zero Carbon Is a Highly Unlikely Outcome.* Vancouver, BC: The Fraser Institute.
Smil, Vaclav. 2025. *Speed.* London: Viking.
Smith, Anthony, and Richard Paterson. 1998. *Television: An International History.* Oxford: Oxford University Press.
Smith, Cyril S. 1967. Metallurgy: Science and practice before 1900. In: Melvin Kranzberg and Carroll W. Pursell, Jr., eds., *Technology in Western Civilization,* vol. 1. New York: Oxford University Press, pp. 592–636.
Smith, Douglas K., and Robert C. Alexander. 1988. *Fumbling the Future: How Xerox Invented, Then Ignored, the First Personal Computer.* New York: W. Morrow.

Smith, Edward S. 1911. Heavy commercial vehicles. In: *Encyclopaedia Britannica*, 11th ed., H. Chisholm, ed., vol. 18. Cambridge: Cambridge University Press, pp. 921–930.

Smith, G. W. 1954. *Dr. Carl Gustaf Patrick de Laval*. New York: Newcomen Society in North America.

Smith, N. 1980. The origins of the water turbine. *Scientific American* 242(1):138–148.

Smith, Roderick A. 2001. Railway technology: The last 50 years and future prospects. *Japan Railway & Transport Review* 27:16–24.

Smith, William W. 1966. *Midway, Turning Point of the Pacific*. New York: Crowell.

Smock, Robert. 1991. Gas turbine, combined cycle orders continue. *Power Engineering* 95(5):17–22.

Snyder, T. 2011. Hitler vs. Stalin: Who was worse? | The New York Review of Books. www.nybooks.com

Society of Petroleum Engineers. 1991. *Horizontal Drilling*. Richardson, TX: Society of Petroleum Engineers.

Solberg, Carl. 1979. *Conquest of the Skies: A History of Commercial Aviation in America*. Boston: Little, Brown.

Solly, Ray. 2022. *The Development of Crude Oil Tankers: A Historical Miscellany*. Barnsley: Pen and Sword.

Solow, Robert M. 1957. Technical change and the aggregate production function. *Review of Economics and Statistics* 39:312–320.

Solow, Robert M. 1987. Nobel Prize Lecture. In: Robert M. Solow, *Growth Theory: An Exposition*. New York: Oxford University Press, pp. ix–xxvi.

Solow, Robert M. 2001. Addendum, August 2001. Robert M. Solow. Prize Lecture: Growth theory and after. www.nobelprize.org

Solzhenitsyn, Aleksandr. 1973–1975. *Arkhipelag Gulag, 1918–1956*. Paris: YMCA Press.

Sone, Satoru. 2001. Japan's rail technology development from 1945 to the future. *Japan Railway & Transport Review* 27:4–15.

Song, Y. 1966/1673. *Tiangong kaiwu* [*The Creations of Nature and Man*] (E. Sun and S. Sun, trans.). State College: Pennsylvania State University Press.

Sony. 2004. *History: Products and Technology Milestones*. Tokyo: Sony. http://www.sony.net

Soulié, Claude, and Jean Tricoire 2002. *Le grand livre du TGV*. Paris: Vie du rail.

Southward, John. 1890. Typography. In: *Encyclopaedia Britannica*, 9th ed., T.S. Baynes, ed., vol. 23. Chicago: R. S. Peale & Co., pp. 697–710.

Spada, Alfred T. 2003. *In Search of Light-Weight Components: Automotive's Cast Aluminum Conversion*. Schaumburg, IL: American Foundry Society.

Spillman, William J. 1903. Systems of farm management in the United States. In: *Yearbook of the United States Department of Agriculture*. Washington, DC: US Government Printing Office, pp. 343–364.

Splinter, W. E. 1976. Centre-pivot irrigation. *Scientific American* 234(6):90–99.

Sprague, S. 2021. The US productivity slowdown: An economy-wide and industry-level analysis. *Labor Review*, US Bureau of Labor Statistics, April 2021. https://doi.org/10.21916/mlr.20

Stalk, George, and Thomas M. Hout. 1990. *Competing Against Time: How Time-based Competition Is Reshaping Global Markets*. New York: Free Press.

Stanley, William. 1912. Alternating-current development in America. *Journal of the Franklin Institute* 173:561–580.

Starratt, F. Weston. 1960. LD ... in the beginning. *Journal of Metals* 12(7):528–529.

Staudinger, Hermann. 1953. Nobel Lecture. Nobel Foundation. www.NobelPrize.org

Steckel, Richard H., and Roderick Floud, eds. 1997. *Health and Welfare During Industrialization*. Chicago: University of Chicago Press.

Steenberg, B. 1995. Carl David Ekman—Pioneer. *Nordisk Pappershistorisk Tidskrift* 23(4):9–16.

Steinhart, John S., and Carol E. Steinhart. 1974. Energy use in the US food system. *Science* 184:307–316.

Stellman, Jeanne M., et al. 2003. The extent and patterns of usage of Agent Orange and other herbicides in Vietnam. *Nature* 422:681–687.
Sterne, Harold. 2001. *A Catalogue of Nineteenth Century Printing Presses.* New London: British Library.
Stilwell, Charles B. 1889. Paper-bag machinery. 1498400465671058416-00410123
Stoltzenberg, Dietrich. 1994. *Fritz Haber: Chemiker, Nobelpreisträger, Deutscher, Jude.* Weinheim: VCH.
Stone, David. 2012. *Shattered Genius: The Decline and Fall of the German General Staff in World War II.* Havertown, PA: Casemate.
Stone, Richard. 1993. *Introduction to Internal Combustion Engines.* Warrendale, PA: Society of Automotive Engineers.
Stone, Richard. 1999. Nuclear strongholds in peril. *Science* 283:158–164.
Stout, B. A., J. L. Butler, and E. E. Garett. 1984. Energy use and management in US agriculture. In: George Stanhill, ed., *Energy Use and Agriculture.* New York: Springer-Verlag, pp. 175–176.
Strada, Gino. 1996. The horror of land mines. *Scientific American* 274(5):40–45.
Straub, Hans. 1996. *Die Geschichte der Bauingenieurkunst.* Basel: Birkhauser Verlag.
Stross, Randall E. 1996. *The Microsoft Way: The Real Story of How the Company Outsmarts Its Competition.* Reading, MA: Addison-Wesley.
Strunk, Peter. 1999. *Die AEG: Aufstieg und Niedergang einer Industrielegende.* Berlin: Nicolai.
Suga, T. 1998. Mr. Hideo Shima (1901–1998). *Japan Railway & Transport Review* 16:59.
Sugawara, T., et al. 1986. Construction and operation of no. 5 blast furnace, Fukuyama Works, Nippon Kokan KK. *Ironmaking and Steelmaking* 3(5):241–251.
Summers, Lawrence, and Alan Heston. 1991. The Penn World Table (Mark 5): An expanded set of international comparisons, 1950–1988. *Quarterly Journal of Economics* 106:327–368.
Sung, Hsun-chang. 1945. *Thermal Cracking of Petroleum.* Ann Arbor: University of Michigan.
Suplee, Henry H. 1913. Rudolf Diesel: An appreciation. *Scientific American* 109:306.
Sutton, Ernest P. 1999. *From Polymers to Propellants to Rockets: A History of Thiokol.* Paper presented at 35th Joint Propulsion Conference, Los Angeles, 20–24 June, 1999. Edina, MN: ATK.
Sutton, George P. 1992. *Rocket Propulsion Elements: An Introduction to the Engineering of Rockets.* New York: Wiley.
Sutton, George P., and Oscar Biblarz. 2001. *Rocket Propulsion Elements.* New York: Wiley.
Szekely, Julian. 1987. Can advanced technology save the US steel industry? *Scientific American* 257(1):34–41.
Szöllösi-Janze, Margit. 1998. *Fritz Haber: 1868–1934. Eine Biographie.* München: Beck.
TAEP (Thomas A. Edison Papers). 2024. *Edison's US Patents, 1880–1882.* Piscataway, NJ: Rutgers University. https://edison.rutgers.edu/
Takeuchi, Hidemaro, et al. 1994. Production of stainless steel strip by twin-drum strip casting process. *Nippon Steel Technical Report* 61:46–51.
Tanner, A. Heinrich. 1998. *Continuous Casting: A Revolution in Steel.* Fort Lauderdale, FL: Write Stuff Enterprises.
Tarde, Gabriel. 1903. *The Laws of Imitation.* New York: Henry Holt.
Taylor, Charles F. 1984. *The Internal-Combustion Engine in Theory and Practice.* Cambridge, MA: MIT Press.
Taylor, Frederick W. 1907. On the art of cutting metals. *Transactions of ASME* 28:31–350.
Taylor, Frederick W. 1911. *Principles of Scientific Management.* New York: Harper & Brothers.
Taylor, Michael J. H., ed. 1989. *Jane's Encyclopedia of Aviation.* New York: Portland House.
Tegler, Jan. 2000. *B-47 Stratojet: Boeing's Brilliant Bomber.* New York: McGraw Hill.
Temin, Peter. 1964. *Iron and Steel in Nineteenth-Century America.* Cambridge, MA: The MIT Press.
Temin, Peter. 1997. Two views of the British Industrial Revolution. *Journal of Economic History* 57:63–82.
Temple, Robert. 1986. *The Genius of China: 3,000 years of Science, Discovery and Invention.* New York: Simon and Schuster.

Tennekes, Henk. 1997. *The Simple Science of Flight.* Cambridge, MA: MIT Press.

Tenner, Edward. 1996. *Why Things Bite Back: Technology and the Revenge of Unintended Consequences.* New York: A. A. Knopf

Tennies, W. L., et al. 1991. MIDREX ironmaking technology. *Direct from MIDREX* 16(3):4–7.

Termuehlen, Heinz. 2001. *100 Years of Power Plant Development.* New York: ASME Press.

Tesla, Nikola. 1888. Electro-Magnetic Motor. Specification Forming Part of Letters Patent No. 391,968, Dated May 1, 1888. Washington, DC: US Patent Office. http://www.uspto.gov

Tesla, Nikola. 1894. Electro-Magnetic Motor. Specification Forming Part of Letters Patent No. 524,426, Dated August 14, 1894. Washington, DC: US Patent Office. http://www/uspto.gov

Tesla, Nikola. 1900. Apparatus for Transmission of Electrical Energy. Specification Forming Part of Letters Patent No. 649,621, Dated May 1900. Washington, DC: US Patent Office. http://www/uspto.gov

Tevini M, ed. 1993. *UV-B Radiation and Ozone Depletion: Effects on Humans, Animals, Plants, Microorganisms, and Materials.* Boca Raton, FL: Lewis Publishers.

TGVweb. 2000. *Under the Hood of a TGV.* Pisa, Italy: Railfan Europe.

The Times. 1915. History of the War, vol. 5. London: The Times.

Thoman, J. R. 1953. *Statistical Survey of Water Supply and Treatment Practices.* Washington, DC: US Public Health Service.

Thompson, Louis S. 1986. High-speed rail. *Technology Review* 89(3):33–41, 70.

Thompson, Louis S. 1994. High-speed rail (HSR) in the United States: Why isn't there more? *Japan Railway & Transport Review* 3:32–39.

Thompson, Richard C., et al. 2004. Lost at sea: Where is all the plastic? *Science* 304:838.

Thompson, Silvanus P. 1901. *Dynamo-Electric Machinery.* New York: Macmillan.

Thomson, William. 1852. On a universal tendency in nature to the dissipation of mechanical energy. *Proceedings of the Royal Society of Edinburg* 3:139.

Thurston, Robert H. 1878. *A History of the Growth of the Steam-Engine.* New York: D. Appleton & Co.

Tilghman, Benjamin C. 1867. Improved Mode of Treating Vegetable Substances for Making Paper-Pulp. Letters Patent No. 70,485, Dated November 5, 1867; Antedated October 26, 1867. Washington, DC: US Patent Office. http://www.uspto.gov

Tingwall, Eric. 2020. Electronics account for 40 percent of the cost of a new car. *Car and Driver.* www.caranddriver.com

Tobin, James. 2003. *To Conquer the Air: The Wright Brothers and the Great Race for Flight.* New York: Free Press.

Tomlinson, Ray. 2002. The invention of e-mail just seemed like a neat idea. *SAP INFO.* Mannheim, Germany: SAP AG. http://www.sap.info

Top500. 2024. *Top 500 Supercomputer Sites.* Sinsheim: Prometheus. Home - | TOP500

Toynbee, Arnold. 1884. *Lectures on the Industrial Revolution.* London: Rivingtons.

Toyota Motor Company. 2004. *Smoothing the Flow.* Toyota City, Japan: Toyota Motor Company.

Trading Economics. 2024. FedEx. FDX stock price, live quote, historical chart. www.tradingeconomics.com

Trailhead. 2024. Meet the three industrial revolutions. Salesforce. Trailhead.

Transocean. 2003. *Facts and Firsts.* Houston, TX: Transocean.

Troyer, James R. 2001. In the beginning: The multiple discovery of the first hormone herbicides. *Weed Science* 49:290–297.

Turing, Alan. 1936. On computable numbers, with an application to the Entscheidungsproblem. *Proceedings of the London Mathematical Society,* Ser. 2, 42:230–265

Twain, Mark. 1889. *A Connecticut Yankee in King Arthur's Court.* New York: Charles L. Webster & Co.

Tylecote, Andrew. 1992. *The Long Wave in the World Economy: The Present Crisis in Historical Perspective.* London: Mackays of Chatham.

Typewriter Topics. 1924. *The Typewriter: History & Encyclopedia.* New York: Typewriter Topics.

UIG (Universal Industrial Gases). 2003. *Overview of Cryogenic Air Separation.* Easton, PA: UIG.

UNECE (United Nations Economic Commission for Europe). 2001. *World Robotics 2001: Statistics, Market Analysis, Forecasts, Case Studies and Profitability of Robot Investment.* Geneva: UNECE.

UNEP (United Nations Environmental Programme). 1995. *Montreal Protocol on Substances That Deplete the Ozone Layer.* Nairobi: UNEP.

UNESCO (United Nations Educational and Scientific Organization). 1999. *World Communication and Information Report.* Paris: UNESCO.

UNO (United Nations Organization). 1956. World energy requirements in 1975 and 2000. In: *Proceedings of the International Conference on the Peaceful Uses of Atomic Energy,* vol. 1. New York: UNO, pp. 3–33.

UNO (United Nations Organization). 2002. *Energy Statistics Yearbook.* New York: UNO.

Updike, John. 1971. *Rabbit Redux.* New York: Knopf.

Updike, John. 2003. TV. *Atlantic Monthly* 292(2):104.

US Auto Manufacturer Sales Data. 2024. 2023 US Auto Manufacturer Sales Figures. GCBC. www.goodcarbadcar.net

USAF Museum. 2024. History gallery. Wright-Patterson Air Force Base, OH. USAF Air Force Museum Foundation. www.afmuseum.com

USBC (US Bureau of the Census). 1954. *US Census of Manufacturers: 1954.* Washington, DC: USGPO.

USBC (US Bureau of the Census). 1975. *Historical Statistics of the United States: Colonial Times to 1970.* Washington, DC: US Department of Commerce.

USCFC (US Centennial Flight Commission). 2003. The first powered flight-1903. 1903 Wright Flyer. National Air and Space Museum. www.si.edu

USDA (US Department of Agriculture). 1980. *Energy and US Agriculture: 1974 and 1978.* Washington, DC: USDA.

USDA (US Department of Agriculture). 2003. *Agricultural Statistics 2003.* Washington, DC: USDA.

USDC (US Department of Commerce). 2002. *Electric Current Abroad.* Washington, DC: USDC.

USDL (US Department of Labor). 2002. *Trends in Labor Productivity.* Washington, DC: USDL.

USDOE (US Department of Energy). 2022. Biden administration implements new cost-saving energy efficiency standards for light bulbs. Department of Energy.

USDOE. 2024. Manhattan Project. Manhattan Project Background Information and Preservation Work | Department of Energy

USFPC (Federal Power Commission). 1965. *Northeast Power Failure: November 9 and 10, 1965.* Washington, DC: USFPC.

USGS (US Geological Survey). 2024. Mineral commodity summaries.| US Geological Survey. www.usgs.gov

USPTO (US Patent and Trademark Office). 2024. *US Patent Activity: Calendar Years 1790–2001.* Washington, DC: USPTO. http://www.uspto.gov/web/offices/ac/ido/oeip/taf/h_counts.htm

Valenti, Michael. 1991. Combined-cycle plants: Burning cleaner and saving fuel. *Mechanical Engineering* 113(9):46–50.

Valenti, Michael. 1993. Propelling jet turbines to new uses. *Mechanical Engineering* 115(3):68–72.Van Basshuysen, Richard, and F. Schäfer. 2007. *Modern Engine Technology from A-Z.* Warrendale, PA: SAE International.

Van der Spiegel, Jan, et al. 2000. The ENIAC: History, operation and reconstruction in VLSI. In: Raúl Rojas and Ulf Hashagen, eds., *The First Computers: History and Architectures.* Cambridge, MA: MIT Press, pp. 121–178.

Van Duijn, Jacob J. 1983. *The Long Wave in Economic Life.* London: George Allen & Unwin.

Van Dyke, Kate. 2000. *Drilling Fluids.* Austin, TX: Petroleum Extension Service.

van Middelkoop, J. H. 1996. *High-Density Broiler Production: The European Way.* Edmonton, AB: Agricultural Food and Rural Development.

Van Zant, Peter. 2004. *Microchip Fabrication.* New York: McGraw-Hill.

Vasey, Ruth. 1996. The world-wide spread of cinema. In: Geoffrey Nowell-Smith, ed., *The Oxford History of World Cinema.* Oxford: Oxford University Press, pp. 53–61.

Vasko, Tibor, Robert Ayres, and L. Fontvieille, eds. 1990. *Life Cycles and Long Waves.* Berlin: Springer-Verlag.

Veblen, Thorstein. 1902. *The Theory of the Leisure Class.* London: Macmillan.

Vendryes, Georges A. 1977. Superphénix: A full-scale breeder reactor. *Scientific American* 236(3):26–35.

Verhoeven, John D., A. H. Pendray, and W. E. Dauksch. 1998. The key role of impurities in ancient Damascus steel blades. *Journal of Metals* 50(9):58–64.

Vinge, Vernor. 1993. The coming technological singularity. www.accelerating.org

Vinylfacts. 2024. Vinyl. What is vinyl? What is Polyvinyl Chloride (PVC)? Arlington, VA: The Vinyl Institute. www.vinylinfo.org

Vlasic, Bill, and Bradley A. Stertz. 2000. *Taken for a Ride: How Daimler-Benz Drove Off with Chrysler.* New York: W. Morrow.

VÖEST (Vereinigte Osterreichische Eisen- und Stahlwerke). 2003. *Das LD-Verfahren.* Linz, Austria: VÖEST.

Vogel, H. U. 1993. The great well of China. *Scientific American* 268(6):116–121.

von Braun, Wernher, and F. I. Ordway. 1975. *History of Rocketry and Space Travel.* New York: Thomas Y. Crowell.

von Hippel, Frank, et al. 1988. Civilian casualties from counterforce attacks. *Scientific American* 259(3):36–42.

von Hippel, Frank, and Suzanne Jones. 1997. The slow death of the fast breeder. *Bulletin of the Atomic Scientists* 53(5):46–51.

Voss, C. A., ed. 1987. *Just-in-Time Manufacture.* Berlin: Springer-Verlag.

Wade, Louise Carroll. 1987. *Chicago's Pride: The Stockyards, Packingtown, and Environs in the Nineteenth Century.* Urbana: University of Illinois Press.

Waitt, Alden H. 1942. *Gas Warfare: The Chemical Weapon, Its Use, and Protection against It.* New York: Duell, Sloan and Pearce.

Wakelin, David H., ed. 1999. *The Making, Shaping and Treating of Steel, Ironmaking Volume.* Pittsburgh, PA: AISE Steel Foundation.

Waldrop, M. Mitchell. 2001. *The Dream Machine: J. C. R. Licklider and the Revolution That Made Computing Personal.* New York: Viking.

Walker, Bryce. 1983. *Fighting Jets.* New York: Time-Life Books.

Walker, R. D. 1985. *Modern Ironmaking Methods.* Brookfield, VT: Gower Publishing.

Walker, Robert, and F. Whitten Peters. 2002. *The Commission on the Future of the US Aerospace Industry.* Washington, DC: US Federal Government.

Wallace, A. R. 1899. *Wonderful Century.* New York: Dodd, Mead and Company.

Walz, Werner, and Harry Niemann. 1997. *Daimler-Benz: Wo das Auto Anfing.* Konstanz: Verlag Stadler.

Wang, Zhongshu. 1982. *Han Civilization.* New Haven and London: Yale University Press.

Ward, Colin. 1984. *Coal Geology and Coal Technology.* Oxford: Blackwell Scientific.

Ward's Communications. 2000. *Ward's Motor Vehicle Facts & Figures 2000.* Southfield, MI: Ward's Communications.

Warneck, Peter. 2000. *Chemistry of the Natural Atmosphere.* San Diego: Academic Press.

Wärtsilä. 2006. The world's most powerful engine enters service. www.wartsila.com

Watson-Watt, R. A. 1957. *Three Steps to Victory: A Personal Account by Radar's Greatest Pioneer.* London: Odhams Press.

Wattenberg, Albert. 1992. A lovely experiment. *Bulletin of the Atomic Scientists* 48(10):41–43.

Weinberg, Alvin M. 1973. Long-range approaches for resolving the energy crisis. *Mechanical Engineering* 95(6):14–18.

Weinberg, Alvin M. 1972. Social institutions and nuclear energy. *Science* 177:27–34.
Weinberg, Alvin. 1994. *The First Nuclear Era*. Washington, DC: American Institute of Physics.
Wellerstein, A. 2021. An unearthly spectacle. *Bulletin of the Atomic Scientists* October 2021. The untold story of the world's biggest nuclear bomb - Bulletin of the Atomic Scientists
Wells, Herbert George. 1902a. *Anticipations of the Reaction of Mechanical and Scientific Progress upon Human Life and Thought*. New York: Harper.
Wells, Herbert George. 1902b. *The Discovery of the Future: A Discourse Delivered to the Royal Institution on January 24, 1902*. London: T. Fisher Unwin.
Wells, Herbert George. 1905. *A Modern Utopia*. New York: Scribner.
Welsbach, Carl A. von. 1902. History of the invention of incandescent gas-lighting. *Chemical News* 85:254–256.
Wescott, N. P. 1936. *Origins and Early History of the Tetraethyl Lead Business*. Wilmington, DE: Du Pont Corporation.
White House Millennium Council. 2000. *National Medal Winner*. Washington, DC: The White House.
White, Matthew. 2003. Historical atlas of the twentieth century. http://users.erols.com/mwhite28/20centry.htm
White, Thomas H. 2024. United States early radio history. United States Early Radio History
Whitt, Frank R., and David G. Wilson. 1982. *Bicycling Science*. Cambridge, MA: MIT Press.
WHO (World Health Organization). 2002. *The Injury Chart Book*. Geneva: WHO.
WHO (World Health Organization). 2023. World Health Statistics 2023. WHO Data. World health statistics 2023 – Monitoring health for the SDGs
Wiener, Norbert. 1948. *Cybernetics*. New York: Wiley.
Wildi, Theodore. 1981. *Electrical Power Technology*. New York: Wiley.
Williams, D. S. D., ed. 1972. *The Modern Diesel: Development and Design*. London: Newnes-Butterworths.
Williams, Eric. 2003. Forecasting material and economic flows in the global production chain for silicon. *Technological Forecasting & Social Change* 70:341–357.
Williams, Michael. 1982. *Great Tractors*. Poole: Blandford Press.
Williams, Robert H., and Eric D. Larson. 1988. Aeroderivative turbines for stationary power. *Annual Review of Energy* 13:429–489.
Williams, Thomas, et al. 2001. *Sound Coil-Tubing Drilling Practices*. Washington, DC: USDOE.
Williams, Trevor I. 1987. *The History of Invention*. New York: Facts on File.
Wilson, Eugene B. 1902. *Cyanide Processes*. New York: Wiley.
Winchell, Mike. 2019. *The Electric War: Edison, Tesla, Westinghouse, and the Race to Light the World*. New York: Henry Holt.
WNA (World Nuclear Association). 2001. *Chernobyl*. London: WNA. http://www.world-nuclear.org/
Wolfram, Stephen. 2023. How did we get here? The tangled history of the second law of thermodynamics. Stephen Wolfram writings. How Did We Get Here? The Tangled History of the Second Law of Thermodynamics—Stephen Wolfram Writings
Womack, James P., Daniel T. Jones, and Daniel Roos. 1990. *The Machine That Changed the World: The Story of Lean Production*. New York: Harper Perennial.
Wood, Gaby. 2002. *Edison's Eve: A Magical History of the Quest for Mechanical Life*. New York: Knopf.
Wordingham, Charles H. 1901. *Central Electrical Stations*. London: C. Griffin & Co.
World Bank. 2024. GDP (current US$). https://data.worldbank.org/indicator/NY.GDP.MKTP.C
Wright, Orville. 1953. *How We Invented the Airplane*. New York: David McKay.
WSA (World Steel Association) 2024. Total production of crude steel. www.worldsteel.org
Wyatt, J. W. 1911. Paper manufacture. In: *Encyclopaedia Britannica*, 11th ed., H. Chisholm, ed., vol. 20. Cambridge: Cambridge University Press, pp. 727–736.

Wyman, William I. 1913. What are the ten greatest inventions of our time? *Scientific American* 109:337–339.

Yang, Jisheng. 2008. *Tombstone: The Great Chinese Famine, 1958–1962.* New York: Farrar, Straus and Giroux.

Younossi, Obaid, et al. 2002. *Military Jet Engine Acquisition: Technology Basics and Cost-Estimating Methodology.* Santa Monica, CA: Rand Corporation.

Zapffe, Carl A. 1948. *A Brief History of the Alloy Steel.* Cleveland, OH: American Society for Metals.

Zellers, John Adam. 1948. *The Typewriter: A Short History, on Its 75th Anniversary, 1873–1948.* New York: Newcomen Society of England, American Branch.

Zhuravlev, A., and R. Riding. 2000. *The Ecology of the Cambrian Radiation.* New York: Columbia University Press.

Ziegler, Karl. 1963. Consequences and development of an invention. Nobel Lecture, December 12, 1963. In: *The Nobel Prize in Chemistry 1963.* Stockholm: Nobel e-Museum. www.NobelPrize.orgZola, Émile. 1877. *L'Assommoir.* In: Henri Mitterand, ed., *Emile Zola: Oeuvres Completes.* Paris: Cercle du Livre Précieux (1966).

Zola, Émile. 1883/2001. *Au Bonheur des Dames* [The Ladies Delight]. (R. Buss, trans.). Paris: G. Charpentier. London: Penguin Books.

Zulehner, Werner. 2003. Historical overview of silicon crystal pulling development. *Materials Science and Engineering B* 73:7–15.

Index

For the benefit of digital users, indexed terms that span two pages (e.g., 52-53) may, on occasion, appear on only one of those pages.

Note: Figures are indicated by an italic *f* following the page number.

Ader, Clément, 157
aeroderivative gas turbines (AGTs), 415, 416–17
afterburners, 408, 411
Age of Electricity. *See* electricity (Age of Electricity)
Age of Electronics, 229–30, 279–80
Age of Oil, the Automobile, and Mass Production, 26
Age of Steam and Railways, 26
Age of Steel, Electricity, and Heavy Engineering, 26. *See also* electricity (Age of Electricity); steel industry
Age of Synergy
  changes with, 291–304
  communication and, 216–17, 243
  before computers, 279–83
  contemporary perceptions of, 325–35
  high-energy societies, 308–23
  Industrial Revolution and, 19–20, 22
  innovation life cycles, 299–304, 300*f*
  introduction to, 19–30
  makers of new era, 304–8
  mass production, 25, 27, 313–23
  mechanization in, 313–23
  missing contexts, 652–56
  need for advances, 656–60
  traditional *vs.* new contrast, 650–52
  triumphs and tragedies, 283–91
  See also technical advances/innovation
Age of Terror, 30
agricultural industry
  ammonia and, 201–13, 211*f*
  animal foods, 487–89
  chemicals and, 480–87
  contemporary perceptions of, 328
  electrical, 39
  food specialization, 489–91
  mechanized, 28–29, 476–77, 477*f*, 478–91, 481*f*
  nitrogen, 201–13, 483–84, 484*f*, 537, 658

Agricultural Revolution, 11–12
Aiken, Howard, 561
air conditioning, 98–99
airplane/flying industry
  aluminum and, 190, 191*f*
  beginnings of, 154–61, 155*f*, 156*f*, 159*f*
  commercial aviation, 547–52, 549*f*, 552*f*
  contemporary perceptions of, 326–27, 327*f*
  electricity and, 39, 40*f*
  Ely, Eugene and, 160, 299–300, 300*f*
  engines, 363–64, 364*f*
  expansion of, 150, 594–603, 595*f*, 597*f*, 602*f*
  freight movement, 557–58, 558*f*
  gas turbines, 363–64, 364*f*, 403–13
  jet propulsion, 403–13, 407*f*
  technical advances/innovations and, 346–47, 584*f*, 668
  Wright brothers and, 11*f*, 11, 154–60, 156*f*, 159*f*
Alexanderson, Ernst Frederik Werner, 270–71, 284
alternating (AC) current, 89–95, 272, 284, 352
aluminum
  car industry and, 190, 456
  contemporary perceptions of, 331
  duplicate discoveries of, 185–88
  Hall, Charles Martin and, 185–92, 186*f*, 187*f*
  Hall-Héroult process, 9–10, 102, 188–92, 189*f*
  Héroult, Paul Louis Toussaint, 185, 186*f*, 188–92
  introduction to, 184–92
AM frequency, 271–73
ammonia, 201–13, 211*f*
ammonia synthesis, 18–19, 24, 194, 204, 205–6, 208–9, 210, 212–13, 345*f*, 345–46, 450, 483, 660–61, 672–73
Ampère, André Marie, 37–38
Andreessen, Mark, 578
animal foods, 487–89

anti-Semitism, 285–86
Arco, Georg Wilhelm Alexander von, 267
artificial intelligence (AI), 663–64
artistic creativity, 318–19
Aspdin, Joseph, 180–81
assembly lines, 25, 141–43
Atanasoff, John, 561*f*, 561–62
automation and mechanization
    in agriculture, 28–29, 476–77, 477*f*, 478–91, 481*f*
    coal mining, 494–96, 496*f*
    digital automation, 524–25
    factory production advances, 506–29
    flexibility and variety, 524–29, 525*f*
    gas industry, 497–506
    mass production, 25, 27, 313–23, 343, 401–2, 509–15, 610–17, 613*f*, 617*f*
    mineral extraction and processing, 492–96, 493*f*, 495*f*, 496*f*
    oil industry, 497–506, 501*f*, 502*f*
    robots/robotization, 474*f*, 475–78, 515–23, 516*f*, 522*f*

Babbage, Charles, 280
Backus, John, 563
Baekeland, Leo Hendrik, 195, 459–61
Bakelite, 459–60, 460*f*, 528–29
Baker, Benjamin, 175
Bardeen, John, 565–66, 566*f*
Barsanti, Eugenio, 111–12
basic oxygen furnace (BOF), 433–38, 436*f*, 446–47
Bayer, Karl Joseph, 188–89
Bayer, Otto, 461
Beau, Alphonse Eugene (Beau de Rochas), 117–19
Becquerel, Henri, 380
Bell, Alexander Graham, 214*f*, 215, 243–51, 244*f*
Bell, Lowthian, 165–66
Benton, Linn Boyd, 221–22
Benz, Karl, 107–8, 108*f*, 109, 119–25
Berliner, Emile, 253–55, 255*f*
Berners-Lee, Tim, 578
Berry, Clifford, 561–62
Berthier, Pierre, 184–85
Bessemer, Henry, 168–70, 169*f*, 172, 176
bicycles, 120, 136–37, 144*f*, 144–45, 147, 155, 157–58, 163, 176, 190, 217–18, 509, 647
binary inputs, 7–8
biomass energies, 31–32, 163–64, 308–9, 313, 322, 365–66, 476–77, 479, 489–90, 492, 505, 589–90, 590*f*, 646, 650, 671
biosphere's evolution, 4–5
Black, Joseph, 12

blast furnaces, 165–66, 166*f*, 192, 310, 416, 429–33, 430*f*, 432*f*, 437–39, 444–45, 446–47, 450, 453, 465–66, 657–58, 659
Boeing, William, 551–52, 552*f*
Boeing Company
    commercial airplanes, 190, 191*f*, 547–48, 550–52, 558, 668
    engines of planes, 362*f*, 363–64, 364*f*, 406–8, 409–13, 410*f*, 415, 420–21
    foreign production, 659
    global use of planes, 610–11, 615
    military use of planes, 526
Bohlin, Nils Ivar, 540*f*, 540
Bohr, Niels, 25, 380
Bolton, Werner von, 51
Boot, Harry A., 549
Borlaug, Norman, 211–12
Bosch, Carl, 18–19, 204*f*, 204, 205–9, 454
Bosch, Robert, 134–35, 321, 322–23
Bourseul, Charles, 243
Bradley, Charles S., 91–92
Brandenberger, Jacques Edwin, 460–61
Branly, Edouard, 261
Brattain, Walter, 565, 566*f*
Braun, Karl Ferdinand, 267, 356–57, 420*f*
Braun, Wernher von, 402, 419
Brearley, Harry, 177
bridge building, 163–64
Bronowski, Jacob, 646
Brown, Harold, 89–90
Buehler, Ernest, 470*f*
Buick, David D., 147–48
Bullock, William, 218
Bunsen, Robert, 184–85
Burroughs, William Seward, 281*f*, 281–82
Burton, William, 137, 504
Byrn, Edward W., 325, 329–30

Cailletet, Louis-Paul, 192
Cambrian eruption, 8
cameras, 217, 231–36
Čapek, Josef, 475
Čapek, Karel, 518–19
carbon dioxide (CO_2) 451–52, 454, 537, 635–39, 638*f*, 669–70. *See also* decarbonization
carbon electrodes, 41–42, 173
carbon-filament electric lights, 38–39, 43*f*, 43–44, 293*f*
car industry and culture
    agricultural industry, 476–77, 477*f*, 478–87, 481*f*
    aluminum and, 190, 456
    conspicuous consumption of, 317–18

INDEX 719

contemporary perceptions of, 325–26
electric cars, 88, 110–11, 132–33
expansion of, 149–61
Ford, Henry and, 137–43, 139f
hydrocarbons and, 148–49, 537, 599–600
initial decades, 144–49
innovations and inventions, 357
introduction to, 130–49, 131f
on mass scale, 533–42, 535f, 541f
motorcycles, 26–27, 111, 120, 121f, 150, 217–18, 349, 537
racing, 145–46, 146f
robot use in, 476
rubber tire, 136–37, 145, 353, 480–81
steel industry and, 175–76
technical challenges, 132–39
technical transformations, 338f, 343, 347, 353–54
See also internal combustion engines
Carlson, David, 471–72
Carnegie, Andrew, 333–35
Carnot, Sadi, 14–15
Carothers, Wallace Hume, 461
Carrier, Willis Haviland, 98–99
Castan, Pierre, 461
Casti, John, 607
cast iron, 75, 163–64, 165–68, 172, 178, 187–88, 331, 427–28, 456, 508, 657–58
Castner, Hamilton Y., 185
cellular phones, 217, 249–50, 333
cement, 164–65, 179–81, 284, 285, 343, 492, 498, 657–60. See also concrete
cementation, 167–68
Cerf, Vinton, 578
Chadwick, James, 380–82, 381f
Chalmers, William, 461
Chanute, Octave, 11–12
chlorofluorocarbons (CFCs), 377–78, 461–62, 631–32, 637f, 637–38
Cho, Alfred Y., 570–71
Claude, Georges, 52–53, 193–94
Clausius, Rudolf, 15, 23–24
Clerk, Dugald, 150
coal mining, 494–96, 496f
Cockroft, John Douglas, 380
Coignet, François, 181
combustion engines. See internal combustion engines
commercial aviation, 547–52, 549f, 552f
Boeing Company, 190, 191f, 547–48, 550–52, 558, 668
commercial electricity generation, 53–54
commercialization, 12–13, 14–15, 18, 129

communication and information
computer technology and, 575–83, 582f
introduction to, 215–18
printed word, 218–30
satellites, 25, 190, 216, 340f, 347, 391, 402–3, 419, 422–23, 469–71, 471f, 578, 580, 606, 615–16
vacuum tubes, 24, 47, 271, 272–73, 349, 561–62, 563–64, 565–67, 576, 579, 603–4, 662
wireless communication, 261–74, 264f, 266f, 269f, 270f
See also telephone
composing machine, 220–21
Compound Vortex Controlled Combustion (CVCC), 537–38
computer-assisted manufacturing (CAM), 517–18, 526–28
computer-integrated manufacturing (CIM), 527
computer numerical control (CNC), 524–26, 525f
computer technology
communication and, 575–83, 582f
electricity use, 666–68
expansion of, 603–9, 605f
foundations of, 560–63
integrated circuits, 243–51
internet and, 576–78, 667
introduction to, 559–83, 560f
mainframes, 576–78
mass media and, 579–83
microchip transistors, 4, 5f, 563–70, 572f, 574f
microprocessors, 4, 378, 559, 570–75
personal computers (PCs), 225–26, 359–60
supercomputers, 559–60, 576–77, 604, 662–64, 663f
transistors in, 243–51
concrete, 27, 164–65, 177–83, 182f. See also cement
construction industry, 27, 164–65, 177–83, 179f, 180f, 182f, 426
consumer expenditures, 316
continuous casting process, 438–43, 441f, 443f
Coolidge, William David, 51–52, 52f
Corporate Average Fuel Economy (CAFE), 538–39
COVID-19 pandemic, 598–99, 642, 653–54
Crompton, R. E. B., 91
Crookes, William, 203–4, 258, 262, 285, 356–57
Cross, Charles Frederick, 253, 460–61
crude oil, 367f, 367–68, 373
cruise ships, 534f

Curtis, Charles, 69–70, 357
Curtiss, Glenn, 327
cyanide process, 330
cybernetics, 519
Czochralski, Jan, 468

Daguerre, Louis Jacques Mandé, 231
Dahl, Carl F., 228–29
Daimler, Gottlieb, 109, 114, 119–25, 120*f*, 121*f*, 289
Dalton, Francis, 653
Dalton, John, 16–17
Darwin, Charles, 653
Das Kapital (Marx), 24
Davenport, Thomas, 78–79
Davidson, Robert, 79
Davis, Francis Wright, 534–35
Davy, Humphry, 41–42, 184
Debussy, Claude, 33
decarbonization, 364–65, 370–74, 671, 672*f*
Deming, William, 513
Deng Xiaoping, 346
Denison, Edward, 623–24, 625
Derick, Link, 567–68
Deville, Henri Saint-Claire, 184–85
Devol, George, 519–21, 520*f*
dichlorodiphenyl trichloroethane (DDT), 484–85, 485*f*
Dickson, William Kennedy Laurie, 236–37
Diebold, John, 515
Diesel, Rudolf, 109, 125–30, 126*f*, 128*f*, 192–93
diesel engines, 125–30, 128*f*, 153, 154
digital automation, 524–25
digital circuits, 7–8
direct (DC) current, 89–95, 272, 352, 368–70, 438*f*, 658–59
direct reduction of iron (DRI), 444–45
Dochring, Carl, 183
Doll, Henri, 498
Dornberger, Walter, 419
Douglas, Donald, 551–52, 552*f*
Doyle, Arthur Conan, 285, 653
drinking water chlorination, 315
Dumont, Alberto Santos, 160
Durant, William, 147–48
Durrer, Robert, 434–35
Duryea, Charles, 147
Duryea, Frank, 147
dynamos, 55–64, 60*f*, 61*f*

Eastman, George, 217, 233–34, 234*f*
Eckert, John Presper, 561–62
Eckert, Mauchly, 561–62

economics/economic growth, 603–30, 605*f*, 622*f*
Edison, Thomas
  admiration after death, 289–90
  commercial electricity generation, 10–11, 14–15, 18, 53–88
  concrete houses, 182
  dynamos, 55–64, 60*f*, 61*f*
  early electric lights, 18, 41–44, 43*f*, 49*f*
  Edison effect, 258–59
  electric motors, 79
  invention of electric lights, 41–64
  lighting experiments, 44–47, 45*f*
  movies, 236–42, 238*f*
  phonograph, 236–37, 242–43, 251–56, 252*f*, 255*f*, 271, 459
  phrenology and, 653
  standardization of electricity, 89–91
  success and competition, 47–51
  on technical innovation, 650
  telephone, 245–49, 248*f*
  unparalleled record of, 284
Edison effect, 258–59
Edison Electric Illuminating Company, 11, 57
Eickemeyer, Rudolf, 80, 94–95
Einstein, Albert, 8, 25–26, 380
Ekman, Carl Daniel, 227
electric arc furnace (EAF), 433–38
electric cars, 88, 110–11, 132–33
electricity (Age of Electricity)
  advances in, 64–88
  alternating (AC) current, 89–95, 272, 284, 352
  commercialization of, 14–15, 18, 53–88
  computer technology use, 666–68
  contemporary perceptions of, 329
  direct (DC) current, 89–95, 368–70, 438*f*, 658–59
  dynamos, 55–64, 60*f*, 61*f*
  expansion of, 96–104
  generation of, 9, 10–11
  globalization and, 589–93
  hydrocarbons and, 50–51, 84, 309, 592
  hydroelectricity, 58, 77, 96–97, 118, 188, 203–4, 309
  introduction of, 3–4, 8, 37–40
  Parsens, Charles Algernon and, 66–72, 67*f*, 69*f*
  plugs and sockets, 103–4, 104*f*
  production of, 374–79, 376*f*, 379*f*
  relative decarbonization, 370–74
  standardization efforts, 88–104
  steam turbines and, 64–72, 71*f*

INDEX 721

technical transformations, 355, 356f
three-wire distribution system, 61–62
transformers and, 55–57, 72–78, 74f
Electric Lighting Act (1882), 102–3
electric motors
    adoption of, 83–88, 85f, 87f
    development of, 78–83
    Tesla, Nikola, 64–65, 78, 80–83, 81f, 82f, 87–88
electric vacuum cleaners, 100
electromagnetic radiation, 3–4
electromagnetic waves, 8, 215–16, 242, 256–61, 257f, 263, 292, 294–95, 348, 661–62
electronic computing. *See* computer technology
Ely, Eugene, 160, 299–300, 300f
e-Manufacturing Networks, 22–23
embourgeoisement, 341–42, 601, 629
energy production
    biomass, 31–32, 163–64, 308–9, 313, 322, 365–66, 476–77, 479, 489–90, 492, 505, 589–90, 590f, 646, 650, 671
    global supply, 372f, 589–93, 590f, 669–73
    growth and performance, 366–70
    hydrocarbons and, 309, 311–12, 313, 366, 575–76, 639, 658–59
    industrial capacities and, 659
    introduction to, 362f, 363–65, 364f
    natural gas, 31–32, 72, 84, 311–12, 313, 370–71, 373–74, 416, 431–33, 444–45, 454, 557, 590, 639, 672
    nuclear energy, 347, 380–400
    nuclear fission, 32, 33, 340–41, 355, 366, 373–74, 380, 382–84, 389–90, 393–400
    prime movers in, 400–23, 591f
    societal improvements, 589–93
    See also electricity; fossil fuels/fossil fuel era; gas turbines
Engelbart, Douglas, 359–60
engines
    afterburners and, 408, 411
    airplane/flying industry, 363–64, 364f
    Boeing Company and, 362f, 363–64, 364f, 406–8, 409–13, 410f, 415, 420–21
    jet propulsion, 403–13, 407f
    liquid fueled, 108, 112–13, 126f, 127, 309, 311–12, 373, 400, 418f, 418–19, 430–31, 626
    rocket engines, 402, 403f, 417–23, 418f, 420f, 421f, 632
    rotary engines, 15
    See also gas turbines; internal combustion engines; steam engines

environmental advances, 364–65, 370–74, 635–39, 637f, 638f, 671, 672f
evolutionary saltation, 7–12
explosives, 195–213

factory production advances, 506–29
Faraday, Michael, 38, 73–74
Farnsworth, Philo, 356–57
Fawcett, Eric O., 463
Félix, Charles, 79–80
Felt, Dorr Eugene, 280
female workers, 328
Ferguson, Harry, 480–81
Fermi, Enrico, 385–86, 393–95, 394f, 398–99
Fernald, John, 666
Ferranti, Sebastian de, 62
Ferraris, Galileo, 80
fertilizer. *See* nitrogen
Fessenden, Reginald Aubrey, 270f, 270–72, 289
Feuerlein, Otto, 51
Feynman, Richard, 363
films, 231–36
First Industrial Revolution, 665–66
First Law of Thermodynamics, 8, 15
Fleming, John Ambrose, 265, 273f
flexible manufacturing system (FMS), 527–28
Flowers, John, 562
fluorescent lamps, 25, 52–53, 355
FM frequency, 271–73
Fontaine, Hippolyte, 79–80
Ford, Henry, 11, 137–43, 139f, 285–86, 509–13, 510f
Ford Motor Company, 139–40, 511–13, 512f
Forest, Lee De, 285
fossil fuels/fossil fuel era
    automation and mechanization, 493
    beginnings of, 31–32, 365–79
    domination of, 32
    energy from and, 364–79, 592, 672
    growth and performance, 366–70
    relative decarbonization, 370–74
    transition from biomass to, 322
Fourdrinier, Henry, 226
Fourneyron, Benoit, 66–67
Fourth Industrial Revolution, 23
Fowler, John, 175
Fox, Marshall, 47–48
Franklin, Benjamin, 37
Franz, Anselm, 405–6
Franz, John E., 486–87
freight movement, 553–58, 554f
French Revolution, 30
Freyssinet, Eugene, 183

Frisch, Otto, 382–83
Frosch, Carl, 567–68
Fuchs, Klaus, 389–90

Galvani, Luigi, 37–38
Garfield, Lewis E., 498
gas industry, 497–506
gasoline-powered engines, 150–54, 151*f*, 326
gas turbines
    aeroderivative gas turbines, 415, 416–17
    airplane/flying industry, 363–64, 364*f*, 403–13
    combustion process, 400–2, 401*f*
    importance of, 32, 357, 362*f*
    industrial gas turbines, 413–17, 414*f*
Gates, Bill, 614
Gaulard, Lucien H., 74–75
General Electric Company, 48–49, 91, 363–64
General Motors, 99, 147–48, 322–23, 505, 509–10, 514, 521, 526, 610–11, 612
Gibbs, John D., 74–75
Gibson, Reginald O., 463
Gilchrist, Percy Carlyle, 169–70
Gililand, Edwin R., 504
Glenn, John, 419–20
globalization, 321–23, 620–30, 628*f*
Global Positioning Systems, 8–9
Goddard, Robert, 25, 417–19, 418*f*
gold extraction, 330
Gordon, James E., 425
Gordon, Robert, 665–66
Grahame, Kenneth, 149
Gramme, Zénobe-Théophile, 55–56
gramophones, 26–27, 242–43, 611–12
graphical user interface (GUI), 576–77
Gray, Elisha, 243–49, 246*f*, 287
Grey, Henry, 178
gross domestic product (GDP), 620–30, 622*f*, 628*f*
gross world product (GWP), 620–30, 622*f*
Groves, Leslie, 384–85, 385*f*

Haagen-Smit, Arie, 537
Haber, Fritz, 18, 204*f*, 204, 205–9, 345*f*, 345–46
Haber-Bosch process, 9–10, 24, 29, 205–9, 207*f*, 349, 657–58. *See also* ammonia synthesis
Hadfield, Robert A., 77, 176–77
Haggerty, Patrick, 571
Hahn, Otto, 382–84, 383*f*
Hall, Charles Martin, 185–92, 186*f*, 187*f*
Hall-Héroult process, 9–10, 102, 188–92, 189*f*
Han Dynasty (207bce–9ce), 6, 7*f*
Harmsworth, Alfred, 326–27

Heald, James, 508
Heilbroner, Robert, 344
Hellbrügge, Heinrich, 434–35
Helmholtz, Hermann von, 259
Henderson, Charles B., 422
Hennebique, François, 181
Henning, Hans, 201
Henry, Joseph, 37–38
herbicides, 486–87
Héroult, Paul Louis Toussaint, 185, 186*f*, 188–92
Héroult, William, 173
Hertz, Heinrich, 8–9, 16–17, 215–16, 256–61, 260*f*, 289
heterodyning, 271–72
heterotrophic metabolism, 13–14, 16–17
Hewitt, Peter Cooper, 52–53
Higonnet, René Alphonse, 579
Hoe, Richard, 218
Hollerith, Herman, 280, 282–83, 283*f*
Holwein, Ludwig, 124*f*
Hook, Marcus, 504
Hopkinson, John, 61–62, 91, 100–1
hormone herbicides, 486–87
Howard, Samuel B., 96–97
Howe, Elias, 21–22
Hughes, David Edward, 257*f*, 257, 259
Hughes, Howard Robard, 497–98, 499*f*
Huntsman, Benjamin, 167–68
hydrocarbons
    car industry and, 148–49, 537, 599–600
    electricity and, 50–51, 84, 309, 592
    energy use and, 309, 311–12, 313, 366, 575–76, 639, 658–59
    gas turbines and, 416
    as industrial gases, 431–33, 444–45, 450–51, 454–56
    internal combustion engines, 148–49, 413–14
    production platforms, 501*f*, 503–6
    rediscovery of, 477–78, 497, 499–500, 501*f*, 503–6
    transition to, 322
hydroelectricity, 58, 77, 96–97, 118, 188, 203–4, 309
hydrogen, 450–51, 454–55
hysteresis, 77

images
    cameras, 217, 231–36
    films, 231–36
    introduction to, 230–42
    lithography, 57, 220, 230, 572*f*, 573–74, 662–63
    movies, 236–42
    photographs, 231–36

INDEX 723

incandescent lights
　carbon filaments, 38–39, 43f, 43–44, 293f
　development of, 18, 20, 24, 25f
　doing away with, 657f
　early electric lights, 41–44, 43f, 49f
　efficacy of, 302, 303f
　efficiency of, 10, 50f
　gas lights and, 20
　invention of, 41–64
　light bulbs, 18, 24, 25f, 293f, 293–94
　lighting experiments, 44–47
　metallic filaments, 51–53
　success and competition, 47–51
　technical transformations, 355
industrial gases
　ammonia, 201–13
　carbon dioxide, 451–52, 454, 537, 635–39, 638f, 669–70
　hydrogen, 450–51, 454–55
　methane, 373, 454, 557, 637–38, 639, 669–70
　nitrogen, 201–13, 449–53
　overview of, 192–94, 449–55
　oxygen, 449–53, 453f
industrial gas turbines, 413–17, 414f
industrialization, 167–68, 320, 321–22, 332, 370–71, 378–79, 494, 506–7, 588, 659–60
Industrial Revolution
　Age of Synergy and, 19–20, 22
　chemical industry and, 12
　First Industrial Revolution, 665–66
　Fourth Industrial Revolution, 23
　Second Industrial Revolution, 22–23, 27, 665–66
　Third Industrial Revolution, 23, 27
information. *See* communication and information
integrated circuits, 243–51
Intel Corporation, 5f
intercontinental ballistic missiles (ICBMs), 390–91, 422
internal combustion, 11
internal combustion engines
　Age of Synergy and, 20–21
　Beau, Alphonse Eugene and, 117–19
　beginnings of, 25–26, 111–30, 112f
　Benz, Karl and, 107–8, 108f, 109, 119–25
　Daimler, Gottlieb and, 119–25, 120f, 121f
　diesel engines, 125–30, 128f, 153, 154
　efficiencies of, 310
　gasoline-powered, 150–54, 151f
　in Germany, 119–25
　introduction to, 107–11
　Langen, Eugen and, 109, 113–16

　Maybach, Wilhelm and, 109, 119–25, 121f
　Otto, Nicolaus August and, 109, 113f, 113–19, 115f, 118f
　rotary engines, 15
　three-wheel vehicle, 106f, 107, 112–13, 122–23, 134–35
　See also car industry and culture
internet, 576–78, 667. *See also* computer technology
Iron Age, 426–49
iron/ironmaking
　accomplishments and consequences, 445–49
　direct reduction of iron (DRI), 444–45
　Iron Age and, 426–49
　MIDREX process, 444f, 444–45
　new developments, 444–45
　pig iron, 166–71, 167f, 173–74, 174f, 191–92, 310–11, 429, 431–33, 435–36, 446, 511
　spiegel iron, 168–69
　steel from, 6–7
Ives, Frederic Eugene, 236

Jacobi, M. H., 78–79
Jeffrey, Thomas B., 317–18
Jellinek, Emil, 123
Jenne, William, 223–24
jet propulsion, 403–13, 407f
Jewitt, Frank F., 185
Jobs, Steven, 571
Johnson, Edward H., 49–50
Johnson, Harry, 521
Jones, Howard Mumford, 25–26
Joule, James Prescott, 15
Joule, Thomas, 192
Joy, Joseph Francis, 494
Judson, Whitcomb L., 21–22
Junghans, Siegfried, 439–40

Kafka, Frank, 641
Kahn, Robert, 578
Kastenbein, Charles, 219–20
Kekulé, Friedrich August, 17–18
Kellner, Carl, 227, 285
Kellogg, M. W., 212–13
Kelly, Mervin, 565
Kelly, William, 168
Kennedy, John, 419–20
Kettering, Charles, 136, 285, 505
Keyes, Robert W., 574
Keynes, John Maynard, 652
Khrushchev, Nikita, 390
Kilburn, Tom, 562
Kilby, Jack S., 568–70, 569f

Kinetograph/Kinetoscope, 236–42, 238f
Kipling, Rudyard, 163
Kipping, F. S., 467
Klatte, Friedrich Hein-rich, 465–66
knowledge economy, 12–19
Kodak cameras, 217, 233–35, 234f
Koenig, Friedrich, 218
Kondratiev cycle, 26
Korolyov, Sergei, 419
kraft pulping, 228–29
Kurzweil, Ray, 18–19

labor force, 617–20
Lamson, Alexander, 235
Lanchester, Frederick William, 137, 284
Langen, Eugen, 109, 113–16
Langmuir, Irving, 24, 25f, 52
la technique, 643, 644–45
Laval, Carl Gustaf Patrick de, 66–67
Lavoisier, Antoine, 13–14, 16–17, 31f
Lawrence, Ernest O., 387–88
Lebon, Philippe, 111–12
leisure activities, 319f, 319–20
Lemelson-MIT Prize Program, 28
Lenoir, Jean Joseph Etienne, 112–13
Levassor, Emile, 109, 133–35, 134f
Lewis, Warren K., 504
Licklider, Joseph, 577–78
Liebermann, Carl, 205
Liebig, Justus von, 16–17, 17f
Life expectancy, 321, 341, 342f, 593, 594f, 630–31
light bulbs, 18, 24, 25f, 293f, 293–94
Lilienthal, Otto, 154
Limited Test Ban Treaty, 391
Linde, Carl von, 192–94, 193f
linotype machine, 220f, 220–23, 221f, 223f
liquefied natural gas (LNG), 72, 557
liquid fuels, 108, 112–13, 126f, 127, 309, 311–12, 373, 400, 418f, 418–19, 430–31, 626
liquid-metal fast breeder reactors (LMFBR), 396–97
lithography, 57, 220, 230, 572f, 573–74, 662–63
Little, J. B., 468–69
Lodge, Oliver Joseph, 261–63, 285, 286f
logic gates, 7–8
London Electric Supply Corporation, 92, 93f
Lougheed, Allan, 551–52
Lougheed, Malcolm, 551–52
Ludd, Ned, 332
Lüders, Johannes, 129–30
Lumière, Louis, 239–41, 285–86

MacArthur, John S., 330
Maillart, Robert, 182
Malcolmson, Alexander, 139–40
Manhattan Project, 339, 384–85, 385f, 388–90
Mannesmann, Max, 175
Mannesmann, Reinhard, 175
Mao Zedong, 346, 658
Marconi, Guglielmo, 261, 262f, 263–67, 264f, 266f, 287
Marsh, Albert, 177
Martin, Emile, 23, 172
Martin, Thomas, 651–52
Marx, Karl, 24
mass production, 25, 27, 313–23, 343, 401–2, 509–15, 610–17, 613f, 617f
material revolution
  ammonia and, 201–13, 211f
  bridge building, 163–64
  explosives manufacturing, 195–213
  Haber-Bosch process, 9–10, 24, 29, 205–9, 207f
  industrial gases, 192–94
  introduction to, 163–65, 425–26
  Iron Age and, 426–49
  plastics industry, 457–72, 458f
  See also aluminum; industrial gases; iron/ironmaking; steel industry
Mateucci, Felice, 111–12
Maxwell, James Clerk, 15–16, 256–57, 257f
Maybach, Adolf, 289
Maybach, Wilhelm, 109, 114, 116, 119–25, 121f, 137–38, 289
Mayer, Robert, 15
McAfee, Almer M., 504
McLean, Malcolm Purcell, 553, 554–55
mechanization. *See* automation and mechanization
Meitner, Lise, 382–83, 383f
Méliès, Georges, 240–41
Melville, George W., 11
Mendeleev, Dimitrii, 17–18
Mergenthaler Ottmar, 220f, 220–23, 221f, 223f, 289
Mestral, George de, 359–60, 360f
metal bottle caps, 294–95, 295f
metallurgical industry/innovations, 6–7, 9–10, 456–57. *See also* iron/ironmaking; steel industry
methane, 373, 454, 557, 637–38, 639, 669–70
Michelin, Andre, 136–37
Michelin, Edouard, 136–37
microchip transistors, 4, 5f, 563–70, 572f, 574f

microprocessors, 4, 353–54, 378, 559, 570–75, 603–4, 626–27, 662–63
Midgley, Thomas, 505–6, 506f
MIDREX process, 444f, 444–45
Milenkovic, Veljko, 521
military industry
  advances in weapons and conflicts, 632–35, 634f
  Boeing Company and, 526
  Ford Motor Company and, 511–13, 512f
  missiles, 390–91, 422–23
  nuclear weapons, 348, 384–93, 387f
  rocket engines, 419, 422
  thermonuclear weapons, 389–90, 391–92, 559, 587, 604, 631–32
mineral extraction and processing, 492–96, 493f, 495f, 496f
missiles, 390–91, 422–23
Mitscherlich, Alexander, 227
Mittasch, Alwin, 208–9
Monet, Claude, 33
Monier, Joseph, 181
Moore, Charles T., 220
Moore, Gordon E., 5f, 570–72, 572f
Moore's Law, 4, 279, 572f, 573, 604
Morehead, James, 453
Morrison, Philip, 339, 384, 386–87, 389, 392–93
Morse signals, 8–9
Moss, Sanford, 401–2
motorcycles, 26–27, 111, 120, 121f, 150, 217–18, 349, 537
Mouret, Octave, 317
movies, 236–42
Moyroud, Louis M., 579
Muller, Oscar, 92–93
multinational corporations, 322–23
multiple independently targeted reentry vehicles (MIRVs), 390–91
multiple-stage rocket, 25
Mushet, Robert Forester, 168–69, 176
Musschenbroek, Pieter van, 37
Muybridge, Eadweard, 231–33, 232f

Nader, Ralph, 536–37
nanomachines, 23, 493
National Academy of Engineering, 27–28
Natta, Giulio, 462–63
natural gas, 31–32, 72, 84, 311–12, 313, 370–71, 373–74, 416, 431–33, 444–45, 454, 557, 590, 639, 672
Neuman, John von, 562
neutrons, 380–400, 381f, 394f
New Economy, 446, 606

Niepce, Claude, 231–32
Niepce, Joseph Nicephore, 230–31
nitrogen, 201–13, 449–53, 483–84, 484f, 537, 658
Nobel, Alfred, 195, 196f, 197–200, 288–89
Norton, Charles, 508
Noyce, Robert, 568–70, 569f
nuclear energy, 347, 380–400
nuclear fission, 32, 33, 340–41, 355, 366, 373–74, 380, 382–84, 389–90, 393–400
nuclear weapons, 348, 384–93, 628f. *See also* thermonuclear weapons

Oberth, Hermann, 402
Odum, Howard, 489, 585
Ohm, Georg Simon, 14–15
oil industry, 497–506, 501f, 502f
oil wells, 312f, 312–13
Olds, Ransom, 147–48
Ōno, Taiichi, 513–14
open hearth furnaces (OHFs), 433–38, 446–47
open-hearth steelmaking, 170–73
Oppenheimer, Robert, 358–59, 384–85, 385f, 387f
organic compounds, 17–18
Organization of the Petroleum Exporting Countries (OPEC), 27, 28f, 371, 537
Ørsted, Hans Christian, 37–38, 184
Orwell, George, 277–78
Ott, John, 49–50
Otto, Nicolaus August, 109, 113f, 113–19, 115f, 118f, 137–38
Owens, Michael Joseph, 515, 516f, 617–18
oxygen, 449–53, 453f

Pacinotti, Antonio, 56
Page, Charles G., 243
paper clips, 295–96, 296f
papermaking, 216–17, 226–30, 228f
parcel post delivery, 217–18
Parkes, Alexander, 234–35
Parsons, Charles Algernon, 15–16, 66–72, 67f, 69f, 658
Parsons, John T., 22–23, 517, 518f
Pelton, Lester Allen, 96–97
Perret, Auguste, 182
personal computers (PCs), 225–26, 359–60, 607–8
phonograph, 236–37, 242–43, 251–56, 252f, 255f, 271, 459
photographs, 231–36
photolithography, 573–74, 662–63

photovoltaic (PV) cells, 13, 32, 40f, 71, 425, 467, 469–72, 471f, 650, 670
phrenology, 653–55, 654f
Pickard, Greenleaf W., 565
Pictet, Raoul, 192
pig iron, 166–71, 167f, 173–74, 174f, 191–92, 310–11, 429, 431–33, 435–36, 446, 511
plastics industry
  overview of, 457–72, 458f
  polyethylene (PE), 457, 458f, 463–67, 464f
  polyethylene terephthalate (PET), 461–62, 462f
  polystyrene, 461
  polytetrafluoroethylene, 461–62
  polyvinyl chloride (PVC), 457, 458f, 461, 463–67
  silicon, 467–72, 470f
Plunkett, Roy, 461–62
polyethylene (PE), 457, 458f, 463–67, 464f
polyethylene terephthalate (PET), 461–62, 462f
polymers (man-made), 457. See also plastics industry
polyphase systems, 103–4
polystyrene, 461
polytetrafluoroethylene, 461–62
polyvinyl chloride (PVC), 457, 458f, 461, 463–67
Pool, Leonard Parker, 451–52
Popov, Alexander S., 16–17, 263
Porsche, Ferdinand, 535
Poulson, Valdemar, 256
Priestley, Joseph, 16–17
prime movers, 13, 14f, 20–21, 400–23, 591f
printed word, 216–17, 218–30, 228f

quality-of-life, 589–93, 594f
quotidian realities, 5–6, 660–61

Rabi, Isidore, 388–89
race car industry and culture, 145–46, 146f
radio, 580–81
radio frequencies, 3
rail industry
  aluminum, 456
  electricity and, 101–2
  freight movement, 553–58, 554f
  on mass scale, 542–47, 544f, 545f, 547f
Randall, John T., 549
refrigeration, 99, 377f, 377–78
Regnault, Henri Victor, 465–66
Reichenbach, Henry M., 234–35
relative decarbonization, 370–74
Renault, Louis, 145, 146f

Reynolds, Osborne, 9
Rickover, Hyman, 396
Ritchie, Dennis, 563
RMS (Robert Mushet's Special Steel), 176
Robert, Nicolas Louis, 226
robots/robotization, 474f, 475–78, 515–23, 516f, 522f
rocket engines, 402, 403f, 417–23, 418f, 420f, 421f, 632
Röntgen, Wilhelm Conrad, 100
Rossi, Irving, 439
rotary engines, 15
rubber tire, 136–37, 145, 353, 480–81
Rue, William de La, 43–44
Rumbel, Keith, 422
Rutherford, Ernest, 380

satellites, 25, 190, 216, 340f, 347, 391, 402–3, 419, 422–23, 469–71, 471f, 578, 580, 606, 615–16
Sawyer, William Edward, 48, 356–57
Schairer, George, 406–7
Scheele, Wilhelm, 16–17
Scheutz, Edvard, 280
Scheutz, George, 280
Schlumberger, Conrad, 498
Schmidt, Carlo, 22–23
Scott, Floyd L., 498
Second Industrial Revolution, 22–23, 27, 665–66
Second Law of Thermodynamics, 15, 23–24
Selden, George B., 148–49
Seldes, Gilbert, 653
Semon, Waldo L., 465–66
Senefelder, Alis, 230
Serpollet, Leon, 132–33
sewage plants, 315–16
Shailor, Frank, 98f
Shallenberger, Oliver, 91–92
Shannon, Claude, 560–61
Sharp, Walter B., 498
shinkansen trains, 352–55
shipping industry, 530f, 531–33, 553–58, 554f
Shockley, William, 565–66, 567f
Shriver, Andrew, 314
Siemens, Werner, 23, 55f, 55–57, 101
Siemens, William, 20, 23, 56, 73, 170–73, 171f, 290, 436–37
silicon, 467–72, 470f
Singularity, 26, 642–43, 650, 663–64, 666–67, 669–70
Smith, Arthur, 459
Smith, S. L., 147–48

Sobrero, Ascanio, 197
social media, 4, 668–69
Soodak, Harry, 397
sound reproduction
   electromagnetic waves, 8, 215–16, 242, 256–61, 257f, 263
   gramophones, 26–27, 242–43, 611–12
   introduction to, 242–74
   phonograph, 236–37, 242–43, 251–56, 252f, 255f, 271, 459
   radio, 580–81
   technical transformations, 347
   telephone, 215, 243–51, 244f, 246f, 248f, 249f
   wireless communication, 261–74, 264f, 266f, 269f, 270f
Southward, John, 219–20
space technology
   liquid fueled engines, 417–23, 418f, 420f, 421f
   rocket engines, 402, 403f, 417–23, 418f, 420f, 421f, 632
spark plugs, 294
spiegel iron, 168–69
The Stammering Century (Seldes), 653
Stanford, Leland, 231–32
Stanley, Francis, 132–33
Stanley, Freelan, 132–33
Stanley, William, 64–65, 76f, 78
Starley, John Kemp, 144–45
steam engines
   Age of Synergy and, 20–21
   efficiency of, 13–14
   industrial gas turbines and, 413–17, 414f
   in industry, 153–54
   invention and commercialization of turbines, 15–16
steam turbines, 9–10, 64–72, 71f, 357, 369f
steel industry
   accomplishments and consequences, 445–49
   advanced variants, 455–72
   Age of, 26
   Bessemer, Henry and, 168–70, 169f, 172
   in car industry, 343
   in construction, 178–83, 179f, 180f, 182f, 426
   continuous casting, 438–43, 441f, 443f
   in electric industry, 343
   furnaces for, 433–38
   global production, 448f, 659
   in modern society, 173–77, 174f, 424f, 426–49, 427f, 428f
   open-hearth steelmaking, 170–73
   rise of, 165–83
   Siemens, William and, 170–73, 171f
Steinmetz, Charles Proteus, 94–95

Stibitz, Georg, 561
Stokes, George Gabriel, 257–58
streetcars, 101f, 101–2
Stulen, Frank L., 517, 518f
submarine-launched missiles (SLBMs), 390–91
Sundback, Gideon, 21–22
supercomputers, 559–60, 576–77, 604, 662–64, 663f. *See also* computer technology
surveillance techniques, 3
Sutton, William, 144–45
Swan, Joseph Wilson, 44–46
synthetic ammonia, 210–13, 211f
synthetic inorganic chemistry, 165
Szilard, Leo, 382, 386, 393–95, 394f, 397

Talbot, Benjamin, 172
Talbot, William Henry Fox, 231–32
Tarde, Gabriel, 301–2
Taylor, Albert H., 548
Taylor, Charles, 13
Taylor, Frederick Winslow, 507–11
Taylor, Robert W., 577–78
Teal, Gordon K., 468–69, 470f
technical advances/innovation
   acceleration of, 7–12
   artifacts and innovations, 293–99
   car industry and culture, 338f, 343
   categories of risk with, 630–39
   causality limits, 344–48
   changes with, 291–304
   before computers, 279–83
   computer technology, 603–9, 605f
   contemporary perceptions of, 325–35
   debt to great innovators, 350, 351f
   economies and, 609–30, 622f
   environmental changes, 635–39, 637f, 638f
   evolutionary saltation, 7–12
   innovations and inventions, 357–60
   introduction to, 3–7, 5f, 7f, 277–78, 585–87, 650
   knowledge economy, 12–19
   labor force and, 617–20
   legacy of, 30–34
   mass production, 509–15, 610–17, 613f, 617f
   missing contexts, 652–56
   missing perspectives on, 278–91
   modern impact of, 587–609
   need for, 656–60
   new departures, 351–57
   quality-of-life improvements, 589–93, 594f
   shinkansen trains, 352–55
   traditional *vs.* new, 650–52

technical advances/innovation (*cont.*)
  transformations and, 339–44, 340*f*, 342*f*, 351–57
  unprecedented saltation, 349–50
  weapons and conflicts, 632–35, 634*f*
  See also Age of Synergy
technical civilization
  contrasting views on, 642–48
  introduction to, 640*f*, 641–42
  Singularity and, 26, 642–43, 650, 663–64, 666–67, 669–70
  See also computer technology
technical determinism, 343–44, 348
telephone
  cellular phones, 217, 249–50, 333
  communication innovation, 215, 217, 243–51, 244*f*, 246*f*, 248*f*, 249*f*
  computer technology and, 580
  contemporary perceptions of, 329
television, 8–9, 99, 217, 243, 261, 342*f*, 355–57, 579, 581, 582*f*, 584*f*, 603–4, 606, 608
Tesla, Nikola
  admiration after death, 289–90
  electric motors, 64–65, 78, 80–83, 81*f*, 82*f*, 87–88
  patent count, 284
  psychosis of, 288–89
  radio and, 267–68, 269*f*, 270–71
  three- wire electrical distribution, 62
thermodynamics, 8, 12, 13–16, 23–24, 79, 125–26, 205, 404–5
thermonuclear weapons, 389–90, 391–92, 559, 587, 604, 631–32
Third Industrial Revolution, 23, 27
Thomas, Sidney Gilchrist, 169–70
Thompson, Ken, 563
Thompson, William, 192
Thomson, William, 15
Three Mile Island accident, 398–99
three-wheel vehicle, 106*f*, 107, 112–13, 122–23, 134–35
three-wire distribution system, 61–62
Tilghman, Benjamin Chew, 227
titanium, 456–57
Tomlinson, Ray, 577–78
Toyota Motor Company, 511, 513–14
trains. *See* rail industry
tramways, 101
transistors in computer technology, 243–51
transportation industry
  commercial aviation, 547–52, 549*f*, 552*f*
  expansion of, 594–603, 595*f*, 597*f*, 602*f*
  freight movement, 553–58, 554*f*
  introduction to, 530*f*, 531–33
  on mass scale, 533–58
  shipping industry, 530*f*, 531–33, 553–58, 554*f*
  See also car industry and culture; rail industry
Truman, Harry, 389
Tsiolkovsky, Konstantin Eduardovich, 402
turbulent flows, 9
Turing, Alan, 519, 560–61
Turner, Samuel N., 235
Twain, Mark, 24
Twenty Thousand Leagues Under the Sea (Verne), 37
typesetting, 219–26
typewriter, 24
typewriting, 219–26, 225*f*

Updike, John, 579
Upton, Francis R., 46

vaccines, 378
vacuum cleaners, 100
vacuum tubes, 24, 47, 271, 272–73, 349, 561–62, 563–64, 565–67, 576, 579, 603–4, 662
Varley, Alfred, 23
Varley, Samuel, 55–56
velcro fasteners, 359–60, 360*f*
Verne, Jules, 37, 531
Villard, Henry, 54
Volta, Alessandro, 37–38
Volter, Heinrich, 226–27

Wallace, Alfred Russell, 653–54
Walton, Ernest T. S., 380
Wanamaker, Rodman, 327
Watson-Watt, Robert, 548
Watt, James, 13–14, 308–9
Watt's syndrome, 286–87
Wayss, Adolf Gustav, 181
Weil, George, 386
Weinberg, Alvin, 395–96, 397, 398
Wells, H. G., 3, 20–21
Welsbach, Carl Auer von, 51
Westinghouse, George, 62, 69–70, 75–76, 76*f*, 89–90, 90*f*, 284, 302, 350, 351*f*
Westinghouse Electric & Manufacturing Company, 49–50, 75–76, 80, 83, 85–86, 87–94, 103–4
Wheatstone, Charles, 23, 55–56, 243
White, J. Maunsel, 507–8
Whittle, Frank, 403–6, 404*f*
Wiener, Norbert, 519
Wigner, Eugene, 386

Wilbrand, Joseph, 201
Wild, Wilhelm, 454
Wilde, Henry, 55–56
Wilhelm, Carl, 170
Williams, F. C., 562
Wilson, Thomas L., 453
The Wind in the Willows (Grahame), 149
wireless communication, 261–74, 264*f*, 266*f*, 269*f*, 270*f*
Wöhler, Friedrich, 16–17
wood pulp, 23–24, 226–27, 228, 229, 371
work hours, 321
World Wide Web, 4, 576–78, 668–69. *See also* computer technology
Wozniak, Steven, 571
Wright, Frank Lloyd, 183
Wright, Orville, 11*f*, 11, 154–60, 156*f*, 159*f*
Wright, Wilbur, 11*f*, 11, 154–60, 156*f*, 159*f*
Wroblewski, Sigmund von, 192–93
Wronski, Christopher, 471–72
Wyman, William I., 330–31

x-rays, 100, 177

Young, Leo C., 548

Ziegler, Karl, 462–65, 464*f*
zipper patents, 21–22
Zola, Émile, 33
Zuse, Knorad, 561